Applied Statistics (Continued)

HANSEN, HURWITZ, and MADOW · Sample Survey Methods and Theory, Volume I
HOEL · Elementary Statistics
KEMPTHORNE · An Introduction to Genetic Statistics
MEYER · Symposium on Monte Carlo Methods
MUDGETT · Index Numbers
RICE · Control Charts
ROMIG · 50–100 Binomial Tables
SARHAN and GREENBERG · Contributions to Order Statistics
TIPPETT · Technological Applications of Statistics
WILLIAMS · Regression Analysis
WOLD and JURÉEN · Demand Analysis
YOUDEN · Statistical Methods for Chemists

Books of Related Interest

ALLEN and ELY · International Trade Statistics
ARLEY and BUCH · Introduction to the Theory of Probability and Statistics
CHERNOFF and MOSES · Elementary Decision Theory
HAUSER and LEONARD · Government Statistics for Business Use, *Second Edition*
STEPHAN and McCARTHY · Sampling Opinions—An Analysis of Survey Procedures

Stochastic

Processes

Stochastic Processes

J. L. DOOB
Professor of Mathematics
University of Illinois

New York · John Wiley & Sons, Inc.
London

FOURTH PRINTING, OCTOBER, 1962

Library of Congress Catalog Card Number: 52–11857

PRINTED IN THE UNITED STATES OF AMERICA

Preface

A STOCHASTIC PROCESS IS THE MATHEMATICAL ABSTRACTION OF an empirical process whose development is governed by probabilistic laws. The theory of stochastic processes has developed so much in the last twenty years that the need of a systematic account of the subject has been strongly felt by students of probability, and the present book is an attempt to fill this need. The reader is warned that this book does not cover the subject completely, and that it stresses most those parts of the subject which appeal most to me.

Although it would be absurd to write a book on stochastic processes which does not assume a considerable background in probability on the part of the reader, there is unfortunately as yet no single text which can be used as a standard reference. To compensate somewhat for this lack, elementary definitions and theorems are stated in detail, but there has been no attempt to make this book a first text in probability, even for the most advanced mathematical reader.

There has been no compromise with the mathematics of probability. Probability is simply a branch of measure theory, with its own special emphasis and field of application, and no attempt has been made to sugar-coat this fact. Using various ingenious devices, one can drop the interpretation of sample sequences and functions as ordinary sequences and functions, and treat probability theory as the study of systems of distribution functions. For a very short time this was the only known way to treat certain parts of the subject (such as the strong law of large numbers) rigorously. However, such a treatment is no longer necessary and results in a spurious simplification of some parts of the subject, and a genuine distortion of all of it.

There is probably no mathematical subject which shares with probability the features that on the one hand many of its most elementary theorems are based on rather deep mathematics, and that on the other hand many of its most advanced theorems are known and understood by many (statisticians and others) without the mathematical background to understand their proofs. It follows that any rigorous treatment of probability is threatened with the charge that it is absurdly overloaded with extraneous mathematics. Against this charge I have an easy defense. Early versions of the book contained many attempts to evade various mathematical points and thereby make the book easier to read. The readers complained that the evasions only increased the obscurity. The evasions were therefore eliminated, and the reader of this final version can now estimate the clarity of the earlier versions. The additional mathematical discussion made advisable a supplement on measure theory. In the Supplement is included a treatment of various aspects of measure theory with which the ordinary reader may not be familiar.

References to the literature and historical remarks have all been collected in the Appendix. Apologies are offered in advance to those who feel that they have been slighted in this Appendix, together with the assurance that they will probably have lots of company, and that no slight is intentional. The articles and books referred to in the Appendix are collected in a separate Bibliography.

Although much of the material in the book has been reorganized as seemed best to fit the point of view of the book, and other material is new, it is stressed that, even where no reference is given, no result is to be considered new unless it is stated as such in the Appendix. Subsidiary results are usually not credited to anyone.

Chapter XII, on prediction theory, is somewhat out of place in the book, since it discusses a rather specialized problem. It was put in because of the importance of the subject matter, and because of the lack of material on prediction theory, in the usual language of probability, readily available to the American reader. I had the benefit of stimulating conversations with Norbert Wiener on this subject.

I take this opportunity to thank a small group of friendly readers and savage critics. The advice of William Feller was particularly important in shaping the organization of the book. The criticism given by him, by Kai Lai Chung, and by J. L. Snell was always prized even when not accepted. Finally, thanks are due to Kathryn Hollenbach who typed most of the manuscript. The Office of Naval Research partly subsidized me during a year of my work on the preparation of the manuscript.

J. L. Doob

University of Illinois
September, 1952

Contents

CHAPTER I

Introduction and Probability Background

1. Mathematical prerequisites

Although advanced mathematical methods are used in this book, the results will always be stated in probability language, and it is hoped that the book will be accessible to readers thoroughly familiar with the manipulation of random variables conditional probability distributions and conditional expectations. The necessary background will be outlined in this chapter.

There is an unavoidable dilemma confronting the authors of advanced probability books. Probability theory is but one aspect of the theory of measure, with special emphasis on certain problems. An advanced probability book must, therefore, either include a section devoted to measure theory or must assume knowledge of measure theoretic facts as given in scattered papers. The second alternative was chosen originally, to make the book less formidable, but pressure of critics and logic forced a compromise which is as inconsistent as most compromises. It is hoped that the book remains in part at least accessible to statisticians and others who are not professional mathematicians but who are familiar with the formal manipulations of probability theory.

2. The basic space

Now that probability theory has become an acceptable part of mathematics, words like "occurrence," "event," "urn," "die," and so on can be dispensed with. These words and the ideas they represent, however, still add intuitive significance to the subject and suggest analytical methods and new investigations. It is for this reason that probability language is still used even in purely theoretical investigations.

In the applications, probability is concerned with the occurrence of events, such as the turning up of a 5 on a die, a displacement in a given direction of $x(t)$ centimeters in t seconds by a particle in a Brownian movement, and so on. Probability numbers are assigned to such events. For example, the number $^1/_6$ is usually assigned to the turning up of a 5 on a die and $^1/_2$ to the turning up of an even number; in the Brownian movement example, the inequality $x(3) > 7$ is assigned a probability according to rules to be described in detail in VIII. For the die, the purely mathematical analysis goes as follows: Each possible result, that is, each one of the integers $1, \cdots, 6$, is assigned the probability number $^1/_6$; any class of n of these results is assigned the number $n/6$. The statement that the probability of getting an even number is $^1/_2 = {}^3/_6$ is then simply the evaluation for this special case when $n = 3$. In the Brownian movement case the situation is more complicated and its analysis will be deferred for the present, but in this case also the question becomes that of assigning numbers to the mathematical abstractions of events. Throughout this book a point set is taken as the mathematical abstraction of an event.

The theory of probability is concerned with the measure properties of various spaces, and with the mutual relations of measurable functions defined on those spaces. Because of the applications, it is frequently (although not always) appropriate to call these spaces *sample spaces* and their measurable sets *events*, and these terms should be borne in mind in applying the results.

The following is the precise mathematical setting. (See the Supplement at the end of the book for a treatment of the concepts of *field* and *measure* suitable for this book.) It is supposed that there is some basic space Ω, and a certain basic collection of sets of points of Ω. These sets will be called *measurable sets*; it is supposed that the class of measurable sets is a Borel field. It is supposed that there is a function $\mathbf{P}\{\cdot\}$, defined for all measurable sets, which is a probability measure, that is, $\mathbf{P}\{\cdot\}$ is a completely additive non-negative set function, with value 1 on the whole space. The number $\mathbf{P}\{\Lambda\}$ will be called the *probability* or *measure* of Λ. The measurability of an ω function is defined in the usual way in terms of the measurable sets, and the integral with respect to the probability measure of a measurable function φ over a measurable set Λ will be denoted by

$$\int_\Lambda \varphi(\omega)\, d\mathbf{P} \qquad \text{or} \qquad \int_\Lambda \varphi\, d\mathbf{P}.$$

A property true at all points of Ω except at those of a set of probability 0 will be said to be true *almost everywhere*, or true at almost all points ω, or true *with probability* 1.

For example, consider the analysis of the tossing (once) of a die. In this case the simplest suitable space Ω consists of the six points $1, \cdots, 6$, with the identification of the point j as the event *the number j turns up when the die is tossed.* Every Ω set is measurable in this case and is assigned as measure one-sixth the number of its points. In this case the space Ω is certainly appropriately described as sample space, because of the simple correspondence between its points and the possible outcomes of the experiment. Now consider a second mathematical model of this same experiment. In this model the space Ω consists of all real numbers and the Ω points $1, \cdots, 6$ are identified with the same events as above. No other identifications are made. All Ω point sets are again measurable and the measure of any point set is one-sixth the number of the points $1, \cdots, 6$ which lie in the set. This mathematical model is practically the same as the preceding one, but the name "sample space" for Ω is somewhat less appropriate because the correspondence between Ω points and events is less simple. Finally consider a third mathematical model of this same experiment. In this model the space Ω consists of all the numbers in the semi-closed interval $[1, 7)$ and this time the interval $[j, j + 1)$ is identified with the event *the number j turns up when the die is tossed.* The measurable point sets in this case are the intervals $[j, j + 1)$ or unions of these intervals, and the measure of any measurable set is one-sixth the number of these intervals in it. This model is just as usable as the two preceding ones, but the name "sample space" is certainly not appropriate for Ω. The natural objection that the last two models, particularly the last one, are unsuitable, and should be excluded because they are needlessly complicated, is easily refuted. In fact "needlessly complicated" models cannot be excluded, even in practical studies. How a model is set up depends on what parameters are taken as basic. In the first model above the actual outcomes of the toss were taken as basic. If, however, the die is thought of as a cube to be tossed into the air and let fall, the position of the cube in space is determined by six parameters, and the most natural space Ω might well be considered to be the twelve-dimensional space of initial positions and velocities. A point of Ω determines how the die will land. Let Λ_j be the set of those Ω points which give rise to the outcome j. Then the usual hypothesis made is that the assignment of probabilities to Ω sets assigns probability $^1/_6$ to each Λ_j. If we are only interested in the probabilities of the results of the experiment, this is all we need to know about Ω probabilities, and we then have a mathematical model similar to our third one above. The point is that both in practical and in theoretical discussions, even if initially the space Ω is chosen to fit some criterion of simplicity (say that the name "sample space" be appropriate), this criterion is likely to be lost in the course of a discussion in which we got distributions derived in some way from the initial one. In accordance with this fact we have imposed no condition whatever on the space Ω not implicit in the existence of the set function $\mathbf{P}\{\Lambda\}$, and the existence of this set function prevents Ω from being empty but is not otherwise restrictive. In the course of this book Ω is usually simply an abstract space. In some cases, however, Ω will be taken to be the interval $0 \leq x \leq 1$, the space of all finite or infinite valued real functions of $t, -\infty < t < \infty$, the finite plane and many other spaces. The only specifically probabilistic hypothesis imposed on the measure function $\mathbf{P}\{\Lambda\}$ is the normalizing condition $\mathbf{P}\{\Lambda\} = 1$. Outside the field of probability there is frequently no reason to restrict measure functions to be finite-valued, and if finite-valued there is no reason to restrict the value on the whole space to be 1.

We now analyze the tossing of a die still further, to illustrate the importance of product spaces as the basic spaces. In analyzing the tossing of a die an unlimited number of times, the natural basic space Ω is the obvious sample space, defined as follows. Each point ω of Ω is a sequence $\xi_1, \xi_2, \cdots$, where ξ_j is one of the integers $1, \cdots, 6$. That is, each point of Ω is one of the conceptually possible outcomes of an experiment in which a die is tossed infinitely often. If the class of all sequences beginning with j is identified with the event *the number j turns up the first time the die is tossed*, and if this class is assigned the probability $^1/_6$ we have still another mathematical model of a single toss. The advantage of this model over those described above is that this model is adaptable to any number of tosses if the class of all sequences beginning with $j_1, \cdots, j_n$ is identified with the event *the numbers* $j_1, \cdots, j_n$ *turn up in the first n tosses*. With the usual hypotheses, this class of sequences, that is, this Ω set, is assigned the measure $(1/6)^n$. While it is true that in many applications spaces of *infinite* sequences of this sort are unnecessary, because only a fixed finite number of experiments is to be discussed, the infinite sequences cannot be avoided in some problems of very simple character. For example, the analysis of the first time an event occurs (say the first 6 in a succession of throws of a die) cannot be done in a satisfactory manner without the space of sample *infinite* sequences, because the number of trials before the event occurs will be a number which cannot be bounded in advance. This number is an unbounded integral valued function of Ω. In this die example Ω is a product space, one with infinitely many factor spaces each of which contains six points. The natural sample space for a repeated trial is always a product space.

If C represents conditions on points of Ω, the notation $\{C\}$ will be used for the set of points satisfying those conditions. For example, if X is a linear set, and if x is an ω function, $\{x(\omega) \in X\}$ is the ω set on which $x(\omega)$ is a number in the set X.

We have not yet assumed that our basic probability measure is complete (see Supplement). That is, if Λ_0 is measurable, and has measure 0, and if Λ is a subset of Λ_0, then Λ need not be measurable. In the language of probability this means that there may be two events, Λ and Λ_0, with the properties that the occurrence of Λ implies that of Λ_0, and that

$$\mathbf{P}\{\Lambda_0\} = 0,$$

and yet Λ is not assigned any probability. (Clearly if $\mathbf{P}\{\Lambda\}$ is defined it is 0.) This possibility may be somewhat disturbing to a mathematician who wishes to preserve his intuitive concept of probability. One has the choice here of either changing one's intuition or one's mathematics if one insists on a close correspondence between the two. The choice is not very important, but in any event the mathematics is easily modified for the sake of those with stubborn intuitions. In fact (see Supplement §2) the given probability measure can be *completed* in a unique way by a slight enlargement of the class of ω sets whose measures are defined. To avoid fussy details in II §2, *we assume in this book that* **P** *measure is complete.*

3. Random variables and probability distributions

A (real) function x, defined on a space of points ω, will be called a (real) *random variable* if there is a probability measure defined on ω sets, and if for every real number λ the inequality $x(\omega) \leq \lambda$ delimits an ω set whose probability is defined, that is, a measurable ω set. Thus

$$F(\lambda) = \mathbf{P}\{x(\omega) \leq \lambda\} \tag{3.1}$$

is defined for all real λ. In mathematical language a (real) random variable is thus simply a (real) measurable function. A complex random variable is an ω function whose real and imaginary parts are measurable.

Throughout this book, whenever more than one random variable is involved in a discussion, it will always be assumed unless the contrary is explicitly stated that the random variables are all defined on the same ω space.

Random variables were manipulated by probabilists long before it was recognized that the mathematical concept involved was that of a measurable function, and in fact long before measure theory was invented. Thus probabilists developed a specialized language of their own, which it is now possible to translate into the language of measure theory. Probability language has not been dropped because it adds intuitive content to the subject and also makes it more accessible to workers in applied fields.

It is customary in analysis to use the same notation for a function as for its value at a given point of its domain of definition, and the ambiguity has its compensations. It will be necessary to be somewhat more precise when we deal with functions in this book. A function will usually be denoted by a single letter, and the usual functional notation will be reserved for the value of the function at a given point. Thus $x(\omega)$ is the value of the function x at the point ω. As usual, it will sometimes be convenient to denote the function x by $x(\cdot)$.

The function F defined by (3.1) is called the *distribution function of the random variable* x. It is monotone non-decreasing, continuous on the right, and

$$\lim_{\lambda \to -\infty} F(\lambda) = 0, \qquad \lim_{\lambda \to \infty} F(\lambda) = 1. \tag{3.2}$$

Any function F satisfying all these conditions will be called a *distribution function*. A distribution function defines a *distribution*, that is, a probability measure

$$\int_A dF(\lambda)$$

of sets A. This is the usual Lebesgue-Stieltjes measure defined by F.

If F is a distribution function, and if, for some Lebesgue measurable and integrable function f,

$$F(\lambda) = \int_{-\infty}^{\lambda} f(\mu)\, d\mu, \qquad -\infty < \lambda < \infty, \tag{3.3}$$

f is called the density function corresponding to F. We have $F'(\lambda) = f(\lambda)$ for almost all λ (Lebesgue measure). The statement that F has a density will be understood to imply the existence of f satisfying (3.3), that is, there is a density if and only if F is absolutely continuous.

If $x_1, \cdots, x_n$ are real random variables, the function defined by

$$F(\lambda_1, \cdots, \lambda_n) = \mathbf{P}\{x_j(\omega) \le \lambda_j, j = 1, \cdots, n\} \tag{3.4}$$

is called their *multivariate distribution function*. The function F defined by (3.4) is monotone non-decreasing and continuous on the right in each variable, with

$$\lim_{\lambda_j \to -\infty} F(\lambda_1, \cdots, \lambda_n) = 0, \qquad j = 1, \cdots, n$$

$$\lim_{\lambda_1, \cdots, \lambda_n \to \infty} F(\lambda_1, \cdots, \lambda_n) = 1.$$

Moreover if $\lambda_j \le \mu_j$, $j = 1, \cdots, n$, then

$$F(\mu_1, \cdots, \mu_n) - \sum_{j=1}^{n} F(\mu_1, \cdots, \mu_{j-1}, \lambda_j, \mu_{j+1}, \cdots, \mu_n)$$
$$+ \sum_{\substack{j,k=1 \\ j<k}}^{n} F(\mu_1, \cdots, \mu_{j-1}, \lambda_j, \mu_{j+1}, \cdots, \mu_{k-1}, \lambda_k, \mu_{k+1}, \cdots, \mu_n)$$
$$- \cdots + (-1)^n F(\lambda_1, \cdots, \lambda_n) \ge 0 .$$

The quantity on the left is the evaluation in terms of F of

$$\mathbf{P}\{\lambda_j < x_j(\omega) \leq \mu_j, j = 1, \cdots, n\}.$$

Any function F satisfying all these conditions will be called an n-variate distribution function. Such a function defines a probability measure

$$\int_A \cdots \int d_{\lambda_1, \cdots, \lambda_n} F(\lambda_1, \cdots, \lambda_n),$$

Lebesgue-Stieltjes measure in n dimensions. (See Supplement Example 2.2.) If, for some Lebesgue measurable and integrable f,

$$F(\lambda_1, \cdots, \lambda_n) = \int_{-\infty}^{\lambda_1} \cdots \int_{-\infty}^{\lambda_n} f(\mu_1, \cdots, \mu_n) d\mu_1 \cdots d\mu_n \tag{3.5}$$

for all $\lambda_1, \cdots, \lambda_n$, then f is called the corresponding density function.

The real random variables $x_1, \cdots, x_n$ are said to be *mutually independent* if

$$\mathbf{P}\{x_j(\omega) \in X_j, j = 1, \cdots, n\} = \prod_{j=1}^{n} \mathbf{P}\{x_j(\omega) \in X_j\} \tag{3.6}$$

for all linear Borel sets $X_1, \cdots, X_n$. Equivalently one can require (3.6) only for open sets X_j, or intervals, or even semi-infinite intervals $(-\infty, b_j]$. It follows that the random variables are mutually independent if and only if their multivariate distribution function is the product of their individual distribution functions.

If x_j in the preceding paragraph represents the m_j variables $x_{j1}, \cdots, x_{jm_j}$, and if X_j is correspondingly an m_j-dimensional set, the preceding paragraph furnishes the definition of mutual independence of finitely many aggregates each containing finitely many random variables. If the aggregates may contain infinitely many random variables, the aggregates are said to be mutually independent if whenever each infinite aggregate is replaced by a finite subaggregate these subaggregates are mutually independent. Infinitely many aggregates of random variables are said to be mutually independent if, whenever $A_1, A_2, \cdots$ are finitely many of the given aggregates, $A_1, A_2, \cdots$ are mutually independent.

The preceding definitions are applicable to complex random variables if we allow the X_j's in (3.6) to be two-dimensional Borel sets, that is, Borel sets of the complex plane, making corresponding conventions in the later discussion. Or the preceding definitions are applicable directly, if the convention is made that a complex random variable is to be considered a pair of real random variables, its real and imaginary parts.

Let x be a random variable. Then its integral over ω space Ω, if the integral exists, will be denoted by $\mathbf{E}\{x\}$, or $\mathbf{E}\{x(\omega)\}$, and called the *expectation* of x,

$$\mathbf{E}\{x\} = \int_{\Omega} x \, d\mathbf{P}. \tag{3.7}$$

We recall that this integral exists if and only if the integral of $|x|$ is finite.

4. Convergence concepts

Let $x, x_1, x_2, \cdots$ be random variables. If

$$\lim_{n \to \infty} x_n(\omega) = x(\omega)$$

for almost all ω, we say that

$$\lim_{n \to \infty} x_n = x$$

with probability 1. There is such convergence if and only if

$$\lim_{n \to \infty} \mathbf{P}\{\underset{m \geq n}{\text{L.U.B.}} |x_m(\omega) - x_n(\omega)| \geq \varepsilon\} = 0 \tag{4.1}$$

for every $\varepsilon > 0$.

The x_n sequence is said to converge *stochastically* to x if the weaker condition

$$\lim_{n \to \infty} \mathbf{P}\{|x_n(\omega) - x(\omega)| \geq \varepsilon\} = 0 \tag{4.2}$$

is satisfied for every $\varepsilon > 0$. Stochastic convergence, denoted by

$$\underset{n \to \infty}{\text{p lim}}\, x_n = x, \tag{4.3}$$

is also known as *convergence in probability* and as *convergence in measure*. The relations between convergence with probability 1 and convergence in probability are well known, and will not be proved here:

(*a*) convergence with probability 1 implies convergence in probability;

(*b*) $\underset{n \to \infty}{\text{p lim}}\, x_n = x$ if and only if every subsequence $\{x_{a_n}\}$ of the x_j's contains a further subsequence which converges to x with probability 1.

The sequence $\{x_n\}$ is said to *converge to x in the mean* if $\mathbf{E}\{|x_n|^2\} < \infty$ for all n, if $\mathbf{E}\{|x|^2\} < \infty$, and if

$$\lim_{n \to \infty} \mathbf{E}\{|x - x_n|^2\} = 0. \tag{4.4}$$

This is written

$$\underset{n \to \infty}{\text{l.i.m.}}\, x_n = x. \tag{4.5}$$

Convergence in the mean implies convergence in probability. In fact, if (4.5) is true,

$$\lim_{n\to\infty} \mathbf{P}\{|x_n(\omega) - x(\omega)| \geq \varepsilon\} \leq \lim_{n\to\infty} \frac{\mathbf{E}\{|x_n - x|^2\}}{\varepsilon^2} = 0 \tag{4.6}$$

for all $\varepsilon > 0$.

Let $\{F_n\}$ be a sequence of one-dimensional distribution functions. The most appropriate definition of convergence to a distribution function F is the condition that $\lim_{n\to\infty} F_n(\lambda) = F(\lambda)$ at every point of continuity of F. If this condition is satisfied, the convergence will necessarily be uniform in any closed interval of continuity of F, finite or infinite. The definition of distance between two distribution functions G_1 and G_2 that matches this convergence is the following. The graphs of G_1 and G_2 are completed with vertical line segments at points of discontinuity, and the distance between G_1 and G_2 is defined as the maximum distance between points of the two completed graphs measured along lines of slope -1. Under this definition the space of distribution functions is a complete metric space and $\lim_{n\to\infty} F_n(\lambda) = F(\lambda)$ at all points of continuity of F if and only if the distance between F_n and F goes to 0 with $1/n$.

If the sequence $\{F_n\}$ is the sequence of distribution functions of the sequence of random variables $\{x_n\}$, and if the distribution functions converge to a distribution function in the sense just described, the sequence of random variables is sometimes said to converge *in distribution*. This terminology is rather unfortunate, since the sequence of random variables may not then converge in any ordinary sense. For example, whenever the distribution functions are identical and the random variables otherwise arbitrary, say mutually independent, the sequence of random variables converges in distribution.

5. Families of random variables

In some discussions, random variables are defined directly. For example, if ω space is the line, if the measurable ω sets are the Lebesgue measurable sets, and if probability measure is defined by

$$\mathbf{P}\{A\} = \frac{1}{\sqrt{2\pi}} \int_A e^{-\lambda^2/2}\, d\lambda, \tag{5.1}$$

then the sequence of functions $1, \omega, \omega^2, \cdots$ is a sequence of random variables. However, in many discussions the specific ω space involved is irrelevant, and only the existence of a family of random variables with specified properties is required. In this case, the common situation is that the multivariate distributions of finite aggregates of the random variables are specified, and it is understood that there exists a family of random variables with the specified distributions. The present section is devoted to a discussion of this point. In II §2 the possibility of prescribing further distributions will be discussed.

For example, when a theorem has the initial phrase "let $x_1, \cdots, x_n$ be mutually independent random variables with distribution functions $F_1, \cdots, F_n$," the theorem would be trivial without the existence theorem that there are n random variables with the stated properties defined on some space. It will clarify the following general discussion if we outline here a proof of this existence theorem. Let ω space be the n-dimensional space of points ω: $(\xi_1, \cdots, \xi_n)$. A probability measure is determined on the Borel sets of ω space by

$$\mathbf{P}\{A\} = \int \cdots \int_A dF_1(\xi_1) \cdots dF_n(\xi_n)$$

(see §3). Let x_j be the jth coordinate variable, that is, $x_j(\omega) = \xi_j$ if ω is the point $(\xi_1, \cdots, \xi_n)$. The x_j's are then mutually independent random variables with the respective distribution functions $F_1, \cdots, F_n$. We have thus proved that there are random variables $x_1, \cdots, x_n$ with the desired properties, and incidentally that the random variables can be taken as the coordinate functions in n-dimensional space. The point of the following paragraph is the generalization of the method used here to the most general aggregates of random variables.

In general, a class of subscripts T is given, and a class of random variables $\{x_t, t \in T\}$ is to be defined. For every finite t set $t_1, \cdots, t_n$, the multivariate distribution function $F_{t_1, \cdots, t_n}$ of $x_{t_1}, \cdots, x_{t_n}$ is prescribed. It is obvious that the distribution functions prescribed must be mutually consistent in the sense that if $\alpha_1, \cdots, \alpha_n$ is a permutation of $1, \cdots, n$, then

$$F_{t_1, \cdots, t_n}(\lambda_1, \cdots, \lambda_n) \equiv F_{t_{\alpha_1}, \cdots, t_{\alpha_n}}(\lambda_{\alpha_1}, \cdots, \lambda_{\alpha_n}),$$

and that, if $m < n$,

$$F_{t_1, \cdots, t_m}(\lambda_1, \cdots, \lambda_m) \equiv \lim_{\substack{\lambda_j \to \infty \\ j = m+1, \cdots, n}} F_{t_1, \cdots, t_n}(\lambda_1, \cdots, \lambda_n).$$

Kolmogorov proved (see Supplement Example 2.3) that these consistency conditions are the only conditions that need be imposed. The x_t's are defined as follows. The space Ω is taken as the space of points ω: $(\xi_t, t \in T)$, where ξ_t is any real number, that is, Ω is the space of functions of $t \in T$, or, from another point of view, Ω is coordinate space, whose dimensionality is the cardinal number of T. The value of a t function at $t = s$ defines an ω function x_s, if we set $x_s(\omega) = \xi_s$, where ω is the function $\xi_{(\cdot)}$. The ω set

$$\{x_{t_j}(\omega) \leq \lambda_j, j = 1, \cdots, n\}$$

is assigned as measure the given number

$$F_{t_1, \cdots, t_n}(\lambda_1, \cdots, \lambda_n);$$

more generally the ω set

$$\{[x_{t_1}(\omega), \cdots, x_{t_n}(\omega)] \in A\}$$

is assigned as measure (if A is an n-dimensional Borel set)

$$\int_A \cdots \int d_{\xi_1, \cdots, \xi_n} F_{t_1, \cdots, t_n}(\xi_1, \cdots, \xi_n).$$

It is proved that this assignment of measures to ω sets of the specified type determines a probability measure on the Borel field of ω sets generated by the class of sets of this type. The family of ω functions $\{x_t, t \in T\}$ is then a family of random variables with the assigned distributions.

The basic ω space Ω was defined in this discussion to be the space Ω_T of all functions of $t \in T$, so that the family of random variables obtained was the family of coordinate functions of a Cartesian space. Even if the basic space of a family of random variables $\{x_t, t \in T\}$ is not this Cartesian space, however, we shall see in §6 that the family can be replaced by such a family, for many purposes.

Let $\{x_t, t \in T\}$ be a family of random variables. Then for fixed ω the function value $x_t(\omega)$ defines a function of t. The t functions defined in this way will be called the *sample functions* of the family. In particular if Ω is coordinate space Ω_T, and if x_t is the tth coordinate function, the concepts of basic point ω and sample function coincide. In any case it is frequently convenient to use such phrases as "almost all sample functions," involving measure concepts and sample functions, with the convention that the measure concepts are referred back to Ω. For example, the phrase "almost all sample functions" means "almost all ω." If the parameter set T is finite or enumerably infinite, it will usually be more appropriate to speak of a sample sequence than of a sample function, and we shall do so.

Finally we remark that, when a family of distribution functions is used as at the beginning of this section to define a measure of Ω_T sets, it may sometimes be useful to take Ω_T not as the space of all functions of t, $t \in T$ but as this space with the restriction that the values of the function lie in a given set X. (We shall not find it necessary in this book to have X depend on the parameter value t.) All the considerations of the above work remain valid if X is a Borel set of the infinite line $-\infty \leq x \leq \infty$

and if the given distributions restrict each of the random variables to X, with probability 1.

To illustrate the preceding discussion we apply it to the analysis of the repeated tossing of a die. Suppose that n tosses are to be considered. The parameter set T will then be taken as the set of integers $1, \cdots, n$. The space Ω_T is the n-dimensional space of points $(\xi_1, \cdots, \xi_n)$. Any point whose coordinates are all integers between 1 and 6 inclusive is given measure $1/6^n$, and the measure of any set of points is the number of these special points in the set, divided by 6^n. In this way we obtain a mathematical model of n tosses of a (balanced) die (see also the discussion in §2), and the jth coordinate of a point of Ω_T is a random variable, corresponding to the number obtained as a result of the jth toss. It is clear that Ω_T has unnecessarily many points in it. The point $(0, \cdots, 0)$ for example, corresponds to no experimental result and simply clutters up the mathematical model. It is for this reason that the set X was introduced in the preceding paragraph. If we take X as the set of integers 1, 2, 3, 4, 5, 6, the ω space becomes the space of 6^n points in n dimensions whose coordinates are all integers between 1 and 6 inclusive. The measure assigned to any set is the number of its points divided by 6^n.

6. Product space representations

Let x be a real random variable, with distribution function F. Then (see Supplement Example 2.1) F determines a Lebesgue-Stieltjes measure of linear sets defined by

$$\tilde{\mathbf{P}}\{A\} = \int_A dF(\xi).$$

The measurable linear sets are the sets measurable with respect to F. These sets include all the Borel sets. In terms of our probability measure on the line, we can interpret an event determined by a condition on x, say $x(\omega) \in A$, as the linear point set A rather than the ω set $\{x(\omega) \in A\}$, and the linear measure has been defined to make the probability the same for the two interpretations. The formal elaboration of this idea is to set up a measure preserving transformation from ω space to the line. This is done in detail in the Supplement (see especially Example 3.1), but the ideas in this section should be clear even without that detailed discussion. Let $\tilde{x}$ be the coordinate random variable on the line. That is, $\tilde{x}$ is the function which takes on the value ξ at the point with coordinate ξ. Suppose that in some investigation one is only concerned with ω random variables of the form $\Phi(x)$, where Φ is a Baire function of a real variable, or more generally is measurable with respect to F. Then it is possible, and frequently convenient, to replace the original basic ω space by the line, using the F measure defined above. For example, an ω random variable $\Phi(x)$ then becomes a random variable $\Phi(\tilde{x})$ defined on the line, and these two random variables, defined on different spaces, have the same distribution. More generally, if $\Phi_1(x), \cdots, \Phi_n(x)$ are ω random variables of this type, they become random variables on the line, and the multivariate distributions of the two sets of random variables, defined on different spaces, are the same. As one application of this idea, consider the expectation of $\Phi(x)$, where Φ is as above. From the point of view of ω space, $\mathbf{E}\{\Phi(x)\}$ is defined by

$$\int_{\Omega} \Phi[x(\omega)]\, d\mathbf{P}.$$

From the point of view of the new basic space $\tilde{\Omega}$, the ξ axis, the random variable in question is $\Phi(\tilde{x})$, and its expectation is defined by

$$\int_{-\infty}^{\infty} \Phi(\xi)\, dF(\xi).$$

(The equality of these two evaluations is proved in the Supplement, §3.) Thus it is unnecessary to revert to the original basic space to calculate the expectation of the random variable.

Let x, y be real random variables with bivariate distribution function F. Then, just as in the one-dimensional case, F determines a probability measure of plane sets,

$$\mathbf{P}\{A\} = \iint_A d_{\xi,\eta}F(\xi, \eta)$$

for A measurable with respect to F. Suppose that in some investigation one is only concerned with random variables $\Phi(x, y)$, where Φ is a function of two real variables, measurable with respect to F. Then, generalizing the one-dimensional case discussed in the preceding paragraph, it is frequently convenient to replace the original basic space Ω by $\tilde{\Omega}$, the ξ, η plane. The point is that the original ω probabilities induce a probability measure on the space $\tilde{\Omega}$, by a point transformation which is measure preserving, and that for many purposes the new space is more convenient than the old one. As in the preceding paragraph, $\mathbf{E}\{\Phi(x, y)\}$, defined in terms of ω space and ω measure by

$$\int_{\Omega} \Phi[x(\omega), y(\omega)]\, d\mathbf{P},$$

is defined in terms of $\tilde{\omega}$ space and $\tilde{\omega}$ measure by

$$\int_{-\infty}^{\infty} \int_{-\infty}^{\infty} \Phi(\xi, \eta)\, d_{\xi,\eta}F(\xi, \eta),$$

and the two evaluations are equal.

In general, let $\{x_t, t \in T\}$ be any family of random variables, and let $\mathscr{B}(x_t, t \in T)$ be the smallest Borel field of ω sets with respect to which the x_t's are measurable. That is, $\mathscr{B}(x_t, t \in T)$ is the Borel field of sets generated by the class of ω sets of the form $\{x_t(\omega) \in X\}$, where $t \in T$ and X is an interval. Then it is frequently convenient to replace the original basic space Ω by $\tilde{\Omega} = \Omega_T$, the space of functions of $t \in T$. It is shown in the Supplement, §3, that a probability measure can be defined on function space, as described in §5, in such a way that, if $\{\tilde{x}_t, t \in T\}$ is the class of coordinate variables in this space, every finite set of $\tilde{x}_t$'s has the same multivariate distribution as the set of corresponding x_t's. In fact there is a transformation taking the ω random variables measurable with respect to $\mathscr{B}(x_t, t \in T)$ into $\tilde{\omega}$ random variables, in a one to one way (if we consider as identical random variables which are equal with probability 1), and satisfying the following conditions:

(i) the transformation takes x_t into $\tilde{x}_t$, and Baire functions of a finite number of x_t's into the same Baire functions of the corresponding $\tilde{x}_t$'s;

(ii) if x is an ω random variable going into the $\tilde{\omega}$ random variable $\tilde{x}$, then if the expectation of either random variable exists, that of the other exists also, and the two expectations are equal;

(iii) if x is an ω random variable, taking on the value 1 on a set $\Lambda \in \mathscr{B}(x_t, t \in T)$, and 0 on the complement of Λ, then the transformation takes x into an $\tilde{\omega}$ random variable $\tilde{x}$ taking on the value 1 on a measurable $\tilde{\omega}$ set $\tilde{\Lambda}$, and 0 on the complement of this set; the set transformation defined in this way is one to one (if we consider as identical sets which differ by sets of measure 0) and measure preserving.

Thus any problem involving ω random variables measurable with respect to $\mathscr{B}(x_t, t \in T)$, or sets of this Borel field, can be expressed as the corresponding problem in terms of $\tilde{\omega}$ random variables. The class of problems that can be considered can be slightly enlarged by completing the probability measure defined on the sets of $\mathscr{B}(x_t, t \in T)$. We refer to the Supplement for a rigorous treatment of the mapping involved here. The particular cases in which T consists of only a single point, or of exactly two points, have been discussed separately at the beginning of this section. Problems involving n random variables can be reduced to problems involving n-dimensional coordinate space, and problems involving infinitely many random variables to problems involving infinite dimensional coordinate space. In each case the original random variables become the coordinate functions of the new space.

As an example of the application of this idea when T consists of two points, consider the theorem that, if x and y are mutually independent random variables, and have expectations, then $\mathbf{E}\{xy\}$ exists also, and

$$\mathbf{E}\{xy\} = \mathbf{E}\{x\}\mathbf{E}\{y\}. \tag{6.1}$$

At first sight this does not appear to be a standard integration theorem, and in fact it is sometimes treated as a very special theorem of the theory of probability. But note that it is a theorem concerned only with the two random variables x and y, so that a representation on the plane is admissible. In this representation we shall see that the theorem becomes a standard integration theorem. If x, y have distribution functions G, H respectively, and if F is the distribution function of the pair, then

$$F(\xi, \eta) \equiv G(\xi)H(\eta),$$

because the random variables are independent, and (6.1) becomes, in terms of the plane representation,

$$\int_{-\infty}^{\infty}\int_{-\infty}^{\infty} \xi\eta \, dG(\xi)\, dH(\eta) = \int_{-\infty}^{\infty}\int_{-\infty}^{\infty} \xi \, dG(\xi)\, dH(\eta) \int_{-\infty}^{\infty}\int_{-\infty}^{\infty} \eta \, dG(\xi)\, dH(\eta). \tag{6.2}$$

The double integrals on the right reduce to single integrals, so that (6.2) becomes

$$\int_{-\infty}^{\infty}\int_{-\infty}^{\infty} \xi\eta \, dG(\xi)\, dH(\eta) = \int_{-\infty}^{\infty} \xi \, dG(\xi) \int_{-\infty}^{\infty} \eta \, dH(\eta). \tag{6.3}$$

Thus (6.1) becomes the evaluation of a double integral by iterated integration. Any direct proof of (6.1) must be equivalent to a justification of this standard evaluation.

In the preceding discussion we have assumed that the random variables were real. The extension of representation theory to complex valued random variables is obvious, and the details will be omitted.

7. Conditional probabilities and expectations

Let y be a random variable and let M be a measurable ω set. We wish to define the conditional probability of M, and the conditional expectation of y, relative to various specific conditions. Before doing so we consider two special cases.

Case 1 Suppose that a random variable x takes on only a finite or enumerable sequence of values $a_1, a_2, \cdots$. The *conditional probability of* M *if* $x(\omega) = a_j$, which we denote by $\mathbf{P}\{\mathrm{M} \mid x(\omega) = a_j\}$, is defined, whenever $\mathbf{P}\{x(\omega) = a_j\} > 0$, by

$$\mathbf{P}\{\mathrm{M} \mid x(\omega) = a_j\} = \frac{\mathbf{P}\{\omega \in \mathrm{M}, x(\omega) = a_j\}}{\mathbf{P}\{x(\omega) = a_j\}}. \tag{7.1}$$

In particular if y takes on only the values $b_1, b_2, \cdots$, we obtain in this way the conditional distribution of y for $x(\omega) = a_j$,

$$\mathbf{P}\{y(\omega) = b_k \mid x(\omega) = a_j\} = \frac{\mathbf{P}\{y(\omega) = b_k,\, x(\omega) = a_j\}}{\mathbf{P}\{x(\omega) = a_j\}}. \tag{7.2}$$

The conditional probability $\mathbf{P}\{\mathrm{M} \mid x(\omega) = a_j\}$ depends on a_j, that is, it defines a function of the values taken on by the random variable x. If we substitute $x(\omega)$ for its value in the definition of this conditional probability, we obtain a random variable z, defined by

$$z(\omega) = \mathbf{P}\{\mathrm{M} \mid x(\omega) = a_j\} \qquad \text{where } x(\omega) = a_j,$$

if $\mathbf{P}\{x(\omega) = a_j\} > 0$. We define $z(\omega)$ arbitrarily on the ω set $\{x(\omega) = a_j\}$ if this set has probability 0. The random variable z is thus defined uniquely, neglecting values on an ω set of probability 0. The *conditional probability of* M *relative to* x, denoted by $\mathbf{P}\{\mathrm{M} \mid x\}$ is defined as the random variable z, or more precisely as any one of the versions of z. Then

$$\mathbf{P}\{\mathrm{M} \mid x\}\,\big|_{x(\omega)=a_j} = \mathbf{P}\{\mathrm{M} \mid x(\omega) = a_j\}, \tag{7.3}$$

if $\mathbf{P}\{x(\omega) = a_j\} > 0$. Let A be any set of a_j's, and define

$$\Lambda = \{x(\omega) \in A\} = \bigcup_{a_j \in A} \{x(\omega) = a_j\}.$$

Then we observe that

$$\mathbf{P}\{\Lambda\mathrm{M}\} = \sum_{a_j \in A} \mathbf{P}\{\mathrm{M} \mid x(\omega) = a_j\}\, \mathbf{P}\{x(\omega) = a_j\}.$$

This equation can also be written in the form

$$\mathbf{P}\{\Lambda\mathrm{M}\} = \int_\Lambda \mathbf{P}\{\mathrm{M} \mid x\}\, d\mathbf{P}. \tag{7.4}$$

Similarly the *conditional expectation of* y *relative to* x, denoted by $\mathbf{E}\{y \mid x\}$, is defined as a random variable for which

$$\mathbf{E}\{y \mid x\}\big|_{x(\omega)=a_j} = \mathbf{E}\{y \mid x(\omega) = a_j\} = \sum_k b_k\, \mathbf{P}\{y(\omega) = b_k \mid x(\omega) = a_j\}$$

if $\mathbf{P}\{x(\omega) = a_j\} > 0$. The equation

$$\sum_k b_k\, \mathbf{P}\{y(\omega) = b_k,\, x(\omega) \in A\} = \sum_{a_j \in A} \mathbf{E}\{y \mid x(\omega) = a_j\}\, \mathbf{P}\{x(\omega) = a_j\}$$

follows at once from our definitions. This equation can also be written in the form

$$\int_\Lambda y\, d\mathbf{P} = \int_\Lambda \mathbf{E}\{y \mid x\}\, d\mathbf{P}. \tag{7.5}$$

Equations (7.4) and (7.5) inspire the general definitions of conditional probability and expectation to be given below.

Case 2 Suppose that Ω is the ξ, η plane, that the measurable ω sets are the Lebesgue measurable sets, and that the given probability measure is determined by a density function,

$$\mathbf{P}\{\Lambda\} = \iint_{\Lambda} f(\xi, \eta)\, d\xi\, d\eta.$$

Then the abscissa and ordinate variables determine coordinate functions x and y taking on the values ξ, η respectively at the point (ξ, η). These functions are random variables whose joint distribution has density f. We define a new density (in the variable η) by

$$\frac{f(\xi, \eta)}{\int_{-\infty}^{\infty} f(\xi, \zeta)\, d\zeta}$$

for each ξ for which the denominator does not vanish. In analogy with the discussion of Case 1, it appears natural to describe the distribution obtained in this way as the conditional distribution of y for $x(\omega) = \xi$, and to describe as the conditional expectation of y for $x(\omega) = \xi$ the ratio

$$\frac{\int_{-\infty}^{\infty} \eta f(\xi, \eta)\, d\eta}{\int_{-\infty}^{\infty} f(\xi, \eta)\, d\eta}.$$

(We make the assumption $\mathbf{E}\{|y|\} < \infty$.) These descriptions will be consistent with the general definitions to be given below.

General case Instead of defining conditional probabilities and expectations relative to a given random variable, we define slightly more general concepts and particularize. It will be seen that a conditional probability is a special case of a conditional expectation, and we therefore define the latter first.

Let y be a random variable whose expectation exists, and let $\mathscr{F}$ be a Borel field of measurable ω sets. Let $\mathscr{F}' \supset \mathscr{F}$ be the Borel field of those ω sets which are either $\mathscr{F}$ sets or which differ from $\mathscr{F}$ sets by sets of probability 0. We recall (see Supplement, Theorem 2.3) that if a random variable is measurable with respect to $\mathscr{F}'$ it is equal with probability 1 to a random variable measurable with respect to $\mathscr{F}$. The *conditional expectation of* y *relative to* $\mathscr{F}$, denoted by $\mathbf{E}\{y \mid \mathscr{F}\}$, is defined as any ω

function which is measurable with respect to $\mathscr{F}'$, which is integrable, and which satisfies the equation

$$\int_\Lambda \mathbf{E}\{y \mid \mathscr{F}\}\, d\mathbf{P} = \int_\Lambda y\, d\mathbf{P}, \qquad \Lambda \in \mathscr{F}. \tag{7.6}$$

(It will necessarily also satisfy this equation for $\Lambda \in \mathscr{F}'$, because of the relation between $\mathscr{F}$ and $\mathscr{F}'$. Thus the definitions of the conditional expectations $\mathbf{E}\{y \mid \mathscr{F}\}$ and $\mathbf{E}\{y \mid \mathscr{F}'\}$ are identical.) Note that the right side of (7.6) defines a function of $\Lambda \in \mathscr{F}$ which is completely additive and which vanishes when $\mathbf{P}\{\Lambda\} = 0$. Hence, according to the Radon-Nikodym theorem (see Supplement, §2), this function of Λ can be expressed as the integral over Λ of an ω function measurable with respect to $\mathscr{F}$. This ω function is thus one possible version of $\mathbf{E}\{y \mid \mathscr{F}\}$. However, according to our definition, any ω function equal almost everywhere to this one is also a possible version of the conditional expectation. Conversely, according to the Radon-Nikodym theorem, any two versions of the conditional expectation are equal almost everywhere. Thus we have defined $\mathbf{E}\{y \mid \mathscr{F}\}$ as any one of a class of random variables. Any two random variables in the class are equal almost everywhere, and any random variable equal almost everywhere to a member of the class is itself in the class. In any expression involving a conditional expectation it will always be understood, unless the contrary is stated explicitly, that any one of the versions of the indicated conditional expectations can be used in the expression.

Let M be a measurable ω set, and let $\mathscr{F}$ be a Borel field of measurable ω sets. Let y be the random variable defined by

$$\begin{aligned} y(\omega) &= 1, \qquad \omega \in \mathrm{M} \\ &= 0, \qquad \omega \in \Omega - \mathrm{M}. \end{aligned}$$

Then the *conditional probability of* M *relative to* $\mathscr{F}$, *denoted by* $\mathbf{P}\{\mathrm{M} \mid \mathscr{F}\}$, is defined as $\mathbf{E}\{y \mid \mathscr{F}\}$, that is, as any one of the versions of this conditional expectation. The conditional probability is thus any ω function which is either measurable with respect to $\mathscr{F}$, or equal almost everywhere to an ω function which is, which is integrable, and which satisfies the equation

$$\int_\Lambda \mathbf{P}\{\mathrm{M} \mid \mathscr{F}\}\, d\mathbf{P} = \mathbf{P}\{\Lambda \mathrm{M}\}, \qquad \Lambda \in \mathscr{F}. \tag{7.7}$$

The preceding definitions are somewhat simplified if $\mathscr{F}$ includes all sets of measure 0, because in that case $\mathscr{F} = \mathscr{F}'$ and conditional probabilities and expectations relative to $\mathscr{F}$ are necessarily measurable relative to $\mathscr{F}$. However, we have seen that in every case there is a version of $\mathbf{E}\{y \mid \mathscr{F}\}$ which is measurable with respect to $\mathscr{F}$.

Now let $\{x_t,\ t \in T\}$ be any family of random variables. Let $\mathscr{F} = \mathscr{B}(x_t,\ t \in T)$ be the smallest Borel field of ω sets with respect to which the x_t's are measurable (that is, $\mathscr{F}$ is the Borel field generated by the class of sets of the form $\{x_t(\omega) \in A\}$ where A is a Borel set) and let $\mathscr{F}'$ be the Borel field of those ω sets which are either $\mathscr{F}$ sets or which differ from $\mathscr{F}$ sets by sets of probability 0. In this book we shall describe an ω set in $\mathscr{F}'$ as a *measurable set on the sample space of the* x_t's, and we shall describe an ω function measurable with respect to $\mathscr{F}'$, that is, one equal almost everywhere to a function measurable with respect to $\mathscr{F}$, as a *random variable on the space of the* x_t's. In particular, if T consists of the integers $1, \cdots, n$, the ω set Λ is measurable on the sample space of the x_t's if and only if it differs by at most an ω set of measure 0 from one of the form

$$\{[x_1(\omega), \cdots, x_n(\omega)] \in A\},$$

where A is a Borel set (n-dimensional in the real case, $2n$-dimensional in the complex case), and the ω function x is a random variable on the sample space of the x_t's if and only if x is equal almost everywhere to a Baire function of $x_1, \cdots, x_n$. (See Supplement, Theorem 1.5.)

Let y be any random variable whose expectation exists, and let M be a measurable ω set. The *conditional expectation* [*probability*] *of* y [M] *relative to the* x_t's, denoted by

$$\mathbf{E}\{y \mid x_t,\ t \in T\} \qquad [\mathbf{P}\{\mathrm{M} \mid x_t,\ t \in T\}],$$

is defined as

$$\mathbf{E}\{y \mid \mathscr{F}\} \qquad [\mathbf{P}\{\mathrm{M} \mid \mathscr{F}\}]$$

that is, as any version of the latter conditional expectation [probability]. Here $\mathscr{F}$ is defined as in the preceding paragraph. As always, we can replace $\mathscr{F}$ by $\mathscr{F}'$ in this definition. Thus the conditional expectation in question is defined as any random variable which is measurable with respect to the sample space of the x_t's and which has the same integral as y over every set measurable on the sample space of the x_t's. When $\mathscr{F}$ is defined in this way, (7.6) and (7.7) can be put in a slightly more convenient form by restricting the class of ω sets on which the equations are to hold. The left and right sides of these equations define completely additive functions of Λ, and such functions are completely determined by their values on any subfield $\mathscr{F}_0$ of $\mathscr{F}$ which generates $\mathscr{F}$. (See Supplement, Theorem 2.1.) In the present case $\mathscr{F}_0$ can be taken as the class of ω sets which are finite unions of sets of the form

$$\{x_{t_j}(\omega) \in X_j, j = 1, \cdots, n\}$$

where $(t_1, \cdots, t_n)$ is any finite subset of T and the X_j's are Borel sets. Thus it is sufficient if (7.6), or (7.7) as the case may be, is satisfied for Λ

of this type. Since the sides of (7.6) and (7.7) are additive in the integration set, it is sufficient to verify them for the above individual summands. If convenient we can even suppose that the X_j's are right semi-closed intervals (or open intervals, or closed intervals.)

In particular, suppose that T in the preceding paragraph consists of the integers $1, \cdots, k$. Then we have defined

$$\mathbf{E}\{y \mid x_1, \cdots, x_k\}, \qquad \mathbf{P}\{M \mid x_1, \cdots, x_k\}$$

in a way consistent with the earlier discussions of Cases 1 and 2. In fact (7.5), derived in Case 1, became the defining property (7.6) in the general case. Consider a version of the conditional expectation which is measurable with respect to $\mathscr{F} = \mathscr{B}(x_1, \cdots, x_k)$. Then we have seen that this version can be written in the form

$$\mathbf{E}\{y \mid x_1, \cdots, x_k\} = \Phi(x_1, \cdots, x_k),$$

where Φ is a Baire function of k variables (Supplement, Theorem 1.5). If such a version is used, we sometimes write

$$\mathbf{E}\{y \mid x_j(\omega) = \xi_j, \qquad j = 1, \cdots, k\}$$

for

$$\mathbf{E}\{y \mid x_1, \cdots, x_k\}\big|_{x_j(\omega)=\xi_j,\, j=1, \cdots, k} = \Phi(\xi_1, \cdots, \xi_k).$$

In particular, if we use a version of $\mathbf{P}\{M \mid x_1, \cdots, x_k\}$ which is a Baire function of $x_1, \cdots, x_k$, we shall sometimes write

$$\mathbf{P}\{M \mid x_j(\omega) = \xi_j, j = 1, \cdots, k\}$$

for

$$\mathbf{P}\{M \mid x_1, \cdots, x_k\}\big|_{x_j(\omega)=\xi_j,\, j=1, \cdots, k}.$$

The discussion we have given justifies the common description of the random variables

$$\mathbf{E}\{y \mid x_t, t \in T\}, \qquad \mathbf{P}\{M \mid x_t, t \in T\}$$

as the conditional expectation of y and probability of M respectively *for given values of the* x_t's, or *for given* $x_t(\omega)$, $t \in T$.

8. Conditional probabilities and expectations: general properties

Let y be a random variable whose expectation exists, and let $\mathscr{F}$, $\mathscr{G}$ be Borel fields of measurable ω sets. Let $\mathscr{F}'$ [$\mathscr{G}'$] be the Borel field of those ω sets which are either $\mathscr{F}$ [$\mathscr{G}$] sets or which differ from such sets by sets of probability 0. Suppose that $\mathscr{G}' \subset \mathscr{F}'$. Then $\mathbf{E}\{y \mid \mathscr{F}\}$ and $\mathbf{E}\{y \mid \mathscr{G}\}$

are not necessarily equal with probability 1. The second conditional expectation is a coarser averaging than the first. More precisely, the two conditional expectations have the same integrals as y over $\mathscr{G}'$ sets, but the first one is not necessarily measurable with respect to $\mathscr{G}'$. However, if the first one happens to be measurable with respect to $\mathscr{G}'$, the two conditional expectations are equal with probability 1, according to the following theorem.

THEOREM 8.1 *Suppose that $\mathscr{G}' \subset \mathscr{F}'$ and that some (and therefore every) version of* $\mathbf{E}\{y \mid \mathscr{F}\}$ *is measurable with respect to $\mathscr{G}'$. Then*

$$\mathbf{E}\{y \mid \mathscr{F}\} = \mathbf{E}\{y \mid \mathscr{G}\} \tag{8.1}$$

with probability 1.

To prove (8.1) we need only remark that $\mathbf{E}\{y \mid \mathscr{F}\}$ is measurable with respect to $\mathscr{G}'$ by hypothesis, and has the same integral over $\mathscr{G}'$ sets as y, since it even has the same integral over $\mathscr{F}'$ sets as y. Thus $\mathbf{E}\{y \mid \mathscr{F}\}$ satisfies the conditions determining $\mathbf{E}\{y \mid \mathscr{G}\}$.

THEOREM 8.2 *Let $\{x_t, t \in T\}$ be a family of random variables, and suppose that T is non-denumerable. Then, if y is a random variable whose expectation exists, there is a denumerable subaggregate $\{t_n, n \geq 1\}$ of T (depending on y) such that*

$$\mathbf{E}\{y \mid x_t, t \in T\} = \mathbf{E}\{y \mid x_{t_1}, x_{t_2}, \cdots\} \tag{8.2}$$

with probability 1.

If S is any subset of T, let $\mathscr{F}_S = \mathscr{B}(x_t, t \in S)$ be the smallest Borel field with respect to which the x_t's with $t \in S$ are measurable. The left side of (8.2) is by definition a version of $\mathbf{E}\{y \mid \mathscr{F}_T\}$. From now on we shall denote by $\mathbf{E}\{y \mid \mathscr{F}_T\}$ a particular version of this conditional expectation which is measurable with respect to $\mathscr{F}_T$. By Theorem 1.6 of the Supplement, since this conditional expectation is measurable with respect to $\mathscr{F}_T$, there is a denumerable subset S of T such that $\mathbf{E}\{y \mid \mathscr{F}_T\}$ is measurable with respect to $\mathscr{F}_S \subset \mathscr{F}_T$. Then by Theorem 8.1

$$\mathbf{E}\{y \mid \mathscr{F}_T\} = \mathbf{E}\{y \mid \mathscr{F}_S\}$$

with probability 1, as was to be proved.

If $\mathscr{F}$ is a Borel field of measurable ω sets, if z is an ω function measurable with respect to $\mathscr{F}$, and if $\mathbf{E}\{|z|\} < \infty$, then

$$\mathbf{E}\{z \mid \mathscr{F}\} = z$$

with probability 1. In fact, z has the defining properties of the indicated conditional expectation. More generally, we prove the following theorem.

THEOREM 8.3 *If y is a random variable, if z is an ω function measurable with respect to the Borel field $\mathscr{F}$ of measurable ω sets, and if*

$$\mathbf{E}\{|y|\} < \infty, \qquad \mathbf{E}\{|zy|\} < \infty,$$

then

$$\mathbf{E}\{zy \mid \mathscr{F}\} = z\mathbf{E}\{y \mid \mathscr{F}\} \tag{8.3}$$

with probability 1, *and*

$$\mathbf{E}\{[y - \mathbf{E}\{y \mid \mathscr{F}\}]z\} = 0. \tag{8.3'}$$

Equation (8.3′) is a trivial consequence of (8.3). Note that its validity for z taking on the values 0 and 1 is the defining property of the conditional expectation $\mathbf{E}\{y \mid \mathscr{F}\}$. To prove (8.3), we have only to prove, according to the definition of conditional expectation, that

$$\int_{\Lambda} zy\, d\mathbf{P} = \int_{\Lambda} z\mathbf{E}\{y \mid \mathscr{F}\}\, d\mathbf{P}, \qquad \Lambda \in \mathscr{F}. \tag{8.4}$$

Now if $z(\omega)$ is 1 or 0 according as ω is or is not a point of $M \in \mathscr{F}$, this equation becomes

$$\int_{\Lambda M} y\, d\mathbf{P} = \int_{\Lambda M} \mathbf{E}\{y \mid \mathscr{F}\}\, d\mathbf{P},$$

and this equation is true by the definition of the integrand on the right. It follows at once that (8.4) is true if z is a linear combination of functions of the type just considered, that is, if z takes on only a finite number of values, each on an $\mathscr{F}$ set, and the general case follows using the usual approximation procedure.

The most useful particular case of (8.3) is

$$\mathbf{E}\{\Phi(x)y \mid x\} = \Phi(x)\mathbf{E}\{y \mid x\} \tag{8.3''}$$

with probability 1, where x and y are random variables, Φ is a Baire function, and

$$\mathbf{E}\{|y|\} < \infty, \qquad \mathbf{E}\{|\Phi(x)y|\} < \infty.$$

In §7 the conditional expectation of y was defined as the integral of y with respect to a conditional probability measure in two simple cases (Cases 1 and 2). Although, as we shall see, this definition is not always possible, because $\mathbf{P}\{M \mid \mathscr{F}\}$ cannot always be considered a probability measure in M for fixed ω, nevertheless $\mathbf{E}\{y \mid \mathscr{F}\}$ considered as a functional of y has many of the properties of an integral. The following facts illustrate this assertion. (The y's are random variables and $\mathscr{F}$ is a Borel field of ω sets throughout.)

CE_1 $\qquad \mathbf{E}\{1 \mid \mathscr{F}\} = 1$ with probability 1.

CE_2 If $y \geq 0$, then $\mathbf{E}\{y \mid \mathscr{F}\} \geq 0$ with probability 1.

CE_3 If $c_1, \cdots, c_n$ are constants,

$$\mathbf{E}\{\sum_{j=1}^{n} c_j y_j \mid \mathscr{F}\} = \sum_{j=1}^{n} c_j \mathbf{E}\{y_j \mid \mathscr{F}\},$$

with probability 1.

CE_4

$$|\mathbf{E}\{y \mid \mathscr{F}\}| \leq \mathbf{E}\{|y| \mid \mathscr{F}\}$$

with probability 1.

CE_5 If $\lim_{n\to\infty} y_n = y$ with probability 1, and if there is a random variable $x \geq 0$, with $\mathbf{E}\{x\} < \infty$, such that

$$|y_n(\omega)| \leq x(\omega)$$

with probability 1, then

$$\lim_{n\to\infty} \mathbf{E}\{y_n \mid \mathscr{F}\} = \mathbf{E}\{y \mid \mathscr{F}\}$$

with probability 1.

In §9 will be shown how to derive these results from the corresponding integration theorems, using representation theory. It may be instructive to derive them directly here, however. Properties CE_1, CE_2, CE_3 are immediate consequences of the conditional expectation definition. To prove CE_4 we suppose first that y is real. Then, using CE_2,

$$\mathbf{E}\{|y| - y \mid \mathscr{F}\} \geq 0,$$
$$\mathbf{E}\{|y| + y \mid \mathscr{F}\} \geq 0,$$

with probability 1. Hence, using CE_3,

$$\mathbf{E}\{y \mid \mathscr{F}\} \leq \mathbf{E}\{|y| \mid \mathscr{F}\}$$
$$-\mathbf{E}\{y \mid \mathscr{F}\} = \mathbf{E}\{-y \mid \mathscr{F}\} \leq \mathbf{E}\{|y| \mid \mathscr{F}\}$$

with probability 1, so that CE_4 is true. If y is complex-valued, choose the random variable z to satisfy

$$z(\omega) = 0, \qquad \text{if } \mathbf{E}\{y \mid \mathscr{F}\} = 0,$$
$$z(\omega)\mathbf{E}\{y \mid \mathscr{F}\} = |\mathbf{E}\{y \mid \mathscr{F}\}|, \qquad \text{if } \mathbf{E}\{y \mid \mathscr{F}\} \neq 0.$$

(Here $\mathbf{E}\{y \mid \mathscr{F}\}$ is a particular version of this conditional expectation, held fast in the following.) Let z_1 be the real part of zy. Then, using the fact that z is measurable with respect to $\mathscr{F}$,

$$|\mathbf{E}\{y \mid \mathscr{F}\}| = z\mathbf{E}\{y \mid \mathscr{F}\} = \mathbf{E}\{zy \mid \mathscr{F}\} = \mathbf{E}\{z_1 \mid \mathscr{F}\}$$

with probability 1. Since z_1 is real, we can apply the real case of CE_4, already proved, to continue this inequality, obtaining, since

$$|z_1(\omega)| \leq |y(\omega)|,$$

the desired inequality

$$|\mathbf{E}\{y \mid \mathscr{F}\}| \leq \mathbf{E}\{|z_1| \mid \mathscr{F}\} \leq \mathbf{E}\{|y| \mid \mathscr{F}\}$$

(with probability 1). To prove CE_5, define $\hat{y}_n$ by

$$\hat{y}_n(\omega) = \underset{j \geq n}{\text{L.U.B.}}\ |y_j(\omega) - y(\omega)|.$$

Then

$$\hat{y}_1(\omega) \geq \hat{y}_2(\omega) \geq \cdots \geq 0, \qquad \hat{y}_n(\omega) \leq 2x(\omega),$$

with probability 1, and

$$\lim_{n \to \infty} \hat{y}_n(\omega) = 0$$

with probability 1. Because of CE_2 and CE_3,

$$\begin{aligned} |\mathbf{E}\{y \mid \mathscr{F}\} - \mathbf{E}\{y_n \mid \mathscr{F}\}| = |\mathbf{E}\{y - y_n \mid \mathscr{F}\}| &\leq \mathbf{E}\{|y - y_n| \mid \mathscr{F}\} \\ &\leq \mathbf{E}\{\hat{y}_n \mid \mathscr{F}\} \end{aligned}$$

with probability 1. Hence it is sufficient to prove that

$$\lim_{n \to \infty} \mathbf{E}\{\hat{y}_n \mid \mathscr{F}\} = 0$$

with probability 1. We have, from CE_2 and CE_3,

$$\mathbf{E}\{\hat{y}_1 \mid \mathscr{F}\} \geq \mathbf{E}\{\hat{y}_2 \mid \mathscr{F}\} \geq \cdots \geq 0$$

with probability 1, so that there must be convergence,

$$\lim_{n \to \infty} \mathbf{E}\{\hat{y}_n \mid \mathscr{F}\} = w$$

with probability 1. Moreover, by definition of conditional expectation [(7.6) with $\Lambda = \Omega$],

$$\mathbf{E}\{w\} \leq \mathbf{E}\{\mathbf{E}\{\hat{y}_n \mid \mathscr{F}\}\} = \mathbf{E}\{\hat{y}_n\},$$

and the right side goes to 0 when $n \to \infty$ because it is the integral of $\hat{y}_n$, where $\hat{y}_n$ is dominated by $2x$, and goes to 0 with probability 1 when $n \to \infty$. Thus $\mathbf{E}\{w\} = 0$, so that $w = 0$ with probability 1, as was to be proved.

The listed properties of conditional expectations imply the following properties of conditional probabilities. (The M's are measurable ω sets and $\mathscr{F}$ is a Borel field of measurable ω sets.)

CP_1
$$0 \leq \mathbf{P}\{M \mid \mathscr{F}\} \leq 1$$
with probability 1.

CP_2 If $\mathbf{P}\{M\} = 0$, then $\mathbf{P}\{M \mid \mathscr{F}\} = 0$ with probability 1;

if $\mathbf{P}\{M\} = 1$, then $\mathbf{P}\{M \mid \mathscr{F}\} = 1$ with probability 1.

CP_3 If either
$$M_1 \supset M_2 \supset \cdots, \quad \bigcap_n M_n = M$$
or
$$M_1 \subset M_2 \subset \cdots, \quad \bigcup_n M_n = M,$$
then
$$\lim_{n\to\infty} \mathbf{P}\{M_n \mid \mathscr{F}\} = \mathbf{P}\{M \mid \mathscr{F}\}$$
with probability 1.

CP_4 If $M_1, M_2, \cdots$ are disjunct, and finite or denumerably infinite in number,
$$\mathbf{P}\{\bigcup_n M_n \mid \mathscr{F}\} = \sum_n \mathbf{P}\{M_n \mid \mathscr{F}\}$$
with probability 1.

With the help of the listed properties of conditional probabilities and expectations, we can derive a suggestive evaluation of a conditional expectation.

THEOREM 8.4 *Let y be a random variable whose expectation exists, let $\mathscr{F}$ be a Borel field of measurable ω sets, and let δ be a positive number. Then the sum*
$$x_\delta = \sum_{-\infty}^{\infty} (j+1)\delta \mathbf{P}\{j\delta < y(\omega) \leq (j+1)\delta \mid \mathscr{F}\}$$
is absolutely convergent, with probability 1, *and if* $\lim_{k\to\infty} \delta_k = 0$ *it follows that*
$$\lim_{k\to\infty} x_{\delta_k} = \mathbf{E}\{y \mid \mathscr{F}\}$$
with probability 1.

Note that this theorem does *not* state that conditional probabilities relative to $\mathscr{F}$ determine a conditional probability measure for fixed ω, and that $\mathbf{E}\{y \mid \mathscr{F}\}$ is the integral of y with respect to the corresponding probability measure. The possibility of such an interpretation is discussed in §9.

To prove the theorem define y_δ and $y_{n,\delta}$ by

$$\begin{aligned} y_\delta(\omega) &= j\delta && \text{where } j\delta < y(\omega) \leq (j+1)\delta, \quad j = 0, \pm 1, \cdots \\ y_{n,\delta}(\omega) &= y_\delta(\omega) && \text{where } -n\delta < y(\omega) \leq (n+1)\delta \\ &= 0 && \text{otherwise.} \end{aligned}$$

Then

$$\lim_{n\to\infty} y_{n,\delta} = y_\delta, \qquad \lim_{\delta\to 0} y_\delta = y,$$

$$|y_{n,\delta}(\omega)| \leq y_\delta(\omega)| \leq |y(\omega)| + \delta.$$

According to CE_5,

$$\lim_{n\to\infty} \mathbf{E}\{y_{n,\delta} \mid \mathscr{F}\} = \mathbf{E}\{y_\delta \mid \mathscr{F}\}$$

with probability 1. Since

$$\mathbf{E}\{y_{n,\delta} \mid \mathscr{F}\} = \sum_{-n}^{n} j\delta\, \mathbf{P}\{j\delta < y(\omega) \leq (j+1)\delta \mid \mathscr{F}\}$$

with probability 1, we have proved that

$$\lim_{n\to\infty} \sum_{-n}^{n} j\delta\, \mathbf{P}\{j\delta < y(\omega) \leq (j+1)\delta \mid \mathscr{F}\} = \mathbf{E}\{y_\delta \mid \mathscr{F}\}$$

with probability 1. Applying this result to $|y|$, we deduce that the infinite series whose nth partial sum appears on the left in the preceding equation converges absolutely, with probability 1, and that therefore

$$\sum_{-\infty}^{\infty} j\delta\, \mathbf{P}\{j\delta < y(\omega) \leq (j+1)\delta \mid \mathscr{F}\} = \mathbf{E}\{y_\delta \mid \mathscr{F}\},$$

with probability 1. The last assertion of Theorem 8.4 is now an immediate consequence of CE_5 applied to the sequence $\{y_{\delta_k}\}$.

9. Conditional probability distributions

It would be very convenient in the theory of probability if corresponding to each Borel field $\mathscr{F}$ of measurable ω sets there were a function defined for every measurable ω set M and point ω, with value $P(\mathrm{M}, \omega)$ at M, ω, such that

CD_1 For each ω, $P(\mathrm{M}, \omega)$ defines a probability measure of M, and, for each M, $P(\mathrm{M}, \omega)$ defines an ω function equal almost everywhere to one measurable with respect to $\mathscr{F}$.

CD_2 For each M,

$$P(\mathrm{M}, \omega) = \mathbf{P}\{\mathrm{M} \mid \mathscr{F}\},$$

with probability 1.

These two properties imply CP_1–CP_5 of §8, but are stronger.

If there is such an M, ω function, it is not necessarily uniquely determined, but any two such functions are equal with probability 1 for fixed M. When there is a function satisfying CD_1 and CD_2, it means that the conditional probabilities (conditioned by $\mathscr{F}$) can be defined in such a way that they determine a new probability measure for each ω, and this probability measure, a function of the parameter ω, is called the *conditional probability distribution relative to* $\mathscr{F}$. Unfortunately there may be no such conditional probability distribution.

As an example of a case where there is such a distribution, suppose that $\Lambda_1, \cdots, \Lambda_n$ are disjunct measurable ω sets with

$$\mathbf{P}\{\Lambda_j\} > 0, \qquad \bigcup_1^n \Lambda_j = \Omega.$$

Let $\mathscr{F}$ be the class of sets which are unions of Λ_j's. Then one possible version of $\mathbf{P}\{M \mid \mathscr{F}\}$ is given by

$$P\{M, \omega\} = \frac{\mathbf{P}\{M\Lambda_j\}}{\mathbf{P}\{\Lambda_j\}}, \qquad \omega \in \Lambda_j, \quad j = 1, \cdots, n.$$

So defined, $P(M, \omega)$ satisfies CD_1 and CD_2.

THEOREM 9.1 *If there is a conditional distribution relative to* $\mathscr{F}$, *then if* y *is any random variable whose expectation exists, one version of* $\mathbf{E}\{y \mid \mathscr{F}\}$ *is given by the integral of* y *with respect to this conditional distribution* (*as a function of* ω); *making the obvious notational conventions*

$$\mathbf{E}\{y \mid \mathscr{F}\} = \int_\Omega y(\omega')\, \mathbf{P}(d\omega', \omega).$$

Let $\mathscr{H}$ be the class of random variables y for which this assertion is true. Then $\mathscr{H}$ includes each random variable which is 1 on a measurable ω set M and vanishes otherwise, according to CD_2. Evidently $\mathscr{H}$ is a linear class, that is, it includes all linear combinations of its elements. Then $\mathscr{H}$ includes the random variables which only take on finitely many values. Finally, with the help of CE_5 of §8, we deduce that $\mathscr{H}$ includes each random variable that is the limit of a sequence of random variables $y_1, y_2, \cdots$ in $\mathscr{H}$, with the property that there is a random variable x such that

$$|y_n(\omega)| \le x(\omega), \qquad \mathbf{E}\{x\} < \infty.$$

Then $\mathscr{H}$ includes every random variable y whose expectation exists, as was to be proved.

Let $y_1, \cdots, y_n$ be random variables, and let $\mathscr{F}_y = \mathscr{B}(y_1, \cdots, y_n)$ be the smallest Borel field of ω sets with respect to which $y_1, \cdots, y_n$ are measurable. Suppose that there is an M, ω function defined for $M \in \mathscr{F}_y$, and all ω, and satisfying CD$_1$ and CD$_2$ for $M \in \mathscr{F}_y$. If Y is an n-dimensional Borel set ($2n$-dimensional if the y_j's are complex), define

$$p(Y, \omega) = P(\mathrm{M}, \omega), \qquad \mathrm{M} = \{[y_1(\omega), \cdots, y_n(\omega)] \in Y\}.$$

Then $p(Y, \omega)$ defines a probability measure of Borel sets, depending on the parameter ω. The probability measure of $\mathscr{F}_y$ sets determined by $P(\mathrm{M}, \omega)$ is called the *conditional probability distribution of the y_j's relative to $\mathscr{F}$.*

THEOREM 9.2 *Suppose that there is a conditional distribution of $y_1, \cdots, y_n$ relative to $\mathscr{F}$, and let Φ be a Baire function of n variables with $\mathbf{E}\{|\Phi(y_1, \cdots, y_n)|\} < \infty$. Then one version of $\mathbf{E}\{\Phi(y_1, \cdots, y_n) \mid \mathscr{F}\}$ is given by the ω integral of $\Phi(y_1, \cdots, y_n)$ with respect to the conditional probability distribution of the y_j's relative to $\mathscr{F}$.*

To prove this theorem let $\mathscr{H}$ be the class of Baire functions Φ for which the assertion of the theorem is true. Then, by definition of the conditional distribution, $\mathscr{H}$ includes each function which is 1 on a Borel set in n dimensions ($2n$ dimensions if the y_j's are complex) and 0 otherwise, and the proof is then carried through like that of the preceding theorem. The particular case $n = 1$, $\Phi(x) \equiv x$, is also trivially deducible from Theorem 8.4.

THEOREM 9.3 *If $P_1(M, \omega)$, $P_2(M, \omega)$ define conditional probability distributions of $y_1, \cdots, y_n$ relative to $\mathscr{F}$, and if $p_i(Y, \omega)$ is defined as above by*

$$p_i(Y, \omega) = P_i(M, \omega),$$

then there is an ω set Λ_0 (which does not depend on Y), of probability 0, such that

$$p_1(Y, \omega) = p_2(Y, \omega), \qquad \omega \notin \Lambda_0.$$

If Y is a Borel set,

$$\mathbf{P}\{[y_1(\omega), \cdots, y_n(\omega)] \in Y \mid \mathscr{F}\} = p_1(Y, \omega) = p_2(Y, \omega)$$

with probability 1. There is therefore an ω set $\Lambda(Y)$, of probability 0, such that

$$p_1(Y, \omega) = p_2(Y, \omega), \qquad \omega \notin \Lambda(Y).$$

Now let $Y_1, Y_2, \cdots$ be an enumeration of the intervals with rational vertices (n-dimensional intervals if the y_j's are real, $2n$-dimensional in the complex case). Define $\Lambda_0 = \bigcup_1^\infty \Lambda(Y_n)$. Then

$$p_1(Y, \omega) = p_2(Y, \omega), \qquad \omega \notin \Lambda_0$$

if Y is a Y_j, and therefore if Y is any interval. The theorem then follows from the fact that two measures of Borel sets are identical if they are equal when the sets are intervals.

We have not yet discussed conditions which insure the existence of a conditional probability distribution of the random variables $y_1, \cdots y_n$ relative to a Borel field. We shall first obtain a preliminary result. Let $\mathscr{F}_y$ be the smallest Borel field of ω sets with respect to which the y_j's are measurable. In the following, Y will denote a Borel set in n dimensions (or in $2n$ dimensions if the y_j's are complex-valued). We shall call a Y, ω function p a *conditional probability distribution of the y_j's in the wide sense, relative to $\mathscr{F}$* if p defines an ω function equal almost everywhere to one measurable with respect to $\mathscr{F}$ when Y is fixed, and a probability measure of Y when ω is fixed, and if, for each Y,

$$\mathbf{P}\{[y_1(\omega), \cdots, y_n(\omega)] \in Y \mid \mathscr{F}\} = p(Y, \omega) \tag{9.1}$$

with probability 1. Since the ω set involved in the conditional probability on the left does not uniquely determine Y, the function p does not define for each ω a probability measure of $\mathscr{F}_y$ sets. That is, the existence of p does not guarantee the existence of a conditional distribution of $y_1, \cdots, y_n$ relative to $\mathscr{F}$. However, if p exists, and if Φ is any Baire function of n variables, with

$$\mathbf{E}\{|\Phi(y_1, \cdots, y_n)|\} < \infty,$$

then

$$\mathbf{E}\{\Phi(y_1, \cdots, y_n) \mid \mathscr{F}\} = \int_{-\infty}^{\infty} \cdots \int_{-\infty}^{\infty} \Phi(\eta_1, \cdots, \eta_n)\, p(d\eta, \omega), \tag{9.2}$$

with probability 1, where the integral on the right is, for each ω, an ordinary integral in n dimensions. The proof is exactly the same as that of Theorem 9.2. The present result differs from that of Theorem 9.2 in that the earlier result involves integration in ω space. We shall see that the existence of a conditional distribution of $y_1, \cdots, y_n$ in the wide sense is almost as useful as that of a conditional distribution. Like the latter, the former is not uniquely determined, but if p_1 and p_2 are both conditional distributions in the wide sense there is an ω set of probability 0 such that is ω is not in this set

$$p_1(Y, \omega) = p_2(Y, \omega)$$

for all Y. The proof follows that of Theorem 9.3.

THEOREM 9.4 *Let $y_1, \cdots, y_n$ be any random variables, and let $\mathscr{F}$ be any Borel field of measurable ω sets. Then there is a conditional distribution p of $y_1, \cdots, y_n$ in the wide sense, relative to $\mathscr{F}$, such that $p(Y, \omega)$, for fixed Y, defines a function measurable relative to $\mathscr{F}$.*

Note that, according to this theorem, if $x_1, \cdots, x_m$ are any random variables, the conditional distribution relative to $x_1, \cdots, x_m$ can be taken as a function of Y and the x_j's which determines a Baire function of the latter variables when Y is fixed. This follows from the theorem, if we take $\mathscr{F}$ to be the smallest Borel field of ω sets with respect to which the x_j's are measurable, so that the functions measurable with respect to $\mathscr{F}$ are the Baire functions of the x_j's.

For definiteness, we give the proof for real y_j's. It will be obvious that the theorem can be formulated, and proved similarly, for enumerably many y_j's. In the following, F will be any distribution function in n dimensions; it will remain fixed throughout the discussion. We must define the function value $p(Y, \omega)$ for Y an n-dimensional Borel set and $\omega \in \Omega$. The definition is given first for Y an interval of the form

$$A_{\lambda_1, \cdots, \lambda_n} = \{-\infty < \xi_j \leq \lambda_j, \quad j = 1, \cdots, n\}.$$

For each n-tuple of rational numbers $(\lambda_1, \cdots, \lambda_n)$ choose a version of the conditional probability $\mathbf{P}\{y_j(\omega) \leq \lambda_j, j = 1, \cdots, n \mid \mathscr{F}\}$ which is measurable with respect to $\mathscr{F}$. It is easily seen, using CP_1–CP_4 of §8, that there is an ω set $\Lambda \in \mathscr{F}$ of probability 0 such that, if $\omega \notin \Lambda$, the function of (rational) $\lambda_1, \cdots, \lambda_n$ determined by this conditional probability coincides on the set of rational points in n dimensions with some distribution function in n dimensions. That is, for $\omega \notin \Lambda$ this function of rational $\lambda_1, \cdots, \lambda_n$ is monotone non-decreasing and continuous on the right in each variable and so on (see §3). Define

$$\begin{aligned} p(A_{\lambda_1, \cdots, \lambda_n}, \omega) &= \mathbf{P}\{y_j(\omega) \leq \lambda_j, j = 1, \cdots, n \mid \mathscr{F}\} \qquad \omega \notin \Lambda, \\ &= F(\lambda_1, \cdots, \lambda_n), \qquad \omega \in \Lambda, \end{aligned}$$

for rational λ_j's, using the above versions of the conditional probabilities. If the λ_j's are not all rational, define

$$p(A_{\lambda_1, \cdots, \lambda_n}, \omega) = \lim_{\mu_j \downarrow \lambda_j, j = 1, \cdots, n} p(A_{\mu_1, \cdots, \mu_n}, \omega)$$

(where the μ_j's are rational). Then $p(A_{\lambda_1, \cdots, \lambda_n}, \omega)$ for fixed ω determines a distribution function in $\lambda_1, \cdots, \lambda_n$, and thereby determines a probability measure of n-dimensional Borel sets Y. Let $p(Y, \omega)$ be the measure assigned to Y in this way. We conclude the proof of the theorem by proving that, for each Y, $p(Y, \omega)$ defines an ω function measurable with respect to $\mathscr{F}$, and that (9.1) is then true, with probability 1. [The exceptional ω set in (9.1) will depend on Y and on the choice of the conditional probability.] The assertion is true, by definition of p, if Y is an interval $A_{\lambda_1, \cdots, \lambda_n}$. Hence, applying CP_4 of §8, the assertion is true if Y is a right semi-closed interval, finite or infinite, and therefore if Y is a finite union of such intervals. Finally, the class of sets Y for which the assertion is true includes, applying CP_3 of §8, the limits of monotone sequences of sets in the class. Hence the class includes all Borel sets Y, as was to be proved.

We now prove that, under rather wide conditions, there is actually a conditional probability distribution of $y_1, \cdots, y_n$ relative to $\mathscr{F}$. The condition is one on the range R of $y_1, \cdots, y_n$, the n-dimensional set ($2n$-dimensional in the complex case) of points $[y_1(\omega), \cdots, y_n(\omega)]$ as ω ranges through its whole space.

THEOREM 9.5 *Let $y_1, \cdots, y_n$ be random variables, and let $\mathscr{F}$ be a Borel field of measurable ω sets. Then, if the range of $y_1, \cdots, y_n$ is a Borel set, there is a conditional distribution of $y_1, \cdots, y_n$ relative to $\mathscr{F}$, such that $P(\mathrm{M}, \omega)$, for fixed* M, *defines a function measurable relative to $\mathscr{F}$.*

We give the proof for real y_j's. In the following, $\mathscr{F}_y$ is as above, and q is any probability measure of $\mathscr{F}_y$ sets, fixed throughout the discussion. The fact that there is such a q will appear in the course of the proof. Let p be a conditional distribution (wide sense) of $y_1, \cdots, y_n$ relative to $\mathscr{F}$ as in Theorem 9.4. Then, applying (9.1), we find that

$$1 = \mathbf{P}\{\Omega \mid \mathscr{F}\} = \mathbf{P}\{[y_1(\omega), \cdots, y_n(\omega)] \in R \mid \mathscr{F}\} = p(R, \omega), \tag{9.3}$$

with probability 1. If $\mathrm{M} \in \mathscr{F}_y$, and if there are two Borel sets Y_1, Y_2 such that

$$\mathrm{M} = \{[y_1(\omega), \cdots, y_n(\omega)] \in Y_1\} = \{[y_1(\omega), \cdots, y_n(\omega)] \in Y_2\}$$

then $Y_1 - Y_1Y_2$ and $Y_2 - Y_1Y_2$ are subsets of the complement of R. Hence

$$p(Y_1, \omega) = p(Y_2, \omega) = p(Y_1Y_2, \omega), \qquad \text{if } p(R, \omega) = 1.$$

Thus the following definition is unique:

$$P(\mathrm{M}, \omega) = p(Y, \omega), \qquad \mathrm{M} = \{[y_1, (\omega), \cdots, y_n(\omega)] \in Y\}, \qquad \text{if } p(R, \omega) = 1. \tag{9.4}$$

This definition makes $P(\mathrm{M}, \omega)$ define a probability measure in M for fixed ω, showing incidentally that such a probability measure in $\mathrm{M} \in \mathscr{F}_y$ exists. Finally set

$$P(\mathrm{M}, \omega) = q(\mathrm{M}) \qquad \text{if } p(R, \omega) < 1. \tag{9.5}$$

The function P defined in this way is the desired conditional probability distribution.

The condition of Theorem 9.5, that the range of $y_1, \cdots, y_n$ be a Borel set, is a useful condition. It is always satisfied, for example, if $y_1, \cdots, y_n$ are discrete random variables, that is, if R is a finite or enumerable set. (In this case, of course the proof can be simplified to the point of triviality.) At the other extreme, it is always satisfied if R is the whole n-dimensional space ($2n$-dimensional space in the complex case). This is the most important special case and is especially useful because, when the representation theory of §6 is used, the random variables under discussion all become coordinate variables in a multidimensional coordinate space, so that R is the whole space. In order to make full use of Theorem 9.5 in this connection we now investigate the transformation of conditional probabilities and expectations in going from random variables to their representations. Since conditional probabilities are special cases of conditional expectations, we shall discuss only the latter. Suppose that y, x_t are random variables, for $t \in T$, and that these random variables are represented by coordinate variables $\tilde{y}$, $\tilde{x}_t$ of a coordinate space. It is supposed that $\mathbf{E}\{|y|\} < \infty$. If T consists of the integers $1, \cdots, n$ we thus represent $y, x_1, \cdots, x_n$ by the coordinate variables $\tilde{y}, \tilde{x}_1, \cdots, \tilde{x}_n$ of $(n+1)$-dimensional space, or $(2n+2)$-dimensional space in the complex case. We have seen that the conditional expectation $\mathbf{E}\{y \mid x_1, \cdots, x_n\}$ can be taken as a certain Baire function Φ of $x_1, \cdots, x_n$. The random variable $\tilde{y}$ defined on $\tilde{\omega}$ space has the same distribution as y. Hence we have $\tilde{\mathbf{E}}\{|\tilde{y}|\} < \infty$, and we can consider the conditional expectation $\tilde{\mathbf{E}}\{\tilde{y} \mid \tilde{x}_1, \cdots, \tilde{x}_n\}$. Let

$$\mathscr{F} = \mathscr{B}(x_1, \cdots, x_n) \qquad [\tilde{\mathscr{F}} = \mathscr{B}(\tilde{x}_1, \cdots, \tilde{x}_n)]$$

be the smallest Borel field of ω [$\tilde{\omega}$] sets with respect to which the x_j's [$\tilde{x}_j$'s] are measurable. Now

$$\int_{\Lambda} y \, d\mathbf{P} = \int_{\Lambda} \Phi(x_1, \cdots, x_n) \, d\mathbf{P}, \qquad \Lambda \in \mathscr{F},$$

by definition of conditional expectation. Then

$$\int_{\tilde{\Lambda}} \tilde{y} \, d\tilde{\mathbf{P}} = \int_{\tilde{\Lambda}} \Phi(\tilde{x}_1, \cdots, \tilde{x}_n) \, d\tilde{\mathbf{P}}, \qquad \tilde{\Lambda} \in \tilde{\mathscr{F}}$$

in view of the properties of the ω to $\tilde{\omega}$ transformation listed in §6, that is, $\Phi(\tilde{x}_1, \cdots, \tilde{x}_n)$ is one version of $\tilde{\mathbf{E}}\{\tilde{y} \mid \tilde{x}_1, \cdots, \tilde{x}_n\}$. More generally, for arbitrary T, each ω random variable corresponds to an $\tilde{\omega}$ random variable in such a way that corresponding random variables have the same integrals in their respective spaces, and that x_t corresponds to $\tilde{x}_t$. Then if $\mathscr{F}$ [$\tilde{\mathscr{F}}$] is the smallest Borel field with respect to which the x_t's [$\tilde{x}_t$'s] are measurable, the $\mathscr{F}$ sets correspond to the $\tilde{\mathscr{F}}$ sets. The functions $\mathbf{E}\{y \mid x_t, t \in T\}$, $\tilde{\mathbf{E}}\{\tilde{y} \mid \tilde{x}_t, t \in T\}$ are determined by the equations

$$\int_{\Lambda} y \, d\mathbf{P} = \int_{\Lambda} \mathbf{E}\{y \mid x_t, t \in T\} \, d\mathbf{P}, \qquad \Lambda \in \mathscr{F},$$

$$\int_{\tilde{\Lambda}} \tilde{y} \, d\tilde{\mathbf{P}} = \int_{\tilde{\Lambda}} \tilde{\mathbf{E}}\{\tilde{y} \mid \tilde{x}_t, t \in T\} \, d\tilde{\mathbf{P}}, \qquad \tilde{\Lambda} \in \tilde{\mathscr{F}}.$$

Since the ω integral of an ω function on an ω set is equal to the $\tilde{\omega}$ integral of the corresponding $\tilde{\omega}$ function on the corresponding $\tilde{\omega}$ set, it follows that the two conditional expectations correspond to each other in the ω $\tilde{\omega}$ correspondence.

If y is a random variable whose expectation exists, and if $\mathscr{F}$ is an arbitrary Borel field of ω sets, we have not yet explained how to apply representation theory to the study of $\mathbf{E}\{y \mid \mathscr{F}\}$ unless, as above, $\mathscr{F}$ is the smallest Borel field of ω sets with respect to which the random variables of some family are measurable. However, the general $\mathscr{F}$ is easily expressed in this way; the general random variable of the family can be taken to be the function which is 1 on an $\mathscr{F}$ set and 0 otherwise.

We exhibit the application of the preceding discussion to the proof of an important inequality. Let y be a real random variable, and suppose that f is a continuous convex function of a single real variable, defined in an interval. Then, according to Jensen's inequality,

$$f[\mathbf{E}\{y\}] \leq \mathbf{E}\{f(y)\} \tag{9.6}$$

if the right side exists. Now let $\mathscr{F}$ be any Borel field of measurable ω sets. Then, if conditional expectations behave like ordinary expectations,

$$f[\mathbf{E}\{y \mid \mathscr{F}\}] \leq \mathbf{E}\{f(y) \mid \mathscr{F}\} \tag{9.7}$$

with probability 1, if y and $f(y)$ have expectations. Although this inequality can be proved directly from the definition of conditional expectations, it is instructive to use the methods we have just developed. We suppose in the following that f is defined in an interval I containing the range of y. It is then trivial to prove that $\mathbf{E}\{y \mid \mathscr{F}\}$ will also lie in I, with probability 1. The simplest proof of the desired inequality uses the conditional distribution p_0 of y, in the wide sense, relative to $\mathscr{F}$. In terms of p_0, (9.7) becomes

$$f\Big[\int_I \eta p_0(d\eta, \omega)\Big] \leq \int_I f(\eta) p_0(d\eta, \omega),$$

for all ω with $p_0(I, \omega) = 1$, and this means for almost all ω. We have thus reduced (9.7) to Jensen's inequality as applied to the wide sense conditional distribution. We also give a proof of (9.7), by means of representation theory, to show how in discussions of this type one can either use wide sense conditional distributions, as we have just done, or true conditional distributions, after applying representation theory. Apply representation theory to obtain representations of y and $\mathscr{F}$ on a coordinate space. Then the following are three pairs of corresponding functions:

$$\begin{array}{ll} \mathbf{E}\{y \mid \mathscr{F}\}, & \tilde{\mathbf{E}}\{\tilde{y} \mid \tilde{\mathscr{F}}\} \\ \mathbf{E}\{f(y) \mid \mathscr{F}\}, & \tilde{\mathbf{E}}\{f(\tilde{y}) \mid \tilde{\mathscr{F}}\} \\ f[\mathbf{E}\{y \mid \mathscr{F}\}], & f[\tilde{\mathbf{E}}\{\tilde{y} \mid \tilde{\mathscr{F}}\}]. \end{array}$$

Since inequalities are preserved in the correspondence, (9.7) is equivalent to

$$f[\tilde{\mathbf{E}}\{\tilde{y} \mid \tilde{\mathscr{F}}\}] \leq \tilde{\mathbf{E}}\{f(\tilde{y}) \mid \tilde{\mathscr{F}}\} \tag{9.8}$$

(to hold with probability 1). However, since $\tilde{y}$ is a coordinate variable with range the whole line, the conditional expectations in (9.8) can be evaluated as integrals, with respect to conditional distributions, according to Theorems 9.2 and 9.5. Thus (9.8) reduces to Jensen's inequality applied to the conditional probability measures.

This argument has been given in detail to illustrate the fact that, even though pathological counterexamples make it impossible to assume that conditional probabilities can always be used to define conditional probability distributions, and that conditional expectations can then be evaluated as ordinary integrals, over ω space, nevertheless, for most purposes, conditional probabilities and expectations can be manipulated as if these examples did not exist.

It remains true, however, that some theorems on conditional probabilities and expectations can be derived just as easily directly as by the use of representation theory. This theory in such cases merely makes it obvious a priori that the theorems are true. The following theorem is an example of this possibility.

Let $\mathscr{G}_0$ be a field of ω sets, and let $\mathscr{G}$ be the Borel field generated by $\mathscr{G}_0$. Then two measures of $\mathscr{G}$ sets which are identical for $\mathscr{G}_0$ sets are identical for $\mathscr{G}$ sets (Supplement, Theorem 2.1), and therefore determine the same integrals of ω functions. This fact is less obvious, but true, for conditional probabilities. The formal statement is as follows.

THEOREM 9.6 *Let $\mathscr{F}_1$, $\mathscr{F}_2$ be Borel fields of ω sets, let $\mathscr{G}_0$ be a field of ω sets, and let $\mathscr{G}$ be the Borel field generated by $\mathscr{G}_0$. Then, if, whenever* $M \in \mathscr{G}_0$,

$$\mathbf{P}\{M \mid \mathscr{F}_1\} = \mathbf{P}\{M \mid \mathscr{F}_2\} \tag{9.9}$$

with probability 1, *it follows that whenever y is an ω function measurable with respect to $\mathscr{G}$, or equal with probability* 1 *to such a function, with* $\mathbf{E}\{|y|\} < \infty$, *then*

$$\mathbf{E}\{y \mid \mathscr{F}_1\} = \mathbf{E}\{y \mid \mathscr{F}_2\} \tag{9.10}$$

with probability 1.

The class of measurable sets M for which (9.9) is true with probability 1 includes $\mathscr{G}_0$, and includes limits of monotone sequences of sets in the class, by CP_3 of §8. Hence, according to the Supplement, Theorem 1.2, the class includes $\mathscr{G}$. The class then must include sets differing from $\mathscr{G}$ sets by sets of probability 0. Finally, (9.10) is now an immediate consequence of the evaluation of a conditional expectation in terms of conditional probabilities given in Theorem 8.4.

10. Iterated conditional expectations and probabilities

The following identity is the particular case of (7.6) obtained by taking $\Lambda = \Omega$,

$$\mathbf{E}\{\mathbf{E}\{y \mid \mathscr{F}\}\} = \mathbf{E}\{y\}. \tag{10.1}$$

In particular, if $y(\omega)$ is 1 on the ω set M, and 0 otherwise, that is, if $\Lambda = \Omega$ in (7.7), we obtain

$$\mathbf{E}\{\mathbf{P}\{\mathrm{M} \mid \mathscr{F}\}\} = \mathbf{P}\{\mathrm{M}\}. \tag{10.2}$$

If $\mathscr{F}$ is the field of sets measurable on the sample space of a single random variable x, the conditional expectation and probability are relative to x. Then (10.1) states that the expected value of a random variable y is (roughly) the probability that $x(\omega)$ takes on a value, multiplied by the expectation of y for $x(\omega)$ given that value, summed over those values.

It is instructive to analyze the meaning of conditional expectations and probabilities when the basic distributions are themselves conditional. We illustrate the situation by the following example. Suppose that all probabilities are conditional, relative to x, and that there is actually a conditional distribution, in the sense of the preceding section. It will be convenient for the moment to denote the conditional probabilities and expectations by $\mathbf{P}_x\{-\}$ and $\mathbf{E}_x\{-\}$ rather than by $\mathbf{P}\{- \mid x\}$ and $\mathbf{E}\{- \mid x\}$. Now suppose that, for $x(\omega)$ fixed, one is interested in the conditional expectation of a random variable y or in the conditional probability of an ω set M relative to a random variable z, that is, in

$$\mathbf{E}_x\{y \mid z\}, \qquad \mathbf{P}_x\{\mathrm{M} \mid z\}.$$

Since y and z are considered given random variables, with given (though conditional) distributions, these doubly conditioned quantities need no new definitions. It is easy to see, however, that this complicated notation is unnecessary, and in fact that we can take

$$\begin{aligned} \mathbf{E}_x\{y \mid z\} &= \mathbf{E}\{y \mid x, z\} \\ \mathbf{P}_x\{\mathrm{M} \mid z\} &= \mathbf{P}\{\mathrm{M} \mid x, z\}. \end{aligned} \tag{10.3}$$

Since we are not attempting to give a rigorous discussion here, we shall not formalize this assertion, but shall verify the second equation in the particular case when Ω is three-dimensional space, the measurable ω sets are the Lebesgue measurable sets, the given probability measure is determined by a density function f, and x, y, z are the coordinate functions of Ω. Then the joint distribution of x, y, z has density f. Finally, we suppose that M is an ω set of the form $\{z(\omega) \in Z\}$, where Z is a Lebesgue measurable set. In this case one version of the bivariate conditional distribution of y, z for $x(\omega) = \xi$ has density

$$g_\xi(\eta, \zeta) = \frac{f(\xi, \eta, \zeta)}{\int_{-\infty}^{\infty} \int_{-\infty}^{\infty} f(\xi, \eta', \zeta')\, d\eta'\, d\zeta'}.$$

Considering this density as defining a y, z distribution, the conditional distribution of y for $z(\omega) = \zeta$ has density

$$\frac{g_\xi(\eta, \zeta)}{\int_{-\infty}^{\infty} g_\xi(\eta, \zeta')\, d\zeta'} \tag{10.4}$$

On the other hand, the conditional distribution of y for $x(\omega) = \xi$ and $z(\omega) = \zeta$ has density

$$\frac{f(\xi, \eta, \zeta)}{\int_{-\infty}^{\infty} f(\xi, \eta', \zeta)\, d\eta'}. \tag{10.5}$$

The equality of (10.4) and (10.5) is exactly the second equation in (10.3) expressed in terms of densities.

The relations between probabilities and expectations obey a certain consistency law. Suppose that all expectations and probabilities are conditional, relative to a conditioning variable x_1. Then the rules of combination of

$$\mathbf{E}_{x_1}\{-\} = \mathbf{E}\{- \mid x_1\}, \qquad \mathbf{E}_{x_1}\{- \mid x_2\} = \mathbf{E}\{- \mid x_1, x_2\}$$
$$\mathbf{P}_{x_1}\{-\} = \mathbf{P}\{- \mid x_1\}, \qquad \mathbf{P}_{x_1}\{- \mid x_2\} = \mathbf{P}\{- \mid x_1, x_2\}$$

must be, respectively, the same as those of

$$\mathbf{E}\{-\}, \qquad \mathbf{E}\{- \mid x_2\},$$
$$\mathbf{P}\{-\} \qquad \mathbf{P}\{- \mid x_2\}.$$

If this were not so, the knowledge that certain parameters in a problem may really be random variables would entirely change the formal handling of the problem.

This consistency principle suggests the following equations. Corresponding to (10.1) and (10.2) we have

$$\mathbf{E}\Big\{\mathbf{E}\{y \mid x_1, x_2\} \Big| x_1\Big\} = \mathbf{E}\{y \mid x_1\} \tag{10.6}$$

and

$$\mathbf{E}\Big\{\mathbf{P}\{M \mid x_1, x_2\} \Big| x_1\Big\} = \mathbf{P}\{M \mid x_1\} \tag{10.7}$$

respectively, with probability 1. These equations reduce to (10.1) and (10.2) if the conditioning variable x_1 is ignored. The equations are to be interpreted as follows. If x_1, x_2, y are random variables, with $\mathbf{E}\{|y|\} < \infty$, and if M is a measurable ω set, then, no matter which versions of the indicated conditional probabilities and expectations are used, the equations are true with probability 1. This is another way of saying that, if the proper versions are used, the equations are true for all ω. Equation (10.7) is a special case of (10.6). Both are almost immediate consequences of the definitions of conditional probabilities and expectations. Before proving them we generalize them as follows. Let $\mathscr{F}_1, \mathscr{F}_2$ be Borel fields of measurable ω sets with $\mathscr{F}_1 \subset \mathscr{F}_2$. Then

$$\mathbf{E}\Big\{\mathbf{E}\{y \mid \mathscr{F}_2\}\Big| \mathscr{F}_1\Big\} = \mathbf{E}\{y \mid \mathscr{F}_1\} \tag{10.8}$$

and

$$\mathbf{E}\Big\{\mathbf{P}\{M \mid \mathscr{F}_2\}\Big| \mathscr{F}_1\Big\} = \mathbf{P}\{M \mid \mathscr{F}_1\} \tag{10.9}$$

with probability 1. Evidently (10.6) and (10.7) are special cases of (10.8) and (10.9), respectively. To prove (10.8), which contains them all, we need only remark that (aside from the measurability conditions, which are obviously satisfied) it states that

$$\int_\Lambda \mathbf{E}\{y \mid \mathscr{F}_2\}\, d\mathbf{P} = \int_\Lambda y\, d\mathbf{P}, \qquad \Lambda \in \mathscr{F}_1,$$

and this equation is true, by definition of the integrand on the left, even for $\Lambda \in \mathscr{F}_2 \supset \mathscr{F}_1$. There are many useful special cases of (10.8) and (10.9). For example, if $y, x_1, x_2, \cdots$ are random variables, with $\mathbf{E}\{|y|\} < \infty$, then according to (10.8),

$$\mathbf{E}\Big\{\mathbf{E}\{y \mid x_1, x_2, \cdots\}\Big| x_2, x_4, \cdots\Big\} = \mathbf{E}\{y \mid x_2, x_4, \cdots\}$$

with probability 1.

11. Characteristic functions

If x is any random variable, with distribution function F, its characteristic function is defined by

$$\Phi(t) = \mathbf{E}\{e^{itx}\} = \int_{-\infty}^{\infty} e^{it\lambda}\, dF(\lambda). \tag{11.1}$$

The function Φ is uniquely determined by F and is accordingly also called the characteristic function of this distribution function. The following fundamental properties of characteristic functions will be used.

(A) Φ is continuous for all t, and

$$|\Phi(t)| \leq \Phi(0) = 1. \tag{11.2}$$

If $\mathbf{E}\{|x|^n\} < \infty$ for some positive integer n, Φ has n continuous derivatives, and

$$\Phi^{(n)}(t) = \int_{-\infty}^{\infty} (i\lambda)^n e^{it\lambda}\, dF(\lambda)$$

(11.3)

$$|\Phi^{(n)}(t)| \le \int_{-\infty}^{\infty} |\lambda|^n\, dF(\lambda),$$

so that

$$\begin{aligned}
\text{(11.4)} \qquad \Phi(t) &= \sum_{j=0}^{n-1} \frac{i^j \mathbf{E}\{x^j\} t^j}{j!} + \int_0^t \Phi^{(n)}(s) \frac{(t-s)^{n-1}}{(n-1)!}\, ds \\
&= \sum_{j=0}^{n} \frac{i^j \mathbf{E}\{x^j\} t^j}{j!} + \int_0^t [\Phi^{(n)}(s) - \Phi^{(n)}(0)] \frac{(t-s)^{n-1}}{(n-1)!}\, ds \\
&= \sum_{j=0}^{n-1} \frac{i^j \mathbf{E}\{x^j\} t^j}{j!} + O(|t|^n), \qquad |O(|t|^n)| \le \frac{\mathbf{E}\{|x|^n\}|t|^n}{n!} \\
&= \sum_{j=0}^{n} \frac{i^j \mathbf{E}\{x^j\} t^j}{j!} + o(|t|^n).
\end{aligned}$$

If $0 < \delta \le 1$ and if $\mathbf{E}\{|x|^{n+\delta}\} < \infty$, we have

$$\text{(11.4}'\text{)} \qquad \Phi(t) = \sum_{j=0}^{n} \frac{i^j \mathbf{E}\{x^j\} t^j}{j!} + O(|t|^{n+\delta}),$$

$$O(|t|^{n+\delta}) \le \frac{2^{1-\delta}\mathbf{E}\{|x|^{n+\delta}\}|t|^{n+\delta}}{(1+\delta)(2+\delta)\cdots(n+\delta)}.$$

To see this we need only find a majorant for the integral in the second line of (11.4), and in fact

$$\begin{aligned}
\left| \int_0^t [\Phi^{(n)}(s) - \Phi^{(n)}(0)] \frac{(t-s)^{n-1}}{(n-1)!}\, ds \right| &= \left| \int_0^t \frac{(t-s)^{n-1}}{(n-1)!}\, ds \int_{-\infty}^{\infty} (i\lambda)^n (e^{is\lambda} - 1)\, dF(\lambda) \right| \\
&\le \int_0^t \frac{(t-s)^{n-1}}{(n-1)!}\, ds \int_{-\infty}^{\infty} 2^{1-\delta} |\lambda|^n |s\lambda|^\delta\, dF(\lambda) \\
&= \frac{2^{1-\delta}\mathbf{E}\{|x|^{n+\delta}\}|t|^{n+\delta}}{(1+\delta)(2+\delta)\cdots(n+\delta)}.
\end{aligned}$$

(B) F is determined by Φ in the sense that (Lévy's formula)

$$\frac{F(\mu)+F(\mu-)}{2}-\frac{F(\lambda)+F(\lambda-)}{2}=\lim_{c\to\infty}\int_{-c}^{c}\frac{e^{-it\mu}-e^{-it\lambda}}{-2\pi it}\Phi(t)\,dt. \tag{11.5}$$

(C) If $x_1, \cdots, x_n$ are mutually independent random variables, with characteristic functions $\Phi_1, \cdots, \Phi_n$, the characteristic function Φ of $x_1+\cdots+x_n$ is

$$\Phi=\prod_{j=1}^{n}\Phi_j.$$

(D) If $F_1, F_2, \cdots$ is a sequence of distribution functions, with $\lim_{n\to\infty} F_n(\lambda)=F(\lambda)$ at all continuity points of a monotone function F, and if F is a distribution function, the sequence of characteristic functions of the F_j's converges to the characteristic function of F uniformly in every finite interval.

We shall have occasion to use the related theorem that, if the functions F_1, F_2 are only supposed monotone non-decreasing and bounded, with $\lim_{n\to\infty} F_n(\lambda)=F(\lambda)$ at all continuity points of a bounded monotone function F, and if g is continuous and bounded, then

$$\lim_{n\to\infty}\int_{-\infty}^{\infty} g(\lambda)\,dF_n(\lambda)=\int_{-\infty}^{\infty} g(\lambda)\,dF(\lambda) \tag{11.6}$$

if

$$F(\infty)-F(-\infty)=\lim_{n\to\infty}[F_n(\infty)-F_n(-\infty)].$$

The latter condition is equivalent to the condition that

$$\lim_{\lambda\to\pm\infty} F_n(\lambda)=F_n(\pm\infty)$$

uniformly in n. If in addition $g(\lambda)=g(t,\lambda)$ depends on a parameter t and if $g(t,\lambda)$ is bounded and continuous in (t,λ), then the convergence in (11.6) is uniform for t in every finite interval. The theorem on characteristic functions stated in the preceding paragraph is a particular case of this theorem.

(E) Conversely, if $F_1, F_2, \cdots$ is a sequence of distribution functions whose corresponding sequence of characteristic functions $\Phi_1, \Phi_2, \cdots$ converges to a characteristic function Φ, and if F is the distribution function corresponding to Φ, then $\lim_{n\to\infty} F_n(\lambda)=F(\lambda)$ at every continuity point of F. If it is only supposed that $\lim_{n\to\infty}\Phi_n(t)$ exists for all t, the limit function will necessarily be a characteristic function if the limit is uniform in some interval containing $t = 0$.

The following consequence of (D) and (E) will be used frequently. If $x_1, x_2, \cdots$ is a sequence of random variables, with characteristic functions $\Phi_1, \Phi_2, \cdots$, $\operatorname{p}\lim_{n\to\infty} x_n = 0$ if and only if $\lim_{n\to\infty} \Phi_n(t) = 1$ uniformly in every finite t interval. In fact, if $F(\lambda)$ is defined as 0 for $\lambda < 0$ and 1 for $\lambda \geq 0$, $\operatorname{p}\lim_{n\to\infty} x_n = 0$ if and only if

$$\lim_{n\to\infty} \mathbf{P}\{x_n(\omega) \leq \lambda\} = F(\lambda), \qquad \lambda \neq 0, \tag{11.7}$$

and the condition for (11.7) in terms of characteristic functions given by (D) and (E) is precisely the stated one. As a matter of fact, if $\lim_{n\to\infty} \Phi_n(t) = 1$ *uniformly in some interval containing* $t = 0$, the same will be true in every finite interval, because of the following inequality in the real part of a characteristic function Φ:

$$\begin{aligned}\Re[1 - \Phi(t)] &= \int_{-\infty}^{\infty} [1 - \cos t\lambda]\, dF(\lambda) \geq \int_{-\infty}^{\infty} \frac{\sin^2 t\lambda}{2}\, dF(\lambda)\\ &= \tfrac{1}{4}\Re[1 - \Phi(2t)].\end{aligned}$$

The following inequalities will be used frequently. They illustrate the simple way in which the characteristic function of a distribution limits the probabilities of large values, and show the simplicity gained by centering the distribution at a median value or at a truncated expectation. Throughout the following discussion, x is a random variable with distribution function F and characteristic function Φ; a, α, μ are positive numbers, and A is a Lebesgue measurable subset of the t interval $[0, a]$, of measure $\rho > 0$. The function L_j is a positive function of whatever variables are indicated, but not depending, for example, on the distribution function F except by way of the indicated variables. Let m be a median value of x,

$$\mathbf{P}\{x(\omega) \leq m\} \geq \tfrac{1}{2}, \qquad \mathbf{P}\{x(\omega) \geq m\} \geq \tfrac{1}{2},$$

define x' as x truncated to m when it is beyond $m \pm \alpha$,

$$\begin{aligned} x'(\omega) &= x(\omega), \qquad &|x(\omega) - m| \leq \alpha\\ &= m, \qquad &|x(\omega) - m| > \alpha,\end{aligned}$$

and define

$$\hat{m} = \mathbf{E}\{x'\}, \qquad \hat{\sigma}^2 = \mathbf{E}\{(x' - \hat{m})^2\},$$

$$\hat{\Phi}(t) = \mathbf{E}\{e^{it(x - \hat{m})}\}.$$

Then we shall prove that there is an $L_1(\mu, \rho, a)$ *such that*

$$\mathbf{P}\{|x(\omega)| \geq \mu\} \leq L_1(\mu, \rho, a) \int_A \Re[1 - \Phi(t)]\, dt. \tag{11.8}$$

Centering at m will yield the more useful

$$\begin{aligned} \mathbf{P}\{|x(\omega) - m| \geq \mu\} &\leq 2L_1(\mu, \rho, a) \int_A [1 - |\Phi(t)|^2]\, dt \\ &\leq -4L_1(\mu, \rho, a) \int_A \log |\Phi(t)|\, dt. \end{aligned} \tag{11.8'}$$

Centering at $\hat{m}$ *will yield an inequality of the same type,*

$$\begin{aligned} \mathbf{P}\{|x(\omega) - \hat{m}| \geq \mu\} &\leq L_2(\alpha, \mu, \rho, a) \int_A [1 - |\Phi(t)|^2]\, dt \\ &\leq -2L_2(\alpha, \mu, \rho, a) \int_A \log |\Phi(t)|\, dt. \end{aligned} \tag{11.$\hat{8}$}$$

Proceeding to study the second moments, we shall prove

$$\int_{-\mu-}^{\mu} \lambda^2\, dF(\lambda) \leq L_3(\mu, \rho, a) \int_A \Re[1 - \Phi(t)]\, dt. \tag{11.9}$$

If we denote the variance of the F distribution truncated to m when it is beyond $m \pm \mu$ by $\hat{\sigma}(\mu)^2$,

$$\hat{\sigma}(\mu)^2 = \int_{m-\mu-}^{m+\mu} (\lambda - m)^2\, dF(\lambda) - \Big[\int_{m-\mu-}^{m+\mu} (\lambda - m)\, dF(\lambda)\Big]^2,$$

then $\hat{\sigma}(\alpha)^2 = \hat{\sigma}^2$, *and* (11.9) *will yield, for some* $L_4(\mu, \rho, a)$,

$$\begin{aligned} \hat{\sigma}(\mu)^2 &\leq L_4(\mu, \rho, a) \int_A [1 - |\Phi(t)|^2]\, dt \\ &\leq -2L_4(\mu, \rho, a) \int_A \log |\Phi(t)|\, dt. \end{aligned} \tag{11.$\hat{9}$}$$

Finally, we shall prove that, for some $L_5(T, \alpha, \rho, a)$,

$$\begin{aligned} |1 - \hat{\Phi}(t)| &\leq L_5(T, \alpha, \rho, a) \int_A [1 - |\Phi(t)|^2]\, dt \\ &\leq -2L_5(T, \alpha, \rho, a) \int_A \log |\Phi(t)|\, dt, \qquad |t| \leq T. \end{aligned} \tag{11.10}$$

The following elementary inequalities will be needed in the proofs of the above inequalities. In the first place the function

$$s - \sin s - \frac{s^3}{\pi^2}$$

vanishes at $s = 0$, π, is positive for small s, and its derivative vanishes at a single point between 0 and π. Hence

$$s - \sin s \geq \frac{s^3}{\pi^2}, \qquad 0 \leq s \leq \pi. \tag{11.11}$$

In the second place, if $\lambda > 0$,

$$\int_A (1\text{-}\cos \lambda t)\, dt \geq \frac{\rho^3}{(a + 2\pi/\lambda)^2}. \tag{11.12}$$

To prove this let a_1 be the smallest number $\geq a$ for which $\lambda a_1/2\pi$ is an integer. Then

$$a_1 \leq a + \frac{2\pi}{\lambda},$$

and $A \subset [0, a_1]$. We minimize the integral in (11.12) by replacing A by $\lambda a_1/\pi$ non-overlapping intervals in $[0, a_1]$, each of length $\pi\rho/a_1\lambda$, each with an endpoint at a point $t = 0, 2\pi/\lambda, \cdots, a_1$ which makes the integrand 0:

$$\int_A (1 - \cos \lambda\, t)\, dt \geq \frac{\lambda a_1}{\pi} \int_0^{\pi\rho/a_1\lambda} (1 - \cos \lambda\, t)\, dt = \rho - \frac{a_1}{\pi} \sin(\pi\rho/a_1).$$

Then by (11.11)

$$\int_A (1 - \cos \lambda t)\, dt \geq \frac{\rho^3}{a_1^2} > \frac{\rho^3}{(a + 2\pi/\lambda)^2},$$

proving (11.12).

To prove (11.8) integrate the inequality

$$\int_{|\lambda| \geq \mu} (1 - \cos \lambda t)\, dF(\lambda) \leq \Re[1 - \Phi(t)]$$

over A to obtain

$$\int_{|\lambda| \geq \mu} dF(\lambda) \int_A (1 - \cos \lambda t)\, dF(\lambda) \leq \int_A \Re[1 - \Phi(t)]\, dt.$$

This inequality implies (11.8) with

$$L_1(\mu, \rho, a) = \frac{(a + 2\pi/\mu)^2}{\rho^3}$$

in view of (11.12).

To prove (11.8′) apply (11.8) to $x - x^*$, where x^* is independent of x and has the same distribution as x. We obtain

$$\mathbf{P}\{|x(\omega) - x^*(\omega)| \geq \mu\} \leq L_1(\mu, \rho, a) \int_A [1 - |\Phi(t)|^2]\, dt,$$

since $x - x^*$ has the characteristic function $|\Phi|^2$. Now

$$\begin{aligned}\mathbf{P}\{|x(\omega) - x^*(\omega)| \geq \mu\} &\geq \mathbf{P}\{x(\omega) - m \geq \mu, x^*(\omega) - m \leq 0\} \\ &\quad + \mathbf{P}\{x(\omega) - m \leq -\mu, x^*(\omega) - m \geq 0\} \\ &\geq \tfrac{1}{2}\mathbf{P}\{x(\omega) - m \geq \mu\} + \tfrac{1}{2}\mathbf{P}\{x(\omega) - m \leq -\mu\} \\ &= \tfrac{1}{2}\mathbf{P}\{|x(\omega) - m| \geq \mu\},\end{aligned}$$

and this inequality combined with the preceding one yields the first half of (11.8′). The other half follows from the inequality

$$1 - a \leq -\log a, \qquad 0 \leq a \leq 1.$$

We defer the proof of (11.$\hat{8}$) until that of (11.10) has been given. To prove (11.9) we use (11.12) to obtain

$$\begin{aligned}\int_A \Re[1 - \Phi(t)]\, dt &\geq \int_{-\mu-}^{\mu} dF(\lambda) \int_A (1 - \cos \lambda t)\, dt \\ &\geq \int_{-\mu-}^{\mu} \frac{\rho^3}{(a + 2\pi/|\lambda|)^2}\, dF(\lambda), \\ &\geq \frac{\rho^3}{(a\mu + 2\pi)^2} \int_{-\mu-}^{\mu} \lambda^2\, dF(\lambda),\end{aligned}$$

so that (11.9) is true with

$$L_3(\mu, \rho, a) = \frac{(a\mu + 2\pi)^2}{\rho^3}.$$

To prove (11.$\hat{9}$), (11.9) with 2μ instead of μ is applied to $x - x^*$ to obtain

$$\begin{aligned}&\int_{-2\mu-}^{2\mu} \lambda^2\, d\mathbf{P}\{x(\omega) - x^*(\omega) \leq \lambda\} \\ &\qquad = \int_{\{|x(\omega) - x^*(\omega)| \leq 2\mu\}} (x - x^*)^2\, d\mathbf{P} \leq L_3(2\mu, \rho, a) \int_A [1 - |\Phi(t)|^2]\, dt.\end{aligned}$$

If we combine this inequality with

$$\begin{aligned}&\int_{\{|x(\omega) - x^*(\omega)| \leq 2\mu\}} (x - x^*)^2\, d\mathbf{P} \geq \int_{\left\{\substack{|x(\omega) - m| \leq \mu \\ |x^*(\omega) - m| \leq \mu}\right\}} (x - x^*)^2\, d\mathbf{P} \\ &\quad = 2\mathbf{P}\{|x(\omega) - m| \leq \mu\} \int_{m-\mu-}^{m+\mu} (\lambda - m)^2\, dF(\lambda) - 2\left[\int_{m-\mu-}^{m+\mu} (\lambda - m)\, dF(\lambda)\right]^2 \\ &\quad \geq 2\hat{\sigma}(\mu)^2 - 2\mu^2 \mathbf{P}\{|x(\omega) - m| \geq \mu\},\end{aligned}$$

we obtain

$$\hat{\sigma}(\mu)^2 \leq \mu^2 \mathbf{P}\{|x(\omega) - m| \geq \mu\} + \tfrac{1}{2} L_3(2\mu, \rho, a) \int_A [1 - |\Phi(t)|^2]\, dt.$$

This implies (11.$\hat{9}$) with

$$\begin{aligned} L_4(\mu, \rho, a) &= 2\mu^2 L_1(\mu, \rho, a) + \tfrac{1}{2} L_3(2\mu, \rho, a) \\ &= \frac{2(a\mu + 2\pi)^2}{\rho^3} + \frac{(2a\mu + 2\pi)^2}{2\rho^3} \\ &< \frac{4(a\mu + 2\pi)^2}{\rho^3}. \end{aligned}$$

To prove (11.10) we note that (using the definition of $\hat{m}$)

$$\begin{aligned} (11.13)\quad |1 - \hat{\Phi}(t)| &\leq \left| \int_{m-\alpha-}^{m+\alpha} (e^{it(\lambda - \hat{m})} - 1)\, dF(\lambda) \right| + 2\mathbf{P}\{|x(\omega) - m| \geq \alpha\} \\ &\leq \left| \int_{m-\alpha-}^{m+\alpha} [(e^{it(\lambda-\hat{m})} - 1 - it(\lambda - \hat{m})]\, dF(\lambda) \right| \\ &\quad + 2\mathbf{P}\{|x(\omega) - m| \geq \alpha\} \\ &\quad + |t(\hat{m} - m)|\mathbf{P}\{|x(\omega) - m| \geq \alpha\} \\ &\leq \frac{t^2}{2} \int_{m-\alpha-}^{m+\alpha} (\lambda - \hat{m})^2\, dF(\lambda) \\ &\quad + [2 + |t(\hat{m} - m)|]\mathbf{P}\{|x(\omega) - m| \geq \alpha\} \\ &= \frac{t^2\hat{\sigma}^2}{2} + [2 + |t(\hat{m} - m)| - \frac{t^2}{2}(\hat{m} - m)^2] \\ &\quad \cdot \mathbf{P}\{|x(\omega) - m| \geq \alpha\} \\ &\leq \frac{t^2\hat{\sigma}^2}{2} + \frac{5}{2}\mathbf{P}\{|x(\omega) - m| \geq \alpha\}. \end{aligned}$$

Hence (11.10) is true with

$$\begin{aligned} L_5(T, \alpha, \rho, a) &= \frac{T^2}{2} L_4(\alpha, \rho, a) + 5L_1(\alpha, \rho, a) \\ &< \frac{(a + 2\pi/\alpha)(2T^2\alpha^2 + 5)}{\rho^3}. \end{aligned}$$

Finally, we prove (11.$\hat{8}$) by applying (11.8) to $x - \hat{m}$, obtaining

$$\mathbf{P}\{|x(\omega) - \hat{m}| \geq \mu\} \leq L_1(\mu, \rho, a) \int_A \Re[1 - \hat{\Phi}(t)]\, dt$$

$$\leq L_1(\mu, \rho, a) \int_A |1 - \hat{\Phi}(t)|\, dt,$$

so that (11.$\hat{8}$) is true, with

$$L_2(\alpha, \mu, \rho, a) = aL_1(\mu, \rho, a)L_5(a, \alpha, \rho, a)$$

$$< a\rho^{-6}\left(a + \frac{2\pi}{\mu}\right)^2 \left(a + \frac{2\pi}{\alpha}\right)^2 (2a^2\alpha^2 + 5).$$

If x is a bounded random variable, (11.9) yields a majorant of the variance of x, if α is sufficiently large, but it is more enlightening to make a direct analysis. We shall prove that *if* $|x| \leq M$, *and if* x *has variance* σ^2, *and characteristic function* Φ, *then*

$$(11.14) \qquad -\log |\Phi(t)| \leq \sigma^2 t^2 \leq -3 \log |\Phi(t)|, \qquad |t| \leq \frac{1}{4M}.$$

To prove this, suppose first that $\mathbf{E}\{x\} = 0$. Then by (11.4) with $n = 2$,

$$(11.15) \qquad |1 - \Phi(t)| \leq \frac{\sigma^2 t^2}{2}$$

and by (11.4′) with $n = 2$, $\delta = 1$,

$$(11.16) \qquad \left|1 - \Phi(t) - \frac{\sigma^2 t^2}{2}\right| \leq \frac{M\sigma^2 t^3}{6}.$$

Now, if z is a complex number, and if $|1 - z| < 1$,

$$|\log z + 1 - z| = \left|\int_z^1 \left(\frac{1}{\zeta} - 1\right) d\zeta\right| \leq \frac{|1 - z|^2}{|z|}$$

(integrating along a line segment). Hence, using (11.4),

$$|\log \Phi + 1 - \Phi| \leq \frac{|1 - \Phi|^2}{|\Phi|} \leq \frac{\sigma^4 t^4}{4(1 - \sigma^2 t^2/2)}, \qquad M^2 t^2 < 2,$$

so that, combining this inequality with (11.16),

$$\left|-\log \Phi - \frac{\sigma^2 t^2}{2}\right| \leq \frac{\sigma^4 t^4}{4(1 - \sigma^2 t^2/2)} + \frac{M\sigma^2 t^3}{6} \leq \frac{\sigma^2 t^2}{6}, \qquad M|t| \leq 1/2.$$

Taking real parts, we find that (11.14) is true for $|t| \leq 1/2M$. If we now drop the restriction that $\mathbf{E}\{x\} = 0$, we apply the inequality to $x - \mathbf{E}\{x\}$ to obtain (11.14) as stated.

CHAPTER II

Definition of a Stochastic Process—Principal Classes

1. Definition of a stochastic process

From the non-mathematical point of view a stochastic process is any probability process, that is, any process running along in time and controlled by probabilistic laws. Numerical observations made as the process continues indicate its evolution. With this background to guide us *we define a stochastic process as any family of random variables* $\{x_t, t \in T\}$. Here x_t is in practice the observation at time t, and T is the time range involved.

Most classical problems in probability involve only finitely many random variables, that is, the corresponding t range T is a finite set. In the following chapters, however, we shall almost always consider processes involving infinitely many random variables, and the term *stochastic process* has usually been applied only in this case. The two most important cases are the following:

(a) T is an infinite sequence; $\{x_t, t \in T\}$ becomes

$$x_m, x_{m+1}, \cdots$$

or

$$\cdots, x_{m-1}, x_m$$

or

$$\cdots, x_0, x_1, \cdots .$$

This type of process is called a discrete parameter process.

(b) *T is an interval;* $\{x_t, t \in T\}$ becomes a continuous parameter family, and the process is called a continuous parameter process. In this case the general sample is a function of t defined on an interval, whereas in the discrete parameter case the general sample is a sequence. More generally we shall call any process whose parameter set is finite or enumerable a discrete parameter process, and any process with a non-denumerable parameter set a continuous parameter process.

Our definition of a stochastic process is historically conditioned and has obvious defects. In the first place there is no mathematical reason for restricting T to be a set of real numbers, and in fact interesting work has already been done in other cases. (Of course, the interpretation of t as time must then be dropped.) In the second place there is no mathematical reason for restricting the value assumed by the x_t's to be numbers. However, in this book we shall consider only processes with T a linear point set, and the random variables will almost always be real- or complex-valued. We allow the defining real random variables of processes to have the values $+\infty$ and $-\infty$, but only with zero probability.

Our definition of a stochastic process is embarrassingly inclusive in that there are not many problems in probability that cannot be formulated as problems in families of random variables. Historically, however, the term stochastic process has been reserved for families (usually infinite) of random variables with some simple relationship between the variables, and this book is devoted to the most important examples of such families. A basic problem is to devise suitable relationships, that is, to discover new types of stochastic processes which are useful, or mathematically elegant, or which conform otherwise to the investigator's criterion of importance.

Let $\{x_t, t \in T\}$ be a stochastic process. A function of $t \in T$ obtained by fixing ω in $x_t(\omega)$ and letting t vary is called a *sample function* of the process. (If T is finite or denumerably infinite, the sample functions are of course *sample sequences*.)

Let $t_1, \cdots, t_n$ be any finite set of parameter values of the process $\{x_t, t \in T\}$. The multivariate distribution of the random variables $x_{t_1}, \cdots, x_{t_n}$ is called a *finite dimensional distribution* of the process. The finite dimensional distributions of a process are the basic distributions of the processes we shall consider in this book, and the processes we shall consider are classified according to their finite dimensional distributions.

Let $\mathscr{F}_T = \mathscr{B}(x_t, t \in T)$ be the Borel field of ω sets generated by the class of sets of the form $\{x_t(\omega) \in A\}$, where $t \in T$, and A is any Borel set (one-dimensional if the x_t's are real, two-dimensional if they are complex). Then $\mathscr{F}_T$ is the smallest Borel field of ω sets with respect to which all the x_t's are measurable. The Borel sets A in the definition can be supposed in a somewhat more restricted class, without changing $\mathscr{F}_T$. For example, the A's can be taken as right semi-closed intervals.

Let $\mathscr{F}_0$ be the field of ω sets of the form

$$\{[x_{t_1}(\omega), \cdots, x_{t_n}(\omega)] \in A\}$$

where $(t_1, \cdots, t_n)$ is any finite parameter set and A is a right semi-closed interval (n-dimensional if the x_t's are real, $2n$-dimensional in the complex case). Then $\mathscr{F}_T$ is the Borel field generated by $\mathscr{F}_0$. According to Theorem 2.3 of the Supplement, if Λ is an ω set which is a measurable set on the sample space of the x_t's, (see I §7 for the definition of this concept) and if $\varepsilon > 0$, there is an $\mathscr{F}_0$ set Λ_ε such that

$$\mathbf{P}\{\Lambda(\Omega - \Lambda_\varepsilon) \cup (\Omega - \Lambda)\Lambda_\varepsilon\} < \varepsilon.$$

According to the same theorem, if x is a random variable measurable on the sample space of the x_t's, and if $\varepsilon > 0$, there is an ω function x_ε which takes on only finitely many values, each on an $\mathscr{F}_0$ set, such that

$$\mathbf{P}\{|x(\omega) - x_\varepsilon(\omega)| > \varepsilon\} < \varepsilon.$$

The given probability measure is also a measure of $\mathscr{F}_T$ sets. Let $\mathscr{F}_T'$ be the domain of definition of this restricted measure after completion (see Supplement, §2), that is, $\mathscr{F}_T'$ consists of the $\mathscr{F}_T$ sets and also of the sets which differ from $\mathscr{F}_T$ sets by sets which are subsets of $\mathscr{F}_T$ sets of probability 0. All the $\mathscr{F}_T'$ sets need not be measurable if the given probability measure is not complete, but in any event there is only one way in which their probabilities can be defined which is compatible with the finite dimensional distributions. In fact the finite dimensional distributions furnish the probabilities of $\mathscr{F}_0$ sets, the probability measure of $\mathscr{F}_0$ sets can be extended in only one way to $\mathscr{F}_T$ sets (Supplement, Theorem 2.1), and the probability measure of $\mathscr{F}_T$ sets can be completed by extension to $\mathscr{F}_T'$ sets in only one way.

If ω sets not in $\mathscr{F}_T'$ are measurable, that is, have probabilities assigned to them, these probabilities are to a certain extent incidental. Some additional principle is necessary if these probabilities are to be determined in terms of the basic probabilities of $\mathscr{F}_0$ sets, that is, in terms of the finite dimensional distributions. Before investigating this question systematically, we illustrate it by an almost trivial example. Let Ω be the semi-closed interval $(0, 1]$. The measurable sets are to be any unions of the six semi-closed intervals $\left(\frac{j-1}{6}, \frac{j}{6}\right]$, $j = 1, \cdots, 6$. The given probability measure is to be length. The stochastic process is to contain only a single random variable x, defined as j on the jth semi-closed interval above. This random variable provides a mathematical model for the result of tossing a balanced die once. The fields $\mathscr{F}_T$ and $\mathscr{F}_0$ are the same and consist of all the measurable sets. It is obvious that the measures of other sets can be defined in many ways which are compatible with the measures already assigned. For example, the set containing the single point $\frac{6}{7}$ is not now measurable. It can be assigned the probability

$\frac{1}{6}$, which means that the rest of the interval $(\frac{5}{6}, 1]$ must be assigned the probability 0. On the other hand, an entirely different assignment of measures compatible with the given ones would be the assignment, to every Lebesgue measurable subset of (0, 1], of its Lebesgue measure. With this assignment, the set containing the single point $\frac{6}{7}$ is measurable, but has probability 0. The distribution of the random variable x is unaffected by these new assignments of probabilities.

We now investigate these measure questions systematically, using the representations of families of random variables discussed in §6 of I. Suppose that $\{x_t, t \in T\}$ is a stochastic process. Assume that the process is real; the obvious modifications are made in the complex case. Let $\tilde{\omega}$ be a real function of $t \in T$; the function values $\pm\infty$ are admissible. Let $\tilde{\Omega}$ be the space of all points $\tilde{\omega}$, that is, of all functions of $t \in T$. Then $\tilde{\Omega}$ is a coordinate space whose dimensionality is the cardinal number of T. Let $\tilde{x}_s(\tilde{\omega})$ be the sth coordinate of $\tilde{\omega}$, that is, the value of the function $\tilde{\omega}$ of t when $t = s$. Then $\tilde{x}_t$ is a representation of x_t based on function space.

If T is finite, containing say the n points $t_1, \cdots, t_n$, $\tilde{\Omega}$ becomes the n-dimensional space of points

$$\tilde{\omega}: [\xi(t_1), \cdots, \xi(t_n)],$$

and a measure is defined on the Borel sets of this space by assigning to the point set

$$\{[\xi(t_1), \cdots, \xi(t_n)] \in A\} = \{[\tilde{x}_{t_1}(\tilde{\omega}), \cdots, \tilde{x}_{t_n}(\tilde{\omega})] \in A\} \tag{1.1}$$

the probability

$$\mathbf{P}\{[x_{t_1}(\omega), \cdots, x_{t_n}(\omega)] \in A\}, \tag{1.2}$$

for every n-dimensional Borel set A. Consistently with our previous notation we denote by $\tilde{\mathscr{F}}_T$ the field of n-dimensional Borel sets. Completing the measure of Borel sets obtained in this way, we obtain an n-dimensional Lebesgue-Stieltjes measure.

If T is infinite, $\tilde{\Omega}$ becomes infinite dimensional Cartesian space. Let $\tilde{\mathscr{F}}_T = \mathscr{B}(\tilde{x}_t, t \in T)$ be the Borel field generated by the class of $\tilde{\omega}$ sets of the form (1.1). Then a measure of $\tilde{\mathscr{F}}_T$ sets is obtained by assigning the probability (1.2) to the "finite dimensional" $\tilde{\omega}$ set (1.1), and more generally every $\tilde{\mathscr{F}}_T$ set $\tilde{\Lambda}$ is assigned as probability the probability of the ω set Λ which determines sample functions in $\tilde{\Lambda}$, that is, the ω set Λ such that if $\omega \in \Lambda$, then $x_t(\omega)$ defines a function of t which is an element of $\tilde{\Lambda}$. (See Supplement, Example 3.2, and I §6.) For each $t \in T$ the $\tilde{\omega}$ function $\tilde{x}_t$ is now a random variable, and for every finite t set $t_1, \cdots, t_n$ the $\tilde{\omega}$ random variables $\tilde{x}_{t_1}, \cdots, \tilde{x}_{t_n}$ have the same multivariate distribution as the ω random variables $x_{t_1}, \cdots, x_{t_n}$. The family $\{\tilde{x}_t, t \in T\}$ was called a representation of the family $\{x_t, t \in T\}$ in I §6. Every family of random variables thus has a representation whose basic space is function space.

On the other hand we have seen in I §5 that, if T is any aggregate and $\tilde{\Omega}$ the space of all functions of $t \in T$, it is possible to define a probability measure on the Borel field $\tilde{\mathscr{F}}_T$ by assigning (finite dimensional) measures to $\tilde{\omega}$ sets of the form (1.1) in a consistent way, and then (if T is infinite) extending the definition to the remaining sets of $\tilde{\mathscr{F}}_T$. Thus an $\tilde{\omega}$ measure of $\tilde{\mathscr{F}}_T$ sets can be defined either as the one induced by a given family of random variables, or as one induced by a given (consistent) family of finite dimensional probability distributions. In either case we finally obtain a family of random variables $\{\tilde{x}_t, t \in T\}$.

The advantage of dealing with the family $\{\tilde{x}_t, t \in T\}$ is that we have complete control over the class of measurable sets. We have already remarked that for a general family of random variables $\{x_t, t \in T\}$ probabilities of sets not in $\mathscr{F}_T'$ may be defined, but if defined these probabilities are to a certain extent arbitrary, not solely determined by the probabilities of $\mathscr{F}_0$ sets and the properties of measure in general. For certain purposes to be discussed in §2, it is necessary to define probabilities of sets not in $\mathscr{F}_T'$, and these must be defined in a specific way to obtain a fruitful theory. Since they may already be defined in some other way, the theory runs into real difficulties here. The two best ways of overcoming the difficulty seem to be the following: (a) one can modify the random variables of a given family in such a way that the finite dimensional distributions are unaltered, and that the sets whose measurability is desired turn out to be in $\mathscr{F}_T'$; (b) one can base the theory on the space $\tilde{\Omega}$ and field $\tilde{\mathscr{F}}_T'$ of measurable sets, defining probabilities of sets not in $\tilde{\mathscr{F}}_T'$ in any way desired that is compatible with the general properties of measure. The first method will be adopted in this book in most of the discussion. Both possibilities will be discussed in §2.

2. The scope of probability measure

Let $\{x_n, n \geq 1\}$ be a stochastic process. The arithmetical operations performed on the x_n's and those involved in finding bounds and limits of x_n's always lead to functions of ω which are random variables, that is, which are measurable. For example, $\operatorname{L.U.B.}_n |x_n|$ is a random variable, because the set equality

$$\{\operatorname*{L.U.B.}_{n} |x_n(\omega)| > \lambda\} = \bigcup_{1}^{\infty} \{|x_n(\omega)| > \lambda\} \tag{2.1}$$

shows that the ω set on the left is measurable, as the union of a sequence of measurable sets. (The usual conventions are made if the least upper bound is infinite.)

The situation is more complicated, however, if one deals with a non-denumerable family, say $\{x_t, t \in I\}$, where I is an interval. In this case (2.1) becomes

$$\{\underset{t \in I}{\text{L.U.B.}}\, |x_t(\omega)| > \lambda\} = \bigcup_{t \in I} \{|x_t(\omega)| > \lambda\}. \tag{2.1'}$$

Since the union on the right involves non-denumerably many sets, the equality does not imply the measurability of the set on the left, and in fact it is not measurable in general. Continuing this investigation, we find that the probabilities that the sample functions are bounded, are continuous, are measurable, are integrable, and so on, may not be defined because the corresponding ω sets may not be measurable. What is worse is that even if these sets are measurable their measures are not uniquely determined by the finite dimensional distributions, and in fact for given finite dimensional distributions there may be considerable latitude in the measures of these sets. This means that the choice that happens to have been made in the original definition of ω measure may not be the one which leads to a fruitful theory.

As an example consider the following very simple case. The parameter set T is the infinite line, $-\infty < t < \infty$. The finite dimensional distributions are determined by

$$\mathbf{P}\{x_t(\omega) = 0\} = 1, \qquad -\infty < t < \infty.$$

Then, if S is any finite or enumerably infinite set (and only in that case), it follows that

$$\mathbf{P}\{x_t(\omega) \equiv 0, t \in S\} = 1.$$

It is desirable for many purposes to assert that

$$\mathbf{P}\{x_t(\omega) \equiv 0, -\infty < t < \infty\} = 1.$$

However, if M is the ω set determined by the condition in the brace, examples will be given at the end of this section in which M is not measurable, in which M is measurable and $\mathbf{P}\{M\} = 1$, in which it is measurable and $\mathbf{P}\{M\} = 0$. The middle case is of course the desirable one, but the others cannot be excluded without introducing some new criterion. The criterion of separability introduced in the next paragraph appears to be the simplest appropriate criterion.

Let $\{x_t, t \in T\}$ be a real stochastic process with linear parameter set T. Let $\mathscr{A}$ be a system of linear Borel sets. The process will be called *separable relative to* $\mathscr{A}$ if there is a sequence $\{t_j\}$ of parameter values and an ω set Λ of probability 0 such that, if $A \in \mathscr{A}$, and if I is any open interval, the ω sets

$$\{x_t(\omega) \in A, t \in IT\}, \qquad \{x_{t_j}(\omega) \in A, t_j \in IT\}$$

differ by at most a subset of Λ. The second of these two ω sets is obviously a measurable ω set which contains the first. The first set is then necessarily also measurable, under the separability hypothesis. The first set is not necessarily measurable in general unless IT is at most denumerable. The two most important special cases are:

(i) $\mathscr{A}$ is the class $\mathscr{A}_1$ of all (finite or infinite) closed intervals. This is the smallest class $\mathscr{A}$ for which the concept of separability is useful, and we shall accordingly write *separable* instead of *separable relative to* $\mathscr{A}_1$.

(ii) $\mathscr{A}$ is the class of all closed sets.

Obviously separability relative to a class of sets implies separability relative to any smaller class. For the purposes of this book we shall need only separability relative to the class $\mathscr{A}_1$ of closed intervals, but for some questions separability relative to a larger class is required. Examples will be given below. In the following we shall call a sequence $\{t_n\}$ a sequence *satisfying the conditions of the separability definition* (relative to a specified class $\mathscr{A}$, or relative to $\mathscr{A}_1$ if none is specified), if there is an ω set Λ which in conjunction with the sequence has the properties stated in the separability definition.

It is obvious how the concept of separability is extended to abstract-valued random variables. In particular, for complex-valued random variables, the above definition need be changed only by making the sets of the class $\mathscr{A}$ two-dimensional Borel sets. We shall say that the process is separable if the processes determined by the real and imaginary parts of the process random variables are separable (relative to $\mathscr{A}_1$). In other words, in the complex case the minimal class corresponding to $\mathscr{A}_1$ in the real case is the class of closed rectangles with sides parallel to the coordinate axes.

Going back to the case of real valued random variables: according to the definition of separability relative to the class of closed intervals, if Λ has the properties stated in the definition, and if $\omega \notin \Lambda$, then

$$\begin{aligned} \underset{t \in IT}{\text{G.L.B.}}\, x_t(\omega) &= \underset{t_j \in IT}{\text{G.L.B.}}\, x_{t_j}(\omega), \\ \underset{t \in IT}{\text{L.U.B.}}\, x_t(\omega) &= \underset{t_j \in IT}{\text{L.U.B.}}\, x_{t_j}(\omega), \end{aligned} \tag{2.2}$$

for every open interval I. Conversely, if there is an ω set Λ with $\mathbf{P}\{\Lambda\} = 0$, such that if $\omega \notin \Lambda$ it follows that (2.2) is true for every open interval I, then the x_t process is obviously separable. Note that for all ω

$$\begin{aligned} \underset{t \in IT}{\text{G.L.B.}}\, x_t(\omega) &\le \underset{t_j \in IT}{\text{G.L.B.}}\, x_{t_j}(\omega), \\ \underset{t \in IT}{\text{L.U.B.}}\, x_t(\omega) &\ge \underset{t_j \in IT}{\text{L.U.B.}}\, x_{t_j}(\omega), \end{aligned}$$

so that if (2.2) is true it will remain true if more values of t are added to the sequence $\{t_j\}$. Hence the content of the statement is that for almost all ω (2.2) is true for a *sufficiently large* denumerable set $\{t_j\}$. We can evidently replace (2.2) by

$$(2.2') \qquad \lim_{n\to\infty} \underset{|t_j-t|<1/n}{\text{G.L.B.}}\, x_{t_j}(\omega) \le x_t(\omega) \le \lim_{n\to\infty} \underset{|t_j-t|<1/n}{\text{L.U.B.}}\, x_{t_j}(\omega), \qquad t \in T.$$

Separability implies that if I is an open interval

$$\underset{t\in IT}{\text{L.U.B.}}\, x_t(\omega), \qquad \underset{t\in IT}{\text{G.L.B.}}\, x_t(\omega), \qquad \limsup_{t\to\tau} x_t(\omega), \qquad \liminf_{t\to\tau} x_t(\omega)$$

are all (finite- or infinite-valued) random variables, that is, measurable functions, since the right sides of (2.2) are random variables. In connection with an example discussed above, we remark that if for each parameter value t

$$\mathbf{P}\{x_t(\omega) = 0\} = 1,$$

then, if $\{t_j\}$ is any sequence of parameter values,

$$\mathbf{P}\{x_{t_j}(\omega) = 0, j \ge 1\} = 1.$$

In particular, if the process is separable, and if $\{t_j\}$ is a sequence satisfying the conditions of the separability definition, the preceding equation implies that

$$\mathbf{P}\{x_t(\omega) = 0, t \in T\} = 1.$$

We have already remarked that this conclusion is not correct without some hypothesis like separability.

Before discussing the existence of separable processes, we prove three theorems for later reference.

THEOREM 2.1 *Let $\{x_t, t \in T\}$ be a real stochastic process.*

(i) *If there is a sequence $\{t_j\}$ of parameter values for which* (2.2) *is true unless $\omega \in \Lambda(I)$, where $\mathbf{P}\{\Lambda(I)\} = 0$, for every I, then the x_t process is separable, and in fact the sequence $\{t_j\}$ satisfies the conditions of the separability definition.*

(ii) *If the x_t process is separable, and if there is a sequence $\{t_j\}$ of parameter values for which* (2.2′) *is true unless $\omega \in \Lambda_t$, where $\mathbf{P}\{\Lambda_t\} = 0$ for every $t \in T$, then the sequence $\{t_j\}$ satisfies the conditions of the separability definition.*

To prove (i) define $\mathrm{M} = \bigcup_r \Lambda(I_r)$, where the I_r's are the intervals with rational endpoints. Then $\mathbf{P}\{\mathrm{M}\} = 0$ and (2.2) is true if $\omega \notin \mathrm{M}$ for every open interval I. The sequence $\{t_j\}$ thus satisfies the conditions of the separability definition. To prove (ii) let $\{\tau_j\}$ be a sequence of parameter values satisfying the conditions of the separability definition, so that

(2.2) is true with the τ_j's unless $\omega \in \mathrm{N}$, where $\mathbf{P}\{\mathrm{N}\} = 0$. Then, unless $\omega \in \mathrm{N} \cup \bigcup_j \Lambda_{\tau_j}$,

$$\underset{t_j \in IT}{\text{G.L.B.}}\, x_{t_j}(\omega) \leq \underset{\tau_j \in IT}{\text{G.L.B.}}\, x_{\tau_j}(\omega) = \underset{t \in IT}{\text{G.L.B.}}\, x_t(\omega)$$

$$\underset{t_j \in IT}{\text{L.U.B.}}\, x_{t_j}(\omega) \geq \underset{\tau_j \in IT}{\text{L.U.B.}}\, x_{\tau_j}(\omega) = \underset{t \in IT}{\text{L.U.B.}}\, x_t(\omega).$$

The inequalities must be equalities since the third term in the first (second) line is certainly not larger (smaller) than the first term. Hence the sequence $\{\tau_k\}$ satisfies the conditions of the separability definition.

THEOREM 2.2 *Let $\{x_t,\ t \in T\}$ be a separable stochastic process, and suppose that, for every $\tau \in T$,*

$$\underset{t\to\tau}{\text{p lim}}\, x_t = x_\tau.$$

(i) *If $\{t_j\}$ is any sequence of parameter values dense in T, this sequence satisfies the conditions of the separability definition.*

(ii) *Let I: $[a, b]$ be any finite closed interval containing points of T, and suppose that, for each n, $a \leq s_0^{(n)} < \cdots < s_{a_n}^{(n)} \leq b$, with $s_j^{(n)} \in T$,*

$$\lim_{n\to\infty} \underset{t \in IT}{\text{L.U.B.}}\, \underset{j \leq a_n}{\text{Min}}\, |t - s_j^{(n)}| = 0,$$

(that is, the $s_j^{(n)}$'s become dense in $[a, b]T$ when $n \to \infty$). Then it follows that, if the process is real,

$$\lim_{n\to\infty} \underset{j}{\text{Min}}\, x_{s_j^{(n)}}(\omega) = \underset{t \in IT}{\text{G.L.B.}}\, x_t(\omega)$$

$$\lim_{n\to\infty} \underset{j}{\text{Max}}\, x_{s_j^{(n)}}(\omega) = \underset{t \in IT}{\text{L.U.B.}}\, x_t(\omega)$$

with probability 1.

The truth of (i) is an obvious consequence of Theorem 2.1 (ii). We prove the second of the two limit relations in (ii) in the real case. Let $\{t_j\}$ be a sequence of parameter values satisfying the conditions of the separability definition. We can suppose that, if a or b is in T, the endpoint is also a t_j.

Fix the parameter value $t \in [a, b]$, and for each n choose j in such a way that $s_n = s_j^{(n)} \to t\ (n \to \infty)$. Then by hypothesis

$$x_t = \underset{n\to\infty}{\text{p lim}}\, x_{s_n}.$$

Hence there is probability 1 convergence for some subsequence of values of n. This fact implies the truth with probability 1 of the far weaker inequality

$$x_t(\omega) \leq \liminf_{n\to\infty} \underset{k \leq a_n}{\text{Max}}\, x_{s_k^{(n)}}(\omega).$$

But then, letting t run through the points of any sequence $\{t_j\}$ satisfying the conditions of the separability definition,

$$\underset{t \in IT}{\text{L.U.B.}}\, x_t(\omega) = \underset{t_j \in IT}{\text{L.U.B.}}\, x_{t_j}(\omega) \leq \liminf_{n\to\infty} \underset{k \leq a_n}{\text{Max}}\, x_{s_k^{(n)}}(\omega)$$

with probability 1. Since the reverse inequality is obvious, we have now derived the desired result.

In the following we shall frequently find it convenient to simplify the typography by using the notation $x(t, \omega)$ and $x(t)$ instead of $x_t(\omega)$ and x_t.

THEOREM 2.3 *Let $\{x(t),\ t \in T\}$ be a real separable stochastic process, and suppose that τ is a limit point of the t set $T\{t > \tau\}$. There is then a sequence $\{\tau_n\}$ in T such that*

$$\tau_1 > \tau_2 > \cdots,\ \tau_n \downarrow \tau,$$

$$\limsup_{n\to\infty} x(\tau_n) = \limsup_{t\downarrow\tau} x(t)$$

$$\liminf_{n\to\infty} x(\tau_n) = \liminf_{t\downarrow\tau} x(t),$$

with probability 1.

This theorem implies for example that, if $\lim_{n\to\infty} x(s_n)$ exists with probability 1 whenever $s_n \downarrow \tau$, then, even if the exceptional ω set may depend on the sequence $\{s_n\}$ in the hypotheses, nevertheless $\lim_{t\downarrow\tau} x(t)$ exists with probability 1, that is, almost all sample functions have limits when $t \downarrow \tau$. In proving the theorem we shall suppose that the superior and inferior limits involved are finite; if this is not true initially, it will be true if $x(t)$ is replaced by arctan $x(t)$. Now choose $t_1, t_2, \cdots$ to satisfy the conditions of the separability definition, so that (2.2) is satisfied with probability 1, for all open intervals I. Then for each n choose a finite number of t_j's, say $s_1^{(n)}, s_2^{(n)}, \cdots$, to satisfy

$$\tau < s_j^{(n)} < \tau + \frac{1}{n}$$

$$\mathbf{P}\left\{\underset{\tau<t<\tau+1/n}{\text{L.U.B.}}\, x(t, \omega) - \underset{j}{\text{Max}}\, x(s_j^{(n)}, \omega) > \frac{1}{n}\right\} < \frac{1}{n}$$

$$\mathbf{P}\left\{\underset{\tau<t<\tau+1/n}{\text{G.L.B.}}\, x(t, \omega) - \underset{j}{\text{Min}}\, x(s_j^{(n)}, \omega) < -\frac{1}{n}\right\} < \frac{1}{n}.$$

If $\{\tau_n\}$ is defined as the $s_j^{(n)}$'s ordered into a monotone sequence, $\tau_n \downarrow \tau$ and

$$\underset{n\to\infty}{\text{p lim}}\ [\underset{\tau<t<\tau+1/n}{\text{L.U.B.}}\, x(t) - \underset{\tau_j<\tau+1/n}{\text{L.U.B.}}\, x(\tau_j)] = 0$$

$$\underset{n\to\infty}{\text{p lim}}\ [\underset{\tau<t<\tau+1/n}{\text{G.L.B.}}\, x(t) - \underset{\tau_j<\tau+1/n}{\text{G.L.B.}}\, x(\tau_j)] = 0.$$

These limit equations imply the truth of the theorem, since the least upper and greatest lower bounds involved converge to the corresponding superior and inferior limits.

The preceding two theorems show the fundamental importance of the concept of separability of a process. It has not yet been explained how much of a restriction it is on a process $\{x_t, t \in T\}$ to suppose that it is separable. Obviously separability is no restriction at all if the parameter set is denumerable, because in that case the parameter set is itself a sequence $\{t_j\}$ which satisfies the conditions of the separability condition (relative to every class $\mathscr{A}$). The following lemma is fundamental. Note that the proof does not use our standard assumption that the parameter space T is linear. The following discussion can thus be generalized to abstract parameter sets as well as to abstract-valued random variables.

LEMMA 2.1 *Let $\{x_t, t \in T\}$ be a stochastic process. To each linear Borel set A there corresponds a finite or enumerable sequence $\{t_n\}$ such that*

$$\mathbf{P}\{x_{t_n}(\omega) \in A, n \geq 1;\ x_t(\omega) \notin A\} = 0, \qquad t \in T. \tag{2.3}$$

More generally, let $\mathscr{A}_0$ be an at most enumerable class of linear Borel sets, and let $\mathscr{A}$ be the class of sets which are intersections of sequences of $\mathscr{A}_0$ sets. Then there is a finite or enumerable sequence $\{t_n\}$ such that to each $t \in T$ corresponds an ω set Λ_t with $\mathbf{P}\{\Lambda_t\} = 0$ and

$$\{x_{t_n}(\omega) \in A, n \geq 1;\ x_t(\omega) \notin A\} \subset \Lambda_t, \qquad A \in \mathscr{A}. \tag{2.4}$$

We prove first that the truth of the first part of the lemma implies that of the second apparently more general part. In fact, if the first part is true, to each $A \in \mathscr{A}_0$ there corresponds a certain parameter sequence for which (2.3) is true, and then (2.3) is true for all $A \in \mathscr{A}_0$ if $\{t_n\}$ is the union of all these parameter sequences. Moreover, with this definition of $\{t_n\}$, if the ω set in (2.3) is denoted by $\Lambda_t(A)$, define

$$\Lambda_t = \bigcup_{A \in \mathscr{A}_0} \Lambda_t(A).$$

Then, if $A \in \mathscr{A}$, if $A_0 \in \mathscr{A}_0$, and if $A \subset A_0$,

$$\{x_{t_n}(\omega) \in A, n \geq 1;\ x_t(\omega) \notin A_0\} \subset \{x_{t_n}(\omega) \in A_0, n \geq 1;\ x_t(\omega) \notin A_0\} \subset \Lambda_t,$$

and the truth of (2.4) follows from the hypothesis that A is the intersection of a sequence of $\mathscr{A}_0$ sets. To prove the truth of the first statement of the lemma, let t_1 be any point of T. If $t_1, \cdots, t_k$ have already been chosen, define

$$\rho_k = \underset{t \in T}{\text{L.U.B.}}\ \mathbf{P}\{x_{t_n}(\omega) \in A, n \leq k;\ x_t(\omega) \notin A\}.$$

Then $\rho_1 \geq \rho_2 \geq \cdots$. If $\rho_k = 0$, then $t_1, \cdots, t_k$ is the desired sequence. If $\rho_k > 0$, choose t_{k+1} as any value of t such that the probability on the right exceeds $\rho_k(1 - 1/k)$. Then, if $\rho_k > 0$ for all k,

$$\mathbf{P}\{x_{t_n}(\omega) \in A, n \geq 1;\ x_t(\omega) \notin A\} \leq \lim_{k\to\infty} \rho_k, \qquad t \in T.$$

Since the ω sets

$$\{x_{t_n}(\omega) \in A, n \leq k;\ x_{t_{k+1}}(\omega) \notin A\}, \qquad k \geq 1$$

are disjunct, their probabilities form a convergent series, so that

$$\lim_{k\to\infty} \rho_k = \lim_{k\to\infty} \rho_k\left(1 - \frac{1}{k}\right) \leq \lim_{k\to\infty} \mathbf{P}\{x_{t_n}(\omega) \in A, n \leq k;\ x_{t_{k+1}}(\omega) \notin A\}$$
$$= 0.$$

This equality, combined with the preceding inequality, implies the truth of the first part of the lemma.

The following theorem shows that separability relative to the class of closed sets is no restriction on the finite dimensional distributions of a stochastic process $\{x_t, t \in T\}$, that is, on the joint distributions of finite aggregates of the x_t's. In the language of §1 this means that separability relative to the class of closed sets is no restriction on the probabilities of sets in the field $\mathscr{F}_T$. In other words, the condition is a restriction only on a probability involving non-denumerably many x_t's. This result is the best that could be hoped for.

THEOREM 2.4 *Let $\{x_t, t \in T\}$ be a stochastic process with linear parameter set T. There is then a stochastic process $\{\tilde{x}_t, t \in T\}$, defined on the same ω space, separable relative to the class of closed sets, with the property that*

$$\mathbf{P}\{\tilde{x}_t(\omega) = x_t(\omega)\} = 1, \qquad t \in T. \tag{2.5}$$

(The $\tilde{x}_t$'s may take on the values $\pm\infty$.)

Note that the joint distribution of finitely many of the $\tilde{x}_t$'s is exactly the same as that of the corresponding x_t's. The ω set $\{\tilde{x}_t(\omega) = x_t(\omega)\}$ is of probability 0 for each t, but this ω set may vary with t. If the union of all these ω sets, as t varies, has probability 0, the x_t process is itself separable relative to the closed sets.

We give the proof of Theorem 2.4 in the real case. The trivial changes to cover the complex case will be obvious. Let $\mathscr{A}_0$ be the class of linear sets which are finite unions of open or closed intervals with rational or infinite endpoints, and let $\mathscr{A}$ be the class of sets which are intersections of sequences of $\mathscr{A}_0$ sets. Then $\mathscr{A}$ includes the closed sets. Let I be an open interval with rational or infinite endpoints. We apply the preceding lemma to the stochastic process $\{x_t, t \in IT\}$, with $\mathscr{A}_0$, $\mathscr{A}$ as just defined.

According to the lemma, there is an at most enumerable set $T(I) \subset IT$, and an ω set $\Lambda_{t,I}$, such that

$$\mathbf{P}\{\Lambda_{t,I}\} = 0, \qquad t \in IT$$

$$\{x_s(\omega) \in A,\ s \in T(I);\ x_t(\omega) \notin A\} \subset \Lambda_{t,I}, \qquad A \in \mathscr{A}.$$

Define

$$S = \bigcup_I T(I), \qquad \Lambda_t = \bigcup_I \Lambda_{t,I}.$$

Let $A(I, \omega)$ be the closure of the set of values of $x_s(\omega)$ for fixed ω and s varying in IS. The set $A(I, \omega)$ may include the values $\pm\infty$. It is closed, non-empty, and

$$x_t(\omega) \in A(I, \omega) \qquad \text{if} \quad t \in IT,\ \omega \notin \Lambda_t.$$

Hence, if the set $A(t, \omega)$ is defined by

$$A(t, \omega) = \bigcap_{I \ni t} A(I, \omega),$$

it follows that this set is closed, non-empty, and

$$x_t(\omega) \in A(t, \omega) \qquad \text{if} \quad t \in T,\ \omega \notin \Lambda_t.$$

For each t, ω we now define $\tilde{x}_t(\omega)$ as follows:

$$\begin{aligned} \tilde{x}_t(\omega) &= x_t(\omega), \qquad && t \in S \\ &= x_t(\omega), && t \notin S,\ \omega \notin \Lambda_t, \end{aligned}$$

and define $\tilde{x}_t(\omega)$ as any value in $A(t, \omega)$ if $t \notin S$ and $\omega \in \Lambda_t$. With this definition we proceed to show that the $\tilde{x}_t$ process satisfies the conditions of the theorem. The condition (2.5) is obviously satisfied. Let A be a closed set. Suppose that I is an open interval with rational or infinite endpoints, and that ω has the property that

$$\tilde{x}_s(\omega) \in A, \qquad s \in SI,$$

that is, $A(I, \omega) \subset A$. It follows from our definition of $\tilde{x}_t(\omega)$ that if $t \in IT$ then

$$\begin{aligned} \tilde{x}_t(\omega) = x_t(\omega) \quad &\in A(I, \omega) \qquad && \text{if } t \in S \\ & && \text{or} \quad \text{if } t \notin S,\ \omega \notin \Lambda_t, \\ &\in A(t, \omega) \subset A(I, \omega) \subset A && \text{if } t \notin S,\ \omega \in \Lambda_t. \end{aligned}$$

Thus

$$\{\tilde{x}_s(\omega) \in A,\ s \in IS\} = \{\tilde{x}_t(\omega) \in A,\ t \in IT\},$$

if A is closed and if I has rational or infinite endpoints. If I' is any open interval, it can be expressed in the form

$$I' = \bigcup_n I_n,$$

where I_n has rational or infinite endpoints. Since the above set equality is true for $I = I_n$, it is also true (taking the intersection in n) for $I = I'$. This completes the proof of the theorem. We observe that we cannot exclude infinite values for $\tilde{x}_t(\omega)$, since the set $A(t, \omega)$ above may contain no finite values. In general, if there is a Borel set X to which the values of each x_t are restricted with probability 1, more precisely if

$$\mathbf{P}\{x_t(\omega) \in X\} = 1, \qquad t \in T,$$

it is possible to define $\tilde{x}_t(\omega)$ to take on values only in the closure of X on the infinite line (here the infinite line is supposed made compact by the adjunction of the points $\pm\infty$). For example, if X is a finite closed interval, this means that the $\tilde{x}_t$'s can be supposed to have values only in this same interval, and are therefore necessarily finite-valued. If X is the set of positive integers, the range of values of each $\tilde{x}_t$ will be the set of positive integers, and also the point ∞, unless the latter value can be excluded by further information on the actual distributions involved. In any event, of course, for each t, $\tilde{x}_t$ is finite-valued with probability 1.

We conclude our discussion of separability with a few remarks on Theorems 2.1 and 2.2. In those theorems we considered only separability relative to the class $\mathscr{A}_1$ of closed intervals, and a few remarks on the extension of the theorems to separability relative to the class of closed sets are now in order. We omit the details since we shall not use the results in this book. Let I be an open interval, let $\{x_t, t \in T\}$ be a stochastic process, and let S be an at most enumerable subset of T. Let $A(I, \omega)$ be the closure of the set of values of $x_s(\omega)$ for $s \in IS$, let $A(t, \omega) = \bigcap_{I \ni t} A(I, \omega)$, and let $A'(I, \omega)$, $A'(t, \omega)$ be defined in the same way except that in the definition of $A'(I, \omega)$ s is not restricted to lie in S. By definition, S satisfies the conditions of the separability definition relative to the class of closed sets if there is an ω set Λ with $\mathbf{P}\{\Lambda\} = 0$, such that for every open interval I and closed set A the ω sets

$$\{x_t(\omega) \in A,\ t \in IS\}, \qquad \{x_t(\omega) \in A,\ t \in IT\}$$

differ by a subset of Λ, that is, if

$$A(I, \omega) = A'(I, \omega), \qquad \omega \notin \Lambda.$$

Moreover [cf. Theorem 2.1. (i)] it is even sufficient if the condition is (apparently) weakened by allowing Λ to depend on I. If the x_t process is known to be separable relative to the class of closed sets, and if

$$\mathbf{P}\{x_t(\omega) \in A(t, \omega)\} = 1, \qquad t \in T,$$

then [cf. Theorem 2.1 (ii)] the set S satisfies the conditions of the separability definition relative to the class of closed sets. This fact implies that Theorem 2.2 (i) is true for separability relative to the class of closed sets.

Let $\{x_t, t \in T\}$ be a stochastic process. In order to take full advantage of the apparatus of measure theory, it is necessary to suppose for some purposes that $x_t(\omega)$ defines a measurable function of the pair of variables (t, ω). Here t measure is taken as Lebesgue measure in T, ω measure as the given probability measure, and (t, ω) measure as the usual product of the two measures, supposed independent. The choice of Lebesgue measure on the t axis, rather than some other extension of a measure of Borel sets, is made because in the applications to be made one is usually interested in ordinary Lebesgue integrals of sample functions and in properties of stochastic processes invariant under translations of the parameter axis. We therefore make the following definition: *the stochastic process $\{x_t, t \in T\}$ is measurable if the parameter set T is Lebesgue measurable and if $x_t(\omega)$ defines a function measurable in the pair of variables (t, ω).*

THEOREM 2.5 *Let $\{x_t, t \in T\}$ be a separable process, with a Lebesgue measurable parameter set T. Suppose that there is a t set T_1 of Lebesgue measure 0 such that*

$$\mathbf{P}\{\lim_{s \to t} x_s(\omega) = x_t(\omega)\} = 1, \qquad t \in T - T_1.$$

Then the x_t process is measurable.

Define $U_t^{(n)}(\omega)$ by

$$U_t^{(n)}(\omega) = \underset{\substack{\frac{j}{n} \le s < \frac{j+1}{n} \\ s \in T}}{\text{L.U.B.}} x_s(\omega), \qquad \underset{t \in T}{\frac{j}{n} \le t < \frac{j+1}{n}}, \qquad j = 0, \pm 1, \cdots.$$

Then $\{U_t^{(n)}, t \in T\}$ is a family of random variables, and as we have defined this family $U_t^{(n)}$ is a single random variable, the above L.U.B., in the part of each t interval $\left[\frac{j}{n}, \frac{j+1}{n}\right)$ lying in T. Then $U_t^{(n)}(\omega)$ is (t, ω) measurable. We define $L_t^{(n)}(\omega)$ in the same way except that L.U.B. is replaced by G.L.B. Then

$$L_t^{(n)}(\omega) \le x_t(\omega) \le U_t^{(n)}(\omega).$$

According to the hypotheses of the theorem, for each $t \in T - T_1$, the extreme terms of this inequality converge to the middle term with probability 1, when $n \to \infty$. Since these extreme terms are (t, ω) measurable, it follows (Fubini's theorem) that the extreme terms have a common (t, ω) measurable limit for almost all (t, ω). Since this common limit must be $x_t(\omega)$, it follows that $x_t(\omega)$ must define a (t, ω) measurable function, as was to be proved.

The following theorem is general enough to cover all the specific stochastic processes discussed in this book. Its significance will be discussed below.

THEOREM 2.6 *Let $\{x_t, t \in T\}$ be a process with a Lebesgue measurable parameter set T. Suppose that there is a t set T_1 of Lebesgue measure 0 such that*

$$\operatorname*{p\,lim}_{s\to t} x_s = x_t, \qquad t \in T - T_1. \tag{2.6}$$

There is then a process $\{\tilde{x}_t, t \in T\}$, defined on the same ω space, which is separable relative to the closed sets, measurable, and for which

$$\mathbf{P}\{\tilde{x}_t(\omega) = x_t(\omega)\} = 1, \qquad t \in T.$$

(The $\tilde{x}_t$'s may take on the values $\pm\infty$.)

According to Theorem 2.4, it is no restriction to assume that the x_t process is separable relative to the closed sets, and we shall do so. We shall also assume that T is a bounded set, and that $|x_t(\omega)| \leq 1$ for all t, ω, since the general case can be reduced to this one by simple transformations. If I is an open interval, and if ω is fixed, let $A(I, \omega)$ be the closure of the range of values of $x_t(\omega)$ for $t \in IT$, and define

$$A(t, \omega) = \bigcap_{I \ni t} A(I, \omega).$$

Let $\{t_n\}$ be a parameter sequence satisfying the conditions of the separability (relative to the closed sets) condition. We observe that $x_t(\omega) \in A(t, \omega)$, and that any process $\{\tilde{x}_t, t \in T\}$ satisfying the conditions

$$\begin{aligned} \mathbf{P}\{\tilde{x}_{t_j}(\omega) = x_{t_j}(\omega)\} &= 1, & j &\geq 1, \\ \tilde{x}_t(\omega) &\in A(t, \omega), & \text{all } t&, \omega, \end{aligned}$$

is necessarily separable relative to the closed sets. For each positive integer n let $s_1^{(n)}, \cdots, s_n^{(n)}$ be the values $t_1, \cdots t_n$ arranged in increasing order, and define $s_0^{(n)} = -\infty$,

$$f_n(t, \omega) = x_{s_j^{(n)}}(\omega), \qquad \text{if} \quad s_{j-1}^{(n)} \leq t < s_j^{(n)}, \quad j \geq 1.$$

Then f_n is (t, ω) measurable, and according to (2.6)

$$\operatorname*{p\,lim}_{n\to\infty} f_n(t, \cdot) = x_t, \qquad t \in T - T_1.$$

This limit equation implies that

$$\lim_{m,\, n\to\infty} \mathbf{E}\{|f_n(t, \omega) - f_m(t, \omega)|\} = 0, \qquad t \in T - T_1,$$

and hence

$$\lim_{m,\, n\to\infty} \int_T \mathbf{E}\{|f_n(t, \omega) - f_m(t, \omega)|\}\, dt = 0.$$

It follows that the sequence $\{f_n\}$ converges in (t, ω) measure, so some subsequence $\{f_{n_j}\}$ converges for almost all (t, ω), to a (t, ω) measurable limit function f, defined only at the points of convergence of this subsequence. By Fubini's theorem, there is a subset T_0 of T, of Lebesgue measure 0, such that the sequence of ω functions $\{f_{n_j}(t, \cdot)\}$ converges with probability 1 if $t \in T - T_0$. We can suppose that $T_0 \supset T_1 \cup \{t_j\}$. Then, by (2.6),

$$\mathbf{P}\{x_t(\omega) = f(t, \omega)\} = 1, \qquad t \in T - T_0.$$

Note that the function $f(t, \cdot)$ is not necessarily defined for all ω. Now define $\tilde{x}_t(\omega)$ as follows. If $f(t, \omega)$ is defined, and if $t \in T - T_0$, define $\tilde{x}_t(\omega) = f_t(\omega)$. If $f(t, \omega)$ is not defined, or if $t \in T_0$, define $\tilde{x}_t(\omega) = x_t(\omega)$. According to this definition,

$$\mathbf{P}\{\tilde{x}_t(\omega) = x_t(\omega)\} = 1, \qquad t \in T.$$

Now $\tilde{x}_{t_j} = x_{t_j}$, since $t_j \in T_0$, and furthermore $\tilde{x}_t(\omega) \in A(t, \omega)$ for all t, ω, so that, according to a remark made at the beginning of the proof, the process $\{\tilde{x}_t, t \in T\}$ is separable relative to the closed sets. Finally, this process is measurable, because $\tilde{x}_t(\omega) = f(t, \omega)$ for almost all t, ω.

The proof just given uses only the fact that (2.6) is true when $s \to t$ from above, and the hypotheses of the theorem can be correspondingly weakened. However, the theorem is not really made stronger in this way. In fact, if for each t in a parameter set either

$$\underset{s \uparrow t}{\text{p lim}}\, x_s \qquad \text{or} \qquad \underset{s \downarrow t}{\text{p lim}}\, x_s$$

exists, then it is easily seen that (VII, Theorem 11.1) except for an at most enumerable subset of this parameter set both exist and are x_t.

The following theorem is typical of the application of the concept of measurability of a stochastic process. The theorem justifies all integrals of sample functions used in this book.

THEOREM 2.7 *Let $\{x_t, t \in T\}$ be a measurable stochastic process. Then almost all sample functions of the process are Lebesgue measurable functions of t. If $\mathbf{E}\{x_t(\omega)\}$ exists for $t \in T$, it defines a Lebesgue measurable function of t. If A is a Lebesgue measurable parameter set and if*

$$\int_A \mathbf{E}\{|x_t(\omega|\}\, dt < \infty,$$

then almost all sample functions are Lebesgue integrable over A.

By hypothesis $x_{\cdot}(\cdot)$ is (t, ω) measurable. It follows (Fubini's theorem) that $x_{\cdot}(\omega)$ is Lebesgue measurable in t for almost all ω, that is, almost all sample functions are Lebesgue measurable, and that $\mathbf{E}\{x_t(\omega)\}$ defines a measurable function of t, if this expectation exists. In particular,

$\mathbf{E}\{|x_t(\omega)|\}$ defines a Lebesgue measurable function of t, not necessarily finite-valued. The second hypothesis of the theorem is that the iterated integral of $|x_t(\omega)|$, first in ω and then in $t \in A$, is finite. The iterated integral in the reverse order is then finite, and the integral

$$\int_A |x_t(\omega)|\, dt$$

is therefore finite for almost all ω, that is, almost all sample functions are Lebesgue integrable over T, as was to be proved. Since the value of an absolutely convergent iterated integral is independent of the order of integration,

$$\mathbf{E}\Big\{\int_A x_t(\omega)\, dt\Big\} = \int_A \mathbf{E}\{x_t(\omega)\}\, dt.$$

Suppose that f is a Lebesgue measurable and integrable function defined on the finite interval $[a, b]$. It will be important for some purposes to approximate the integral of f by a Riemann sum

$$R[f, s_1, \cdots, s_n] = \sum_0^n f(s_j)(s_{j+1} - s_j), \quad s_0 = a < \cdots < s_{n+1} = b. \tag{2.7}$$

It is clear that R may not be a good approximation to the integral even if $\delta = \operatorname{Max}(s_{j+1} - s_j)$ is small, since we have not supposed that f is Riemann integrable, but it will be shown that R is a good approximation for properly chosen s_j's. If $0 \leq t < b - a$, and if we translate $s_1, \cdots, s_n$ to $s_1 + t, \cdots, s_n + t$ (where we suppose that $s_j + t \neq b$ and decrease $s_j + t$ to $s_j + t - b + a$ whenever $s_j + t > b$), we obtain a new Riemann sum $R_t[f, s_1, \cdots, s_n]$, and we shall show that for most values of t, in a sense made precise below, R_t is a good approximation to the integral of f if δ is small. To show this it is convenient to define

$$\begin{aligned} f(t) &= f(t - b + a), \qquad b < t < 2b - a, \\ R_t'[f, s_1, \cdots, s_n] &= \sum_1^{n-1} f(s_j + t)(s_{j+1} - s_j). \end{aligned} \tag{2.7'}$$

Then it is trivial to verify that R_t and R_t' differ by at most three summands in the sum defining R_t and one in the sum defining R_t', so that

$$\lim_{\delta\to 0} \int_0^{b-a} |R_t - R_t'|\, dt = 0. \tag{2.8}$$

We now prove that

$$\lim_{\delta\to 0} \int_0^{b-a} \Big|R_t[f, s_1, \cdots, s_n] - \int_a^b f(s)\, ds\Big|\, dt = 0. \tag{2.9}$$

This equation shows, among other things, that for small δ the measure of the t set for which the translated s_j's do not give a Riemann sum which is a good approximation to the Lebesgue integral is small. To prove (2.9) let ε be a positive number, and let f_ε be a continuous function on $[a, 2b-a]$, with

$$\int_a^{2b-a} |f(t)-f_\varepsilon(t)|\,dt \le \varepsilon.$$

Then

$$\int_0^{b-a} \Big|R_t'[f, s_1, \cdots, s_n] - \int_a^b f(s)\,ds\Big|\,dt$$

$$= \int_0^{b-a} \Big|R_t'[f, s_1, \cdots, s_n] - \int_a^b f(s+t)\,ds\Big|\,dt$$

$$\le \sum_1^{n-1} \int_{s_j}^{s_{j+1}} ds \int_0^{b-a} |f(s_j+t)-f(s+t)|\,dt$$

$$\le \sum_1^{n-1} \int_{s_j}^{s_{j+1}} ds \int_0^{b-a} |f_\varepsilon(s_j+t)-f_\varepsilon(s+t)|\,dt + 2\varepsilon(b-a)$$

$$\to 2\varepsilon(b-a), \qquad (\delta \to 0).$$

Since ε can be taken arbitrarily near 0, (2.9) is true if R_t is replaced by R_t', and therefore is true as written, in view of (2.8). The following theorem applies this result to stochastic processes.

THEOREM 2.8 *Let $\{x_t,\ a \le t \le b\}$ be a measurable stochastic process (a, b finite). Then in the notation of the preceding paragraph*

$$\lim_{\delta\to 0} \int_0^{b-a} \Big|R_t[x_.(\omega), s_1, \cdots, s_n] - \int_a^b x_s(\omega)\,ds\Big|\,dt = 0 \tag{2.10}$$

for all Lebesgue measurable and integrable sample functions. If almost all sample functions are Lebesgue measurable and integrable, to every $\varepsilon > 0$ corresponds a choice of the s_j's, say $s_j = t_j$, with

$$\delta < \varepsilon, \qquad \mathbf{P}\Big\{\Big|R[x_.(\omega), t_1, \cdots, t_n] - \int_a^b x_s(\omega)\,ds\Big| \ge \varepsilon\Big\} \le \varepsilon. \tag{2.11}$$

The first statement of the theorem is a trivial application of what we have proved. To prove the second statement let $u(t, \omega)$ be the absolute value in (2.10). Then (2.10) implies that, if $\varepsilon > 0$, the Lebesgue measure of the t set where $u \geq \varepsilon$ goes to 0 with δ for almost all ω. It follows that the (t, ω) measure of the (t, ω) set where $u(t, \omega) \geq \varepsilon$ goes to 0, that is,

$$\lim_{\delta \to 0} \int_0^{b-a} \mathbf{P}\{u(t, \omega) \geq \varepsilon\}\, dt = 0.$$

Then, if δ is sufficiently small,

$$\mathbf{P}\{u(t, \omega) \geq \varepsilon\} \leq \varepsilon$$

on a t set of positive measure. Then this inequality is true for some s_j's with $\delta \leq \varepsilon$, and some t. The set of t_j's of (2.11) can be taken as the set of s_j's translated through t.

In particular if

$$\int_a^b \mathbf{E}\{|x_t(\omega)|\}\, dt < \infty \tag{2.12}$$

it is easy to show that

$$\lim_{\delta \to 0} \mathbf{E}\left\{\int_0^{b-a} \left|R_t[x_.(\omega), s_1, \cdots, s_n] - \int_a^b x_s(\omega)\, ds\right| dt\right\} = 0. \tag{2.13}$$

In fact we need only observe that we have just proved that the integrand of this (t, ω) integral goes to 0 with δ in (t, ω) measure, and it is easy to verify that the integrands are uniformly integrable, so that integration to the limit is legitimate.

Finally we observe that, if we suppose that (2.12) is true, and also assume

$$\lim_{t \to s} \mathbf{E}\{|x_t - x_s|\} = 0, \qquad a \leq s \leq b, \tag{2.14}$$

then the averaging integration in t in (2.10) and (2.13) becomes unnecessary, because it can be shown that with these hypotheses

$$\lim_{\delta \to 0} \mathbf{E}\left\{\left|R[x_.(\omega), s_1, \cdots, s_n] - \int_a^b x_s(\omega)\, ds\right|\right\} = 0.$$

The limit equations we have discussed make it possible to define $\int_a^b x_t(\omega)\, dt$ as a random variable which is a limit in a suitable sense of Riemann sums. However, the use of Riemann sums to define the integral loses the useful literal interpretation of the integral as the ordinary integral of the general sample function.

Let $\{x_t, t \in T\}$ be a stochastic process. In §1 we defined a field $\mathscr{F}_T{}'$ of measurable ω sets, the smallest Borel field of sets with respect to which all the x_t's are ω measurable and which contains every subset of any one of its sets of probability 0. The finite dimensional distributions determine the measures of $\mathscr{F}_T'$ sets, but do not determine the measure of any measurable set not in $\mathscr{F}_T'$. Let $\{\tilde{x}_t, t \in T\}$ be a stochastic process with

$$\mathbf{P}\{\tilde{x}_t(\omega) = x_t(\omega)\} = 1, \qquad t \in T,$$

and such that for each t the ω set where $\tilde{x}_t(\omega) \neq x_t(\omega)$ is an $\mathscr{F}_T'$ set. The x_t process will be called a *standard modification* of the x_t process in the following discussion. We have not made a point of it in the statements of the theorems, but a glance at their proofs shows that the $\tilde{x}_t$ processes in Theorems 2.4 and 2.6 are standard modifications of the x_t processes.

For many purposes it is legitimate to replace a stochastic process by a standard modification. This change does not affect the finite dimensional distributions, but may decrease the field $\mathscr{F}_T'$.

If $\{x_t, t \in T\}$ is a stochastic process, with standard measurable modifications $\{x_t^{(i)}, t \in T\}$, $i = 1, 2$, then

$$\mathbf{P}\{x_t^{(1)}(\omega) = x_t^{(2)}(\omega)\} = 1, \qquad t \in T,$$

and it follows (Fubini's theorem) that, for almost all ω,

$$x_t^{(1)}(\omega) = x_t^{(2)}(\omega)$$

for almost all t, that is, corresponding sample functions are equal for almost all t, with probability 1. Thus we can *define*

$$\int_A x_t(\omega)\, dt$$

as the corresponding integral for a measurable standard modification of the given process, obtaining a unique random variable, if values on ω sets of probability 0 are neglected. This will sometimes be convenient below.

According to Theorem 2.3 every process has a standard modification which is separable. (Examples of non-separable processes will be given below.) Separability is thus not a restriction on the finite dimensional distributions. On the other hand, if $\{x_t, t \in T\}$ is a process with Lebesgue measurable parameter set, the process is necessarily measurable if T has Lebesgue measure 0, whereas, if T has positive Lebesgue measure, the measurability of the process is a restriction on the finite dimensional distributions. The following is a trivial example of a process (with a Lebesgue measurable parameter set) which is not measurable and which has no standard modification which is measurable. We impose no restriction on Ω, but suppose that T has positive Lebesgue measure.

There is then a bounded function $a(\cdot)$ of $t \in T$, which is not Lebesgue measurable. Suppose that the finite dimensional distributions are determined by

$$\mathbf{P}\{x_t(\omega) = a(t)\} = 1, \qquad t \in T.$$

With this definition, if $\{\tilde{x}_t, t \in T\}$ is any standard modification of the x_t process, $\mathbf{E}\{\tilde{x}_t(\omega)\} = \mathbf{E}\{x_t(\omega)\} = a(t)$ and the $\tilde{x}_t$ process cannot be measurable, or $\mathbf{E}\{\tilde{x}_t(\omega)\}$ would be measurable by Theorem 2.7. Less trivial examples of processes with no standard modifications which are measurable processes will be given below.

In discussing the measurability of a process $\{x_t, t \in T\}$ with Lebesgue measurable parameter set T it is no restriction to assume that T is the infinite line $(-\infty, \infty)$. In fact, if T is not the infinite line, define $x_t(\omega) = 0$ for $t \notin T$ to obtain a new process with an enlarged parameter range. The new process is measurable if and only if the old one is, and there is a measurable standard modification of the new one if and only if there is a measurable standard modification of the old one.

Suppose then that $\{x_t, -\infty < t < \infty\}$ is a stochastic process, and define for each c

$$x_t^{(n)}(c, \omega) = x_{c+j/2^n}(\omega), \qquad \frac{j-1}{2^n} < t \leq \frac{j}{2^n},$$

$$j = 0, \pm 1, \cdots,$$

$$n = 0, 1, 2, \cdots.$$

It can be shown that the x_t process has a measurable standard modification if and only if there is a value of c for which

$$\mathbf{P}\{\lim_{n \to \infty} x_t^{(n)}(c, \omega) = x_{t+c}(\omega)\}$$

for each t not in some set of Lebesgue measure 0. (See the Appendix for references to equivalent results.) Moreover, it can be shown that there is a separable (relative to the closed sets) measurable standard modification whenever there is a measurable standard modification. Note that the stated necessary and sufficient condition is a condition on the measures of $\mathscr{F}_T$ sets, that is, on the finite dimensional distributions. This condition implies, for example, that if the random variables of the process are mutually independent and have a common distribution (not concentrated at a single point) the process has no measurable standard modification.

Now let $\{\tilde{x}_t, t \in T\}$ be a stochastic process in which $\tilde{\omega}$ space $\tilde{\Omega}$ is function space and $\tilde{x}_t$ is the tth coordinate function, as discussed in §1. We shall call such a process one of *function space type*. Processes of this type are the ones most commonly considered in the literature, since, as described in I §5, they can be defined simply by prescribing a mutually consistent collection of finite dimensional distributions. These processes have the simple property that the general basic point $\tilde{\omega}$ coincides with the general sample function of the process. In discussing stochastic processes it is not possible to suppose that all processes encountered are of function space type. This fact is illustrated in the following example. Suppose that one wishes to consider besides the basic random variables of the process, the $\tilde{x}_t$'s, some functions of the $\tilde{x}_t$'s, say the squares. In other words, one wishes to consider the process $\{\tilde{x}_t^2, t \in T\}$. This process no longer has the property that $\tilde{\omega}$ is the sample function of the process, it is no longer of function space type, and thus even in this trivial case it is necessary to use other processes.

In spite of the preceding example of the impossibility of considering only processes of function space type, it is enlightening to see how the concepts of separability and measurability can be handled in this case. The following discussion, which is given without proofs because we shall not use this point of view, shows how these concepts can be treated without going to other types of processes. Standard modifications cannot be used for this purpose, because a standard modification of a process of function space type need not be of function space type.

In our discussion of processes of function space type we shall use a notation which exhibits the range of values of the functions and the class of measurable sets. The process $\{\tilde{x}_t, t \in T, X, \tilde{\mathscr{F}}\}$ is the process of function space type in which $\tilde{\Omega}$ is the space of all functions with domain T and range X, and $\tilde{\mathscr{F}}$ is the class of measurable sets. In agreement with previous notation, $\tilde{\mathscr{F}}_T$ will denote the smallest Borel field of $\tilde{\omega}$ sets with respect to which the $\tilde{x}_t$'s are measurable, and $\tilde{\mathscr{F}}_T'$ will denote the smallest Borel field of $\tilde{\omega}$ sets such that $\tilde{\mathscr{F}}_T' \supset \tilde{\mathscr{F}}_T$ and that $\tilde{\mathscr{F}}_T'$ contains all subsets of its sets of probability 0. It will sometimes be necessary to suppose that X is a closed point set of the infinite line (or plane in the complex case), and in speaking of such a closed set we shall consider that the finite line has been closed at both ends, by the two points $-\infty, +\infty$, and that the plane is the direct product of the coordinate axes, closed in this way. The problem we discuss is how to enlarge the Borel field $\tilde{\mathscr{F}}_T$ to obtain a class $\tilde{\mathscr{F}}_1$ of measurable $\tilde{\omega}$ sets which will make the process separable and measurable. It has been shown that, if X is the finite line (and the proof is even applicable if X contains at least two points), if c is any real number, and if I is any non-denumerable parameter set, then the $\tilde{\omega}$ set

$$\{\tilde{x}_t(\tilde{\omega}) \leq c, t \in I\}$$

has probability 0 if it is in the class $\tilde{\mathscr{F}}_T'$. It follows that, if the process $\{\tilde{x}_t, a < t < b, X, \tilde{\mathscr{F}}_T'\}$ (with X containing at least two points) is separable, if $\{t_n\}$ is a sequence of parameter values satisfying the conditions of the separability definition, and if I is an open interval of parameter values, then

$$\mathbf{P}\{\underset{t \in I}{\text{L.U.B.}}\, \tilde{x}_t(\tilde{\omega}) = \infty\} = \mathbf{P}\{\underset{t_n \in I}{\text{L.U.B.}}\, \tilde{x}_{t_n}(\tilde{\omega}) = \infty\} = 1.$$

This may happen, and in fact the process under consideration will be separable for certain assignments of finite dimensional probability distributions, but this does not happen in the most interesting cases. It has been shown, if X is the finite line (and again the proof is applicable whenever X contains at least two points), that the process $\{\tilde{x}_t, a < t < b, X, \tilde{\mathscr{F}}_T'\}$ is not measurable for any assignment of finite dimensional probability distributions. These facts make clear the advisability of enlarging the field of measurable sets. The method of enlargement is described in the next paragraph.

Let $\tilde{\Gamma}$ be an $\tilde{\omega}$ set of outer measure 1 relative to the given field $\tilde{\mathscr{F}}_T'$ of measurable sets, that is, we suppose that the relation $\tilde{\Gamma} \subset \tilde{\mathrm{M}} \in \tilde{\mathscr{F}}_T$ implies that $\mathbf{P}\{\tilde{\mathrm{M}}\} = 1$. Define the Borel field $\tilde{\mathscr{F}}_1$ as the class of all $\tilde{\omega}$ sets $\tilde{\Lambda}$ of the form

$$\tilde{\Lambda} = \tilde{\Lambda}_1\tilde{\Gamma} + \tilde{\Lambda}_2(\tilde{\Omega} - \tilde{\Gamma}), \qquad \tilde{\Lambda}_i \in \tilde{\mathscr{F}}_T'$$

and for each such $\tilde{\Lambda}$ define

$$\mathbf{P}_1\{\tilde{\Lambda}\} = \mathbf{P}\{\tilde{\Lambda}_1\}.$$

It is easily shown that $\mathbf{P}_1\{\tilde{\Lambda}\}$ is uniquely defined in this way, and is a probability measure, that $\tilde{\mathscr{F}}_T' \subset \tilde{\mathscr{F}}_1$, and that

$$\mathbf{P}_1\{\tilde{\Lambda}\} = \mathbf{P}\{\tilde{\Lambda}\}, \qquad \tilde{\Lambda} \in \tilde{\mathscr{F}}_T'.$$

The stochastic process $\{\tilde{x}_t, t \in T\}$ with $\tilde{\mathscr{F}}_T'$ replaced by $\tilde{\mathscr{F}}_1$ and $\mathbf{P}\{\cdot\}$ by $\mathbf{P}_1\{\cdot\}$ will be called a *standard extension* of the given one. The standard extension depends on the choice of $\tilde{\Gamma}$. Note that a standard extension of the $\tilde{x}_t$ process still is of function space type. The only difference between it and the given process is that more classes of t functions $\tilde{\omega}$ have been assigned probabilities. In this approach the counterpart of Theorem 2.4 is:

THEOREM 2.4′ *Let $\{\tilde{x}_t, t \in T, X, \tilde{\mathscr{F}}_T'\}$ be a process of function space type, with X a closed set of the infinite line (or plane in the complex case). There is then a standard extension of the process which is separable relative to the closed sets.*

We omit the proof of this theorem. (It is essentially the same theorem as Theorem 2.4.)

The problem of measurability is solved in an equally satisfactory manner. In fact the following theorem can be proved.

THEOREM 2.9 *Let $\{\tilde{x}_t, t \in T, X, \tilde{\mathscr{F}}\}$ be a process of function space type, with X a closed set of the infinite line (or plane in the complex case), and T a Lebesgue measurable set. There is a standard extension of the process which is separable and measurable if and only if there is a measurable standard modification.*

The proof will be omitted.

We conclude the discussion of processes of function space type by applying the results to a rather trivial example already considered in this and the previous section. Let $\{\tilde{x}_t, -\infty < t < \infty, X, \tilde{\mathscr{F}}_T'\}$ be a process of function space type, with $\tilde{\mathscr{F}}_T'$ as defined in the above discussion. Suppose that

$$\mathbf{P}\{\tilde{x}_t(\tilde{\omega}) = 0\} = 1, \qquad -\infty < t < \infty.$$

If the range X of the function values is the single point 0, the function space $\tilde{\Omega}$ contains only a single function, the identically vanishing one. In this case

$$\mathbf{P}\{\tilde{x}_t(\tilde{\omega}) = 0, -\infty < t < \infty\} = 1,$$

and the $\tilde{x}_t$ process is separable and measurable. If X contains a second point, the $\tilde{x}_t$ process is neither separable nor measurable, and the probability

$$\mathbf{P}\{\tilde{x}_t(\tilde{\omega}) = 0, -\infty < t < \infty\}$$

is not defined. This probability will be defined, and will be 1, for every separable standard extension of the process. It is easy to see that the $\tilde{\omega}$ set $\tilde{\Gamma}$ in terms of which the standard extension is defined can be taken as the set containing only the identically vanishing function. This extension makes the process separable. Every separable standard extension is measurable, by Theorem 2.5. On the other hand, a second standard extension can be defined in which the set Γ just used is replaced by its complement. With this definition

$$\mathbf{P}\{\tilde{x}_t(\tilde{\omega}) = 0, -\infty < t < \infty\} = 0,$$

but this standard extension is neither separable nor measurable. This example shows the rather arbitrary nature of the standard extensions, and the necessity of a new criterion, such as separability, to aid in determining the choice of the extension. The situation in this respect is the same as for standard modifications of processes. A new criterion like separability must be used to determine the choice of standard modification also.

Let Ω be a space on sets of which a probability measure $\mathbf{P}$ is defined. Let x be a random variable with a distribution function F. All we shall require of F is that it does not define a distribution confined to a single value. Define $y(\omega) \equiv 0$. Then we can think of the random variable y as a function of the random variable x. This is, of course, trivial. It is not true in all cases, however, if y is the identically vanishing random variable on some Ω, that we can write $y(\omega) = f[x(\omega)]$, where x has the distribution function F, for the simple reason that there may be no such random variable x defined on Ω. For example, if Ω consists of exactly one point (to which considered as a point set the probability 1 is assigned), it is clear that the distribution of every random variable is confined to a single value. In order to overcome the difficulties that are encountered in situations like this, in which Ω and the probability measure $\mathbf{P}\{\Lambda\}$ are of too simple a structure for the problems considered, we define *extension by adjunction*, as follows. Let $\Omega^{(i)}$ be a space on sets of which a probability measure $\mathbf{P}^{(i)}$ is defined, $i = 1, 2$. Define $\Omega^{(12)}$ as the space of pairs $\omega^{(12)}$: $(\omega^{(1)}, \omega^{(2)})$, $\omega^{(i)} \in \Omega^{(i)}$, and define a probability measure $\mathbf{P}$ on $\Omega^{(12)}$ in the usual way as the product of the $\Omega^{(1)}$ and $\Omega^{(2)}$ measures considered as independent, so that, if $\Lambda^{(1)}$ is $\omega^{(1)}$ measurable, and $\Lambda^{(2)}$ is $\omega^{(2)}$ measurable,

$$\mathbf{P}\{\Lambda\} = \mathbf{P}^{(1)}\{\Lambda^{(1)}\}\mathbf{P}^{(2)}\{\Lambda^{(2)}\}, \qquad \Lambda = \{\omega^{(1)} \in \Lambda^{(2)}, \omega^{(2)} \in \Lambda^{(2)}\}.$$

Every $\omega^{(1)} \in \Omega^{(1)}$ then corresponds to an $\omega^{(12)}$ set, the set of all points $(\omega^{(1)}, \omega^{(2)})$ with the given first coordinate. This correspondence makes a measurable $\omega^{(1)}$ set go into a measurable $\omega^{(12)}$ set of the same probability. A random variable x on $\Omega^{(1)}$ can also be considered a random variable on $\Omega^{(12)}$, and will have the same distribution function. If a stochastic process is defined using the space $\Omega^{(1)}$, we obtain in this way a stochastic process with the same finite dimensional distributions, separability and measurability properties, and so on, defined on the space $\Omega^{(12)}$. The new process will be said to be obtained by *adjoining* $\Omega^{(2)}$ to $\Omega^{(1)}$. This adjunction procedure is, of course, the procedure used in discussing independent trials, and results in a space and probability measure with a finer structure than the initial one. For example, if $\Omega^{(2)}$ is the unit interval, and if $\Omega^{(2)}$ probability measure is Lebesgue measure, there are $\Omega^{(2)}$ random variables with all distribution functions, and therefore $\Omega^{(12)}$ random variables with all distribution functions.

3. Gaussian processes—strict and wide sense concepts

A stochastic process $\{x_t, t \in T\}$ is called *Gaussian* if the joint distribution of every finite set of the x_t's is Gaussian. These processes are of great importance theoretically as well as in applied work, and hence deserve separate consideration. The following theorem shows how the most general Gaussian process is determined.

THEOREM 3.1 *Let T be any finite or infinite aggregate. Let $\mu(\cdot)$ be any function of $t \in T$, and let $r(\cdot, \cdot)$ be a function of s and t ($s, t \in T$), satisfying the conditions*

(a) $$r(s, t) = \overline{r(t, s)},$$

(b) *if $t_1, \cdots, t_N$ is any finite T set, the matrix $[r(t_m, t_n)]$ is non-negative definite.*

There is then a Gaussian stochastic process $\{x_t, t \in T\}$ for which

$$\mathbf{E}\{x_t\} = \mu(t)$$
$$\mathbf{E}\{x_s\bar{x}_t\} - \mu(s)\overline{\mu(t)} = r(s, t). \tag{3.1}$$

If the functions $\mu(\cdot)$ and $r(\cdot, \cdot)$ are real, there is a real Gaussian process $\{x_t, t \in T\}$ satisfying (3.1). *In any case there is a complex Gaussian process $\{x_t, t \in T\}$ satisfying* (3.1) *and also*

$$\mathbf{E}\{x_s x_t\} = \mu(s)\mu(t). \tag{3.2}$$

Suppose first that $\mu(t)$ is real and that $r(s, t)$ is real and satisfies (a) and (b) of the theorem. If $t_1, \cdots, t_N$ is any finite t set there is then an N variate Gaussian distribution with mean values $\mu(t_1), \cdots, \mu(t_N)$ and covariance matrix $[r(t_m, t_n)]$. This is the distribution function with characteristic function

$$e^{-\frac{1}{2}\sum_{m,n=1}^{N} r(t_m, t_n)\lambda_m\lambda_n + i\sum_{m=1}^{N}\lambda_m\mu(t_m)} \tag{3.3}$$

If the matrix $[r(t_m, t_n)]$ is non-singular, this distribution has density

$$\frac{|a_{mn}|^{1/2}}{(2\pi)^{N/2}} e^{-\frac{1}{2}\sum_{m,n=1}^{N} a_{mn}(x_m - \mu(t_m))(x_n - \mu(t_n))}, \tag{3.4}$$

where the matrix $[a_{mn}]$, with determinant $|a_{mn}|$, is the inverse of the matrix $[r(t_m, t_n)]$. Moreover, if $M \le N$ the distribution defined by the characteristic function (3.3) assigns to $x_{t_1}, \cdots, x_{t_M}$ a Gaussian distribution with means $\mu(t_1), \cdots, \mu(t_M)$ and the covariance matrix $[r(t_m, t_n)]$ with $m, n \le M$. Thus if $M < N$ the marginal distribution of $x_{t_1}, \cdots, x_{t_M}$ is the same as that assigned to $x_{t_1}, \cdots, x_{t_M}$. Consequently, the consistency conditions of Kolmogorov (I §5) are satisfied, and there is a real Gaussian process satisfying (3.1), with ω space a certain coordinate space. The distributions involved are determined uniquely by the assigned means and covariances.

We complete the proof by dropping the hypothesis that the process is necessarily real, and defining a Gaussian process which satisfies (3.1) and (3.2). The process is obviously uniquely determined by these conditions, or, more exactly, the finite dimensional distributions involved are uniquely determined by these conditions. Note that according to these conditions

$$\mathbf{E}\{[x_t - \mu(t)]^2\} \equiv 0.$$

The process will therefore only be real in trivial cases, in fact only when $\mu(t)$ is real and $r(s, t) \equiv 0$, in which case

$$\mathbf{P}\{x_t(\omega) = \mu(t)\} \equiv 1, \qquad t \in T.$$

We shall define real random variables ξ_t, η_t to satisfy†

$$\begin{aligned} \mathbf{E}\{\xi_t\} &= \Re\{\mu(t)\} \qquad \mathbf{E}\{\eta_t\} = \Im\{\mu(t)\} \\ \mathbf{E}\{\xi_s\xi_t\} - \mathbf{E}\{\xi_s\}\mathbf{E}\{\xi_t\} &= \mathbf{E}\{\eta_s\eta_t\} - \mathbf{E}\{\eta_s\}\mathbf{E}\{\eta_t\} \\ &= \tfrac{1}{2}\Re\{r(s, t)\} \\ \mathbf{E}\{\xi_s\eta_t\} - \mathbf{E}\{\xi_s\}\mathbf{E}\{\eta_t\} &= -\tfrac{1}{2}\Im\{r(s, t)\}. \end{aligned} \tag{3.5}$$

The relations (3.5) imply (3.1) and (3.2) if $x_t = \xi_t + i\eta_t$. To show that the ξ_t, η_t process really exists, we apply the criterion for real processes already proved. That is we show that the covariance matrix of any finite number of ξ_t's and η_t's, as defined by (3.5), is symmetric and non-negative definite. It is no restriction to take corresponding pairs of ξ_t's and η_t's, that is, we take

$$\xi_{t_1}, \cdots, \xi_{t_N}, \eta_{t_1}, \cdots, \eta_{t_N}$$

and investigate the corresponding $2N$-dimensional covariance matrix $[\rho_{mn}]$ defined by (3.5). The matrix is obviously symmetric. Moreover, if $\lambda_1, \cdots, \lambda_{2N}$ are real,

$$\begin{aligned} \sum_{m,n=1}^{2N} \rho_{mn}\lambda_m\lambda_n &= \frac{1}{2} \sum_{m,n=1}^{N} \Re\{r(t_m, t_n)\}(\lambda_m\lambda_n + \lambda_{N+m}\lambda_{N+n}) \\ &\quad + \frac{1}{2} \sum_{m,n=1}^{N} \Im\{r(t_m, t_n)\}(\lambda_{N+m}\lambda_n - \lambda_m\lambda_{N+n}) \\ &= \frac{1}{2} \sum_{m,n=1}^{N} r(t_m, t_n)(\lambda_m\lambda_n + \lambda_{N+m}\lambda_{N+n}) \\ &\quad - \frac{i}{2} \sum_{m,n=1}^{N} r(t_m, t_n)(\lambda_{N+m}\lambda_n - \lambda_m\lambda_{N+n}) \\ &= \frac{1}{2} \sum_{m,n=1}^{N} r(t_m, t_n)(\lambda_m - i\lambda_{N+m})(\lambda_n + i\lambda_{N+n}) \\ &\geq 0 \end{aligned} \tag{3.6}$$

† Throughout this book, $\Re$ and $\Im$ will be used to denote "real part" and "imaginary part," respectively.

since the matrix $[r(t_m, t_n)]$ is non-negative definite. Hence the matrix $[\rho_{m,n}]$ is non-negative definite, as was to be proved.

This theorem can be used to obtain Gaussian processes intimately related to given processes, as far as first and second moments are concerned. For example, according to the following theorem, if a y_t process is given, a corresponding x_t Gaussian process exists with the same means and covariances. This means that in a discussion involving only means and covariances of the y_t's, it is no restriction to assume that the y_t process is Gaussian. If the y_t process is real, the corresponding x_t process can also be supposed real. This means that, whenever two y_t's are uncorrelated, the corresponding x_t's are independent. Whether the y_t's are real or not, the x_t process can be chosen to satisfy (3.2), and uncorrelated y_t's will then correspond to independent x_t's. If the y_t process is not real, the x_t process is not real either, and the assignment of the x_t means and covariances do not determine the x_t distributions. One token of this fact is that it is possible to superimpose the arbitrary condition (3.2).

THEOREM 3.2 *Let $\{y_t, t \in T\}$ be any stochastic process, with*

$$\mathbf{E}\{|y_t|^2\} < \infty, \qquad t \in T.$$

The parameter set T may be any finite or infinite aggregate. There is then a corresponding Gaussian process, with the same range of the parameter, but defined on a different measure space, whose random variables $\{x_t\}$ satisfy the equations

$$\begin{aligned} &\mathbf{E}\{x_t\} = 0, \\ &\mathbf{E}\{x_s\bar{x}_t\} = \mathbf{E}\{y_s\bar{y}_t\} \end{aligned} \qquad s, t \in T. \tag{3.7}$$

If the y_t process is real, there is a real x_t process satisfying (3.7). *In any case there is a complex Gaussian process satisfying* (3.7) *and the further condition*

$$\mathbf{E}\{x_s x_t\} = 0. \tag{3.8}$$

If the y_t and x_t processes are both real, and satisfy (3.7) *or if both* (3.7) *and* (3.8) *are satisfied, the orthogonality of two y_t's*† *implies independence of the corresponding x_t's.*

If y_t is replaced by $y_t - \mathbf{E}\{y_t\}$ in the above, zero correlation of two y_t's implies independence of the corresponding x_t's.

More generally, if v_1, v_2 are any two random variables in the closed linear

† The random variables u, v are said to be orthogonal if $\mathbf{E}\{u\bar{v}\} = 0$.

manifold† *of the* y_t's *and if* u_1, u_2 *are the corresponding random variables in the closed linear manifold of the* x_t's, *then* (3.7) (*for all* y_s, y_t) *implies*

$$\mathbf{E}\{u_t\} = 0 \tag{3.7'}$$

$$\mathbf{E}\{u_s\bar{u}_t\} = \mathbf{E}\{v_s\bar{v}_t\}$$

and (3.8) (*for all* x_s, x_t) *implies*

$$\mathbf{E}\{u_s u_t\} = 0. \tag{3.8'}$$

Applying this theorem to the real and imaginary parts of the random variables of a given complex process, it is clear that there is a complex Gaussian process whose real and imaginary parts have the same covariance relations as the real and imaginary parts of the given process. Then (3.7) is certainly satisfied, and also

$$\mathbf{E}\{x_s x_t\} = \mathbf{E}\{y_s y_t\}. \tag{3.9}$$

The exact covariance correspondence afforded by (3.9) in conjunction with (3.7) is not useful, however, and destroys the correspondence between orthogonality and independence.

The proof of Theorem 3.2 is accomplished simply by noting that, if $r(s, t) = \mathbf{E}\{y_s\bar{y}_t\}$, then $r(s, t) = \overline{r(t, s)}$, and, if $z_1, \cdots, z_N$ are any complex numbers,

$$\sum_{m,n=1}^{N} r(t_m, t_n)z_m\bar{z}_n = \sum_{m,n}^{N} \mathbf{E}\{y_{t_m}\bar{y}_{t_n}\}z_m\bar{z}_n = \mathbf{E}\{|\sum_{1}^{N} z_m y_{t_m}|^2\} \geq 0,$$

so that the hypotheses of Theorem 3.1 are certainly satisfied, if we take $\mu(t) \equiv 0$. The extension of the previous theorem involved in (3.7′) and (3.8′) is trivial, since the latter relations are obvious when v_s and v_t are linear combinations of the y_t's; the general case is proved by going to the limit. The theorem states in abstract language that there is a linear transformation taking the closed linear manifold of the y_t's into that of the x_t's, taking y_t into x_t for each t, which leaves the inner product $\mathbf{E}\{v_1\bar{v}_2\}$ invariant. (The transformation is unitary.)

One of the reasons why Gaussian processes are important is the simplification the Gaussian hypothesis brings to the theory of least squares approximation. Let the random variables x, y have a bivariate Gaussian distribution, with zero expectations. We suppose that x and y

† The closed linear manifold determined by a set of random variables is the collection of random variables which are finite linear combinations of the given random variables or limits in the mean of sequences of such linear combinations.

are either real or, if not, satisfy the relations $\mathbf{E}\{xy\} = 0$. Then the difference

$$(3.10) \qquad y - ax, \qquad a = \frac{\mathbf{E}\{y\bar{x}\}}{\mathbf{E}\{|x|^2\}},$$

has zero expectation and is orthogonal to and uncorrelated with and therefore independent of x. It follows that the conditional distribution of y for given x is Gaussian, with expectation ax and variance

$$(3.11) \qquad \mathbf{E}\{|y - ax|^2\} = \mathbf{E}\{|y|^2\}\left[1 - \frac{|\mathbf{E}\{y\bar{x}\}|^2}{\mathbf{E}\{|x|^2\}\mathbf{E}\{|y|^2\}}\right].$$

The bracket is between 0 and 1 by Schwarz's inequality. More generally, if the random variables $x_1, \cdots, x_n, y$ have a multivariate Gaussian distribution with zero expectations and if the variables are either real or, if not, satisfy the relations

$$\mathbf{E}\{x_j x_k\} = \mathbf{E}\{x_j y\} = 0, \qquad i, k = 1, \cdots, n,$$

the difference,

$$(3.10') \qquad y - \sum_1^n a_j x_j,$$

has zero expectation and is uncorrelated with and therefore independent of $x_1, \cdots, x_n$ for properly chosen $a_1, \cdots, a_n$ (see IV §3). It follows that the conditional distribution of y for given $x_1, \cdots, x_n$ is Gaussian, with

$$(3.12) \qquad \mathbf{E}\{y \mid x_1, \cdots, x_n\} = \sum_1^n a_j x_j.$$

The difference in (3.10′) is independent of and therefore orthogonal to every function f measurable on the sample space of $x_1, \cdots, x_n$ (in particular to every Baire function of these variables) whose square is integrable. Hence

$$(3.13) \qquad \begin{aligned} \mathbf{E}\{|y - f|^2\} &= \mathbf{E}\{|(y - \sum_1^n a_j x_j) + (\sum_1^n a_j x_j - f)|^2\} \\ &= \mathbf{E}\{|y - \sum_1^n a_j x_j|^2\} + \mathbf{E}\{|\sum_1^n a_j x_j - f|^2\}. \end{aligned}$$

This equality shows that the problem of minimizing the left side of (3.13), the problem of *least squares approximation*, is solved by setting f equal to the conditional expectation in (3.12), and that this solution is uniquely determined aside from the usual ambiguity in the definition of a conditional expectation. The Gaussian character of the distributions made it possible to infer independence from zero correlation. If we only suppose that $x_1, \cdots, x_n, y$ are any random variables with

$$\mathbf{E}\{|x_j|^2\} < \infty, \qquad \mathbf{E}\{|y|^2\} < \infty$$

the difference (3.10′) with the a_j's calculated in the same way is still orthogonal to $x_1, \cdots, x_n$, from which it follows only that (3.13) is true if f is a linear combination of the x_j's. Thus the random variable $\sum_1^n a_j x_j$, which we shall denote by $\hat{\mathbf{E}}\{y \mid x_1, \cdots, x_n\}$, solves the problem of minimizing the left side of (3.13) for all linear combinations f of the x_j's, the problem of *linear least squares approximation.*

It will be necessary to extend the above discussion to the case of infinitely many conditioning variables in later chapters.

Many concepts which are used in the theory of stochastic processes can be formulated in two ways, in a "strict sense" or in a "wide sense." The general principle is the following: Suppose a y_t process has a certain property P expressed in terms of variances and covariances. Suppose the corresponding complex Gaussian process given by Theorem 3.2 satisfying (3.7) and (3.8) has the corresponding but stronger property P'. Then P' is called a strict sense property and P a wide sense property. (If the y_t process is real, the corresponding Gaussian process can be taken as the uniquely defined corresponding real Gaussian process given by Theorem 3.2, satisfying (3.7).) The point is that a wide sense property implies much more when it is also supposed that the underlying distributions are Gaussian. We shall see that a great many theorems have strict sense and wide sense versions, and that the distinction serves to organize and clarify the theory of stochastic processes.

Example 1 Let $y_1, y_2, \cdots$ be mutually orthogonal random variables. We take this as a wide sense characterization, and find the corresponding strict sense characterization by noting that, if an x_j process is determined as in Theorem 3.2, the x_j's will be mutually independent. Thus a process with mutually orthogonal random variables (or mutually uncorrelated random variables if $y_j - \mathbf{E}\{y_j\}$ is considered instead of y_j) is a process with independent random variables in the wide sense.

Example 2 The least squares analysis made above shows that the best linear least squares approximation $\hat{\mathbf{E}}\{y \mid x_1, \cdots, x_n\}$ is the wide sense version of the best least squares approximation. We carry the analysis further by evaluating the best least squares approximation in the non-Gaussian case. Suppose then that $x_1, \cdots x_n, y$ are any random variables with $\mathbf{E}\{|y_n|^2\} < \infty$, and define

$$y_0 = \mathbf{E}\{y \mid x_1, \cdots, x_n\}.$$

Then

$$\mathbf{E}\{|y_0|^2\} \leq \mathbf{E}\{\mathbf{E}\{|y|^2 \mid x_1, \cdots, x_n\}\} = \mathbf{E}\{|y|^2\},$$

and it follows from I, Theorem 8.3, that $y - y_0$ is orthogonal to every function f which is measurable on the sample space of $x_1, \cdots, x_n$ and whose square is integrable. Hence

(3.13′)

$$\mathbf{E}\{|y-f|^2\} = \mathbf{E}\{|(y-y_0)+(y_0-f)|^2\} = \mathbf{E}\{y-y_0|^2\} + \mathbf{E}\{|y_0-f|^2\},$$

so that the problem of minimizing the left side of (3.13′) is solved by setting $f = y_0$. Thus the strict sense version of $\hat{\mathbf{E}}\{y \mid x_1, \cdots, x_n\}$ is $\mathbf{E}\{y \mid x_1, \cdots, x_n\}$. The latter symbol is of course defined for any family of approximating random variables, finite or infinite, and solves the corresponding least squares approximation problem, since (3.13′) needs no change in the general case. The symbol $\hat{\mathbf{E}}\{y \mid -\}$ will be defined for infinite families of approximating random variables in IV. It obeys the same rules of combination as the symbol $\mathbf{E}\{y \mid -\}$ since the two become the same in the real Gaussian case (and in the complex Gaussian case if the random variables all satisfy the condition that the expected value of the product of any pair vanishes).

Aside from theorems which are strict sense versions of wide sense theorems, very few facts specifically true of Gaussian processes are known. For this reason no later chapter will be devoted to Gaussian processes as such, although many types of Gaussian processes will be treated.

4. Processes with mutually independent random variables

These processes are discussed only in the discrete parameter case, since the sample functions in the continuous parameter case are too irregular to arise in practice. The processes under discussion are thus simply sequences of mutually independent random variables. Most studies of the "foundations of probability," which commonly identify the mathematical with the philosophical foundations, to the detriment of the former, have as their purpose the study of a sequence of mutually independent random variables with a common distribution function.

We have already remarked that, if $F_1, F_2, \cdots$ is any sequence of distribution functions, there is a sequence of mutually independent random variables with those distribution functions. Because of the independence hypothesis, the joint distribution functions of the x_j's are simply products of the F_j's. It is quite feasible to study these processes without mentioning the word probability. For example, if $s_n = x_1 + \cdots + x_n$, the distribution function of s_n is the convolution of those of $x_1, \cdots, x_n$, that is,

(4.1)

$$\mathbf{P}\{s_n(\omega) \le \lambda\} = \int_{-\infty}^{\infty} dF_1(\xi_1) \cdots \int_{-\infty}^{\infty} F_n(\lambda - \xi_1 - \cdots - \xi_{n-1})\, dF_{n-1}(\xi_{n-1}).$$

The study of many of the properties of the partial sums of the series $\sum_1^\infty x_n$ can thus be carried on as a study of iterated convolutions, with no mention of probability. To prove (4.1) we observe that, if $x_1, \cdots, x_n$ are the coordinate functions of the n-dimensional space of points $(\xi_1, \cdots, \xi_n)$, the probabilities of ω sets are probabilities of $(\xi_1, \cdots, \xi_n)$ sets, with

$$\mathbf{P}\{\Lambda\} = \int \cdots \int_\Lambda dF_1(\xi_1) \cdots dF_n(\xi_n).$$

Then (4.1) is simply the evaluation by iterated integration of $\mathbf{P}\{\Lambda\}$ for Λ defined by the inequality $\sum_1^n \xi_j \le \lambda$. Finally, using the representation theory of I §6, it is no restriction in proving (4.1) to assume that the x_j's are the stated coordinate functions. The proof given also proves the following slightly more general equation, which we shall have occasion to use:

$$\text{(4.2)}\quad \mathbf{P}\{x_j(\omega) \le \lambda_j, j = 1, \cdots, n-1, s_n(\omega) \le \lambda\}$$
$$= \int_{-\infty}^{\lambda_1} dF_1(\xi_1) \cdots \int_{-\infty}^{\lambda_{n-1}} F_n(\lambda - \xi_1 - \cdots - \xi_{n-1})\, dF_{n-1}(\xi_{n-1}).$$

5. Processes with uncorrelated or orthogonal random variables

A process with uncorrelated random variables is one whose random variables are uncorrelated in pairs, that is, it is supposed that

$$\text{(5.1)}\qquad \mathbf{E}\{|x_t|^2\} < \infty$$

and that

$$\text{(5.2)}\qquad \mathbf{E}\{x_s\bar{x}_t\} = \mathbf{E}\{x_s\}\mathbf{E}\{\bar{x}_t\}, \qquad s \ne t.$$

A closely related type of process is a process with orthogonal random variables, that is, one for which (5.1) is true and for which

$$\text{(5.3)}\qquad \mathbf{E}\{x_s\bar{x}_t\} = 0, \qquad s \ne t.$$

If an x_t process is a process with uncorrelated random variables, the y_t process determined by

$$y_t = x_t - \mathbf{E}\{x_t\}$$

is a process with orthogonal random variables. Thus, if x is a real random variable, uniformly distributed in the interval $(0, \pi)$, the random variables

$$\sin x, \sin 2x, \cdots$$

constitute both a process with uncorrelated random variables and a process with orthogonal random variables (with zero means). On the other hand, the random variables

$$1 + \sin x, 2 + \sin 2x, \cdots$$

constitute a process with uncorrelated random variables but not one with orthogonal random variables.

As already remarked in the introduction, the theory of probability is but one aspect of the theory of measure. This is particularly clearly exhibited at the present stage. The theory of orthogonal functions and series has occupied mathematicians for over 100 years, but has never been considered a part of probability. From the present point of view, a sequence of orthogonal functions is a sequence of orthogonal random variables, and theorems like the measure theoretic Riesz-Fischer theorem become probability theorems. The probability approach, however, imposes the rather pointless condition that the basic measure space have total measure 1. Up to the present time this condition has been imposed by the physical interpretation of probability, although many of the ideas and definitions used in probability theory do not intrinsically presuppose the condition. Certain physical situations (such as a free particle in quantum mechanics which is equidistributed throughout space) even suggest the usefulness of a probability measure with a value $+\infty$ for the whole space. For these reasons and because of later applications in this book, the hypothesis of finiteness of measure is dropped in IV, where orthogonal series are discussed, and to avoid confusion the customary language of measure theory is used.

As already noted in §3, the processes of this section are the wide sense versions of processes with independent random variables. Although this terminology is not ordinarily used, the point of view will prove illuminating.

6. Markov processes

(*a*) *Strict sense* In the following we shall consider only real processes; the changes to be made in going to the complex case will be obvious. A (strict sense) *Markov process* is a process $\{x_t, t \in T\}$ satisfying the following condition: *for any integer* $n \geq 1$, if $t_1 < \cdots < t_n$ *are parameter values, the conditional* x_{t_n} *probabilities relative to* $x_{t_1}, \cdots, x_{t_{n-1}}$ *are the same as those relative to* $x_{t_{n-1}}$ *in the sense that for each* λ

$$\mathbf{P}\{x_{t_n}(\omega) \leq \lambda \mid x_{t_1}, \cdots, x_{t_{n-1}}\} = \mathbf{P}\{x_{t_n}(\omega) \leq \lambda \mid x_{t_{n-1}}\} \tag{6.1}$$

with probability 1. Then, if $T_1 \subset T$, the process $\{x_t, t \in T_1\}$ is also a Markov process. When a process is called a Markov process, "strict sense" is always understood.

Condition (6.1) is at first glance weaker than

$$\mathbf{P}\{x_{t_n}(\omega) \leq \lambda \mid x_{t_1}, \cdots, x_{t_{n-1}}\} = f(t_n, x_{t_{n-1}}) \tag{6.1'}$$

(to hold with probability 1), where the random variable on the right is a Baire function of $x_{t_{n-1}}$ or more generally is measurable on the sample space of $x_{t_{n-1}}$. Actually, however, f is then necessarily given by the right side of (6.1). In fact, if the operation $\mathbf{E}\{- \mid x_{t_{n-1}}\}$ is performed on both sides of (6.1′), the right side is unchanged, and the left side becomes the right side of (6.1), by I (10.9). Because of the equivalence of (6.1) and (6.1′), a Markov process is sometimes loosely described as one in which the conditional probability on the left in (6.1) depends only on $x_{t_{n-1}}(\omega)$.

The condition (6.1) is also equivalent to the condition that, if $s_1 < s_2$ and if λ is arbitrary, then

$$\mathbf{P}\{x_{s_2}(\omega) \leq \lambda \mid x_t, t \leq s_1\} = \mathbf{P}\{x_{s_2}(\omega) \leq \lambda \mid x_{s_1}\} \tag{6.2}$$

(also with probability 1). In fact, if (6.2) is true with probability 1, and if $t_1 < \cdots < t_{n-1} = s_1 < t_n = s_2$, we can perform the operation $\mathbf{E}\{- \mid x_{t_1}, \cdots, x_{t_{n-1}}\}$ to both sides of (6.2) to obtain (6.1) (with probability 1). Conversely, if (6.1) is true with probability 1, then

$$\int_\Lambda \mathbf{P}\{x_{s_2}(\omega) \leq \lambda \mid x_{s_1}\}\, d\mathbf{P} = \mathbf{P}\{x_{s_2}(\omega) \leq \lambda, \omega \in \Lambda\}$$

whenever Λ is measurable on the sample space of finitely many x_t's with $t \leq s$, and we have seen in I §7 that it is sufficient to verify this equality for these sets Λ to be able to identify the integrand with a version of the left side of (6.2).

We shall now prove that, if $\{x_t, t \in T\}$ is a Markov process, if $s \in T$, if y is an ω function measurable on the sample space of the x_t's with $t \geq s$, and if $\mathbf{E}\{|y|\} < \infty$, then

$$\mathbf{E}\{y \mid x_t, t \leq s\} = \mathbf{E}\{y \mid x_s\} \tag{6.3}$$

with probability 1. This equality contains (6.2) as a special case, and thus, as we have seen, implies (6.1). We shall call the validity of (6.3) the *Markov property*. The Markov property will be derived first under the assumption that T contains only finitely many points $\geq s$. Let these points be $s = u_0 < u_1 < \cdots < u_m$. Let $\lambda_0, \cdots, \lambda_m$ be arbitrary numbers. Then, if y is defined by

$$\begin{aligned} y(\omega) &= 1 \qquad \text{if } x_{u_j}(\omega) \leq \lambda_j, j = 0, \cdots, m \\ &= 0 \text{ otherwise,} \end{aligned}$$

(6.3) reduces to

$$\begin{aligned} \mathbf{P}\{x_{u_j}(\omega) \leq \lambda_j, j = 0, \cdots, m \mid x_t, t \leq s\} \\ = \mathbf{P}\{x_{u_j}(\omega) \leq \lambda_j, j = 0, \cdots, m \mid x_s\}. \end{aligned} \tag{6.3'}$$

Conversely, if (6.3′) is true with probability 1 for arbitrary $\lambda_0, \cdots, \lambda_m$, then (6.3) holds with probability 1 if y is measurable on the sample space of $x_{u_0}, \cdots, x_{u_m}$, and if $\mathbf{E}\{|y|\} < \infty$. In fact, under the stated hypothesis, if the ω set M is defined by

$$\mathrm{M} = \{[x_{u_0}(\omega), \cdots, x_{u_m}(\omega)] \in A\},$$

it follows trivially that

$$\mathbf{P}\{\mathrm{M} \mid x_t, t \leq s\} = \mathbf{P}\{\mathrm{M} \mid x_s\} \tag{6.4}$$

with probability 1, if A is an $m + 1$-dimensional right semi-closed interval. Since the class of sets M of this type, together with the ω sets of probability 0, generates the class of ω sets measurable on the sample space of $x_{u_0}, \cdots, x_{u_m}$, it follows (I, Theorem 9.6) that (6.3) is true with probability 1 whenever y is as described. We now proceed to prove (6.3) by induction in m. If $m = 0$, (6.3) is trivial, since both sides of the equation reduce to y, with probability 1. If $m = 1$, (6.3′)—and therefore (6.3)—is true with probability 1 because (with probability 1)

$$\begin{aligned}\mathbf{P}\{x_{u_0}(\omega) \leq \lambda_0, x_{u_1}(\omega) \leq \lambda_1 \mid x_t, t \leq s\}& \\ = \mathbf{P}\{x_{u_0}(\omega) \leq \lambda_0, x_{u_1}(\omega) \leq \lambda_1 \mid x_s\} = 0 &\qquad \text{if } x_s(\omega) > \lambda_0\end{aligned}$$

and, using (6.2),

$$\begin{aligned}\mathbf{P}\{x_{u_0}(\omega) \leq \lambda_0, x_{u_1}(\omega) \leq \lambda_1 \mid x_t, t \leq s\}& \\ = \mathbf{P}\{x_{u_1}(\omega) \leq \lambda_1 \mid x_t, t \leq s\}& \\ = \mathbf{P}\{x_{u_1}(\omega) \leq \lambda_1 \mid x_s\}& \\ = \mathbf{P}\{x_{u_0}(\omega) \leq \lambda_0, x_{u_1}(\omega) \leq \lambda_1 \mid x_s\} &\qquad \text{if } x_s(\omega) \leq \lambda_0.\end{aligned}$$

Now assume that (6.3′) is true with probability 1 for some $m \geq 1$. We then show that it is true with probability 1 for m replaced by $m + 1$. Let λ_0, λ_1 be any real numbers, and define y, z by

$$\begin{aligned} y(\omega) &= 1 \quad \text{if } x_{u_0}(\omega) \leq \lambda_0 & z(\omega) &= 1 \quad \text{if } x_{u_j}(\omega) \leq \lambda_j, j = 1, \cdots, m + 1 \\ &= 0 \quad \text{otherwise} & &= 0 \quad \text{otherwise.}\end{aligned}$$

Then, applying the induction hypothesis to the x_t process, and also to this process with all parameter points $< u_1$ deleted except the point $s = u_0$,

$$\mathbf{E}\{z \mid x_t, t \leq u_1\} = \mathbf{E}\{z \mid x_{u_1}\} = \mathbf{E}\{z \mid x_s, x_{u_1}\} \tag{6.5}$$

with probability 1. Now

$$(6.6)\quad \mathbf{E}\Big\{y\mathbf{E}\{z \mid x_t, t \le u_1\}\Big| x_t, t \le s\Big\}$$
$$= \mathbf{E}\Big\{\mathbf{E}\{yz \mid x_t, t \le u_1\}\Big| x_t, t \le s\Big\}$$
$$= \mathbf{E}\{yz \mid x_t, t \le s\}$$
$$= \mathbf{P}\{x_{u_j}(\omega) \le \lambda_j, j = 0, \cdot\cdot\cdot, m+1 \mid x_t, t \le s\}$$

with probability 1, and, using the induction hypothesis with $m = 1$,

$$(6.7)\quad \mathbf{E}\Big\{y\mathbf{E}\{z \mid x_s, x_{u_1}\} \mid x_t, t \le s\Big\}$$
$$= \mathbf{E}\Big\{\mathbf{E}\{yz \mid x_s, x_{u_1}\}\Big| x_t, t \le s\Big\}$$
$$= \mathbf{E}\Big\{\mathbf{E}\{yz \mid x_s, x_{u_1}\}\Big| x_s\Big\}$$
$$= \mathbf{E}\{yz \mid x_s\}$$
$$= \mathbf{P}\{x_{u_j}(\omega) \le \lambda_j, j = 0, \cdot\cdot\cdot, m+1 \mid x_s\}$$

with probability 1. Combining (6.6) and (6.7) with (6.5), we obtain (6.3′)—and hence also (6.3)—with m replaced by $m + 1$. Finally, suppose that T contains infinitely many points $\ge s$. Then the result just obtained implies that (6.4) is true with probability 1 if M is measurable on the sample space of some finite aggregate of x_t's with $t \ge s$. It then follows (I, Theorem 9.6) that (6.3) is true with probability 1 if $\mathbf{E}\{|y|\} < \infty$, and if y is equal with probability 1 to an ω function measurable with respect to the Borel field of ω sets generated by the class of sets M, that is, if y is measurable on the sample space of the x_t's with $t \ge s$, as was to be proved.

The definition of a Markov process given above is one-sided in t. An alternative definition, stated in the form of a necessary and sufficient condition, is the following, which, because it is two-sided, shows that, if the process $\{x_t, t \in T\}$ is a Markov process, that with variables $\{x_{-t}\}$ is also a Markov process. *A process is a Markov process if and only if, for any positive integers, m, n, and real numbers* $\lambda_1, \cdot\cdot\cdot, \lambda_m, \mu_1, \cdot\cdot\cdot, \mu_n$, *if* $s_1 < \cdot\cdot\cdot < s_m < t < t_1 < \cdot\cdot\cdot < t_n$ *are parameter values, then*

$$(6.8)\quad \mathbf{P}\{x_{s_j}(\omega) \le \lambda_j, j = 1, \cdot\cdot\cdot, m;\ x_{t_k}(\omega) \le \mu_k, k = 1, \cdot\cdot\cdot, n \mid x_t\}$$
$$= \mathbf{P}\{x_{s_j}(\omega) \le \lambda_j, j = 1, \cdot\cdot\cdot, m \mid x_t\}\, \mathbf{P}\{x_{t_k}(\omega) \le \mu_k, k = 1, \cdot\cdot\cdot, n \mid x_t\}$$

with probability 1. It is tempting to restate (6.8) by saying that for fixed $x_t(\omega)$ the sets of random variables $x_{s_1}, \cdot\cdot\cdot, x_{s_m}$ and $x_{t_1}, \cdot\cdot\cdot, x_{t_n}$ are mutually independent; that is, for $x_t(\omega)$ (the present) known, the past and future are independent of each other. However, the ambiguity in the definition of conditional probabilities makes the interpretation of such statements rather delicate. The following is an intuitive and suggestive proof of the equivalence of (6.8) and the Markov property, in the spirit of the time interpretation of (6.8). A rigorous proof will also be given. For given $x_t(\omega)$, (6.8) states that a certain past event and a certain future event are mutually independent, in view of the multiplying of the corresponding probabilities (all conditioned by the knowledge of the present). An alternative condition for independence is that certain conditional probabilities do not actually involve the conditioning variables; specifically, if $\mathbf{P}^{(t)}$ denotes probabilities conditioned by x_t, this version of the independence condition becomes

$$(6.8') \quad \mathbf{P}^{(t)}\{x_{t_k}(\omega) \leq \mu_k, k = 1, \cdots, n \mid x_{s_1}, \cdots, x_{s_m}\} = \mathbf{P}^{(t)}\{x_{t_k}(\omega) \leq \mu_k, k = 1, \cdots, n\}$$

with $\mathbf{P}^{(t)}$ probability 1. We have seen in I §10 that this equation can be written in the form

$$(6.8'') \quad \mathbf{P}\{x_{t_k}(\omega) \leq \mu_k, k = 1, \cdots, n \mid x_{s_1}, \cdots, x_{s_m}, x_t\} = \mathbf{P}\{x_{t_k}(\omega) \leq \mu_k, k = 1, \cdots, n \mid x_t\}.$$

If $n = 1$, this equation reduces to (6.1); for arbitrary n it is still a special case of the Markov property.

The following is a rigorous discussion of the condition (6.8). Define y, z by

$$\begin{aligned} y(\omega) &= 1 \quad && \text{if } x_{s_j}(\omega) \leq \lambda_j, j = 1, \cdots, m \\ &= 0 && \text{otherwise;} \\ z(\omega) &= 1 && \text{if } x_{t_k}(\omega) \leq \mu_k, k = 1, \cdots, n, \\ &= 0 && \text{otherwise.} \end{aligned}$$

Then, using I, Theorem 8.3, we find that

$$(6.9) \qquad \mathbf{E}\{y \mid x_t\}\mathbf{E}\{z \mid x_t\} = \mathbf{E}\Big\{y\mathbf{E}\{z \mid x_t\} \Big| x_t\Big\}$$

with probability 1. If the process is a Markov process, the right side becomes, using I, Theorem 8.3, once more,

$$(6.10) \qquad \mathbf{E}\Big\{y\mathbf{E}\{z \mid x_s, s \leq t\} \Big| x_t\Big\} = \mathbf{E}\Big\{\mathbf{E}\{yz \mid x_s, s \leq t\} \Big| x_t\Big\} = \mathbf{E}\{yz \mid x_t\}$$

with probability 1. Combining (6.9) and (6.10), we find that

$$(6.11) \qquad \mathbf{E}\{y \mid x_t\}\mathbf{E}\{z \mid x_t\} = \mathbf{E}\{yz \mid x_t\}$$

with probability 1, and this is precisely (6.8). Conversely, if (6.11) is true with probability 1, it follows that

$$\mathbf{E}\{yz \mid x_t\} = \mathbf{E}\{y \mid x_t\}\mathbf{E}\{z \mid x_t\} = \mathbf{E}\Big\{y\mathbf{E}\{z \mid x_t\} \Big| x_t\Big\}$$

with probability 1, so that

$$\int_\Lambda yz \, d\mathbf{P} = \int_\Lambda y\mathbf{E}\{z \mid x_t\} \, d\mathbf{P}$$

if Λ is a set measurable on the sample space of x_t. But then

$$\int_{\{y(\omega)=1\}\Lambda} z \, d\mathbf{P} = \int_{\{y(\omega)=1\}\Lambda} \mathbf{E}\{z \mid x_t\} \, d\mathbf{P},$$

so that

$$\int_M z \, d\mathbf{P} = \int_M \mathbf{E}\{z \mid x_t\} \, d\mathbf{P} \tag{6.12}$$

if M is an ω set of the form $\{x_{s_j}(\omega) \le \lambda_j, j = 1, \cdots, m, x_t(\omega) \le \lambda\}$, and we have seen in I §7 that equality in (6.12) for these sets M is sufficient to identify the integrand on the right with $\mathbf{E}\{z \mid x_{s_1}, \cdots, x_{s_m}, x_t\}$, that is,

$$\mathbf{P}\{x_{t_k}(\omega) \le \mu_k, k = 1, \cdots, n \mid x_{s_1}, \cdots, x_{s_m}, x_t\}$$
$$= \mathbf{P}\{x_{t_k}(\omega) \le \mu_k, k = 1, \cdots, n \mid x_t\}$$

with probability 1. This is (6.1) in a slightly different notation.

Example 1 Let $y_1, y_2, \cdots$ be mutually independent random variables. Then the process $\{y_n, n \ge 1\}$ is a Markov process, and in fact $\mathbf{P}\{y_n(\omega) \le \lambda\}$ is one version of both $\mathbf{P}\{y_n(\omega) \le \lambda \mid y_1, \cdots, y_{n-1}\}$ and $\mathbf{P}\{y_n(\omega) \le \lambda \mid y_{n-1}\}$. Thus in this case the conditional distribution of y_n relative to $y_1, \cdots, y_{n-1}$ exists in the technical sense of I §9.

Example 2 Let the y_j's be as in Example 1, and define $x_n = \sum_1^n y_j$. Let F_{mn} be the distribution function of $\sum_{m+1}^n y_j$. Then the x_n process is a Markov process, and in fact, if $s < t$, one version of both the conditional distribution of x_t relative to x_s and that relative to $x_1, \cdots, x_s$ is given by

$$\mathbf{P}\{x_t(\omega) \le \lambda \mid x_1, \cdots, x_s\} = F_{st}(\lambda - x_s).$$

This is intuitively clear, but it may be instructive to give a formal proof. We must show that, if Λ is an ω set measurable on the sample space of $x_1, \cdots, x_s$, or, which amounts to the same thing, measurable on the sample space of $y_1, \cdots, y_s$, then

$$\int_\Lambda F_{st}(\lambda - x_s) \, d\mathbf{P} = \mathbf{P}\{\omega \in \Lambda, x_t(\omega) \le \lambda\}. \tag{6.13}$$

To derive this equation it will be convenient to assume that the $s+1$ random variables

$$y_1, \cdots, y_s, \sum_{s+1}^{t} y_j$$

are the coordinate functions in the $s+1$-dimensional space of points $(\eta_1, \cdots, \eta_{s+1})$ with

$$\mathbf{P}\{(\eta_1, \cdots, \eta_{s+1}) \in A\} = \int \cdots \int_A dF_1(\eta_1) \cdots dF_{s+1}(\eta_{s+1})$$

where

$$F_j = F_{j-1j}, \qquad j = 1, \cdots, s$$

$$F_{s+1} = F_{st}$$

for every $s+1$-dimensional Borel set A. According to the representation theory of I §6 it is no restriction to make this assumption. Then, translating (6.13), it is sufficient to prove that

$$\int dF_1(\eta_1) \cdots \int_\Lambda dF_{s-1}(\eta_{s-1}) \int F_{s+1}(\lambda - \sum_1^s \eta_j)\, dF_s(\eta_s)$$
$$= \mathbf{P}\{(\eta_1, \cdots, \eta_s) \in \Lambda, x_t(\omega) \leq \lambda\}$$

for Λ an s-dimensional Borel set in $(\eta_1, \cdots, \eta_s)$ space. It is sufficient to prove the equality for Λ an s-dimensional interval determined by the inequalities

$$-\infty < \eta_j \leq \lambda_j, \qquad j \leq s,$$

and the equality then becomes a special case of (4.2).

Example 3 Let $P_1, P_2, \cdots$ be functions of ξ, A, where ξ is a real number and A a linear Borel set, with the following properties:

(i) $P_j(\xi, A)$ defines a probability measure of A for fixed ξ;

(ii) $P_j(\xi, A)$ defines a Baire function of ξ for fixed A.

Let P be a probability measure of linear Borel sets. Then there is a Markov process $\{x_n, n \geq 1\}$, such that $P_n(x_n, A)$ is one version of $\mathbf{P}\{x_{n+1}(\omega) \in A \mid x_n\}$ and that

$$\mathbf{P}\{x_1(\omega) \in A\} = P(A).$$

To see this we observe first that, for every m, with the obvious conventions,

$$\int P(d\xi_1) \int P_1(\xi_1, d\xi_2) \cdots \int_{A_m} P_{m-1}(\xi_{m-1}, d\xi_m)$$

defines a probability measure of m-dimensional Borel sets A_m in the space of points $(\xi_1, \cdots, \xi_m)$. We define a sequence of random variables $\{x_n, n \geq 1\}$ such that the distribution of $x_1, \cdots, x_m$ is given by the above multiple integral. This is possible, with the x_j's the coordinate functions of infinite dimensional space, if the above measures are mutually consistent (see I §5), and it is trivial to verify that they are. It is then also trivial to verify that the x_n process obtained in this way is a Markov process, with

$$\mathbf{P}\{x_{n+1}(\omega) \in A \mid x_1, \cdots, x_n\} = P_n(x_n, A)$$

with probability 1. In particular, if P and the P_j's are given by densities, so that

$$P(A) = \int_A p(\eta)\, d\eta, \; P_j(\xi, A) = \int_A p_j(\xi, \eta)\, d\eta,$$

the distribution of $x_1, \cdots, x_m$ has density

$$p(\xi_1)p_1(\xi_1, \xi_2) \cdots p_{m-1}(\xi_{m-1}, \xi_m).$$

In this form the significance of the Markov property lies in the fact that p_j does not involve $\xi_1, \cdots, \xi_{j-1}$. We have seen that the x_n process reversed in time is also a Markov process. In the density case just described, the reversed transition probabilities can be exhibited directly in an elegant form. The conditional distribution of x_n for $x_{n+1}(\omega) = \xi_{n+1}$ has density (in ξ_n)

(6.14)

$$\frac{\int_{-\infty}^{\infty} \cdots \int_{-\infty}^{\infty} p(\eta_1)p_1(\eta_1, \eta_2) \cdots p_{n-1}(\eta_{n-1}, \xi_n)\, d\eta_1 \cdots d\eta_{n-1}}{\int_{-\infty}^{\infty} \cdots \int_{-\infty}^{\infty} p(\eta_1)p_1(\eta_1, \eta_2) \cdots p_n(\eta_n, \xi_{n+1})\, d\eta_1 \cdots d\eta_n} p_n(\xi_n, \xi_{n+1}).$$

(Here we have assumed that the p_j's are Baire functions of their pairs of variables. .Actually it is easily seen that these densities can always be chosen to have this property.) Note that the forward transition probability density function p_n can be chosen quite independently of the initial probability density function p, so that a change of p does not force a change of p_n. Once the p_n's are chosen, however, the reverse transition probability density function (6.14) will in general depend on the choice of p.

Now let $\{x_n, n \geq 1\}$ be any Markov process with the indicated parameter set. We have seen in I §9 that there is a function P_n, of a linear Borel set A and a real variable, the wide sense conditional distribution of x_{n+1} relative to x_n, which determines a Baire function when A is fixed and determines a probability measure when the variable is fixed, such that

$$\mathbf{P}\{x_{n+1}(\omega) \in A \mid x_n\} = P_n(x_n, A)$$

with probability 1, for each n, A. Let P be the distribution of x_1,

$$P(A) = \mathbf{P}\{x_1(\omega) \in A\}.$$

We show (cf. Example 3) that, if x is a Baire function of $x_1, \cdots, x_m$, then

$$(6.15) \qquad \mathbf{E}\{x\} = \int_{-\infty}^{\infty} P(d\xi_1) \int_{-\infty}^{\infty} P_1(\xi_1, d\xi_2) \cdots \int_{-\infty}^{\infty} x P_{m-1}(\xi_{m-1}, d\xi_m).$$

Using the Markov property, the first integration on the right yields a version of $\mathbf{E}\{x \mid x_1, \cdots, x_{m-1}\}$ which is a Baire function of $x_1, \cdots, x_{m-1}$, according to our work in I §9. The next integration yields

$$\mathbf{E}\Big\{\mathbf{E}\{x \mid x_1, \cdots, x_{m-1}\} \Big| x_1, \cdots, x_{m-2}\Big\} = \mathbf{E}\{x \mid x_1, \cdots, x_{m-2}\},$$

and proceeding in this way the last integral yields

$$\mathbf{E}\Big\{\mathbf{E}\{x \mid x_1\}\Big\} = \mathbf{E}\{x\},$$

as was to be proved. Thus, if $P, P_1, P_2, \cdots$ are used as in Example 3 to define a Markov process whose variables are coordinate variables, the Markov process so obtained is a representation of the given process.

Now let $\{x_t, t \in T\}$ be any Markov process. Then, if $s < \tau < t$, the transition probability satisfies the equation

$$(6.16) \qquad \mathbf{P}\{x_t(\omega) \in A \mid x_s\} = \mathbf{E}\Big\{\mathbf{P}\{x_t(\omega) \in A \mid x_\tau\} \Big| x_s\Big\}$$

with probability 1. In fact, the conditional probability on the right is also

$$\mathbf{P}\{x_t(\omega) \in A \mid x_s, x_\tau\},$$

in view of the Markov property, so that (6.16) is a consequence of the general theorems on conditional expectations and their iterations; see I (10.9). This equation is known as the Chapman-Kolmogorov equation, or, in particular cases, as the Smoluchovski equation. Generalizing the sequence of functions $P_1, P_2, \cdots$ of the preceding paragraph, we now observe that there is a function P of ξ, s, A, t, with $s < t$, which determines a probability measure of the linear Borel set A when ξ, s, t are fixed, and determines a Baire function of ξ when s, A, t are fixed, such that

$$(6.17) \qquad P(x_s, s;\ A, t) = \mathbf{P}\{x_t(\omega) \in A \mid x_s\}$$

with probability 1, for each s, A, t. Then (6.16) can also be written in the form

$$(6.18) \qquad P(\xi, s;\ A, t) = \int_{-\infty}^{\infty} P(\zeta, \tau;\ A, t) P(\xi, s;\ d\zeta, \tau).$$

This equation holds for ξ not in some Borel set B (depending on s, t, τ, A) with

$$\mathbf{P}\{x_s(\omega) \in B\} = 0.$$

In the applications, one is usually given not a Markov process but transition probabilities in terms of which the process is to be constructed. Specifically, it is usually supposed that T has a minimum value t_0, and that a function P is given, satisfying (6.18) identically in the free variables. If an initial distribution is given at t_0, a Markov process with the transition probability function P is defined as follows. If $t_0 < \cdots < t_n$, the random variables $x_{t_0}, \cdots, x_{t_n}$ are to have a joint distribution determined by the preassigned x_{t_0} distribution and the preassigned transition probabilities as in (6.15) (with x defined as 1 on an m-dimensional Borel set, and 0 otherwise, that is, as in Example 3). We have already remarked that it is then trivial to verify that this definition will make the random variables $x_{t_0}, \cdots, x_{t_n}$ a Markov process. The distributions assigned in this way are mutually consistent, that is, the Kolmogorov consistency conditions are satisfied, so that there is actually a (Markov) process with the assigned initial and transition probabilities (I §5). The transition function P is frequently supposed to be given by a density,

$$P(\xi, s;\ A, t) = \int_A p(\xi, s;\ \eta, t)\, d\eta,$$

and in this case (6.18) reduces to

$$p(\xi, s;\ \eta, t) = \int_{-\infty}^{\infty} p(\zeta, \tau;\ \eta, t) p(\xi, s;\ \zeta, \tau)\, d\zeta. \tag{6.19}$$

Equation (6.19) is frequently justified intuitively by the statement that the probability of a transition from ξ at time s to η at time t is the probability of a transition to ζ at the intermediate time τ multiplied by the probability of a transition from ζ at τ to η at t, summed over all values of ζ. There is nothing wrong with this statement, which is simply an imprecise paraphrase of (6.19), but it is imprecise enough to have led some unwary students to believe that (6.19) is true of all stochastic processes. Note, however, that without the Markov property the first factor under the integral sign in (6.19) would depend on ξ and s, in general.

A sequence of random variables $\{x_n\}$ is said to constitute a *multiple Markov process* if there is an integer ν such that for each λ and each n

$$\mathbf{P}\{x_n(\omega) \le \lambda \mid x_{n-1}, x_{n-2}, \cdots\} = \mathbf{P}\{x_n(\omega) \le \lambda \mid x_{n-1}, \cdots, x_{n-\nu}\} \tag{6.1''}$$

with probability 1. If $\nu = 1$ the process is then a Markov process (sometimes also called a *simple* Markov process). The generalization is not very significant, because the (vector) process with random variables $\{\hat{x}_n\}$, $\hat{x}_n = (x_n, \cdots, x_{n+\nu-1})$ has the Markov property [defined for vector processes by the obvious modification of (6.1)]. Thus multiple Markov processes can be reduced to simple ones at the small expense of going to vector-valued random variables. In particular, in the important case of random variables $\{x_n\}$ which take on only a fixed finite set of values (Markov chains), the $\hat{x}_n$ process will be one of the same type; the variables $\{\hat{x}_n\}$ need not be considered vector variables but simply variables taking on N^ν values, if the x_n's take on N values.

(*b*) *Wide sense* Let $\{x_t, t \in T\}$ be a Gaussian process with zero means, $\mathbf{E}\{x_t\} \equiv 0$, and which either is real or, if not, satisfies the equation

$$\mathbf{E}\{x_s x_t\} \equiv 0$$

(cf. Theorem 3.2). To define a Markov process in the wide sense we must discover what apparently weaker property of the x_t process, defined in terms of variances and covariances, is equivalent to the Markov property, or at least to an important special case of this property. We have already remarked (§3) that if $t_1 < \cdots < t_n$, one version of $\mathbf{E}\{x_{t_n} \mid x_{t_1}, \cdots, x_{t_{n-1}}\}$ is that linear combination of $x_{t_1}, \cdots, x_{t_{n-1}}$, $\hat{\mathbf{E}}\{x_{t_n} \mid x_{t_1}, \cdots, x_{t_{n-1}}\}$ in the notation of §3, which is the closest to x_{t_n} in the sense of minimizing

$$\mathbf{E}\{|x_{t_n} - \sum_{1}^{n-1} a_j x_{t_j}|^2\} = \sum_{j,k=1}^{n} \mathbf{E}\{x_{t_j}\bar{x}_{t_k}\} a_j \bar{a}_k, \qquad (a_n = -1). \tag{6.20}$$

Now the Markov property implies that

$$\mathbf{E}\{x_{t_n} \mid x_{t_1}, \cdots, x_{t_{n-1}}\} = \mathbf{E}\{x_{t_n} \mid x_{t_{n-1}}\} \tag{6.21}$$

with probability 1, and in fact considerably more, since the Markov property applies to the conditional distribution of x_{t_n}, not merely to its conditional expectation. However, the condition (6.1) is equivalent to (6.21) in the present case, as we shall see in a moment. Hence the condition that the x_t process be a Markov process can be written in the form

$$\hat{\mathbf{E}}\{x_{t_n} \mid x_{t_1}, \cdots, x_{t_{n-1}}\} = \hat{\mathbf{E}}\{x_{t_n} \mid x_{t_{n-1}}\}, \tag{6.21'}$$

with probability 1. This is a condition involving only variances and covariances, according to (6.20). Hence we shall say that *any process, Gaussian or not, is a Markov process in the wide sense if* $\mathbf{E}\{|x_t|^2\} < \infty$ *for all t, and if, whenever* $t_1 < \cdots < t_n$, (6.21′) *is satisfied with probability* 1. We still must justify, for the original Gaussian x_t process our statement that (6.21) implies (6.1). For this process, the difference

$$x_{t_n} - \mathbf{E}\{x_{t_n} \mid x_{t_1}, \cdots, x_{t_{n-1}}\}$$

is a Gaussian variable, with mean 0, and is orthogonal to and therefore independent of $x_{t_1}, \cdots, x_{t_{n-1}}$. Hence the conditional distribution of x_{t_n}, given that

$$x_{t_1}(\omega) = \xi_1, \cdots, x_{t_{n-1}}(\omega) = \xi_{n-1},$$

is that of

$$y = x_{t_n} - \mathbf{E}\{x_{t_n} \mid x_{t_1}(\omega) = \xi_1, \cdots, x_{t_{n-1}}(\omega) = \xi_{n-1}\} \tag{6.22}$$

which is Gaussian, with mean a certain linear combination of $\xi_1, \cdots, \xi_{n-1}$ [the last term in (6.22)], and variance that of y. The conditional distribution of x_{t_n} is thus entirely determined by the conditional expectation, and (6.21) implies (6.1), in the present case, as was to be proved.

In V §8 a simple condition will be derived which is necessary and sufficient that a process be a Markov process in the wide sense. Note that a process which is a Markov process in the strict sense is not necessarily one in the wide sense, even if the expectations of the squares of the variables involved exist.

7. Martingales

A stochastic process $\{x_t, t \in T\}$ is called a *martingale* if $\mathbf{E}\{|x_t|\} < \infty$ for all t and if, whenever $n \geq 1$ and $t_1 < \cdots < t_{n+1}$,

$$\mathbf{E}\{x_{t_{n+1}} \mid x_{t_1}, \cdots, x_{t_n}\} = x_{t_n} \tag{7.1}$$

with probability 1. This is a strict sense definition, and *martingale* will always mean martingale in this strict sense. A stochastic process with variables $\{x_t\}$ is called a *martingale in the wide sense* if $\mathbf{E}\{|x_t|^2\} < \infty$ for all t and if, whenever $n \geq 1$ and $t_1 < \cdots < t_{n+1}$,

$$\hat{\mathbf{E}}\{x_{t_{n+1}} \mid x_{t_1}, \cdots, x_{t_n}\} = x_{t_n} \tag{7.1'}$$

with probability 1. Applying the rules of combination of $\mathbf{E}$ and $\hat{\mathbf{E}}$ (cf. IV §3 for the derivation of these rules for $\hat{\mathbf{E}}$), it is seen that a sequence of random variables $x_1, x_2, \cdots$ is a martingale if and only if

$$\mathbf{E}\{|x_n|\} < \infty, n \geq 1, \qquad \mathbf{E}\{x_{n+1} \mid x_1, \cdots, x_n\} = x_n \tag{7.2}$$

with probability 1, a martingale in the wide sense if and only if

$$\mathbf{E}\{|x_n|^2\} < \infty, n \geq 1, \qquad \hat{\mathbf{E}}\{x_{n+1} \mid x_1, \cdots, x_n\} = x_n \tag{7.2'}$$

with probability 1.

If $y_1, y_2, \cdots$ are defined as

$$y_1 = x_1, \qquad y_2 = x_2 - x_1, \cdots$$

then, if the x_n process is a martingale,

$$\mathbf{E}\{|y_n|\} < \infty, \qquad \mathbf{E}\{y_{n+1} \mid y_1, \cdots, y_n\} = 0, \qquad n \geq 1, \tag{7.3}$$

with probability 1. The x_n's are thus partial sums of the series $\sum_n y_n$, where the y_n's satisfy (7.3). Conversely, the partial sums of any such series constitute a martingale. If the x_n process is a martingale in the wide sense, the corresponding property of the y_j's is that these random variables are mutually orthogonal.

A y_n process satisfying (7.3) has independent interest and merits further examination. The condition (7.3) lies between zero correlation and independence of the y_n's. In fact, the condition of zero correlation of the y_n's is

$$\mathbf{E}\{y_m\bar{y}_n\} = \mathbf{E}\{y_m\}\mathbf{E}\{\bar{y}_n\} \qquad m \neq n.$$

The condition (7.3) can be reformulated as follows: if $\Phi(y_1, \cdots y_n)$ is any bounded Baire function of the indicated variables,

$$\mathbf{E}\{y_{n+1}\bar{\Phi}(y_1, \cdots, y_n)\} = 0, \qquad n \geq 1. \tag{7.4}$$

This is, of course, stronger than zero correlation. [Note that $\mathbf{E}\{y_n\} = 0$ if (7.3) is true.] On the other hand, the condition of independence of the y_n's is equivalent to the still stronger condition that for every Φ as above and every bounded Baire function $\Psi(y_{n+1})$ of y_{n+1}.

$$\mathbf{E}\{\Psi(y_{n+1})\bar{\Phi}(y_1, \cdots, y_n)\} = \mathbf{E}\{\Psi(y_{n+1})\}\mathbf{E}\{\bar{\Phi}(y_1, \cdots, y_n)\} \tag{7.5}$$

A Markov process involves a stronger restriction, in one sense, than a martingale, since the Markov property involves distributions rather than expectations; on the other hand, a Markov process need not be a martingale.

Example 1 Let $\eta, \xi_1, \xi_2, \cdots$ be any random variables with

$$\mathbf{E}\{|\eta|\} < \infty.$$

Then, if x_n is defined by

$$x_n = \mathbf{E}\{\eta \mid \xi_1, \cdots, \xi_n\}, \tag{7.6}$$

the x_n process is a martingale. In fact,

$$\begin{aligned} \mathbf{E}\{x_{n+1} \mid \xi_1, \cdots, \xi_n\} &= \mathbf{E}\Big\{\mathbf{E}\{\eta \mid \xi_1, \cdots, \xi_{n+1}\} \Big| \xi_1, \cdots, \xi_n\Big\} \\ &= \mathbf{E}\{\eta \mid \xi_1, \cdots, \xi_n\} = x_n, \end{aligned} \tag{7.7}$$

with probability 1. Hence, since $x_1, \cdots, x_n$ are random variables on the sample space of $\xi_1, \cdots, \xi_n$,

$$\mathbf{E}\{x_{n+1} \mid x_1, \cdots, x_n, \xi_1, \cdots, \xi_n\} = \mathbf{E}\{x_{n+1} \mid \xi_1, \cdots, \xi_n\} = x_n, \tag{7.8}$$

with probability 1. Taking the conditional expectation of both sides of (7.8) with $x_1, \cdots, x_n$ fixed gives (7.2).

The corresponding continuous parameter example of a martingale is given by

$$x_t = \mathbf{E}\{\eta \mid \xi_\tau, \tau \leq t\} \tag{7.9}$$

where the ξ_τ's and η are arbitrary random variables except that $\mathbf{E}\{|\eta|\} < \infty$.

If $\mathbf{E}$ is replaced by $\hat{\mathbf{E}}$ in this example, the processes derived are martingales in the wide sense. The proofs are still valid.

Example 2 If the random variables $x_1, x_2, \cdots$ can be put in the form

$$(7.10) \qquad x_n = y_1 + \cdots + y_n, n \geq 1, \qquad \mathbf{E}\{|y_j|\} < \infty, \qquad n \geq 1,$$

where the y_j's are mutually independent, and $\mathbf{E}\{y_m\} = 0$, $m > 1$, then the x_n process is a martingale (also a martingale in the wide sense if $\mathbf{E}\{|y_n|^2\} < \infty$.) This is a special case of the general form of a discrete parameter martingale already discussed. The continuous parameter version of this example will be defined in §9.

Example 3 Let $y_1, y_2, \cdots$ be any random variables, and suppose that the distribution of $y_1, \cdots, y_n$ is given by a Baire density function p_n in n-dimensional space. In this way a sequence of density functions $p_1, p_2, \cdots$ is defined. Let $q_1, q_2, \cdots$ be a second such sequence; in the following we shall consider the y_j process distributions as determined by the p_j's. Define the random variable x_n by

$$(7.11) \qquad x_n = \frac{q_n(y_1, \cdots, y_n)}{p_n(y_1, \cdots, y_n)}.$$

Note that the denominator vanishes with probability 0. We shall prove that the x_n process is a martingale if we suppose that $q_n = 0$ whenever $p_n = 0$. To prove this assertion we shall assume, as we can for the purposes of the proof according to the representation theory of I §6, that $y_1, y_2, \cdots$ are the coordinate variables in infinite dimensional space. Then there is a conditional probability density of y_{n+1} for given $y_1, \cdots, y_n$,

$$\frac{p_{n+1}(y_1, \cdots, y_n, \lambda)}{p_n(y_1, \cdots, y_n)},$$

so that

$$(7.12) \quad \mathbf{E}\{x_{n+1} \mid y_1, \cdots, y_n\}$$

$$= \int_{-\infty}^{\infty} x_{n+1}\Big|_{y_{n+1}=\lambda} \frac{p_{n+1}(y_1, \cdots, y_n, \lambda)}{p_n(y_1, \cdots, y_n)}\, d\lambda$$

$$= \int_{-\infty}^{\infty} \frac{q_{n+1}(y_1, \cdots, y_n, \lambda)}{p_n(y_1, \cdots, y_n)}\, d\lambda = \frac{q_n(y_1, \cdots, y_n)}{p_n(y_1, \cdots, y_n)} = x_n.$$

Taking the conditional expectation of both sides of (7.12) relative to $x_1, \cdots, x_n$ gives (7.2).

The martingale defined in (7.11) has important statistical applications. In statistics the ratio defining x_n is called a "likelihood ratio." Note that, if $p_n(\lambda_1, \cdots, \lambda_n)$ and $q_n(\lambda_1, \cdots, \lambda_n)$ are not interpreted as densities but as the probability that a set of discrete-valued random variables take on the values $\lambda_1, \cdots, \lambda_n$, the x_n process defined by (7.11) is still a martingale. [The integrals in (7.12) become sums.] A more general type of martingale is defined as follows. Let $y_1, y_2, \cdots$ and $z_1, z_2, \cdots$ be sequences of random variables. (The z_n sequence is not necessarily defined on the same ω space as the y_n sequence.) Let P_n, Q_n be the measures of n-dimensional Borel sets A defined by

$$P_n(A) = \mathbf{P}\{[y_1(\omega), \cdots, y_n(\omega)] \in A\},$$
$$Q_n(A) = \mathbf{P}\{[z_1(\omega), \cdots, z_n(\omega)] \in A\}.$$

Suppose that $Q_n(A) = 0$ if $P_n(A) = 0$, that is, that Q_n measure is absolutely continuous with respect to P_n measure. There is then a relative density, according to the Radon-Nikodym theorem, that is, there is a Baire function Φ_n of n variables, such that

$$Q_n(A) = \int_A \Phi_n \, dP_n.$$

Define x_n by

$$x_n = \Phi_n(y_1, \cdots, y_n).$$

Then the x_n process is a martingale. The proof can be carried through as in the preceding special case or by direct appeal to the definition of the conditional expectations involved. Likelihood ratios will be examined in more detail from the martingale point of view in VII.

8. Stationary stochastic processes

(*a*) *Strictly stationary processes* A strictly stationary stochastic process $\{x_t, t \in T\}$ is one whose distributions remain the same as time passes; that is, the multivariate distribution of the random variables $x_{t_1+h}, \cdots, x_{t_n+h}$ is independent of h. Here $t_1, \cdots, t_n$ is any finite set of parameter values, and h is chosen so that the translated parameter values are also parameter values.

Example 1 Let $\cdots, x_0, x_1, \cdots$ be mutually independent random variables with a common distribution function. Then the x_j process is strictly stationary, as is any $\tilde{x}_n$ process, where

$$\tilde{x}_n = \sum_{-\infty}^{\infty} a_{n+m} x_m,$$

and the a_j's are chosen to make the series converge in probability. (We shall see in III that convergence in probability of a series of mutually independent random variables implies convergence with probability 1.)

A strictly stationary process is subject to the strong law of large numbers: *if the parameter is integral-valued and if* $\mathbf{E}\{|x_0|\} < \infty$, *then* (X, Theorem 2.1)

$$\lim_{n\to\infty} \frac{x_0 + \cdots + x_n}{n+1} = \hat{x}$$

exists with probability 1, *that is, for almost all sample sequences.* The limit $\hat{x}$ is identically constant (with probability 1) in many important special cases. For example, if the x_j's are mutually independent it will be seen that

$$\hat{x} = \mathbf{E}\{x_1\}$$

with probability 1.

In the continuous parameter case the strong law of large numbers for strictly stationary processes becomes: *if* $\mathbf{E}\{|x_0|\} < \infty$ *and if the process is measurable, then* (XI, Theorem 2.1)

$$\lim_{t\to\infty} \frac{1}{t} \int_0^t x_s\, ds = \hat{x}$$

exists with probability 1, *that is, for almost all sample functions.*

(*b*) *Wide sense stationary processes* The process $\{x_t,\ t \in T\}$ is called *stationary in the wide sense* if $\mathbf{E}\{|x_t|^2\} < \infty$ for $t \in T$, and if $\mathbf{E}\{x_{s+t}\bar{x}_s\} = R(t)$ does not depend on s. The function R is called the *covariance function* of the process. Usually the added condition that $\mathbf{E}\{x_s\}$ does not depend on s is imposed. This condition is unnatural mathematically, and has nothing to do with the essential properties of interest in these processes, and we shall therefore not impose it. When the added condition is satisfied, however,

$$\mathbf{E}\{[x_{s+t} - \mathbf{E}\{x_{s+t}\}]\,\overline{[x_s - \mathbf{E}\{x_s\}]}\} = \mathbf{E}\{x_{s+t}\bar{x}_s\} - \mathbf{E}\{x_s\}\mathbf{E}\{\bar{x}_{s+t}\}$$

is also independent of t and the process with variables $\{x_t - \mathbf{E}\{x_t\}\}$ is used rather than the original process. If this is done the random variables determining the process have zero expectations and $R(t)$ is a true covariance.

If a real Gaussian process is stationary in the wide sense, and if $\mathbf{E}\{x_s\}$ does not depend on s, the process is strictly stationary because the determining parameters of Gaussian distributions are the means and covariances. If a complex Gaussian process is stationary, if $\mathbf{E}\{x_t\} \equiv 0$, and if (see Theorem 3.2) $\mathbf{E}\{x_s x_t\} = 0$ for all s, t, the process is strictly stationary for the same reason. Thus the definition given is a proper wide sense definition to match the strict sense definition of a stationary process.

A strictly stationary process is stationary in the wide sense if $\mathbf{E}\{|x_t|^2\} < \infty$ for all t.

In this book, both strictly stationary and wide sense stationary processes will be referred to as stationary processes. In the literature, however, "stationary processes" sometimes means "strictly stationary processes." "Temporally homogeneous" has been used as a synonym for "stationary" but is now uncommon.

Example 2 Let $\cdots, x_0, x_1, \cdots$ be mutually independent random variables with

$$\mathbf{E}\{x_n\} = 0, \qquad \mathbf{E}\{|x_n|^2\} = \sigma^2 > 0.$$

Then the x_n process is stationary in the wide sense with

$$\begin{aligned} R(n) &= 0 \qquad n \neq 0 \\ &= \sigma^2 \qquad n = 0. \end{aligned}$$

The process is strictly stationary if and only if the x_n's have a common distribution function.

A process stationary in the wide sense is subject to the law of large numbers, and in fact the theorems stated in this connection for strictly stationary processes remain true if "limit with probability 1" is replaced by "l.i.m."; see X, Theorem 6.1, and XI, Theorem 6.1.

9. Processes with independent increments

A process with independent increments is one whose random variables $\{x_t\}$ have the property that, if $t_1 < \cdots < t_n$ ($n \geq 3$), the differences

$$x_{t_2} - x_{t_1}, \cdots, x_{t_n} - x_{t_{n-1}}$$

are mutually independent. If $x_0, x_1, \cdots$ constitute such a process this means that $x_1 - x_0, x_2 - x_1, \cdots$ are mutually independent; $x_n - x_0$ is the nth partial sum of the series $\sum_1^\infty (x_m - x_{m-1})$ of mutually independent random variables. Conversely, if $y_1, y_2, \cdots$ are mutually independent random variables, if x_0 is an arbitrary random variable, and if x_n for $n \geq 1$ is defined by $x_n = x_0 + y_1 + \cdots + y_n$, the x_n process is a process with independent increments. In practice the term "process with independent increments" is used only in the continuous parameter case.

A continuous parameter process $\{x_t, 0 \leq t < \infty\}$ with independent increments and $\mathbf{P}\{x_0(\omega) = 0\} = 1$ is for $t > 0$ the continuous parameter version of a discrete parameter process whose random variables are the partial sums of a series of mutually independent random variables. Just as the latter process is one example of a discrete parameter Markov process and also (if the expectations of the summands all vanish) an example of a discrete parameter martingale, so the former process is a continuous parameter Markov process and, if $\mathbf{E}\{x_t - x_s\} = 0$, is also a martingale.

If the distribution of $x_t - x_s$ depends only on $t - s$, a process with independent increments will be said to have stationary (strict sense) increments.

Examples of processes with independent increments are specified by the distribution of $x_t - x_s$. If $s_1 < s_2 < s_3$, the random variables

$$y_1 = x_{s_2} - x_{s_1}, \qquad y_2 = x_{s_3} - x_{s_2}, \qquad y_3 = x_{s_3} - x_{s_1}$$

will thereby be assigned distributions. Since $y_3 = y_1 + y_2$, and since y_1 and y_2 are mutually independent, the distribution assigned to y_3 must be that of the sum of two independent random variables, one with the distribution of y_1, the other with that of y_2. This consistency condition is easily checked in the examples to be given, and serves to insure the truth of the Kolmogorov consistency conditions required to insure the possibility of setting up the stochastic process. Note that it is not necessary to assign distributions to the x_t's themselves, since only the differences are usually involved, and in fact the procedure used to set up a function space measure is applicable without the x_t distributions. This will mean, however, that x_t will not be a random variable, and that the random variables of the process are really the differences $x_t - x_s$. This situation is usually avoided by choosing some parameter point t_0 and considering the process $\{x_t - x_{t_0}, t \in T\}$, that is, the variables of the process are normalized to vanish at t_0.

Example 1 *Brownian motion process* In this case it is supposed that $x_t - x_s$ is real and normally distributed, with

$$(9.1) \qquad \begin{aligned} \mathbf{E}\{x_t - x_s\} &\equiv 0 \\ \mathbf{E}\{[x_t - x_s]^2\} &= \sigma^2|t - s|, \end{aligned}$$

where $\sigma > 0$ is a fixed parameter. The parameter set T is usually taken as either the whole t-axis or the half-axis $[0, \infty)$, and in the latter case x_0 is usually defined to be 0 with probability 1, that is, we consider the differences $x_t - x_0$ as explained above. This process was first discussed by Bachelier, and later more rigorously by Wiener. It is sometimes called the Wiener process. For fixed t_0 the differences $\{x_t - x_{t_0}, t \geq t_0\}$ constitute a Markov process and also a martingale.

If microscopic particles are observed in a fluid, they are seen to move in an irregular fashion under the impacts of the fluid molecules. The motion is called Brownian motion after the English botanist Brown who reported the phenomenon in 1826. Einstein and Smoluchovski showed that to a first approximation $x(t)$, the x coordinate of a Brownian particle at time t, defines a function of t for each particle motion which can be identified with a sample function of a Brownian motion stochastic process. The constant σ^2 depends on the mass of the particle and on the viscosity of the fluid.

It will be shown in VIII §2 that the sample functions of a separable Brownian motion process are almost all continuous functions, and in fact (VIII §7) this is essentially the only process with independent increments having this property.

Example 2 *Poisson process* Here it is supposed that for every pair $s < t$, $x_t(\omega) - x_s(\omega)$ is integral-valued, with

$$(9.2) \quad \mathbf{P}\{x_t(\omega) - x_s(\omega) = \nu\} = \frac{e^{-c(t-s)}c^\nu(t-s)^\nu}{\nu!}, \qquad \nu = 0, 1, 2, \cdots,$$

where $c \geq 0$ is a fixed parameter. A Poisson process has stationary increments (strict sense). For fixed t_0 the differences $\{x_t - x_{t_0},\ t \geq t_0\}$ constitute a Markov process, and the differences $\{x_t - x_{t_0} - ct,\ t \geq t_0\}$ constitute a process which is both a Markov process and a martingale.

It will be shown in VIII §4 that the sample functions of a separable Poisson process are (almost all) monotone non-decreasing, increasing in isolated jumps of unit magnitude. The points where these jumps occur can be considered the times when some sort of random event occurs, and $x_t - x_{t_0}$ is then the number of events that have occurred by time t, beginning the count at time t_0. The constant c is the expected rate of occurrence. With this interpretation it is shown that the conditional distribution of events in an interval (s, t), if it is known that n have occurred in the interval, is that of n points chosen independently in (s, t), each choice uniformly distributed over (s, t). The Poisson process has been found a good approximation to the process governing the times of emissions of radioactive material. Each sample function can also be considered as defining an infinite sequence of values of t (positive and negative if the condition $t \geq 0$ is dropped), the points where the jumps occur, and it is such sequences that are usually in the minds of those who speak of a uniform distribution of points over an infinite interval or of a "purely random" sequence of points on such an interval.

Almost all sample functions of a separable Poisson process are continuous except at the points of an at most enumerable parameter set, and even at the points of discontinuity left- and right-hand finite limits exist. This assertion is true of the general process with independent increments after a suitable centering has been accomplished by replacing x_t by $x_t - f(t)$, where f is a function of t but not of ω.

10. Processes with uncorrelated or orthogonal increments

A process $\{x_t, t \in T\}$ is said to have uncorrelated increments if

$$\mathbf{E}\{|x_t - x_s|^2\} < \infty, \qquad s, t \in T, \tag{10.1}$$

and, if whenever parameter values satisfy the inequality $s_1 < t_1 \leq s_2 < t_2$ the increments $x_{t_1} - x_{s_1}$ and $x_{t_2} - x_{s_2}$ are uncorrelated with each other,

$$\mathbf{E}\{(x_{t_2} - x_{s_2})\overline{(x_{t_1} - x_{s_1})}\} = \mathbf{E}\{x_{t_2} - x_{s_2}\}\mathbf{E}\{\overline{x_{t_1} - x_{s_1}}\}. \tag{10.2}$$

A closely related type of process is a process with orthogonal increments, that is, one for which (10.1) is true, and for which (10.2) is replaced by

$$\mathbf{E}\{(x_{t_2} - x_{s_2})\overline{(x_{t_1} - x_{s_1})}\} = 0. \tag{10.3}$$

If an x_t process has uncorrelated increments, the y_t process determined by

$$y_t = x_t - \mathbf{E}\{x_t\}$$

is a process with uncorrelated and orthogonal increments.

If $\{x_n, n \geq 0\}$ is a process with uncorrelated [orthogonal] increments, $x_n - x_0$ is the nth partial sum of the series $\sum_1^\infty (x_m - x_{m-1})$ of mutually uncorrelated [orthogonal] random variables. Conversely, if $y_1, y_2, \cdots$ are mutually uncorrelated [orthogonal] random variables, if x_0 is an arbitrary random variable, and if x_n for $n \geq 1$ is defined by

$$x_n = x_0 + y_1 + \cdots + y_n,$$

then the x_n process is a process with uncorrelated [orthogonal] increments. In practice the term "process with uncorrelated [orthogonal] increments" is used only in the continuous parameter case.

If the increments of a process satisfying (10.1) are stationary as far as second moments are concerned, in the sense that

$$\mathbf{E}\{|x_t - x_s|^2\}$$

depends only on $t - s$, the process is said to have *stationary (wide sense) increments*. An elementary calculation shows that in this case the expectation

$$\mathbf{E}\{(x_s - x_t)\overline{(x_u - x_v)}\}$$

depends only on the differences $t - s$, $u - t$, $v - t$.

If a process with uncorrelated or orthogonal increments is Gaussian, with

$$\mathbf{E}\{x_t\} \equiv 0,$$

and if it is real, or if (see Theorem 3.2)

$$\mathbf{E}\{x_s x_t\} \equiv 0,$$

then the process has independent increments. Thus the processes with uncorrelated or orthogonal increments are processes with independent increments in the wide sense.

If a process has independent increments, and if (10.1) is true, the process also has uncorrelated increments. Thus the Brownian movement and Poisson processes defined in §9 are processes with uncorrelated increments, and in fact with stationary (strict as well as wide sense) increments.

Let $\{x_t, t \in T\}$ be a process with orthogonal increments. Then $F(t)$ can be defined to satisfy

$$\mathbf{E}\{|x_t - x_s|^2\} = F(t) - F(s), \qquad s < t. \tag{10.4}$$

For example, we can take any $t_0 \in T$ and define

$$\begin{aligned} F(t) &= \mathbf{E}\{|x_t - x_{t_0}|^2\}, & t \geq t_0, \\ &= -\mathbf{E}\{|x_t - x_{t_0}|^2\}, & t < t_0. \end{aligned}$$

The function F is monotone non-decreasing, and is determined by (10.4) up to an additive constant. The process has stationary (wide sense) increments if and only if the difference $F(t) - F(s)$ depends only on $t - s$. In the stationary case denote this difference by $F_1(t - s)$, $s < t$. Then

$$F_1(t_3 - t_1) = F_1(t_3 - t_2) + F_1(t_2 - t_1), \qquad t_1 \leq t_2 \leq t_3.$$

Let T_1 be the set of differences $t - s$ with $s, t \in T$ and $s \leq t$. Then F_1 is a monotone non-decreasing function, defined on T_1, and satisfying the functional equation

$$F_1(u + v) = F_1(u) + F_1(v), \qquad u, v \in T_1.$$

The only monotone solution of this functional equation, if T_1 is an interval, is

$$F_1(s) = \text{const. } s.$$

Thus, if T_1 is an interval, the x_t process has stationary (wide sense) increments if and only if

$$F(t) = \text{const.} + \sigma^2 t$$

for some $\sigma \geq 0$.

If the x_t process has uncorrelated increments, $m(t)$ can be defined to satisfy

$$\mathbf{E}\{x_t - x_s\} = m(t) - m(s)$$

and the $x_t - m(t)$ process will then have orthogonal increments.

The Brownian movement process defined in §9 is a process with stationary uncorrelated (and orthogonal) increments. The mean and variance functions m and F are, if the arbitrary additive constants are chosen properly,

$$\mathbf{E}\{x_t - x_0\} = m(t) = 0$$

$$\mathbf{E}\{(x_t - x_0)^2\} = F(t) = \sigma^2 t.$$

The Poisson process defined in §9 is a process with stationary uncorrelated (but not orthogonal) increments, with

$$\mathbf{E}\{x_t - x_0\} = m(t) = ct$$

$$\mathbf{E}\{|x_t - x_0 - ct|^2\} = ct.$$

Processes with uncorrelated and orthogonal increments are essential tools in the study of stationary processes, and will, therefore, be studied in IX before stationary processes are taken up in X and XI.

It will sometimes be convenient to write (10.4) symbolically in the form

$$\mathbf{E}\{|dx_t|^2\} = dF(t).$$

CHAPTER III

Processes with Mutually Independent Random Variables

1. General remarks

In this chapter the random variables will be real-valued. The extension of the results to complex-valued random variables will however be obvious.

As already remarked in II §4, processes with mutually independent random variables are only useful in the discrete parameter case. In other words the useful case is typically a sequence $x_1, x_2, \cdots$ of mutually independent random variables. The process is characterized by the distribution functions of the individual variables, because of the independence property.

One of the most striking properties of these processes is the zero-one law, which will be applied frequently in the following sections.

THEOREM 1.1 (*Zero-one law*) *Let $x_1, x_2, \cdots$ be mutually independent random variables. Then, if Λ is an ω set which is measurable on the sample space of $x_n, x_{n+1}, \cdots$ for every n, it follows that $\mathbf{P}\{\Lambda\} = 0$ or $\mathbf{P}\{\Lambda\} = 1$, and, if y is a random variable measurable on the sample space of $x_n, x_{n+1}, \cdots$ for every n, it follows that there is a constant c such that $\mathbf{P}\{y(\omega) = c\} = 1$.*

This theorem is usually stated somewhat loosely as follows: let $x_1, x_2, \cdots$ be a sequence of mutually independent random variables. Then, if Λ is an event dependent on $x_n, x_{n+1}, \cdots$ for every n, Λ has probability either 0 or 1, and, if y is equal with probability 1 to a function of these same variables for every n, then $y(\omega) \equiv$ const., with probability 1.

The second assertion of the theorem follows readily from the first, using the fact that, if y is a random variable satisfying the hypotheses of the second assertion, the ω set $\{y(\omega) \in A\}$ satisfies those of the first, for every Borel set A. We therefore discuss only the first assertion. Suppose then that Λ is an ω set with the stated properties, and let $\mathscr{G}$ be the class of measurable ω sets M with the property that

$$\mathbf{P}\{\Lambda \mathrm{M}\} = \mathbf{P}\{\Lambda\}\mathbf{P}\{\mathrm{M}\}. \tag{1.1}$$

Then, according to our hypotheses on Λ, for each n the class $\mathscr{G}$ includes every ω set measurable on the sample space of $x_1, \cdots, x_n$. Let $\mathscr{F}_0$ be the class of ω sets measurable on the sample space of $x_1, \cdots, x_n$ for some n. Then $\mathscr{F}_0$ is a field, and we have just shown that $\mathscr{F}_0 \subset \mathscr{G}$. The class $\mathscr{G}$ obviously includes the limits of every monotone sequence of $\mathscr{G}$ sets, and therefore (Supplement, Theorem 1.2) includes the Borel field generated by $\mathscr{F}_0$. This Borel field is the class of ω sets measurable on the sample space of $x_1, x_2, \cdots$. Hence $\Lambda \in \mathscr{G}$, so that

$$\mathbf{P}\{\Lambda\} = \mathbf{P}\{\Lambda\}^2, \tag{1.2}$$

and it follows that $\mathbf{P}\{\Lambda\} = 0$ or $\mathbf{P}\{\Lambda\} = 1$, as was to be proved.

The hypothesis of mutual independence of the x_n's served only to insure the truth of (1.1) for $\mathrm{M} \in \mathscr{F}_0$. If we drop this independence hypothesis, and in fact only suppose that Λ is a measurable ω set with the property that for each n

$$P\{\Lambda \mid x_1, \cdots, x_n\} = \mathbf{P}\{\Lambda\}$$

with probability 1, it is still true that (1.1) follows, for $\mathrm{M} \in \mathscr{F}_0$, and the proof of (1.2) then goes through as in the above special case. However, even in this more general form the theorem is a trivial corollary of a martingale convergence theorem (VII, Theorem 4.3).

Before applying the zero-one law we give an example to show the possibilities if the x_n's are not mutually independent. Suppose that

$$x_1(\omega) = x_2(\omega) = \cdots$$

with probability 1. Then $y = x_1$ satisfies the condition of Theorem 1.1, but this random variable need not be a constant, since it is completely unrestricted.

As a first application of the zero-one law, consider the following problem. Let $\mathrm{M}_1, \mathrm{M}_2, \cdots$ be measurable ω sets, and let Λ be the ω set of those points in infinitely many M_j's,

$$\Lambda = \bigcap_{n=1}^{\infty} \bigcup_{j=n}^{\infty} \mathrm{M}_j.$$

The problem is to evaluate $\mathbf{P}\{\Lambda\}$. In the less prosaic language of events, the problem is to evaluate the probability that infinitely many events (represented by the M_j's) occur. Define x_j by

$$\begin{aligned} x_j(\omega) &= 1, \qquad && \omega \in \mathrm{M}_j \\ &= 0 && \text{otherwise.} \end{aligned}$$

Then obviously Λ has the property presupposed in Theorem 1.1. If the M_j's do not represent independent events, that is to say, if the x_j's are not mutually independent, then $\mathbf{P}\{\Lambda\}$ may be any number between 0 and 1. However, according to the zero-one law, if these events are mutually independent then $\mathbf{P}\{\Lambda\}$ must be 0 or 1. The following theorem, usually called the Borel-Cantelli lemma, gives a criterion for each case.

THEOREM 1.2 *Let* $M_1, M_2, \cdots$ *be measurable* ω *sets, and let* Λ *be the set of points in infinitely many* M_j's. *If* $\Sigma\mathbf{P}\{M_j\} < \infty$, *then* $\mathbf{P}\{\Lambda\} = 0$. *Conversely, if* $\Sigma\mathbf{P}\{M_j\} = \infty$ *and if the* M_j's *are mutually independent, then* $\mathbf{P}\{\Lambda\} = 1$.

If $\Sigma\mathbf{P}\{M_j\} < \infty$, then

$$\mathbf{P}\{\Lambda\} \leq \mathbf{P}\{\bigcup_n^\infty M_j\} \leq \sum_n^\infty \mathbf{P}\{M_j\} \to 0 \qquad (n \to \infty), \tag{1.3}$$

so that $\mathbf{P}\{\Lambda\} = 0$. Now suppose that the M_j events are independent. Since the contrary of infinitely many events occurring is that none occurs after the nth for sufficiently large n,

$$1 - \mathbf{P}\{\Lambda\} = \lim_{n\to\infty} \mathbf{P}\{\bigcap_{n+1}^\infty M_j'\} = \lim_{n\to\infty} \prod_n^\infty [1 - \mathbf{P}\{M_j\}], \tag{1.4}$$

where M_j' is the complement of M_j. If $\Sigma\,\mathbf{P}\{M_j\} = \infty$, the infinite product

$$\prod [1 - \mathbf{P}\{M_j\}]$$

must diverge to 0, which implies that the limit on the right in (1.4) is 0, so that $\mathbf{P}\{\Lambda\} = 1$.

An extension of the zero-one law and the Borel-Cantelli lemma, due to Lévy, will be discussed in VII.

A related result is the following. Let $M_1, M_2, \cdots$ be independent ω sets in the above sense. Let N be the expected number of M_j events that occur, that is, N is the expectation of a random variable which at each ω is defined as the number of M_j's containing ω,

$$N = \Sigma\,\mathbf{P}\{M_j\}.$$

We wish to compare N with the probability P that at least one M_j event occurs,

$$P = \mathbf{P}\{\bigcup M_j\}.$$

According to the following inequality N and P approach 0 together. In fact, since $p_j = \mathbf{P}\{M_j\} \leq P$,

$$\begin{aligned} P \leq N = \sum_j p_j &= -\sum_j [\log(1 - p_j) + O(p_j^2)] \\ &\leqq -\log \prod_1^\infty (1 - p_j) + NO\,(\text{L.U.B. } p_j) \\ &\leqq -\log(1 - P) + NO(P) = P + O(P^2) + NO(P), \end{aligned}$$

so that

$$P \leq N \leq P + O(P^2) = O(P).$$

2. Series

The key to the study of sums of mutually independent random variables is the fact that, roughly speaking, partial sums cannot be large unless the total sum is. The following two theorems both express this fact.

THEOREM 2.1 *Let $y_1, \cdots, y_n$ be random variables with*

$$\begin{aligned} \mathbf{E}\{y_j\} &= 0, \qquad & \mathbf{E}\{y_j y_k\} &= 0 \qquad & j &\neq k \\ & & &= \sigma_k^2 & j &= k, \\ x_j &= y_1 + \cdots + y_j, & \mathbf{E}\{x_j^2\} &= \sigma_1^2 + \cdots + \sigma_j^2 = \hat{\sigma}_j^2. \end{aligned}$$

Then, if $\varepsilon > 0$,

$$\mathbf{P}\{|x_n(\omega)| \geq \varepsilon\} \leq \frac{\hat{\sigma}_n^2}{\varepsilon^2} \tag{2.1}$$

and if in addition

$$\mathbf{E}\{y_j \mid y_1, \cdots, y_{j-1}\} = 0, \qquad j > 1, \tag{2.2}$$

with probability 1, *then*

$$\mathbf{P}\{\operatorname*{Max}_{j \leq n} |x_j(\omega)| \geq \varepsilon\} \leq \frac{\hat{\sigma}_n^2}{\varepsilon^2}. \tag{2.1'}$$

The first part of the theorem is simply an application of Chebyshev's inequality, which requires no proof here. The extra hypothesis (2.2), which is that the x_j process is a martingale (see II §7), is certainly satisfied if the y_j's are mutually independent, and Kolmogorov's proof of (2.1′) in the latter case goes through without change in the general case, as follows. Let $|x_\nu(\omega)|$ be the first $|x_j(\omega)|$, if any, which is $\geq \varepsilon$. Then

$$\hat{\sigma}_n^2 = \int_\Omega x_n^2 \, d\mathbf{P} \geq \sum_{j=1}^n \int_{\{\nu(\omega)=j\}} [x_j^2 + 2(x_n - x_j)x_j + (x_n - x_j)^2] \, d\mathbf{P}. \tag{2.3}$$

Now if z_j is defined by

$$\begin{aligned} z_j(\omega) &= 2x_j(\omega), \qquad & \nu(\omega) &= j, \\ &= 0, & \nu(\omega) &\neq j, \end{aligned}$$

the integral of the second term in the bracket above becomes $\mathbf{E}\{z_j(x_n - x_j)\}$ and, since z_j depends only on $y_1, \cdots, y_j$,

$$\begin{aligned} \mathbf{E}\{z_j(x_n - x_j)\} &= \mathbf{E}\Big\{\mathbf{E}\{z_j(x_n - x_j) \mid y_1, \cdots, y_j\}\Big\} \\ &= \mathbf{E}\Big\{z_j \mathbf{E}\{y_{j+1} + \cdots + y_n \mid y_1, \cdots, y_j\}\Big\} = 0, \end{aligned}$$

by (2.2). Hence

$$\hat{\sigma}_n^{\,2} \geq \sum_{j=1}^{n} \int_{\{\nu(\omega)=j\}} x_\nu^{\,2}\, d\mathbf{P} \geq \varepsilon^2 \sum_{j=1}^{n} \mathbf{P}\{\nu(\omega) = j\}$$

$$= \varepsilon^2 \mathbf{P}\{\operatorname*{Max}_{j \leqq n} |x_j(\omega)| \geq \varepsilon\},$$

giving the desired inequality. See VII, Theorem 3.2, for a generalization of this result.

THEOREM 2.2 *Let $y_1, \cdots, y_n$ be mutually independent random variables, and let $x_j = y_1 + \cdots + y_j$. Then, if $x_n - x_1$, $x_n - x_2, \cdots$ have symmetric distributions,*

$$(2.4)\quad 2\mathbf{P}\{x_n(\omega) \geq \lambda + 2\varepsilon\} - 2 \sum_{j=1}^{n} \mathbf{P}\{y_j(\omega) \geq \varepsilon\} \leq \mathbf{P}\{\operatorname*{Max}_{j \leqq n} x_j(\omega) \geq \lambda\}$$

$$\leq 2\mathbf{P}\{x_n(\omega) \geq \lambda\},$$

for every $\lambda > 0$, and every $\varepsilon > 0$. The right-hand half of (2.4) remains valid if each $x_n - x_k$ is supposed to have zero median, but not necessarily to be symmetrically distributed.

Clearly

$$(2.5)\quad \mathbf{P}\{\operatorname*{Max}_{j \leqq n} x_j(\omega) \geq \lambda, x_n(\omega) \geq \lambda\} = \mathbf{P}\{x_n(\omega) \geq \lambda\}.$$

On the other hand, using the hypotheses of symmetry and independence, if $x_\nu(\omega)$ is the first $x_j(\omega)$, if any, which is $\geq \lambda$,

$$(2.6)\quad \mathbf{P}\{\operatorname*{Max}_{j \leqq n} x_j(\omega) \geq \lambda, x_n(\omega) < \lambda\}$$

$$= \sum_{k=1}^{n-1} \mathbf{P}\{\nu(\omega) = k, x_n(\omega) < \lambda\}$$

$$\leq \sum_{k=1}^{n-1} \mathbf{P}\{\nu(\omega) = k, x_n(\omega) - x_k(\omega) < 0\}$$

$$= \sum_{k=1}^{n-1} \mathbf{P}\{\nu(\omega) = k\}\mathbf{P}\{x_n(\omega) - x_k(\omega) < 0\}$$

$$= \sum_{k=1}^{n-1} \mathbf{P}\{\nu(\omega) = k, x_n(\omega) - x_k(\omega) > 0\}$$

$$\leq \sum_{k=1}^{n-1} \mathbf{P}\{\nu(\omega) = k, x_n(\omega) \geq \lambda\}$$

$$\leq \mathbf{P}\{x_n(\omega) \geq \lambda\}.$$

Add (2.5) to (2.6) to get the right-hand half of (2.4). Note that even if $x_n - x_k$ is not symmetrically distributed, but if at least

$$\mathbf{P}\{x_n(\omega) - x_k(\omega) \geq 0\} \geq \tfrac{1}{2}, \qquad k = 1, \cdots, n-1,$$

so that

$$\mathbf{P}\{x_n(\omega) - x_k(\omega) \geq 0\} \geq \mathbf{P}\{x_n(\omega) - x_k(\omega) < 0\}, \qquad k = 1, \cdots, n-1,$$

then (2.6) remains true except that in the fifth line the (first) equality sign must be replaced by "$\leq$" and the "$>$" must be replaced by "$\geq$." In particular, if every difference $x_n - x_k$ has zero median, (2.6), and therefore also the right-hand half of (2.4), remains true. To derive the other half of (2.4), note that, if λ, $\varepsilon > 0$, and if each $x_n - x_k$ is symmetrically distributed, then

$$\begin{aligned}
(2.7)\quad & \mathbf{P}\{\operatorname*{Max}_{j \leq n} x_j(\omega) \geq \lambda, x_n(\omega) < \lambda\} \\
&= \sum_{k=1}^{n-1} \mathbf{P}\{\nu(\omega) = k, x_n(\omega) < \lambda\} \\
&\geq \sum_{k=1}^{n-1} \mathbf{P}\{\nu(\omega) = k, x_n(\omega) - x_k(\omega) < -\varepsilon\} - \sum_{k=1}^{n-1} \mathbf{P}\{y_k(\omega) \geq \varepsilon\} \\
&= \sum_{k=1}^{n-1} \mathbf{P}\{\nu(\omega) = k, x_n(\omega) - x_k(\omega) > \varepsilon\} - \sum_{k=1}^{n-1} \mathbf{P}\{y_k(\omega) \geq \varepsilon\} \\
&\geq \sum_{k=1}^{n-1} \mathbf{P}\{\nu(\omega) = k, x_n(\omega) \geq \lambda + 2\varepsilon\} - 2\sum_{k=1}^{n-1} \mathbf{P}\{y_k(\omega) \geq \varepsilon\} \\
&\geq \mathbf{P}\{x_n(\omega) \geq \lambda + 2\varepsilon\} - \mathbf{P}\{y_n(\omega) \geq \varepsilon\} - 2\sum_{k=1}^{n-1} \mathbf{P}\{y_k(\omega) \geq \varepsilon\}.
\end{aligned}$$

Add (2.5) to (2.7) to get the left-hand half of (2.4).

The right-hand half of (2.4) is the most important one; the left-hand half will be used in this book only in VIII. There is one important case in which a more precise evaluation can be obtained. If the y_j's only take on the values ± 1, with probability $1/2$ for each, and if N is an integer,

$$(2.8)\qquad \mathbf{P}\{\operatorname*{Max}_{j \leq n} x_j(\omega) \geq N\} = 2\mathbf{P}\{x_n(\omega) \geq N\} - \mathbf{P}\{x_n(\omega) = N\}.$$

This can be seen by the appropriate modification of (2.6), summarized as follows (reflection principle of D. André): if $\operatorname*{Max}_{j \leq n} x_j(\omega) \geq N$, there is a first $x_j(\omega)$, say $x_\nu(\omega)$, which reaches N. From there on, any succession of $x_j(\omega)$ values culminating in an $x_n(\omega) < N$ has the same probability as the succession reflected in the line $x(\omega) = N$, which culminates in an $x_n(\omega) > N$; that is,

$$\mathbf{P}\{\operatorname*{Max}_{j\leq n} x_j(\omega) \geq N, x_n(\omega) < N\} = \mathbf{P}\{\operatorname*{Max}_{j\leq n} x_j(\omega) \geq N, x_n(\omega) > N\}$$
$$(= \mathbf{P}\{x_n(\omega) > N\})$$

and this, combined with (2.5), gives (2.8). This evaluation is used in the problem of ruin. The method is also applicable of course if the y_j's take on only the values $\pm\, \varepsilon$ (instead of $\pm\, 1$), and from this it is but a step to get a precise evaluation in the limiting case when the parameter j is continuous, when x_j is replaced by x_t and the x_t process is the Brownian movement process. The essential point is that there must be a first value of the parameter at which the value λ is actually attained. This is true if the $x_t(\omega)$'s only change in integral multiples of a given number ε, as above, choosing λ as a multiple of ε, or, if the parameter t is continuous, it is true if the sample functions are continuous functions of t (cf. VIII §2).

THEOREM 2.3 *Let $y_1, y_2, \cdots$ be mutually independent random variables, with variances $\sigma_1^2, \sigma_2^2, \cdots$. Then, if $\sum_1^\infty \sigma_n^2 = \sigma^2 < \infty$ and if $\sum_1^\infty \mathbf{E}\{y_n\}$ converges, $\sum_1^\infty y_n = x$ is convergent with probability* 1 *and also in the mean. Moreover*

$$\mathbf{E}\{x\} = \sum_{n=1}^\infty \mathbf{E}\{y_n\}, \qquad \mathbf{E}\{x^2\} - \mathbf{E}\{x\}^2 = \sigma^2 \tag{2.9}$$

and if $x_n = \sum_1^n [y_j - \mathbf{E}\{y_j\}]$

$$\mathbf{P}\{\operatorname*{L.U.B.}_{n} |x_n(\omega)| \geq \varepsilon\} \leq \frac{\sigma^2}{\varepsilon^2}. \tag{2.10}$$

Conversely, if $\operatorname*{l.i.m.}_{n\to\infty} \sum_1^n y_j = x$ *exists, the series* $\sum_1^\infty \sigma_j^2$ *and* $\sum_1^\infty \mathbf{E}\{y_j\}$ *converge.*

By Theorem 2.1, if m is fixed,

$$\mathbf{P}\{\operatorname*{Max}_{m<n\leq m+r} |x_n(\omega) - x_m(\omega)| \geq \varepsilon\} \leq \frac{1}{\varepsilon^2} \sum_{m+1}^{m+r} \sigma_j^2 \leq \frac{1}{\varepsilon^2} \sum_{m+1}^{\infty} \sigma_j^2. \tag{2.11}$$

When $r \to \infty$ this becomes, if ε is a point of continuity of the distribution function of $\operatorname*{L.U.B.}_{n>m} |x_n - x_m|$,

$$\mathbf{P}\{\operatorname*{L.U.B.}_{n>m} |x_n(\omega) - x_m(\omega)| \geq \varepsilon\{ \leq \frac{1}{\varepsilon^2} \sum_{m+1}^{\infty} \sigma_j^2; \tag{2.12}$$

this inequality is then true for all $\varepsilon > 0$, by a continuity argument. Hence ($\varepsilon \to \infty$) the inferior and superior limits of the sequence $\{x_n\}$ are finite, with probability 1, and

$$(2.13)\qquad \mathbf{P}\{\limsup_{n\to\infty} x_n(\omega) - \liminf_{n\to\infty} x_n(\omega) \geq 2\varepsilon\} \leq \frac{1}{\varepsilon^2}\sum_{m+1}^{\infty}\sigma_j^2.$$

Thus ($m \to \infty$) the inferior and superior limits differ by at most 2ε, with probability 1, for every $\varepsilon > 0$ and are consequently equal with probability 1; that is to say, the series $\sum_1^\infty [y_j - \mathbf{E}\{y_j\}]$ converges with probability 1, as does $\sum_1^\infty y_j$, since $\sum_1^\infty \mathbf{E}\{y_j\}$ converges. Inequality (2.10) is a special case of (2.12) with $m = 0$, defining $x_0 = 0$. There is convergence in the mean of $\sum_1^\infty [y_j - \mathbf{E}\{y_j\}]$ since

$$\lim_{m,\,n\to\infty} \mathbf{E}\{|x_n - x_m|^2\} = \lim_{m,\,n\to\infty} \sum_{m+1}^{n} \sigma_j^2 = 0,$$

and this together with the convergence of $\sum_1^\infty \mathbf{E}\{y_j\}$ implies the convergence in the mean of $\sum_1^\infty y_j$. It is a general property of convergence in the mean that $\text{l.i.m.}_{n\to\infty}\, z_n = z$ implies that $\lim_{n\to\infty} \mathbf{E}\{z_n\} = \mathbf{E}\{z\}$ and $\lim_{n\to\infty} \mathbf{E}\{z_n^2\} = \mathbf{E}\{z^2\}$, and this general property gives (2.9). Conversely, if $\sum_1^\infty y_n = x$, where the series converges in the mean, $\sum_1^\infty \mathbf{E}\{y_n\}$ converges to $\mathbf{E}\{x\}$ in accordance with this same property. Hence $\sum_1^\infty [y_n - \mathbf{E}\{y_n\}]$ also converges in the mean, that is, $\text{l.i.m.}_{n\to\infty}\, x_n$ exists. Then the evaluation just given of $\mathbf{E}\{|x_n - x_m|^2\}$ shows that $\sum_1^\infty \sigma_j^2 < \infty$.

Since the differences $\{y_j - \mathbf{E}\{y_j\}\}$ are mutually orthogonal, this theorem can be considered a theorem on a very special type of orthogonal series; it is the strict sense version of IV, Theorem 4.1.

Theorem 2.1 can be interpreted to state that convergence in the mean of $\sum_1^\infty y_j$ implies convergence with probability 1. The basic fact about series of mutually independent random variables is that almost any limitation on the spread of the partial sums, such as convergence in the mean, implies convergence with probability 1. Before elaborating on this point we note that the set of sample sequences of the x_j's for which there is convergence is defined by conditions on the x_j's for large j. In other words, the zero-one law (Theorem 1.1) is applicable, and states that there is convergence (to a finite limit) with either probability 1 or probability 0.

Let $y_1, y_2, \cdots$ be a sequence of mutually independent random variables. If there are constants $c_1, c_2, \cdots$ such that $\sum_1^\infty (y_n - c_n)$ converges with probability 1, the series $\sum_1^\infty y_n$ will be said to converge with probability 1 when centered, and $c_1, c_2, \cdots$ will be called *centering constants.* If $c_1', c_2', \cdots$ is another sequence of constants, the c_j''s will be centering constants if and only if $\sum_1^\infty (c_n - c_n')$ converges. If $\sum_1^\infty y_n$ converges with probability 1 when centered and if there are centering constants $\hat{c}_1, \hat{c}_2, \cdots$ with the property that $\sum_1^\infty (y_n - \hat{c}_n)$ converges with probability 1, for any ordering of the terms in the sum, $\hat{c}_1, \hat{c}_2, \cdots$ will be called *absolute centering constants.* If $\hat{c}_1', \hat{c}_2', \cdots$ is another sequence of constants, the $\hat{c}_j'$'s will be absolute centering constants if and only if $\sum_1^\infty |\hat{c}_n - \hat{c}_n'|$ converges (since $\sum_1^\infty (\hat{c}_n - \hat{c}_n')$ must converge for any ordering of the terms). It will be shown in Theorem 2.6 that when there are centering constants there are always absolute centering constants (and that the sums using these constants are independent of the order of summation). As an example suppose that y_n has finite variance σ_n^2, and that $\sum_1^\infty \sigma_n^2 < \infty$. Then according to Theorem 2.3 $\mathbf{E}\{y_1\}, \mathbf{E}\{y_2\}, \cdots$ are appropriate centering constants. Since $\sum_1^\infty \sigma_n^2$ converges with any ordering of the terms, the same is true of

$$\sum_1^\infty [y_n - \mathbf{E}\{y_n\}].$$

This means that $\mathbf{E}\{y_1\}, \mathbf{E}\{y_2\}, \cdots$ are absolute centering constants. On the other hand, $0, 0, \cdots$ are centering constants for the series if and only if $\sum_1^\infty \mathbf{E}\{y_n\}$ converges, and are absolute centering constants if and only if $\sum_1^\infty |\mathbf{E}\{y_n\}|$ converges.

Even if the variances are not finite, appropriate centering constants can always be written down explicitly. For example, an evaluation of the

nth centering constant given in Theorem 2.6 makes it depend only on the nth summand.

The following theorem is a weakened converse to Theorem 2.3 in so far as the latter pertained to probability 1 convergence.

THEOREM 2.4 *Let $y_1, y_2, \cdots$ be mutually independent random variables which are uniformly bounded, $|y_n| \leq c$, with variances $\sigma_1^2, \sigma_2^2, \cdots$. Then, if $\sum_1^\infty y_n$ converges with probability* 1, *the series $\sum_1^\infty \sigma_n^2$ and $\sum_1^\infty \mathbf{E}\{y_n\}$ converge.*

Let Φ_n and Φ be, respectively, the characteristic functions of y_n and $\sum_1^\infty y_n = x$. Since the distribution of $\sum_1^n y_j$ converges to that of x, $\prod_1^n \Phi_j(t) \to \Phi(t)$ uniformly in every finite t interval. Then, applying I (11.14),

$$\sum_1^\infty \sigma_n^2 \leq -\frac{3}{t^2} \log |\Phi(t)|, \qquad |t| \leq \frac{1}{4c},$$

and the right side is finite if t is sufficiently small. Then by Theorem 2.3 $\sum_1^\infty [y_n - \mathbf{E}\{y_n\}]$ converges with probability 1. Since $\sum_1^\infty y_n$ also converges with probability 1, $\sum_1^\infty \mathbf{E}\{y_n\}$ must converge.

THEOREM 2.5 (*Three series theorem*) *Let $y_1, y_2, \cdots$ be mutually independent random variables, and let $\{a_n\}$, $\{b_n\}$ be sequences of numbers, with*

$$0 < \liminf_{n\to\infty} \begin{cases} a_n \\ b_n \end{cases} \leq \limsup_{n\to\infty} \begin{cases} a_n \\ b_n \end{cases} < \infty. \tag{2.14}$$

Define y_n' by

$$\begin{aligned} y_n'(\omega) &= y_n(\omega), & -a_n \leq y_n(\omega) \leq b_n, \\ &= 0, & \text{otherwise}. \end{aligned}$$

Then $\sum_1^\infty y_n$ converges with probability 1 *if and only if the series*

$$\sum_1^\infty \mathbf{P}\{y_n(\omega) \neq y_n'(\omega)\}, \qquad \sum_1^\infty \mathbf{E}\{y_n'\}, \qquad \sum_1^\infty [\mathbf{E}\{y_n'^2\} - \mathbf{E}\{y_n'\}^2] \tag{2.15}$$

converge.

If $\sum_1^\infty y_n$ converges, $\lim_{n\to\infty} y_n = 0$, so that $y_n(\omega) = y_n'(\omega)$ for large n. If these statements are true with probability 1, the first series in (2.15) converges, by the Borel-Cantelli lemma, Theorem 1.2, and the convergence of the other two series in (2.15) is proved by applying Theorem 2.4 to the series $\sum_1^\infty y_n'$, which also converges with probability 1. Conversely, suppose that the three series in (2.15) converge. Then, by Theorem 1.2, $y_n(\omega) = y_n'(\omega)$ for large n, with probability 1, and the convergence of $\sum_1^\infty y_n'$ with probability 1, which is implied by the convergence of means and variances, implies that of $\sum_1^\infty y_n$.

If $\sum_1^\infty y_n$ is a series of mutually independent random variables which converges with probability 1 when centered, it is not obvious that a series of the same summands in a different order also converges with probability 1 when centered. This implication is correct, however, and can be seen in various ways. For example, we shall show (Theorem 2.7) that the series $\sum_1^\infty y_n$ converges with probability 1 when centered if and only if the infinite product whose factors are the absolute values of the characteristic functions of the y_n's is everywhere convergent. Since this convergence is independent of the order of the factors, the stated property of the series $\sum_1^\infty y_n$ is independent of the order of summation. Another way of proving this same fact is to show that there is a set of absolute centering constants whenever there is a set of centering constants. This fact is contained in the following theorem.

THEOREM 2.6 *Let $y_1, y_2, \cdots$ be mutually independent random variables and suppose that $\sum_1^\infty y_n$ converges with probability* 1, *when centered.*

(i) *There are always absolute centering constants, $\hat{c}_1, \hat{c}_2, \cdots$. For example, if m_n is a median value of y_n, if $\alpha > 0$, and if*

$$\begin{aligned} y_n'(\omega) &= y_n(\omega), & |y_n(\omega) - m_n| &\leq \alpha, \\ &= m_n, & |y_n(\omega) - m_n| &> \alpha, \end{aligned}$$

then we can take $\hat{c}_n = \mathbf{E}\{y_n'\}$. If $y_1, y_2, \cdots$ are symmetrically distributed, 0, 0, $\cdots$ *are absolute centering constants.*

(ii) *If $\hat{c}_1, \hat{c}_2, \cdots$ are absolute centering constants, the sum $\sum_1^\infty (y_n - \hat{c}_n)$ is independent of the order of the terms, neglecting zero probabilities, and any subseries $\sum_1^\infty y_{\nu_n}$ has $\hat{c}_{\nu_1}, \hat{c}_{\nu_2}, \cdots$ as absolute centering constants.*

Proof of (i) Let $c_1, c_2, \cdots$ be centering constants for the y_n's, so that $\sum_1^\infty (y_n - c_n)$ converges with probability 1. Then $\lim_{n\to\infty} (m_n - c_n) = 0$.

The definition of y_n' amounts to cutting off $(y_n - c_n)$ α units below and above its median value $m_n - c_n$, that is, for n large very nearly α units below and above 0. This is a legitimate cutoff of $y_n - c_n$ in accordance with the conditions of Theorem 2.5, and yields the truncated expectation $\mathbf{E}\{y_n'\} - c_n$. Hence by Theorem 2.5 $\sum_1^\infty [\mathbf{E}\{y_n'\} - c_n]$ converges. Therefore $\mathbf{E}\{y_1'\}$, $\mathbf{E}\{y_2'\}$, $\cdots$ are centering constants for the series, and $\sum_1^\infty [y_n - \mathbf{E}\{y_n'\}]$ converges with probability 1. We shall use below the fact, implied by this, that $m_n - \mathbf{E}\{y_n'\} \to 0$. Now define $\hat{y}_n = y_n - \mathbf{E}\{y_n'\}$, so that $\sum_1^\infty \hat{y}_n$ converges with probability 1, and apply Theorem 2.5 to this series, truncating $\hat{y}_n$ outside the closed interval $[m_n - \mathbf{E}\{y_n'\} - \alpha, m_n - \mathbf{E}\{y_n'\} + \alpha]$ to get $\hat{y}_n' = y_n' - \mathbf{E}\{y_n'\}$. This truncation satisfies the conditions on a_n, b_n in Theorem 2.5 because $m_n - \mathbf{E}\{y_n'\} \to 0$. According to Theorem 2.5,

$$\sum_1^\infty \mathbf{P}\{\hat{y}_n(\omega) \neq \hat{y}_n'(\omega)\} < \infty, \qquad \sum_1^\infty \mathbf{E}\{\hat{y}_n'^2\} < \infty,$$

and we have also $\mathbf{E}\{\hat{y}_n'\} = 0$. Since the series written here are absolutely convergent, and since the convergence of these series gives sufficient conditions in Theorem 2.5, the series

$$\sum_1^\infty \hat{y}_n = \sum_1^\infty [y_n - \mathbf{E}\{y_n'\}]$$

converges with probability 1, regardless of the order of the terms in the sum. The constants $\mathbf{E}\{y_1'\}$, $\mathbf{E}\{y_2'\}$, $\cdots$ are therefore absolute centering constants. In particular, if the y_n's are symmetrically distributed, $\mathbf{E}\{y_n'\} = 0$ so that $0, 0, \cdots$ are absolute centering constants. Note that, since every subseries of either of the above two series of constants converges absolutely, every subseries of $\sum_1^\infty [y_n - \mathbf{E}\{y_n'\}]$ converges with probability 1, regardless of the order of the terms in the sum. Thus each subseries of $\sum_1^\infty y_n$ has the corresponding subsequence of $\{\mathbf{E}\{y_n'\}\}$ as sequence of absolute centering constants.

Proof of (ii) If $\hat{c}_1, \hat{c}_2, \cdots$ are absolute centering constants, $\sum_1^\infty |\hat{c}_n - \mathbf{E}\{y_n'\}| < \infty$. Any subseries $\sum_1^\infty y_{\nu_n}$ has $\hat{c}_{\nu_1}, \hat{c}_{\nu_2}, \cdots$ as absolute centering constants because $\sum_1^\infty |\hat{c}_{\nu_n} - \mathbf{E}\{y_{\nu_n}'\}| < \infty$ and because we have seen in (i) that $\mathbf{E}\{y_{\nu_1}'\}$, $\mathbf{E}\{y_{\nu_2}'\}$, $\cdots$ are absolute centering constants for the subseries. Finally, to prove that the sum $\sum_1^\infty (y_n - \hat{c}_n)$ is independent of the order of the terms, it is sufficient to take $\hat{c}_n = \mathbf{E}\{y_n'\}$ because the sum of the absolutely convergent series $\sum_1^\infty [\hat{c}_n - \mathbf{E}\{y_n'\}]$ is independent of the order of the terms. Thus we must prove that, if $n_1, n_2, \cdots$ is any permutation of the natural numbers,

$$\sum_1^\infty \hat{y}_j = \sum_1^\infty \hat{y}_{n_j} \tag{2.16}$$

with probability 1. Since the interchange of a finite number of terms of a series does not change the sum, we can modify the above series, if convenient, so that $n_j = j, j = 1, \cdots, N$, where N is arbitrary. We prove first that

$$\sum_1^\infty \hat{y}_j' = \sum_1^\infty \hat{y}_{n_j}' \tag{2.16'}$$

with probability 1. To prove this it is sufficient to prove that the series have the same limit in the mean sums, which follows from the evaluation, (applying (2.9) to the y'_j's)

$$\mathbf{E}\{|\sum_1^\infty \hat{y}_j' - \sum_1^r \hat{y}_{n_j}'|^2\} = \sum_1^\infty \mathbf{E}\{\hat{y}_j'^2\} - \sum_1^r \mathbf{E}\{\hat{y}_{n_j}'^2\} \to 0, \qquad r \to \infty.$$

Then (2.16′) is true, and (2.16) follows from

$$\begin{aligned}\mathbf{P}\{\sum_1^\infty \hat{y}_j(\omega) \neq \sum_1^\infty \hat{y}_{n_j}(\omega)\} &\leq \sum_{N+1}^\infty \mathbf{P}\{\hat{y}_j(\omega) \neq \hat{y}_j'(\omega)\} + \sum_{N+1}^\infty \mathbf{P}\{\hat{y}_{n_j}(\omega) \neq \hat{y}_{n_j}'(\omega)\} \\ &= 2\sum_{N+1}^\infty \mathbf{P}\{\hat{y}_j(\omega) \neq \hat{y}_j'(\omega)\} \to 0 \qquad (N \to \infty),\end{aligned}$$

where we have assumed, as we have already remarked we can, that the first N of the n_j's have been changed into $1, \cdots, N$.

Note that in proving this theorem we have *not* proved that the series $\sum_1^\infty (y_n - \hat{c}_n)$ converges absolutely with probability 1, and in fact this may be false. The following theorem distinguishes between the different possibilities in terms of the characteristic functions of the y_n's. The conditions are stated in terms of certain infinite products. We recall that, if an infinite product is convergent, its value is 0 if and only if one of the factors vanishes. This means, for example, that if $\{\Phi_n\}$ is a sequence of characteristic functions, the infinite product $\prod_1^\infty |\Phi_n(t)|$ has a value for every t, even though the product may diverge for some values of t. The value of the product will be 0 at all points of divergence, and at any further points where one of the factors vanishes.

THEOREM 2.7 *Let $y_1, y_2, \cdots$ be mutually independent random variables, with characteristic functions $\Phi_1, \Phi_2, \cdots$.*

(i) *If $\sum_1^\infty y_n$ converges with probability 1 when centered, then $\prod_1^\infty |\Phi_n|$ is continuous, is 1 when $t = 0$, and this infinite product converges uniformly in every finite t interval. Conversely, if $\prod_1^\infty |\Phi_n(t)| > 0$ on a t set of positive Lebesgue measure (or slightly more generally if this infinite product converges on a t set of positive Lebesgue measure), the series $\sum_1^\infty y_n$ converges with probability 1 when centered.*

(ii) *If $\sum_1^\infty y_n$ converges with probability 1 (that is if $0, 0, \cdots$ are appropriate centering constants), $\prod_1^\infty \Phi_n$ converges uniformly in every finite t interval. Conversely, if this infinite product converges on a t set of positive Lebesgue measure, $\sum_1^\infty y_n$ converges with probability 1.*

(iii) *If $\sum_1^\infty y_n$ converges with probability 1 regardless of the order of summation (that is, if $0, 0, \cdots$ are appropriate absolute centering constants), $\sum_1^\infty |\Phi_n - 1|$ converges uniformly in every finite t interval and the Weierstrass M test criterion is applicable. Conversely, if $\sum_1^\infty |\Phi_n - 1|$ converges on a t set of positive Lebesgue measure, $\sum_1^\infty y_n$ converges with probability 1, regardless of the order of summation.*

Proof of (i) If $\sum_1^\infty y_n$ converges with probability 1 when centered, let $c_1, c_2, \cdots$ be centering constants. Then $\sum_1^\infty (y_n - c_n)$ converges with probability 1, so that

$$\lim_{N\to\infty} \prod_1^N \Phi_n(t)\, e^{-ic_nt}$$

exists uniformly in every finite interval, and defines the characteristic function of $\sum_1^\infty (y_n - c_n)$. Then $\lim_{N\to\infty} \prod_1^N |\Phi_n|$ exists uniformly in every finite interval, and $\prod_1^\infty |\Phi_n|$ is continuous, and is 1 when $t = 0$. Since

$$\lim_{N\to\infty} \sum_N^\infty (y_n - c_n) = 0,$$

with probability 1, it follows that

$$\lim_{N\to\infty} \prod_N^\infty \Phi_n(t)\, e^{-ic_n t} = 1$$

uniformly in every finite t interval, so that the infinite product $\prod_1^\infty |\Phi_n|$ converges uniformly in every finite t interval. Conversely, suppose that the t-set A_1, where $\prod_1^\infty |\Phi_n(t)| > 0$, has positive Lebesgue measure. Since $\Phi_n(-t) = \bar{\Phi}_n(t)$, A_1 is symmetric in $t = 0$. There is then a bounded set A of Lebesgue measure $\rho > 0$, and a positive number a such that

$$\underset{t \in A}{\text{G.L.B.}} \prod_1^\infty |\Phi_n(t)| > 0, \qquad t \in A \subset [0, a].$$

Suppose that $\alpha > 0$, and define y_n' by

$$\begin{aligned} y_n'(\omega) &= y_n(\omega), & |y_n(\omega) - m_n| &\le \alpha \\ &= m_n, & |y_n(\omega) - m_n| &> \alpha, \end{aligned}$$

where m_n is a median value of y_n. Then, by I (11.8′),

$$\begin{aligned} \sum_1^\infty \mathbf{P}\{y_n(\omega) \ne y_n'(\omega)\} &= \sum_1^\infty \mathbf{P}\{|y_n(\omega) - m_n| > \alpha\} \\ &\le -4L_1(\alpha, \rho, a) \int_A \log \prod_1^\infty |\Phi_n(t)|\, dt < \infty, \end{aligned}$$

and, by I (11.9̂), if y_n' has variance σ_n^2,

$$\sum_1^\infty \sigma_n^2 \le -2L_4(\alpha, \rho, a) \int_A \log \prod_1^\infty |\Phi_n(t)|\, dt < \infty.$$

The three series theorem (Theorem 2.5) then states that the series $\sum_1^\infty [y_n - \mathbf{E}\{y_n'\}]$ converges with probability 1, that is, the series $\sum_1^\infty y_n$

converges with probability 1 when centered, as was to be proved. More generally, suppose that the infinite product $\prod_1^\infty |\Phi_n|$ is convergent (rather than divergent to 0) on a t set A of positive Lebesgue measure. If there is such a set A,

$$\lim_{\nu\to\infty} \prod_\nu^\infty |\Phi_n(t)| = 1, \qquad t \in A$$

so that there is a ν for which $\prod_\nu^\infty |\Phi_n(t)| > 0$ on a t set of positive Lebesgue measure. But then, according to what we have just proved, $\sum_\nu^\infty y_n$ converges with probability 1 when centered, so the same must be true of $\sum_1^\infty y_n$.

Proof of (ii) If $\sum_1^\infty y_n$ converges with probability 1, $\lim_{N\to\infty} \prod_1^N \Phi_n$ exists uniformly in every finite interval (and is the characteristic function of $\sum_1^\infty y_n$). Moreover, since $\lim_{N\to\infty} \sum_N^\infty y_n = 0$ with probability 1, it follows that $\lim_{N\to\infty} \prod_N^\infty \Phi_n = 1$ uniformly in every finite t interval. Then the infinite product $\prod_1^\infty \Phi_n$ is uniformly convergent in every finite t interval. Conversely, if this product converges on a t set A of positive Lebesgue measure, the same is true of $\prod_1^\infty |\Phi_n|$. There is then, according to (i), a sequence of constants $\{c_n\}$ such that $\sum_1^\infty (y_n - c_n)$ converges with probability 1. It follows from what we have just proved that $\prod_1^\infty \Phi_n(t)e^{-ic_nt}$ is a convergent product for all t. On the other hand, $\prod_1^\infty \Phi_n(t)$ converges, by hypothesis, when $t \in A$. These two facts taken together imply that $\sum_1^\infty c_n$ converges. Then $\sum_1^\infty y_n$ converges with probability 1, as was to be proved.

Proof of (iii) If $\sum_1^\infty y_n$ converges with probability 1, define $y_1', y_2', \cdots$ as in (i). We have seen that the truncated expectations $\mathbf{E}\{y_1'\}, \mathbf{E}\{y_2'\}, \cdots$ are always absolute centering constants. If we suppose that $0, 0, \cdots$ are also absolute centering constants, it follows that

$$\sum_1^\infty |\mathbf{E}\{y_n'\}| < \infty.$$

We write $c_n = \mathbf{E}\{y_n'\}$ and $\hat{\Phi}_n(t) = \Phi_n(t)e^{-ic_nt}$. Then by, I (11.10),

$$|\Phi_n(t) - 1| = |\hat{\Phi}_n(t) - e^{-itc_n}| \leq |\hat{\Phi}_n(t) - 1| + |e^{-itc_n} - 1|$$

$$\leq M_n(T) + T|c_n|, \qquad |t| \leq T,$$

where

$$M_n(T) = -2L_5(T, \alpha, a, a) \int_0^a \log |\Phi_n(s)|\, ds$$

and a is taken so small that $\prod_1^\infty |\Phi_n(t)| > \frac{1}{2}$ for $|t| \leq a$. Since

$$\sum_1^\infty [M_n(T) + |c_n|] < \infty,$$

the series $\sum_1^\infty |\Phi_n - 1|$ converges uniformly in every finite interval, and the Weierstrass M test is applicable. Conversely, if $\sum_1^\infty |\Phi_n - 1|$ converges on some t set of positive Lebesgue measure, the infinite product $\prod_1^\infty \Phi_n$ also converges on this t set, so the series $\sum_1^\infty y_n$ converges with probability 1, by (ii). Since $\sum_1^\infty |\Phi_n - 1|$ converges on this set regardless of the order of the summands, $\sum_1^\infty y_n$ converges with probability 1 regardless of the order of the summands, as was to be proved.

The following two corollaries, of which the second is the important one, are now easily derived. The first is proved only because it will be used in VIII.

COROLLARY 1 *Let* $y_1, y_2, \cdots$ *be mutually independent random variables. Suppose that* $\sum_1^\infty y_j$ *converges with probability* 1, *regardless of the order of summation. Let* $A_1, A_2, \cdots$ *be disjunct sets of natural numbers, with* $A = \bigcup_1^\infty A_j$, *and define* $\Sigma_n = \sum_{j \in A_n} y_j$. *Then* $\Sigma_1, \Sigma_2, \cdots$ *are mutually independent random variables, and* $\sum_1^\infty \Sigma_n = \sum_{j \in A}^\infty y_j$ *with probability* 1, *where the sums involved converge with probability* 1, *regardless of the order of summation.*

The series for Σ_n converges with probability 1 by Theorem 2.6 (ii), and $\Sigma_1, \Sigma_2, \cdots$ are obviously mutually independent. Let Φ_j be the characteristic function of y_j. According to Theorem 2.7 (iii),

$$\sum_1^\infty |\Phi_n(t) - 1| < \infty$$

for all t. Hence (see the Appendix for a proof of the inequality used here)

$$\sum_{n=1}^\infty \Big| \prod_{j \in A_n} \Phi_j(t) - 1 \Big| \leq \sum_{n=1}^\infty \sum_{j \in A_n} |\Phi_j(t) - 1| < \infty,$$

for all t, and since the nth product on the left is the characteristic function of Σ_n, the inequality implies, again by Theorem 2.7 (iii), that $\sum_1^\infty \Sigma_n$ converges with probability 1 regardless of the order of summation. The difference between this sum and the sum $\sum_{j \in A} y_j$ is evidently independent of $y_1, \cdots, y_n$ for every n, and therefore is identically constant with probability 1, by the zero-one law. The constant must be 0 because the two sums have the same characteristic function $\prod_{j \in A} \Phi_j$. The order of the factors in the latter product does not affect the value of the product, since $\sum_1^\infty |\Phi_j(t) - 1| < \infty$.

COROLLARY 2 *Let* $y_1, y_2, \cdots$ *be mutually independent random variables. Then, if the partial sums of* $\sum_1^\infty y_n$ *converge in distribution or in probability, the series converges with probability* 1.

Since convergence in distribution of the partial sums means that $\prod_1^n \Phi_j$ converges uniformly in every finite t interval $(n \to \infty)$, and since convergence in probability implies convergence in distribution, this statement is a special case of part (ii) of the theorem.

The hypotheses of the following theorem seem somewhat artificial at first, but they are frequently true (cf. the application of this theorem in VIII §6).

THEOREM 2.8 *Let* $y_1, y_2, \cdots$ *be mutually independent random variables, and suppose that there is a random variable* y *for which, if* Δ_n *is defined by*

$$y_1 + \cdots + y_n + \Delta_n = y,$$

Δ_n is independent of $y_1, \cdots, y_n$. Then $\sum_1^\infty y_j$ converges with probability 1 when centered.

In fact, if Φ_j is the characteristic function of y_j, Ψ_j that of Δ_j, and Φ that of y,

$$\prod_1^n |\Phi_j(t)| \geq [\prod_1^n |\Phi_j(t)|]|\Psi_n(t)| = |\Phi(t)|.$$

Hence $\prod_1^\infty |\Phi_j(t)| \geq |\Phi(t)| > \frac{1}{2}$ for small t, and Theorem 2.7 (i) can be applied to give the stated result.

In discussing the convergence of a series $\sum_1^\infty y_j$ of mutually independent random variables it is sometimes convenient to use a symmetrization procedure. Let $y_1^*, y_2^*, \cdots$ be random variables defined in such a way that y_j and y_j^* have the same distribution and that $y_1, y_1^*, y_2, y_2^*, \cdots$ are mutually independent. (If the given ω space is not complex enough to support such a sequence $\{y_j^*\}$, the ω space may be adjusted by the adjunction of a space with such a sequence, as described in II §2.) Then the characteristic function of $(y_j - y_j^*)$ is $|\Phi_j|^2$. According to Theorem 2.7 the series $\sum_1^\infty y_j$ converges with probability 1 when centered if and only if $\prod_1^\infty |\Phi_j|$ converges uniformly in every finite t interval, and $\sum_1^\infty (y_j - y_j^*)$ converges with probability 1 if and only if $\prod_1^\infty |\Phi_j|^2$ converges uniformly in every finite t interval. Then $\sum_1^\infty y_j$ converges with probability 1 when centered if and only if $\sum_1^\infty (y_j - y_j^*)$ converges with probability 1. The conditions of Theorem 2.7 for convergence with probability 1 when centered, and for convergence with probability 1, coincide for the series $\sum_1^\infty (y_j - y_j^*)$. The introduction of y_j^* makes it possible to reduce the discussion of the convergence of $\sum_1^\infty y_j$ when centered to the special case when the characteristic functions of the summands are real and ≥ 0.

The fact that the convergence of $\sum_1^\infty (y_j - y_j^*)$ with probability 1 implies the convergence of $\sum_1^\infty y_j$ when centered can also be obtained as a consequence of Theorem 2.8, with

$$y = \sum_1^\infty (y_j - y_j^*), \qquad \Delta_n = \sum_1^n y_j^* + \sum_{n+1}^\infty (y_j - y_j^*).$$

THEOREM 2.9 *Let $y_1, y_2, \cdots$ be mutually independent random variables, and suppose that for some $K > 0$*

$$\limsup_{n\to\infty} \mathbf{P}\{|\sum_1^n y_j(\omega)| \le K\} > 0.$$

Then $\sum_1^\infty y_j$ converges with probability 1 when centered.

By hypothesis there is an $\varepsilon > 0$ and an increasing sequence $\{n_m\}$ of positive integers such that

$$\mathbf{P}\{|\sum_1^n y_j(\omega)| \le K\} > \varepsilon, \qquad n = n_m.$$

Let $y_1^*, y_2^*, \cdots$ be symmetrizing random variables as in the above discussion. Then

$$\mathbf{P}\{|\sum_1^n [y_j(\omega) - y_j^*(\omega)]| \le 2K\} \ge \mathbf{P}\{|\sum_1^n y_j(\omega)| \le K, |\sum_1^n y_j^*(\omega)| \le K\}$$
$$> \varepsilon^2, \qquad n = n_m.$$

If Φ_j is the characteristic function of y_j, we have

$$\frac{1}{2\delta}\int_{-\delta}^{\delta} \prod_1^n |\Phi_j(t)|^2\, dt = \frac{1}{2\delta}\int_{-\delta}^{\delta} dt \int_{-\infty}^{\infty} e^{it\lambda}\, d\mathbf{P}\{\sum_1^n [y_j(\omega) - y_j^*(\omega)] \le \lambda\}$$
$$= \int_{-\infty}^{\infty} \frac{\sin \lambda\delta}{\lambda\delta}\, d\mathbf{P}\{\sum_1^n [y_j(\omega) - y_j^*(\omega)] \le \lambda\}.$$

By Helly's theorem there is a subsequence of the distribution functions of $y_{n_1} - y_{n_1}^*, y_{n_2} - y_{n_2}^*, \cdots$ which converges for all λ to some bounded monotone function G. Then, since the last integrand in the preceding equation vanishes at $\pm\infty$,

$$\frac{1}{2\delta}\int_{-\delta}^{\delta} \prod_1^\infty |\Phi_n(t)|^2\, dt = \int_{-\infty}^{\infty} \frac{\sin \lambda\delta}{\lambda\delta}\, dG(\lambda). \tag{2.17}$$

If the series $\sum_1^\infty y_j$ does not converge with probability 1 when centered, the integral on the left vanishes, by Theorem 2.7 (i). But then

$$0 = \int_{-\infty}^{\infty} \frac{\sin \lambda\delta}{\lambda\delta}\, dG(\lambda) = G(\infty) - G(-\infty) \qquad (\delta \to 0),$$
$$\ge G(K) - G(-K) \ge \varepsilon^2,$$

and this contradiction implies the truth of the theorem.

This theorem shows that the distribution of $\sum_1^n y_j$ goes out to infinity if $\sum_1^\infty y_j$ does not converge with probability 1 when centered. In this case, then,

$$\lim_{n\to\infty} \mathbf{P}\{|\sum_1^n y_j(\omega) - c_j)| \leq K\} = 0$$

for all choices of centering constants $c_1, c_2, \cdots$ and all $K > 0$. Another way of writing this is

$$\lim_{n\to\infty} \underset{-\infty<d<\infty}{\text{L.U.B.}} \mathbf{P}\{|[\sum_1^n y_j(\omega)] - d| \leq K\} = 0.$$

The nth least upper bound in this equation is a function of K which measures the concentration of the distribution of $\sum_1^n y_j$. The study of series $\sum_1^\infty y_j$ of mutually independent random variables has been carried through by Lévy in a way which bases itself on such functions of concentration, and Kawata has based the theory on averages of these functions.

3. The law of large numbers

Let $y_1, y_2, \cdots$ be random variables. If for some constants $a_1, b_1, a_2, b_2, \cdots$

$$\lim_{n\to\infty} \frac{1}{b_n} \sum_1^n (y_j - a_j) \tag{3.1}$$

exists in some sense of convergence, the sequence $y_1, y_2, \cdots$ is said to be subject to the law of large numbers (relative to the centering constants $a_1, a_2, \cdots$ and scaling constants $b_1, b_2, \cdots$). The law is described as the weak law if the convergence in (3.1) is convergence in probability, the strong law if the convergence in (3.1) is convergence with probability 1. It is of course always possible to choose the b_j's so large that the limit in (3.1) exists with probability 1 and is 0.

In the present section the problem is simplified by the hypothesis that the y_j's are mutually independent. As a first example of the significance of this hypothesis note that, if $b_n \to \infty$ in (3.1) (or even if $\limsup_{n\to\infty} b_n = \infty$) and if the limit x in (3.1) is a limit in probability, then x is a random variable which is unaffected by changes in the values of a finite number of y_j's. Hence, according to the zero-one law (Theorem 1.1), x is identically a constant, with probability 1. This constant can be absorbed in the centering constants $\{a_j\}$ if desired.

THEOREM 3.1 *Let $y_1, y_2, \cdots$ be mutually independent random variables with characteristic functions $\Phi_1, \Phi_2, \cdots$ and let $b_1, b_2, \cdots$ be any non-zero constants. There are constants $a_1, a_2, \cdots$, for which*

$$\operatorname{p}\lim_{n\to\infty} \frac{1}{b_n} \sum_1^n (y_j - a_j) = 0 \tag{3.2}$$

if and only if

$$\lim_{n\to\infty} \prod_1^n |\Phi_j(t/b_n)| = 1 \tag{3.3}$$

uniformly in every finite t-interval.

If (3.2) is true, the distributions of the quotients converge to the distribution concentrated at 0; hence

$$\lim_{n\to\infty} \prod_{j=1}^n \Phi_j(t/b_n)e^{-ita_j/b_n} = 1$$

uniformly in every finite t-interval, which implies (3.3). Conversely, suppose that (3.3) is true. Choose c_n as a median value of $\sum_{j=1}^n y_j/b_n$, and define $a_1, a_2, \cdots$ by

$$a_1 = b_1c_1$$

$$a_j = b_jc_j - b_{j-1}c_{j-1}, \qquad j > 1.$$

Then by I (11.8′), if $\mu > 0$,

$$\mathbf{P}\left\{\left|\frac{1}{b_n}\sum_1^n [y_j(\omega) - a_j)]\right| \geq \mu\right\} = \mathbf{P}\left\{\left|\frac{1}{b_n}[\sum_1^n y_j(\omega)] - c_n\right| \geq \mu\right\}$$

$$\leq -4L_1(\mu, a, a)\int_0^a \log \prod_{j=1}^n \left|\Phi_j\left(\frac{t}{b_n}\right)\right| dt,$$

and the right side goes to 0 with $1/n$, by (3.3). Note that the hypothesis of uniformity in (3.3) was not used in this sufficiency proof. When $b_n = n$ we shall use the following fact without special mention in each application: if $\lim_{n\to\infty} \frac{1}{n}\sum_1^n y_j$ exists, then $\lim_{n\to\infty} y_n/n = 0$. In fact, then

$$0 = \lim_{n\to\infty} \frac{1}{n}\sum_1^n y_j - \lim_{n\to\infty} \frac{1}{n-1}\sum_1^{n-1} y_j = \lim_{n\to\infty} \frac{1}{n}\sum_1^n y_j - \lim_{n\to\infty} \frac{1}{n}\sum_1^{n-1} y_j$$

$$= \lim_{n\to\infty} \frac{y_n}{n}.$$

If the y_j's are random variables, "lim" can be interpreted as "p lim" or "l.i.m." in this equation.

In the most common case, $b_n = n$, $a_j = \mathbf{E}\{x_j\}$, and the limit in (3.1) is 0. If this is so, and if the limit

$$\lim_{n\to\infty} \frac{1}{n}\sum_{1}^{n} \mathbf{E}\{y_j\} \tag{3.4}$$

exists, we can take $a_j = 0$, and the limit in (3.1) will be that in (3.4). One problem of the theory is to find conditions under which the law of large numbers holds with these constants. For example, suppose that $y_1, y_2, \cdots$ are mutually independent with variances $\sigma_1^2, \sigma_2^2, \cdots$. Then the average $x_n = \frac{1}{n}\sum_{1}^{n} y_j$ has expectation and variance

$$\mathbf{E}\{x_n\} = \frac{1}{n}\sum_{1}^{n} \mathbf{E}\{y_j\} \tag{3.5}$$

$$\mathbf{E}\{[x_n - \mathbf{E}\{x_n\}]^2\} = \frac{1}{n^2}\sum_{1}^{n} \sigma_j^2.$$

Hence, if

$$\lim_{n\to\infty} \frac{1}{n^2}\sum_{1}^{n} \sigma_j^2 = 0, \tag{3.6}$$

which will be true for example if $\sigma_j^2 \leq \text{const.}$, $j \geq 1$, the variance of x_n goes to 0, that is,

$$\underset{n\to\infty}{\text{l.i.m.}}\ \frac{1}{n}\sum_{1}^{n} [y_j - \mathbf{E}\{y_j\}] = 0. \tag{3.7}$$

Thus in this case the law of large numbers holds with $a_n = \mathbf{E}\{y_n\}$, $b_n = n$ in the sense of convergence in the mean (which implies convergence in probability) and conversely (3.7) implies (3.6). These considerations have barely used the fact of mutual independence of the y_j's and have in fact resulted in no stronger result than the corresponding weak sense Theorem 5.1 of IV, in which the y_j's are only supposed orthogonal. However, far stronger theorems can be proved. We first prove a partial converse, restating the direct part, for completeness.

THEOREM 3.2 *Let $y_1, y_2, \cdots$ be mutually independent random variables with finite variances $\sigma_1^2, \sigma_2^2, \cdots$. Then* (3.7) *is true if and only if* (3.6) *is true. Moreover*

$$\underset{n\to\infty}{\text{l.i.m.}}\ \frac{1}{n}\sum_{1}^{n} (y_j - a_j) = 0 \tag{3.8}$$

if and only if (3.6) *and*

$$\lim_{n\to\infty} \frac{1}{n^2}\sum_{1}^{n} [\mathbf{E}\{y_j\} - a_j]^2 = 0 \tag{3.9}$$

are true. If, for some constants $c_1, c_2, \cdots$, $\limsup_{n\to\infty} \frac{c_n}{n} < \infty$ *and* $|y_n| \le c_n$, *then*

$$\text{p}\lim_{n\to\infty} \frac{1}{n} \sum_1^n (y_j - a_j) = 0 \tag{3.8'}$$

is true if and only if (3.8) *is true.*

The equivalence of (3.6) and (3.7) has already been noted. The equivalence of (3.8) and (3.6) combined with (3.9) follows from

$$\mathbf{E}\left\{\left[\frac{1}{n}\sum_1^n (y_j - a_j)\right]^2\right\} = \frac{1}{n^2}\sum_1^n [\sigma_j^2 + \mathbf{E}\{y_j\} - a_j)^2].$$

If (3.8) is true, the weaker (3.8′) is always true. If (3.8′) is true and if $|y_j| \le c_j, j \ge 1$, let Φ_j be the characteristic function of y_j and let m_j be a median value of y_j. Then (3.8′) implies by (3.3) that

$$\lim_{n\to\infty} \prod_1^n |\Phi_j(t/n)| = 1$$

uniformly in every finite t-interval, and, according to I (11.14) applied to $(y_j - m_j)/n$,

$$\frac{1}{n^2}\sum_1^n \sigma_j^2 t^2 \le - 3 \log \prod_1^n |\Phi_j(t/n)| \to 0,$$

$$|t| \le \left[4 \operatorname{Max}_{j\le n} \frac{c_j}{n}\right]^{-1}.$$

Thus (3.6) is true and consequently (3.7) is also true. Since (3.7) implies convergence in probability, it can be combined with (3.8′) to give

$$\lim_{n\to\infty} \frac{1}{n}\sum_1^n [\mathbf{E}\{y_j\} - a_j] = 0,$$

and this equation in turn combines with (3.7) to give the desired (3.8).

THEOREM 3.3 *Let* $y_1, y_2, \cdots$ *be mutually independent random variables and let* c *be any positive constant. If*

$$\lim_{n\to\infty} \sum_{j=1}^n \mathbf{P}\{|y_j(\omega)| > cn\} = 0, \tag{3.10}$$

and if, when y_{nj} *is defined by*

$$\begin{aligned} y_{nj}(\omega) &= y_j(\omega), \qquad & |y_j(\omega)| \le cn \\ &= 0 & |y_j(\omega)| > cn \end{aligned}$$

and has variance σ_{nj}^2,

$$\lim_{n\to\infty}\frac{1}{n^2}\sum_{j=1}^{n}\sigma_{nj}^2=0, \tag{3.11}$$

then

$$\operatorname*{p\,lim}_{n\to\infty}\frac{1}{n}\sum_{j=1}^{n}(y_j-a_j)=0 \tag{3.12}$$

for some constants $a_1, a_2, \cdots$ *satisfying*

$$\limsup_{n\to\infty}\frac{|a_n|}{n}\le c. \tag{3.13}$$

Conversely, if

$$\operatorname*{p\,lim}_{n\to\infty}\frac{1}{n}\sum_{j=1}^{n}y_j=0, \tag{3.12'}$$

then (3.10) *and* (3.11) *are also true, for every* $c>0$.

To prove the direct half, note that

$$\mathbf{P}\{y_{nj}(\omega)=y_j(\omega), j\le n\}\ge 1-\sum_{j=1}^{n}\mathbf{P}\{|y_j(\omega)|>cn\},$$

so that by (3.10)

$$\operatorname*{p\,lim}_{n\to\infty}\frac{1}{n}\sum_{j=1}^{n}(y_j-y_{nj})=0. \tag{3.14}$$

Now (Theorem 3.2, (3.11) implies

$$\operatorname*{p\,lim}_{n\to\infty}\frac{1}{n}\sum_{j=1}^{n}(y_{nj}-\mathbf{E}\{y_{nj}\})=0$$

so that, combining the last two equations,

$$\operatorname*{p\,lim}_{n\to\infty}\frac{1}{n}\sum_{j=1}^{n}(y_j-\mathbf{E}\{y_{nj}\})=0,$$

which implies (3.12) for properly chosen a_j's. This equation in turn implies

$$\operatorname*{p\,lim}_{n\to\infty}\frac{y_n-a_n}{n}=0,$$

whereas, by (3.10), $\mathbf{P}\{|y_n(\omega)|>cn\}\to 0$. Hence (3.13) is true.

Conversely, suppose that (3.12′) is true. Then, if Φ_j is the characteristic function of y_j,

$$\lim_{n\to\infty}\prod_{1}^{n}\Phi_j(t/n)=1 \tag{3.15}$$

uniformly in every finite interval. Let a_n' be a median value of y_n. Then, since by (3.12′) $\operatorname{p\,lim}_{n\to\infty} \frac{y_n}{n} = 0$, it follows that

$$\lim_{n\to\infty} \frac{a_n'}{n} = 0.$$

By I (11.8′), and by (3.15)

$$\sum_1^n \mathbf{P}\left\{\left|\frac{y_j(\omega) - a_j'}{n}\right| \geq \mu\right\} \leq -4L_1(\mu, \mu, \mu) \int_0^{\mu} \log \prod_1^n |\Phi_j(t/n)|\, dt \to 0,$$

and this combined with the preceding equation yields (3.10) if $\mu < c$. Now, if y_{nj} is defined as in the statement of the theorem, we have seen above that (3.10) implies (3.14), and (3.12′) together with (3.14) imply

$$\operatorname{p\,lim}_{n\to\infty} \frac{1}{n} \sum_{j=1}^n y_{nj} = 0.$$

The reasoning which in the proof of Theorem 3.2 led from convergence in probability (3.8′) to convergence in the mean (3.8) is applicable without change here, leading to (3.11).

If the condition (3.6) is strengthened slightly, the conclusion of convergence in the mean and in probability can be strengthened to convergence with probability 1. To simplify the notation we take $\mathbf{E}\{y_j\} = 0$.

THEOREM 3.4 *Let $y_1, y_2, \cdots$ be mutually independent random variables with*

$$\mathbf{E}\{y_j\} = 0, \qquad \mathbf{E}\{y_j^2\} = \sigma_j^2 < \infty.$$

Then, if

$$\sum_1^\infty \frac{\sigma_j^2}{j^2} < \infty,$$

$$\operatorname{l.i.m.}_{n\to\infty} \frac{y_1 + \cdots + y_n}{n} = \lim_{n\to\infty} \frac{y_1 + \cdots + y_n}{n} = 0 \tag{3.16}$$

with probability 1.

According to Theorem 2.3 the present hypotheses imply that the series

$$\sum_1^\infty \frac{y_j}{j} = S$$

is convergent in the mean and with probability 1. Now, if $S_n = \sum_1^n \frac{y_j}{j}$,

$$\begin{aligned} \frac{1}{n}\sum_1^n y_j &= \frac{1}{n}\Big[\sum_1^n jS_j - \sum_1^{n-1} (j+1)S_j\Big] \\ &= -\frac{1}{n}\sum_1^n S_j + \frac{n+1}{n} S_n. \end{aligned} \tag{3.17}$$

When $n \to \infty$ the last term on the right converges to S both in the mean and with probability 1. The same is true of the average on the right since convergence in the mean and ordinary convergence imply the same type of convergence to the same limit when averages are taken (Cesàro summability). Hence the limit on the left exists both in the mean and with probability 1, and is 0.

Although this theorem has made use of the hypothesis of mutual independence of the y_j's, the weak sense version (IV, Theorem 5.2), in which the only qualitative hypothesis is that the y_j's are mutually orthogonal, has only slightly stronger conditions on the sequence $\mathbf{E}\{y_1^2\}$, $\mathbf{E}\{y_2^2\}, \cdots$

It is sufficient for the convergence of the critical series if $\sigma_n^2 \leq$ const., $n \geq 1$, and the classical examples of the law of large numbers are special cases of this. For example, suppose that

$$\mathbf{P}\{y_j(\omega) = 1\} = p_j, \qquad \mathbf{P}\{y_j(\omega) = 0\} = q_j = 1 - p_j.$$

Then

$$\mathbf{E}\{y_j\} = p_j, \qquad \sigma_j^2 = p_j q_j \leq \tfrac{1}{4}$$

so that, by Theorem 3.4,

$$\lim_{n\to\infty} \frac{1}{n} \sum_1^n [y_j - p_j] = 0$$

with probability 1. If the p_j's are all equal, $p_j = p$ (Bernoulli), we have

$$\lim_{n\to\infty} \frac{1}{n} \sum_1^n y_j = p;$$

if $\lim_{n\to\infty} \frac{1}{n} \sum_1^n p_j = p$ exists (Poisson), the same equation holds, with the new interpretation of p. The left side is in the usual language the "success ratio," the number of successful trials out of n divided by n, and the theorem in the Bernoulli case states that the success ratio approaches with probability 1 the probability of success in each trial.

4. Infinitely divisible distributions and the central limit theorem

Let x be a random variable, and suppose that it can be expressed in the form

$$x = y_1 + \cdots + y_n \tag{4.1}$$

where $y_1, \cdots, y_n$ are mutually independent. This is no restriction on the distribution of x, since we can take $y_1 = x$, $y_2 = \cdots = y_n = 0$. If it is supposed in addition that the y_j's are small, however, (4.1) is a real restriction on x. To avoid discussing the somewhat extraneous issue of the complexity of the relevant ω spaces we shall formulate conditions in terms of the distributions or their characteristic functions rather than in terms of corresponding random variables.

A distribution function F is called *infinitely divisible in the generalized sense* if, for each $\eta > 0$, F can be written as a convolution of distribution functions $F_1, \cdots, F_n$,

$$F(\lambda) = \int_{-\infty}^{\lambda} dF_1(\lambda_1) \cdots \int_{-\infty}^{\lambda-\lambda_1-\cdots-\lambda_{n-2}} F_n(\lambda - \lambda_1 - \cdots - \lambda_{n-1})\, dF_{n-1}(\lambda_{n-1}) \tag{4.1'}$$

with

$$1 - F_j(\eta -) + F_j(-\eta) \leq \eta, \qquad j \leq n.$$

If $y_1, \cdots, y_n$ are mutually independent random variables with distribution functions $F_1, \cdots, F_n$, their sum x will then have distribution function F and $\mathbf{P}\{|y_j(\omega)| \geq \eta\} \leq \eta$, $j \leq n$. Note that both n and the F_j's will depend on η except in trivial special cases.

Evidently Φ is the characteristic function of an infinitely divisible law (generalized sense) if and only if, for each $\varepsilon > 0$, Φ can be written in the form

$$\Phi = \prod_{1}^{n} \Phi_j \tag{4.1''}$$

where $\Phi_1, \cdots, \Phi_n$ are characteristic functions with $|1 - \Phi_j(t)| \leq \varepsilon$ for $|t| \leq 1/\varepsilon$; n and the Φ_j's will depend on ε except in trivial special cases.

A distribution function F is called *infinitely divisible* if for each n it can be written in the form (4.1′) with $F_1 \equiv \cdots \equiv F_n$, that is, if for each n its characteristic function Φ is the nth power of a characteristic function, $\Phi = \Psi_n{}^n$. If $\delta > 0$ is chosen so small that $\Phi(t) \neq 0$ for $|t| \leq \delta$, it is clear that $\lim_{n\to\infty} \Psi_n(t) = 1$ uniformly for $|t| \leq \delta$, and (cf. I §11) this implies that $\lim_{n\to\infty} \Psi_n(t) = 1$ uniformly in every finite t-interval. Thus Φ must be the characteristic function of a distribution that is infinitely divisible in the generalized sense. It will be shown below that conversely, if Φ is infinitely divisible in the generalized sense, it is also infinitely divisible. In other words, it will be shown that the added hypothesis that the F_j's in (4.1′) or the Φ_j's in (4.1″) are identical is no further restriction on the given distribution; the phrase "in the generalized sense" will then be superfluous and will be omitted thereafter.

Before proceeding we give some simple examples of infinitely divisible distributions, obtaining the decomposition (4.1′) or (4.1″) in each case.

(a) Let the distribution be concentrated at a single point γ, so that

$$F(\lambda) = 0 \qquad \lambda < \gamma$$

$$= 1 \qquad \lambda \geq \gamma, \tag{4.2}$$

$$\Phi(t) = e^{it\gamma}.$$

Then, for every n, F is the convolution of n distributions each concentrated at γ/n, $\Phi(t) = (e^{it\gamma/n})^n$.

(b) Let the distribution be Gaussian, with expectation γ and variance σ^2. For each n this is the convolution of n Gaussian distributions, with expectation γ/n and variance σ^2/n;

$$\Phi(t) = [e^{it\gamma/n}e^{-(1/2)\sigma^2/n}]^n.$$

(c) Let the distribution be a Poisson distribution:

$$\mathbf{P}\{x(\omega) = n\} = e^{-c}\frac{c^n}{n!}, \qquad n = 0, 1, \cdots, c > 0. \tag{4.3}$$

The characteristic function Φ of this distribution is given by

$$\log \Phi(t) = c(e^{it} - 1) \tag{4.4}$$

and $\Phi^{1/n}$ is then the characteristic function of the same distribution except that c is replaced by c/n. More generally, if x has the distribution (4.3), the characteristic function of $ax + b$ has logarithm

$$itb + c(e^{iat} - 1) \tag{4.5}$$

and the (properly chosen) nth root of the characteristic function is the characteristic function of the same distribution with a, b, c replaced by a, b/n, c/n.

We shall show below that every infinitely divisible distribution in the generalized sense has characteristic function Φ given by Lévy's formula, stated at this point in Khintchine's form,

$$\log \Phi(t) = i\gamma t + \int_{-\infty}^{\infty} \left(e^{it\lambda} - 1 - \frac{it\lambda}{1 + \lambda^2}\right) \frac{1 + \lambda^2}{\lambda^2} \, dG(\lambda), \tag{4.6}$$

where G is monotone non-decreasing and bounded, and the integrand is defined as its limiting value $-t^2/2$ when $\lambda = 0$. Before proving this we shall examine the functions Φ given by (4.6) in various special cases:

$$G(\lambda) \equiv 0. \tag{a$'$}$$

Then $\Phi(t) = e^{it\gamma}$, and Φ is the characteristic function of the infinitely divisible distribution discussed under (a) above.

(b′)
$$\begin{aligned} G(\lambda) &= 0, \qquad \lambda < 0, \\ &= \sigma^2, \qquad \lambda \geq 0. \end{aligned}$$

Then
$$\log \Phi(t) = i\gamma t - \frac{\sigma^2}{2} t$$

and Φ is the characteristic function of the Gaussian distribution with mean γ and variance σ^2 discussed under (b) above.

(c′)
$$\begin{aligned} G(\lambda) &= 0, \qquad \lambda < \lambda_0, \\ &= c_0, \qquad \lambda \geq \lambda_0, \qquad \text{where } c_0 > 0, \lambda_0 \neq 0. \end{aligned}$$

Then
$$\begin{aligned} \log \Phi(t) &= i\gamma t + \left(e^{it\lambda_0} - 1 - \frac{it\lambda_0}{1 + \lambda_0^2}\right) \frac{1 + \lambda_0^2}{\lambda_0^2} c_0 \\ &= i\left(\gamma - \frac{1}{\lambda_0}\right) t + c_0 \frac{1 + \lambda_0^2}{\lambda_0^2} (e^{it\lambda_0} - 1), \end{aligned}$$

which is the Poisson case discussed under (c) above. Note that λ_0, the λ at the jump of G, is the magnitude of the increment in the Poisson distribution.

Now, since the integrand is bounded and continuous, the integral in (4.6) always exists. Moreover, if $-N = \lambda_0 < \cdots < \lambda_n = N$ and if $|t| \leq T$,

$$\begin{aligned} &\left|\log \Phi(t) - \left\{i\gamma t + \sum_1^n \left[e^{it\lambda_j} - 1 - \frac{it\lambda_j}{1 + \lambda_j^2}\right] \frac{1 + \lambda_j^2}{\lambda_j^2} [G(\lambda_j) - G(\lambda_{j-1})]\right\}\right| \\ &\qquad \leq K[G(\infty) - G(N) + G(-N) - G(-\infty)] + \eta[G(N) - G(-N)], \end{aligned}$$

where K is the L.U.B. of the absolute value of the integrand in (4.6) for all λ, $|t| \leq T$, and η is the maximum oscillation of the integrand in the n intervals $(\lambda_0, \lambda_1), \cdots, (\lambda_{n-1}, \lambda_n)$ for $|t| \leq T$. The right-hand side can be made arbitrarily small for fixed T by choosing N large and then choosing $\operatorname{Max}_j (\lambda_j - \lambda_{j-1})$ small. Hence the function in (4.6) can be expressed as the limit of a sequence of functions obtained by replacing the integral by properly chosen Riemann-Stieltjes sums, and the limit is uniform in every finite t-interval. Since the expression with sums is the logarithm of a characteristic function—it is the characteristic function of a convolution of Gaussian and Poisson distributions discussed under (b) and (c)—log Φ as defined by (4.6) must also be the logarithm of a characteristic function. The corresponding distribution must be infinitely divisible since $(1/n) \log \Phi$ is given by the same formula with γ, G replaced by γ/n, G/n.

THEOREM 4.1 *An infinitely divisible distribution is infinitely divisible in the generalized sense and conversely. A distribution is infinitely divisible if and only if its characteristic function* Φ *never vanishes and is given by* (4.6), *where* G *is monotone non-decreasing and bounded, and* γ *is real.*

It will be sufficient to prove that the characteristic function of an infinitely divisible distribution in the generalized sense can be written in the form (4.6). In fact, we have already remarked that this form always gives infinitely divisible distributions. By hypothesis, for every $\varepsilon > 0$ we can write Φ in the form

$$\Phi = \prod_j \Phi_{\varepsilon j}$$

where $\Phi_{\varepsilon j}$ is a characteristic function, and

$$|1 - \Phi_{\varepsilon j}(t)| \leq \varepsilon \quad \text{for} \quad |t| \leq \frac{1}{\varepsilon}.$$

The minimum number of factors n depends on ε. Then, if $\varepsilon < 1$, $\Phi(t) \neq 0$ for $|t| \leq 1/\varepsilon$, so that $\Phi(t)$ never vanishes. We center the distribution corresponding to $\Phi_{\varepsilon j}$ as we have done repeatedly, by means of a truncated expectation; if $F_{\varepsilon j}$ is the distribution function we subtract the centering constant

$$m_{0\varepsilon j} = \int_{m_{\varepsilon j}-\alpha-}^{m_{\varepsilon j}+\alpha} \lambda \, dF_{\varepsilon j}(\lambda) + m_{\varepsilon j}[1 - F_{\varepsilon j}(m_{\varepsilon j} + \alpha) + F_{\varepsilon j}(m_{\varepsilon j} - \alpha -)],$$

where $m_{\varepsilon j}$ is a median value of the $F_{\varepsilon j}$ distribution, getting the new distribution function $\hat{F}_{\varepsilon j}$, with $\hat{F}_{\varepsilon j}(\lambda) = F_{\varepsilon j}(\lambda + m_{0\varepsilon j})$. Then if $\hat{\Phi}_{\varepsilon j}$ is the characteristic function of the centered distribution we can write

$$\Phi(t) = e^{it\gamma_\varepsilon} \prod_j \hat{\Phi}_{\varepsilon j}(t),$$

where γ_ε is the sum of the centering constants. The positive constant α will be fixed throughout the following discussion. Then it is clear that as $\varepsilon \to 0$ the individual centering constants go to 0 uniformly so that, changing the notation if necessary, we can suppose that

$$|1 - \hat{\Phi}_{\varepsilon j}(t)| \leq \varepsilon, \qquad |t| \leq \frac{1}{\varepsilon}.$$

We shall use o O notation referring to $\varepsilon \to 0$ with t restricted to a finite interval, and there will always be uniformity in t with this restriction. Now

$$\begin{aligned}\log \Phi(t) &= it\gamma_\varepsilon + \sum_j \log \hat{\Phi}_{\varepsilon j}(t) \\ &= it\gamma_\varepsilon + \sum_j [(\hat{\Phi}_{\varepsilon j}(t) - 1) + O|\hat{\Phi}_{\varepsilon j}(t) - 1|^2]\end{aligned}$$

and by I (11.10), if $\varepsilon < 1/T$,

$$\begin{aligned}\sum_j |\hat{\Phi}_{\varepsilon j}(t) - 1|^2 &\leq \varepsilon \sum_j |\hat{\Phi}_{\varepsilon j}(t) - 1| \\ &\leq -2\varepsilon L_5(T, \alpha, 1, 1) \int_0^1 \log |\Phi(s)|\, ds, \qquad |t| \leq T, \\ &= o(1)\end{aligned}$$

so that

$$\log \Phi(t) = it\gamma_\varepsilon + \sum_j [\hat{\Phi}_{\varepsilon j}(t) - 1] + o(1).$$

Now by I (11.8) and (11.9), using our evaluation of L_1, and the approximation of $\log \Phi$ just obtained, if $\hat{F}_\varepsilon = \sum_j \hat{F}_{\varepsilon j}$,

$$\begin{aligned}(4.7)\quad \hat{F}_\varepsilon(\infty) - \hat{F}_\varepsilon(\mu -) + \hat{F}_\varepsilon(-\mu) &\leq L_1\left(\mu, \frac{1}{\mu}, \frac{1}{\mu}\right) \int_0^{1/\mu} \sum_j \Re[1 - \hat{\Phi}_{\varepsilon j}(t)]\, dt \\ &\leq (1 + 2\pi)^2 \mu \int_0^{1/\mu} \sum_j \Re[1 - \hat{\Phi}_{\varepsilon j}(t)]\, dt \\ &= -(1 + 2\pi)^2 \mu \int_0^{1/\mu} \log |\Phi(t)|\, dt + o(1)\end{aligned}$$

and

$$\begin{aligned}(4.8)\qquad \int_{-\mu-}^{\mu} \lambda^2\, d\hat{F}_\varepsilon(\lambda) &\leq L_3(\mu, 1, 1) \int_0^1 \sum_j \Re[1 - \hat{\Phi}_{\varepsilon j}(t)]\, dt \\ &= -L_3(\mu, 1, 1) \int_0^1 \log |\Phi(t)|\, dt + o(1);\end{aligned}$$

here $o(1)$ is uniform in t, μ, for $|t| \leq T$, $1/\mu \leq T$, for any fixed T, in the first inequality, and for $|t| \leq T$, fixed μ, in the second inequality. Then if G_ε is defined by

$$G_\varepsilon(\lambda) = \int_{-\infty}^{\lambda} \frac{s^2}{1 + s^2}\, d\hat{F}_\varepsilon(s)$$

G_ε is monotone in λ, uniformly bounded when $\varepsilon \to 0$, according to (4.7) and (4.8), and, according to (4.7),

$$\lim_{\lambda \to \pm\infty} G_\varepsilon(\lambda) = G_\varepsilon(\pm \infty)$$

uniformly in ε when $\varepsilon \to 0$. By Helly's theorem a decreasing sequence of values $\varepsilon_1, \varepsilon_2, \cdots$ of ε can be found along which $G_\varepsilon(\lambda) \to G(\lambda)$ for all λ, where G is monotone and bounded; and $G(-\infty) = 0$, $G(\infty) = \lim_{k\to\infty} G_{\varepsilon_k}(\infty)$ according to the remark on uniformity just made. Now, replacing $\hat{F}_\varepsilon$ by G_ε, $\log \Phi$ can be put in the form

$$\log \Phi(t) = it\gamma_\varepsilon + \sum_j [\hat{\Phi}_{\varepsilon j}(t) - 1] + o(1) \tag{4.9}$$

$$= it\gamma_\varepsilon + \int_{-\infty}^{\infty} (e^{it\lambda} - 1) \frac{1+\lambda^2}{\lambda^2}\, dG_\varepsilon(\lambda) + o(1)$$

$$= it\gamma_\varepsilon' + \int_{-\infty}^{\infty} \left(e^{it\lambda} - 1 - \frac{it\lambda}{1+\lambda^2}\right) \frac{1+\lambda^2}{\lambda^2}\, dG_\varepsilon(\lambda) + o(1)$$

where

$$\gamma_\varepsilon' = \gamma_\varepsilon + \int_{-\infty}^{\infty} \frac{dG_\varepsilon(\lambda)}{\lambda} = \gamma_\varepsilon + \int_{-\infty}^{\infty} \frac{\lambda}{1+\lambda^2}\, d\hat{F}_\varepsilon(\lambda).$$

When $\varepsilon \to 0$ along the sequence $\{\varepsilon_k\}$, the integral in the last line of (4.9) goes to that in (4.6) uniformly in every finite t-interval (cf. the discussion of Stieltjes integration to the limit in I §11), and, since the left side of (4.9) does not depend on ε, γ_{ε_k}' must converge to some limit γ. This finishes the proof.

It will be useful below to know that Φ in (4.6) determines γ and G uniquely (assuming, say, that G is normalized by supposing it continuous on the right and 0 at $-\infty$). To show this note that

$$\log \Phi(t) - \tfrac{1}{2} \int_{t-1}^{t+1} \log \Phi(s)\, ds = \int_{-\infty}^{\infty} e^{it\lambda} \left(1 - \frac{\sin \lambda}{\lambda}\right) \frac{1+\lambda^2}{\lambda^2}\, dG(\lambda) \tag{4.10}$$

$$= \int_{-\infty}^{\infty} e^{it\lambda}\, dH(\lambda),$$

where H is monotone non-decreasing and bounded. Thus H is the "distribution" function corresponding to the "characteristic" function log Φ and is, together with G, therefore determined by Φ. The constant is then determined uniquely as the difference between log Φ and the integral in (4.6).

Note that in the derivation of (4.6) there appears to be an ambiguity in the definition of G since it was defined as the limit of some sequence of functions $\{G_{\varepsilon_k}\}$, and at that stage it was conceivable that a different sequence would give a different limit. Since G has now been shown to be uniquely determined at all its continuity points by Φ, if $G(-\infty) = 0$, it follows that G is the only possible limit, that is, $\lim_{\varepsilon\to 0} G_\varepsilon(\lambda) = G(\lambda)$ at all continuity points of G; similarly $\lim_{\varepsilon\to 0} \gamma_\varepsilon' = \gamma$.

In the most important particular case G is constant except for a jump when $\lambda = 0$, so that the given infinitely divisible distribution is Gaussian. The following corollary gives a sufficient condition for this.

COROLLARY 1 *If for every* $\eta > 0$ *a distribution function* F *can be written in the form* (4.1′) *with*

$$\sum_j [1 - F_j(\eta -) + F_j(-\eta)] \leq \eta,$$

then the distribution function is Gaussian.

The condition on F is much stronger than that of infinite divisibility. To prove the corollary we follow through the proof of the theorem, using the added hypothesis of the corollary. Then the F_j of the corollary corresponds to $F_{\varepsilon j}$ in the proof of the theorem, that is, in the proof of the theorem we can suppose that, for any $\eta > 0$,

$$\sum_j [1 - F_{\varepsilon j}(\eta -) + F_{\varepsilon j}(-\eta)] \leq \eta,$$

if ε is sufficiently small. If $\eta < \frac{1}{2}$, $|m_{\varepsilon j}| \leq \eta$ and $|m_{0\varepsilon j}| < \eta(2 + \alpha)$. Then the condition on the $F_{\varepsilon j}$'s implies the condition

$$1 - \hat{F}_\varepsilon(\eta' -) + \hat{F}_\varepsilon(-\eta') = \sum_j [1 - \hat{F}_{\varepsilon j}(\eta' -) + \hat{F}_{\varepsilon j}(-\eta')] \leq \eta,$$
$$\eta' = \eta(3 + \alpha),$$

on $\hat{F}_\varepsilon$. Thus the $\hat{F}_\varepsilon$ "distribution" becomes concentrated at the origin. The same is then true of the G_ε "distribution," so that G must be constant except for a possible jump at the origin. Then the original distribution is Gaussian. (We identify a random variable that is identically constant with a Gaussian variable having variance 0.) This corollary is not vacuous, since, if x is a Gaussian random variable with

$$\mathbf{E}\{x\} = 0, \qquad \mathbf{E}\{x^2\} = \sigma^2 > 0,$$

and if we write x in the form $x = \sum_1^n y_j$, where $y_1, \cdots, y_n$ are mutually independent with

$$\mathbf{E}\{y_j\} = 0, \qquad \mathbf{E}\{y_j^2\} = \frac{\sigma^2}{n},$$

then

$$\sum_1^n \mathbf{P}\{|y_j(\omega)| \geq \eta\} = n\mathbf{P}\{|y_1(\omega)| \geq \eta\}$$

$$= \frac{2n}{\sqrt{2\pi}} \int_{\eta\sqrt{n}}^{\infty} e^{-\lambda^2/2\sigma^2}\, d\lambda \to 0, \qquad n \to \infty.$$

In terms of random variables this corollary states that, if a random variable x can be represented as the sum of mutually independent random variables, $x = \sum_1^n y_j$, with $\mathbf{P}\{\operatorname{Max}_j |y_j(\omega)| \geq \eta\} \leq \eta$ for arbitrarily small η, then x is Gaussian. In fact the conditions

$$\sum_j \mathbf{P}\{|y_j(\omega)| \geq \eta\} \leq \eta, \qquad \mathbf{P}\{\operatorname{Max}_j |y_j(\omega)| \geq \eta\} \leq \eta$$

are equivalent in accordance with the fact (see §1) that the probability of at least one of a number of independent events (in this case $|y_j(\omega)| \geq \eta$) happening goes to zero together with the expected number happening.

In the above discussion we took a given distribution and analyzed the restriction imposed on it by the hypothesis that it was a convolution of distributions concentrated near the origin. A slightly more general question is the following. Suppose that $y_1, \cdots, y_n$ are mutually independent and small,

$$\mathbf{P}\{|y_j(\omega)| \geq \varepsilon\} \leq \varepsilon, \qquad j = 1, \cdots, n.$$

Then what can one say of the asymptotic character of the distribution of their sum x? If $\Phi_{\varepsilon j}$ is the characteristic function of the y_j distribution and if Φ_ε is that of the x distribution,

$$\Phi_\varepsilon = \prod_j \Phi_{\varepsilon j}.$$

The difference between this question and the one treated already lies in the fact that the left side depends on the y_j's. We can use the notation and methods of the proof of Theorem 4.1, however. This proof shows that, if when $\varepsilon \to 0$ the y_j's are chosen so that x has a limiting distribution, then this limiting distribution must be given by (4.6), and thus be infinitely divisible. In particular, if a given distribution can be approximated by infinitely divisible distributions it can be approximated by convolutions of small y_j's, so that it must be infinitely divisible, that is, we have the following corollary:

COROLLARY 2 *Any limiting distribution of infinitely divisible distributions is infinitely divisible.*

It is easy to show, using (4.10), that the G's of the approximating distributions—cf. (4.6)—converge to that of the limit distribution, at the points of continuity of the latter, if all are normalized to be 0 at $-\infty$.

The most natural theorem from this point of view is that the distribution of the sum of a large number of small mutually independent random variables is near some infinitely divisible distribution whose γ and G in (4.6) are expressible in terms of the summands. In proving such a theorem one can proceed as in the proof of Theorem 4.1, but there is an additional difficulty. In Theorem 4.1 the distribution of the sum was given. Thus in inequalities (4.7) and (4.8), involving on the right the characteristic function of this distribution, the right sides are fixed. But from the present point of view the Φ in those inequalities is not fixed, and in fact it is to be proved that this Φ has some asymptotic character when ε is small. Thus the hypotheses of the theorem must themselves limit the left sides of (4.7) and (4.8). We shall restrict ourselves to the case of a Gaussian limiting distribution, that is, to the case of a G in (4.6) which is constant except for a jump at $\lambda = 0$. We shall thus be proving one form of the *central limit theorem*, a generic name applied to any theorem which states that the sum of small random variables, under appropriate restrictions, is nearly normally distributed.

THEOREM 4.2 *Let $x = y_1 + \cdots + y_n$ be a sum of mutually independent random variables. Let η, α, b be positive numbers, with $1/b < \alpha < b$, and define*

$$\gamma_j = \int_{-\alpha-}^{\alpha} \lambda \, d\mathbf{P}\{y_j(\omega) \le \lambda\}, \qquad \gamma = \sum_j \gamma_j,$$

$$\sigma_j^2 = \int_{-\alpha-}^{\alpha} \lambda^2 \, d\mathbf{P}\{y_j(\omega) \le \lambda\} - \gamma_j^2, \qquad \sigma^2 = \sum_j \sigma_j^2.$$

Suppose that $\sigma^2 \le b$ and that

$$\sum_j \mathbf{P}\{|y_j(\omega)| \ge \eta\} \le \eta. \tag{4.11}$$

Then there is an η_1 depending only on η, b, and going to 0 with η for fixed b, such that

$$\left| \mathbf{P}\{x(\omega) - \gamma \le \lambda\} - \frac{1}{\sigma\sqrt{2\pi}} \int_{-\infty}^{\lambda} e^{-\mu^2/2\sigma^2} \, d\mu \right| \le \eta_1, \qquad -\infty < \lambda < \infty. \tag{4.12}$$

Conversely, if $x = y_1 + \cdots + y_n$ *is a sum of mutually independent random variables with zero medians, if* η, α, b *are positive numbers with* $1/b < \alpha < b$, *if* γ, σ *are defined as above, if*

$$\mathbf{P}\{|y_j(\omega)| \geq \eta\} \leq \eta, \qquad j \geq 1, \tag{4.13}$$

and if

$$\left| \mathbf{P}\{x(\omega) - \tilde{\gamma} \leq \lambda\} - \frac{1}{\tilde{\sigma}\sqrt{2\pi}} \int_{-\infty}^{\lambda} e^{-\mu^2/2\tilde{\sigma}^2}\, d\mu \right| \leq \eta, \qquad -\infty < \lambda < \infty, \tag{4.12'}$$

with $|\tilde{\gamma}| \leq b$, $\tilde{\sigma}^2 \leq b$, *then there is an* η' *depending only on* b, η, *such that* η' *goes to* 0 *with* η *for fixed* b, *and that*

$$\sum_j \mathbf{P}\{|y_j(\omega)| \geq \eta'\} \leq \eta', \qquad |\sigma^2 - \tilde{\sigma}^2| \leq \eta', \qquad |\gamma - \tilde{\gamma}| \leq \eta'. \tag{4.14}$$

The condition (4.11) can be replaced by the asymptotically equivalent condition

$$\mathbf{P}\{\underset{j \leq n}{\mathrm{Max}}\, |y_j(\omega)| \geq \eta\} \leq \eta. \tag{4.11'}$$

To show the real simplicity of the direct half of this theorem, we prove it without recourse to the methods of Theorem 4.1. Define y_j' by

$$\begin{aligned} y_j'(\omega) &= y_j(\omega), \qquad & |y_j(\omega)| \leq \alpha \\ &= 0, & |y_j(\omega)| > \alpha, \end{aligned}$$

so that

$$\gamma_j = \mathbf{E}\{y_j'\}, \qquad \sigma_j^2 = \mathbf{E}\{(y_j' - \gamma_j)^2\}.$$

From now on we shall assume that η is so small that

$$\eta \leq \frac{1/b}{2(1 + 1/b)} \leq \frac{\alpha}{2(1 + \alpha)} \leq \frac{b}{2(1 + b)}. \tag{4.15}$$

Then

$$|\gamma_j| \leq \eta + \mathbf{P}\{|y_j(\omega)| > \eta\} \leq \eta(1 + b),$$

$$\sigma_j^2 \leq \mathbf{E}\{y_j'^2\} \leq \eta^2 + 4\alpha^2\mathbf{P}\{|y_j(\omega)| > \eta\} \leq \eta(1 + 4b^2),$$

$$\begin{aligned} \mathbf{E}\{|y_j' - \gamma_j|^3\} &\leq \eta(2 + b)\sigma_j^2 + 8\alpha^3\mathbf{P}\{|y_j'(\omega) - \gamma_j| > \eta(2 + b)\} \\ &\leq \eta(2 + b)\sigma_j^2 + 8\alpha^3\mathbf{P}\{|y_j'(\omega)| > \eta\}, \end{aligned}$$

so that

$$\sum_j \mathbf{E}\{|y_j' - \gamma_j|^3\} \leq \eta(2 + b)\sigma^2 + 8\alpha^3\eta \leq \eta[(2 + b)b^2 + 8b^3] = b_1\eta.$$

Now according to I (11.4) with $n = 3$, if Φ_j is the characteristic function of $y_j' - \gamma_j$,

$$\Phi_j(t) = 1 - \frac{\sigma_j^2 t^2}{2} + u_j, \qquad |u_j| \le \mathbf{E}\{|y_j' - \gamma_j|^3\} \frac{|t|^3}{6}.$$

If $T > 0$,

$$\operatorname*{Max}_j \left| -\frac{\sigma_j^2 t^2}{2} + u_j \right| < \frac{1}{2}, \qquad |t| \le T,$$

for sufficiently small η, and we can write

$$\log \Phi_j(t) = -\frac{\sigma_j^2 t^2}{2} + u_j + v_j, \qquad |t| \le T,$$

where

$$|v_j| \le \left(\frac{\sigma_j^2 t^2}{2} + u_j\right)^2.$$

Then if $\hat{\Phi}$ is the characteristic function of $\sum_j (y_j' - \gamma_j)$,

$$\log \hat{\Phi}(t) = -\frac{\sigma^2 t^2}{2} + \sum_j (u_j + v_j), \qquad |t| \le T, \tag{4.16}$$

and

$$\begin{aligned} \left|\sum_j (u_j + v_j)\right| &\le \sum_j |u_j| + \operatorname*{Max}_j \left(\frac{\sigma_j^2 t^2}{2} + u_j\right) \sum_j \left(\frac{\sigma_j^2 t^2}{2} + |u_j|\right) \\ &\le \frac{b_1 \eta T^3}{6} + \left[\frac{\eta(1 + 4b^2)T^2}{2} + \frac{b_1 \eta T^3}{6}\right]\left(\frac{bT^2}{2} + \frac{b_1 \eta T^3}{6}\right), \\ &\qquad\qquad |t| \le T, \end{aligned}$$

so that the sum in (4.16) goes to 0 with η uniformly in t for $|t| \le T$. Thus $\sum_j (y_j' - \gamma_j)$ is asymptotically normal with mean 0 and variance σ^2, as $\eta \to 0$. Since

$$\mathbf{P}\{y_j(\omega) = y_j'(\omega), j \ge 1\} > 1 - \eta,$$

the distribution of $\sum_j (y_j - \gamma_j)$ is asymptotically $(\eta \to 0)$ the same as that of $\sum_j (y_j' - \gamma_j)$, and this finishes the proof of the direct half of Theorem 4.2.

Suppose now that the hypotheses of the converse are true. We then assume that $\eta \to 0$ along a sequence of values, that for each η value there is a sum x as described, that η' is given, and we prove that (4.14) is true for η sufficiently small. If Φ is the characteristic function of x, the characteristic function $\Phi e^{-it\tilde{\gamma}}$ is by hypothesis asymptotically $e^{-\tilde{\sigma}^2 t^2/2}$. On the other hand, the proof of Theorem 4.1 carried through here makes $\Phi e^{-it\tilde{\gamma}}$ asymptotically a characteristic function given by (4.6). (We can assume that $\tilde{\gamma}$ and $\tilde{\sigma}^2$ are asymptotically near finite limiting values.) Then the function G in (4.6) must be constant except for a jump at the origin. According to the derivation of (4.6), this means that

$$\sum_j \mathbf{P}\{|y_j(\omega) - \gamma_j| \geq \eta'\}$$

goes to 0 with η. Since $|\gamma_j| \leq \eta(1 + b)$, it follows that if η is so small that $\eta(1 + b) < \eta'/2$ then

$$\sum_j \mathbf{P}\{y_j(\omega)| \geq \eta'\} \leq \sum_j \mathbf{P}\{|y_j(\omega) - \gamma_j| \geq \eta'/2\} \to 0, \qquad (\eta \to 0).$$

The direct half of the theorem now asserts that $\sum_j (y_j - \gamma_j)$ is asymptotically Gaussian with mean 0 and variance σ^2, so that γ, σ^2 must be asymptotically $\tilde{\gamma}$, $\tilde{\sigma}^2$, respectively, as was to be proved.

Two useful special cases of this theorem will be proved. To illustrate the method of proving these theorems the first special case (Theorem 4.3) will be deduced independently and the second one (Theorem 4.4) will be derived from Theorem 4.2.

THEOREM 4.3 *Let $y_1, y_2, \cdots$ be independent random variables with a common distribution function, having finite variance σ^2. Then*

$$\frac{1}{\sqrt{n}} \sum_1^n [y_j - \mathbf{E}\{y_j\}]$$

is asymptotically Gaussian with expectation 0 and variance σ^2,

$$\lim_{n\to\infty} \mathbf{P}\left\{\frac{1}{\sqrt{n}} \sum_{j=1}^n [y_j(\omega) - \mathbf{E}\{y_j\}] < \sigma\lambda\right\} = \frac{1}{\sqrt{2\pi}} \int_{-\infty}^{\lambda} e^{-\mu^2/2}\, d\mu,$$

uniformly in λ.

In fact, if Φ is the characteristic function of $y_j - \mathbf{E}\{y_j\}$, we have, using I (11.4),

$$\Phi(t) = 1 - \frac{\sigma^2}{2} t^2 + o(t^2) \qquad (t \to 0)$$

so that

$$\log \Phi(t) = -\frac{\sigma^2 t^2}{2} + o(t^2)$$

if t is sufficiently small. The characteristic function of $\frac{1}{\sqrt{n}} \sum_{j=1}^n [y_j - \mathbf{E}\{y_j\}]$ is $\Phi(t/\sqrt{n})^n$, and

$$\log \Phi\left(\frac{t}{\sqrt{n}}\right)^n = -\frac{\sigma^2 t^2}{2} + n\, o\left(\frac{t^2}{n}\right).$$

Since this converges to $-\sigma^2/2$ uniformly in every finite interval the theorem is true.

We conclude with a famous version of the central limit theorem due to Liapounov, which has a wide range of applicability.

THEOREM 4.4 *Let $y_1, \cdots, y_n$ be mutually independent random variables, and suppose that $\mathbf{E}\{y_j\} = 0$, $\mathbf{E}\{|y_j|^{2+\delta}\} = c_j < \infty$, for some $\delta > 0$.*

Define B_n, C_n by

$$B_n = \mathbf{E}\{(\sum_1^n y_j)^2\} = \sum_1^n \mathbf{E}\{y_j^2\}.$$

$$C_n = \sum_1^n c_j.$$

Then, if $B_n > 0$, and if

$$\frac{C_n}{B_n^{1+\delta/2}} \leq \varepsilon,$$

it follows that

$$\mathbf{P}\{\sum_1^n y_j(\omega) \leq \lambda \sqrt{B_n}\} = \frac{1}{\sqrt{2\pi}} \int_{-\infty}^{\lambda} e^{-\mu^2/2}\, d\mu + o(1)$$

where $o(1)$ refers to $\varepsilon \to 0$ and is uniform in λ and in the y_j distributions involved.

Note that

$$s_j^2 = \mathbf{E}\{y_j^2\} \leq c_j^{2/(2+\delta)} \leq \varepsilon^{2/(2+\delta)} B_n \qquad j = 1, \cdots, n,$$

so that (summing over j)

$$B_n \leq n\varepsilon^{2/(2+\delta)} B_n, \qquad n \geq \varepsilon^{-2/(2+\delta)}.$$

This means that for ε small the number of y_j's is large.

Theorem 4.4 is easily proved directly just as Theorem 4.3 was, using I (11.4′). However, it is instructive to derive it from the general result of Theorem 4.2. We apply Theorem 4.2 to the variables $y_1 B_n^{-1/2}, \cdots, y_n B_n^{-1/2}$. In the first place, if $\eta > 0$, the obvious generalization of Chebyshev's inequality to the exponent $2 + \delta$ yields

$$\sum_{j=1}^n \mathbf{P}\{|y_j(\omega)|\, B_n^{-1/2} \geq \eta\} \leq \frac{C_n}{\eta^{2+\delta} B_n^{1+\delta/2}} < \frac{\varepsilon}{\eta^{2+\delta}} \leq \eta$$

if $\varepsilon \leq \eta^{3+\delta}$. Thus (4.11) is satisfied in the present case if ε is sufficiently small. It remains to calculate the truncated expectations and variances used in Theorem 4.2. We have

$$\gamma_j = \int_{\{|\lambda| \leq \alpha\}} \lambda\, d\mathbf{P}\{y_j(\omega) B_n^{-1/2} \leq \lambda\} = -B_n^{-1/2} \int_{\{|\lambda| > \alpha B_n^{1/2}\}} \lambda\, d\mathbf{P}\{y_j(\omega) \leq \lambda\}$$

so that

$$\sum_j |\gamma_j| \leq B_n^{-1/2} \sum_j (\alpha B_n^{1/2})^{-1-\delta} \int_{\{|\lambda| > \alpha B_n^{1/2}\}} |\lambda|^{2+\delta}\, d\mathbf{P}\{y_j(\omega) \leq \lambda\}$$

$$\leq \frac{C_n}{\alpha^{1+\delta} B_n^{1+\delta/2}} \leq \frac{\varepsilon}{\alpha^{1+\delta}}.$$

Similarly

$$\sigma_j^2 = \int_{\{|\lambda| \leq \alpha\}} \lambda^2 \, d\mathbf{P}\{y(\omega)B_n^{-1/2} \leq \lambda\} - \gamma_j^2 = B_n^{-1} \int_{\{|\lambda| \leq \alpha B_n^{1/2}\}} \lambda^2 \, d\mathbf{P}\{y_j(\omega) \leq \lambda\} - \gamma_j^2$$

$$= B_n^{-1} \left[s_j^2 - \int_{\{|\lambda| > \alpha B_n^{1/2}\}} \lambda^2 \, d\mathbf{P}\{y_j(\omega) \leq \lambda\}\right] - \gamma_j^2.$$

Then (using the fact that $|\gamma_j| \leq \alpha$)

$$1 - \sum_j \sigma_j^2 = B_n^{-1} \sum_j \int_{\{|\lambda| > \alpha B_n^{1/2}\}} \lambda^2 \, d\mathbf{P}\{y_j(\omega) \leq \lambda\} + \sum_j \gamma_j^2$$

$$\leq B_n^{-1} \sum_j (\alpha B_n^{1/2})^{-\delta} \int_{\{|\lambda| > \alpha B_n^{1/2}\}} |\lambda|^{2+\delta} \, d\mathbf{P}\{y_j(\omega) \leq \lambda\} + \alpha \sum_j |\gamma_j|$$

$$\leq \frac{2C_n}{\alpha^\delta B_n^{1+\delta/2}} \leq 2\frac{\varepsilon}{\alpha^\delta}.$$

Thus this theorem follows from Theorem 4.2.

As an example of the applicability of this theorem note that, if $y_1, y_2, \cdots$ are mutually independent random variables with $\mathbf{E}\{y_j\} = 0$, $\mathbf{E}\{|y_j|^{2+\delta}\} \leq K$ for some constant K, then, if the variance of y_j is bounded away from 0, $\mathbf{E}\{y_j^2\} \geq \sigma^2 > 0$,

$$\frac{\sum_1^n y_j}{\left[\sum_1^n \mathbf{E}\{y_j^2\}\right]^{1/2}}$$

is asymptotically normal, $n \to \infty$, with mean 0 and variance 1. In fact, in this case we have $c_j \leq K$ and the critical ratio goes to zero,

$$\frac{C_n}{B_n^{1+\delta/2}} \leq \frac{nK}{n^{1+\delta/2}\sigma^{2+\delta}} \to 0.$$

5. Stationary case

In this section we shall suppose that the random variables $x_1, x_2, \cdots$ are mutually independent and have a common distribution function. The first theorem we prove is the form of the strong law of large numbers applicable to this case:

THEOREM 5.1 *If $x_1, x_2, \cdots$ are mutually independent, and have a common distribution function, with $\mathbf{E}\{|x_j|\} < \infty$, then*

$$\lim_{n\to\infty} \frac{x_1 + \cdots + x_n}{n} = \mathbf{E}\{x_1\} \tag{5.1}$$

with probability 1.

Before proving the theorem we note that, if $\mathbf{E}\{x_j^2\} < \infty$, this theorem is a special case of Theorem 3.4, with $\sigma_1^2 = \sigma_2^2 = \cdots$. To prove the theorem, define x_n' by

$$\begin{aligned} x_n'(\omega) &= x_n(\omega), \qquad & |x_n(\omega)| &\le n \\ &= 0, & |x_n(\omega)| &> n. \end{aligned}$$

We prove first that $x_j'(\omega) = x_j(\omega)$ for all large j, with probability 1. In fact (Borel-Cantelli lemma), this follows from

$$\begin{aligned} \sum_j \mathbf{P}\{x_j'(\omega) \ne x_j(\omega)\} = \sum_j \mathbf{P}\{|x_1(\omega)| > j\} &= \sum_2^\infty j\,\mathbf{P}\{j-1 < |x_1(\omega)| \le j\} \\ &\le \mathbf{E}\{|x_1| + 1\}. \end{aligned}$$

Hence

$$\lim_{n\to\infty} \frac{1}{n} \sum_1^n (x_j - x_j') = 0$$

with probability 1, and we shall therefore discuss first the x_j' averages. Now

$$\begin{aligned} \sum_2^\infty \frac{\mathbf{E}\{[x_n' - \mathbf{E}\{x_n'\}]^2\}}{n^2} \le \sum_2^\infty \frac{\mathbf{E}\{x_n'^2\}}{n^2} &= \sum_2^\infty \int_0^n \frac{\lambda^2}{n^2}\, d\mathbf{P}\{|x_1(\omega)| \le \lambda\} \\ &\le \int_2^\infty \frac{d\mu}{(\mu-1)^2} \int_0^\mu \lambda^2\, d\mathbf{P}\{|x_1(\omega)| \le \lambda\} \\ &\le \int_2^\infty \frac{\lambda^2}{\lambda - 1}\, d\mathbf{P}\{|x_1(\omega)| \le \lambda\} + 4 \\ &\le 2\mathbf{E}\{|x_1|\} + 4. \end{aligned}$$

It now follows from Theorem 3.4 that

$$\lim_{n\to\infty} \frac{1}{n} \sum_1^n [x_j' - \mathbf{E}\{x_j'\}] = 0,$$

with probability 1, so that

$$\lim_{n\to\infty} \frac{1}{n} \sum_1^n [x_j - \mathbf{E}\{x_j'\}] = 0$$

with probability 1. Since $\lim_{n\to\infty} \mathbf{E}\{x_n'\} = \mathbf{E}\{x_1\}$,

$$\lim_{n\to\infty} \frac{1}{n} \sum_1^n \mathbf{E}\{x_j'\} = \mathbf{E}\{x_1\}.$$

Adding this to the previous equation gives the desired result.

A sequence of mutually independent random variables with a common distribution function corresponds to the physical picture of repeated trials of an experiment (giving numerical answers). The theory of probability was originally invented to deal with this situation, and discussions of the foundations of probability usually restrict themselves to it. Now two facts are very striking to anyone who actually performs repeated trials, and any mathematical analysis must contain theorems which correspond to these facts.

(A) As n increases, the sample ratios $\{[x_1(\omega) + \cdots + x_n(\omega)]/n\}$ are observed to vary only slightly. The mathematical version of this fact is Theorem 5.1.

The existence of the limit in (5.1) cannot be verified in practice, of course, because infinitely many trials are involved. In fact, the experimenter's observations cannot a priori be exact enough for him to demand of the theoretician that in the latter's theoretical structure the limit in (5.1) exist with (mathematical) probability 1. It is gratifying that the limit exists in this strong sense, but it would not be disturbing if it only existed in some weaker sense, say as a limit in probability.

(B) The experimenter also notices that the sample ratios cluster around the same value as before if the results of some trials are not counted at all. For example, if the experimenter goes out to lunch, leaving his apparatus going, but with no recordings being taken, the noon hour x_j's will be irretrievably lost, but, if they are simply ignored, and the x_j's recorded only when the experimenter is present and interested, the cluster value is unaffected. More generally, the cluster value is unaffected if the experimenter, instead of merely ignoring certain trials because of the pangs of hunger or boredom or love or because of other irrelevant reasons independent of the trials, actually ignores some because of the results of past trials, say out of disgust with past results. In other words, the criterion of acceptance or rejection may depend on past results. This can be expressed formally as follows: Let $x_1(\omega), x_2(\omega), \cdots$ be the full sequence of (original) sample values, and let $x_1'(\omega), x_2'(\omega), \cdots$ be the $x_j(\omega)$'s that are actually recorded. It is supposed, then, that there are integers $n_1 < n_2 < \cdots$ and $x_j' = x_{n_j}$; the n_j's may be random variables. Now the experimenter may choose the trials to record, that is, the n_j's, on the basis of his lack of appetite, his hunch that the experiment is going well, the results of previous trials, or general cussedness. However, we shall not allow him to be clairvoyant. This is interpreted by prescribing that after he has recorded $x_{k-1}'(\omega) = x_{n_{k-1}(\omega)}(\omega)$, his choice of the next $x_j(\omega)$ to record is to be based only on past events; that is, the condition $n_j(\omega) = \nu$ is a condition on $x_1, \cdots, x_{\nu-1}$ alone. In mathematical terms $x_1, x_2, \cdots$ are random variables, $n_1, n_2, \cdots$ are integral-valued random variables, and the ω set $\{n_j(\omega) = \nu\}$ differs by at most an ω set of probability 0 from an ω set of the form $\{[x_1(\omega), \cdots, x_\nu(\omega)] \in A\}$, where A is a ν-dimensional Borel set. The question is thus reduced to the relation between the old random variables $x_1, x_2, \cdots$ and the new ones $x_1', x_2', \cdots$, where $x_j' = x_{n_j}$. In the following we shall assume that in this way $N \leq \infty$ new random variables $x_1', x_2', \cdots$ have been defined, and we shall simplify the argument unessentially by assuming that these random variables are certainly defined, that is, each n_j is defined with probability 1 or not at all.

In the time-honored language of gambling, the x_j's can be considered the numbers turned up in some sort of game in which a gambling system is applied to choose which plays to bet on and which to reject. For example, if x_j can take on two values 1 or 0 corresponding to red or black in roulette x_1' might be the first number after the first 1, x_2' the first after a run of two 1's, and in general x_j' the first after a run of j 1's. This gambling system, in which the usual gambler would bet that the x_j''s have more chance of being 0 than the x_j's, has an important advantage; there is a longer and longer wait between bets, that is, between plays that are accepted, and there is thus more and more time available to the gambler to think and reform before he loses his money, and to study probability rather than gambling.† The disadvantage, or rather lack of advantage, of this system, is that like all other systems it leaves the gambler's chances entirely unaffected. This is the substance of the following theorem. Note again that this theorem specifically excludes the hypothesis of clairvoyance. Naturally any prophet can make money at a gambling casino. The fact that those recently discovered do not do so shows that their high principles nullify their supernatural advantages over mathematically limited (by Theorem 5.2) mortals.

THEOREM 5.2 *The random variables $x_1', x_2', \cdots$ have the same probability properties as $x_1, x_2, \cdots$; that is, $x_1', x_2', \cdots$ are mutually independent, with a common distribution function, that of x_1. Moreover (defining $x_j = 0$ for $j \leq 0$) for each j, the sets of random variables $\{\cdots, x_{n_j-2}, x_{n_j-1}\}$ $\{x_{n_j} = x_j', x_{j+1}', \cdots\}$ are mutually independent.*

Note that, according to Theorem 5.1, applied to $x_1', x_2', \cdots$ (assuming the truth of Theorem 5.2),

$$\lim_{n\to\infty} \frac{x_1' + \cdots + x_n'}{n} = \mathbf{E}\{x_1'\} = \mathbf{E}\{x_1\}$$

with probability 1. It is usually this property, that limiting ratios are not altered by the use of a system, that is used in the foundation studies,

† This formulation ignores those who would find the first less interesting than the second; in fact it considers only those who are trying to make gambling profitable rather than merely fascinating.

rather than the more general theorem as stated, and this is of course the mathematical version of (B) above.

To prove Theorem 5.2 we note that for any intervals $I_1, \cdots, I_m$

$$\begin{aligned}
&\mathbf{P}\{x_1'(\omega) \in I_1, \cdots, x_m'(\omega) \in I_m\} \\
&\quad = \sum_{a_1 < \cdots < a_m} \mathbf{P}\{n_j(\omega) = a_j, x_{a_j}(\omega) \in I_j, j = 1, \cdots, m-1, \\
&\qquad\qquad\qquad n_m(\omega) = a_m, x_{a_m}(\omega) \in I_m\} \\
&\quad = \sum_{a_1 < \cdots < a_m} \mathbf{P}\{n_j(\omega) = a_j, x_{a_j}(\omega) \in I_j, j = 1, \cdots, m-1, \\
&\qquad\qquad\qquad n_m(\omega) = a_m\} \mathbf{P}\{x_1(\omega) \in I_m\},
\end{aligned}$$

since the conditions imposed on $n_1, \cdots, n_m, x_{a_1}, \cdots, x_{a_{m-1}}$ involve only x_j's for $j < a_m$, so that the factor

$$\mathbf{P}\{x_{a_m}(\omega) \in I_m\} = \mathbf{P}\{x_1(\omega) \in I_m\}$$

can be separated out. Summing first over a_m we now obtain, since by hypothesis n_m is certainly finite,

$$\sum_{a_1 < \cdots < a_{m-1}} \mathbf{P}\{n_j(\omega) = a_j, x_{a_j}(\omega) \in I_j, j = 1, \cdots, m-1\} \mathbf{P}\{x_1(\omega) \in I_m\}.$$

Repeating this procedure $m - 1$ times, we finally evaluate the multiple sum, finding it to be

$$\prod_{j=1}^{m} \mathbf{P}\{x_1(\omega) \in I_j\},$$

so that the x_j''s have the same distribution as the x_j's. If now we evaluate

$$\mathbf{P}\{x_{n_j+k-1}(\omega) \in I_k, k = -m, \cdots, m\},$$

the same reasoning reduces this to

$$\mathbf{P}\{x_{n_j+k-1}(\omega) \in I_k, k = -m, \cdots, 0\} \mathbf{P}\{x_1(\omega) \in I_1\} \cdots \mathbf{P}\{x_1(\omega) \in I_m\},$$

which proves the last statement of the theorem.

Theorem 5.2 is commonly used implicitly in probability discussions. For example, let $x_1, x_2, \cdots$ be mutually independent random variables, with

$$\mathbf{P}\{x_j(\omega) = 1\} = p, \qquad \mathbf{P}\{x_j(\omega) = 0\} = 1 - p.$$

Let t_0 be the number of 0's of the x_j's before the first 1, so that

$$\mathbf{P}\{t_0(\omega) = m\} = (1 - p)^m p.$$

Now consider the successive times $t_1, t_2, \cdots$ between groups of 1's; more precisely define $a_1, a_2, \cdots, t_1, t_2, \cdots$ by:

$$a_1(\omega) = \text{Min } n \text{ such that } x_n(\omega) = 1, x_{n+1}(\omega) = 0$$

$$a_1(\omega) + t_1(\omega) = \text{Min } n > a_1(\omega) \text{ such that } x_n(\omega) = 0, x_{n+1}(\omega) = 1$$

$$a_2(\omega) = \text{Min } n > a_1(\omega) + t_1(\omega) \text{ such that } x_n(\omega) = 1, x_{n+1}(\omega) = 0$$

$$\cdot \quad \cdot \quad \cdot \quad \cdot \quad \cdot \quad \cdot \quad \cdot$$

It is commonly accepted without proof that the random variables $t_0 + 1, t_1, \cdots$ are mutually independent and have a common distribution. The fact that this is true as well as intuitively obvious is deduced from Theorem 5.2 as follows. In accordance with this theorem, $x_{a_j+2}, x_{a_j+3}, \cdots$ are mutually independent with the same distribution as x_1. The number of 0's, before the first 1 for this sequence is $t_j - 1$. Hence t_j has the same distribution as $t_0 + 1$. Moreover according to this theorem x_{a_j+2}, $x_{a_j+3}, \cdots$ form a sequence independent of $\cdots, x_{a_j}, x_{a_j+1}$. Hence t_j is independent of the latter variables, and so of $t_0, \cdots, t_{j-1}$.

CHAPTER IV

Processes with Mutually Uncorrelated or Orthogonal Random Variables

1. General remarks

In this chapter the random variables will be complex-valued, and two random variables equal almost everywhere will be considered identical.

Let x and y be random variables, with

$$\mathbf{E}\{|x|^2\} < \infty, \qquad \mathbf{E}\{|y|^2\} < \infty.$$

Then if

$$\mathbf{E}\{x\bar{y}\} = \mathbf{E}\{x\}\,\mathbf{E}\{\bar{y}\}$$

the random variables are said to be uncorrelated; if

$$\mathbf{E}\{x\bar{y}\} = 0$$

they are said to be orthogonal. If x and y are uncorrelated $x - \mathbf{E}\{x\}$ and $y - \mathbf{E}\{y\}$ are orthogonal. In most discussions of uncorrelated random variables the variables are centered by subtraction of expectations, obtaining orthogonal random variables, and it is for this reason that throughout the present chapter we shall discuss only processes with mutually orthogonal variables. We shall not, however, suppose that their expectations vanish, since this hypothesis is usually irrelevant although frequently satisfied. Only the discrete parameter case is of interest here, just as in the case of processes with mutually independent random variables and for the same reason; in the continuous parameter case the sample functions are too discontinuous for the processes to be useful.

It is always difficult to distinguish between probability theory and measure theory; it is pointless even on historical grounds to do so in studying orthogonality. Random variables are measurable functions and "expectation" is a semantical obfuscation of "integral," but here the deception ceases; two functions f, g are called orthogonal if the integral of $f\bar{g}$ vanishes. It is, however, somewhat embarrassing, and needlessly restrictive, to suppose that the total measure is 1, or even that it is finite.

In fact, throughout this chapter we shall discuss only formal properties of orthogonality which do not involve the measure of the space itself. However, if the measure of the space is infinite, we suppose that the space is the union of enumerably many measurable sets of finite measure, to avoid integration difficulties. It is to be understood, then, that in the present chapter "random variable" may be interpreted as "measurable function," defined on a measure space with no restriction other than the one just stated. It is merely for the sake of consistency with the point of view of the rest of the book that the theorems are stated in probability rather than in measure terminology. In several subsequent chapters the theorems on orthogonal series will be applied to cases in which the total measure need not be 1.

There are many good and readily available treatments of orthogonal functions, and the treatment in this chapter is therefore somewhat sketchy. The sketchiness of the treatment should not, however, lead the reader to believe that this chapter is not a part of probability. The law of large numbers is a probability theorem whether the random variables involved are orthogonal or whether they are mutually independent, and it is only historical tradition that has relegated the corresponding theorems to different books. It is true, however, that the probabilist is less interested in particular sequences of orthogonal functions, like Legendre polynomials, trigonometric functions, etc., than the analyst. The former is usually more interested in the properties shared by all such sequences than in those possessed by particular ones.

2. Geometrical considerations

Throughout this chapter all the random variables will have finite second moments and the "distance" between two random variables x and y will be the root mean square distance

$$[\mathbf{E}\{|x-y|^2\}]^{1/2}.$$

A collection of random variables will be called a *linear manifold* if, whenever any finite number of them, say $x_1, \cdots, x_n$, are in the collection, any linear combination $\sum_1^n a_j x_j$ is also in the collection. (In the following, "linear combination" always means "finite linear combination.") The manifold is said to be closed if, whenever $x_1, x_2, \cdots$ are random variables in the manifold and $\underset{n\to\infty}{\text{l.i.m.}}\, x_n = x$, x is also in the manifold. This is in conformity with the usual concept of closure, using the root mean square distance.

All possible linear combinations of the random variables of a given collection make up a new (in general larger) collection. The new collection is a linear manifold, the smallest linear manifold containing the given collection; it is called the *linear manifold generated by the given collection*. If in addition all possible limits in the mean of sequences of random variables in this linear manifold are adjoined to the manifold, a new (in general larger) linear manifold is obtained. This new linear manifold is closed, and is the smallest closed linear manifold containing the given collection; it is called the *closed linear manifold generated by the given collection*.

Let $\mathfrak{M}$ be any linear manifold and let x be orthogonal to every random variable in it. The variable x is then said to be orthogonal to $\mathfrak{M}$. The variable x will then automatically be orthogonal to the closed linear manifold generated by $\mathfrak{M}$. In fact, if $y = \operatorname*{l.i.m.}_{n\to\infty} y_n$ and if y_n is in $\mathfrak{M}$,

$$|\mathbf{E}\{x\bar{y}\}|^2 = |\mathbf{E}\{x(\bar{y} - \bar{y}_n)\}|^2 \leq \mathbf{E}\{|x|^2\}\mathbf{E}\{|y - y_n|^2\}$$

and the second factor goes to 0 when $n \to \infty$.

If x is orthogonal to all the random variables of a certain collection, it is obviously orthogonal to the linear manifold (and hence to the closed linear manifold) they generate.

Two linear manifolds are said to be orthogonal if every random variable in one is orthogonal to every random variable in the other.

3. General definition of projection

Let $y_1, y_2, \cdots$ be a finite or infinite sequence of random variables. If

$$\mathbf{E}\{y_j\bar{y}_k\} = \delta_{jk} \qquad \begin{pmatrix} = 0 & j \neq k \\ 1 & j = k \end{pmatrix}$$

the variables are said to form an *orthonormal sequence*. If $x = \sum_j a_j y_j$ and if term by term integration can be justified, $\mathbf{E}\{x\bar{y}_k\} = a_k$. It is therefore natural to consider for any random variable x the series $\sum_j a_j y_j$ with a_j's defined in this way. We write

$$x \sim \sum_j a_j y_j \qquad a_j = \mathbf{E}\{x\bar{y}_j\}. \tag{3.1}$$

The coefficients $a_1, a_2, \cdots$ are called the *Fourier coefficients of* x, and the sign $\sim$ means merely that x corresponds to the indicated series, called the *Fourier series* of x with respect to the given orthonormal sequence. We have already observed that, if any series $\sum_j b_j y_j$ converges to x, then, if the series is well behaved (for example, if it is only a finite series), the b_j's must necessarily be the Fourier coefficients. The Fourier series of x may converge, however, without converging to x. For example, if probability is defined by a uniform distribution (that is, constant density) on the interval $0 \leq \xi \leq 2\pi$,

$$\sqrt{2} \cos \xi, \quad \sqrt{2} \sin \xi, \quad \sqrt{2} \cos 2\xi, \quad \sqrt{2} \sin 2\xi, \quad \cdots$$

is an orthonormal sequence. The function $x = 1$ is a random variable all of whose Fourier coefficients vanish,

$$1 \sim \sum_j 0 = 0.$$

Let $w_1, w_2, \cdots$ be a finite or infinite sequence of random variables, (not all vanishing almost everywhere). It is possible to *orthogonalize* this sequence, that is, to find an orthonormal sequence $y_1, y_2, \cdots$ of random variables with the property that each y_j is a linear combination of w_j's and conversely. Then the linear manifold generated by the y_j's is the same as that generated by the w_j's. The orthogonalization will be sketched briefly. Dropping any w_j which is a linear combination of preceding w_j's, we can suppose that the w_j's are linearly independent. Define $\hat{y}_j$ by

$$\hat{y}_j = \begin{vmatrix} w_j \cdots\cdots\cdots w_1 \\ \mathbf{E}\{w_j\bar{w}_1\} \cdots\cdots \mathbf{E}\{w_1\bar{w}_1\} \\ \cdots\cdots\cdots\cdots\cdots\cdots \\ \mathbf{E}\{w_j\bar{w}_{j-1}\} \cdots\cdots \mathbf{E}\{w_1\bar{w}_{j-1}\} \end{vmatrix}$$

It is easy to verify that $\hat{y}_j$ is a linear combination of $w_1, \cdots, w_j$, orthogonal to $w_1, \cdots, w_{j-1}$ and therefore to $\hat{y}_1, \cdots, \hat{y}_{j-1}$, and that, because $w_1, \cdots, w_j$ are linearly independent, $\mathbf{E}\{|\hat{y}_j|^2\} \neq 0$; we set $y_j = \hat{y}_j/[\mathbf{E}\{|\hat{y}_j|^2\}]^{1/2}$.

Let $w_1, \cdots, w_n$ be n random variables, not all vanishing almost everywhere, and let $\mathfrak{M}$ be the linear manifold generated by the w_j's, consisting of all linear combinations $\sum_1^n c_j w_j$. Orthogonalizing the w_j's, $m \leq n$ random variables $y_1, \cdots, y_m$ are found in $\mathfrak{M}$ forming an orthonormal sequence which generates $\mathfrak{M}$. If the w_j's are linearly independent, $m = n$. Any random variable x in $\mathfrak{M}$ can be written in the form

$$x = \sum_j a_j y_j, \tag{3.2}$$

and we have seen that the a_j's must be the Fourier coefficients of x. If x' is a second random variable in this manifold,

$$x' = \sum_j a_j' y_j,$$

we have

$$\mathbf{E}\{|x - x'|^2\} = \sum_j |a_j - a_j'|^2. \tag{3.3}$$

Thus each random variable x corresponds to a vector $(a_1, \cdots, a_m)$ with m complex components, and distance between random variables is the usual Euclidean distance between the endpoints of the corresponding vectors. Linear combinations of random variables go over into the same linear combinations of the corresponding vectors, so that a linear manifold of random variables corresponds to a linear manifold of vectors, geometrically, to a plane through the origin. Closed linear manifolds correspond to closed linear manifolds. In particular, the manifold $\mathfrak{M}$ itself corresponds to the whole vector space, and is therefore a closed linear manifold. Moreover

$$\mathbf{E}\{x\bar{x}'\} = \sum_j a_j \bar{a}_j', \tag{3.4}$$

so that orthogonal random variables correspond to orthogonal vectors. The random variables $y_1, \cdots, y_m$ correspond to the coordinate vectors

$$(1, 0, \cdots, 0), \cdots, (0, \cdots, 0, 1).$$

If x is any random variable in $\mathfrak{M}$, $a_j = \mathbf{E}\{x\bar{y}_j\}$ is, in terms of the vector picture, the component of the vector corresponding to x on the jth coordinate vector, and $a_j y_j$ goes into the projection of the former vector on the latter. More generally, if $\mathfrak{M}_1$ is the linear manifold of random variables generated by $y_1, \cdots, y_k$, the Fourier series for x in terms of these y_j's goes into the projection of the vector corresponding to x on the plane corresponding to $\mathfrak{M}_1$. This geometric picture clarifies the following remarks, in which we assume $\mathfrak{M}$ and the y_j's as in this paragraph, but also consider random variables not in $\mathfrak{M}$.

If x is any random variable, and if

$$x \sim \sum_j a_j y_j = x_1, \qquad a_j = \mathbf{E}\{x\bar{y}_j\},$$

then $x = x_1$ with probability 1 if x is in $\mathfrak{M}$, and in any case the following relations hold:

$$\mathbf{E}\{|x_1|^2\} = \sum_j |a_j|^2. \tag{a}$$

This is verified by direct calculation.

$$\mathbf{E}\{|x|^2\} = \sum_j |a_j|^2 + \mathbf{E}\{|x - x_1|^2\}. \tag{b}$$

This is verified by evaluating $\mathbf{E}\{|x - x_1|^2\}$, and, in the m-dimensional picture, is simply the Pythagorean theorem. In particular,

$$\mathbf{E}\{|x|^2\} \geq \sum_j |a_j|^2.$$

This inequality is known as *Bessel's inequality*.

(c) The random variables x and x_1 have the same Fourier series. Hence $x - x_1$ is orthogonal to every y_j and consequently to every random variable in $\mathfrak{M}$. Conversely, if x_2 is in $\mathfrak{M}$ and if $x - x_2$ is orthogonal to every y_j, $x_2 = x_1$ with probability 1. In fact, then x and x_2 have the same Fourier coefficients so that x_2 (in $\mathfrak{M}$) is the sum of the same Fourier series as x_1.

(d) The random variable x_1 is in $\mathfrak{M}$ and is closer to x than any other random variable in $\mathfrak{M}$ (counting random variables as the same if they are equal with probability 1). In fact, if x_2 is also in $\mathfrak{M}$, $x_1 - x_2$ is in $\mathfrak{M}$ and therefore orthogonal to $x - x_1$. Hence

$$\begin{aligned}\mathbf{E}\{|x - x_2|^2\} &= \mathbf{E}\{|(x - x_1) + (x_1 - x_2)|^2\} \\ &= \mathbf{E}\{|x - x_1|^2\} + \mathbf{E}\{|x_1 - x_2|^2\} \geq \mathbf{E}\{|x - x_1|^2\}\end{aligned}$$

and there is equality if and only if $x_1 = x_2$ with probability 1.

In II we have called x_1 the wide sense conditional expectation of x relative to $y_1, \cdots, y_m$ (or $w_1, \cdots, w_n$),

$$x_1 = \hat{\mathbf{E}}\{x \mid y_1, \cdots, y_m\} = \hat{\mathbf{E}}\{x \mid w_1, \cdots, w_n\}.$$

We shall sometimes write

$$x_1 = \hat{\mathbf{E}}\{x \mid \mathfrak{M}\},$$

or we may even put for $\mathfrak{M}$ any collection of random variables which generates the closed linear manifold $\mathfrak{M}$.

Two trivial cases have been excluded above. If $\mathfrak{M}$ contains no random variables, or only random variables which vanish almost everywhere, we define

$$\hat{\mathbf{E}}\{x \mid \mathfrak{M}\} = 0.$$

Let $w_1, w_2, \cdots$ be an infinite sequence of random variables, generating the closed linear manifold $\mathfrak{M}$. By orthogonalizing the w_j's an orthonormal sequence $y_1, y_2, \cdots$ can be found, in the linear manifold generated by the w_j's, which also generate $\mathfrak{M}$. We shall suppose that there are infinitely many linearly independent w_j's so that there are infinitely many y_j's; otherwise the present case would be the same as the preceding one. We wish to show that all the considerations of the preceding case carry over to the present one. To show this we remark that, if $x \sim \sum_1^\infty a_j y_j$, Bessel's inequality

$$\mathbf{E}\{|x|^2\} \geq \sum_j |a_j|^2$$

still holds, since it has been seen to hold for any finite number of the y_j's. We shall now show that every random variable x in $\mathfrak{M}$ can be written in the form

$$x = \sum_j a_j y_j, \qquad a_j = \mathbf{E}\{x\bar{y}_j\}, \tag{3.2'}$$

where $\sum_j |a_j|^2 < \infty$ and the sums converges in the mean, that is,

$$x = \operatorname*{l.i.m.}_{n\to\infty} \sum_1^n a_j y_j,$$

and we shall show that conversely, if $\sum_j |a_j|^2 < \infty$, $\sum_j a_j y_j$ converges in the mean to a random variable in $\mathfrak{M}$, whose Fourier coefficients are the a_j's. We show first that, if $\sum_j |a_j|^2 < \infty$, the series $\sum_j a_j y_j$ converges in the mean, by verifying that the Cauchy condition for convergence in the mean is satisfied. In fact,

$$\mathbf{E}\{|\sum_1^n a_j y_j - \sum_1^m a_j y_j|^2\} = \mathbf{E}\{|\sum_{m+1}^n a_j y_j|^2\} = \sum_{m+1}^n |a_j|^2$$

and the last sum goes to 0 when $m, n \to \infty$. The series $\sum_j a_j y_j$ thus has a sum, which is certainly in $\mathfrak{M}$. Conversely, if x is in $\mathfrak{M}$ and if $x \sim \sum_j a_j y_j$, then $\sum_j |a_j|^2 < \infty$ by Bessel's inequality, so that the Fourier series converges (in the mean) to some random variable x_1 in $\mathfrak{M}$. Moreover x_1 has the same Fourier coefficients as x, since, if $n \geq k$,

$$\begin{aligned}\mathbf{E}\{x_1 \bar{y}_k\} &= \mathbf{E}\{(\sum_1^n a_j y_j)\bar{y}_k\} + \mathbf{E}\{(\sum_{n+1}^\infty a_j y_j)\bar{y}_k\} \\ &= a_k + \mathbf{E}\{(\sum_{n+1}^\infty a_j y_j)\bar{y}_k\};\end{aligned}$$

here the last term is 0 since y_k is orthogonal to $y_{n+1}, y_{n+2}, \cdots$ and therefore to the sum $\sum_{n+1}^\infty y_j$ which is in the closed linear manifold generated by these y_j's. We have now proved that $x - x_1$ has Fourier series 0, so that $x - x_1$ is orthogonal to the y_j's, and therefore to $\mathfrak{M}$. In particular, $x - x_1$ is orthogonal to itself,

$$\mathbf{E}\{|x - x_1|^2\} = 0,$$

that is, $x = x_1$ almost everywhere. This proves that if x is in $\mathfrak{M}$ it is the sum of its Fourier series (convergence in the mean).

Thus (3.2) goes over to the case of infinite series, and statements (a), (b), (c), (d) are true also for infinite series; since they really depended on (3.2) their proofs need no change.

We have already defined the symbol $\hat{\mathbf{E}}\{x \mid \mathfrak{M}\}$ if $\mathfrak{M}$ is generated by a finite number of random variables. In the same way, if $\mathfrak{M}$ cannot be so generated but is generated by a denumerably infinite sequence of random variables, we define $\hat{\mathbf{E}}\{x \mid \mathfrak{M}\}$ as x_1, the random variable in $\mathfrak{M}$ closest to x.

(Note that this geometrical property implies that x_1 does not depend on the particular orthonormal sequence used to define it.) The random variable $\hat{\mathbf{E}}\{x \mid \mathfrak{M}\}$ is sometimes called the *projection of x on $\mathfrak{M}$*. We have seen that from the point of view of this book it is the *wide sense conditional expectation of x relative to the variables of $\mathfrak{M}$*. The above treatment is easily extended to include the possibility that $\mathfrak{M}$ may not be generated by any denumerable sequence of random variables. It will sometimes be convenient to replace $\mathfrak{M}$ in the notation $\hat{\mathbf{E}}\{x \mid \mathfrak{M}\}$ by any collection of random variables which generate $\mathfrak{M}$. This agrees with our earlier definition of $\hat{\mathbf{E}}\{x \mid w_1, \cdots, w_n\}$ as a certain linear combination of $w_1, \cdots, w_n$.

Finally we show that $\hat{\mathbf{E}}\{- \mid -\}$ and $\mathbf{E}\{- \mid -\}$ satisfy analogous functional equations. (In the following it is to be understood that the equations are to be true with probability 1, not identically.) In the first place

$$\hat{\mathbf{E}}\{cx \mid \mathfrak{M}\} = c\hat{\mathbf{E}}\{x \mid \mathfrak{M}\}$$

$$\hat{\mathbf{E}}\{x_1 + x_2 \mid \mathfrak{M}\} = \hat{\mathbf{E}}\{x_1 \mid \mathfrak{M}\} + \hat{\mathbf{E}}\{x_2 \mid \mathfrak{M}\}.$$

These equations follow, for example, from the linearity property of Fourier coefficients. In the second place, to prove the analogue of I, (10.8) we prove that if $\mathfrak{M}_1$ and $\mathfrak{M}_2$ are any collections of random variables (with finite second moments), with $\mathfrak{M}_1 \subset \mathfrak{M}_2$,

$$\hat{\mathbf{E}}\left\{\hat{\mathbf{E}}\{x \mid \mathfrak{M}_2\} \mid \mathfrak{M}_1\right\} = \hat{\mathbf{E}}\{x \mid \mathfrak{M}_1\}.$$

To prove this we prove that

$$\hat{\mathbf{E}}\left\{x - \hat{\mathbf{E}}\{x \mid \mathfrak{M}_2\} \mid \mathfrak{M}_1\right\} = 0.$$

Now $x - \hat{\mathbf{E}}\{x \mid \mathfrak{M}_2\}$ is a random variable orthogonal to the closed linear manifold generated by $\mathfrak{M}_2$. Hence its projection on $\mathfrak{M}_1$ is 0, as was to be proved.

Finally we remark that, if $\mathfrak{M}_1$ is orthogonal to $\mathfrak{M}_2$,

$$\hat{\mathbf{E}}\{x \mid \mathfrak{M}_1, \mathfrak{M}_2\} = \hat{\mathbf{E}}\{x \mid \mathfrak{M}_1\} + \hat{\mathbf{E}}\{x \mid \mathfrak{M}_2\}.$$

This fact implies that $\hat{\mathbf{E}}\{x \mid \mathfrak{M}\}$ is unchanged if random variables orthogonal to x are adjoined to $\mathfrak{M}$ or if any collection $\mathfrak{M}_0$ of random variables in $\mathfrak{M}$ is replaced by the single random variable $\hat{\mathbf{E}}\{x \mid \mathfrak{M}_0\}$.

4. Series of orthogonal random variables

THEOREM 4.1 *Let $y_1, y_2, \cdots$ be mutually orthogonal random variables, with $\mathbf{E}\{|y_n|^2\} = \sigma_n^2$. Then the series $\sum_1^\infty y_n$ converges in the mean if and only if $\sum_1^\infty \sigma_n^2 < \infty$.*

The theorem is a consequence of the evaluation

$$\mathbf{E}\{|\sum_1^n y_j - \sum_1^m y_j|^2\} = \mathbf{E}\{|\sum_{m+1}^n y_j|^2\} = \sum_{m+1}^n \sigma_j^2.$$

Alternatively the theorem can be reduced to theorems on Fourier series by the remark that $y_1/\sigma_1, y_2/\sigma_2, \cdots$ is an orthonormal sequence (deleting the terms for which $\sigma_j = 0$). This theorem is the weak sense version of III, Theorem 2.3.

For many purposes it is desirable to have conditions under which $\sum_j y_j$ converges with probability 1. Theorem 4.2 gives one set of such conditions. Its proof depends on a preliminary lemma.

Lemma 4.1 *If $y_1 + \cdots + y_n$ are mutually orthogonal, with*

$$\mathbf{E}\{|y_j|^2\} = \sigma_j^2,$$

then

$$\mathbf{E}\{\operatorname{Max}_{j \leq n} |y_1 + \cdots + y_j|^2\} \leq \left(\frac{\log 4n}{\log 2}\right)^2 (\sigma_1^2 + \cdots + \sigma_n^2). \tag{4.1}$$

Note that if the left side were replaced by $\mathbf{E}\{|y_1 + \cdots + y_n|^2\}$ it would be equal to $(\sigma_1^2 + \cdots + \sigma_n^2)$. The Max is made possible by the factor $(\log 4n/\log 2)^2$. If $n = 1$, (4.1) is true with the factor decreased to 1. If $n > 1$, define the integer $r \geq 0$ by

$$2^r < n \leq 2^{r+1} = N$$

and define $y_j = 0$ for $n < j < N$. Let s be the sum of all partial sums of the form

$$|y_{\alpha+1} + \cdots + y_\beta|^2, \qquad \alpha = \mu 2^\nu \qquad \beta = (\mu + 1)2^\nu$$
$$\nu = 0, \cdots, r + 1 \quad \mu = 0, \cdots, 2^{r+1-\nu} - 1.$$

Then for each ν the sums in s for that value of ν have total expectation $(\sigma_1^2 + \cdots + \sigma_n^2)$, so that

$$\mathbf{E}\{s\} = (r + 2)(\sigma_1^2 + \cdots + \sigma_n^2\}.$$

Now we can separate $y_1 + \cdots + y_j$ into partial sums of the type $y_{\alpha+1} + \cdots + y_\beta$, with α, β as above, say

$$y_1 + \cdots + y_j = \eta_1 + \cdots + \eta_k$$

where η_j contains 2^{r_j} terms, $r + 1 \geq r_1 > r_2 > \cdots > r_k \geq 0$. Then by Schwarz's inequality

$$|y_1 + \cdots + y_j|^2 \leq k \sum_1^k |\eta_i|^2 \leq (r + 2)s.$$

Hence

$$\mathbf{E}\{\operatorname*{Max}_{j\le n} |y_1+\cdots+y_j|^2\} \le (r+2)\mathbf{E}\{s\} \le (r+2)^2(\sigma_1^2+\cdots+\sigma_n^2)$$
$$< \left(\frac{\log n}{\log 2}+2\right)^2(\sigma_1^2+\cdots+\sigma_n^2),$$

as was to be proved.

THEOREM 4.2 *Let $y_1, y_2, \cdots$ be mutually orthogonal random variables, with $\mathbf{E}\{|y_n|^2\} = \sigma_n^2$. Then, if $\sum_1^\infty \sigma_n^2 \log^2 n < \infty$, the series $\sum_1^\infty y_n$ converges in the mean and with probability* 1.

Convergence in the mean is assured by Theorem 4.1. Let the sum be x and let $x_n = \sum_{j=1}^{n} y_j$. Then

$$\mathbf{E}\{|x-x_n|^2\} = \sum_{j=n+1}^{\infty} \sigma_j^2 \le \frac{c}{\log^2 n}, \qquad c = \sum_{j=1}^{\infty} \sigma_j^2 \log^2 j.$$

It follows that, if $n = 2^r$,

$$\sum_{r=1}^{\infty} \mathbf{P}\{|x(\omega)-x_n(\omega)| \ge (\log n)^{-1/3}\} \le \sum_{r=1}^{\infty} \frac{c(r\log 2)^{2/3}}{(r\log 2)^2} < \infty$$

so that (Borel-Cantelli lemma)

$$|x(\omega)-x_n(\omega)| < (\log n)^{-1/3} \qquad n = 2^r \tag{4.2}$$

for sufficiently large n, with probability 1. Moreover, according to Lemma 4.1,

$$\mathbf{E}\{\operatorname*{Max}_{n<m<2n} |x_m - x_n|^2\} \le \left(\frac{\log 4(n-1)}{\log 2}\right)^2 \sum_{j=n+1}^{2n-1} \sigma_j^2 \tag{4.3}$$
$$\le \left(\frac{\log 4(n-1)}{\log 2 \log(n+1)}\right)^2 \sum_{j=n+1}^{2n-1} \sigma_j^2 \log^2 j$$
$$< \frac{4}{\log^2 2} \sum_{j=n+1}^{2n-1} \sigma_j^2 \log^2 j = \varepsilon_n,$$

for $n > 1$. Here, if $n = 2^r$, $\sum_{r=1}^{\infty} \varepsilon_n < \infty$. Choose $\delta_2, \delta_4, \cdots$ so that

$$\delta_n > 0, \qquad \lim_{n\to\infty} \delta_n = 0, \qquad \sum_{r=1}^{\infty} \frac{\varepsilon_n}{\delta_n} < \infty \quad (n = 2^r).$$

Then (4.3) yields

$$\sum_{r=1}^{\infty} \mathbf{P}\{\operatorname*{Max}_{n<m<2n} |x_m(\omega)-x_n(\omega)| \ge \delta_n^{1/2}\} \le \sum_{r=1}^{\infty} \frac{\varepsilon_n}{\delta_n} < \infty.$$

Hence (Borel-Cantelli lemma again)

$$(4.4)\qquad |x_m(\omega) - x_n(\omega)| < \delta_n^{1/2}, \qquad n = 2^r < m < 2n$$

for sufficiently large n, with probability 1. The combination of (4.2) with (4.4) gives convergence of $\sum_{n=1}^{\infty} y_n$ with probability 1.

Note that if the y_n's are independent and have zero expectations the condition of Theorem 4.2 can be weakened to $\sum_1^\infty \sigma_n^2 < \infty$ (III, Theorem 2.3).

5. The law of large numbers

We shall see in this section how little the quantitative hypotheses of some of the theorems on the law of large numbers for mutually independent random variables (III §3) need be strengthened if only mutual orthogonality is presupposed.

THEOREM 5.1 *If $y_1, y_2, \cdots$ are mutually orthogonal random variables, with $\mathbf{E}\{|y_j|^2\} = \sigma_j^2$,*

$$(5.1)\qquad \operatorname*{l.i.m.}_{n\to\infty} \frac{y_1 + \cdots + y_n}{n} = 0$$

if and only if

$$(5.2)\qquad \lim_{n\to\infty} \frac{\sigma_1^2 + \cdots + \sigma_n^2}{n^2} = 0.$$

The theorem is obvious from the evaluation

$$\mathbf{E}\left\{\left|\frac{y_1 + \cdots + y_n}{n}\right|^2\right\} = \frac{\sigma_1^2 + \cdots + \sigma_n^2}{n^2};$$

it is the weak sense version of III, Theorem 3.2. The part of the latter theorem connecting III (3.8) and (3.9) is still true, but is of less interest when the y_j's are supposed merely mutually orthogonal.

According to III, Theorem 3.4, if the y_j's have zero expectations and are mutually independent, and if $\sum_1^\infty \sigma_n^2/n^2 < \infty$, the strong law of large numbers holds. Under the present hypothesis of mutual orthogonality somewhat more severe restrictions must be imposed on the σ_n^2's.

THEOREM 5.2 *If $y_1, y_2, \cdots$ are mutually orthogonal random variables, with $\mathbf{E}\{|y_j|^2\} = \sigma_j^2$, and if*

$$\sum_1^\infty \frac{\sigma_n^2}{n^2} \log^2 n < \infty,$$

then

$$(5.3)\qquad \operatorname*{l.i.m.}_{n\to\infty} \frac{y_1 + \cdots + y_n}{n} = \lim_{n\to\infty} \frac{y_1 + \cdots + y_n}{n} = 0$$

with probability 1.

As in the proof of Theorem 3.4 of III, it is sufficient to prove that $\sum_1^n \frac{y_n}{n}$ converges in the mean and with probability 1, and this convergence is assured by Theorem 4.2.

This theorem is less well known than its counterpart for independent random variables. We therefore point out a special case: if $y_1, y_2, \cdots$ are random variables, with

$$\mathbf{E}\{y_n\} = 0, \qquad \mathbf{E}\{|y_n|^2\} \leq K, \qquad n \geq 1$$

the strong law of large numbers (5.3) holds not only if the variables are mutually independent (III, Theorem 3.4) but even if they are only mutually uncorrelated (that is, mutually orthogonal since the expectations vanish). Note, however, that the condition that $\mathbf{E}\{y_n\} = 0$ does not appear in Theorem 5.2 and is quite irrelevant. *Orthogonality* is the essential condition and $\mathbf{E}\{y_n\} = 0$ is only important in that, together with zero correlation between pairs of y_j's, it implies orthogonality of the pairs.

6. Series $\sum_0^\infty a_j e^{2\pi ij\lambda}$ of power series type

In the present section we suppose that probabilities are defined as length (Lebesgue measure) on the interval $-\frac{1}{2} \leq \lambda \leq \frac{1}{2}$, so that

$$1, e^{2\pi i\lambda}, e^{-2\pi i\lambda}, \cdots$$

is an orthonormal sequence. In particular we are interested in characterization of those functions of λ whose Fourier series contain no negative powers of $e^{2\pi i\lambda}$. The results are sketched, and will be used only in XII, in the theory of least squares prediction.

Note that, since the orthogonal sequence under discussion consists of bounded functions, any integrable function has a Fourier series, whether the second moment exists or not.

The following theorem is stated for ready reference. Its proof will be omitted.

THEOREM 6.1 *If* $f(z) = \sum_0^\infty \gamma_n z^n$ *is analytic for* $|z| < 1$ *then* $\int_{-1/2}^{1/2} |f(re^{2\pi i\lambda})|\,d\lambda$ *is monotone non-decreasing in r. Suppose that this integral has a finite limit when* $r \to 1$. *Then*

$$\lim_{r\to 1} f(re^{2\pi i\lambda}) = f(e^{2\pi i\lambda}) \tag{6.1}$$

exists for almost all values of λ. *Moreover* $f(e^{2\pi i\lambda})$ *is integrable,*

$$\lim_{r\to 1} \int_{-1/2}^{1/2} |f(e^{2\pi i\lambda}) - f(re^{2\pi i\lambda})|\,d\lambda = 0 \tag{6.2}$$

and

$$f(e^{2\pi i\lambda}) \sim \sum_{0}^{\infty} \gamma_n e^{2\pi i n\lambda}; \tag{6.3}$$

f(z) is given by the Cauchy integral formula

$$f(z) = \frac{1}{2\pi i} \int_{|\zeta|=1} \frac{f(\zeta)\, d\zeta}{\zeta - z}. \tag{6.4}$$

Conversely, if $f(e^{2\pi i\lambda})$ is any integrable function of λ, whose Fourier series has the form (6.3), *the function $f(z) = \sum_{0}^{\infty} \gamma_n z^n$ defined for $|z| < 1$ has all the above stated properties. Finally, in this case, $f(e^{2\pi i\lambda}) = 0$ only on a set of Lebesgue measure* 0, *or on a set of Lebesgue measure* 1 *(when $\gamma_0 = \gamma_1 = \cdots = 0$).*

If $g(e^{2\pi i\lambda})$ is a real integrable function of λ, with Fourier series

$$g(e^{2\pi i\lambda}) \sim \sum_{-\infty}^{\infty} \gamma_n e^{2\pi i n\lambda}, \tag{6.3'}$$

then $g(re^{2\pi i\lambda})$, given by

$$\begin{aligned} g(re^{2\pi i\lambda}) &= \int_{-1/2}^{1/2} g(e^{2\pi i\mu}) \frac{1 - r^2}{1 - 2r\cos 2\pi(\lambda - \mu) + r^2}\, d\mu \\ &= \sum_{-\infty}^{\infty} \gamma_n r^{|n|} e^{2\pi i n\lambda}, \end{aligned} \tag{6.4'}$$

defines a harmonic function of $re^{2\pi i\lambda}$ for $r < 1$, and

$$\lim_{r\to 1} g(re^{2\pi i\lambda}) = g(e^{2\pi i\lambda}) \tag{6.1'}$$

for almost all values of λ.

The next theorem provides the factoring of a spectral density, as required for prediction theory.

THEOREM 6.2 *If $f(\lambda)$ (≥ 0) is Lebesgue integrable over the interval $-\frac{1}{2} \leq \lambda \leq \frac{1}{2}$, if $f = 0$ at most on a set of measure* 0, *and if*

$$\int_{-1/2}^{1/2} \log f(\lambda)\, d\lambda > -\infty, \tag{6.5}$$

there is a uniquely determined sequence of numbers $\gamma_0, \gamma_1, \cdots$ with γ_0 real and positive, $\sum_{0}^{\infty} |\gamma_n^2| < \infty$,

$$\sum_{0}^{\infty} \gamma_n z^n \neq 0, \qquad |z| < 1, \tag{6.6}$$

$$\log \gamma_0 = \tfrac{1}{2} \int_{-1/2}^{1/2} \log f(\lambda)\, d\lambda, \tag{6.7}$$

and

$$f(\lambda) = |\sum_{0}^{\infty} \gamma_n e^{2\pi i n\lambda}|^2. \tag{6.8}$$

(The series converges in the mean.) The γ_n*'s are determined by the relations*

$$\tfrac{1}{2} \log f(\lambda) \sim \sum_{-\infty}^{\infty} a_n e^{2\pi i n\lambda}$$

$$\sum_{0}^{\infty} \gamma_n z^n = e^{a_0 + 2\sum_{1}^{\infty} a_n z^n} \qquad |z| < 1. \tag{6.9}$$

Conversely suppose that $\gamma_0, \gamma_1, \cdots$ *are any numbers with* $0 < \sum_{0}^{\infty} |\gamma_n|^2 < \infty$, *and that* (6.7) *is true. Then, if* $f(\lambda)$ *is defined by* (6.8), $f = 0$ *at most on a set of measure* 0, (6.5) *is true, and in fact*

$$\log |\gamma_0| \le \tfrac{1}{2} \int_{-1/2}^{1/2} \log f(\lambda)\, d\lambda. \tag{6.7'}$$

Note that, for $f > 1$, $\log f \le f$. Hence if f is integrable $\log f$ is integrable unless f is too small, that is, the integral of $\log f$ is finite or $-\infty$.

If $|f_1|^2 = f$, f_1 has a Fourier series $\sum_{-\infty}^{\infty} \gamma_n e^{2\pi n i\lambda}$, with $\sum_{-\infty}^{\infty} |\gamma_n|^2 < \infty$. Thus

$$f(\lambda) = |\sum_{-\infty}^{\infty} \gamma_n e^{2\pi i n\lambda}|^2.$$

The point of the theorem is that f_1 can be chosen so that $\gamma_n = 0$ for $n < 0$ if (6.5) is true.

The proof of the theorem relies on Theorem 6.1. If (6.5) is true, $\log \sqrt{f}$ has a Fourier series,

$$\log \sqrt{f(\lambda)} \sim \sum_{-\infty}^{\infty} a_n e^{2\pi i n\lambda}, \qquad a_{-n} = \bar{a}_n.$$

Let g be the function

$$g(z) = a_0 + 2 \sum_{1}^{\infty} a_n z^n,$$

analytic for $|z| < 1$, whose real part

$$\Re g(re^{2\pi i\lambda}) = \sum_{-\infty}^{\infty} a_n r^{|n|} e^{2\pi n i\lambda}$$

is the harmonic function with boundary function $\log \sqrt{f(\lambda)}$ (see Theorem 6.1). Finally define $\gamma(z) = e^{g(z)}$. Then $\gamma(z) \neq 0$ for $|z| < 1$ and

$$\text{(6.10)} \qquad \lim_{r \to 1} |\gamma(re^{2\pi i\lambda})| = \lim_{r \to 1} e^{\Re g(re^{2\pi i\lambda})} = e^{\log \sqrt{f(\lambda)}} = \sqrt{f(\lambda)}$$

for almost all values of λ. The function $\gamma(z)$ has a power series development

$$\text{(6.11)} \qquad \gamma(z) = \sum_0^\infty \gamma_n z^n = e^{a_0 + 2\sum_1^\infty a_n z^n}, \qquad |z| < 1.$$

To show that this can be extended to $|z| = 1$, in the sense that $\gamma(e^{2\pi i\lambda})$ is defined as the radial limit of $\gamma(z)$ and has the Fourier series $\sum_0^\infty \gamma_n e^{2\pi in\lambda}$, we show that Theorem 6.1 is applicable:

$$\text{(6.12)} \quad \int_{-1/2}^{1/2} |\gamma(re^{2\pi i\lambda})| \, d\lambda = \int_{-1/2}^{1/2} e^{\Re g(re^{2\pi i\lambda})} \, d\lambda$$

$$= \int_{-1/2}^{1/2} e^{\int_{-1/2}^{1/2} \log \sqrt{f(\mu)} \frac{1-r^2}{1-2r\cos 2\pi(\lambda-\mu)+r^2} d\mu} \, d\lambda$$

$$\leq \int_{-1/2}^{1/2} d\lambda \int_{-1/2}^{1/2} \sqrt{f(\mu)} \frac{1-r^2}{1-2r\cos 2\pi(\lambda-\mu)+r^2} \, d\mu$$

$$= \int_{-1/2}^{1/2} \sqrt{f(\mu)} \, d\mu \int_{-1/2}^{1/2} \frac{1-r^2}{1-2r\cos 2\pi(\lambda-\mu)+r^2} \, d\lambda$$

$$= \int_{-1/2}^{1/2} \sqrt{f(\mu)} \, d\mu.$$

Thus the integral on the left remains bounded when $r \to 1$, which means according to Theorem 6.1 that $\gamma(z)$ has a boundary function $\lim_{r \to 1} \gamma(re^{2\pi i\lambda}) = \gamma(e^{2\pi i\lambda})$ for almost all values of λ, and that $\gamma(e^{2\pi i\lambda})$ has the Fourier series $\sum_0^\infty \gamma_n e^{2\pi in\lambda}$. Since, as we have already seen, $|\gamma(e^{2\pi i\lambda})| = \sqrt{f(\lambda)}$, it follows that $|\gamma(e^{2\pi i\lambda})|^2$ is integrable, so that $\sum_0^\infty |\gamma_n|^2 < \infty$, and thus we have obtained the desired representation [in which (6.7) follows from (6.9) with $z = 0$]

$$f(\lambda) = |\gamma(e^{2\pi i\lambda})|^2 = |\sum_0^\infty \gamma_n e^{2\pi ni\lambda}|^2.$$

The uniqueness of the γ_n's will be proved after the proof of the converse.

Conversely, suppose that f is given by (6.8), where $0 < \sum_0^\infty |\gamma_n|^2 < \infty$, and suppose that $\gamma(z) = \sum_0^\infty \gamma_n z^n$ does not vanish for $|z| < 1$. Then $\log |\gamma(z)|$ is harmonic and

$$\log |\gamma(0)| = \log |\gamma_0| = \int_{-1/2}^{1/2} \log |\gamma(re^{2\pi i\lambda})| \, d\lambda, \qquad r < 1. \tag{6.13}$$

Now according to Theorem 6.1 $\gamma(e^{2\pi i\lambda})$ is defined for almost all λ in terms of radial approach to $|z| = 1$, and is the function with Fourier series $\sum_0^\infty \gamma_n e^{2\pi in\lambda}$. Let $h_b(x) = \text{Max}\,[x, b]$. Then

$$\lim_{r\to 1} \int_{-1/2}^{1/2} h_b\,[\log |\gamma(re^{2\pi i\lambda})|]\, d\lambda = \int_{-1/2}^{1/2} h_b\,[\log |\gamma(e^{2\pi i\lambda})|]\, d\lambda,$$

because the square of the integrand on the left is less than $|\gamma|$ for large values of $|\gamma|$, and the integral of $|\gamma|$ is bounded when $r \to 1$. Hence

$$\log |\gamma_0| \leq \int_{-1/2}^{1/2} h_b\,[\log |\gamma(e^{2\pi i\lambda})|]\, d\lambda.$$

When $b \to -\infty$ in this inequality, it yields (6.7′). Note that a trivial variation of this discussion shows that $\log |\gamma(z)| \leq a(z)$, where $a(z)$ is the value at z of the harmonic function determined by the Poisson integral with boundary function $\log |\gamma(e^{2\pi i\lambda})|$. The discussion treated the case $z = 0$. Since the difference $a(z) - \log |\gamma(z)|$ defines a non-negative harmonic function, this difference either never vanishes or vanishes identically. Thus, if (6.7′) is known to be true, so that this difference vanishes when $z = 0$, we have now shown that $\log |\gamma(re^{2\pi i\lambda})|$ can be represented by the Poisson integral with boundary function $\log |\gamma(e^{2\pi i\lambda})|$. This fact is used in the following uniqueness proof.

To prove the uniqueness statement of the first part of the theorem, we must show that the hypotheses that $f(\lambda)$ is given by (6.8) with $\gamma_n = \gamma_n'$

and also with $\gamma_n = \gamma_n''$, and that (6.6) and (6.7) are satisfied by both representations of $f(\lambda)$ imply that $\gamma_n' = c\gamma_n''$, $n \geq 0$, where $|c| = 1$. Define

$$\gamma_1(z) = \sum_0^\infty \gamma_n' z^n$$

$$\gamma_2(z) = \sum_0^\infty \gamma_n'' z^n.$$

It will be sufficient to show that these functions are proportional. Now $\log \gamma_1(z)$, $\log \gamma_2(z)$ are analytic for $|z| < 1$, and we have seen above that $\log |\gamma_1(z)|$ is determined by its boundary function $\log |\sum_0^\infty \gamma_n' e^{2\pi i n\lambda}|$; that is, $\log |\gamma_1(z)| = \log |\gamma_2(z)|$. Then $\log \gamma_1(z)$ and $\log \gamma_2(z)$ can differ at most by an imaginary constant, $i\alpha$, so that

$$\gamma_2(z) = e^{i\alpha}\gamma_1(z),$$

as was to be proved.

7. Martingales in the wide sense

Martingales (wide sense) were defined in II §7 as processes whose random variables $\{x_t\}$ have finite second moments and satisfy the equation

$$\hat{\mathbf{E}}\{x_{t_{n+1}} \mid x_{t_1}, \cdots, x_{t_n}\} = x_{t_n} \tag{7.1}$$

(with probability 1) whenever $t_1 < \cdots < t_{n+1}$; here n is an arbitrary positive integer. This is true if and only if $x_{t_{n+1}} - x_{t_n}$ is orthogonal to x_{t_j} for $j \leq n$, and, since these t_j's are arbitrary, $x_{t_{n+1}} - x_{t_n}$ must be orthogonal to every x_t for $t \leq t_n$. In other words, (7.1) is equivalent to the condition

$$\hat{\mathbf{E}}\{x_t \mid x_r, r \leq s\} = x_s \tag{7.1$'$}$$

with probability 1 whenever $s < t$. In the present section t will range through the integers, or some of the integers, sometimes including $\pm\infty$. If t is integral it is sufficient if (7.1′) is replaced by

$$\hat{\mathbf{E}}\{x_n \mid x_1, \cdots, x_{n-1}\} = x_{n-1}. \tag{7.1$''$}$$

Note that the defining properties (7.1) and (7.1′) are meaningful whenever the range of t is an ordered set, and that any set of random variables of a martingale (ordered as before) also constitutes a martingale.

THEOREM 7.1 *If the random variables $\{x_t\}$ constitute a martingale (wide sense) then, if $t_1 < t_2$,*

$$\mathbf{E}\{|x_{t_1}|^2\} \leq \mathbf{E}\{|x_{t_2}|^2\} \tag{7.2}$$

and there is equality only if $x_{t_1} = x_{t_2}$ with probability 1.

In fact, according to the martingale property $x_{t_2} - x_{t_1}$ is orthogonal to x_{t_1}, so that

$$\begin{aligned}\mathbf{E}\{|x_{t_2}|\}^2 &= \mathbf{E}\{|(x_{t_2} - x_{t_1}) + x_{t_1}|^2\}\\ &= \mathbf{E}\{|x_{t_2} - x_{t_1}|^2\} + \mathbf{E}\{|x_{t_1}|^2\} \geq \mathbf{E}\{|x_{t_1}|^2\}\end{aligned}$$

and there is equality only if $x_{t_2} = x_{t_1}$ with probability 1.

If the random variables $x_1, x_2, \cdots$ constitute a martingale, and if y_j is defined by

$$\begin{aligned} y_j &= x_j - x_{j-1} \qquad j > 1\\ &= x_1 \qquad\qquad\quad j = 1\end{aligned}$$

then the martingale property implies that the y_j's are mutually orthogonal, and

$$x_n = \sum_1^n y_j.$$

Conversely, if the y_j's are any mutually orthogonal random variables and if x_n is defined in this way, the x_n process is a wide sense martingale. This would make it appear that the whole subject of martingales in the wide sense might well be forgotten since the variables are simply the partial sums of a series of orthogonal random variables. However, the natural way they arise in Hilbert space considerations and the light they shed on problems of least square approximation (XII) and on martingales in the strict sense (VII) makes it worth while to study their properties briefly. The most interesting case is the continuous parameter case (IX).

THEOREM 7.2 *Let $x_1, x_2, \cdots$ be random variables constituting a martingale (wide sense). Then*

$$\mathbf{E}\{|x_1|^2\} \leq \mathbf{E}\{|x_2|^2\} \leq \cdots \tag{7.3}$$

If $\lim_{n\to\infty} \mathbf{E}\{|x_n|^2\} = l < \infty$, $\operatorname*{l.i.m.}_{n\to\infty} x_n = x_\infty$ *exists and the random variables* $x_1, x_2, \cdots, x_\infty$ *constitute a martingale (wide sense).*

The strict sense version of this theorem is VII, Theorem 4.1 (i), (ii), (iii). To prove the theorem we need only remark that, if x_n is written in the above form with mutually orthogonal y_j's,

$$\mathbf{E}\{|x_n|^2\} = \sum_1^n \mathbf{E}\{|y_j|^2\}.$$

Hence (Theorem 4.1), if $l < \infty$, $\operatorname*{l.i.m.}_{n\to\infty} x_n = x_\infty$ exists. To prove the last statement of the theorem we must prove that $x_\infty - x_n$ is orthogonal to x_j for $j \leq n$. The difference $x_m - x_n$ $(m > n)$ has this property, and when $m \to \infty$ we obtain the desired result.

THEOREM 7.3 *Let* $\cdots, x_{-2}, x_{-1}$ *be random variables constituting a martingale (wide sense), let* $\mathfrak{M}_n$ *be the closed linear manifold generated by* $\cdots, x_{n-1}, x_n$, *and let* $\mathfrak{M} = \bigcap_{-\infty}^{-1} \mathfrak{M}_n$. *Then*

$$\text{(7.4)} \qquad \underset{n\to\infty}{\text{l.i.m.}}\ x_n = \hat{\mathbf{E}}\{x_{-1} \mid \mathfrak{M}\} \qquad (= x_{-\infty})$$

and the random variables $x_{-\infty}, \cdots, x_{-2}, x_{-1}$ *constitute a martingale (wide sense).*

The strict sense version of this theorem is VII, Theorem 4.2. Let $y_n = x_n - x_{n-1}$. Then the y_j's are mutually orthogonal and (Theorem 4.1) the existence of the limit in the mean $x_{-\infty}$ will follow from the convergence of the series $\sum_1^\infty \mathbf{E}\{|y_{-j}|^2\}$. This convergence is a consequence of

$$\text{(7.5)} \qquad \begin{aligned} \sum_1^n \mathbf{E}\{|y_{-j}|^2\} &= \mathbf{E}\{|x_{-1} - x_{-n-1}|^2\} \\ &\leq 2\mathbf{E}\{|x_{-1}|^2\} + 2\mathbf{E}\{|x_{-n-1}|^2\} \\ &\leq 4\mathbf{E}\{|x_{-1}|^2\}. \end{aligned}$$

To show that $x_{-\infty}$ is the projection indicated in (7.4) we note that the projection is characterized by two conditions: it is in $\mathfrak{M}$ (as is $x_{-\infty}$ since $x_{-\infty}$ is in every $\mathfrak{M}_n$), and $x_{-1} - \hat{\mathbf{E}}\{x_{-1} \mid \mathfrak{M}\}$ is orthogonal to $\mathfrak{M}$. To show that $x_{-1} - x_{-\infty}$ is orthogonal to $\mathfrak{M}$ we need only remark that $x_{-1} - x_n$ is orthogonal to $\mathfrak{M}_n \supset \mathfrak{M}$ and hence to $\mathfrak{M}$; going to the limit, $n \to -\infty$ gives the desired result. (Of course, $x_{-\infty} = \hat{\mathbf{E}}\{x_k \mid \mathfrak{M}\}$ for every k.) To prove the last statement of the theorem we need only prove that if $m < n$ then

$$\hat{\mathbf{E}}\{x_n \mid x_{-\infty}, \cdots, x_m\} = x_m$$

with probability 1, that is, $x_n - x_m$ is orthogonal to $x_{-\infty}, \cdots, x_m$. By hypothesis $x_n - x_m$ is orthogonal to the random variables $\cdots, x_{m-1}, x_m$ and to the closed linear manifold $\mathfrak{M}_m$ they generate, which includes $x_{-\infty}$, and the proof is now complete.

According to Theorem 7.3 if the random variables $\cdots, x_{-1}, x_0, \cdots$ constitute a martingale (wide sense) they always can be written in the form

$$x_n = \sum_{j \leq n} y_j,$$

where the y_j's are mutually orthogonal, $j \geq -\infty$, and the series converges in the mean. In fact we can take

$$y_{-\infty} = x_{-\infty}, \qquad y_j = x_j - x_{j-1}, \qquad j > -\infty.$$

Conversely any sums of this form define a martingale (wide sense).

THEOREM 7.4 *Let z be a random variable with a finite second moment, and let* $\cdots \subset \mathfrak{M}_1 \subset \mathfrak{M}_2 \subset \cdots$ *be closed linear manifolds of random variables. Let* $\mathfrak{M}_{-\infty} = \bigcap_n \mathfrak{M}_n$ *and let* $\mathfrak{M}_\infty$ *be the closed linear manifold generated by the random variables in* $\bigcup_n \mathfrak{M}_n$. *Then the random variables*

$$\hat{\mathbf{E}}\{z \mid \mathfrak{M}_{-\infty}\}, \cdots, \hat{\mathbf{E}}\{z \mid \mathfrak{M}_1\}, \hat{\mathbf{E}}\{z \mid \mathfrak{M}_2\}, \cdots, \hat{\mathbf{E}}\{z \mid \mathfrak{M}_\infty\}$$

constitute a martingale (*wide sense*), *and*

$$(7.6) \qquad \begin{aligned} \underset{n\to-\infty}{\text{l.i.m.}}\ \hat{\mathbf{E}}\{z \mid \mathfrak{M}_n\} &= \hat{\mathbf{E}}\{z \mid \mathfrak{M}_{-\infty}\} \\ \underset{n\to\infty}{\text{l.i.m.}}\ \hat{\mathbf{E}}\{z \mid \mathfrak{M}_n\} &= \hat{\mathbf{E}}\{z \mid \mathfrak{M}_\infty\} \end{aligned}$$

with probability 1.

Define

$$x_n = \hat{\mathbf{E}}\{z \mid \mathfrak{M}_n\}, \qquad -\infty \leq n \leq \infty.$$

Then x_n is a random variable in the manifold $\mathfrak{M}_n$, and, if $m < n$, $x_n - x_m$ is orthogonal to $\mathfrak{M}_m$, and therefore to every x_j with $j \leq m$. In other words

$$x_m = \hat{\mathbf{E}}\{x_n \mid \cdots, x_{m-1}, x_m\}$$

with probability 1, that is, the x_j's constitute a martingale in the wide sense. By Theorem 7.3 the first limit in (7.6) exists. To show that the limit x is the projection $x_{-\infty}$ we note that the projection $x_{-\infty}$ is characterized by two conditions: it is in $\mathfrak{M}_{-\infty}$ (as is x, since $x_n \in \mathfrak{M}_n$), and $z - x_{-\infty}$ is orthogonal to $\mathfrak{M}_{-\infty}$ (as is $z - x$ since $z - x_n$ is orthogonal to $\mathfrak{M}_n$ and therefore to $\mathfrak{M}_{-\infty}$). This completes the proof of the first line in (7.6). To prove the second line we note that, by Theorem 7.1,

$$\cdots \leq \hat{\mathbf{E}}\{|x_1|^2\} \leq \cdots \leq \hat{\mathbf{E}}\{|x_\infty|^2\},$$

so that the mean limit in the second line exists, by Theorem 7.2. Just as in the preceding proof, the limit is then shown to have the properties characterizing the projection x_∞.

COROLLARY 1 *Let* $z, y_1, y_2, \cdots$ *be any random variables with finite second moments. Then if* $\mathfrak{N}_n$ *is the closed linear manifold generated by the* y_j's, *with* $j \geq n$,

$$(7.6') \qquad \begin{aligned} \underset{n\to\infty}{\text{l.i.m.}}\ \hat{\mathbf{E}}\{z \mid y_n, y_{n+1}, \cdots\} &= \hat{\mathbf{E}}\{z \mid \bigcap_1^\infty \mathfrak{N}_n\} \\ \underset{n\to\infty}{\text{l.i.m.}}\ \hat{\mathbf{E}}\{z \mid y_1, \cdots, y_n\} &= \hat{\mathbf{E}}\{z \mid y_1, y_2, \cdots\} \end{aligned}$$

with probability 1. *In particular if z is a random variable in the closed linear manifold generated by the* y_j's, *the second limit can be replaced by* z.

To reduce the first of these equations to the first in (7.6) identify $\mathfrak{N}_n$ with $\mathfrak{M}_{-n}$. To reduce the second to the second in (7.6) identify with $\mathfrak{M}_n$ in (7.6) the closed linear manifold generated by $y_1, \cdots, y_n$.

The strict sense versions of Theorem 7.4 and its corollary are VII, Theorem 4.3, and its corollary. The study of projections that we have made in this section is really a study of one aspect of the geometry of Hilbert space. Since we have proved all the results to be used, we stop the study at this point.

Equation (7.6′) is of fundamental importance. As one application suppose that z is actually in the closed linear manifold generated by the y_j's. Then, according to (7.6′), if one approximates z in the sense of least squares, in terms of linear combinations of $y_1, \cdots, y_n$, the (least squares) error approaches 0 when $n \to \infty$.

As an instructive example we apply (7.6′) to a very special case. Suppose that the sequence $\{y_n'\}$ is an orthonormal sequence of random variables. We apply (7.6′) with the following identifications:

$$z = y_1'$$

$$y_n = y_1' + \cdots + y_n'.$$

Now the closed linear manifold $\mathfrak{N}_n$ generated by $y_n, y_{n+1}, \cdots$ is the same as that generated by $y_1' + \cdots + y_n', y_{n+1}', y_{n+2}', \cdots$. These random variables are mutually orthogonal, so any random variable in $\mathfrak{N}_n$ can be written in the form

$$a(y_1' + \cdots + y_n') + \sum_{n+1}^{\infty} a_j y_j', \qquad \sum_{n+1}^{\infty} |a_j|^2 < \infty,$$

where the series converges in the mean. If the random variable is in $\mathfrak{N} = \bigcap_1^{\infty} \mathfrak{N}_n$ it is in every $\mathfrak{N}_n$, so we must have $a_j = a$ for all j, and this is incompatible with the convergence of $\sum_j |a_j|^2$ unless the a_j's vanish. In other words, $\mathfrak{N}$ contains only the null random variable. We therefore deduce from (7.6′) that

$$\operatorname*{l.i.m.}_{n\to\infty} \hat{\mathbf{E}}\{y_1' \mid \mathfrak{N}_n\} = \operatorname*{l.i.m.}_{n\to\infty} \hat{\mathbf{E}}\{y_1' \mid y_n, y_{n+1}, \cdots\} = 0. \tag{7.7}$$

Now

$$\begin{aligned}\hat{\mathbf{E}}\{y_1' \mid \mathfrak{N}_n\} &= \hat{\mathbf{E}}\{y_1' \mid y_1' + \cdots + y_n', y_{n+1}', y_{n+2}', \cdots\} \\ &= \hat{\mathbf{E}}\{y_1' \mid y_1' + \cdots + y_n'\}\end{aligned}$$

because $y_{n+1}', y_{n+2}', \cdots$ are orthogonal to y_1' and $y_1' + \cdots + y_n'$ and can therefore be ignored in the calculation. Finally by symmetry

$$\begin{aligned}\hat{\mathbf{E}}\{y_1' \mid y_1' + \cdots + y_n'\} &= \hat{\mathbf{E}}\{y_j' \mid y_1' + \cdots + y_n'\} \qquad j \leq n \\ &= \frac{1}{n}\sum_1^n \hat{\mathbf{E}}\{y_j' \mid y_1' + \cdots + y_n'\} \\ &= \hat{\mathbf{E}}\left\{\frac{1}{n}\sum_{j=1}^n y_j' \mid y_1' + \cdots + y_n'\right\} \\ &= \frac{y_1' + \cdots + y_n'}{n}.\end{aligned}$$

This projection can, of course, also be evaluated by a simple evaluation of the appropriate Fourier coefficients. Combining this evaluation with (7.7), we find that

$$\operatorname*{l.i.m.}_{n\to\infty} \frac{y_1' + \cdots + y_n'}{n} = 0,$$

a result that is hardly surprising since

$$\mathbf{E}\left\{\left|\frac{y_1' + \cdots + y_n'}{n}\right|^2\right\} = \frac{1}{n},$$

and the limit equation simply states that this mean square expectation goes to 0 when $n \to \infty$. This rather ridiculous proof of a trivial case of the law of large numbers has been given for two reasons. In the first place it illustrates the application of the limit theorems proved in this section. In the second place the strong sense version of this theorem is an important and non-trivial theorem (III, Theorem 5.1), an important case of the strong law of large numbers for strictly stationary processes, which can be proved by giving the strong sense version of the above proof. (See VII §6.)

CHAPTER V

Markov Processes—Discrete Parameter

1. Markov chains—definitions

A Markov chain is defined as a Markov process (II §6) whose random variables can (with probability 1) only assume values in a certain finite or denumerably infinite set. This set is usually taken, for convenience, to be the integers $1, \cdots, N$ (finite case) or the integers $1, 2, \cdots$ (infinite case). We shall assume that these choices have been made. Physically one speaks of a system which evolves through numbered states in accordance with probabilistic laws satisfying the Markov hypothesis. Let the random variables of the chain be $x_1, x_2, \cdots$. If $\mathbf{P}\{x_m(\omega) = i\} > 0$, define ${}_mp_{ij}$ by

$$ {}_mp_{ij} = \mathbf{P}\{x_{m+1}(\omega) = j \mid x_m(\omega) = i\} = \frac{\mathbf{P}\{x_m(\omega) = i, x_{m+1}(\omega) = j\}}{\mathbf{P}\{x_m(\omega) = i\}}, $$

and let ${}_mP$ be the matrix $[{}_mp_{ij}]$. Then, if $\mathbf{P}\{x_m(\omega) = i\} > 0$,

$$ {}_mp_{ij} \geq 0, \qquad \sum_j {}_mp_{ij} = 1, \tag{1.1} $$

and

$$ \mathbf{P}\{x_{m+2}(\omega) = j \mid x_m(\omega) = i\} = \sum_k {}_mp_{ik}\, {}_{m+1}p_{kj}. \tag{1.2} $$

Since the process has the Markov property, the second factors in the sum do not depend on i. [In this equation it is understood that if ${}_{m+1}p_{kj}$ is undefined, because $\mathbf{P}\{x_{m+1}(\omega) = k\} = 0$, the corresponding summand ${}_mp_{ik}\, {}_{m+1}p_{kj}$ is to be taken as 0; we observe that, for such a value of k, ${}_mp_{ik} = 0$.] Thus the matrix of probabilities of transitions from i at time m to j at time $m + 2$, that is, from i at time m to j in two steps, is the product ${}_mP\, {}_{m+1}P$ of the matrices of one-step transitions. In many applications ${}_mP$ is independent of m so that we write P: $[p_{ij}]$ for

${}_mP : [{}_mp_{ij}]$ and speak of *stationary transition probabilities.* In this case, if $p_{ij}^{(n)}$ is the probability of a transition from i to j in n steps,

$$[p_{ij}^{(n)}] = P^n,$$

so that

$$p_{ij}^{(m+n)} = \sum_k p_{ik}^{(m)} p_{kj}^{(n)}, \qquad (P^{m+n} = P^m P^n). \tag{1.3}$$

In practice matrices $[{}_mp_{ij}]$ are given, satisfying (1.1), and a Markov process is defined which has these matrices as transition probability matrices. This process is defined by choosing initial probabilities $\mathbf{P}\{x_1(\omega) = i\}$ and defining

$$\mathbf{P}\{x_1(\omega) = a_1, \cdots, x_n(\omega) = a_n\} = \mathbf{P}\{x_1(\omega) = a_1\}\, {}_1p_{a_1a_2}\, {}_2p_{a_2a_3} \cdots {}_{n-1}p_{a_{n-1}a_n}.$$

With this definition of $x_1, x_2, \cdots$ probabilities, the x_n process is actually a Markov process with the given transition probabilities. More precisely, if with this definition $\mathbf{P}\{x_m(\omega) = i\} > 0$, then ${}_mp_{ij}$ is the transition probability

$$\mathbf{P}\{x_{m+1}(\omega) = j \mid x_m(\omega) = i\}.$$

Whether $\mathbf{P}\{x_m(\omega) = i\}$ is positive or not depends both on the initial probabilities and on the given transition probabilities.

According to II §6, any Markov process reversed in time is still a Markov process. In the case of a chain, the reverse one-step transition probabilities are easily calculated, and we find

$$\begin{aligned}\mathbf{P}\{x_m(\omega) = j \mid x_{m+1}(\omega) = i\} &= \frac{\mathbf{P}\{x_m(\omega) = j\}\, {}_mp_{ij}}{\mathbf{P}\{x_{m+1}(\omega) = i\}} \\ &= \frac{\mathbf{P}\{x_m(\omega) = j\}\, {}_mp_{ij}}{\sum_k \mathbf{P}\{x_m(\omega) = k\}\, {}_mp_{ki}},\end{aligned}$$

where

$$\mathbf{P}\{x_m(\omega) = j\} = \sum_{a_1, \cdots, a_{m-1}} \mathbf{P}\{x_1(\omega) = a_1\}\, {}_1p_{a_1a_2}\, {}_2p_{a_2a_3} \cdots {}_{m-1}p_{a_{m-1}j}.$$

Note that the reverse transition probabilities depend on the choice of the initial probabilities.

According to the previous discussion any matrix $[p_{ij}]$ satisfying the conditions

$$p_{ij} \geq 0 \qquad \sum_j p_{ij} = 1 \tag{1.4}$$

can be used, together with initial probabilities $\{p_i\}$ satisfying

$$p_i \geq 0, \qquad \sum_i p_i = 1, \tag{1.5}$$

to define a Markov chain with stationary transition probabilities. A matrix $[p_{ij}]$ satisfying (1.4) is called a *stochastic matrix.* The n-step transition probabilities, which also determine stochastic matrices, are defined successively by

$$p_{ij}^{(1)} = p_{ij}$$

$$(1.6) \qquad p_{ij}^{(n+1)} = \sum_k p_{ik}^{(n)} p_{kj} = \sum_k p_{ik} p_{kj}^{(n)}$$

and satisfy (1.3). The probability $\mathbf{P}\{x_n(\omega) = j\}$, the probability of state j at time n, is given recursively by

$$p_j^{(1)} = p_j$$

$$(1.7) \qquad p_j^{(n)} = \sum_i p_i p_{ij}^{(n-1)}, \qquad n > 1.$$

The quantities $\{p_j^{(n)}\}$ are called the *absolute probabilities* of the process. In particular, if $p_j^{(1)} = p_j^{(2)} = p_j$, for all j, so that

$$(1.8) \qquad p_j = \sum_i p_i p_{ij},$$

which implies $p_j^{(1)} = p_j^{(2)} = \cdots$, the p_j's are called *stationary absolute probabilities.* It will be shown below that in the finite dimensional case (that is, when there are finitely many states, so that the matrices involved are finite dimensional) there is always at least one set of stationary absolute probabilities. There may be none in the infinite dimensional case.

2. Finite dimensional Markov chains with stationary transition probabilities

In this section we shall consider finite dimensional chains with $N < \infty$ states. The transition probabilities will be stationary, and the problem to be discussed is, in the notation of §1, the determination of the asymptotic character of $p_{ij}^{(n)}$ for $n \to \infty$, that is, the determination of the characteristic properties of the system under examination after a long lapse of time. In particular, an analysis of the dependence of the long-time characteristics of the system on the initial states is desired.

The problem will be considered in various special cases leading up to the general case (d), and the results will then be specialized to further special cases.

Case (a) $p_{ij} = p_j$ *is independent of* i.

The condition imposed means that the random variables $x_1, x_2, \cdots$ of a Markov chain with these transition probabilities are mutually independent, that is, the system assumes its successive states as a result of independent trials of some experiment. For example, the system may be a die, with

$$p_1 = \cdots = p_6 = \tfrac{1}{6}.$$

The probability of a 4 on the $(m+1)$th toss is $\frac{1}{6} = p_4$, regardless of the results of previous tosses. In Case (a) it is intuitively clear, and trivially verifiable, that

$$p_{ij}^{(n)} = p_{ij} = p_j, \qquad n = 1, 2, \cdots,$$

and the problem of this section is thus trivial. Moreover, no matter how $(p_j^{(1)})$ is chosen,

$$p_j^{(n)} = \sum_{i=1}^{N} p_i^{(1)} p_{ij}^{(m-1)} = \sum_{i=1}^{N} p_i^{(1)} p_j = p_j, \qquad n > 1.$$

The process is entirely unaffected by the initial conditions after the initial step has been taken.

Case (b) *There is an integer $\nu \geq 1$ and a set J of $N_1 \geq 1$ values of j such that*

$$\min_{\substack{1 \leq i \leq N \\ j \in J}} p_{ij}^{(\nu)} = \delta > 0.$$

In this case there are numbers $p_1, \cdots, p_N$ such that

$$\lim_{n \to \infty} p_{ij}^{(n)} = p_j, \qquad i, j = 1, \cdots, N, \qquad p_j \geq \delta, j \in J; \tag{2.1}$$

$p_1, \cdots, p_N$ form a set of stationary absolute probabilities. Moreover,

$$\left|p_{ij}^{(n)} - p_j\right| \leq (1 - N_1\delta)^{(n/\nu)-1}. \tag{2.2}$$

Thus in Case (b) the long-run properties of the transitions make the process like that of Case (a) in the sense that no matter what the initial distribution there is probability approximately p_j that the system will be in state j after a large number of transitions.

Proof Define $m_j^{(r)}, M_j^{(r)}$ by

$$m_j^{(r)} = \min_i p_{ij}^{(r)}, \qquad M_j^{(r)} = \max_i p_{ij}^{(r)}$$

Then

$$\begin{aligned} m_j^{(r+1)} &= \min_i \sum_k p_{ik} p_{kj}^{(r)} \geq \min_i \sum_k p_{ik} m_j^{(r)} = m_j^{(r)} \\ M_j^{(r+1)} &= \max_i \sum_k p_{ik} p_{kj}^{(r)} \leq \max_i \sum_k p_{ik} M_j^{(r)} = M_j^{(r)} \end{aligned} \tag{2.3}$$

so that

$$m_j^{(1)} \leq m_j^{(2)} \leq \cdots \leq M_j^{(2)} \leq M_j^{(1)}. \tag{2.4}$$

If, for fixed α, β, $\sum_k^+$ and $\sum_k^-$ denote summation over values of k for which $p_{\alpha k}^{(\nu)} \geq p_{\beta k}^{(\nu)}$ and for which $p_{\alpha k}^{(\nu)} < p_{\beta k}^{(\nu)}$ respectively, then

$$\sum_k^+ (p_{\alpha k}^{(\nu)} - p_{\beta k}^{(\nu)}) + \sum_k^- (p_{\alpha k}^{(\nu)} - p_{\beta k}^{(\nu)}) = 1 - 1 = 0, \tag{2.5}$$

and, if there are s values of $k \in J$ involved in $\sum^+$,

$$\sum_k^+ (p_{\alpha k}^{(\nu)} - p_{\beta k}^{(\nu)}) \leq 1 - (N_1 - s)\delta - s\delta = 1 - N_1\delta. \tag{2.6}$$

Using these two facts, we find that, if $n > 0$,

$$\begin{aligned} M_j^{(n+\nu)} - m_j^{(n+\nu)} &= \operatorname*{Max}_{\alpha, \beta} \sum_k (p_{\alpha k}^{(\nu)} - p_{\beta k}^{(\nu)}) p_{kj}^{(n)} \\ &\leq \operatorname*{Max}_{\alpha, \beta} \{\sum_k^+ (p_{\alpha k}^{(\nu)} - p_{\beta k}^{(\nu)}) M_j^{(n)} + \sum_k^- (p_{\alpha k}^{(\nu)} - p_{\beta k}^{(\nu)}) m_j^{(n)}\} \\ &= \operatorname*{Max}_{\alpha, \beta} \sum_k^+ (p_{\alpha k}^{(\nu)} - p_{\beta k}^{(\nu)})(M_j^{(n)} - m_j^{(n)}) \\ &\leq (1 - N_1\delta)(M_j^{(n)} - m_j^{(n)}). \end{aligned} \tag{2.7}$$

Moreover,

$$\begin{aligned} M_j^{(\nu)} - m_j^{(\nu)} &= \operatorname*{Max}_{\alpha, \beta} (p_{\alpha j}^{(\nu)} - p_{\beta j}^{(\nu)}) \\ &\leq \operatorname*{Max}_{\alpha, \beta} \sum_k^+ (p_{\alpha k}^{(\nu)} - p_{\beta k}^{(\nu)}) \leq 1 - N_1\delta. \end{aligned} \tag{2.8}$$

Therefore, combining the two preceding inequalities, we find that

$$M_j^{(k\nu)} - m_j^{(k\nu)} \leq (1 - N_1\delta)^k. \tag{2.9}$$

Then [cf. (2.4)] $M_j^{(n)}$ and $m_j^{(n)}$ must have a common limit p_j when $n \to \infty$, and

$$|p_{ij}^{(n)} - p_j| \leq M_j^{(n)} - m_j^{(n)} \leq (1 - N_1\delta)^{(n/\nu)-1}. \tag{2.10}$$

If $j \in J$, $0 < \delta \leq m_j^{(\nu)} \leq p_j$. When $n \to \infty$ in (1.6), the resulting equation (taken together with the fact that the p_j's are non-negative and have sum 1) is precisely the condition that $p_1, \cdots, p_N$ be a set of stationary absolute probabilities. This concludes the discussion of Case (b).

There are many methods which can be used to carry the study of Markov chains further. We shall use one based on an analysis of the actual transitions. Before doing this we shall, however, give an example of the possibilities of a purely analytic method. The theorem we prove in this way will appear again as a consequence of more detailed results.

THEOREM 2.1 *If P: $[p_{ij}]$ is a stochastic matrix, $i, j = 1, \cdots, N$, there is a stochastic matrix Q: $[q_{ij}]$ such that*

$$\lim_{n\to\infty} \frac{1}{n} \sum_{m=1}^{n} p_{ij}^{(m)} = q_{ij} \qquad i, j = 1, \cdots, N. \tag{2.11}$$

Moreover, $QP = PQ = Q$ and $Q^2 = Q$.

Note that this is true with no restrictions on the stochastic matrix $[p_{ij}]$. This theorem states that there is a long-time average probability that a system starting in state i will be in state j. In the following, if A_n: $[a_{ij}(n)]$ is a matrix for each n, $A_n \to A$: $[a_{ij}]$ means $a_{ij}(n) \to a_{ij}$ $(n \to \infty)$ for all i, j. If A: $[a_{ij}]$ and B: $[b_{ij}]$ are matrices $A + B$ is the matrix $[a_{ij} + b_{ij}]$, and if λ is a constant λA is the matrix $[\lambda a_{ij}]$. Then (2.11) becomes

$$\lim_{n\to\infty} \frac{1}{n} \sum_{m=1}^{n} P^m = Q. \tag{2.11'}$$

Now the matrices

$$\frac{1}{n} \sum_{m=1}^{n} P^m, \qquad n = 1, 2, \cdots,$$

are stochastic matrices. Since the elements of stochastic matrices are bounded by 1, every sequence of stochastic matrices contains a convergent subsequence. Hence to prove the existence of the limit Q in (2.11') it is sufficient to show that the convergent subsequences of the averages on the left all have the same limit. Suppose then that A is the limit of some convergent subsequence, that is, for some sequence of integers $n_1 < n_2 < \cdots$

$$\lim_{j\to\infty} \frac{1}{n_j} \sum_{m=1}^{n_j} P^m = A.$$

Multiplying by P on the left and right, we find that

$$\lim_{n\to\infty} \frac{1}{n_j} \sum_{m=2}^{n_j+1} P^m = AP = PA.$$

Now the two averages in the preceding equations have the same limit when $j \to \infty$ since these averages differ by the two terms

$$\frac{P^{n_j+1}}{n_j}, \qquad \frac{P}{n_j}$$

which go to zero when $j \to \infty$. Thus

$$A = AP = PA.$$

This implies that, for every n,

$$A = AP^n = P^nA = A\left(\frac{1}{n}\sum_{m=1}^{n} P^m\right) = \left(\frac{1}{n}\sum_{m=1}^{n} P^m\right)A.$$

Hence, if B is any limit matrix of a subsequence of the averages in (2.11′),

$$A = AB = BA.$$

Since A and B are arbitrary limit matrices here, they can be interchanged, giving

$$B = BA = AB.$$

Then $A = B$, that is, there is only one limit matrix, as was to be proved. Finally, according to the preceding equations, $Q = QP = PQ$ and the limit Q is idempotent, $Q^2 = Q$. The study of idempotent stochastic matrices leads to more details on the asymptotic character of $p_{ij}^{(n)}$ for large n. These details will be obtained by other methods below.

We shall use the following lemma.

LEMMA 2.2 *Let S be a set of positive integers, with highest common factor d. Then, if $m + n \in S$ whenever both $m \in S$ and $n \in S$, all sufficiently large multiples of d are in S.*

To prove this we note first that there are finitely many integers $m_1, \cdots, m_\mu$ in S whose highest common factor is d. Then there are integers $n_1, \cdots, n_\mu$ such that

$$n_1m_1 + \cdots + n_\mu m_\mu = d.$$

This equation states (putting the terms with negative n_j's on the right) that there are two integers in S of the form $p, p + d$. Then for every positive integer m the integers

$$mp, mp + d, \cdots, mp + md$$

are in S, that is, if $p_1 = p/d$, the integers

$$mp + nd = d(mp_1 + n), \qquad 0 \le n \le m$$

are in S. These integers obviously include all multiples of d larger than p_1^2d.

In the following j will be called a *consequent* of i (of order n), relative to a given stochastic matrix, if $p_{ij}^{(n)} > 0$. This means, if $n = 1$, that $p_{ij} > 0$, or, if $n > 1$, that there are intermediate states $\alpha_1, \cdots, \alpha_{n-1}$ with $p_{i\alpha_1} \cdots p_{\alpha_{n-1}j} > 0$. If j is a consequent of i, it may be a consequent of several orders, but *there is always a smallest order* $\le N$, where N is the dimension of the matrix. In fact, if two α_j's are equal (and here we take $i = \alpha_0, j = \alpha_n$), say $\alpha_a = \alpha_b$, the states $\alpha_a, \cdots, \alpha_{b-1}$ can be dropped from the succession; dropping all possible intermediate states in this way, a succession of distinct α_j's is finally obtained (except that $\alpha_0 = \alpha_n$ is possible) with $n \leq N$.

Case (c) *For every i and j, j is a consequent of i.*

Let d be the highest common factor of the orders in which 1 is a consequent of itself. Since

$$p_{11}^{(m+n)} \geq p_{11}^{(m)} p_{11}^{(n)}$$

these orders include the number $m + n$ if m and n are included. Hence by the lemma these orders include all large multiples of d. Now

$$p_{11}^{(l+m+n)} \geq p_{1i}^{(l)} \, p_{ij}^{(m)} \, p_{j1}^{(n)},$$

and l and n can be chosen to make the first and last factors on the right positive. Then this inequality shows that $p_{ij}^{(m)} = 0$ except possibly for values of m differing by certain multiples of d. Then if C_α ($\alpha = 1, \cdots, d$) is defined as the set of integers which are consequents of 1 of order $nd + \alpha$ for some n, the C_α's are disjunct and exhaust $1, \cdots, N$. Evidently $C_{\alpha+1}$ consists of the integers which are consequents of those in C_α of order

$[p_{ij}]$:	C_1	C_2	C_3
C_1	0	P_{12}	0
C_2	0	0	P_{23}
C_3	P_{31}	0	0

$[p_{ij}^{(2)}]$:	C_1	C_2	C_3
C_1	0	0	$P_{12}P_{23}$
C_2	$P_{23}P_{31}$	0	0
C_3	0	$P_{31}P_{12}$	0

$[p_{ij}^{(3)}]$:	C_1	C_2	C_3
C_1	$P_{12}P_{23}P_{31}$	0	0
C_2	0	$P_{23}P_{31}P_{12}$	0
C_3	0	0	$P_{31}P_{12}P_{23}$

CYCLICALLY MOVING CLASSES

$nd + 1$ for some n (interpreting C_{d+1} as C_1), and so on. Hence, if the states are renumbered so that those in C_1 come first, then those in $C_2, \cdots$, the matrix $[p_{ij}]$ and its successive powers take the indicated forms ($d = 3$), running through these forms cyclically. Only the elements in a submatrix P_{ij} can be positive. The system runs cyclically through the sets of states $C_1, C_2, \cdots, C_d, C_1, \cdots$. Then $[p_{ij}^{(d)}]$ is a matrix whose elements vanish except for those in d square submatrices down the main diagonal.

We have seen that there is a ν so large that $p_{11}^{(md)} > 0$ if $m \geq \nu$. For each i, j

$$p_{ij}^{(l+md+n)} \geq p_{i1}^{(l)} p_{11}^{(md)} p_{1j}^{(n)}$$

and we can choose l and n so that the first and third factors are positive, and so that $l \leq N$, $n \leq N$. The second factor is positive if $m \geq \nu$. If i, j are in the same C_α, d will be a factor of $l + n$. Hence

$$p_{ij}^{(rd)} > 0 \quad \text{if} \quad r \geq \frac{2N}{d} + \nu, \qquad i, j \in C_\alpha.$$

The matrix $[p_{ij}^{(d)}]$ with $i, j \in C_\alpha$ is thus a stochastic matrix satisfying the hypotheses of Case (b) above with J the class of all integers in C_α. Therefore

$$\lim_{n\to\infty} p_{ij}^{(nd)} = \pi_{\alpha j} > 0, \qquad i, j \in C_\alpha,$$

$$\sum_{j \in C_\alpha} \pi_{\alpha j} = 1.$$

More generally, if $i \in C_\alpha$, and if $j \in C_\beta$

$$\begin{aligned}(2.12) \qquad \lim_{n\to\infty} p_{ij}^{(nd+m)} &= \lim_{n\to\infty} \sum_{k \in C_\beta} p_{ik}^{(m)} p_{kj}^{(nd)} = \sum_{k \in C_\beta} p_{ik}^{(m)} \pi_{\beta j} \\ &= \pi_{\beta j} \qquad \beta \equiv \alpha + m \pmod{d} \\ &= 0 \qquad \text{otherwise.}\end{aligned}$$

This implies that

$$(2.13) \qquad \lim_{n\to\infty} \frac{1}{n} \sum_{m=1}^{n} p_{ij}^{(m)} = \frac{\pi_{\beta j}}{d}, \qquad j \in C_\beta,$$

which is independent of i. Thus, regardless of the initial state i, there is asymptotically an average probability $\pi_{\beta j}/d$ of being in state j after a large number of transitions; more exactly, there is probability 0 or nearly $\pi_{\beta j}$ of being in state j after a large number n of transitions, depending on whether (if $i \in C_\alpha$, $j \in C_\beta$) $\alpha + n \equiv \beta \pmod{d}$ or not. For large n the non-zero boxes of $p_{ij}^{(n)}$ contain only positive elements.

Case (d) *General case* The integers $1, \cdots, N$ can be divided into two classes as follows: a *transient integer* is one which has a consequent of which it is not itself a consequent; a *non-transient integer* is one which is a consequent of every one of its consequents. If j is a consequent of i, a non-transient integer, j is itself non-transient. In fact, if k is a consequent of j it is also one of i, $i \to j \to k$, and hence i must be a consequent of k (i is non-transient), $i \to j \to k \to i$. Then j is a consequent of k, $k \to i \to j$, as was to be proved. In the following F will be the class of transient integers. The non-transient integers are further divided into *ergodic classes*, $E_1, E_2, \cdots$, by putting two integers in the same class if they are consequents of each other. Then, if $i \in E_a$, $p_{ij} = 0$ for $j \notin E_a$.

	E_1	E_2	E_3	F
E_1	R_{11}	0	0	0
E_2	0	R_{22}	0	0
E_3	0	0	R_{33}	0
F	R_{41}	R_{42}	R_{43}	R_{44}

Ergodic Classes

If the states are numbered so that those in E_1 are first, then those in $E_2, \cdots$ and finally those in F, the matrix $[p_{ij}]$ takes the indicated form (three ergodic classes are shown). Only the elements in a submatrix R_{ij} can be positive. The powers of $[p_{ij}]$ have this same form. Now suppose that $i \in F$. We show that at least one of its consequents lies in an ergodic class, that is, at least one of its consequents is non-transient. Let Γ_i be the set of consequents of i. Then Γ_i certainly includes the consequents of all the integers in it, that is, if $j \in \Gamma_i$ then $\Gamma_j \subset \Gamma_i$; let Γ_k be the Γ_j, $j \in \Gamma_i$, with the smallest number of elements. Because of this minimal property, Γ_k must be the class of consequents of each of its own members. Then Γ_k consists of non-transient integers, and in fact is an ergodic class; the integers of this ergodic class are all consequents of i. This means that in the first place there must actually be non-transient integers and in the second place each of the F rows of $[p_{ij}^{(n)}]$ must, for some n, contain a positive element in an E_a column. For each a, E_a determines its own stochastic matrix, the ath submatrix down the main diagonal of $[p_{ij}]$, and the powers of this matrix are the corresponding submatrices in the powers of $[p_{ij}]$. Moreover, each of these stochastic matrices comes under Case (c). Thus we know the character of $p_{ij}^{(n)}$ for large n if $i \in E_a$. If d_a is the number of cyclically moving classes ${}_aC_1, \cdots, {}_aC_{d_a}$ in E_a,

$$(2.12') \quad \lim_{n\to\infty} p_{ij}^{(nd_a+m)} = {}_a\pi_{\beta j} > 0, \qquad \beta \equiv \alpha + m \pmod{d_a}, \qquad \begin{matrix} i \in {}_aC_\alpha \\ j \in {}_aC_\beta \end{matrix}$$

$$= 0 \quad \text{otherwise},$$

where

$$\sum_{j \in {}_aC_\beta} {}_a\pi_{\beta j} = 1.$$

Then

$$(2.13') \qquad \lim_{n\to\infty} \frac{1}{n} \sum_{m=1}^{n} p_{ij}^{(m)} = \frac{{}_a\pi_{\beta j}}{d_a}, \qquad i \in E_a, \quad j \in {}_aC_\beta.$$

We have seen that the convergence is exponentially fast in (2.12′).

If $i \in F$, $\sum_{j \in F} p_{ij}^{(n)}$ cannot increase when n increases; it is the probability of going from i into F in n transitions, that is, of remaining in F throughout the transitions, since once out of F the system stays out. Moreover, we have seen that the sum is < 1 for large n. Now choose μ so large that all these row sums are < 1 for $n \geq \mu$,

$$\sum_{j \in F} p_{ij}^{(n)} \leq \eta < 1 \qquad n \geq \mu, \quad i \in F.$$

Then

$$\sum_{j \in F} p_{ij}^{((m+1)\mu)} = \sum_{j, k \in F} p_{ik}^{(m\mu)} p_{kj}^{(\mu)} \leq \eta \sum_{k \in F} p_{ik}^{(m\mu)}, \qquad i \in F,$$

so that

$$\sum_{j \in F} p_{ij}^{(m\mu)} \leq \eta^m, \qquad i \in F.$$

Since the sum $\sum_{j \in F} p_{ij}^{(n)}$ is monotone in n,

$$\sum_{j \in F} p_{ij}^{(n)} \leq \eta^{(n/\mu)-1} \to 0 \quad (n \to \infty), \qquad i \in F.$$

Now the sum on the left is the probability of being in F after n transitions, and the sum of this over n converges. Hence we have the following theorem:

THEOREM 2.3 *The system will remain among the transient states through only a finite number of transitions, with probability* 1.

The system thus goes into the non-transient states in the long run, even if it is initially in a transient state. Let $\rho_i(E_a)$ be the probability that the system will finally be in E_a, if it was initially in state i,

$$\rho_i(E_a) = \lim_{n\to\infty} \sum_{j \in E_a} p_{ij}^{(n)}.$$

(The sum is monotone non-decreasing in n, since once in E_a the system stays there, so that the nth sum is the probability of entry into E_a at some time up to the nth transition.) If $i \in E_a$, $\rho_i(E_a) = 1$; if $i \in E_b$, $b \neq a$, $\rho_i(E_a) = 0$. Now if $j \in E_a$

$$p_{ij}^{(m+n)} = \sum_{k \in E_a} p_{ik}^{(m)} p_{kj}^{(n)} + \sum_{k \in F} p_{ik}^{(m)} p_{kj}^{(n)}.$$

The last sum is $\leq \sum_{k \in F} p_{ik}^{(m)}$, and we have seen that this goes to 0 when $m \to \infty$. Then, if $m, n \to \infty$ in this equation and if $d_a = 1$, we find that

$$\lim_{l\to\infty} p_{ij}^{(l)} = \rho_i(E_a)\, {}_a\pi_{1j} \qquad j \in E_a.$$

More generally, if $d_a \geq 1$ let $\rho_i({}_aC_\beta)$ be the probability that the system will finally be in ${}_aC_\beta$, if it was initially in state i, in the course of transitions of orders d_a, $2d_a, \cdots$,

$$\rho_i({}_aC_\beta) = \lim_{n\to\infty} \sum_{j \in {}_aC_\beta} p_{ij}^{(nd_a)}.$$

Then

$$\rho_i(E_a) = \sum_\beta \rho_i({}_aC_\beta).$$

If $j \in {}_aC_\beta$,

$$p_{ij}^{(m_1d_a+m_2d_a+m)} = \sum_{k \in {}_aC_{\beta-m}} p_{ik}^{(m_1d_a)} p_{kj}^{(m_2d_a+m)} + \sum_{k \in F} p_{ik}^{(m_1d_a)} p_{kj}^{(m_2d_a+m)},$$

where $\beta - \alpha$ is to be interpreted mod d_a. When $m_1, m_2 \to \infty$ in this equation we find that

$$\lim_{n\to\infty} p_{ij}^{(nd_a+m)} = \rho_i({}_aC_{\beta-m})\, {}_a\pi_{\beta j}, \qquad j \in {}_aC_\beta. \tag{2.12''}$$

[Equation (2.12′) is the special case with $i \notin F$.] Then

$$\lim_{n\to\infty} \frac{1}{n} \sum_{m=1}^{n} p_{ij}^{(m)} = \frac{\sum_\alpha \rho_i({}_aC_\alpha)\, {}_a\pi_{\beta j}}{d_a} = \frac{\rho_i(E_a)\, {}_a\pi_{\beta j}}{d_a}, \qquad j \in {}_aC_\beta. \tag{2.13''}$$

We have thus proved Theorem 2.1 once again, in a proof giving insight into the transitions of the process. A detailed nomenclature has been used to distinguish between the various important types of stochastic matrices which we shall now discuss, but this nomenclature will not be given here. It is doubtful that the nomenclature would have been invented if the full results and their connotation in ergodic theory had been known earlier.

The above diagram for $[p_{ij}]$ exhibiting the effect of the division into ergodic classes is also valid for $[q_{ij}]$, the limit in (2.11) and (2.13″); however, the $F \times F$ box is now also a zero box. Each $E_a \times E_a$ box has identical rows,

$$q_{ij} = {}_a\pi_{\beta j}/d_a > 0, \qquad i \in E_a, \quad j \in {}_aC_\beta.$$

The part of any F row below this box is proportional to the rows of this box, with constant of proportionality $\rho_i(E_a) \geq 0$.

Case (e) *The limit q_{ij} of Theorem* 2.1 *is independent of i* if and only if there is only a single ergodic class. This case is usually described by the statement that the system's state becomes independent of the initial conditions as the number of transitions becomes infinite.

Case (f) *The limit q_{ij} of Theorem* 2.1 *can be taken as an ordinary limit*

$$\lim_{n\to\infty} p_{ij}^{(n)} = q_{ij}$$

if and only if there are no cyclically transferring subclasses in any ergodic class.

Case (g) *The limit q_{ij} of Theorem* 2.1 *is positive for all i, j* if and only if there are no transient states and only one ergodic class. The elements q_{ij} are then independent of i.

Case (b) above, in which, for some ν, $p_{ij}^{(\nu)} > 0$ whenever $j \in J$, can now be characterized as follows: there are no transient states in J; there is only one ergodic class, and this class contains no cyclically moving subclasses.

Case (h) *If $p_{ii} > 0$ for all i*, the same will be true no matter how the states are renumbered. Then under this condition there can be no cyclically moving classes of states, so that by (f)

$$\lim_{n\to\infty} p_{ij}^{(n)} = q_{ij}$$

for all i, j. Thus in this case the Cesàro limit in Theorem 2.1 can be replaced by an ordinary limit. The condition used is unnecessarily strong, however.

Case (i) *If $p_{ij} = p_{ji}$*, the matrix $[p_{ij}]$, all its powers, and therefore also the limit matrix $[q_{ij}]$ are symmetric. Moreover, this is true no matter how the states are renumbered. There can, therefore, be no cyclically moving classes of states, and no transient states (both of which cases require unsymmetric matrices). Hence we have

$$\lim_{n\to\infty} p_{ij}^{(n)} = {}_a\pi_{1j} = q_{ij}, \qquad i \in E_a.$$

Since $[q_{ij}]$ is symmetric, this becomes

$$\lim_{n\to\infty} p_{ij}^{(n)} = \frac{1}{N_a}, \qquad i \in E_a,$$

where N_a is the number of states in E_a. In particular, if there is only one ergodic class, this becomes

$$\lim_{n\to\infty} p_{ij}^{(n)} = \frac{1}{N}, \qquad i, j = 1, \cdot\cdot\cdot, N.$$

Case (j) In some applications *the basic stochastic matrix satisfies the condition*

$$\sum_{i=1}^{N} p_{ij} = 1, \qquad j = 1, \cdot\cdot\cdot, N.$$

This condition, somewhat more general than that of symmetry, will be satisfied no matter how the states are renumbered, and will also be satisfied by the iterates $[p_{ij}^{(n)}]$, and by the limit $[q_{ij}]$. Consequently there can be no transient states (which give rise to zero columns in $[q_{ij}]$). Moreover, (2.11) becomes

$$\lim_{n\to\infty} \frac{1}{n} \sum_{m=1}^{n} p_{ij}^{(m)} = \frac{1}{N_a},$$

where N_a is the number of states in E_a. The Cesàro limit cannot in general be replaced by an ordinary limit.

The possible sets of stationary absolute probabilities are fully described in the following theorem.

THEOREM 2.4 *For each* i, $q_{i1}, \cdots, q_{iN}$ *[defined by* (2.11)*] is a set of stationary absolute probabilities. Conversely, any set of stationary absolute probabilities is a linear combination (non-negative coefficients with sum* 1) *of these* N *sets, and even of the sets for* $i \notin F$. *More generally, every solution of*

$$\sum_{i=1}^{N} \xi_i p_{ij} = \xi_j, \qquad j = 1, \cdots, N \tag{2.14}$$

is a linear combination of these sets, for $i \notin F$.

Note that according to this theorem each transient state has probability 0 in a *stationary* Markov chain with a finite number of states. If $i \in E_a$, $(q_{i1}, \cdots, q_{iN})$ is independent of i. According to the theorem each ergodic class thus defines a single set of stationary absolute probabilities. Since the one defined by E_a assigns probability 0 to all states outside E_a, the sets of stationary absolute probabilities defined in this way are linearly independent. We have already remarked that, if $i \in F$ and $j \in E_a$, then q_{ij} is proportional to q_{ij} for $i \in E_a$ and $j \in E_a$, so that a row $(q_{i1}, \cdots, q_{iN})$ with $i \in F$ defines a set of stationary absolute probabilities linearly dependent on those already described.

To prove the theorem, we remark that according to Theorem 2.1, $QP = Q$, that is,

$$\sum_{j=1}^{N} q_{ij} p_{jk} = q_{ik}, \qquad i, k = 1, \cdots, N,$$

and according to this equation $q_{i1}, \cdots, q_{iN}$ is a set of stationary absolute probabilities. Conversely, if $p_1, \cdots, p_N$ is a set of stationary absolute probabilities, then for all j and n

$$\sum_{i=1}^{N} p_i p_{ij}^{(n)} = p_j$$

and therefore also

$$\sum_{i=1}^{N} p_i \frac{1}{n} \left(\sum_{m=1}^{n} p_{ij}^{(m)} \right) = p_j.$$

Hence ($n \to \infty$)

$$\sum_{i=1}^{N} p_i q_{ij} = p_j \qquad j = 1, \cdots, N,$$

and these equations exhibit $p_1, \cdots, p_N$ as a linear combination, with coefficients $p_1, \cdots, p_N$, of the rows of Q, as was to be proved. More generally, this argument is applicable if $p_1, \cdots, p_N$ are replaced by $\xi_1, \cdots, \xi_N$ satisfying (2.14).

As an application consider the following rather crude idealization of a diffusion process. Two urns, U_1 and U_2, contain white and black balls. A ball is taken at random from each urn and put in the other, and this procedure is continued indefinitely. The problem is to find the asymptotic distribution of balls in the two urns. Suppose that there are N balls in each urn and that there are N white balls and N black balls altogether. Let x_n be the number of white balls in U_1 after the nth interchange. The x_n process is evidently a Markov chain, with stochastic transition matrix $[p_{ij}]$ $i, j = 0, \cdots, N$ given by

$$p_{ii} = 2\frac{i(N-i)}{N^2} \qquad 0 \le i \le N$$

$$p_{ii+1} = \left(\frac{N-i}{N}\right)^2 \qquad i < N$$

$$p_{ii-1} = \left(\frac{i}{N}\right)^2 \qquad 0 < i$$

$$p_{ij} = 0 \qquad \text{otherwise.}$$

In this case $p_{ij}^{(N)} > 0$ for all i, j so that there are no transient states, only one ergodic class, and this has no cyclically moving subclasses. Hence

$$\lim_{n\to\infty} p_{ij}^{(n)} = p_j > 0$$

exists for all i, j; $p_0, \cdots, p_N$ is the uniquely defined set of absolute probabilities. To evaluate the p_j's one could solve the equations they satisfy,

$$\sum_{i=0}^{N} p_i p_{ij} = p_j \qquad \sum_{0}^{N} p_j = 1,$$

but it is simpler to guess at a solution and check the validity of the guess. In fact, it is intuitively reasonable that an initial random distribution of the balls in the urns, obtained by choosing N balls at random out of the $2N$ for U_1, would be preserved. This suggests that

$$p_j = \frac{(N!)^4}{(j!)^2[(N-j)!]^2(2N)!},$$

and in fact this evaluation satisfies the equations for the p_j's. Therefore

$$\lim_{n\to\infty} p_{ij}^{(n)} = \frac{(N!)^4}{(j!)^2[(N-j)!]^2(2N)!},$$

and we have proved that no matter what the initial distribution of black and white balls in the two urns, the distribution after a large number of interchanges is asymptotically that obtained in filling the urns by choosing the balls for each at random.

3. Multiple Markov chains

We have already remarked, in II §6, that a multiple Markov process, that is, a process in which, roughly speaking, conditional probabilities of future states depend on $\nu \geq 1$ past states, can be reduced to Markov processes whose random variables are vectors. In the case of a finite dimensional Markov chain this means that the multiple processes can be reduced to finite dimensional Markov chains with a larger number of possible states. If the given transition probabilities are stationary, the only case we shall consider, we are given the transition probabilities

$$\mathbf{P}\{x_n(\omega) = j \mid x_{n-\nu}(\omega) = i_1, \cdots, x_{n-1}(\omega) = i_\nu\}$$

from which the other transition probabilities can be evaluated. If now $\hat{x}_n$ is defined as the ν-dimensional vector $(x_{n-\nu+1}, \cdots x_n)$, the $\hat{x}_n$ process is a Markov process with stationary transition probabilities $p_{i_1, \cdots i_\nu j_1 \cdots j_\nu}$. If each x_n can assume N values, $p_{i_1, \cdots i_\nu j_1 \cdots j_\nu}$ defines a stochastic matrix with N^ν rows and columns, corresponding to the N^ν values which each $\hat{x}_n$ can assume. However, most of the elements of the stochastic matrix vanish, since obviously

$$p_{i_1 \cdots i_\nu j_1 \cdots j_\nu} = 0 \qquad \text{unless} \quad j_1 = i_2, \cdots, j_{\nu-1} = i_\nu;$$

thus all but $N^{\nu+1}$ of the transition probabilities necessarily vanish. If the N^ν possible sets of values $i_1, \cdots, i_\nu$ are enumerated in some way, the $\hat{x}_n$ process can be considered an ordinary Markov chain with stationary transition probabilities, whose random variables take on the values $1, \cdots, N^\nu$; $p_{i_1 \cdots i_\nu j_1 \cdots j_\nu}$ becomes p_{ij} with $i, j = 1, \cdots, N^\nu$. As already remarked, most of the p_{ij}'s vanish. The results of the previous section can now be applied, and the conditions of the previous section become conditions on the $\hat{x}_n$ process which can be translated into conditions on the x_n process. For example, by Theorem 2.1,

$$\lim_{n\to\infty} \frac{1}{n} \sum_{m=1}^{n} p^{(m)}{}_{i_1 \cdots i_\nu j_1 \cdots j_\nu}$$

exists for all $i_1, \cdots, i_\nu, j_1, \cdots, j_\nu$. Summing over $j_1, \cdots, j_{\nu-1}$, we find that

$$\lim_{n\to\infty} \frac{1}{n} \sum_{m=\nu}^{n} \mathbf{P}\{x_m(\omega) = j \mid x_1(\omega) = i_1, \cdots, x_\nu(\omega) = i_\nu\}$$

exists for all $i_1, \cdots, i_\nu, j$. In particular, suppose that

$$\mathbf{P}\{x_{\nu+1}(\omega) = j \mid x_1(\omega) = i_1, \cdots, x_\nu(\omega) = i_\nu\} > 0$$

for all $i_1, \cdots, i_\nu, j$. Then it follows that

$$p^{(\nu)}{}_{i_1 \cdots i_\nu j_1 \cdots j_\nu} > 0$$

for all $i_1, \cdots, i_\nu, j_1, \cdots, j_\nu$, which implies [Case (b) of the preceding section] that

$$\lim_{n\to\infty} p^{(n)}{}_{i_1 \cdots i_\nu j_1 \cdots j_\nu}$$

exists and is independent of $i_1, \cdots, i_\nu$, and accordingly that

$$\lim_{n\to\infty} \mathbf{P}\{x_n(\omega) = j \mid x_1(\omega) = i_1, \cdots, x_\nu = i_\nu\}$$

exists and is independent of $i_1, \cdots, i_\nu$.

4. Application to card mixing

The problem of card mixing is a good example of the application of Markov chains. It is supposed that M cards, labeled $1, \cdots, M$, are shuffled repeatedly. Between shuffles their order in the deck is examined. The problem is to determine reasonable hypotheses to be imposed on the shuffling process that imply thorough mixing.

A shuffle is simply a permutation of the order in the deck; that is, a shuffle corresponds to a permutation

$$\begin{matrix} 1 \ldots M \\ a_1 \ldots a_M \end{matrix}$$

of the integers $1, \cdots, M$; the jth card from the top of the deck goes into the a_jth card from the top, $j = 1, \cdots, M$. A succession of shuffles thus corresponds to a succession of the $M!$ possible permutations. The thoroughness of the shuffling is dependent on how the permutations vary from shuffle to shuffle. For example, if the permutations are identical, each card will run cyclically through a succession of positions in the deck, and its position can be determined from a knowledge of the number of shuffles. The order in the deck, taking all the cards together, will also vary in this cyclic manner. This would not be consistent with any reasonable definition of thorough shuffling. On the other hand, the same cyclic character of changes of position would be true if the permutations ran successively over the $M!$ possibilities, repeating the succession indefinitely.

Neither of these examples exhibits the generally accepted idea of thorough mixing, which includes unpredictability on the basis of the initial positions and number of shuffles. It is necessary to introduce probabilistic ideas. To do so let X_n, the permutation induced by the nth shuffle, be a random variable. We number the $m!$ permutations, so that X_n can take on the values $1, \cdots, M!$ The character of the shuffling then depends on the properties of the X_n process, the *shuffling process.*

In examining the transitions of individual cards or groups of cards it is appropriate to analyze certain processes derived from the shuffling process rather than the shuffling process itself. Let $i^{(1)}, \cdots, i^{(m)}$ be any m ($\leq M$) distinct natural numbers, and let

$$x_n^{(m)} = (i_n^{(1)}, \cdots, i_n^{(m)})$$

be the m-tuple of positions of the cards, after n shuffles, that were initially in positions $i^{(1)}, \cdots, i^{(m)}$ respectively ("position" meaning, say, the number of a card counted from the top of the deck). The $x_n^{(m)}$ process is determined completely by the initial positions $i^{(1)}, \cdots, i^{(m)}$ and by the X_n process. The $x_n^{(m)}$ process is the natural tool in the study of the transitions of m-tuples of cards.

The shuffling process will be said to induce thorough mixing if in the long run the $M!$ different orders of the cards become equally likely, regardless of the initial shuffles, in the sense that for every positive integer s

$$(4.1)\quad \lim_{n\to\infty} \mathbf{P}\{x_n^{(M)}(\omega) = \xi \mid x_1^{(M)}(\omega) = \xi_1, \cdots, x_s^{(M)}(\omega) = \xi_s\} = \frac{1}{M!}$$

for all sets of the $s + 1$ M-tuples $\xi, \xi_1, \cdots, \xi_s$ of distinct positive integers between 1 and M. For some purposes a less thorough mixing is as useful as this. For example, if a magician (now known as a telepath) tells a victim to shuffle a deck and "take a card," the victim need not be sure of thorough mixing to test the magician's powers. The relevant condition here would be the far weaker one that, for every positive integer s and every integer k between 1 and M,

$$\lim_{n\to\infty} \mathbf{P}\{x_n^{(1)}(\omega) = k \mid x_1^{(1)}(\omega) = j_1, \cdots, x_s^{(1)}(\omega) = j_s\} = \frac{1}{M}$$

for all integers $j_1, \cdots, j_s$. If this is true, the M different possible positions of any card become equally likely in the long run, regardless of the first s positions. More generally, the shuffling process will be said to induce thorough μ-tuple mixing if, for every positive integer s,

$$(4.2)\quad \lim_{n\to\infty} \mathbf{P}\{x_n^{(\mu)}(\omega) = \xi \mid x_1^{(\mu)}(\omega) = \xi_1, \cdots, x_s^{(\mu)}(\omega) = \xi_s\} = \frac{(M-\mu)!}{M!}$$

for all sets of the $s + 1$ μ-tuples $\xi, \xi_1, \cdots, \xi_s$ of distinct integers between 1 and M. Thorough mixing is thus thorough M-tuple mixing. It is easily verified that thorough μ-tuple mixing implies thorough μ_0-tuple mixing if $\mu_0 < \mu$.

HYPOTHESIS (A) *The variables $X_1, X_2, \cdots$ of the shuffling process are mutually independent and have a common distribution.*

This hypothesis, that the X_j's correspond to repeated independent trials, is probably the most natural one. Let

$$P_j = \mathbf{P}\{X_n(\omega) = j\},$$

and let

$$p_{i^{(1)} \cdots i^{(\mu)} j^{(1)} \cdots j^{(\mu)}}$$

be the probability that cards in positions $i^{(1)}, \cdots, i^{(\mu)}$ go into positions $j^{(1)}, \cdots, j^{(\mu)}$ by means of the nth shuffle. Then

$$p_{i^{(1)} \cdots i^{(\mu)} j^{(1)} \cdots j^{(\mu)}} = \sum P_k$$

where the sum is over all those permutations taking the first μ-tuple into the second. The $x_n^{(\mu)}$ process is clearly a Markov chain, and we have just evaluated the transition probability matrix. Moreover, when either the i's or the j's run through all possible μ-tuples the permutations involved in the above sum run through all possible permutations, that is,

$$\sum_{i^{(1)} \cdots i^{(\mu)}} p_{i^{(1)} \cdots i^{(\mu)} j^{(1)} \cdots j^{(\mu)}} = \sum_{j^{(1)} \cdots j^{(\mu)}} p_{i^{(1)} \cdots i^{(\mu)} j^{(1)} \cdots j^{(\mu)}}$$
$$= \sum_{k=1}^{M!} P_k = 1.$$

Thus the $x_n^{(\mu)}$ process is a Markov chain with stationary transition probabilities, whose stochastic matrix has column sums (as well as row sums) 1. The implications of this have been discussed in §2, Case (j).

Since the $x_n^{(\mu)}$ process is a Markov chain, condition (4.2) for thorough mixing becomes

$$\lim_{n \to \infty} \mathbf{P}\{x_n^{(\mu)}(\omega) = \xi \mid x_s^{(\mu)}(\omega) = \xi_s\} = \frac{(M - \mu)!}{M!}, \tag{4.2$'$}$$

and since there are stationary transition probabilities it is no restriction to take $s = 1$.

We continue by making more specific hypotheses on the shuffling process. These are certainly necessary to obtain interesting results, since, for example, Hypothesis (A) does not exclude the possibility that all the shuffles are identical.

HYPOTHESIS (A_μ) *Hypothesis* (A) *is satisfied, and it is not possible to divide the μ-tuples of card positions into two groups in such a way that no permutation with positive P_k takes a μ-tuple of one group into one of the other.*

Obviously (A_μ) implies (A_{μ_0}) if $\mu_0 < \mu$.

If (A_μ) is satisfied the $x_n^{(\mu)}$ chain has only a single ergodic class, so that [cf. §2, Case (e)]

$$\lim_{n\to\infty} \frac{1}{n} \sum_{m=1}^{n} p^{(m)}{}_{i^{(1)}\ldots i^{(\mu)} j^{(1)}\ldots j^{(\mu)}} = \frac{(M-\mu)!}{M!},$$

that is to say, there is thorough μ-tuple mixing in an average sense. It would contradict hypothesis (A_μ) if all the permutations with positive P_k left the top μ cards of the deck on the top. On the other hand, suppose that every permutation with positive P_k permutes the top and bottom halves of the deck (the example is modified slightly if M is odd), permuting among themselves the cards in each half. Then hypotheses (A_1), $\cdots$, (A_M) may be satisfied but none of the following hypotheses will be.

HYPOTHESIS (A'_μ) *Hypothesis* (A_μ) *is satisfied and for all large s at least one μ-tuple of card positions has positive probability of returning to itself after s shuffles.*

Hypothesis (A'_μ) implies (A'_{μ_0}) if $\mu_0 < \mu$.

If (A'_μ) is satisfied, the $x_n^{(\mu)}$ process has only one ergodic class, and there are no cyclically moving subclasses. Hence

$$\lim_{n\to\infty} p^{(n)}{}_{i^{(1)}\ldots i^{(\mu)} j^{(1)}\ldots j^{(\mu)}} = \frac{(M-\mu)!}{M!},$$

and this is precisely condition (4.2′) for thorough μ-tuple mixing.

Hypotheses (A_μ) and (A'_μ) are hypotheses on the zeros of the P_k's. If all the P_k's are positive, that is, if all permutations are possible in the shuffling process, Hypothesis (A'_μ) will be satisfied for all μ and there will be thorough mixing. In that case the stochastic matrices will have no zero elements.

Hypotheses (A_μ) and (A'_μ) assume a particularly simple form if $\mu = M$: (A_M) is equivalent to the statement that (A) is satisfied and that it is possible to go from any ordering of the deck to any other by way of permutations with $P_r > 0$. (That is, the group of all permutations is generated by those with positive probability.) Hypothesis (A'_M) adds the condition that for all large s it is possible to go from any ordering of the deck to any other (or back to the same) by way of s permutations of positive probability. This additional condition will be satisfied, for example, if the identical permutation (corresponding to not shuffling the deck at all) has positive probability.

It is natural to attempt to extend Hypothesis (A). One way would be to make hypotheses on the subsidiary $x_n^{(\mu)}$ processes to insure thorough μ-tuple mixing and to derive the corresponding hypotheses for the X_n process. Suppose, for example, that the $x_n^{(\mu)}$ process is a Markov process with stationary transition probabilities. This implies, if $\mu = M$, that the probability P_r of the rth permutation at the nth shuffle is unaffected by the results of previous shuffles. In other words, for $\mu = M$ the hypothesis is simply Hypothesis (A). If $\mu < M$ the hypothesis becomes weaker, but its formulation in terms of the X_n process is not particularly enlightening.

Another way to weaken Hypothesis (A) would be to suppose only that the X_n process is a Markov chain with stationary transition probabilities. This hypothesis allows the shuffling process a certain memory, which is reasonable. In general if the X_n process is a multiple Markov chain of multiplicity ν, the $x_n^{(M)}$ process will be a multiple Markov chain of multiplicity $\nu + 1$, and the general Markov chain theorems can be applied to analyze the mixing (cf. §3). The subsidiary $x_n^{(\mu)}$ processes with $\mu < M$ will not necessarily be multiple Markov chains, however.

5. Generalization of §2 to general state spaces

The following is the natural generalization of a stochastic matrix $[p_{ij}]$. Let X be a space of points ξ, and let $\mathscr{F}_X$ be a Borel field of X sets. A function $p(\cdot, \cdot)$ of $\xi \in X$ and $A \in \mathscr{F}_X$ will be called a *stochastic transition function* if it has the following properties:

(i) $p(\xi, A)$ for fixed ξ determines a probability measure in A;

(ii) $p(\xi, A)$ for fixed A determines a ξ function measurable with respect to the field $\mathscr{F}_X$.

If an initial probability distribution $p(\cdot)$ of sets $A \in \mathscr{F}_X$ is chosen, a probability measure can be defined in the space Ω of sequences $\omega: (\xi_1, \xi_2, \cdots)$, $\xi_j \in X$, as follows. Let x_n be the nth coordinate function, so that $x_n(\omega) = \xi_n$ if ω is the point $(\xi_1, \xi_2, \cdots)$. Define

$$\mathbf{P}\{x_1(\omega) \in A_1\} = p(A_1), \qquad A_1 \in \mathscr{F}_X,$$

and in general define

$$\mathbf{P}\{x_1(\omega) \in A_1, \cdots, x_n(\omega) \in A_n\} = \int_{A_1} p(d\xi_1) \int_{A_2} p(\xi_1, d\xi_2) \cdots \int_{A_n} p(\xi_{n-1}, d\xi_n), \qquad A_j \in \mathscr{F}_X. \tag{5.1}$$

This definition determines a probability measure of $\mathscr{F}$ sets, where $\mathscr{F} = \mathscr{F}_X \times \mathscr{F}_X \times \cdots$ is the Borel field of ω sets generated by the class of those of the form

$$\{x_j(\omega) \in A\}, \qquad j \geq 1, \qquad A \in \mathscr{F}_X,$$

(see Supplement, §2).

In particular, if the initial probability distribution is concentrated at the point ξ, that is, if

$$\begin{aligned} p(A) &= 1, \quad \xi \in A \\ &= 0, \quad \xi \in X - A \end{aligned} \qquad A \in \mathscr{F}_X,$$

we shall denote the probability of an ω set Λ by

$$\mathbf{P}\{\Lambda \mid x_1(\omega) = \xi\},$$

and the expectation of a random variable x by

$$\mathbf{E}\{x \mid x_1(\omega) = \xi\}.$$

Here "random variable" means as usual a measurable ω function, and the use of conditional probability and expectation notation will be justified in the next paragraph. Replacing sequences $(\xi_1, \xi_2, \cdots)$ by sequences $(\xi_n, \xi_{n+1}, \cdots)$, we find, in the obvious way, definitions of

$$\mathbf{P}\{\Lambda \mid x_n(\omega) = \xi\}, \qquad \mathbf{E}\{x \mid x_n(\omega) = \xi\}.$$

If X is the set of real numbers, and if $\mathscr{F}_X$ is the field of linear Borel sets, a stochastic process $\{x_n, n \geq 1\}$ is defined by the above procedure. For any initial distribution the process is a Markov process, and the quantities defined above become versions of the conditional probabilities and expectations indicated by the notation. These versions will always be the ones used in the following sections. The restriction of X and $\mathscr{F}_X$ to the line and the class of Borel sets will not be made, however. In other words, in this section we consider Markov processes whose random variables are not necessarily numerically valued. The reader made uneasy by this generality may prefer to restrict X and $\mathscr{F}_X$, but will not thereby simplify the discussion.

The n-step transition probabilities are easily calculated [cf. (1.6)],

$$\begin{aligned} p^{(1)}(\xi, A) &= p(\xi, A), \\ p^{(n+1)}(\xi, A) &= \int_X p^{(n)}(\eta, A) p(\xi, d\eta), \end{aligned} \tag{5.2}$$

and are also stochastic transition functions. The probability of being in A at time n is given recursively by [cf. (1.7)]

$$\begin{aligned} \mathbf{P}\{x_n(\omega) \in A\} &= p(A) && n = 1 \\ &= \int_X p^{(n-1)}(\xi, A) p(d\xi) && n > 1. \end{aligned}$$

If this probability is independent of n the x_n process is strictly stationary and $p(\cdot)$ is called a stationary absolute probability distribution.

Example 1 Let X space consist of exactly N points labeled $1, \cdots, N$ and let $\mathscr{F}_X$ consist of X and all its subsets. Let $[p_{ij}]$ be any N-dimensional stochastic matrix, and define $p(\xi, A)$ by

$$p(\xi, A) = \sum_{\eta \in A} p_{\xi\eta}$$

Then $p(\cdot, \cdot)$ is a stochastic transition function, and every stochastic transition function with this X and $\mathscr{F}_X$ can be generated in this way. Moreover, the iterated stochastic transition functions determined by (5.2) are generated in this way by the powers of $[p_{ij}]$. Thus the study of $p^{(n)}(\xi, A)$ for large n becomes that of $p_{ij}^{(n)}$ for large n; this study was carried out in §2. The results obtained in that case go over almost completely in the general case if rather weak hypotheses are imposed on $p(\cdot, \cdot)$. Example 4 below shows, however, that these results do not always hold.

For each ξ, if A is small, $p(\xi, A)$ is small, roughly speaking, since $p(\xi, A) \to 0$ when $A \to 0$. The hypotheses made on $p(\xi, A)$ usually impose some kind of uniformity in ξ on the smallness of $p(\xi, A)$ for small A. In the following we shall use the hypothesis of Doeblin, somewhat generalized over his original formulation.

Hypothesis (D) *There is a (finite-valued) measure φ of sets $A \in \mathscr{F}_X$ with $\varphi(X) > 0$, an integer $\nu \geq 1$, and a positive ε, such that*

$$p^{(\nu)}(\xi, A) \leq 1 - \varepsilon \qquad \text{if} \quad \varphi(A) \leq \varepsilon.$$

We observe that, if (D) is satisfied, for some φ, ν, and ε, it is satisfied for every $n > \nu$ with the same φ and ε, since, if $\varphi(A) \leq \varepsilon$,

$$p^{(\nu+m)}(\xi, A) = \int_X p^{(\nu)}(\eta, A) p^{(m)}(\xi, d\eta) \leq (1 - \varepsilon) p^{(m)}(\xi, X) = 1 - \varepsilon.$$

Example 1 (*continued*) Hypothesis (D) is always satisfied in Example 1. In fact, if $\varphi(A)$ is defined as the number of points (integers) in A, then, if $\varphi(A) < 1$, A is the null set so that $p(\xi, A) = 0$. Then (D) is satisfied if $0 < \varepsilon < 1$. Thus Hypothesis (D) imposes no restriction on finite dimensional stochastic matrices. Now suppose that the space X has denumerably infinitely many points denoted by $1, 2, \cdots$, so that the corresponding stochastic matrix is infinite dimensional. Any φ must be determined as follows. Let $c_1, c_2, \cdots$ be any non-negative numbers with $\sum_i c_i < \infty$ and define $\varphi(A)$ by

$$\varphi(A) = \sum_{i \in A} c_i.$$

Then (D) is certainly satisfied if, for example, the series $\sum_j p_{ij}$ converges uniformly in i; but uniformity of convergence is an unnecessarily strong condition. If $[p_{ij}]$ is the identity matrix, condition (D) is not satisfied for any choice of φ, that is, for any choice of the c_i's.

Example 2 Let X be a Borel set in m-dimensional Euclidean space, and let $\mathscr{F}_X$ consist of the Borel subsets of X. Let $p_0(\cdot, \cdot)$ be a Baire function of ξ, η for which

$$p_0(\xi, \eta) \geq 0, \qquad \int_X p_0(\xi, \eta)\, d\eta = 1,$$

where ξ, η are points in m dimensions and the integration is with respect to Lebesgue measure in m dimensions. Then the function defined by

$$p(\xi, A) = \int_A p_0(\xi, \eta)\, d\eta$$

is a stochastic transition function. If we define $p_0^{(n)}(\xi, \eta)$ recursively by

$$\begin{aligned} p_0^{(n)}(\xi, \eta) &= p_0(\xi, \eta) & n &= 1 \\ &= \int_X p_0^{(n-1)}(\xi, \zeta) p_0(\zeta, \eta)\, d\zeta, & n &> 1, \end{aligned}$$

then

$$p^{(n)}(\xi, A) = \int_A p_0^{(n)}(\xi, \eta)\, d\eta, \qquad n \geq 1.$$

In this important case $p(\cdot, \cdot)$ is the integral of a *stochastic transition density function* $p_0(\cdot, \cdot)$, and the iterate $p^{(n)}(\cdot, \cdot)$ is the integral of the iterated density function $p_0^{(n)}(\cdot, \cdot)$. This case can be generalized in the obvious way by dropping the hypothesis that X is a set in a Euclidean space and that the basic measure of X sets is Lebesgue measure. Without going to this generalization we remark that, if $\varphi(A)$ is defined as the Lebesgue measure of A, if $\varphi(X) < \infty$, and if $p_0(\cdot, \cdot)$ is bounded, say, $p_0(\xi, \eta) \leq K$, then

$$p(\xi, A) \leq K\varphi(A)$$

so that Hypothesis (D) is satisfied if $\varepsilon \leq 1/(K + 1)$. More generally, if $\varphi(X) < \infty$ and if $p_0(\xi, \eta)$ is supposed uniformly (in ξ) integrable in η Hypothesis (D) is still satisfied. However, even this condition, for the given φ measure, is far stronger than (D) since it implies that, if $\varphi(A)$ is near 0, $p(\xi, A)$ is near 0 uniformly in ξ, whereas (D) only requires that under those hypotheses $p(\xi, A)$ be < 1 uniformly in ξ.

Example 3 If $p(\cdot, \cdot)$ is a stochastic transition function and if ψ is some finite-valued measure of $\mathscr{F}_X$ sets with the property that for some ν

$$p^{(\nu)}(\xi, A) \leq \psi(A),$$

then (D) is satisfied if $0 < \varepsilon \leq \frac{1}{2}$ and if $\varphi = \psi$. The following stronger condition is sometimes useful:

$$p^{(\nu)}(\xi, A) \leq Kp^{(\nu)}(\eta, A)$$

for all ξ, η, A, for some integer ν and constant K. This condition is far stronger than the first, since it implies that if, for some ξ and some sequence $A_1, A_2, \cdots$,

$$\lim_{n\to\infty} p^{(\nu)}(\xi, A_n) = 0$$

then this limit equation is true for all ξ and is uniform in ξ.

In the following we denote by $\tilde{p}^{(n)}(\xi, A)$ the conditional probability that (from initial point ξ) the system will be in a state of A at some time during the first n transitions, that is,

$$\tilde{p}^{(n)}(\xi, A) = \mathbf{P}\{\bigcup_{j=2}^{n+1} [x_j(\omega) \in A] \mid x_1(\omega) = \xi\}$$

The quantity $\tilde{p}^{(n)}(\xi, A)$ is non-decreasing in n.

LEMMA 5.1 *Under Hypothesis* (D), *if a set* $A \in \mathscr{F}_X$ *has the property that*

$$\lim_{n\to\infty} \tilde{p}^{(n)}(\xi, A) = \text{L.U.B.}_{n}\, \tilde{p}^{(n)}(\xi, A) > 0$$

for all ξ, *then there is positive integer* μ *and a positive* $\rho < 1$ *for which*

$$\tilde{p}^{(n)}(\xi, A) \geq 1 - \rho^{(n/\mu)-1}, \qquad \xi \in X.$$

Since $\tilde{p}^{(n)}(\xi, A)$ is monotone non-decreasing in n, and positive for sufficiently large n (depending on ξ), there is a $\delta > 0$, a positive integer α, and a set $B \in \mathscr{F}_X$ for which

$$\varphi(B) \leq \varepsilon,$$

$$\tilde{p}^{(\alpha)}(\xi, A) \geq \delta, \qquad \xi \in X - B.$$

Then, according to Hypothesis (D),

$$p^{(\nu)}(\xi, B) \leq 1 - \varepsilon, \qquad \xi \in X. \tag{5.3}$$

To prove the lemma we take $\mu = \alpha + \nu$ and prove

$$1 - \tilde{p}^{(m\mu)}(\xi, A) \leq (1 - \delta\varepsilon)^m, \tag{5.4}$$

which implies the desired statement, with $\rho = 1 - \delta\varepsilon$. When $m = 1$, (5.4) becomes

$$\tilde{p}^{(\alpha+\nu)}(\xi, A) \geq \delta\varepsilon, \qquad \xi \in X$$

which follows [using (5.3)] from

$$\tilde{p}^{(\alpha+\nu)}(\xi, A) \geq \int_{X-B} \tilde{p}^{(\alpha)}(\eta, A) p^{(\nu)}(\xi, d\eta)$$

$$\geq \delta p^{(\nu)}(\xi, X - B) \geq \delta\varepsilon.$$

To prove (5.4) in general we suppose it is true for m and prove it for $m + 1$. Let $\pi^{(n)}(\xi, E)$ be the conditional probability, starting at ξ, of going into E after $n\mu$ transitions, always remaining out of A. Then

$$1 - \tilde{p}^{(n\mu)}(\xi, A) = \pi^{(n)}(\xi, X - A)$$

and assuming (5.4) for some m (and also for $m = 1$, in which case it has already been proved)

$$\begin{aligned}\pi^{(m+1)}(\xi, X - A) &= \int_{X-A} \pi^{(m)}(\eta, X - A)\pi^{(1)}(\xi, d\eta)\\ &\leq (1 - \delta\varepsilon)^m \pi^{(1)}(\xi, X - A)\\ &\leq (1 - \delta\varepsilon)^{m+1}\end{aligned}$$

as was to be proved.

The study of the asymptotic properties of $p^{(n)}(\xi, A)$ as $n \to \infty$ will go step by step, taking in succession cases which are generalized versions of those in §2. If $p(\xi, A)$ is the integral of a density function which is reasonably positive, the proofs go over with no change. The difficulty is to show how Hypothesis (D) implies the existence and good behavior of a density function. To handle these density functions we introduce a "two-dimensional" measure of (ξ, η) sets, $\xi \in X$, $\eta \in X$ in the usual way. If E and E' are sets in the class $\mathscr{F}_X$, the ξ, η "rectangle" $\hat{E}$ defined by the conditions $\xi \in E$, $\eta \in E'$ is assigned the measure

$$\hat{\varphi}(\hat{E}) = \varphi(E)\varphi(E'),$$

where φ is the measure in Hypothesis (D). This measure is extended in the usual way to the sets of the Borel field $\hat{\mathscr{F}}_X = \mathscr{F}_X \times \mathscr{F}_X$ of the (ξ, η) sets generated by the class of rectangles.

For each ξ and each n, $p^{(n)}(\xi, \cdot)$ is a measure of $\mathscr{F}_X$ sets, and therefore has an absolutely continuous and a singular component with respect to the measure φ (see Supplement, §2). That is, we can write

$$p^{(n)}(\xi, E) = \int_E p_0^{(n)}(\xi, \eta)\varphi(d\eta) + \Delta^{(n)}(\xi, E), \tag{5.5}$$

where $p_0^{(n)}(\xi, \cdot)$ is an η function measurable with respect to $\mathscr{F}_X$, and $\Delta^{(n)}(\xi, \cdot)$ is a measure of $\mathscr{F}_X$ sets with maximum value on a set (depending on n and ξ), of measure 0. *We shall denote by Condition* (Σ) *the condition that this representation of* $p^{(n)}$ *is possible for all* n *with* $p_0^{(n)}$ *measurable in the pair* ξ, η, *that is, measurable with respect to* $\hat{\mathscr{F}}_X$, and when Condition (Σ) is satisfied we shall always suppose that the $p_0^{(n)}$ used has this property. Attention will be called to every use of this condition, which will be used only to obtain certain preliminary results. The final results will not require the validity of this condition. Note that, if Condition (Σ) is satisfied, $\Delta^{(n)}(\cdot, A)$ is a ξ function measurable with respect to $\mathscr{F}_X$. Moreover, under this condition,

$$p^{(m+n)}(\xi, A) = \int_X p^{(m)}(\zeta, A)p^{(n)}(\xi, d\zeta)$$

$$\geq \int_A \varphi(d\eta) \int_X p_0^{(m)}(\zeta, \eta)p_0^{(n)}(\xi, \zeta)\varphi(d\zeta),$$

so that, if A is a subset of the complement of the singular set of the measure $\Delta^{(m+n)}(\xi, \cdot)$,

$$\int_A p_0^{(m+n)}(\xi, \eta)\varphi(d\eta) \geq \int_A \varphi(d\eta) \int_X p_0^{(m)}(\zeta, \eta)p_0^{(n)}(\xi, \zeta)\varphi(d\zeta).$$

Then this inequality holds for all A, since the singular set in question has φ measure 0, and it follows that

$$p_0^{(m+n)}(\xi, \eta) \geq \int_X p_0^{(m)}(\zeta, \eta)p_0^{(n)}(\xi, \zeta)\varphi(d\zeta)$$

for almost all η (φ measure). *We shall assume from now on, if Condition* (Σ) *is satisfied, that the* $p_0^{(k)}$*'s have been chosen in such a way that this inequality is true for all* η. To justify this assumption, we must show that the $p_0^{(k)}$'s can be chosen to satisfy this condition. Such a choice can be made as follows. Assume first any choice of the $p_0^{(k)}$'s. We then choose $\hat{p}_0^{(k)}$'s, new versions of the $p_0^{(k)}$'s, which satisfy the stated condition. Choose $\hat{p}_0^{(1)} = p_0^{(1)}$. If $\hat{p}_0^{(j)}$ has been chosen already, for $j < k$ $(k > 1)$ we have seen that for each ξ

$$p_0^{(k)}(\xi, \eta) \geq \operatorname*{Max}_{1 \leq m < k} \int_X \hat{p}_0^{(m)}(\zeta, \eta)\hat{p}_0^{(k-m)}(\xi, \zeta)\varphi(d\zeta),$$

for almost all η. Define $\hat{p}_0^{(k)}(\xi, \eta)$ as the maximum of the left and right sides of the inequality, for all ξ, η. It is immediately verified that this choice satisfies the stated conditions.

According to Example 2.7 of the Supplement, Condition (Σ) is satisfied if the following condition is satisfied.

CONDITION (Σ') *There is a sequence of* $\mathscr{F}_X$ *sets, generating the Borel field* $\mathscr{G}$, *such that to every* $\mathscr{F}_X$ *set corresponds a* $\mathscr{G}$ *set differing from it by a set of* φ *measure* 0.

Condition (Σ') is satisfied in most cases of interest. For example, if X is a Borel set in k-dimensional Cartesian space, if $\mathscr{F}_X$ is the class of

k-dimensional Borel subsets of X, and if φ is a measure of Borel sets, the sequence of $\mathscr{F}_X$ sets of Condition (Σ') can be taken as the intersections with X of the k-dimensional open intervals with rational vertices.

We now discuss various cases, always under (D) and sometimes under the additional condition (Σ).

Case (a) *$p(\xi, E) = p(E)$ is independent of ξ.* In this case,

$$p^{(n)}(\xi, E) = p(E)$$

for $n > 1$, the random variables of the Markov process are mutually independent, and the discussion of §2, Case (a), needs no modification.

Case (b) *Suppose that* (D) *is strengthened to* (D'): *there is a measure φ of $\mathscr{F}_X$ sets, with $0 < \varphi(X) < \infty$, an integer $\nu \geq 1$, a $\delta > 0$, and an $\mathscr{F}_X$ set C for which*

$$\varphi(C) > 0,$$

$$p_0^{(\nu)}(\xi, \eta) \geq \delta, \qquad \xi \in X, \eta \in C.$$

Here $p_0^{(\nu)}(\xi, \cdot)$ is, as above, the density of the absolutely continuous component of $p^{(\nu)}(\xi, \cdot)$ with respect to φ, but we do not suppose that Condition (Σ) is satisfied. Then there is a stationary absolute probability distribution $p(\cdot)$, with $p(C_1) \geq \delta\varphi(C_1)$ when $C_1 \subset C$, for which

$$|p^{(n)}(\xi, A) - p(A)| \leq [1 - \delta\varphi(C)]^{(n/\nu)-1}, \qquad n = 1, 2, \cdots. \tag{5.6}$$

We observe that (D') implies (D), since if (D') is true and if $\varphi(A) \leq \varphi(C)/2$,

$$p^{(\nu)}(\xi, A) \leq 1 - \int_{(X-A)C} p_0^{(\nu)}(\xi, \eta)\varphi(d\eta) \leq 1 - \frac{\delta\varphi(C)}{2},$$

so that (D) is true with the same ν, φ, and with ε the smaller of the numbers $\varphi(C)/2$, $\delta\varphi(C)/2$.

Case (b) is the obvious analogue of §2, Case (b), and the proof of (5.6) will only be sketched since it follows that of §2 (2.7). Define $m_E^{(r)}$, $M_E^{(r)}$ by

$$m_E^{(r)} = \underset{\xi}{\text{G.L.B.}}\, p^{(r)}(\xi, E), \qquad M_E^{(r)} = \underset{\xi}{\text{L.U.B.}}\, p^{(r)}(\xi, E).$$

Then just as in §2, cf. (2.4),

$$m_E^{(1)} \leq m_E^{(2)} \leq \cdots \leq \cdots M_E^{(2)} \leq M_E^{(1)}.$$

For fixed ξ and η the set function defined by

$$\psi(E) = p^{(\nu)}(\xi, E) - p^{(\nu)}(\eta, E)$$

is completely additive. There is therefore a certain set S^+ (that on which $\psi(E)$ is a maximum) on every $\mathscr{F}_X$ subset of which $\psi(E) \geq 0$ and on every $\mathscr{F}_X$ subset of whose complement S^-, $\psi(E) \leq 0$. Then

$$\psi(S^+) + \psi(S^-) = \psi(X) = 0,$$

$$\begin{aligned}\psi(S^+) &= p^{(\nu)}(\xi, S^+) - p^{(\nu)}(\eta, S^+)\\ &\leq 1 - \int_{S^-} p_0^{(\nu)}(\xi, \zeta)\varphi(d\zeta) - \int_{S^+} p_0^{(\nu)}(\eta, \zeta)\varphi(d\zeta)\\ &\leq 1 - \delta\varphi(C).\end{aligned}$$

Using these two facts, we find that

$$\begin{aligned}M_E^{(n+\nu)} - m_E^{(n+\nu)} &= \underset{\xi,\eta}{\text{L.U.B.}} \int_X p^{(n)}(\zeta, E)[p^{(\nu)}(\xi, d\zeta) - p^{(\nu)}(\eta, d\zeta)]\\ &\leq \underset{\xi,\eta}{\text{L.U.B.}} \left[\int_{S^+} M_E^{(n)}\psi(d\zeta) + \int_{S^-} m_E^{(n)}\psi(d\zeta)\right]\\ &= \underset{\xi,\eta}{\text{L.U.B.}}\ \psi(S^+)[M_E^{(n)} - m_E^{(n)}]\\ &\leq [1 - \delta\varphi(C)](M_E^{(n)} - m_E^{(n)}).\end{aligned}$$

Then as in §2 we find that

$$M_E^{(k\nu)} - m_E^{(k\nu)} \leq [1 - \delta\varphi(C)]^k, \qquad k \geq 1,$$

that $M_E^{(n)}$, $m_E^{(n)}$ must have a common limit $p(E)$ when $n \to \infty$, and that

$$|p^{(n)}(\xi, E) - p(E)| \leq M_E^{(n)} - m_E^{(n)} \leq [1 - \delta\varphi(C)]^{(n/\nu)-1}.$$

If $C_1 \subset C$,

$$p(C_1) \geq m_{C_1}^{(\nu)} \geq \delta\varphi(C_1).$$

The set function $p(\cdot)$ is non-negative, with $p(X) = 1$. As the uniform limit of completely additive set functions it is itself completely additive. That is, $p(\cdot)$ is a probability measure. From (5.2),

$$p^{(m+n)}(\xi, E) = \int_X p^{(n)}(\eta, E)p^{(m)}(\xi, d\eta),$$

and this becomes, when $m \to \infty$,

$$p(E) = \int_X p^{(n)}(\eta, E)p(d\eta).$$

This equation identifies $p(\cdot)$ as a stationary absolute probability distribution, and ends the discussion of Case (c).

We shall see that under Hypothesis (D) there is a $q(\xi, E)$ for which

$$\lim_{n\to\infty} \frac{1}{n} \sum_{m=1}^{n} p^{(m)}(\xi, E) = q(\xi, E)$$

exists uniformly in ξ and E. We omit the proof corresponding to that of Theorem 2.1 because it involves ideas of compactness in Banach spaces, and abstract ergodic theorems which would take us too far afield. The result will be obtained by a detailed discussion of the actual transitions, following Doeblin.

Case (c) *Suppose that the* φ *of* (D) *has the property that, whenever* $\varphi(E) > 0$,

$$\text{L.U.B.}_{n}\, p^{(n)}(\xi, E) > 0$$

for all ξ. This condition is obviously equivalent to the condition that

$$\text{L.U.B.}_{n}\, \tilde{p}^{(n)}(\xi, E) > 0$$

for all ξ, which implies, by Lemma 5.1, that $\lim_{n\to\infty} \tilde{p}_n(\xi, E) = 1$ uniformly in ξ. In other words, in this natural generalization of Case (c) of §2 we suppose that every ξ has positive probability of entering (at some time) every set E of positive φ measure. The next two lemmas will make it possible to use the arguments of §2 to treat this case.

LEMMA 5.2 *Let* $\hat{H}$ *be a* (ξ, η) *set in* $\hat{\mathscr{F}}_X$. *There is then a sequence of decompositions of* X,

$$X = \bigcup_{j=1}^{a_n} H_j^{(n)}, \qquad n = 1, 2, \cdots, \quad H_j^{(n)} \in \mathscr{F}_X, \quad H_j^{(n)} H_k^{(n)} = 0\,(j \neq k),$$

with the property that, if for each n *the subscripts* i, j *are chosen, as functions of* ξ, η, *so that* $\xi \in H_i^{(n)}$, $\eta \in H_j^{(n)}$, *and if* $\hat{H}_{ij}^{(n)}$ *is the* (ξ, η) *set* $\{\xi \in H_i^{(n)}, \eta \in H_j^{(n)}\}$ *then*

$$\lim_{n\to\infty} \frac{\hat{\varphi}(\hat{H} \cdot \hat{H}_{ij}^{(n)})}{\hat{\varphi}(\hat{H}_{ij}^{(n)})} = 1 \tag{5.7}$$

for almost all (ξ, η) *in* $\hat{H}$ *(*$\hat{\varphi}$ *measure).*

This lemma states, in more geometric language, that the $H_i^{(n)}$'s can be chosen in such a way that almost all points of $\hat{H}$ have density 1 in $\hat{H}$ relative to the net of the $\hat{H}_{ij}^{(n)}$'s. If we define the function $\hat{g}$ of $\hat{\mathscr{F}}_X$ sets by

$$\hat{g}(\hat{A}) = \varphi(\hat{A}\hat{H}),$$

then

$$\hat{g}(\hat{A}) = \int_{\hat{A}} y(\hat{\xi})\hat{\varphi}(d\hat{\xi}),$$

where y is the characteristic function of $\hat{H}$, that is,

$$y(\hat{\xi}) = 1, \qquad \xi \in \hat{H}$$
$$= 0 \qquad \text{otherwise.}$$

Thus y is the density of the absolutely continuous set function $\hat{g}$ relative to $\hat{\varphi}$. The ratio in (5.7) is the nth generalized difference quotient at (ξ, η) of $\hat{g}$, on the net of $\hat{H}_{ij}{}^{(n)}$'s, relative to $\hat{\varphi}$. If every $\hat{H}_{rs}{}^{(n+1)}$ is a subset of some $\hat{H}_{ij}{}^{(n)}$, this nth generalized difference quotient converges for almost all (ξ, η) when $n \to \infty$ to the derivative of $\hat{g}$ with respect to $\hat{\varphi}$ relative to the net, according to Theorem 2.4 of the Supplement. According to this same theorem, this derivative is equal to the density y almost everywhere if y is measurable with respect to the Borel field generated by the $\hat{H}_{ij}{}^{(n)}$'s, that is, if $\hat{H}$ is in this Borel field. Now (Supplement, Theorem 2.5) there is a sequence $\{B_n\}$ of $\mathscr{F}_X$ sets such that, if $\mathscr{F}_X{}'$ is the Borel field generated by $\{B_n\}$, then $\hat{H} \in \mathscr{F}_X{}' \times \mathscr{F}_X{}'$. For each n define $H_1{}^{(n)}, \cdots, H_{2^n}{}^{(n)}$ as the intersections of the form $\bigcap_1^n A_j$, where A_j is either B_j or $X - B_j$. These $H_j{}^{(n)}$'s are disjunct with union X, for each n, each $H_j{}^{(n+1)}$ is a subset of some $H_i{}^{(n)}$, and the class of $H_i{}^{(n)}$'s, $i, n \geq 1$, generates the Borel field $\mathscr{F}_X{}'$. Then the $H_i{}^{(n)}$'s satisfy the requirements of the lemma.

LEMMA 5.3 *Under Conditions* (D) *and* (Σ) *there are sets* $A \in \mathscr{F}_X$, $B \in \mathscr{F}_X$, *for which*

$$\varphi(A) > 0, \qquad \varphi(B) > 0$$
$$\underset{\substack{\xi \in A \\ \eta \in B}}{\text{G.L.B.}}\, p_0{}^{(2\nu)}(\xi, \eta) > 0$$

To prove this let $\hat{H}$ be the (ξ, η) set where

$$\frac{\varepsilon}{2\varphi(X)} \leq p_0{}^{(\nu)}(\xi, \eta) \leq \frac{1}{\varepsilon}.$$

Here ε is as described in (D). Let A_ξ be the η set for which $(\xi, \eta) \in \hat{H}$ and let B_η be the ξ set for which $(\xi, \eta) \in \hat{H}$. We prove first that

$$\varphi(A_\xi) \geq \frac{\varepsilon^2}{2}. \tag{5.8}$$

In fact,

$$\begin{aligned}\frac{\varphi(A_\xi)}{\varepsilon} &\geq \int_{A_\xi} p_0{}^{(\nu)}(\xi, \eta)\varphi(d\eta) \\ &= 1 - \int_{\{p_0{}^{(\nu)}(\xi,\eta) < \varepsilon/[2\phi(X)]\}} p_0{}^{(\nu)}(\xi, \eta)\varphi(d\eta) \\ &\qquad - \int_{\{p_0{}^{(\nu)}(\xi,\eta) > 1/\varepsilon\}} p_0{}^{(\nu)}(\xi, \eta)\varphi(d\eta) - \Delta^{(\nu)}(\xi, X).\end{aligned}$$

Now the first integral on the right side of this inequality is at most $\varepsilon/2$. The domain of integration of the second has φ measure at most ε (since the value of the integral is at most 1). Hence the second integral and $\Delta^{(\nu)}(\xi, X)$, which is the contribution to $p^{(\nu)}(\xi, X)$ from a set of φ measure 0 give a total contribution of at most $1 - \varepsilon$, by condition (D). Thus

$$\frac{\varphi(A_\xi)}{\varepsilon} \geq 1 - \frac{\varepsilon}{2} - (1 - \varepsilon) \geq \frac{\varepsilon}{2},$$

and we have proved (5.8). We now apply Lemma 5.2 to $\hat{H}$, and show that *there are two points* (ξ_0, η_0) *and* (ξ_1, η_1) *in* $\hat{H}$ *where there is a limit* 1 *in* (5.7) *and for which* $\eta_0 = \xi_1$. According to Lemma 5.2, applying Fubini's theorem, there is a ξ set X' with $\varphi(X - X') = 0$ and for each $\xi \in X'$ a φ measurable set $A_\xi' \subset A_\xi$ with $\varphi(A_\xi - A_\xi') = 0$ such that the limit in (5.7) is 1 if $\xi \in X'$, $\eta \in A_\xi'$. Choose $\xi_0, \eta_0, \xi_1, \eta_1$ to satisfy the following conditions:

$$\xi_0 \in X'$$

$$\eta_0 = \xi_1 \in A_{\xi_0}'X'$$

$$\eta_1 \in A_{\xi_1}'.$$

This is possible since $\varphi(X') = \varphi(X) > 0$ and, by (5.8), $\varphi(A_{\xi_0}'X') = \varphi(A_{\xi_0}) \geq \varepsilon^2/2$, and $\varphi(A_{\xi_1}') = \varphi(A_{\xi_1}) \geq \varepsilon^2/2$. The two points (ξ_0, η_0), (ξ_1, η_1) have the desired properties. Lemma 5.3 will now be proved. For each n choose i, j, k so that

$$\xi_0 \in H_i^{(n)}, \ \eta_0 = \xi_1 \in H_j^{(n)}, \ \eta_1 \in H_k^{(n)}$$

(using the notation of Lemma 5.2). Denote by $\hat{H}_{rs}^{(n)}$ the $\hat{X}$ set of points ξ, η where $\xi \in H_r^{(n)}$, $\eta \in H_s^{(n)}$. By Lemma 5.2 n can be chosen so large that

$$\int_{H_i^{(n)}} \varphi(A_\xi H_j^{(n)})\varphi(d\xi) = \hat{\varphi}(\hat{H}\,\hat{H}_{ij}^{(n)}) > \tfrac{3}{4}\hat{\varphi}(\hat{H}_{ij}^{(n)}) = \tfrac{3}{4}\varphi(H_i^{(n)})\varphi(H_j^{(n)})$$

$$\int_{H_k^{(n)}} \varphi(B_\eta H_j^{(n)})\varphi(d\eta) = \hat{\varphi}(\hat{H}\,\hat{H}_{jk}^{(n)}) > \tfrac{3}{4}\hat{\varphi}(\hat{H}_{jk}^{(n)}) = \tfrac{3}{4}\varphi(H_j^{(n)})\varphi(H_k^{(n)}).$$

In the following n will be chosen in this way and held fast. There are then sets A and B in $\mathscr{F}_X$, for which

$$A \subset H_i^{(n)}, \quad \varphi(A) > 0, \quad \varphi(A_\xi H_j^{(n)}) > \tfrac{3}{4}\varphi(H_j^{(n)}), \quad (\xi \in A),$$

$$B \subset H_k^{(n)}, \quad \varphi(B) > 0, \quad \varphi(B_\eta H_j^{(n)}) > \tfrac{3}{4}\varphi(H_j^{(n)}), \quad (\eta \in B).$$

It follows that $\varphi(A_\xi B_\eta H_j^{(n)}) > \frac{1}{2}\varphi(H_j^{(n)})$ so that, if $\xi \in A$ and $\eta \in B$,

$$p_0^{(2\nu)}(\xi, \eta) \geq \int_{A_\xi B_\eta H_j^{(n)}} p_0^{(\nu)}(\xi, \zeta) p_0^{(\nu)}(\zeta, \eta)\varphi(d\zeta) \geq \left[\frac{\varepsilon}{2\varphi(X)}\right]^2 \frac{1}{2}\varphi(H_j^{(n)}),$$

which proves the lemma.

LEMMA 5.4 *In Case* (c), *under Conditions* (D) *and* (Σ), *there is a set* C, *with* $\varphi(C) > 0$, *for which*

$$\underset{\xi, \eta \in C}{\text{G.L.B.}}\, p_0^{(\alpha)}(\xi, \eta) > 0$$

for some α.

In fact, let A and B be as in Lemma 5.3. In Case (c) there must be a set $C \in \mathscr{F}_X$, a positive δ, and a positive integer β for which

$$C \subset B, \quad \varphi(C) > 0, \quad p^{(\beta)}(\xi, A) \geq \delta, \qquad (\xi \in C).$$

Then, taking into account transitions from C to A to C, we find that

$$p_0^{(\beta+2\nu)}(\xi, \eta) \geq \delta \underset{\substack{\xi \in A \\ \eta \in B}}{\text{G.L.B.}}\, p_0^{(2\nu)}(\xi, \eta).$$

This proves the lemma, with $\alpha = \beta + 2\nu$.

We proceed to the study of the transitions in Case (c). For each C with the properties described in Lemma 5.4 let $I(C)$ be the class of integers n for which

$$\underset{\xi, \eta \in C}{\text{G.L.B.}}\, p_0^{(n)}(\xi, \eta) > 0.$$

The class $I(C)$ contains $n_1 + n_2$ if it contains n_1 and n_2, and therefore, according to Lemma 2.2, contains all large multiples of $d(C)$, the highest common factor of its members. Moreover, we now prove that $d(C)$ is independent of C. In fact, if $n_1 \in I(C_1)$, if $n_2 \in I(C_2)$, if n_{12} is chosen so that $p^{(n_{12})}(\xi, C_2) > 0$ on a subset of C_1 of positive φ measure [hypothesis of Case (c)], and if n_{21} is chosen so that $p^{(n_{21})}(\xi, C_1) > 0$ on a subset of C_2 of positive φ measure, the succession of transitions C_1 to C_1 to C_2 to C_2 to C_1 to C_1 corresponds to the fact that

$$n_1 + n_{12} + n_2 + n_{21} + n_1 \in I(C_1).$$

Since this is true for all $n_2 \in I(C_2)$, $d(C_1)$ is a factor of every difference between two values of n_2; in particular, $d(C_1)$ is a factor of $d(C_2)$. Hence $d(C_1) = d(C_2)$ because C_1 and C_2 can be interchanged in this argument. In the following we shall write d for $d(C)$.

Suppose now that $d = 1$; the general case will be reduced to this one. Let C have the properties described in Lemma 5.4. By Lemma 5.1 there is a $\delta > 0$ and an integer μ for which

$$\tilde{p}^{(\mu)}(\xi, C) \geq \delta, \qquad \xi \in X.$$

Then, since

$$\tilde{p}^{(\mu)}(\xi, C) \leq \sum_{j=1}^{\mu} p^{(j)}(\xi, C),$$

there is a $\beta = \beta(\xi) \leq \mu$ for which $p^{(\beta)}(\xi, C) \geq \delta/\mu$. The hypothesis $d = 1$ means that $I(C)$ contains all sufficiently large integers, say all $\geq N$. It follows that, if $\eta \in C$ and ξ is arbitrary,

$$\begin{aligned} p_0^{(N+\mu)}(\xi, \eta) &\geq \int_C p_0^{(N+\mu-\beta)}(\zeta, \eta) p^{(\beta)}(\xi, d\zeta) \\ &\geq \frac{\delta}{\mu} \operatorname*{G.L.B.}_{\eta, \zeta \in C} p_0^{(N+\mu-\beta)}(\zeta, \eta) \\ &\geq \frac{\delta}{\mu} \operatorname*{Min}_{N \leq n \leq N+\mu-1} \operatorname*{G.L.B.}_{\eta, \zeta \in C} p_0^{(n)}(\zeta, \eta) > 0, \end{aligned}$$

and this is precisely the hypothesis of Case (b) with the ν of that hypothesis identified with $N + \mu$ here. Then, according to the discussion under Case (b), there is a stationary absolute probability distribution π such that

$$\lim_{n \to \infty} p^{(n)}(\xi, E) = \pi(E)$$

exists uniformly in ξ and E. The limit is approached with exponential speed. Moreover, if $C_1 \subset C$ and if $\varphi(C_1) > 0$, then $\pi(C_1) > 0$. We shall prove that in the present case $\varphi(E) > 0$ implies that $\pi(E) > 0$ for all E (in $\mathscr{F}_X$). In fact (stationarity)

$$\begin{aligned} \pi(E) = \int_X p^{(m)}(\xi, E)\pi(d\xi) &= \frac{1}{n} \sum_1^n \int_X p^{(m)}(\xi, E)\pi(d\xi) \\ &\geq \frac{1}{n} \int_X \tilde{p}^{(n)}(\xi, E)\pi(d\xi). \end{aligned}$$

Now in Case (c), if $\varphi(E) > 0$, $\tilde{p}^{(n)}(\xi, E)$ is positive for large n (and in fact converges uniformly to 1 when $n \to \infty$, by Lemma 5.1). Then $\pi(E) > 0$ if $\varphi(E) > 0$.

Now suppose that $d > 1$, and define $\tilde{C}_j$, $j = 1, \cdots, d$, as the ξ set for which $p^{(nd-j)}(\xi, C) > 0$, for some integer $n \geq 1$. Then $X = \bigcup_{j=1}^{d} \tilde{C}_j$, by the hypothesis of Case (c). The $\tilde{C}_j$'s are not necessarily disjunct.

LEMMA 5.5 *For all n, under Conditions* (D) *and* (Σ),

$$p^{(n)}(\xi, \tilde{C}_j\tilde{C}_k) = 0, \qquad j \neq k$$

except perhaps on a ξ set of φ measure 0, *and*

$$\varphi(\tilde{C}_j\tilde{C}_k) = 0.$$

In fact, if $p^{(\alpha)}(\xi, \tilde{C}_j\tilde{C}_k) > 0$ on a set of positive φ measure, there are integers m and n for which the inequalities

$$p^{(\alpha+md-j)}(\xi, C) > 0, \qquad p^{(\alpha+nd-k)}(\xi, C) > 0$$

are simultaneously true on a set of positive φ measure and (decreasing the latter set slightly to another set E of positive φ measure) for which both inequalities are true with 0 replaced on the right-hand sides by a positive number. By the hypothesis of Case (c) there is a $C' \subset C$, with $\varphi(C') > 0$ and a β, such that

$$\underset{\xi \in C'}{\text{G.L.B.}}\, p^{(\beta)}(\xi, E) > 0.$$

Hence if $i \in I(C)$ the transitions C to C' to E to C to C show that

$$i + \beta + \alpha + md - j + i \in I(C)$$

$$i + \beta + \alpha + nd - k + i \in I(C).$$

Then d is a factor of both left-hand sums, and therefore of their difference $|k-j|$. This implies that $k = j$, and finishes the proof of the first part of the lemma. To prove the second part we note that if $\varphi(\tilde{C}_j\tilde{C}_k) > 0$ there is an n for which $p^{(n)}(\xi, \tilde{C}_j\tilde{C}_k) > 0$ on a ξ set of positive φ measure by the hypothesis of Case (c), which means, by the first part of the lemma, that $j = k$.

We shall use the following *reduction* procedure below. Suppose that $F \in \mathscr{F}_X$ and that $p(\xi, F) = 0$ if $\xi \in X - F$. Then if $X - F$ is not empty the *reduced function* $p(\cdot, \cdot)$ of ξ and E, considered only for $\xi \in X - F$ and $E \subset X - F$, is a stochastic transition function with X replaced by $X - F$. If a point is ever in $X - F$ it stays there, so that

$$\tilde{p}^{(n)}(\xi, X - F) = p^{(n)}(\xi, X - F), \qquad \xi \in X.$$

For example, if $F_0 \in \mathscr{F}_X$ and if F_n is the ξ set defined by the condition $p^{(n)}(\xi, F_0) > 0$, then $F = \bigcup_{n=0}^{\infty} F_n$ has the property demanded of F above. Finally we remark that in every case, *if Condition* (D) *is satisfied by* $p(\xi, E)$, *it is also satisfied by the reduced* $p(\xi, E)$, *and in particular if* $\varphi(F) = 0$,

$$\lim_{n\to\infty} p^{(n)}(\xi, F) = 0$$

uniformly in $\xi \in X$. The first part of this statement is obvious; the second follows from Lemma 5.1 since [Condition (D)]

$$p^{(\nu)}(\xi, X - F) = \tilde{p}^{(\nu)}(\xi, X - F) \geq \varepsilon, \qquad \xi \in X,$$

if $\varphi(F) = 0$. The convergence to 0 will be exponentially fast, according to Lemma 5.1.

Now, if the $\tilde{C}_j$'s of Lemma 5.5 were actually disjunct, we should have

$$p(\xi, \tilde{C}_{j+1}) = 1, \qquad \xi \in \tilde{C}_j$$

(interpreting $\tilde{C}_{d+1}$ as $\tilde{C}_1$). The system would run cyclically through states in $\tilde{C}_1, \tilde{C}_2, \cdots, \tilde{C}_d, \tilde{C}_1, \tilde{C}_2, \cdots$. Moreover, the function $p^{(d)}(\xi, E)$, for $\xi \in \tilde{C}_j$, $E \subset \tilde{C}_j$, would be a stochastic transition function with the properties the original $p(\xi, E)$ would have in case $d = 1$. It would then follow from the preceding work that

$$\lim_{n\to\infty} p^{(nd)}(\xi, E) \qquad (\xi \in \tilde{C}_j)$$

exists uniformly in ξ and E. Although in general the $\tilde{C}_j$'s are not disjunct, we can apply the reduction procedure described above to obtain a reduced stochastic transition function for which they are disjunct. We delete from X the set $F_0 = \bigcup_{j\neq k} \tilde{C}_j\tilde{C}_k$ augmented by the F_n's as described in the reduction procedure. According to Lemma 5.5, $\varphi(F_0) = 0$. If $\varphi(F_n) > 0$ for some n, there would be positive probability that a point in $X - F_0$ would go into F_n in some number (say m) of transitions, according to the hypothesis of Case (c). Then that point in $X - F_0$ would go into F_0 in $m + n$ transitions, with positive probability, contradicting Lemma 5.5. Thus $\varphi(F_n) = 0$ for all n, and it follows that $\varphi(F) = 0$. For the reduced stochastic transition function the $\tilde{C}_j$'s become the disjunct sets

$$C_1, \cdots, C_d, \qquad C_j = \tilde{C}_j(X - F).$$

Then the remarks made above relevant to disjunct C_j's apply. Taking into account the fact (proved in discussing the reduction procedure) that

$$\lim_{n\to\infty} p^{(n)}(\xi, F) = 0 \qquad \xi \in X$$

uniformly (exponentially fast) in ξ, we find that

$$\lim_{n\to\infty} p^{(nd)}(\xi, E) = \pi_\alpha(E), \qquad \xi \in C_\alpha$$

uniformly in ξ and $E \in \mathscr{F}_X$. Here π_α is a probability measure of sets E, with

$$\pi_\alpha(C_\alpha) = 1, \qquad \pi_\alpha(E) > 0 \quad \text{if} \quad \varphi(EC_\alpha) > 0.$$

More generally, as in §2,

$$p^{(nd+m)}(\xi, E) = p^{(nd+m)}(\xi, EC_\beta), \qquad \xi \in C_\alpha$$
$$\beta \equiv \alpha + m \pmod d$$

and

$$\lim_{n\to\infty} p^{(nd+m)}(\xi, E) = \pi_\beta(E), \qquad \xi \in C_\alpha$$
$$\beta \equiv \alpha + m \pmod d.$$

The convergence is uniform and exponentially fast. This equation implies that

$$\lim_{n\to\infty} \frac{1}{n} \sum_{m=1}^{n} p^{(m)}(\xi, E) = \frac{\sum_{\alpha=1}^{d} \pi_\alpha(E)}{d}, \qquad \xi \in \bigcup_1^d C_\alpha.$$

The asymptotic character of $p^{(n)}(\xi, E)$ for $\xi \in F$ will be investigated in the next (more general) case.

Case (d) *General case under Hypothesis* (D). A set E will be called a *consequent* set if, for some ξ_0, $p^{(n)}(\xi_0, E) = 1$ for all n, and in this case E will be called a *consequent* of ξ_0. By (D), if E is a consequent set $\varphi(E) \geq \varepsilon$. A set which is a consequent of every one of its points will be called an *invariant set*. An invariant set is then either empty or has φ measure $\geq \varepsilon$.

If E is a consequent of ξ_0 and if F_n is the set of points ξ of E for which $p^{(n)}(\xi, E) < 1$, then $E - F_n$ is a consequent of ξ_0, since otherwise $p^{(m)}(\xi_0, F_n) > 0$ for some m, and then

$$\begin{aligned} 1 = p^{(m+n)}(\xi_0, E) &= \int_{F_n} p^{(n)}(\xi, E) p^{(m)}(\xi_0, d\xi) \\ &\quad + \int_{E-F_n} p^{(n)}(\xi, E) p^{(m)}(\xi_0, d\xi) \\ &< p^{(m)}(\xi_0, F_n) + p^{(m)}(\xi_0, E - F_n) = p^{(m)}(\xi_0, E) = 1. \end{aligned}$$

Thus if E is a consequent set it contains the non-null invariant set $\bigcap_1^\infty (E - F_n)$.

Now suppose that E is an invariant set. If it contains no (non-null) invariant set of smaller φ measure it will be called *minimal*. Every consequent set E contains a non-null minimal invariant set, which can be obtained as follows. We can assume that E is invariant (decreasing E if necessary to make this true). If E contains no point ξ_1 with consequent E_1 for which $\varphi(E_1) < \varphi(E)$, then E is minimal. If there is such a pair, ξ_1, E_1, E_1 can be supposed invariant. Repeating the argument, we find a finite or infinite sequence $E_1, E_2, \cdots$ of invariant sets, with

$$E_1 \supset E_2 \supset \cdots$$

$$\varphi(E_1) > \varphi(E_2) > \cdots \geq \varepsilon.$$

Let $\rho_n =$ G.L.B. $\varphi(F)$ for F non-null, invariant, and a subset of E_n. Then we can suppose that if E_n is not minimal E_{n+1} is chosen so that $\varphi(E_{n+1}) < \rho_n + 1/n$. If there are only finitely many E_j's, the last is the desired minimal invariant set; if there are infinitely many, $\bigcap_1^\infty E_n$ is the desired set, since it is non-null, is invariant, and has φ measure $\lim_{n\to\infty} \rho_n$.

If E_1 and E_2 are invariant, E_1E_2 must also be invariant. In particular, if E_1 is minimal, E_1E_2 is an invariant subset of a minimal invariant set. It follows that $\varphi(E_1) = \varphi(E_1E_2)$ unless E_1E_2 is the null set. Hence (if E_2 is also minimal) *two minimal invariant sets are either disjunct or differ by at most a set of* φ *measure* 0. Let $E_1, E_2, \cdots$ be an enumeration of the essentially distinct non-null minimal invariant sets. That is, $E_jE_k = 0$ $(j \neq k)$, $\varphi(E_j) \geq \varepsilon$, and if E is any non-null minimal invariant set it differs from some E_j by at most a set of φ measure 0. There are at most $\varphi(X)/\varepsilon$ E_j's.

For every ξ_0, $\underset{n}{\text{L.U.B.}}\, p^{(n)}(\xi_0, \bigcup_j E_j) > 0$ or $X - \bigcup_j E_j$ would be a consequent of ξ_0 and would, therefore, contain a non-null minimal invariant subset which should have appeared among the E_j's. Thus every point of X will finally enter the E_j's; once in a point stays in, so that

$$p^{(n)}(X, \bigcup_j E_j) = \tilde{p}^{(n)}(\xi, \bigcup_j E_j).$$

THEOREM 5.6 (cf. Theorem 2.3) *For every* ξ,

$$\lim_{n\to\infty} p^{(n)}(\xi, \bigcup_j E_j) = 1,$$

and in fact for some $\rho < 1$

$$1 - p^{(n)}(\xi, \bigcup_j E_j) \leq \text{const. } \rho^n \qquad n = 1, 2, \cdots$$

uniformly in ξ. *Hence, each* ξ *will remain* (*with probability* 1) *outside the* E_j*'s only a finite number of times in its transitions.*

This theorem follows at once from Lemma 5.1 and the Borel-Cantelli lemma.

For each a, $p(\xi, E)$ with $\xi \in E_a$ and $E \subset E_a$ defines a stochastic transition function for which the hypotheses of Case (c) are satisfied. Hence, if we suppose for the moment that Condition (Σ) is satisfied, and if ${}_aC_1, \cdots, {}_aC_{d_a}$ are the cyclically moving classes in E_a, as discussed under Case (c), with E_a reduced as in Case (c), to make $E_a = \bigcup_\alpha {}_aC_\alpha$,

$$(5.9) \quad p^{(nd_a+m)}(\xi, E) = p^{(nd_a+m)}(\xi, E\,{}_aC_\beta), \qquad \xi \in {}_aC_\alpha, \quad \beta \equiv \alpha + m \pmod{d_a},$$

and

$$(5.10) \quad \lim_{n\to\infty} p^{(nd_a+m)}(\xi, E) = {}_a\pi_\beta(E), \qquad \xi \in {}_aC_\alpha, \quad \beta \equiv \alpha + m \pmod{d_a}.$$

The convergence is uniform and exponentially fast. Here ${}_a\pi_\beta$ is a probability measure of sets E, with

$${}_a\pi_\beta({}_aC_\beta) = 1, \qquad {}_a\pi_\beta(E) > 0 \quad \text{if} \quad \varphi(E\,{}_aC_\beta) > 0.$$

The conditional probability that the system, initially at ξ, will finally be in E_a, is

$$\rho(\xi, E_a) = \lim_{n\to\infty} p^{(n)}(\xi, E_a), \tag{5.11}$$

where $p^{(n)}(\xi, E_a)$ is non-decreasing in n since once the system is in E_a it remains there. If $\xi \in E_a$, $\rho(\xi, E_a) = 1$ and in general, according to Theorem 5.6, $\sum_a \rho(\xi, E_a) = 1$. Considering only positions after $d_a, 2d_a, \cdots$ transitions, the conditional probability that the system, initially at ξ, will finally be in ${}_aC_\alpha$ is

$$\rho(\xi, {}_aC_\alpha) = \lim_{n\to\infty} p^{(nd_a)}(\xi, {}_aC_\alpha). \tag{5.12}$$

If $\xi \in {}_aC_\alpha$, $\rho(\xi, {}_aC_\alpha) = 1$. Then

$$\rho(\xi, E_a) = \sum_{\alpha=1}^{d_a} \rho(\xi, {}_aC_\alpha), \qquad \xi \in {}_aC_\alpha,$$

and, as in §2,

$$\lim_{n\to\infty} p^{(nd_a+m)}(\xi, E) = \sum_{a,\alpha} \rho(\xi, {}_aC_\alpha)\, {}_a\pi_{\alpha+m}(E), \tag{5.13}$$

for all ξ, where $\alpha + m$ is to be reduced mod d_a in the subscript. If ξ is in an ergodic set, this reduces to the limit equation already obtained. The convergence is again uniform and exponentially fast. Finally this limit equation implies that, for all $E \in \mathscr{F}_X$,

$$\begin{aligned} \lim_{n\to\infty} \frac{1}{n} \sum_{m=1}^{n} p^{(m)}(\xi, E) &= \sum_{a,\alpha} \sum_{m=1}^{d_a} \frac{\rho(\xi, {}_aC_\alpha)\, {}_a\pi_{\alpha+m}(E)}{d_a} \\ &= \sum_{a,\alpha} p(\xi, {}_aC_\alpha)\, {}_a\pi(E) \\ &= \sum_{a} \rho(\xi, E_a)\, {}_a\pi(E) \end{aligned} \tag{5.14}$$

where

$$ {}_a\pi(E) = \sum_{\alpha=1}^{d_a} \frac{{}_a\pi_\alpha(E)}{d_a}. \tag{5.15}$$

As so defined, ${}_a\pi$ is a probability measure of sets E, with

$$\begin{aligned} {}_a\pi(E_a) &= 1 \\ {}_a\pi(E) &> 0 \qquad \text{if} \quad \varphi(EE_a) > 0. \end{aligned}$$

These equations hold for all ξ and E, and the limits are uniform in ξ and E; the limits are approached exponentially fast where Cesaro sums are not involved.

The results (5.9)–(5.15) have been based on the validity of Condition (Σ) as well as Condition (D). We now show how to obtain these results without Condition (Σ). We shall call a Borel field $\mathscr{F}_X' \subset \mathscr{F}_X$ *admissible* if the ξ function $p(\cdot, A)$ is measurable with respect to $\mathscr{F}_X'$ when $A \in \mathscr{F}_X'$, and if Condition (Σ'), which is stronger than (Σ), is satisfied for $\mathscr{F}_X'$. For example, the Borel field containing only the null set and X is admissible. Then the following statements are true.

(i) *If* $\{B_n\}$ *is a finite or infinite sequence of* $\mathscr{F}_X$ *sets, there is an admissible Borel field* $\mathscr{F}_X'$ *with* $\{B_n\} \subset \mathscr{F}_X'$.

(ii) *If* $\{\mathscr{F}_X^{(n)}\}$ *is a finite or infinite sequence of admissible Borel fields, there is an admissible Borel field* $\mathscr{F}_X'$ *with* $\bigcup_n \mathscr{F}_X^{(n)} \subset \mathscr{F}_X'$.

The first statement implies the second, because, if (i) is true, we can take as the B_n's in (ii) all the $\mathscr{F}_X$ sets in the sequences of sets involved in the (Σ') condition for $\mathscr{F}_X^{(1)}, \mathscr{F}_X^{(2)}, \cdots$. To prove (i), define $\mathscr{F}(A)$ for any $A \in \mathscr{F}_X$ as the denumerable class of ξ sets of the form

$$\{p(\xi, A) \leq r\} \qquad r \text{ rational.}$$

Then define $\mathscr{F}_X'$ as the Borel field of ξ sets generated by $\mathscr{G}_\infty = \bigcup_n \mathscr{G}_n$, where $\mathscr{G}_0 = \{B_n\}$, and, if $\mathscr{G}_0, \cdots, \mathscr{G}_n$ have already been defined,

$$\mathscr{G}_{n+1} = \bigcup_0^n \mathscr{G}_j \cup \mathscr{G}_n', \qquad \mathscr{G}_n' = \bigcup_A{}' \mathscr{F}(A),$$

and the prime on the union symbol means that the union is over all sets A which are finite unions of finite intersections of sets which are $\mathscr{G}_n$ sets or complements of $\mathscr{G}_n$ sets. Then $\mathscr{G}_\infty$ is denumerable. By definition of $\mathscr{F}_X'$, $\{B_n\} \subset \mathscr{F}_X'$. Let $\mathscr{G}$ be the class of $\mathscr{F}_X$ sets A with the property that $p(\cdot, A)$ is a ξ function measurable with respect to $\mathscr{F}_X'$. We show that $\mathscr{F}_X'$ is admissible by showing that $\mathscr{G} \supset \mathscr{F}_X'$. Class $\mathscr{G}_\infty$ is a field of sets, since by definition $\mathscr{G}_{n+1}$ includes the field generated by $\mathscr{G}_n$. Moreover $\mathscr{G} \supset \mathscr{G}_\infty$ because, if $A \in \mathscr{G}_n$, $p(\cdot, A)$ is a ξ function measurable with respect to the Borel field generated by $\mathscr{G}_{n+1}$. Finally $\mathscr{G}$ obviously includes limits of monotone sequences of $\mathscr{G}$ sets. Hence (Supplement, Theorem 1.2), $\mathscr{G} \supset \mathscr{F}_X'$, as was to be proved.

Since all the results of this section are applicable if $\mathscr{F}_X$ is replaced by an admissible Borel field, we can use the properties (i), (ii) of these admissible Borel fields to obtain results for $\mathscr{F}_X$ itself. For example, if $A \in \mathscr{F}_X$ and if $\mathscr{F}_X$ satisfies Condition (Σ), we have proved that

$$\lim_{n\to\infty} \frac{1}{n} \sum_1^n p^{(j)}(\xi, A)$$

exists, uniformly in ξ and A. If $\mathscr{F}_X$ does not satisfy Condition (Σ) and if $A \in \mathscr{F}_X$, replace $\mathscr{F}_X$ by an admissible Borel field containing A to get that the above limit exists uniformly in ξ. To get the full results of this section, we proceed as follows. We defined ergodic classes $E_1, E_2, \cdots$ without the use of Condition (Σ). Let $\mathscr{F}_X'$ be any admissible Borel field of ξ sets, such that $\{E_n\} \subset \mathscr{F}_X'$. There is such an admissible Borel field by (i). Then, replacing $\mathscr{F}_X$ by $\mathscr{F}_X'$, we find cyclically moving classes ${}_aC_1, \cdots, {}_aC_{d_a}$, $d_a \geq 1$ in E_a, relative to $\mathscr{F}_X'$. If $\mathscr{F}_X''$ is another admissible Borel field, with $\mathscr{F}_X' \subset \mathscr{F}_X''$, the cyclically moving classes relative to $\mathscr{F}_X''$ are for each a the same ${}_aC_j$'s, or sets obtained from these by a finer decomposition of E_a with d_a a multiple of its previous value. Since $d_a \leq \varphi(X)/\varepsilon$, there must be for each a an admissible $\mathscr{F}_X'$ maximizing d_a. Then, using (ii), there must be an admissible Borel field $\mathscr{G}_X$ maximizing d_a for all a. Now suppose that $\mathscr{F}_X'$ is any admissible Borel field of ξ sets, with $\mathscr{G}_X \subset \mathscr{F}_X'$. Then the cyclically moving classes relative to $\mathscr{F}_X'$ can be taken the same as those relative to $\mathscr{G}_X$, and all the constants which govern the speed of convergence will be the same as those relative to $\mathscr{G}_X$. Since $\mathscr{F}_X'$ can be taken to include any $\mathscr{F}_X$ set, by combining (i) and (ii), we have proved that all our results, including the uniformity results, summarized in (5.9)–(5.15), are true even without Condition (Σ).

If a stochastic transition function $p(\cdot, \cdot)$ satisfies Condition (D) with the triple φ, ν, ε, the triple will be called a (D) triple (for the given transition function). If a stochastic transition function has one (D) triple, it has infinitely many. This fact, together with the fact that even for a given (D) triple the decomposition of X obtained above is not uniquely determined, obscures to some extent the significance of the condition and the results obtained. The following discussion is given to clarify the choice of (D) triples, and the decompositions of X obtained.

In the following, if F is a set for which

$$\lim_{n\to\infty} p^{(n)}(\xi, F) = 0, \qquad \xi \in X,$$

the set F will be called a transient set. Suppose that there is a decomposition of X into disjunct invariant sets $E_1, E_2, \cdots$ and a transient set $F = X - \bigcup_a E_a$, and that to each E_a corresponds a probability measure ${}_a\pi$ of sets $E \in \mathscr{F}_X$ such that

$${}_a\pi(E_a) = 1, \qquad \lim_{n\to\infty} \frac{1}{n} \sum_{m=1}^{n} p^{(m)}(\xi, E) = {}_a\pi(E), \qquad \xi \in E_a.$$

Then the E_a's will be called *ergodic sets*. Suppose furthermore that E_a can be decomposed into $d_a > 1$ disjunct sets ${}_aC_1, \cdots, {}_aC_{d_a}$ such that

$$p(\xi, {}_aC_{\alpha+1}) = 1, \qquad \xi \in {}_aC_\alpha, \qquad \alpha = 1, \cdots, d_a$$

(where ${}_aC_{d_a+1}$ is interpreted as ${}_aC_1$), and that to each ${}_aC_\alpha$ corresponds a probability measure ${}_a\pi_\alpha$ of $\mathscr{F}_X$ sets such that

$${}_a\pi_\alpha({}_aC_\alpha) = 1, \qquad \lim_{n\to\infty} p^{(nd_a+m)}(\xi, E) = {}_a\pi_\beta(E), \qquad \xi \in {}_aC_\alpha,$$
$$\beta \equiv \alpha + m \pmod{d_a}.$$

Then the sets ${}_aC_1, \cdots, {}_aC_{d_a}$ will be called *cyclically moving subsets* of the ergodic set E_a. If the above decompositions are possible, $\rho(\xi, E_a)$ and $\rho(\xi, {}_aC_\alpha)$ can be defined as above, and (5.13), (5.14), and (5.15) will be true. We have proved that these decompositions are possible if Condition (D) is satisfied. We now show that, if a decomposition into ergodic and transient sets is possible, the set functions ${}_1\pi, {}_2\pi, \cdots$ are uniquely determined (aside from order). To see this suppose that

$$E_1, E_2, \cdots, F, \qquad {}_1\pi, {}_2\pi, \cdots$$

and

$$E_1', E_2', \cdots, F', \qquad {}_1\pi'(E), {}_2\pi'(E), \cdots$$

are two decompositions into ergodic and transient sets with the corresponding set functions. Clearly no $E_j(E_j')$ can be entirely contained in $F'(F)$, because the ergodic sets are invariant, whereas the points of a transient set finally go into its complement. Set E_1 must therefore have a non-null intersection with some E_a', say with E_1'. Then, if $\xi \in E_1E_1'$, (5.14) implies that ${}_1\pi(E) \equiv {}_1\pi'(E)$. Since no two set functions ${}_a\pi', {}_b\pi'$ with $a \neq b$ can be identical, E_1 cannot have points in common with any other E_a'. An obvious elaboration of the argument we have given then shows that the E_a's and E'_a's can be numbered in such a way that

$$E_a \subset E_a' + F' \qquad E_aE_a' \neq 0$$
$$E_a' \subset E_a + F$$
$${}_a\pi(E) \equiv {}_a\pi'(E), \qquad a = 1, 2, \cdots.$$

Thus we have proved that the set functions ${}_1\pi, {}_2\pi, \cdots$ are uniquely determined (aside from order), and that the ergodic sets are uniquely determined neglecting points in transient sets. A similar argument proves that if there are cyclically moving sets ${}_aC_1, \cdots, {}_aC_{d_a}$, with corresponding set functions ${}_a\pi_1, \cdots, {}_a\pi_{d_a}$, these set functions are uniquely determined (aside from order), and the cyclically moving sets are uniquely determined (aside from order), points in transient sets being neglected. The number of ergodic sets in a given decomposition into ergodic and transient sets is then uniquely determined. If there are n we shall simply say there are n ergodic sets. Similarly we shall say that an ergodic set has d_a cyclically moving subsets if this is the number in any decomposition of the ergodic set into such subsets. In the following we shall suppose that a decomposition into cyclically moving sets and a transient set is possible and that some ordering is used, so that the set functions $\{{}_a\pi_\alpha\}, \{{}_a\pi\}$ are uniquely determined. The function values $\rho(\xi, E_a)$, $\rho(\xi, {}_aC_\alpha)$ are also uniquely determined (that is, they depend only on ξ and a, and on ξ, a, and α respectively), as is obvious from their definition.

We can define a decomposition of X into ergodic and transient sets using only the set functions ${}_1\pi$, ${}_2\pi$, $\cdots$ by setting $\bar{\bar{E}}_a$ as the ξ set where

$$\lim_{n\to\infty} \frac{1}{n}\sum_{m=1}^{n} p^{(m)}(\xi, E) = {}_a\pi(E), \qquad E \in \mathscr{F}_X.$$

(This set $\bar{\bar{E}}_a$ is the set where $\rho(\xi, E_a) = 1$.) Similarly we can define a decomposition into cyclically moving sets by setting ${}_a\bar{C}_\alpha$ as the ξ set where

$$\lim_{n\to\infty} p^{(nd_a)}(\xi, E) = {}_a\pi_\alpha(E), \qquad E \in \mathscr{F}_X.$$

(The set ${}_a\bar{C}_\alpha$ is the set where $\rho(\xi, {}_aC_\alpha) = 1$.) If we define

$$\bar{E}_a = \bigcup_\alpha {}_a\bar{C}_\alpha,$$

the sets $\bar{E}_1, \bar{E}_2, \cdots, X - \bigcup_a \bar{E}_a$ provide still another decomposition into ergodic and transient sets. If $E_1, E_2, \cdots, X - \bigcup_a E_a$ is any decomposition into ergodic and transient sets,

$$E_a \subset \bar{\bar{E}}_a, \qquad a \geq 1,$$

if the ergodic sets are ordered properly. If ${}_aC_1, {}_aC_2, \cdots$ is a further decomposition into cyclically moving sets,

$${}_aC_\alpha \subset {}_a\bar{C}_\alpha,$$

if the cyclically moving sets are ordered properly.

Now suppose that Condition (D) is satisfied, so that ergodic and transient sets $\{E_a\}$, F, and cyclically moving sets $\{{}_aC_\alpha\}$ exist, as proved above. Define $\bar{\varphi}(E)$ by

$$\bar{\varphi}(E) = \sum_a {}_a\pi(E).$$

Then, assuming proper ordering of the sets involved,

$$E_a \subset \bar{\bar{E}}_a \subset E_a + F$$

$${}_aC_\alpha \subset {}_a\bar{C}_\alpha \subset {}_aC_\alpha + F$$

$$\bar{\varphi}(F) = 0.$$

Now we can write the function φ of Condition (D) in the form

$$\varphi(E) = \int_E f(\xi)\bar{\varphi}(d\xi) + \psi(E)$$

where $f(\xi) \geq 0$ and ψ, a finite-valued measure, is the singular component of φ with respect to $\bar{\varphi}$, so that there is a set of $\bar{\varphi}$ measure 0 on which ψ takes its maximum value, $\psi(X)$. Since ${}_a\pi(E) > 0$ whenever $\varphi(EE_a) > 0$, $\bar{\varphi}(E) = 0$ implies that $\varphi(E \bigcup_a E_a) = 0$. Then

$$\psi(\bigcup_a E_a) = 0.$$

In other words, the ψ measure is confined to F. Since

$$p^{(nd_a)}(\xi, {}_aC_\alpha) = 1, \qquad \xi \in {}_aC_\alpha,$$

Condition (D) implies that $\varphi({}_aC_\alpha) \geq \varepsilon$. Then

$$\int_{{}_aC_\alpha} f(\xi)\bar{\varphi}(d\xi) \geq \varepsilon$$

and hence $f(\xi)$ must be > 0 on a set of positive $\bar{\varphi}$ measure on each ${}_aC_\alpha \subset {}_a\bar{C}_\alpha$. Conversely, if Condition (D) is satisfied, and if f_1 is any non-negative function measurable with respect to $\mathscr{F}_X$, and > 0 on a subset of each ${}_a\bar{C}_\alpha$ of positive $\bar{\varphi}$ measure, and if ψ_1 is any finite-valued measure of sets $E \in \mathscr{F}_X$ which is singular with respect to $\bar{\varphi}$ measure, define $\varphi_1(E)$ by

$$\varphi_1(E) = \int_E f_1(\xi)\,\bar{\varphi}(d\xi) + \psi_1(E).$$

Then it is easy to verify that Condition (D) is satisfied with φ_1, ν_1, ε_1, for some ν_1, ε_1. We have thus characterized the class of φ's occurring in (D) triples [assuming that there is at least one (D) triple]. The simplest choice of φ is $\bar{\varphi}$ itself, obtained by setting $f_1 = 1$, $\psi_1 = 0$. With this choice a far stronger condition than (D) is satisfied. It is easily shown, using the intimate relation between $\bar{\varphi}$ and the transition probabilities, that given any $\varepsilon_1 > 0$ there is an $\varepsilon_2 > 0$ and an integer ν such that if $\bar{\varphi}(E) \leq \varepsilon_2$ then $p^{(\nu)}(\xi, E) \leq \varepsilon_1$ for all ξ. (If ε_1 is very small, ν will be very large.) In practice, of course, the existence of $\bar{\varphi}$ is not known until it has been verified that there is some (D) triple, and the φ of this (D) triple will not usually be $\bar{\varphi}$. However, this remark shows that Condition (D) is a posteriori equivalent to much stronger conditions.

The preceding discussion implies that if the $E_a^{(i)}$'s and ${}_aC_\alpha^{(i)}$'s are choices of the ergodic and cyclically moving sets determined by a stochastic transition function satisfying Condition (D), for $i = 1, 2$, then (with proper ordering of the sets)

$$\begin{aligned} &\bar{\varphi}(E_a^{(i)} - E_a^{(1)}\,E_a^{(2)}) = 0 \\ &\bar{\varphi}({}_aC_\alpha^{(i)} - {}_aC_\alpha^{(1)}\,{}_aC_\alpha^{(2)}) = 0 \end{aligned} \qquad i = 1, 2.$$

Moreover, if $\varphi_1, \nu_1, \varepsilon_1$ is a (D) triple for this stochastic transition function and if the $E_a^{(1)}$'s are minimal invariant sets for this (D) triple, as defined above,

$$\varphi_1(E_a^{(1)} - E_a^{(1)} E_a^{(2)}) = 0$$

$$\varphi_1({}_aC_\alpha^{(1)} - {}_aC_\alpha^{(1)} {}_aC_\alpha^{(2)}) = 0.$$

Then, if the $E_a^{(2)}$'s are minimal invariant sets for the (D) triple $\varphi_2, \nu_2, \varepsilon_2$, $E_a^{(1)}$ differs from $E_a^{(2)}$ (and ${}_aC_\alpha^{(1)}$ from ${}_aC_\alpha^{(2)}$) by at most the union of a set of φ_1 measure 0 and a set of φ_2 measure 0.

The stationary absolute probability distributions are fully described in the following theorem.

THEOREM 5.7 *Under Condition* (D), $q(\xi, E)$ [*defined as the limit in* (5.14)] *defines for each* ξ *a stationary absolute probability distribution*; *if* $\xi \in E_a$, $q(\cdot, E)$ *is independent of* ξ, $q(\xi, E) = {}_a\pi(E)$. *Conversely, every stationary absolute probability distribution is a linear combination* (*non-negative coefficients with sum* 1) *of the* ${}_a\pi$*'s. More generally, every solution of*

$$\int_X p(\xi, E)\psi(d\xi) = \psi(E)$$

(ψ *finite-valued and completely additive*) *is a linear combination of the* ${}_a\pi$*'s.*

The proof is exactly the same as in the stochastic matrix case (Theorem 2.4) and will be omitted. This theorem implies that the number of ergodic sets depends only on the given stochastic transition function, not on the (D) triple involved in Condition (D). We have already derived this fact above. It can be shown that the stationary absolute probability distributions form a convex set in a suitably defined linear space, and that ${}_1\pi, {}_2\pi, \cdots$ are the vertices of this set. This gives a characterization of the stationary absolute probability distributions independent of particular (D) triples.

The important classes of stochastic transition functions satisfying Condition (D) can be treated in exactly the same way as stochastic matrices (see §2), and the discussion will be given without proofs. It is supposed throughout that there is a (D) triple $\varphi, \nu, \varepsilon$, and that ergodic sets, a transient set, and cyclically moving sets (if any) have been chosen in some definite way.

Case (e) *The limit* $q(\xi, E)$ *defined by* (5.14) *is independent of* ξ *if and only if there is only a single ergodic set.*

Case (f) *The limit* $q(\xi, E)$ *exists as an ordinary* (*rather than Cesàro*) *limit* if and only if there are no cyclically moving subsets in any ergodic set ($d_a = 1$).

Case (g) *The limit* $q(\xi, E)$ *is positive for all* ξ *whenever* $\varphi(E) > 0$ if and only if the transient set has φ measure 0 and there is only one ergodic set.

Case (h) *Under Condition* (Σ) *if for every* $\delta > 0$ *there are sets* $S_1, S_2, \cdots$ *in* $\mathscr{F}_X$ *with* $\bigcup_j S_j = X$, $\varphi(S_j) \leq \delta$, *and* $p_0(\xi, \eta) > 0$ *for every* $\xi, \eta \in S_j$, *for each* j, then there can be no cyclically moving subsets of ergodic sets. This will be true, for example, if X is a Euclidean space, $\mathscr{F}_X$ the field of Borel sets, and if $p(\cdot, \cdot)$ is continuous with $p_0(\xi, \xi) > 0$ for all ξ.

Case (i) *Under Condition* (Σ) *if* $p(\xi, \cdot)$ *is absolutely continuous with respect to* $\varphi(\cdot)$, *with density* $p_0(\xi, \cdot)$, *and if* $p_0(\xi, \eta) = p_0(\eta, \xi)$, then $p^{(n)}(\xi, \cdot)$ will also have a symmetric density function. There will then be no cyclically moving subsets of ergodic sets, and the transient set will have φ measure 0. The limit $q(\xi, E)$ will be determined by the symmetric density $q_0(\cdot, \cdot)$ (cf. the discussion below of Examples 2 and 3), which satisfies the equations

$$
\begin{aligned}
q_0(\xi, \eta) &= \frac{1}{\varphi(E_a)} \qquad && \xi \in E_a, \quad \eta \in E_a \\
&= 0 && \xi \in E_a, \quad \eta \notin E_a, \\
&= 0 && \eta \in F.
\end{aligned}
$$

In particular, if there is only one ergodic set,

$$
\begin{aligned}
q_0(\xi, \eta) &= \frac{1}{\varphi(X)}, \qquad && \eta \notin F \\
&= 0 && \eta \in F.
\end{aligned}
$$

Case (j) *If*

$$\int_X p(\xi, E)\varphi(d\xi) = \varphi(E), \qquad E \in \mathscr{F}_X$$

(that is, if $\varphi(\cdot)/\varphi(X)$ is a stationary absolute probability distribution), then $p^{(n)}(\cdot, \cdot)$ and therefore $q(\cdot, \cdot)$ satisfy this same equation. Hence (substituting F for E) the transient set F must have φ measure 0. Moreover, since $q(\xi, E) = {}_a\pi(E)$ if $\xi \in E_a$,

$$q(\xi, E) = {}_a\pi(E) = \frac{\varphi(E)}{\varphi(E_a)}, \qquad \xi \in E_a.$$

Thus, as in Case (i), $q(\xi, E)$ is given by a constant density $1/\varphi(E_a)$ for $\xi \in E_a$. Since there may be cyclically moving classes in case (j) [although they cannot be present in Case (i)], the Cesàro limit in (5.14) cannot be replaced by an ordinary limit.

Examples 2,3 (*continued*) Suppose that Condition (Σ) is satisfied and that the stochastic transition function has the property that

$$p^{(\nu)}(\xi, E) \leq \varphi(E)$$

for all ξ and $E \in \mathscr{F}_X$, where ν is some integer and φ is some finite-valued measure. Then we have seen that Condition (D) is satisfied with the given φ and ν if $\varepsilon \leq \frac{1}{2}$. The transition function $p^{(\nu)}(\cdot, \cdot)$ has a density function,

$$p^{(\nu)}(\xi, E) = \int_E p_0^{(\nu)}(\xi, \eta)\varphi(d\eta),$$

and, more generally, $p_0^{(n)}(\xi, \cdot)$, the density of $p^{(n)}(\xi, \cdot)$, is defined for $n > \nu$ by

$$p_0^{(n)}(\xi, \eta) = \int_X p_0^{(\nu)}(\zeta, \eta)p^{(n-\nu)}(\xi, d\zeta).$$

We observe that, although $p_0^{(\nu)}(\cdot, \cdot)$ is not uniquely determined by $p^{(\nu)}(\cdot, \cdot)$, $p_0^{(\nu)}(\xi, \cdot)$ is determined up to an η set of φ measure 0, for each ξ. We can assume that $p_0^{(\nu)}(\cdot, \cdot)$ satisfies the conditions

$$\begin{aligned} &0 \leq p_0^{(\nu)}(\xi, \eta) \leq 1 \\ &p_0^{(\nu)}(\xi, \eta) = 0 \qquad (\xi \in E_a, \quad \eta \notin E_a), \\ &p_0^{(\nu)}(\xi, \eta) = 0 \qquad (\xi \in {}_aC_\alpha, \quad \eta \notin {}_aC_{\alpha+\nu}). \end{aligned}$$

If $p_0^{(\nu)}(\xi, \eta)$ is chosen to satisfy these conditions, $p_0^{(n)}(\xi, \eta)$, uniquely defined as above, satisfies the same conditions with ν replaced by $n > \nu$. [As usual the forward subscript of ${}_aC_\alpha$ is to be interpreted (mod d_a).] Since $p^{(n)}(\xi, E) = 0$ if $n \geq \nu$ whenever $\varphi(E) = 0$, the limits $q(\xi, E)$, ${}_a\pi(E)$, ${}_a\pi_\alpha(E)$ must also vanish whenever $\varphi(E) = 0$, so that ${}_a\pi$ and ${}_a\pi_\alpha$ are absolutely continuous with respect to φ, and they determine density functions. We shall show that the limit theorems proved above for the set functions imply the obvious corresponding limit theorems for the densities.

For example, since

$$p^{(n)}(\xi, F) \leq \gamma\rho^n$$

for some constants γ, ρ, with $0 < \rho < 1$, uniformly in ξ and n, it follows that

$$p_0^{(n)}(\xi, \eta) \leq \gamma\rho^{n-\nu} \qquad \eta \in F.$$

In fact,

$$\begin{aligned} p_0^{(n)}(\xi, \eta) = \int_F p_0^{(\nu)}(\zeta, \eta)p^{(n-\nu)}(\xi, d\zeta) &\leq p^{(n-\nu)}(\xi, F) \\ &\leq \gamma\rho^{n-\nu}, \qquad \eta \in F. \end{aligned}$$

The other limit theorems go over in the same way; the densities approach the limit densities uniformly with the same speed as the distributions approach their limits. To show the principle involved, we take the case in which there is only one ergodic set, which has no cyclically moving subsets, and prove that, from the inequality we have already proved in this case

$$|p^{(n)}(\xi, E) - p(E)| \leq \gamma\rho^n,$$

where $p = {}_1\pi$ is the (unique) stationary absolute probability distribution, $\gamma > 0$, and $0 < \rho < 1$, it follows that

$$|p_0^{(n)}(\xi, \eta) - p_0(\eta)| \leq 2\gamma\rho^n$$

for all ξ, η. Here $p_0(\cdot)$ is the density of the $p(\cdot)$ distribution,

$$\int_E p_0(\eta)\varphi(d\eta) = p(E)$$

for all $E \in \mathscr{F}_X$. The condition that $p_0(\cdot)$ be the $p(\cdot)$ density only fixes it up to a set of φ measure 0. The function is defined uniquely as follows: Since $p(\cdot)$ is a stationary absolute probability distribution,

$$\int_X p^{(n)}(\xi, E)p(d\xi) = p(E),$$

and this implies that

$$\int_X p_0^{(n)}(\xi, \eta)p(d\xi) = p_0(\eta), \qquad n \geq \nu$$

for almost all η (φ measure). We can, therefore, define $p_0(\eta)$ uniquely as the left side of this equation for $n = \nu$. It follows that the equation is true for all $n \geq \nu$. With this definition we have

$$\begin{aligned}|p_0^{(n)}(\xi, \eta) - p_0(\eta)| &= \left| \int_X p_0^{(\nu)}(\zeta, \eta)[p^{(n-\nu)}(\xi, d\eta) - p(d\zeta)]\right| \\ &\leq 2\gamma\rho^{n-\nu},\end{aligned}$$

as was to be proved. If the above conventions had not been made on $p_0^{(n)}(\xi, \eta)$ and $p_0(\eta)$, these results would hold for each ξ up to an η set of φ measure 0.

Finally we discuss the implications of the stronger of the two conditions of Example 3,

$$p^{(\nu)}(\xi, E) \leq Kp^{(\nu)}(\eta, E)$$

for some integer ν and constant K; the condition is to hold for all ξ, η, $E \in \mathscr{F}_X$. It implies

$$\frac{1}{K}p^{(\nu)}(\eta, E) \leq p^{(\nu)}(\xi, E) \leq Kp^{(\nu)}(\eta, E).$$

Hence, if we define $\varphi(E) = p^{(\nu)}(\eta, E)$ for some η, we see that the previous discussion is applicable and moreover that the density $p_0^{(\nu)}(\xi, \eta)$ can be defined to satisfy

$$\frac{1}{K} \leq p_0^{(\nu)}(\xi, \eta) \leq K.$$

Under this condition then the hypotheses of Case (b) are satisfied. (It is actually obvious from the original condition, without the introduction of densities, that there is only one ergodic class and that there are no cyclically moving subclasses.) This case was first discussed by Kolmogorov, who proved the conclusions drawn in Case (b), using essentially the proof given (without the introduction of densities, which is quite superfluous in this case).

Finally we give a simple example in which some of the conclusions drawn in this section hold, but in which Condition (D) is not satisfied.

Example 4 Let $x_1, x_2, \cdots$ be real-valued random variables constituting a Gaussian process with

$$\mathbf{E}\{x_n\} = 0$$

$$\mathbf{E}\{x_n x_m\} = \rho^{|n-m|}.$$

Here ρ is a real parameter, $0 < \rho < 1$. The x_n process is a stationary Markov process (see §8 and X, §4). The conditional distribution of x_{n+1} for given x_1 is Gaussian with expectation $\rho^n x_1$ and variance $1 - \rho^{2n}$; that is,

$$p^{(n)}(\xi, E) = [2\pi(1 - \rho^{2n})]^{-1/2} \int_E e^{-\frac{(\eta - \rho^n \xi)^2}{2(1-\rho^{2n})}} \, d\eta.$$

Then

$$\lim_{n\to\infty} p^{(n)}(\xi, E) = \frac{1}{\sqrt{2\pi}} \int_E e^{\frac{-\eta^2}{2}} \, d\eta.$$

If E is a finite interval this convergence is *not* uniform in ξ, because, for each n, $p^{(n)}(\xi, E)$ can be made arbitrarily small by making ξ large. Thus, simple as this case is, (D) is not satisfied. The iterated stochastic transition functions converge (but not uniformly) to a stationary absolute probability distribution function.

6. The law of large numbers

In this section we shall prove a version of the law of large numbers applicable to the processes studied in §5. We suppose first only that $p(\cdot, \cdot)$ is a stochastic transition function as defined in §5, and that $x_1, x_2, \cdots$ are random variables (not necessarily numerically valued) constituting a Markov process with the given $p(\cdot, \cdot)$ as transition probability distribution (from x_n to x_{n+1}).

THEOREM 6.1 *If the x_1 distribution $p(\cdot)$ is a stationary absolute probability distribution, and if f is a function of $\xi \in X$, measurable with respect to $\mathscr{F}_X$, with*

$$\mathbf{E}\{|f(x_1)|\} = \int_X |f(\xi)|\,p(d\xi) < \infty,$$

then

$$\lim_{n\to\infty} \frac{1}{n} \sum_{m=1}^{n} f(x_m) \tag{6.1}$$

exists with probability 1. *In particular, under Hypothesis* (D), *if there is only one ergodic set,*

$$\lim_{n\to\infty} \frac{1}{n} \sum_{m=1}^{n} f(x_m) = \int_X f(\xi)p(d\xi)\ (= \mathbf{E}\{f(x_1)\}) \tag{6.1'}$$

with probability 1.

The first statement of the theorem is simply an application of the strong law of large numbers for strictly stationary processes (ergodic theorem) to be proved in X (X, Theorem 2.1). It is remarked in X, §1, that if the hypotheses of the second part of Theorem 6.1 are satisfied the process is metrically transitive, and in that case the strong law of large numbers prescribes the limit in (6.1′).

As a simple application suppose that the process is a Markov chain with a finite number of states, as discussed in §2, and suppose that there is only one ergodic class, with no cyclically moving subclasses. According to §2 there is then one and only one set of stationary absolute probabilities; suppose that these are used to determine a stationary process. Then, if j_n is the number of the first n states of the system which are j,

$$\lim_{n\to\infty} \frac{j_n}{n} = p_j = \mathbf{P}\{x_1(\omega) = j\}$$

with probability 1. In fact, if $f(\xi)$ is defined as 1 if $\xi = j$ and 0 otherwise,

$$\frac{1}{n} \sum_{m=1}^{n} f(x_m) = \frac{j_n}{n}$$

and Theorem 6.1 therefore implies the desired statement. The extension of this result to general state spaces is obvious.

If Condition (D) is satisfied, and if there is more than one ergodic class, the most general stationary absolute probability distribution has the form

$\sum_a q_a \, {}_a\pi$, where $0 \leq q_a$, $\sum_a q_a = 1$, and we use here and in the following the notation of §5. The limit in (6.1) under these conditions is clearly

$$\int_{E_a} f(\xi) \, {}_a\pi(d\xi) = \int_X f(\xi) \, {}_a\pi(d\xi), \qquad x_1(\omega) \in E_a,$$

with probability 1. We observe that

$$\mathbf{P}\{x_1(\omega) \in F = X - \bigcup_a E_a\} = 0.$$

The limit may thus depend on $x_1(\omega)$ and will, in general, if there is more than one ergodic set.

Probabilities were defined in Theorem 6.1 and in the extension just made by the given stochastic transition function, together with a stationary initial distribution of x_1. More generally, we shall now allow any initial distribution π of x_1 sets. The relevant expectations and probabilities will be denoted by $\mathbf{E}_\pi\{-\}$, $\mathbf{P}_\pi\{-\}$ to stress the dependence on the initial distribution π. The most important initial distributions are the stationary ones and those concentrated at a single point. In the latter case, if the point is ξ, expectations and probabilities become the conditional expectations and probabilities $\mathbf{E}\{- \mid x_1(\omega) = \xi\}$, $\mathbf{P}\{- \mid x_1(\omega) = \xi\}$.

THEOREM 6.2 *Let f be a function of $\xi \in X$, measurable with respect to $\mathscr{F}_X$, with*

$$\int_{E_a} |f(\xi)| \, {}_a\pi(d\xi) < \infty \qquad a = 1, 2, \cdots.$$

Then, under Hypothesis (D), *for any initial distribution of probabilities,*

$$\lim_{n \to \infty} \frac{1}{n} \sum_{m=1}^{n} f(x_m) \tag{6.2}$$

exists with probability 1, *and is* $\int_{E_a} f(\xi) \, {}_a\pi(d\xi)$ *if* $x_1(\omega) \in E_a$, *with probability* 1.

According to Theorem 6.1, as trivially generalized after its proof, the present theorem is true if the initial distribution π is a stationary absolute probability distribution. The choice $\pi = 1/l \sum_a {}_a\pi$, where l is the number of ergodic classes, then shows that if $x_1(\omega) = \xi_1 \in \bigcup_a E_a$, but if ξ_1 does not belong to some set A_0 with $\sum_a {}_a\pi(A_0) = 0$, the limit (6.2) will exist (and have the stated value if $\xi_1 \in E_a$) with probability 1 when the π distribution is concentrated at the single point ξ_1. It follows that the limit (6.2) will exist and have the stated value with probability 1 when the π distribution is unrestricted except that $\pi(A_0) = 0$. To finish the proof it will thus suffice to show that A_0 is the null set. According to §5, for any ξ_1, if $x_1(\omega) = \xi_1$, $x_n(\omega)$ will finally be in an ergodic set, and in fact will remain in whichever one it enters first. More than that, $x_n(\omega)$ will at some time be in

$$G = \bigcup_a E_a - A_0 \bigcup_a E_a$$

because ${}_a\pi(A_0) = 0$ for all a implies by (5.11) that

$$\lim_{n \to \infty} p^{(n)}(\xi_1, G) = 1.$$

Consider the sample sequences for which $x_N(\omega)$ is the first $x_n(\omega)$ in G. Then for these sample sequences the averages in (6.2) have the same limits (if any) as the averages

$$\frac{1}{n} \sum_{m=N}^{n} f[x_m(\omega)], \qquad n = N, N+1, \cdots.$$

But these averages behave like the averages

$$\frac{1}{n} \sum_{m=1}^{n-N+1} f[x_m(\omega)], \qquad n = N, N+1, \cdots$$

with the initial x_1 distribution confined to G. Hence these averages approach

$$\int_{E_a} f(\xi)\, {}_a\pi(d\xi)$$

when $x_N(\omega) \in E_a$, with probability 1. The limit in (6.2) thus exists with probability 1 for the $\pi(A)$ distribution concentrated at $x_1(\omega) = \xi_1$. In particular, if $\xi_1 \in E_a$, the limit is as stated in the theorem. Thus A_0 is empty, as was to be proved.

7. The central limit theorem

In this section we shall discuss the applicability of the central limit theorem to Markov processes. We shall assume (D_0):

(a) *Condition* (D) *is satisfied;*

(b) *there is only a single ergodic set and this set contains no cyclically moving subsets.*

We have proved in §5 that under Hypothesis (D_0) there are positive constants γ, and ρ, $\rho < 1$, and a (unique) stationary absolute probability distribution $p(\cdot)$, such that

$$|p^{(n)}(\xi, E) - p(E)| \leq \gamma\rho^n \qquad n = 1, 2, \cdots. \tag{7.1}$$

The distribution $p(\cdot)$ taken as initial distribution together with the stochastic transition function determines a stationary Markov process, and in this section the expectations and probabilities involved will always be based on this process unless the contrary is stated explicitly. Thus

$$\mathbf{E}\{f(x_n)\} = \int_X f(\xi)p(d\xi)$$

$$\mathbf{E}\{f(x_m)g(x_{m+n})\} = \int_X f(\xi)p(d\xi) \int_X g(\eta)p^{(n)}(\xi, d\eta)$$

and so on.

Define $S(\xi, A)$ by

$$S(\xi, A) = \sum_{n=1}^{\infty} [p^{(n)}(\xi, A) - p(A)],$$

and $V_n(\xi, A)$ by

$$V_n(\xi, A) = \operatorname*{Max}_{B \subset A} [p^{(n)}(\xi, B) - p(B)] - \operatorname*{Min}_{B \subset A} [p^{(n)}(\xi, B) - p(B)].$$

Then $V_n(\xi, A)$ is the variation of the nth summand of $S(\xi, \cdot)$ on A, for fixed ξ; the first term on the right is the positive variation, the second is the negative variation. There is a set A_ξ on all subsets of which

$$p^{(n)}(\xi, A) - p(A) \geq 0$$

and on all subsets of whose complement

$$p^{(n)}(\xi, A) - p(A) \leq 0,$$

so that the defining equation of $V_n(\xi, A)$ becomes

$$V_n(\xi, A) = [p^{(n)}(\xi, A A_\xi) - p(A A_\xi)] \\ - [p^{(n)}(\xi, A(X - A_\xi)) - p(A(X - A_\xi))].$$

The variation $V_n(\xi, A)$ is completely additive in A for fixed ξ, and

$$\begin{aligned} &V_n(\xi, A) \leq 2\gamma\rho^n \\ (7.2)\qquad &V_n(\xi, A) \leq p^{(n)}(\xi, A) + p(A). \end{aligned}$$

We shall prove several lemmas needed for the central limit theorem. In each lemma it is supposed that Hypothesis (D_0) is satisfied, and the lemmas all express in various ways the fact that then x_m and x_{m+n} are nearly independent if n is large.

LEMMA 7.1 *Under Hypothesis* (D_0), *if f is a random variable on $x_1, \cdots x_m$ sample space, and g one on $x_{m+k}, x_{m+k+1}, \cdots$ sample space, and if for some $r > 1$, $s > 1$ with $1/r + 1/s = 1$,*

$$\mathbf{E}\{|f|^r\} < \infty, \qquad \mathbf{E}\{|g|^s\} < \infty,$$

then

$$(7.3)\qquad |\mathbf{E}\{fg\} - \mathbf{E}\{f\}\mathbf{E}\{g\}| \leq 2\gamma^{1/r}\rho^{k/r}\, \mathbf{E}^{1/r}\{|f|^r\}\, \mathbf{E}^{1/s}\{|g|^s\}, \qquad k \geq 1.$$

In particular, if f and g are functions of $\xi \in X$, measurable with respect to $\mathscr{F}_X$, and if, with r, s as above,

$$\mathbf{E}\{|f(x_1)|^r\} < \infty, \qquad \mathbf{E}\{|g(x_1)|^s\} < \infty,$$

then

$$(7.4) \quad \sum_{k=1}^{\infty} [\mathbf{E}\{f(x_1)g(x_{k+1})\} - \mathbf{E}\{f(x_1)\}\mathbf{E}\{g(x_1)\}] = \int_X f(\xi)p(d\xi) \int_X g(\eta)S(\xi, d\eta).$$

In fact, applying Holder's inequality repeatedly,

$$\begin{aligned}
&|\mathbf{E}\{fg\} - \mathbf{E}\{f\}\mathbf{E}\{g\}| \\
&= |\mathbf{E}\{f[\mathbf{E}\{g \mid x_m\} - \mathbf{E}\{g\}]\}| \\
&= \left|\mathbf{E}\left\{f \int_X \mathbf{E}\{g \mid x_{m+k}(\omega) = \eta\}[p^{(k)}(x_m, d\eta) - p(d\eta)]\right\}\right| \\
&\le \mathbf{E}\left\{|f| \int_X \mathbf{E}\{|g| \mid x_{m+k}(\omega) = \eta\}V_k(x_m, d\eta)\right\} \\
&\le \mathbf{E}^{1/r}\{|f|^r\}\mathbf{E}^{1/s}\left\{\left[\int_X \mathbf{E}\{|g| \mid x_{m+k}(\omega) = \eta\}V_k(x_m, d\eta)\right]^s\right\} \\
&\le (2\gamma\rho^k)^{1/r}\mathbf{E}^{1/r}\{|f|^r\}\mathbf{E}^{1/s}\left\{\int_X \mathbf{E}^s\{|g| \mid x_{m+k}(\omega) = \eta\}V_k(x_m, d\eta)\right\} \\
&\le (2\gamma\rho^k)^{1/r}\mathbf{E}^{1/r}\{|f|^r\}\mathbf{E}^{1/s}\left\{\int_X \mathbf{E}\{|g|^s \mid x_{m+k}(\omega) = \eta\}[p^{(k)}(x_m, d\eta) + p(d\eta)]\right\} \\
&= (2\gamma\rho^k)^{1/r}\mathbf{E}^{1/r}\{|f|^r\}[2\mathbf{E}\{|g|^s\}]^{1/s} \\
&= 2\gamma^{1/r}\rho^{k/r}\mathbf{E}^{1/r}\{|f|^r\}\mathbf{E}^{1/s}\{|g|^s\},
\end{aligned}$$

which yields (7.3). In particular, if f and g are as described in the second part of the theorem, we can apply the preceding continued inequality to $f(x_1)$, $g(x_{k+1})$, with $m = 1$, obtaining from the fourth and the last lines

$$\int_X |f(\xi)|\, p(d\xi) \int_X |g(\eta)|\, V_k(\xi, d\eta) \le 2\gamma^{1/r}\rho^{k/r}\mathbf{E}^{1/r}\{|f|^r\}\mathbf{E}^{1/s}\{|g|^s\}.$$

Then the integral on the right in (7.4) is absolutely convergent, because

$$|S(\xi, A)| \le \sum_{k=1}^{\infty} V_k(\xi, A),$$

so that the integral is dominated by the convergent series

$$\sum_{k=1}^{\infty} \int_X |f(\xi)|p(d\xi) \int_X |g(\eta)|\, V_k(\xi, d\eta) \le \sum_{k=1}^{\infty} 2\gamma^{1/r}\rho^{k/r}\mathbf{E}^{1/r}\{|f|^r\}\mathbf{E}^{1/s}\{|g|^s\}.$$

The series on the left in (7.4) is absolutely convergent because of (7.3). The equality

$$\sum_{k=1}^{n} [\mathbf{E}\{f(x_1)g(x_{k+1})\} - \mathbf{E}\{f(x_1)\}\mathbf{E}\{g(x_1)\}]$$
$$= \sum_{k=1}^{n} \int_X f(\xi)p(d\xi) \int_X g(\eta)[p^{(k)}(\xi, d\eta) - p(d\eta)]$$

then implies (7.4) when $n \to \infty$.

According to Lemma 7.1 the correlation between $f(x_1)$ and functions of x_k for large k goes to 0 exponentially when $k \to \infty$. If f is bounded, the reason underlying this fact can be expressed in a particularly simple form as follows.

LEMMA 7.2 *Under Condition* (D_0), *if f is a bounded random variable, $|f| \le M$, on $x_{k+1}, x_{k+2}, \cdots$ sample space, then*

$$|\mathbf{E}\{f \mid x_1\} - \mathbf{E}\{f\}| \le 2\gamma M\rho^k. \tag{7.5}$$

If $f(\omega) = 1$ when $x_{k+1}(\omega) \in A$ and $f(\omega) = 0$ otherwise, this inequality reduces to

$$|p^{(k)}(\xi, A) - p(A)| \le 2\gamma\rho^k,$$

and this is true even without the factor 2, by (7.1). The general case is easily treated as follows:

$$|\mathbf{E}\{f \mid x_1(\omega) = \xi\} - \mathbf{E}\{f\}| = \left| \int_X \mathbf{E}\{f \mid x_{k+1}(\omega) = \eta\}[p^{(k)}(\xi, d\eta) - p(d\eta)]\right|$$
$$\le \int_X \mathbf{E}\{|f| \mid x_{k+1}(\omega) = \eta\}V_k(\xi, d\eta)$$
$$\le MV_k(\xi, X) \le 2\gamma M\rho^k.$$

LEMMA 7.3 *Under Condition* (D_0), *let $f(\cdot)$ be a function of $\xi \in X$, measurable with respect to $\mathscr{F}_X$, with*

$$\mathbf{E}\{f(x_1)\} = 0, \qquad \mathbf{E}\{|f(x_1)|^2\} = \sigma^2 < \infty,$$

$$\mathbf{E}\{|f(x_1)|^2\} + 2\Re \int_X f(\xi)p(d\xi) \int_X \overline{f(\eta)}S(\xi, d\eta)$$
$$= \mathbf{E}\{|f(x_1)|^2\} + 2\Re \sum_{k=1}^{\infty} \mathbf{E}\{f(x_1)\overline{f(x_{k+1})}\} = \sigma_1^2.$$

Then

$$\lim_{n\to\infty} \mathbf{E}\left\{\left|\frac{1}{\sqrt{n}} \sum_{j=1}^{n} f(x_j)\right|^2\right\} = \sigma_1^2$$

and in fact

$$\lim_{n\to\infty} [\mathbf{E}\{|\sum_{j=1}^{n} f(x_j)|^2 - n\sigma_1^2] = -2\Re \sum_{k=1}^{\infty} k\mathbf{E}\{f(x_1)\overline{f(x_{k+1})}\}. \tag{7.6}$$

If $\sigma^2 \ne 0$, the limit on the right and σ_1^2 do not vanish simultaneously.

In particular, if the x_j's are mutually independent, the nth bracket on the left in (7.6) vanishes and $\sigma_1{}^2 = \sigma^2$. The lemma is thus trivial in this case. In general this bracket has the value

$$2\Re \sum_{k=1}^{n-1} (n-k)\mathbf{E}\{f(x_1)\overline{f(x_{k+1})}\} - 2n\Re \sum_{k=1}^{\infty} \mathbf{E}\{f(x_1)\overline{f(x_{k+1})}\}$$

$$= - 2\Re \sum_{k=1}^{n-1} k\mathbf{E}\{f(x_1)\overline{f(x_{k+1})}\} - 2n\Re \sum_{k=n}^{\infty} \mathbf{E}\{f(x_1)\overline{f(x_{k+1})}\}.$$

When $n \to \infty$ this becomes the desired equation (7.6), in view of the inequalities of Lemma 7.1, with $g = \bar{f}$, $r = s = 2$. If $\sigma_1 = 0$ and if the right side of (7.6) also vanishes, (7.6) implies

$$\underset{n\to\infty}{\text{l.i.m.}} \sum_{j=1}^{n} f(x_j) = 0.$$

But then $\underset{n\to\infty}{\text{l.i.m.}}\, f(x_n) = 0$, and this is impossible, since $\mathbf{E}\{|f(x_n)|^2\} = \sigma^2$, unless $\sigma = 0$, that is, unless $f(x_1) = 0$ with probability 1.

LEMMA 7.4 *Under Condition* (D_0), *let* $f(\cdot)$ *be a function of* $\xi \in X$, *measurable with respect to* $\mathscr{F}_X$, *with*

$$\mathbf{E}\{f(x_1)\} = 0, \qquad \mathbf{E}\{|f(x_1)|^l\} < \infty$$

for some $l \geq 2$. *Then there is a constant* a, *for which*

$$\mathbf{E}\{|\sum_1^n f(x_j)|^l\} \leq an^{l/2}, \qquad n = 1, 2, \cdots. \tag{7.7}$$

Lemma 7.3 shows that (7.7) is true for $l = 2$. It will therefore be sufficient to assume that (7.7) is true if l is an integer $m \geq 2$ and prove that it is then true if $l = m + \delta$, where $0 < \delta \leq 1$. We thus assume in the following that $\mathbf{E}\{|f(x_1)|^{m+\delta}\} < \infty$ and that the lemma is true if $l = m$.

Let k be a positive integer, to be determined more precisely below, and define s_n, t_n, $\hat{s}_n$, c_n, by

$$s_n = \sum_1^n f(x_j), \qquad t_n = \sum_{n+1}^{n+k} f(x_j), \qquad \hat{s}_n = \sum_{n+k+1}^{2n+k} f(x_j)$$

$$c_n = \mathbf{E}\{|\sum_1^n f(x_j)|^{m+\delta}\}.$$

Then we are to prove

$$c_n \leq an^{\frac{m+\delta}{2}}, \qquad n = 1, 2, \cdots, \tag{7.7'}$$

for proper choice of a. In order to prove this we first prove that, if $\varepsilon_1 > 0$,

$$\mathbf{E}\{|s_n + \hat{s}_n|^{m+\delta}\} \leq (2 + \varepsilon_1)c_n + a_1 n^{(m+\delta)/2}, \qquad n = 1, 2, \cdots, \tag{7.8}$$

for proper choice of a_1 and k. In fact, remembering that s_n and $\hat{s}_n$ have the same distribution,

$$(7.9)\quad \mathbf{E}\{|s_n + \hat{s}_n|^{m+\delta}\} \leq \mathbf{E}\{|s_n + \hat{s}_n|^m(|s_n|^\delta + |\hat{s}_n|^\delta)\}$$

$$\leq 2c_n + \mathbf{E}\left\{\sum_{j=0}^{m-1}\binom{m}{j}|s_n|^{j+\delta}\,|\hat{s}_n|^{m-j} + \sum_{j=1}^{m}\binom{m}{j}|s_n|^j|\hat{s}_n|^{m-j+\delta}\right\}.$$

Now by Lemma 7.1, with

$$f = |s_n|^u, \qquad g = |\hat{s}_n|^v, \qquad u + v = m + \delta$$

$$r = \frac{m+\delta}{u} \qquad s = \frac{m+\delta}{v}$$

$$(7.10)\qquad \mathbf{E}\{|s_n|^u|\hat{s}_n|^v\} \leq 2\gamma^{\frac{u}{m+\delta}}\rho^{\frac{u}{m+\delta}(k+1)}\,c_n + \mathbf{E}\{|s_n|^u\}\mathbf{E}\{|s_n|^v\}.$$

We substitute this in (7.9), giving u and v the appropriate values. In each case $0 < \delta \leq u \leq m$, $0 < \delta \leq v \leq m$. Hence, using Hölder's inequality, the last term in (7.10) is at most

$$\mathbf{E}\{|s_n|^m\}^{u/m}\mathbf{E}\{|s_n|^m\}^{v/m} = \mathbf{E}\{|s_n|^m\}^{(m+\delta)/m}.$$

Now we are supposing that (7.7′) is true (with some a) if $\delta = 0$. Hence the preceding product is at most const. $n^{(m+\delta)/2}$. Combining these results, we find that

$$\mathbf{E}\{|s_n + \hat{s}_n|^{m+\delta}\} \leq (2 + b\rho^{\frac{\delta}{m+\delta}(k+1)})c_n + a_1 n^{\frac{m+\delta}{2}} \qquad n = 1, 2, \cdots,$$

for some constants a_1, b not involving k. To prove (7.8) we need only increase k, if necessary, to make the second term in the parentheses $< \varepsilon_1$. Next we prove that, if $\varepsilon > 0$, there is a constant a_2 and a value of k for which

$$(7.11)\qquad c_{2n} \leq (2 + \varepsilon)c_n + a_2 n^{\frac{m+\delta}{2}}, \qquad n \geq 1.$$

In fact, applying Minkowski's inequality and (7.8), we find that

$$c_{2n} = \mathbf{E}\{|s_n + \hat{s}_n + t_n - \sum_{j=2n+1}^{2n+k} f(x_j)|^{m+\delta}\}$$

$$\leq \left[\mathbf{E}^{1/(m+\delta)}\{|s_n + \hat{s}_n|^{m+\delta}\} + \left(\sum_{j=n+1}^{n+k} + \sum_{j=2n+1}^{2n+k}\right)\mathbf{E}^{1/(m+\delta)}\{|f(x_j)|^{m+\delta}\}\right]^{m+\delta}$$

$$\leq \left[[(2 + \varepsilon_1)c_n + a_1 n^{(m+\delta)/2}]^{1/(m+\delta)} + 2kc_1^{1/(m+\delta)}\right]^{m+\delta}$$

$$\leq \left[(1 + \varepsilon_1)[(2 + \varepsilon_1)c_n + a_1 n^{(m+\delta)/2}]^{1/(m+\delta)}\right]^{m+\delta}$$

if n is sufficiently large. Then

$$c_{2n} \leq (1 + \varepsilon_1)^{m+\delta}[(2 + \varepsilon_1)c_n + a_1 n^{(m+\delta)/2}]$$

if n is sufficiently large. If ε_1 is so small that $(1+\varepsilon_1)^{m+\delta}(2+\varepsilon_1) \leq 2+\varepsilon$, there must be an a_2 for which (7.11) is true. According to (7.11)

$$c_{2^r} \leq (2+\varepsilon)^r c_1 + a_2 [2^{(r-1)\frac{m+\delta}{2}} + (2+\varepsilon)2^{(r-2)\frac{m+\delta}{2}} + \cdots + (2+\varepsilon)^{r-1}]$$

$$< (2+\varepsilon)^r c_1 + a_2 \frac{2^{(r-1)\frac{m+\delta}{2}}}{1 - \dfrac{2+\varepsilon}{2^{\frac{m+\delta}{2}}}}, \qquad r \geq 1,$$

if ε is so small that $2 + \varepsilon < 2^{(m+\delta)/2}$. Then, if ε is chosen in this way,

$$c_{2^r} < a_3 2^{r\frac{m+\delta}{2}}, \qquad r \geq 0, \tag{7.12}$$

where

$$a_3 = c_1 + a_2 \frac{2^{-(m+\delta)/2}}{1 - \dfrac{2+\varepsilon}{2^{\frac{m+\delta}{2}}}}.$$

Finally if n is any positive integer it can be written in the form

$$n = 2^r + \nu_1 2^{r-1} + \cdots + \nu_r \leq 2^r + 2^{r-1} + \cdots + 1$$

where

$$2^r \leq n < 2^{r+1}$$

and each ν_j is either 0 or 1. Then s_n can be written as the sum of $r + 1$ groups of sums containing 2^r, $\nu_1 2^{r-1}, \cdots$ terms and using Minkowski's inequality, (7.12), and the fact that the $f(x_j)$ process is stationary,

$$c_n \leq \Big[\mathbf{E}^{1/(m+\delta)}\{|s_{2^r}|^{m+\delta}\} + \mathbf{E}^{1/(m+\delta)}\{|s_{2^{r-1}}|^{m+\delta}\} + \cdots + \mathbf{E}^{1/(m+\delta)}\{|s_1|^{m+\delta}\}\Big]^{m+\delta}$$

$$\leq a_3 [2^{r/2} + 2^{(r-1)/2} + \cdots + 1]^{m+\delta}$$

$$= a_3 \left[\frac{2^{(r+1)/2} - 1}{2^{\frac{1}{2}} - 1}\right]^{m+\delta} \leq a n^{(m+\delta)/2}$$

for some constant a, as was to be proved.

We are now in a position to discuss the central limit theorem. Let π be a probability measure of $\mathscr{F}_X$ sets, and consider the Markov process obtained using π as an initial probability distribution together with a stochastic transition function satisfying condition (D_0). The relevant expectations and probabilities will be distinguished from those obtained when $\pi = p$ by use of the subscript π so that

$$\mathbf{E}_\pi\{f(x_1)\} = \int_X f(\xi)\pi(d\xi), \mathbf{E}\{f(x_1)\} = \int_X f(\xi)p(d\xi),$$

and so on. We wish to show that, for a wide class of functions f, $\frac{1}{\sqrt{n}} \sum_{m=1}^{n} f(x_m)$ is nearly normally distributed for large n, no matter what initial distribution π is used. The most important cases are $\pi = p$ and π a distribution concentrated at a single point ξ_1 so that probabilities become conditional probabilities under the condition $x_1(\omega) = \xi_1$.

THEOREM 7.5 *Suppose that Condition* (D_0) *is satisfied, and that* $f(\cdot)$ *is a real function of* $\xi \in X$, *measurable with respect to* $\mathscr{F}_X$, *with*

$$\mathbf{E}\{|f(x_1)|^{2+\delta}\} = \int_X |f(\xi)|^{2+\delta}p(d\xi) < \infty$$

for some $\delta > 0$. *Then*

$$\lim_{n\to\infty} \mathbf{E}\left\{\left[\frac{1}{\sqrt{n}} \sum_{m=1}^{n} (f(x_m) - \mathbf{E}\{f(x_m)\})\right]^2\right\} = \sigma_1^2 \tag{7.13}$$

exists; if $\sigma_1^2 > 0$, *and if* π *is any initial distribution (of* x_1),

$$\lim_{n\to\infty} \mathbf{P}_\pi\left\{\frac{1}{\sqrt{n}} \sum_{m=1}^{n} (f(x_m) - \mathbf{E}\{f(x_m)\}) \leq \lambda\right\} \tag{7.14}$$

$$= \frac{1}{\sigma_1\sqrt{2\pi}} \int_{-\infty}^{\lambda} e^{-\mu^2/2\sigma_1^2}\, d\mu$$

uniformly in λ.

The limit equation (7.13) has already been proved (Lemma 7.3) and is only restated here for completeness. It is no restriction to assume that $\mathbf{E}\{f(x_m)\} = 0$, and we shall do so.

Let α, β be positive integers and let $\mu(\alpha + \beta)$ be the largest multiple of $\alpha + \beta$ which is $\leq n$. Define y_m, y_m' by

$$y_m = \sum_{j=(m-1)(\alpha+\beta)+1}^{(m-1)(\alpha+\beta)+\alpha} f(x_j), \qquad m = 1, \cdots, \mu,$$

$$y_m' = \sum_{(m-1)(\alpha+\beta)+\alpha+1}^{m(\alpha+\beta)} f(x_j) \qquad m = 1, \cdots, \mu$$

$$y_{\mu+1}' = \sum_{\mu(\alpha+\beta)+1}^{n} f(x_j).$$

The theorem is proved by choosing β *so large* that the y_m's are nearly mutually independent and therefore nearly subject to the central limit theorem for mutually independent random variables, but *so small* that the contribution of the y_m''s in (7.14) can be neglected when $n \to \infty$. We first consider the case $\pi = p$. In the course of the discussion α, β, and μ will vary with n, becoming infinite when $n \to \infty$ in such a way that (7.15) and (7.18) will be satisfied.

We prove first that, if

$$\lim_{n\to\infty} \frac{\mu\beta}{\alpha} = 0, \tag{7.15}$$

then

$$\operatorname*{p\,lim}_{n\to\infty} \frac{1}{\sqrt{n}} \sum_1^{\mu+1} y_m' = 0. \tag{7.16}$$

In fact, using (7.6) and Minkowski's inequality,

$$\begin{aligned}\mathbf{E}^{1/2}\left\{\left[\frac{1}{\sqrt{n}} \sum_1^{\mu+1} y_m'\right]^2\right\} &\leq \frac{1}{\sqrt{n}} \sum_1^{\mu+1} \mathbf{E}^{1/2}\{y_m'^2\} \\ &\leq \frac{\mu}{\sqrt{n}} [\beta\sigma_1^2 + \text{const.}]^{1/2} + \frac{[(\alpha+\beta)\sigma_1^2 + \text{const.}]^{1/2}}{\sqrt{n}} \\ &\leq \sqrt{\frac{\mu}{\alpha}}(\beta\sigma_1^2 + \text{const.})^{1/2} + \frac{[(\alpha+\beta)\sigma_1^2 + \text{const.}]^{1/2}}{\sqrt{\mu\alpha}}.\end{aligned}$$

The right side of this inequality goes to 0 when $n \to \infty$ if (7.15) is true, and this implies (7.16). Then if (7.15) is true the contribution of the y_m''s in (7.14) can be neglected when $n \to \infty$.

In the second place we prove that, if $\Phi_m(t)$ is defined by

$$\Phi_m(t) = \mathbf{E}\{e^{it\sum_{j=1}^{m} f(x_j)}\},$$

then for each μ

$$\mathbf{E}\{e^{it\sum_1^{\mu} y_j}\} = \Phi_\alpha(t)^\mu + \zeta_\mu, \quad |\zeta_\mu| < 2\gamma\mu\rho^{\beta+1}. \tag{7.17}$$

In fact, the left side of (7.17) is given by

$$\begin{aligned}&\mathbf{E}\left\{e^{it\sum_1^{\mu-1} y_j} \mathbf{E}\{e^{it\,y_\mu} \mid x_1, \cdots, x_{(\mu-2)(\alpha+\beta)+\alpha}\}\right\} \\ &= \mathbf{E}\{e^{it\sum_1^{\mu-1} y_j}[\Phi_\alpha(t) + \eta_1']\} \\ &= \mathbf{E}\{e^{it\sum_1^{\mu-1} y_j}\}\Phi_\alpha(t) + \eta_1, \qquad |\eta_1| < 2\gamma\rho^{\beta+1},\end{aligned}$$

where we have used Lemma 7.2. Repeating this argument, we find that the left side of this equation can be put in the form

$$\Phi_\alpha(t)^\mu + \eta_1 + \cdots + \eta_{\mu-1} \qquad |\eta_j| < 2\gamma\rho^{\beta+1},$$

which implies (7.17). Thus, if we can choose α, β, μ to satisfy (7.15) and also

$$\lim_{n\to\infty} \mu\rho^{\beta} = 0, \tag{7.18}$$

the distribution to be proved asymptotically normal, with mean 0 and variance σ_1^2 is that with characteristic function $\Phi_\alpha(t/\sqrt{n})^\mu$. This is the distribution of $\sum_1^\mu z_m$, where the z_m's are mutually independent and each has the distribution of $y_1/\sqrt{n}$. It is clearly sufficient to prove the stated asymptotic normality for the random variable $\sum_1^\mu z_m$.

Now $\mathbf{E}\{z_m\} = 0$, and, using Lemma 7.3,

$$\mu\mathbf{E}\{z_m^2\} = \frac{\mu}{n}\mathbf{E}\{[\sum_1^\alpha f(x_j)]^2\} \to \sigma_1^2, \qquad (n \to \infty),$$

if (7.15) is true (so that $\alpha\mu/n \to 1$), and, by Lemma 7.4,

$$\mathbf{E}\{|z_m|^{2+\delta}\} = \frac{1}{n^{(2+\delta)/2}}\mathbf{E}\{|\sum_1^\alpha f(x_j)|^{2+\delta}\} \le \frac{a\alpha^{(2+\delta)/2}}{n^{(2+\delta)/2}}.$$

Then, for large n,

$$\frac{\sum_1^\mu \mathbf{E}\{|z_m|^{2+\delta}\}}{[\sum_1^\mu \mathbf{E}\{z_m^2\}]^{(2+\delta)/2}} < \frac{2\mu a(\alpha/n)^{1+\delta/2}}{\sigma_1^{2+\delta}}.$$

This quotient goes to 0 like $(\alpha/n)^{\delta/2}$ when $n \to \infty$, and this implies the stated asymptotic normality (for $\pi = p$) by III, Theorem 4.4. We observe that (7.15) and (7.18) are consistent; for example, we can take β as the largest integer whose fourth power is $\le n$, $\alpha = \beta^3$, and then $\mu = n/(\alpha + \beta) = \beta$ approximately, so that (7.15) and (7.18) are certainly satisfied.

To prove asymptotic normality for an arbitrary initial distribution π we observe that for any m, using Lemma 7.2,

$$\begin{aligned}
&\left| \mathbf{E}_\pi\{e^{\frac{it}{n^{1/2}}\sum_{m+1}^n f(x_j)}\} - \mathbf{E}\{e^{\frac{it}{n^{1/2}}\sum_{m+1}^n f(x_j)}\}\right| \\
&\qquad = \left| \int \pi(d\xi)\left[\mathbf{E}\{e^{\frac{it}{n^{1/2}}\sum_{m+1}^n f(x_j)} \mid x_1(\omega) = \xi\} - \mathbf{E}\{e^{\frac{it}{n^{1/2}}\sum_{m+1}^n f(x_j)}\}\right]\right| \\
&\qquad \le 2\gamma\rho^m.
\end{aligned}$$

Then, if $n > m$,

$$\left| \mathbf{E}_\pi\{e^{\frac{it}{n^{1/2}}\sum_1^n f(x_j)}\} - \mathbf{E}\{e^{\frac{it}{n^{1/2}}\sum_1^n f(x_j)}\}\right|$$

$$\leq \mathbf{E}_\pi\{|e^{\frac{it}{n^{1/2}}\sum_1^m f(x_j)} - 1|\} + \mathbf{E}\{|e^{\frac{it}{n^{1/2}}\sum_1^m f(x_j)} - 1|\} + 2\gamma\rho^m.$$

The third term on the right can be made arbitrarily small by choosing m large. For such a choice of m, the first two terms on the right can then be made arbitrarily small for t in any finite interval by choosing n large. Thus the characteristic functions of $n^{-1/2}\sum_1^n f(x_j)$, $n \geq 1$, are asymptotically the same, as $n \to \infty$, for the two definitions of probability considered here, one with initial distribution p, the other with initial distribution π. It follows that the distribution of $n^{-1/2}\sum_1^n f(x_j)$ is asymptotically normal, with mean 0 and variance σ_1^2. This finishes the proof of the theorem.

Finally we prove a generalization of this theorem which is useful in statistical applications. Suppose that in the theorem the functions depend on more than one x_j. Let $f(\cdot, \ldots, \cdot)$ be a function of $\xi_1, \cdots, \xi_r$, $\xi_j \in X$, measurable with respect to $\mathscr{F}_X \times \cdots \times \mathscr{F}_X$. We consider the random variables

$$\frac{1}{\sqrt{n}}\sum_{m=1}^n [f(x_m, \cdots, x_{m+r-1}) - \mathbf{E}\{f(x_m, \cdots, x_{m+r-1})\}], \qquad n = 1, 2, \cdots,$$

and wish to derive asymptotic normality for $n \to \infty$. Theorem 7.1 covers the case $r = 1$, and we shall now show how to reduce the general case to this one. To do this, replace X by the space $\tilde{X}$ of points $\tilde{\xi}$: $(\xi_1, \cdots, \xi_r)$, $\xi_j \in X$, replace $\mathscr{F}_X$ by the product field $\tilde{\mathscr{F}}_X = \mathscr{F}_X \times \cdots \times \mathscr{F}_X$, and replace the space of points ω: $(\xi_1, \xi_2, \cdots)$, $\xi_j \in X$ by the space of points $\tilde{\omega}$: $(\tilde{\xi}_1, \tilde{\xi}_2, \cdots)$, $\tilde{\xi}_j \in \tilde{X}$. Let $\tilde{x}_j$ be the new jth coordinate function, so that $\tilde{x}_j(\tilde{\omega}) = \tilde{\xi}_j$. Define $\tilde{x}_1, \tilde{x}_2, \cdots \tilde{\omega}$ probabilities to be the same as $\hat{x}_1, \hat{x}_2, \cdots \omega$ probabilities, where $\hat{x}_j$ is the r-tuple $(x_j, \cdots, x_{j+r-1})$. Then the $\tilde{x}_j$ process is a Markov process satisfying Condition (D_0) if the x_j process is such a process. The function f of $(\xi_1, \cdots, \xi_r)$ defines a function $\tilde{f}$ of $\tilde{\xi}_1$, and the ω random variables

$$\{f(x_m, \cdots, x_{m+r-1}), m \geq 1\}$$

have the same joint distributions as the $\tilde{\omega}$ random variables $\{\tilde{f}(\tilde{x}_m), m \geq 1\}$. Thus we have reduced the problem involving f to the corresponding one involving $\tilde{f}$, for which $r = 1$. Applying Theorem 7.5 to the latter problem, we obtain the following.

THEOREM 7.5′ *Suppose that Condition* (D_0) *is satisfied, and that* $f(\cdot, \ldots, \cdot)$ *is a real function of* $\xi_1, \cdots, \xi_r$, *measurable with respect to* $\mathscr{F}_X \times \cdots \times \mathscr{F}_X$, *with*

$$\mathbf{E}\{|f(x_1, \cdots, x_r)|^{2+\delta}\} < \infty$$

for some $\delta > 0$. *Then, if* $f_m = f(x_m, \cdots, x_{m+r-1})$,

$$\lim_{n\to\infty} \mathbf{E}\left\{\left[\frac{1}{n}\sum_{m=1}^{n}(f_m - \mathbf{E}\{f_m\})\right]^2\right\} = \sigma_1^2 \tag{7.13′}$$

exists; if $\sigma_1^2 > 0$ *and if* π *is any initial distribution (of* x_1*),*

$$\lim_{n\to\infty} \mathbf{P}_\pi\left\{\frac{1}{n}\sum_{m=1}^{n}(f_m - \mathbf{E}\{f_m\}) \le \lambda\right\} = \frac{1}{\sigma_1\sqrt{2\pi}}\int_{-\infty}^{\lambda} e^{-\mu^2/2\sigma_1^2}\, d\mu \tag{7.14′}$$

uniformly in λ.

As an example of the application of Theorems 7.5 and 7.5′, let $x_1, x_2, \cdots$ be mutually independent random variables with a common distribution function. Then Condition (D_0) is certainly satisfied. Theorem 7.5 is applicable, but reduces to III, Theorem 4.3 (which has a weaker hypothesis, $\delta = 0$). Theorem 7.5′ yields something new. The limit variance σ_1^2 in (7.13′) is, according to Lemma 7.3,

$$\sigma_1^2 = \mathbf{E}\{f_1^2\} - \mathbf{E}^2\{f_1\} + 2\sum_{1}^{r-1}[\mathbf{E}\{f_1 f_{k+1}\} - \mathbf{E}^2\{f_1\}].$$

[This is also easily evaluated from (7.13′).] As a particular case suppose that the x_j's are real-valued random variables, with

$$\mathbf{E}\{x_1\} = 0, \qquad \mathbf{E}\{x_1^2\} = a_2, \qquad \mathbf{E}\{x_1^4\} = a_4,$$

and that $r = 2, f_1 = (x_2 - x_1)^2$. Then

$$\sigma_1^2 = 4a_4$$

and we find that

$$\frac{1}{\sqrt{n}}\sum_{m=1}^{n}[(x_{m+1} - x_m)^2 - 2a_2]$$

is asymptotically normal with mean 0 and variance $4a_4$, if $\mathbf{E}\{|x_1|^{4+\delta}\} < \infty$ for some $\delta > 0$.

The Condition (D_0) was used to simplify the formal work above. If Condition (D) is satisfied and if there is only one ergodic set but if there may be cyclically moving subsets, Theorems 7.5 and 7.5′ are easily extended to show asymptotic normality of the same sums. If there is more than one ergodic class, however, the limit distributions may be weighted averages of normal distributions.

8. Markov processes in the wide sense

Let $\{x_t, t \in T\}$ be a family of real- or complex-valued random variables and suppose that

$$\mathbf{E}\{|x_t|^2\} < \infty$$

for all t.

Define $R(s, t)$ by

$$\begin{aligned} R(s, t) &= \frac{\mathbf{E}\{x_t \bar{x}_s\}}{\mathbf{E}\{|x_s|^2\}}, \qquad \mathbf{E}\{|x_s|^2\} > 0 \\ &= 0, \qquad \mathbf{E}\{|x_s|^2\} = 0. \end{aligned}$$

Then $x_t - R(s, t)x_s$ is orthogonal to x_s, that is,

$$\hat{\mathbf{E}}\{x_t \mid x_s\} = R(s, t)x_s.$$

According to II §6 the x_t process is a *Markov process in the wide sense* if, whenever $t_1 < \cdots < t_n$,

$$\hat{\mathbf{E}}\{x_{t_n} \mid x_{t_1}, \cdots, x_{t_{n-1}}\} = \hat{\mathbf{E}}\{x_{t_n} \mid x_{t_{n-1}}\}, \tag{8.1}$$

with probability 1.

THEOREM 8.1 *An x_t process is a Markov process in the wide sense if and only if $\mathbf{E}\{|x_t|^2\} < \infty$ and R satisfies the functional equation*

$$R(s, u) = R(s, t)R(t, u), \qquad s \leq t \leq u. \tag{8.2}$$

To prove the theorem define $y(t, u)$ as the difference

$$x_u - R(t, u)x_t = y(t, u).$$

Then $y(t, u)$ is orthogonal to x_t. If the x_t process is a Markov process in the wide sense,

$$\hat{\mathbf{E}}\{x_u \mid x_s, x_t\} = \hat{\mathbf{E}}\{x_u \mid x_t\} = R(t, u)x_t \tag{8.3}$$

so that $y(t, u)$ is orthogonal to x_s also,

$$\mathbf{E}\{x_u \bar{x}_s\} - R(t, u)\mathbf{E}\{x_t \bar{x}_s\} = 0. \tag{8.4}$$

This is equivalent to (8.2). Conversely, if (8.2) is true, (8.4) is also true, and this equation states that $y(t, u)$ is orthogonal to every x_s for $s < t$; that is,

$$R(t, u)x_t = \hat{\mathbf{E}}\{x_u \mid x_t\} = \hat{\mathbf{E}}\{x_u \mid x_{s_1}, \cdots, x_{s_k}, x_t\}$$

with probability 1 if $s_1 < \cdots < s_k < t < u$, and this is equivalent to the condition (8.1) that the process be a Markov process in the wide sense.

In particular, if the x_t process is real and Gaussian and if $\mathbf{E}\{x_t\} = 0$, the condition of the theorem is necessary and sufficient that the x_t process be a Markov process in the strict sense (II §6) in accordance with the general relationship between wide sense and strict sense concepts.

Theorem 8.1 becomes particularly simple if the x_t process is stationary in the wide sense (see II §8). In this case $R(s, t)$ depends only on the difference $t - s$, so that we can write $R(s, t) = R(t - s)$. The condition (8.2) then becomes

$$R(t_1 + t_2) = R(t_1)R(t_2), \qquad t_1, t_2 > 0.$$

If the x_t's are a sequence $x_1, x_2, \cdots$, this means that

$$R(n) = \mathbf{E}\{x_{m+n}\bar{x}_m\} = a^n R(0), \qquad n > 0$$

for some constant a. By Schwarz's inequality

$$|R(n)| \leq R(0),$$

so that $|a| \leq 1$. In the continuous parameter case

$$R(t) = \mathbf{E}\{x_{s+t}\bar{x}_s\} = e^{-ct}R(0), \qquad t > 0$$

where c has non-negative real part, if $R(\cdot)$ is known to be continuous.

Finally we remark that, if the x_t's are a sequence $x_1, x_2, \cdots$, the condition (8.1) is easily shown to be equivalent to the condition

$$\hat{\mathbf{E}}\{x_n \mid x_1, \cdots, x_{n-1}\} = \hat{\mathbf{E}}\{x_n \mid x_{n-1}\} \tag{8.1'}$$

(with probability 1) for all $n > 1$.

CHAPTER VI

Markov Processes—Continuous Parameter

1. Markov chains with finitely many states

This section leans heavily on V §1 and §2, which treat the discrete parameter case. Let $\{x_t, 0 \leq t < \infty\}$ be a Markov process with a finite number of states labeled $1, \cdots, N$. That is, we suppose that

$$\sum_{j=1}^{N} \mathbf{P}\{x_t(\omega) = j\} = 1.$$

(In most applications the random variables of the process actually do not assume other values than $1, \cdots, N$, but this restriction makes no difference in the theory.) If $\mathbf{P}\{x_s(\omega) = i\} > 0$, define $p_{ij}(s, t)$ by

$$p_{ij}(s, t) = \mathbf{P}\{x_t(\omega) = j \mid x_s(\omega) = i\},$$

and let $P(s, t)$ be the matrix $[p_{ij}(s, t)]$. Then, if $\mathbf{P}\{x_s(\omega) = i\} > 0$,

$$p_{ij}(s, t) \geq 0, \qquad \sum_j p_{ij}(s, t) = 1, \tag{1.1}$$

and

$$p_{ik}(s, u) = \sum_j p_{ij}(s, t) p_{jk}(t, u), \qquad 0 \leq s < t < u, \tag{1.2}$$

[where in (1.2) we only sum over values of j for which $p_{jk}(t, u)$ is defined.] In matrix notation the preceding equation, a special case of the Chapman-Kolmogorov equation, becomes

$$P(s, u) = P(s, t)P(t, u), \qquad 0 \leq s < t < u.$$

It is convenient to define $P(t, t)$ as the identity matrix, and we shall use this definition throughout. The Chapman-Kolmogorov equation system (1.2) is then valid for $0 \leq s \leq t \leq u$. The stochastic process is said to have stationary transition probabilities if for each pair i, j, whenever

$\mathbf{P}\{x_s(\omega) = i\} > 0$, the transition probability $p_{ij}(s, t)$ depends only on $t - s$. In this case we write $p_{ij}(t - s)$ for $p_{ij}(s, t)$, and (1.1) and (1.2) specialize to

$$p_{ij}(t) \geq 0, \qquad \sum_j p_{ij}(t) = 1, \qquad t > 0, \tag{1.1'}$$

$$p_{ik}(s + t) = \sum_j p_{ij}(s)p_{jk}(t) \qquad s, t > 0. \tag{1.2'}$$

In matrix notation the preceding equation is

$$P(s + t) = P(s)P(t).$$

The matrix $P(0)$ is by definition the identity matrix.

A matrix function $P(\cdot, \cdot)$ satisfying (1.1) and (1.2) will be called a *Markov transition matrix function.* (To avoid needless complexity we suppose that every element of the matrix function is defined for all values of the arguments.) A matrix function $P(\cdot)$ satisfying (1.1′) and (1.2′) will be called a *stationary Markov transition matrix function.* Given a Markov transition matrix function, there is a corresponding Markov process $\{x_t, 0 \leq t < \infty\}$ obtained by choosing any initial probabilities $\mathbf{P}\{x_0(\omega) = i\}$, $i = 1, \cdots, N$, and, for every finite t set $0 = t_0 < t_1 < \cdots < t_n$, defining

$$\mathbf{P}\{x_{t_0}(\omega) = a_0, \cdots, x_{t_n}(\omega) = a_n\}$$
$$= \mathbf{P}\{x_0(\omega) = a_0\}p_{a_0a_1}(0, t_1) \cdots p_{a_{n-1}a_n}(t_{n-1}, t_n)$$

(see II §6). These basic probabilities determine an x_t process which is a Markov chain with the given initial and transition probabilities. The basic ω space can be taken as the space of all functions of t, $0 \leq t < \infty$ which assume only the integral values $1, \cdots, N$ or as the space of all real functions of t, $0 \leq t < \infty$, without this restriction.

THEOREM 1.1 *If* $[p_{ij}(\cdot)]$ *is a stationary Markov transition matrix function, then* $\lim_{t\to\infty} p_{ij}(t)$ *exists for all* i, j *and the limit is approached exponentially fast.*

Note that this result is simpler than that in the discrete parameter case, because Cesàro limits are not required here. Note also that we have made no continuity or even measurability assumptions whatever. To prove the theorem we first fix t and prove that

$$\lim_{n\to\infty} p_{ij}(nt) = w_{ij}(t) \tag{1.3}$$

exists for all i, j. We shall then show that the matrix $W(t)$: $[w_{ij}(t)]$ does not depend on t, and finally we shall consider the asymptotic behavior of $p_{ij}(t)$ when $t \to \infty$ in an arbitrary manner. We have seen in V §2 that either $\lim_{n\to\infty} p_{ij}(nt_0)$ exists for all i, j [when the stochastic matrix $P(t_0)$ does not determine cyclically moving classes of states] or that for some choice of ν $\lim_{n\to\infty} p_{ij}(n\nu t_0)$ exists for all i, j. Here ν is any integer divisible by certain cycle lengths denoted by $d_1, d_2, \cdots$ in V §2, and $\sum_j d_j \le N$, so that we can certainly set $\nu = N!$ Since t_0 is arbitrary, we can derive (1.3) by setting $t = N!t_0$. Obviously

$$W(s + t) = W(s)W(t)$$

$$W(t)^n = W(nt) = W(t), \qquad n = 1, 2, \cdots.$$

Now suppose that $s < t$. Then

$$W(s + t) = W(2s)W(t - s) = W(s)W(t - s) = W(t),$$

so that $W(u) = W(t)$ for $t \le u \le 2t$. Hence $W(t)$ is independent of t, $W(t) \equiv W$: $[w_{ij}]$, and (1.3) becomes

$$\lim_{n\to\infty} p_{ij}(nt) = w_{ij} \tag{1.3'}$$

for all i, j, t. Here $W = W^2$ and the matrix W has the characteristics of a limit matrix described in V §2: there are (ergodic) classes $E_1, E_2, \cdots$ of states and a class F of transient states such that:

$$\begin{aligned} p_{ij}(t) &= 0, && (i \in E_a, j \notin E_a), \\ w_{ij} &= {}_a\pi_j && i, j \in E_a, \\ &= \rho_i(E_a)\,{}_a\pi_j && (i \in F, j \in E_a), \\ &= 0 && (i \in E_a, j \notin E_a), \\ &= 0 && j \in F, \end{aligned}$$

where

$${}_a\pi_j > 0, \qquad j \in E_a, \qquad \sum_{j \in E_a} {}_a\pi_j = 1,$$

$$\rho_i(E_a) \ge 0, \qquad \sum_a \rho_i(E_a) = 1.$$

Just as in the discrete parameter case of V §2, $\sum_{j \in F} p_{ij}(t)$ is monotone non-increasing in t, for each i. When $t \to \infty$ with t integral-valued, we have the discrete parameter case, and know that this sum approaches 0 exponentially fast. Hence the same is true when $t \to \infty$ with no restriction on the value of t,

$$\lim_{t\to\infty} p_{ij}(t) = 0, \qquad j \in F.$$

To investigate $p_{ij}(t)$ for $i, j \in E_a$, define $m_j^{(t)}$, $M_j^{(t)}$ by

$$m_j^{(t)} = \min_{i \in E_a} p_{ij}(t), \qquad M_j^{(t)} = \max_{i \in E_a} p_{ij}(t) \qquad j \in E_a.$$

Then, just as in the discrete parameter case, see V §2, Case (b), $m_j^{(t)}$ and $M_j^{(t)}$ are respectively monotone non-decreasing and non-increasing. When $t \to \infty$ with t integral-valued, we are in the discrete parameter case, and know that these functions approach ${}_a\pi_j$ exponentially fast; hence the same is true when $t \to \infty$ with no restriction on the values of t, and we have proved that

$$\lim_{t\to\infty} p_{ij}(t) = {}_a\pi_j$$

and that the convergence is exponentially fast. Finally

$$p_{ik}(s+t) = \sum_{j \notin F} p_{ij}(s)p_{jk}(t) + \sum_{j \in F} p_{ij}(s)p_{jk}(t) \qquad \begin{matrix} i \in F \\ k \notin F \end{matrix}$$

so that when $t \to \infty$ the first sum approaches a limit exponentially fast for fixed s, whereas the second is at most $\sum_{j \in F} p_{ij}(s)$, which is (exponentially) small for large s. Therefore $\lim_{u\to\infty} p_{ik}(u)$ exists when $i \in F$, $k \notin F$ and the limit is approached exponentially fast. This finishes the proof of the theorem. As in the discrete parameter case, the rows of the limit matrix are sets of stationary absolute probabilities and every set of stationary absolute probabilities is a linear combination of these rows.

We proceed to a discussion of the sample functions of a Markov chain with a given stationary Markov matrix transition function. We shall always make the assumption that $P(\cdot)$ is continuous when $t = 0$,

$$\lim_{t\to 0} p_{ij}(t) = 1 \quad i = j \qquad (1.4)$$
$$= 0 \quad i \neq j.$$

[The first line implies the second in view of (1.1′).] This condition that the $p_{ij}(\cdot)$'s are continuous when $t = 0$ implies that they are continuous for all t, because

$$\lim_{\varepsilon\downarrow 0} p_{ij}(s+\varepsilon) = \lim_{\varepsilon\downarrow 0} \sum_k p_{ik}(s)p_{kj}(\varepsilon) = p_{ij}(s)$$

$$\begin{aligned} \lim_{\varepsilon\downarrow 0} \, [p_{ij}(s) - p_{ij}(s-\varepsilon)] &= \lim_{\varepsilon\downarrow 0} \, [\sum_k p_{ik}(s-\varepsilon)p_{kj}(\varepsilon) - p_{ij}(s-\varepsilon)] \\ &= \lim_{\varepsilon\downarrow 0} \Big[\sum_{k\neq j} p_{ik}(s-\varepsilon)p_{kj}(\varepsilon) + p_{ij}(s-\varepsilon)[1 - p_{jj}(\varepsilon)]\Big] \\ &= 0. \end{aligned}$$

Since

$$\begin{aligned}\mathbf{P}\{x_t(\omega) \neq x_s(\omega)\} &= \sum_i \mathbf{P}\{x_\tau(\omega) = i\}[1 - p_{ii}(|t-s|)] \\ &\leq \operatorname{Max}_i [1 - p_{ii}(|t-s|)], \qquad \tau = \operatorname{Min}(s, t),\end{aligned}$$

the condition (1.4) is equivalent to the condition that the quantity on the left goes to 0 when $t \to s$, for every initial distribution of probabilities, and all s. That is, (1.4) is true if and only if

$$\operatorname{p\,lim}_{t \to s} x_t = x_s \tag{1.4'}$$

for every initial distribution of probabilities. The stochastic limit (1.4′) will be strengthened to a probability-one limit in Theorem 1.2.

If (1.4) is true $p_{ii}(t) > 0$ for small t. Hence the inequality

$$p_{ii}(s + t) \geq p_{ii}(s)p_{ii}(t)$$

implies that $p_{ii}(t) > 0$ for all $t \geq 0$. Moreover, if $j \neq i$, $p_{ij}(t)$ either vanishes identically or never vanishes except when $t = 0$. In fact, fixing i and j with $i \neq j$, suppose that $p_{ij}(s) > 0$ for some s which is fixed in the following argument. Then we prove that $p_{ij}(t) > 0$ if $t > 0$. Since

$$p_{ij}(t) \geq p_{ij}(u)p_{jj}(t - u), \qquad 0 < u < t$$

and since the second factor on the right is always positive, it is sufficient to prove that $p_{ij}(u) > 0$ for some $u < t$. Now, for every positive integer m, $p_{ij}\left(m \dfrac{s}{m}\right) > 0$, that is, in the language of V §2, j is a consequent of i of order m for the stochastic matrix $P\left(\dfrac{s}{m}\right)$. It was proved in V §2 that j is then a consequent of i of some order $n \leq N$, $p_{ij}\left(n \dfrac{s}{m}\right) > 0$. If $m > Ns/t$, the desired value of u can be taken as ns/m.

The fact that, if (1.4) is true, $p_{ij}(t)$ never vanishes for $t > 0$ unless it vanishes identically shows that the stochastic matrix $P(t)$ (t fixed) cannot determine cyclically moving classes of states. This fact has already been proved above in connection with Theorem 1.1 without the assumption that (1.4) is true.

Suppose now that $p_{ij}(\cdot)$ has a derivative $p_{ij}'(\cdot)$ for all $t \geq 0$ and set

$$\begin{aligned} q_i &= \lim_{t \to 0} \frac{1 - p_{ii}(t)}{t} = -p_{ii}'(0) \\ q_{ij} &= \lim_{t \to 0} \frac{p_{ij}(t)}{t} = p_{ij}'(0). \end{aligned} \tag{1.5}$$

Let Q be the matrix $[q_{ij}]$, where we set $q_{ii} = -q_i$. Using (1.1′), we find that

$$q_i \geq 0, \quad \sum_{j \neq i} q_{ij} = q_i. \tag{1.6}$$

Differentiating the Chapman-Kolmogorov equation (1.2′) with respect to each variable and setting it equal to 0, we obtain two systems of differential equations,

$$p_{ik}'(t) = -q_i p_{ik}(t) + \sum_{j \neq i} q_{ij} p_{jk}(t) \qquad i, k = 1, \cdots, N \tag{1.7}$$

$$p_{ik}'(t) = -p_{ik}(t) q_k + \sum_{j \neq k} p_{ij}(t) q_{jk} \qquad i, k = 1, \cdots, N. \tag{1.7′}$$

The first system is called the backward system, the second the forward system, for reasons to be given below. In each case the initial conditions are given by

$$\begin{aligned} p_{ij}(0) &= 1 \qquad i = j \\ &= 0 \qquad i \neq j. \end{aligned} \tag{1.8}$$

The q_i's and q_{ij}'s determine the $p_{ij}(t)$'s uniquely. We shall investigate this question from two points of view. Firstly, we shall show that the systems (1.7) and (1.7′) with initial conditions (1.8) have a unique solution satisfying (1.1′) and (1.2′) if Q satisfies (1.6). Secondly we shall construct a stochastic process by means of a given Q satisfying (1.6) which will be a Markov chain and whose transition probabilities will satisfy (1.7), (1.7′), (1.8).

We consider first the system (1.7). This system of differential equations is studied most easily in its matrix form

$$P'(t) = QP(t)$$

with the initial condition

$$P(0) = I,$$

where I is the identity. Then we can write a solution [to both (1.7) and (1.7′)]

$$P(t) = e^{Qt},$$

where the exponential of a matrix is defined as the (element by element) sum of the exponential series. It is clear that, for any matrix Q, $P(t)$ as so defined furnishes a solution of (1.7) which satisfies the initial conditions (1.8) and the Chapman-Kolmogorov equations (1.2′). If Q satisfies (1.6), $P(\cdot)$ is a probability solution in the sense that (1.1′) is also true. In fact, firstly, if Q satisfies (1.6) we sum over k in (1.7′) to obtain

$$\sum_k p_{ik}'(t) = 0$$

so that $\sum_k p_{ik}(t) \equiv$ const., and this constant is 1 since it is 1 when $t = 0$. Secondly, the following argument shows that the $p_{ij}(t)$'s are ≥ 0. Suppose that no q_{ij} vanishes. Then $p_{ij}(t) > 0$ for sufficiently small t since, if $j = i$, $p_{ij}(0) = 1$ and, if $j \neq i$, $p_{ij}(0) = 0$, $p_{ij}'(0) = q_{ij} > 0$. Unless the $p_{ij}(t)$'s are always ≥ 0 there is a finite positive δ such that δ is the largest h for which

$$p_{ij}(t) \geq 0, \qquad i, j = 1, \cdots, N, \qquad 0 \leq t \leq h.$$

But the equation $P(\delta + h) = P(\delta)P(h)$ shows that the elements of $P(t)$ are ≥ 0 for $t \leq 2\delta$, in contradiction to the definition of δ. Thus the $p_{ij}(t)$'s are all ≥ 0 if no q_{ij} vanishes. In the general case let Q_n be the matrix Q with every $q_{ii} = -q_i$ replaced by $q_i + (N-1)/n$ and every q_{ij} $(j \neq i)$ replaced by $q_{ij} + 1/n$. Then Q_n satisfies (1.6) and has no vanishing elements, so the elements of $P_n(t)$, defined by

$$P_n(t) = e^{Q_n t},$$

are ≥ 0. When $n \to \infty$ an elementary calculation shows that $P_n(t)$ becomes $P(t)$, so that the elements of $P(t)$ are all ≥ 0. We have not yet shown that the solution we have obtained is unique. However, if

$$P'(t) = QP(t),$$

it follows that

$$P^{(n)}(t) = Q^n P(t).$$

An application of Taylor's theorem with remainder then shows that $P(t)$ must be given by

$$P(t) = e^{Qt}P(0),$$

as was to be proved.

The system (1.7′) is, in matrix form, $P'(t) = P(t)Q$, and can be treated in the same way. It has solution

$$P(t) = P(0)e^{Qt}.$$

Thus the systems (1.7) and (1.7′) have the same solution under the initial condition $P(0) = I$ which we have accepted.

The following two examples are typical of the simple examples that arise in practical applications. In such examples the q_i's and q_{ij}'s are commonly given by theoretical considerations or deduced from experimental data, using the fact that, neglecting second-order terms, $q_{ij}\,dt$ is the probability of a transition from i to j in time dt, and $1 - q_i\,dt$ the probability of no transition.

Example 1 Suppose that only transitions from i to $i + 1$ are possible (and from N to 1), and in fact that

$$\begin{aligned} q_{ij} &= 0 \qquad j \neq i + 1 \qquad (\text{mod } N). \\ &= q \qquad j = i + 1 \end{aligned}$$

Then (1.7) becomes (if we fix k and identify $N+1$ with 1)

$$p_{ik}'(t) = -qp_{ik}(t) + qp_{i+1k}(t) \qquad i = 1, \cdots, N.$$

If $\hat{p}_{ik}(t) = e^{qt}p_{ik}(t)$, we find that

$$\hat{p}_{ik}'(t) = q\hat{p}_{i+1k}(t), \qquad \hat{p}_{ik}(0) = \delta_{ik}.$$

It follows that

$$\hat{p}_{ik}^{(N)}(t) = q^N \hat{p}_{ik}(t),$$

so that $\hat{p}_{ik}(t)$ must have the form

$$\hat{p}_{ik}(t) = \sum_{\nu=1}^{n} c_{ik}^{(\nu)} e^{q\alpha_\nu t}, \qquad \alpha_\nu = e^{\frac{2\pi\sqrt{-1}}{N}\nu},$$

and, substituting in the difference-differential equation for $\hat{p}_{ik}(\cdot)$, we find that we can write $c_{ik}^{(\nu)}$ in the form $c_{ik}^{(\nu)} = \alpha_\nu^{\,i} c_k^{(\nu)}$. Combining this with the initial conditions, we find that

$$\hat{p}_{ik}(t) = \frac{1}{N}\sum_{\nu=1}^{N} \alpha_\nu^{\,i-k} e^{q\alpha_\nu t},$$

so that

$$p_{ik}(t) = \frac{1}{N}\sum_{\nu=1}^{N} \alpha_\nu^{\,i-k} e^{qt(\alpha_\nu - 1)}.$$

If $\nu = N$, $\alpha_\nu - 1 = 0$; otherwise $\alpha_\nu - 1$ has a negative real part. Hence

$$\lim_{t\to\infty} p_{ik}(t) = \frac{1}{N},$$

as was intuitively clear in the first place.

Example 2 In Example 1 there is only one ergodic class of states, and there are no transient states. To exhibit other possibilities we modify the above example by making the system remain in state N after reaching it:

$$\begin{aligned} q_{ij} &= 0 \qquad j \neq i+1 \\ &= q \qquad j = i+1 \end{aligned} \qquad i = 1, \cdots, N-1$$

$$q_N = 0.$$

In this case (1.7) becomes

$$\begin{aligned} p_{ik}'(t) &= -qp_{ik}(t) + qp_{i+1k}(t) \qquad i < N \\ p_{Nk}'(t) &= 0, \end{aligned}$$

with initial conditions $p_{ik}(0) = \delta_{ik}$. Then

$$\begin{aligned} p_{Nk}(t) &\equiv 0 \qquad k \neq N \\ p_{NN}(t) &\equiv 1, \end{aligned}$$

and it is easily verified that the solution is

$$\begin{aligned} p_{ik}(t) &= 0 & k < i \\ &= \frac{(qt)^{k-i}e^{-qt}}{(k-i)!} & i \leq k < N \\ &= e^{-qt}\left[e^{qt} - 1 - qt - \cdots - \frac{(qt)^{N-i-1}}{(N-i-1)!}\right] & k = N. \end{aligned}$$

In this example there is only one ergodic class of states, the class containing the Nth state only. The other states are all transient.

The preceding example illustrates a general procedure. Given any Markov chain as studied in this section, one can choose a state, say the Nth state, and modify the process so that if the system ever reaches that state it remains there. In terms of the q_i's and q_{ij}'s this means simply that q_N and $q_{N1}, \cdots, q_{NN-1}$ are replaced by 0. In terms of the old process and sample functions the new transition probability $\tilde{p}_{ij}(t)$ obtained in this way is the probability that, if $x_0(\omega) = i$, then $x_t(\omega) = j$ and $x_s(\omega) \neq N$ for any $s < t$; $\tilde{p}_{iN}(t)$ is the probability that, if $x_0(\omega) = i$, then $x_s(\omega) = N$ for some $s \leq t$. This procedure can be considered as making the state N an absorbing barrier. Several states can, of course, be made absorbing barriers simultaneously.

We shall now investigate in detail stationary Markov transition matrix functions, under the continuity restriction (1.4). It will be shown that the transition probabilities necessarily have derivatives which satisfy the systems of differential equations (1.7) and (1.7′). Particular attention will be paid to the relation between the properties of the transition probabilities and the properties of the sample functions of the separable Markov processes determined by the matrix transition functions. We shall always adopt a unique definition of a conditional probability of the form $\mathbf{P}\{\Lambda \mid x_{t_0}(\omega) = i\}$, where Λ is defined by restrictions on sample functions for $t \geq t_0$. We take the obvious definition in terms of the given transition probabilities, and accept this whether or not $\mathbf{P}\{x_{t_0}(\omega) = i\}$ happens to be positive.

In the rest of this chapter it will frequently be convenient to denote a random variable depending on the parameter t by $x(t)$, rather than x_t, and to denote its value at ω by $x(t, \omega)$, rather than $x_t(\omega)$.

THEOREM 1.2 *If* $[p_{ij}(\cdot)]$ *is a stationary Markov transition matrix function satisfying* (1.4), *the limit*

$$\lim_{t\to 0} \frac{1 - p_{ii}(t)}{t} = q_i < \infty \tag{1.9}$$

exists for all i.

If $\{x(t), 0 \leq t < \infty\}$ *is a separable process determined by* $[p_{ij}(\cdot)]$ *together with an initial probability distribution, then*

$$(1.10) \quad \mathbf{P}\{x(\tau, \omega) \equiv i, \qquad t_0 \leq \tau \leq t_0 + \alpha \mid x(t_0, \omega) = i\} = e^{-q_i\alpha},$$

and, if $x(t_0, \omega_0) = i$, *then* $x(t, \omega_0) \equiv i$ *in some neighborhood of* t_0 *(whose size depends on* ω_0*), with probability* 1.

We observe that the last clause does not state that almost all sample functions are continuous, but only that at most the sample functions of a collection of sample functions of probability 0 have a discontinuity at any particular value of t.

The existence of the limit in (1.9) is, of course, a purely analytic fact which has of itself nothing to do with the theory of probability; it is implied by the conditions (1.1′) (1.2′), and the continuity condition (1.4). However, it is convenient and somewhat instructive to use probability theory to prove the existence of this limit, and we shall do so. Suppose then that the $x(t)$ process is as described in the second half of the theorem. Then the conditional probability on the left is a function $f(\cdot)$ which satisfies the functional equation

$$f(\alpha + \beta) = f(\alpha)f(\beta), \qquad \alpha, \beta > 0.$$

Moreover, $0 \leq f(\alpha) \leq 1$. Hence $f(\cdot)$ must be monotone non-increasing, and the only solution of the functional equation under this restriction has the form

$$f(\alpha) = e^{-\text{const.}\,\alpha},$$

where the constant is non-negative and may be $+\infty$, in which case we interpret the exponential function as 0. We, therefore, have (1.10) with some constant q_i [not yet identified as the limit in (1.9)], where $0 \leq q_i \leq +\infty$. The following argument excludes the case $q_i = +\infty$. Suppose that $0 = \tau_0 < \cdots < \tau_n = \alpha$ and define ${}_\nu p_{ii}$ by

$$\begin{aligned} {}_\nu p_{ii} &= 1, & \nu &= 0 \\ &= \mathbf{P}\{x(\tau_j, \omega) = i, j = 0, \cdots, \nu \mid x(0, \omega) = i\}, & \nu &\geq 1. \end{aligned}$$

Then, if $\varepsilon > 0$ and if α is so small that $p_{jj}(t) \geq 1 - \varepsilon$ for all j when $t \leq \alpha$,

$$\begin{aligned} 1 - \varepsilon \leq p_{ii}(\alpha) &= {}_n p_{ii} + \sum_{\nu=0}^{n-2} \sum_{j \neq i} {}_\nu p_{ii} p_{ij}(\tau_{\nu+1} - \tau_\nu) p_{ji}(\alpha - \tau_{\nu+1}) \\ &\leq {}_n p_{ii} + \sum_{\nu=0}^{n-2} \sum_{j \neq i} {}_\nu p_{ii} p_{ij}(\tau_{\nu+1} - \tau_\nu)\varepsilon \\ &\leq {}_n p_{ii} + \varepsilon(1 - {}_n p_{ii}). \end{aligned}$$

According to the definition of a separable process (see II §2) ${}_n p_{ii}$ can be made to approximate $f(\alpha)$ arbitrarily closely by choosing the τ_k's properly. Hence the preceding inequality implies

$$1 - \varepsilon \leq f(\alpha) + \varepsilon[1 - f(\alpha)],$$

and ε can be made arbitrary small in this inequality by choosing α small. This inequality excludes the possibility that $f(t) \equiv 0$ if we take $\varepsilon < \frac{1}{2}$. Then $f(\cdot)$ must be the exponential function written above, and we have also proved that, for any $\varepsilon > 0$, if t is sufficiently small,

$$p_{ii}(t) \leq e^{-q_i t} + \varepsilon(1 - e^{-q_i t}).$$

Combining this inequality with the inequality

$$p_{ii}(t) \geq f(t) = e^{-q_i t},$$

we find

$$\frac{1 - e^{-q_i t}}{t}(1 - \varepsilon) \leq \frac{1 - p_{ii}(t)}{t} \leq \frac{1 - e^{-q_i}}{t}$$

and this implies (1.9). Finally we find from (1.9) that, if

$$\mathbf{P}\{x(t_0, \omega) = i\} > 0, \quad h_2 \geq 0, \quad \text{and} \quad 0 \leq h_1 \leq t_0,$$

then

$$\mathbf{P}\{x(\tau, \omega) = i, t_0 - h_1 \leq \tau \leq t_0 + h_2 \mid x(t_0, \omega) = i\}$$
$$= \frac{\mathbf{P}\{x(t_0 - h_1, \omega) = i\}}{\mathbf{P}\{x(t_0, \omega) = i\}} e^{-q_i(h_1 + h_2)}.$$

If $h_1 \downarrow 0$, $h_2 \downarrow 0$, the quantity on the right goes to 1, and this fact means that, if $x(t_0, \omega) = i$, then $x(t, \omega) \equiv i$ for t near t_0 except possibly for a set of sample functions of probability 0. The sample functions are therefore almost all continuous at t_0, as was to be proved.

Note that, if the existence of the limit q_i in (1.9) is made a hypothesis, the probability (1.10) can be evaluated at once, as follows. According to the definition of a separable stochastic process (II §2) there is a sequence $\{t_j\}$ in the interval $[t_0, t_0 + \alpha]$ such that the two ω sets

$$\{x(t_j, \omega) = i, j \geq 1\}, \qquad \{x(\tau, \omega) \equiv i, t_0 \leq \tau \leq t_0 + \alpha\}$$

differ by at most a set of probability 0. Moreover, $\{t_j\}$ can be taken as any sequence dense in the interval $[t_0, t_0 + \alpha]$ because of the continuity condition (1.4′). In particular, if we take $\{t_j\}$ as the sequence of all points of the form $t_0 + k\alpha/2^n$, $k = 0, \cdots, 2^n$, $n = 1, 2, \cdots$, we find that

$$\mathbf{P}\{x(\tau, \omega) \equiv i, t_0 \leq \tau \leq t_0 + \alpha\}$$
$$= \lim_{n\to\infty} \mathbf{P}\{x(t_0 + k\alpha/2^n, \omega) = i, k = 1, \cdots, 2^n\}$$
$$= \mathbf{P}\{x(t_0, \omega) = i\} \lim_{n\to\infty} p_{ii}(\alpha/2^n)^{2^n}$$
$$= \mathbf{P}\{x(t_0, \omega) = i\} e^{-q_i\alpha}$$

and this is equivalent to (1.10).

A function $g(\cdot)$ will be called a *step function* if it has only finitely many points of discontinuity in every finite closed interval, if it is identically constant in every open interval of continuity points, and if, when t_0 is a point of discontinuity,

$$(1.11) \qquad g(t_0-) \le g(t_0) \le g(t_0+) \qquad \text{or} \qquad g(t_0-) \ge g(t_0) \ge g(t_0+).$$

A function $g(\cdot)$ will be said to have a *jump* at a point t_0 if it is discontinuous there, but if the one-sided limits $g(t_0-)$ and $g(t_0+)$ exist and satisfy one of the two preceding inequalities. The discontinuities of a step function in the interior of its interval of definition are jumps. A function of t taking on only finitely many values and continuous except for jumps is a step function.

We shall prove that almost all sample functions of a Markov chain of the type we are considering are step functions (if the process is separable). Theorem 1.2 shows that the probability of a discontinuity at any one point is 0. The following theorem goes considerably further.

THEOREM 1.3 *Let* $[p_{ij}(\cdot)]$ *be a stationary Markov transition matrix function satisfying* (1.4).

(i) *The limits*

$$(1.12) \qquad \lim_{t\to 0} \frac{p_{ij}(t)}{t} = q_{ij}, \qquad i \ne j$$

exist and

$$(1.13) \qquad \sum_{j\ne i} q_{ij} = q_i.$$

(ii) *Let* $\{x(t), 0 \le t < \infty\}$ *be a separable process determined by* $[p_{ij}(\cdot)]$ *together with an initial probability distribution. If* $q_i > 0$ *and if* $x(t_0, \omega) = i$, *there is with probability* 1 *a sample function discontinuity for some* $t > t_0$, *and in fact a first discontinuity, which is a jump; if* $0 < \alpha \le \infty$, *the probability that if there is a discontinuity in the interval* $[t_0, t_0 + \alpha)$ *the first is a jump to* j *is* q_{ij}/q_i.

Let $\{x(t), 0 \le t < \infty\}$ be a separable process as described in (ii). It is no restriction on the transition matrix function to assume the existence of the $x(t)$ process. In proving (ii) it is no restriction to assume that $t_0 = 0$, and we shall do so.

If $q_i = 0$, (1.10) shows that, whenever $x(0, \omega) = i$, $x(t, \omega) \equiv i$ for $t > 0$, so that $p_{ii}(t) \equiv 1$. The theorem is thus true in this case, and we shall assume $q_i > 0$ from now on. Let δ, t be any positive numbers and let $m\delta$ be the smallest multiple of δ which is $\ge t$. The probability $p_{ij}(m\delta)$, $j \ne i$, is at least equal to the probability that, if $x(0, \omega) = i$, then $x(m\delta, \omega) = j$ and the first change of $x(\mu\delta, \omega)$ as the integer μ increases is a transition from i to j; analytically

$$p_{ij}(m\delta) \ge \sum_{\nu=0}^{m-1} p_{ii}(\delta)^\nu p_{ij}(\delta) p_{jj}((m-\nu-1)\delta).$$

If t is so small that $p_{jj}(s) \geq 1 - \varepsilon$ for $s \leq t$, this inequality becomes

$$p_{ij}(m\delta) \geq (1 - \varepsilon) \frac{1 - p_{ii}(\delta)^m}{1 - p_{ii}(\delta)} p_{ij}(\delta).$$

When $\delta \to 0$ this implies, by Theorem 1.2,

$$p_{ij}(t) \geq (1 - \varepsilon) \frac{1 - e^{-q_i t}}{q_i} \limsup_{\delta \to 0} \frac{p_{ij}(\delta)}{\delta}.$$

The limit superior on the right must therefore be finite. We now divide both sides by t and obtain, when $t \to 0$, using the fact that ε can be made arbitrarily small,

$$\liminf_{t \to 0} \frac{p_{ij}(t)}{t} \geq \limsup_{\delta \to 0} \frac{p_{ij}(\delta)}{\delta}.$$

Then the limit q_{ij} in (1.12) exists, and (1.13) follows from the equality

$$\sum_{j \neq i} \frac{p_{ij}(t)}{t} = \frac{1 - p_{ii}(t)}{t}.$$

We prove now that for any $\alpha > 0$ the first discontinuity if any of a sample function for $0 \leq t \leq \alpha$ is a jump. Fix i, take the initial condition $x(0, \omega) = i$, and consider the ω set $\Lambda_{n,\beta}^{(j)}$ for which, for some ν, $2 \leq \nu \leq n$,

$$\left.\begin{aligned} x(\tau, \omega) &= i, \quad 0 \leq \tau \leq \frac{(\nu - 1)\alpha}{n} \\ x(\tau, \omega) &= j, \quad \frac{\nu\alpha}{n} \leq \tau \leq \frac{\nu\alpha}{n} + \beta. \end{aligned}\right\} \quad (j \neq i)$$

Then

$$\begin{aligned} \mathbf{P}\{\Lambda_{n,\beta}^{(j)} \mid x(0, \omega) = i\} &= \sum_{\nu=2}^{n} e^{-\frac{q_i(\nu-1)\alpha}{n}} p_{ij}(s/n) e^{-q_j\beta} \\ &= \frac{e^{-q_i\alpha/n} - e^{-q_i\alpha}}{1 - e^{-q_i\alpha/n}} p_{ij}(\alpha/n) e^{-q_j\beta} \\ &\to (1 - e^{-q_i\alpha}) \frac{q_{ij}}{q_i} e^{-q_i\beta}, \qquad n \to \infty. \end{aligned}$$

It follows that, if $\Lambda_\beta^{(j)}$ is the set of points ω in $\Lambda_{n,\beta}^{(j)}$ for infinitely many values of n, that is,

$$\Lambda_\beta^{(j)} = \bigcap_{k=1}^{\infty} \bigcup_{n=k}^{\infty} \Lambda_{n,\beta}^{(j)},$$

then

$$\mathbf{P}\{\Lambda_\beta^{(j)} \mid x(0, \omega) = i\} \geq (1 - e^{-q_i\alpha}) \frac{q_{ij}}{q_i} e^{-q_j\beta}.$$

The set $\Lambda_\beta^{(j)}$ increases when β decreases. Hence, if $\Lambda^{(j)}$ is the set of points ω in $\Lambda_\beta^{(j)}$ for some $\beta > 0$,

$$\mathbf{P}\{\Lambda^{(j)} \mid x(0, \omega) = i\} \geq (1 - e^{-q_i\alpha}) \frac{q_{ij}}{q_i}.$$

Any sample function corresponding to an ω in $\Lambda^{(j)}$ is identically i in some interval with left-hand endpoint at $t = 0$, has a discontinuity at $\tau_1 < \alpha$, and is identically j in some open interval with left-hand endpoint τ_1. Since

$$\sum_j \mathbf{P}\{\Lambda^{(j)} \mid x(0, \omega) = i\} \geq 1 - e^{-q_i\alpha} = \mathbf{P}\{x(\tau, \omega) \neq i, 0 \leq \tau \leq \alpha\},$$

there must be equality in this inequality, and therefore also in the preceding one. Thus excluding an ω set of probability 0, $x(t, \omega) \equiv i$ for $t < \tau_1$ and $x(t, \omega) \equiv j$ for $\tau_1 < t < \tau_1 + \delta$, where j and δ depend on ω, and, on the assumption that there is a discontinuity in the interval $(0, \alpha)$, the probability that $x[\tau_1(\omega) +, \omega] = j$ is q_{ij}/q_i. If $q_i = 0$, and if $x(0, \omega) = i$, the probability of a discontinuity in the interval $(0, \alpha)$ is 0 for all α, as we have already noted above. If $q_i > 0$, and if $x(0, \omega) = i$, the probability is $1 - e^{-q_i\alpha}$ that there is a discontinuity in the interval $(0,\alpha)$, that is,

$$\mathbf{P}\{\tau_1(\omega) \geq \alpha \mid x(0, \omega) = i\} = 1 - e^{-q_i\alpha} \qquad \alpha \geq 0.$$

We have not yet proved that the discontinuity at $\tau_1(\omega)$ is a jump for almost all ω because we have not verified that the condition (1.11) is satisfied at $\tau_1(\omega)$. This not very important fact is a consequence of the following argument. By the definition of separability of a stochastic process (II §2) there is a denumerable t set with the property that, neglecting an ω set of probability 0, $x(\cdot, \omega)$ has the same least upper and greatest lower bounds on every open t interval as on the part of this denumerable t set in the interval. Since the probability of a discontinuity at any point of this denumerable set is 0, $x[\tau_1(\omega), \omega]$ must lie between $x[\tau_1(\omega) -, \omega]$ and $x[\tau_1(\omega) +, \omega]$ inclusive, that is, the discontinuity at $\tau_1(\omega)$ is a jump, with probability 1. This fact is a more or less accidental result of our definition of separability of a stochastic process. If the process is supposed separable relative to the closed sets, it follows that the sample functions are even almost all continuous on the right or left at $\tau_1(\omega)$.

THEOREM 1.4 *The sample functions of a separable Markov chain* (*with a finite number of states*) *which has stationary transition probabilities satisfying the continuity condition* (1.4), *are almost all step functions.*

We shall suppose in the proof that there is a stationary Markov transition matrix function which together with an initial probability distribution determines the process. This is a slight restriction since the transition probability $\mathbf{P}\{x(t, \omega) = j \mid x(s, \omega) = i\}$ may not be defined for all i, j, but it will be obvious what reservations to make in a few statements to make the proof perfectly general.

Let $\{x(t), 0 \leq t < \infty\}$ be the process in question. We have seen that (neglecting a set of sample functions of probability 0), if $x(0, \omega) = i$ and if $q_i = 0$, then $x(t, \omega) \equiv i$, whereas, if $q_i > 0$, there is a first discontinuity, a jump, at a point $\tau_1(\omega)$. If $q_i = 0$ set $\tau_1(\omega) = \tau_2(\omega) = \cdots = \infty$. Continuing the argument, it is easily proved that, if $x(\tau_1 +, \omega) = j$ and if $q_j = 0$, then $x(t, \omega) \equiv j$ for $t > \tau_1(\omega)$, in which case we define $\tau_2(\omega) = \tau_3(\omega) = \cdots = \infty$, whereas, if $q_j > 0$, there is a first discontinuity after $\tau_1(\omega)$, a jump, at a point $\tau_2(\omega)$, with

$$\mathbf{P}\{\tau_2(\omega) - \tau_1(\omega) \geq \alpha \mid x[\tau_1(\omega) +, \omega] = j\} = e^{-q_j\alpha},$$

and so on. In general we have not only

$$\mathbf{P}\{\tau_{n+1}(\omega) - \tau_n(\omega) \geq \alpha \mid x[\tau_n(\omega) +, \omega] = j\} = e^{-q_j\alpha}$$

(in case $q_j > 0$), but to the condition $x[\tau_n(\omega) +, \omega] = j$ we can add other conditions which involve sample functions at parameter values $\leq \tau_n(\omega)$ without changing the exponential function on the right. This fact explains why the procedure can be continued indefinitely. To prove the theorem we show that $\lim_{n\to\infty} \tau_n = +\infty$ with probability 1. Let $q = \operatorname{Max}_j q_j$. Then $\{\tau_n, n \geq 1\}$ is a sequence of random variables with

$$\begin{aligned}
&\mathbf{P}\{\tau_{n+1}(\omega) - \tau_n(\omega) \geq \alpha\} \\
&= \sum_{j=1}^{N} \mathbf{P}\{\tau_{n+1}(\omega) - \tau_n(\omega) \geq \alpha \mid x[\tau_n(\omega) +, \omega] = j\}\mathbf{P}\{x[\tau_n(\omega) +, \omega] = j\} \\
&= \sum_{j=1}^{N} e^{-q_j\alpha}\, \mathbf{P}\{x[\tau_n(\omega) +, \omega] = j\} \\
&\geq e^{-q\alpha}, \qquad n \geq 0, \qquad \alpha \geq 0.
\end{aligned}$$

Here we have put $\tau_0(\omega) = 0$, and interpret $\infty - c$ as ∞ for $c \leq \infty$. Then infinitely many differences $\tau_{n+1}(\omega) - \tau_n(\omega)$ will be $\geq \alpha$, with probability $\geq e^{-q\alpha}$, so that $\lim_{n\to\infty} \tau_n(\omega) = \infty$ with probability $\geq e^{-q\alpha}$ for every $\alpha > 0$. This probability is then 1, as was to be proved.

Let ${}_1p_{ik}(t)$ be the probability that, if $x(t_0, \omega) = i$, then $x(t_0 + t, \omega) = j$ and that the transition from i to k has been accomplished in a single step. One is tempted to evaluate ${}_1p_{ik}(t)$ by stating that $x(\tau, \omega)$ must be identically i to the jump point s (probability e^{-q_is}), that there must then be a jump to k (probability $q_{ik}\,ds$), and that then $x(\tau, \omega)$ must be identically k to t (probability $e^{-q_k(t-s)}$), so that

$$
\begin{aligned}
{}_1p_{ik}(t) &= \int_0^t e^{-q_is-q_k(t-s)}\,q_{ik}\,ds \qquad && k \neq i \\
&= 0 && k = i.
\end{aligned}
$$

This reasoning will now be justified in full detail, but similar reasoning will be used below without further justification. Take $t_0 = 0$ again, fix i and $k \neq i$, let M_n be the ω set for which, for some ν, $0 < \nu < n - 1$,

$$
\begin{aligned}
x(\tau, \omega) &= i, \qquad && 0 \leq \tau < \frac{\nu}{n}\,t \\
&= k && \frac{\nu + 1}{n}\,t < \tau \leq t,
\end{aligned}
$$

and let M be the ω set for which, for some s, $0 < s < t$,

$$
\begin{aligned}
x(\tau, \omega) &= i, \qquad && 0 \leq \tau < s \\
&= k, && s < \tau \leq t.
\end{aligned}
$$

Then M is the set of points ω in infinitely many M_n's as well as the set of points ω in all but a finite number of M_n's, that is, in the usual language of set theory $\lim_{n\to\infty} \mathrm{M}_n = \mathrm{M}$, and it follows that

$$
\lim_{n\to\infty} \mathbf{P}\{\mathrm{M}_n \mid x(0, \omega) = i\} = \mathbf{P}\{\mathrm{M} \mid x(0, \omega) = i\} = {}_1p_{ik}(t).
$$

On the other hand

$$
\mathbf{P}\{\mathrm{M}_n \mid x(0, \omega) = i\} = \sum_{\nu=1}^{n-2} e^{-q_i\frac{\nu}{n}}\,p_{ik}\left(\frac{t}{n}\right) e^{-q_k\frac{n-\nu-1}{n}t},
$$

and when $n \to \infty$ this sum approaches the integral evaluation tentatively obtained above for ${}_1p_{ik}(t)$, an evaluation which is thereby justified.

Let ${}_np_{ik}(t)$ be the probability that, if $x(t_0, \omega) = i$, than $x(t_0 + t, \omega) = k$, and the transition from i to k has been accomplished in n steps. We have already evaluated ${}_1p_{ik}(t)$. In the same way we derive equations for ${}_{n+1}p_{ik}(t)$ in terms of ${}_np_{ik}(t)$, and thereby obtain an inductive definition of the sequence,

$$
\begin{aligned}
{}_0p_{ik}(t) &= 0, \qquad && k \neq i, \\
&= e^{-q_it}, && k = i; \qquad (1.14)
\end{aligned}
$$

$$
{}_{n+1}p_{ik}(t) = \sum_{j\neq i} \int_0^t e^{-q_is} q_{ij}\,{}_np_{jk}(t - s)\,ds, \qquad n \geq 0,
$$

or alternatively

$$
\begin{aligned}
{}_0p_{ik}(t) &= 0, && k \neq i, \\
&= e^{-q_i t}, && k = i;
\end{aligned}
\tag{1.14'}
$$

$$
{}_{n+1}p_{ik}(t) = \sum_{j \neq k} \int_0^t {}_n p_{ij}(s) q_{jk} e^{-q_k(t-s)}\, ds, \qquad n \geq 0.
$$

Since almost all sample functions are step functions, it follows that

$$
p_{ik}(t) = \sum_{n=0}^{\infty} {}_n p_{ik}(t). \tag{1.15}
$$

If (1.14) and (1.15) are combined, we find

$$
p_{ik}(t) = \delta_{ik} e^{-q_i t} + \sum_{j \neq i} \int_0^t e^{-q_i s} q_{ij} p_{jk}(t-s)\, ds, \tag{1.16}
$$

and, if (1.14′) and (1.15) are combined, we find

$$
p_{ik}(t) = \delta_{ik} e^{-q_k t} + \sum_{j \neq k} \int_0^t p_{ij}(s) q_{jk} e^{-q_k(t-s)}\, ds. \tag{1.16'}
$$

These two integral equations are so important that we derive them independently to exhibit their probabilistic meaning. The first exhibits $p_{ik}(t)$ as the probability that, if $x(0, \omega) = i$, either $x(\tau, \omega)$ remains at i for time t (possible only if $k = i$, in that case probability $e^{-q_i t}$) or else $x(\tau, \omega)$ will remain at i to time s (probability $e^{-q_i s}$), will then jump to some state $j \neq i$ in time ds (probability $q_{ij}\, ds$), and will go to k in time $t - s$ [probability $p_{jk}(t - s)$]. Thus (1.16) depends essentially on the fact that almost all sample functions have a first discontinuity, which is a jump. Similarly (1.16′) depends essentially on the existence of a last discontinuity, which is a jump, in the interval $[0, t]$.

Equations (1.15) and (1.16) show that the elements of a stationary Markov transition matrix function satisfying the continuity condition (1.4) have continuous derivatives. Taking derivatives in these equations, we obtain (1.7) and (1.7′) respectively, obtained earlier by direct differentiation of the Chapman-Kolmogorov equation. The differential equation (1.7) and (1.7′) can thus be interpreted in terms of the continuity properties of the sample functions. This interpretation becomes critically important in its generalized form for the processes to be discussed in §2.

Equations (1.14) or (1.14′), and (1.15), provide an explicit algorithm for calculating $p_{ij}(t)$ in terms of the q_i's and q_{ij}'s which has more probabilistic significance than that given above, $P(t) = e^{tQ}$, in the course of the solution of the systems of differential equations (1.7) and (1.7′).

Suppose now that $q_1, \cdots, q_N, q_{ij}, i, j = 1, \cdots, N(i \neq j)$ are any constants satisfying (1.6). We have already proved that there is a stationary Markov transition matrix function [satisfying the continuity condition (1.4)] with these q_i's and q_{ij}'s, that is, satisfying (1.9) and (1.12). In fact, we showed above that e^{tQ} is such a matrix function. A second method is to define $p_{ij}(t)$ by (1.14) and (1.15), or (1.14′) and (1.15), and it is in fact not difficult to prove directly that the series in (1.15) converges, and defines a stationary Markov transition matrix function. (An indirect proof would consist simply of the remark that there is a unique stationary Markov transition matrix function—satisfying the continuity condition (1.4)—with the given q_i's and q_{ij}'s, namely e^{tQ}, and we have already deduced that (1.15) is then true.) A third method, now to be given, is important in that it defines the matrix function by defining the corresponding Markov chains in terms of their sample functions, and thus proves an existence theorem for the systems of differential equations (1.7) and (1.7′) by purely probabilistic methods which are applicable to the considerably more complicated integral-differential systems considered in §2. To construct a Markov chain with given q_i's and q_{ij}'s we simply adapt the proof of Theorem 1.3, as follows.

Let z_1 be any random variable taking on the values $1, \cdots, N$ only. Let τ_1 be a positive random variable whose joint distribution with z_1 is determined by setting

$$\mathbf{P}\{\tau_1(\omega) > \alpha \mid z_1(\omega) = i\} = e^{-q_i\alpha}, \qquad \alpha > 0.$$

(If $q_i = 0$ the interpretation is the obvious one that $\tau_1(\omega) = \infty$.) If $z_1, \cdots, z_n, \tau_1, \cdots, \tau_n$ have already been defined, z_{n+1} is a random variable assuming the values $1, \cdots, N$ only, whose joint distribution with $z_1, \cdots, z_n, \tau_1, \cdots, \tau_n$ is determined by setting

$$\mathbf{P}\{z_{n+1}(\omega) = j \mid \tau_1, \cdots, \tau_n, z_1, \cdots, z_n\} = \frac{q_{ij}}{q_i} \qquad (z_n(\omega) = i)$$

and τ_{n+1} is a positive random variable whose joint distribution with $z_1, \cdots, z_{n+1}, \tau_1, \cdots, \tau_n$ is determined by

$$\mathbf{P}\{\tau_{n+1}(\omega) - \tau_n(\omega) > \alpha \mid \tau_1, \cdots, \tau_n, z_1, \cdots, z_{n+1}\} = e^{-q_j\alpha}$$
$$(z_{n+1}(\omega) = j,\ \alpha > 0).$$

We have assumed here that, if $z_n(\omega) = i$, then $q_i \neq 0$. To complete the definition we define

$$\mathbf{P}\{z_{n+1}(\omega) = i \mid \tau_1, \cdots, \tau_n, z_1, \cdots, z_n\} = 1 \qquad (z_n(\omega) = i,\quad q_i = 0),$$

$$\mathbf{P}\{\tau_{n+1}(\omega) = \infty \mid \tau_1, \cdots, \tau_n, z_1, \cdots, z_{n+1}\} = 1 \qquad (z_n(\omega) = i,\quad q_i = 0).$$

A sequence of random variables $z_1, \tau_1, z_2, \tau_2, \cdots$ has thus been defined inductively, and this will of course be identified with sequence obtained in the proof of Theorem 1.4 with $x(\tau_n +) = z_{n+1}$. To do this we note first that $\lim_{n\to\infty} \tau_n = \infty$ with probability 1, using the proof of the corresponding fact in Theorem 1.3, so that if we take $\tau_0 = 0$ we can define $x(t)$ by

$$x(t, \omega) = z_n(\omega) \qquad \tau_{n-1}(\omega) \le t < \tau_n(\omega)$$

for $0 \le t < \infty$. The following argument shows that the $x(t)$ process defined in this way is a Markov chain, and that the given q_i's and q_{ij}'s satisfy (1.9) and (1.12). We shall use the fact that, if c is a positive constant and if x is a positive random variable with density $ce^{-c\lambda}(\lambda > 0)$, and if ξ is any positive number,

$$\mathbf{P}\{x(\omega) > \lambda\} = \mathbf{P}\{x(\omega) - \xi > \lambda \mid x(\omega) - \xi > 0\}.$$

This trivial fact means that, if in the procedure described above for constructing the general sample function we choose some $s > 0$ and stop the procedure when a τ_j is reached which exceeds s, so that $x(t)$ is only defined for $t \le s$, and if then the defining procedure is recommenced at $t = s$ in exactly the same way as it was started at $t = 0$, using, however, the $x(s)$ values as the initial values with the probabilities already found for these values, then the following τ_j's and x_j's will have the same distribution as they would have had if the procedure had not been interrupted. But then

(a) the conditional probability distribution of $x(t)$, for prescribed values of $x(\tau)$ when $\tau \le s$, depends only on $x(s)$, that is, the process is a Markov process;

(b) the probability $\mathbf{P}\{x(t, \omega) = j \mid x(s, \omega) = i\}$ is a function of $(t - s)$, that is, the process has stationary transition probabilities.

The continuity condition (1.4) is obviously satisfied and, since the q_i's and q_{ij}'s have exactly the sample function significance of the q_i's and q_{ij}'s of Theorems 1.1 and 1.2, these sets of constants can be identified with each other. This completes the discussion of the construction of a chain with given q_i's and q_{ij}'s. We remark that the corresponding analytical calculations of $p_{ij}(t)$ would be by means of (1.14) and (1.15) or (1.14′) and (1.15), using the fact that

$${}_np_{ik}(t) = \mathbf{P}\{z_{n+1}(\omega) = k, \tau_n(\omega) \le t < \tau_{n+1}(\omega) \mid z_1(\omega) = i\}.$$

The relation between the differential equation system (1.7) and the adjoint system (1.7′) shows up more clearly in the non-stationary case, about which we shall therefore make a few remarks. We shall neither maximize the rigor nor minimize the hypotheses in the following. Suppose that $[p_{ij}(\cdot, \cdot)]$ is a Markov transition matrix function and that there are functions $q_i(\cdot)$, $q_{ij}(\cdot)$ for which

$$\begin{aligned} p_{ij}(s, t) &= 1 - q_i(t)(t - s) + o(t - s) \qquad & i = j \\ &= q_{ij}(t)(t - s) + o(t - s) & i \neq j, \end{aligned}$$

so that, by (1.1),

$$q_i(t) \geq 0, \qquad q_{ij}(t) \geq 0, \qquad q_i(t) = \sum_{j \neq i} q_{ij}(t).$$

Then

$$\begin{aligned} \left.\frac{\partial p_{ij}(s, t)}{\partial s}\right|_{s=t} &= q_i(t) \qquad & i = j \\ &= -\, q_{ij}(t) & i \neq j \end{aligned}$$

and

$$\begin{aligned} \left.\frac{\partial p_{ij}(s, t)}{\partial t}\right|_{t=s} &= -\, q_i(s) \qquad & i = j \\ &= q_{ij}(s) & i \neq j. \end{aligned}$$

Taking partial derivatives in the Chapman-Kolmogorov system of equations (1.2) with respect to s, setting $s = t$, and then finally replacing the pair (t, u) by the pair s, t, we find

$$\frac{\partial p_{ik}(s, t)}{\partial s} = q_i(s) p_{ik}(s, t) - \sum_{j \neq i} q_{ij}(s) p_{jk}(s, t). \tag{1.17}$$

Taking partial derivatives in (1.2) with respect to u and setting $u = t$, we find

$$\frac{\partial p_{ik}(s, t)}{\partial t} = -\, p_{ik}(s, t) q_k(t) + \sum_{j \neq k} p_{ij}(s, t) q_{jk}(t). \tag{1.17$'$}$$

The system (1.17) is called the *backward* system because it involves differentiation with respect to the earlier time s, and the system (1.17′) is called the *forward* system because it involves differentiation with respect to the later time t. In the stationary case we have seen that the backward system (1.7) involves in an essential way the first sample function jump after a given time t_0 (before time $t_0 + t$) while the forward system (1.7′) involves the last sample function jump (after time t_0) before time $t_0 + t$.

We have made essential use throughout this section of the hypothesis that the chains we have considered have only a finite number N of states. The situation is considerably more complicated if there are enumerably many states. In this case, as we shall see in the next section, it is characteristic that for a wide class of cases the backward system of differential equations holds when the forward system does not. This will be related to the study of sample function discontinuities, which may be worse than jumps.

2. Generalization of §1 to a continuous state space

Let X be a linear Borel set. In the present section we consider Markov processes $\{x(t), 0 \leq t < \infty\}$ whose random variables take on values in X. In particular, if X is the set of integers $1, \cdots, N$, the processes are Markov chains with N states, the processes considered in the preceding section. The choice of X is not critical, in the sense that any larger set will serve as well, and the fact that X contains more points than necessary will do no harm. For example, if X is the whole line, but, if

$$\sum_{j=1}^{N} \mathbf{P}\{x(t, \omega) = j\} = 1,$$

the results of the preceding section hold, whether or not the random variables ever actually take on values other than $1, \cdots, N$. If the stochastic process under consideration is to be supposed separable, we recall from II §2 that it may be necessary to suppose that X is closed in the extended infinite line.

The results of this section are valid with no change if X is multidimensional and with some moderate changes even if X is an abstract space (with no topology defined on it), but in the latter case there are certain measure theoretic problems that would affect the type of argument that could be used in the proofs.

Rather than supposing that a Markov process is given we suppose the stochastic transition functions of the process are given, in the following form:

It is supposed that a function $p(\cdot, \cdot;\ \cdot, \cdot)$ of s, ξ, t, A is given, defined for $0 < s < t$, $\xi \in X$, A a Borel subset of X, satisfying:

(*a*) $p(s, \cdot;\ t, A)$ is a Baire function of ξ for fixed s, t, A;
(*b*) $p(s, \xi;\ t, \cdot)$ is a probability measure in A for fixed s, ξ, t;
(*c*) $p(\cdot, \cdot;\ \cdot, \cdot)$ satisfies the Chapman-Kolmogorov equation

$$p(s, \xi;\ u, A) = \int_X p(t, \eta;\ u, A) p(s, \xi;\ t, d\eta), \qquad s < t < u.$$

A function $p(\cdot, \cdot;\ \cdot, \cdot)$ satisfying the above conditions will be called a *Markov transition function.* It is convenient to define $p(t, \xi;\ t, A)$ to be 1 if $\xi \in A$ and 0 otherwise, and we shall use this definition throughout. If $p(\cdot, \cdot;\ \cdot, \cdot)$ is a Markov transition function and if $p(\cdot)$ is any probability distribution of sets A, we have seen in II §6 that there is a Markov process $\{x(t), 0 \leq t < \infty\}$ whose random variables take on values in X, for which

$$\mathbf{P}\{x(0, \omega) \in A\} = p(A)$$

$$\mathbf{P}\{x(t, \omega) \in A \mid x(s)\} = p(s, x(s);\ t, A),$$

with probability 1. The exceptional ω set may depend on s, t, A. More generally we shall say that any Markov stochastic process for which the latter equation is satisfied in the sense just described is one with p as its transition function.

In the case of stationary transition probabilities, when by definition $p(s, \xi;\ t, A)$ depends only on $t - s$, we use the notation

$$p(t - s, \xi, A) = p(s, \xi, t, A),$$

and call $p(\cdot, \cdot, \cdot)$ a *stationary Markov transition function.*

The Chapman-Kolmogorov equation becomes

$$p(s + t, \xi, A) = \int_X p(t, \eta, A)p(s, \xi, d\eta).$$

In the following it will always be assumed that the transition probabilities are stationary unless the contrary is explicitly stated. The results will include those of §1 as special cases, and §1 was written only to illuminate the general case of this section.

We shall say that Doeblin's condition (D) is satisfied if there is a finite valued measure $\varphi(\cdot)$ of Borel subsets of X, an $\varepsilon > 0$, and an $s > 0$, such that $p(s, \xi, A) \leq 1 - \varepsilon$ if $\varphi(A) \leq \varepsilon$. It follows as in the discrete parameter case (see V §5) that then

$$p(t, \xi, A) \leq 1 - \varepsilon \qquad t \geq s$$

if $\varphi(A) \leq \varepsilon$, so that to every t_0 corresponds a ν such that

$$p(\nu t_0, \xi, A) \leq 1 - \varepsilon$$

if $\varphi(A) \leq \varepsilon$. Thus each stochastic transition function $p(t_0, \cdot, \cdot)$ satisfies Doeblin's condition (D) in the sense of V §5.

THEOREM 2.1 *If the stationary Markov transition function* $p(\cdot, \cdot, \cdot)$ *satisfies condition* (D), *then* $\lim_{t\to\infty} p(t, \xi, A)$ *exists for all* ξ *and* A. *The convergence is uniformly exponentially fast.*

The proof of this theorem follows that of Theorem 1.1, which is a special case, and only the beginning will be sketched to show that, if $\varphi(\cdot)$ is the measure function of condition (D), and if N is the largest integer $\leq \varphi(X)/\varepsilon$, then N plays the same role here as the number of states played in the chain case of §1. In fact, we have seen in V §5 that for each t_0

either $\lim_{n\to\infty} p(nt_0, \xi, A)$ exists for all ξ, A or the stochastic transition function $p(t_0, \cdot, \cdot)$ determines cyclically moving classes of states and that because of condition (D) each such class must have φ measure at least ε. If the cycle lengths are $d_1, d_2, \cdots$, we then have $\sum_j d_j\varepsilon \leq \varphi(X)$, so that $\sum_j d_j \leq N$. Moreover, we showed that if ν is an integer divisible by the d_j's, say $\nu = N!$, $\lim_{n\to\infty} p(n\nu t_0, \xi, A)$ exists for all ξ, A. Since t_0 is arbitrary, $\lim_{n\to\infty} p(nt, \xi, A)$ exists for all t, ξ, A and the proof goes on as in the special case Theorem 1.1.

We shall be interested in the remainder of this section in stochastic processes whose sample functions are almost all step functions, and in closely related processes. The appropriate corresponding continuity condition to impose on the stochastic transition function is

$$\lim_{t\to 0} p(t, \xi, \{\xi\}) = 1 \tag{2.1}$$

for all ξ, where $\{\xi\}$ is the point set containing the single point ξ. Among other results Doeblin's result will be proved, that if (2.1) is satisfied uniformly in ξ the sample functions of a separable process will almost all be step functions. Without this added uniformity condition the statement is false in general. If X is finite, then if (2.1) is satisfied at all it is necessarily satisfied uniformly in ξ. In the preceding section we studied the case in which X consisted of the points $1, \cdots, N$ and wrote $p_{ij}(t)$ for $p(t, i, \{j\})$. With this notational change (2.1) becomes (1.4).

According to the Chapman-Kolmogorov equation

$$p(t + \varepsilon, \xi, A) = \int_X p(\varepsilon, \eta, A)p(t, \xi, d\eta) \qquad \varepsilon > 0,$$

and, since when $\varepsilon \to 0$ the integrand converges to 0 if $\eta \notin A$, and converges to 1 if $\eta \in A$, by (2.1), it follows that

$$\lim_{\varepsilon \downarrow 0} p(t + \varepsilon, \xi, A) = p(t, \xi, A),$$

that is, the stochastic transition functions are continuous on the right. Further hypotheses seem to be necessary to obtain continuity on the left. It is certainly sufficient if (2.1) is assumed to hold uniformly in ξ because then each integrand in the equality

$$p(t - \varepsilon, \xi, A) - p(t, \xi, A) = \int_A [1 - p(\varepsilon, \eta, A)]p(t - \varepsilon, \xi, d\eta) - \int_{X-A} p(\varepsilon, \eta, A)p(t - \varepsilon, \xi, d\eta) \qquad \varepsilon > 0,$$

is uniformly small with ε.

In the following it will be convenient to have uniquely defined conditional probabilities of the form

$$\mathbf{P}\{\Lambda \mid x(t_0, \omega) = \zeta\}$$

in which Λ is defined by restrictions on the sample functions for $t \geq t_0$. This conditional probability is defined as follows: Define a Markov process with parameter $t \geq t_0$, defining the $x(t_0)$ distribution to be concentrated at the point ζ, and using the given transition probability function. The probability assigned to Λ in this way is uniquely defined, and the above conditional probability will always be understood as this value. It is clear that this definition is legitimate, and we shall use it without further comment.

THEOREM 2.2 *If* $p(\cdot, \cdot, \cdot)$ *is a stationary Markov transition function satisfying the continuity condition* (2.1), *then the limit*

$$\lim_{t \to 0} \frac{1 - p(t, \xi, \{\xi\})}{t} = q(\xi) \leq \infty \tag{2.2}$$

exists for all ξ. *If* $q(\cdot)$ *is bounded on a set* A, *then the continuity condition* (2.1) *holds uniformly on* A. *If* (2.1) *holds uniformly on* X, $q(\cdot)$ *is a bounded function, and the limit in* (2.2) *is uniform in* ξ.

If $\{x(t), 0 \leq t < \infty\}$ *is a separable Markov process with* $p(\cdot, \cdot, \cdot)$ *as its transition function, then, if* $x(t_0, \omega) = \xi$, *it follows that* $x(t, \omega) \equiv \xi$ *in some interval* $(t_0, t_0 + \tau)$ (*where* τ *depends on* ω) *with probability* 1.

Just as in proving Theorem 1.2, which is a special case, it will be illuminating to prove this theorem by probabilistic methods, although the existence of the limit $q(\xi)$ can of course be proved without explicit use of probability concepts. Suppose then that $\{x(t), 0 \leq t < \infty\}$ is a separable process as described in the second half of the theorem. We have seen in II §2 that there always is such a process if X is a closed set on the line closed at $+\infty$. It appears at first therefore that we have imposed an extra condition here, but we shall show how to get around this point at the end of the proof. Let α and δ be positive numbers, and let $m\delta$ be the smallest multiple of δ, which is $\geq \alpha$. Then (see II §2) since (2.1) obviously implies that

$$\underset{t \downarrow t_0}{\text{p lim}}\, x(t) = x(t_0),$$

we have (since II, Theorem 2.2, is trivially adaptable to the present case)

$$\begin{aligned}
&\mathbf{P}\{x(t, \omega) = \xi, 0 \leq t \leq \alpha \mid x(0, \omega) = \xi\} \\
&= \lim_{\delta \to 0} \mathbf{P}\{x(j\delta, \omega) = \xi, j = 1, \cdots, m - 1, x(\alpha, \omega) = \xi \mid x(0, \omega) = \xi\} \\
&= \lim_{\delta \to 0} \mathbf{P}\{x(j\delta, \omega) = \xi, j = 1, \cdots, m \mid x(0, \omega) = \xi\}.
\end{aligned} \tag{2.3}$$

If the limit is positive, its logarithm is

$$\lim_{\delta \to 0} m \log p(\delta, \xi, \{\xi\}) = - \lim_{\delta \to 0} \alpha \frac{1 - p(\delta, \xi, \{\xi\})}{\delta},$$

so that the existence of $q(\xi)$ in (2.2) is established and

$$\mathbf{P}\{x(t, \omega) = \xi, 0 \leq t \leq \alpha \mid x(0, \omega) = \xi\} = e^{-q(\xi)\alpha}. \tag{2.4}$$

If the limit in (2.3) is 0, then $q(\xi)$ in (2.2) is infinite and (2.4) still holds, with the obvious convention. The evaluation (2.4) implies that

$$e^{-q(\xi)\alpha} \leq p(\alpha, \xi, \{\xi\}).$$

It follows that the continuity condition (2.1) holds uniformly in ξ on any set on which $q(\cdot)$ is bounded. If (2.1) holds uniformly in $\xi \in X$, choose $\varepsilon > 0$ and suppose that α is so small that

$$p(s, \xi, \{\xi\}) \geq 1 - \varepsilon, \qquad s \leq \alpha, \xi \in X.$$

Then, if $0 = \tau_0 < \cdots < \tau_n = \alpha$ and if ${}_\nu p(\xi, \xi)$ is defined by

$$\begin{aligned} {}_\nu p(\xi, \xi) &= 1, & \nu &= 0, \\ &= \mathbf{P}\{x(\tau_j, \omega) = \xi, j = 0, \cdots, \nu \mid x(0, \omega) = \xi\}, & \nu &\geq 1, \end{aligned}$$

it follows that

$$\begin{aligned} 1 - \varepsilon &\leq p(\alpha, \xi, \{\xi\}) \\ &= {}_n p(\xi, \xi) + \sum_{\nu=0}^{n-2} \int_{\{\eta \neq \xi\}} {}_\nu p(\xi, \xi) p(\alpha - \tau_{\nu+1}, \eta, \{\xi\}) p(\tau_{\nu+1} - \tau_\nu, \xi, d\eta) \\ &\leq {}_n p(\xi, \xi) + \varepsilon \sum_{\nu=0}^{n-2} \int_{\{\eta \neq \xi\}} {}_\nu p(\xi, \xi) p(\tau_{\nu+1} - \tau_\nu, \xi, d\eta) \\ &\leq {}_n p(\xi, \xi) + \varepsilon[1 - {}_n p(\xi, \xi)]. \end{aligned}$$

Then

$$(1 - \varepsilon)[1 - {}_n p(\xi, \xi)] \leq 1 - p(\alpha, \xi, \{\xi\}) \leq \varepsilon$$

and, since this is true for all choices of the τ_j's,

$$(1 - \varepsilon)[1 - e^{-q(\xi)\alpha}] \leq 1 - p(\alpha, \xi, \{\xi\}) \leq \varepsilon.$$

If $\varepsilon < \frac{1}{2}$ this inequality shows that $q(\cdot)$ is a bounded function. Moreover, the inequality

$$(1 - \varepsilon) \frac{1 - e^{-q(\xi)\alpha}}{\alpha} \leq \frac{1 - p(\alpha, \xi, \{\xi\})}{\alpha} \leq \frac{1 - e^{-q(\xi)\alpha}}{\alpha},$$

valid uniformly in ξ for sufficiently small α, now shows that the convergence in (2.2) is uniform in ξ, and even that

$$\lim_{\alpha\to 0} \frac{1-p(\alpha, \xi, \{\xi\})}{\alpha q(\xi)} = 1$$

uniformly in ξ (with the obvious conventions when $q(\xi) = 0$).

If $q(\xi) = 0$, and $x(t_0, \omega) = \xi$, then $x(t, \omega) = \xi$ for $t > t_0$ with probability 1, by (2.4) (remembering that the transition probabilities are stationary). If $0 < q(\xi) < \infty$ and if $x(t_0, \omega) = \xi$, (2.4) shows that the distribution function of the length of the maximum interval $(t_0, t_0 + \tau)$ in which $x(t, \omega) \equiv \xi$ is

$$\begin{aligned} \mathbf{P}\{\tau(\omega) \leq \lambda\} &= 1 - e^{-q(\xi)\lambda}, && \lambda \geq 0. \\ &= 0, && \lambda < 0. \end{aligned}$$

Finally we remark that the hypothesis of separability of the process was unnecessary in proving the first part of the theorem. This is intuitively obvious because the statement of this part of the theorem does not involve the process sample functions. However, the proof we have given definitely used separability, without which the conditional probability (2.3) might not be defined. There are two simple ways of avoiding this hypothesis. One way is to replace the left side of (2.3) by

$$\mathbf{P}\{x(t, \omega) = \xi, 0 \leq t \leq \alpha \text{ } (t \text{ rational}) \mid x(0, \omega) = \xi\}.$$

This conditional probability does not involve the concept of separability of the process, and the change in (2.3) would require no change in the proof we have given [except that of course the same change would be made in (2.4)]. A second method if X is closed (as it was in the chain case of §1, where we used this method) is to make the process separable by changing each $x(t)$ on at most an ω set of probability 0 to get a new process which is separable (see II §2). This forces no change in the transition probabilities of the process. If X is not closed, this second method makes it necessary to replace X by its closure $\bar{X}$, but this causes no difficulty if we set $p(t, \xi, A) \equiv 0$ if $\xi \notin A \subset \bar{X} - X$ and $p(t, \xi, \{\xi\}) \equiv 1$ if $\xi \in \bar{X} - X$.

In the following we shall call any Borel subset of X on which $q(\cdot)$ is bounded a *q-bounded* set. Thus any finite set on which $q(\xi) < \infty$ is a q-bounded set.

THEOREM 2.3 *Let $p(\cdot, \cdot, \cdot)$ be a stationary Markov transition function satisfying the continuity condition* (2.1).

(i) *If* $q(\xi) = \infty$, *then*

$$\lim_{t\to 0} \frac{p(t, \xi, A)}{1 - p(t, \xi, \{\xi\})} = \lim_{t\to 0} \frac{p(t, \xi_1, \{\xi\})}{1 - p(t, \xi, \{\xi\})} = 0, \tag{2.5}$$

if A does not contain ξ, *and if the continuity condition* (2.1) *holds uniformly on A* (*in particular if A is q-bounded*), *and if* $\xi_1 \neq \xi$.

(ii) *If* $q(\xi) < \infty$, *then the limits*

$$\lim_{t\to 0} \frac{p(t, \xi, A)}{t} = q(\xi, A) \leq q(\xi) \tag{2.6}$$

$$\lim_{t\to 0} \frac{p(t, \xi_1, \{\xi\})}{t} \quad (< \infty) \tag{2.7}$$

exist, if A and ξ_1 *are restricted as in* (i). *If B is a q-bounded set, then, for each* $\xi \in B$, *the convergence in* (2.6) *is uniform for* $A \subset B - \{\xi\}$, *and* $q(\xi, A)$ *is completely additive in* $A \subset B - \{\xi\}$.

(iii) *Let*

$$\bar{q}(\xi) = \underset{A}{\text{L.U.B.}}\, q(\xi, A) \qquad (A \text{ } q\text{-bounded, } A \subset X - \{\xi\})$$

if $q(\xi) < \infty$. *Then* $\bar{q}(\xi) \leq q(\xi)$ *and if there is equality at some* ξ *the limit* (2.6) *will exist* (*for that* ξ) *uniformly for all Borel subsets A of* $X - \{\xi\}$, *and* $q(\xi, A)$ *is completely additive in A.*

(iv) *Suppose that* $\{x(t), 0 \leq t < \infty\}$ *is a separable Markov process with* $p(\cdot, \cdot, \cdot)$ *as its transition function. If* $q(\xi) = 0$, *and if* $x(t_0, \omega) = \xi$, *then* $x(t, \omega) \equiv \xi$ *for* $t > t_0$, *with probability* 1. *If* $0 < q(\xi) < \infty$ *and if* $x(t_0, \omega) = \xi$, *there is with probability* 1 *a sample function discontinuity for some* $t > t_0$. *Suppose that* $0 < q(\xi) < \infty$ *and that* $A \subset X - \{\xi\}$ *is q-bounded. Then the probability is* $q(\xi, A)/q(\xi)$ *that, if there is a sample function discontinuity in the finite or infinite interval* $(t_0, t_0 + \alpha)$, *there is a first, which is a jump, and there are positive numbers* $\tau_1 < \tau_2$ *and a point* $\xi_2 \in A$, $\xi_2 \neq \xi$, *all three depending on the sample function, such that* $x(t, \omega) = \xi$ *for* $0 \leq t < \tau_1$ *and* $x(t, \omega) = \xi_2$ *for* $\tau_1 < t < \tau_2$. *If* $0 < \bar{q}(\xi) = q(\xi) < \infty$, *there is with probability* 1 *a first discontinuity after* t_0, *of the type just described, and the preceding evaluation* $q(\xi, A)/q(\xi)$ *holds whenever A is a Borel subset of* $X - \{\xi\}$.

Before giving the proofs of the several sections of this theorem we derive some inequalities which will be used throughout. Let B be a q-bounded set. Choose $\varepsilon > 0$, and then $\alpha > 0$ so small that

$$(1 - \varepsilon)\alpha q(\xi) \leq 1 - e^{-q(\xi)\alpha} \leq \varepsilon, \qquad \xi \in B.$$

Let A be a Borel subset of B and let ξ, ξ_1 be points of X, with $\xi_1 \neq \xi \in B - A$. Choose δ, $0 < \delta < \alpha$, and let $m\delta$ be the smallest multiple of δ which is $\geq \alpha$. Then

$$(2.8) \quad p(m\delta, \xi, A) \geq \sum_{\nu=0}^{m-1} p(\delta, \xi, \{\xi\})^\nu \int_A e^{-q(\eta)(m-\nu-1)\delta} p(\delta, \xi, d\eta)$$

$$\geq (1-\varepsilon) \sum_{\nu=0}^{m-1} p(\delta, \xi, \{\xi\})^\nu p(\delta, \xi, A)$$

$$= (1-\varepsilon) \frac{1 - p(\delta, \xi, \{\xi\})^m}{1 - p(\delta, \xi, \{\xi\})} p(\delta, \xi, A).$$

Note that this inequality is also true for sufficiently small α if A is any Borel set, q-bounded or not, on which the continuity condition (2.1) holds uniformly. Moreover,

$$(2.9) \quad p(m\delta, \xi_1, \{\xi\}) \geq \sum_{\nu=0}^{m-1} p((m-\nu-1)\delta, \xi_1, \{\xi_1\}) p(\delta, \xi_1, \{\xi\}) p(\delta, \xi, \{\xi\})^\nu$$

$$\geq (1-\varepsilon) p(\delta, \xi_1, \{\xi\}) \sum_{\nu=0}^{m-1} p(\delta, \xi, \{\xi\})^\nu$$

$$= (1-\varepsilon) p(\delta, \xi_1, \{\xi\}) \frac{1 - p(\delta, \xi, \{\xi\})^m}{1 - p(\delta, \xi, \{\xi\})}.$$

Proof of (i) If $q(\xi) = \infty$, then when $\delta \to 0$ in (2.8) and (2.9) we obtain

$$p(\alpha, \xi, A) \geq (1-\varepsilon) \limsup_{\delta \to 0} \frac{p(\delta, \xi, A)}{1 - p(\delta, \xi, \{\xi\})},$$

$$p(\alpha, \xi_1, \{\xi\}) \geq (1-\varepsilon) \limsup_{\delta \to 0} \frac{p(\delta, \xi_1, \{\xi\}}{1 - p(\delta, \xi, \{\xi\})},$$

and if $\varepsilon < 1$ these inequalities imply (2.5) when $\alpha \to 0$.

Proof of (ii) The case $q(\xi) = 0$ is solved by the evaluation (2.4). If $0 < q(\xi) < \infty$, (2.8) and (2.9) become, when $\delta \to 0$,

$$p(\alpha, \xi, A) \geq (1-\varepsilon) \frac{1 - e^{-q(\xi)\alpha}}{q(\xi)} \limsup_{\delta \to 0} \frac{p(\delta, \xi, A)}{\delta}$$

$$p(\alpha, \xi_1, \{\xi\}) \geq (1-\varepsilon) \frac{1 - e^{-q(\xi)\alpha}}{q(\xi)} \limsup_{\delta \to 0} \frac{p(\delta, \xi_1, \{\xi\})}{\delta}.$$

These inequalities imply the finiteness of the superior limits on the right. Moreover, the first inequality implies that

$$\liminf_{\alpha \to 0} \frac{p(\alpha, \xi, A)}{\alpha} \geq (1-\varepsilon) \limsup_{\delta \to 0} \frac{p(\delta, \xi, A)}{\delta},$$

which implies the existence of the limit in (2.6), and in the same way the second inequality implies the existence of the limit in (2.7). The first inequality now becomes

$$p(\alpha, \xi, A) \geq (1 - \varepsilon) \frac{1 - e^{-q(\xi)\alpha}}{q(\xi)} q(\xi, A) \geq (1 - \varepsilon)^2 \alpha q(\xi, A)$$

$$\geq \alpha q(\xi, A) - 2\alpha\varepsilon q(\xi).$$

If we apply this inequality to $B - A - \{\xi\}$ and combine the two inequalities, we obtain

$$-2\varepsilon q(\xi) \leq \frac{p(\alpha, \xi, A)}{\alpha} - q(\xi, A)$$

$$\leq \frac{p(\alpha, \xi, B - \{\xi\})}{\alpha} - q(\xi, B - \{\xi\}) + 2\varepsilon q(\xi).$$

Then the limit in (2.6) is uniform in $A \subset B - \{\xi\}$ for fixed ξ. The set function $q(\xi, \cdot)$ is obviously additive. The uniformity of convergence we have just proved shows that, for each ξ, $q(\xi, \cdot)$ is completely additive for sets $A \subset B - \{\xi\}$.

Proof of (iii) Let $\bar{q}(\xi) = q(\xi)$; ξ is fixed throughout the following discussion. We shall use repeatedly the fact [see (ii)] that $q(\xi, \cdot)$ is an additive function of q-bounded subsets of $X - \{\xi\}$, and is completely additive on the subsets of a fixed q-bounded set. By definition of $\bar{q}(\xi)$ there is a sequence $C_1, C_2, \cdots$ of q-bounded subsets of $X - \{\xi\}$ such that

$$\lim_{n \to \infty} q(\xi, C_n) = \bar{q}(\xi) = q(\xi),$$

and we can suppose that $C_1 \subset C_2 \subset \cdots$, replacing C_n by $\bigcup_1^n C_j$ if necessary, to achieve this. Let $C = \bigcup_1^\infty C_n$ and let $C' = X - C - \{\xi\}$. Then, if A is q-bounded and if $\xi \notin A$,

$$q(\xi, A) = q(\xi, AC) + q(\xi, AC').$$

Now, if $q(\xi, AC') > 0$, it follows that

$$q(\xi, C_n + AC') = q(\xi, C_n) + q(\xi, AC') > \bar{q}(\xi)$$

for large n, which is impossible. Hence $q(\xi, AC') = 0$ and it follows that we can write $q(\xi, A)$ in the form

$$q(\xi, A) = \lim_{n \to \infty} q(\xi, AC_n),$$

if $A \subset X - \{\xi\}$ and if A is q-bounded. We therefore can define $q(\xi, A)$ for every Borel subset A of $X - \{\xi\}$ as this limit, without any conflict with the previous definition. We now show that, with this definition, (2.6) holds uniformly in the sets A considered, and this will imply that $q(\xi, \cdot)$

(which is obviously additive) is completely additive. Let ε be a positive number and choose n so large that

$$\bar{q}(\xi) = q(\xi) < q(\xi, C_n) + \varepsilon/5$$

With this choice of n choose $\delta > 0$ so small that

$$\left| \frac{1 - p(t, \xi, \{\xi\})}{t} - q(\xi) \right| < \frac{\varepsilon}{5}$$

$$0 < t < \delta$$

$$\left| \frac{p(t, \xi, AC_n)}{t} - q(\xi, AC_n) \right| < \frac{\varepsilon}{5},$$

for all Borel subsets A of $X - \{\xi\}$. The second inequality is feasible by (ii) because AC_n is a subset of the fixed q-bounded set C_n. Then, if $0 < t < \delta$,

$$\begin{aligned} \left| \frac{p(t, \xi, A)}{t} - q(\xi, A) \right| &\leq \left| \frac{p(t, \xi, AC_n)}{t} - q(\xi, AC_n) \right| \\ &+ \frac{1 - p(t, \xi, \{\xi\})}{t} - \frac{p(t, \xi, C_n)}{t} \\ &+ q(\xi) - q(\xi, C_n) \\ &< \varepsilon, \end{aligned}$$

proving the stated uniform convergence.

Proof of (iv) The first two statements of (iv) follow from (2.4). In proving the remaining statements take $t_0 = 0$ as usual. In the following let ξ be fixed, with $q(\xi) < \infty$, and let A be a Borel subset of $X - \{\xi\}$. We suppose that A is q-bounded if $\bar{q}(\xi) < q(\xi)$. Choose $\alpha > 0$, $\beta > 0$ and let $\Lambda_{n,\beta}{}^{(A)}$ be the ω set for which, for some ν, $2 \leq \nu \leq n$,

$$\begin{aligned} x(t, \omega) &\equiv \xi, & 0 \leq \tau \leq \frac{\nu - 1}{n}\alpha \\ &\equiv \eta \in A & \frac{\nu\alpha}{n} \leq t \leq \frac{\nu\alpha}{n} + \beta. \end{aligned}$$

Then, just as in §1,

$$\begin{aligned} \mathbf{P}\{\Lambda_{n,\beta}{}^{(A)} \mid x(0, \omega) = \xi\} &= \sum_{\nu=2}^{n} \int_A e^{-q(\xi)\frac{\nu-1}{n}\alpha - q(\eta)\beta} p\left(\frac{\alpha}{n}, \xi, d\eta\right) \\ &\to \left[\frac{1 - e^{-q(\xi)\alpha}}{q(\xi)}\right] \int_A e^{-q(\eta)\beta} q(\xi, d\eta) \qquad (n \to \infty) \\ &\to \left[1 - e^{-q(\xi)\alpha}\right] \frac{q(\xi, A)}{q(\xi)} \qquad (\beta \to 0), \end{aligned}$$

and the stated results all follow from this limiting equation. (See the corresponding discussion in the proof of Theorem 1.3.)

Example 1 *Markov chains with infinitely many states* As an example of the application of these theorems let X be the set of positive integers, with $p(t, i, \{j\}) = p_{ij}(t)$ as usual. The continuity condition (2.1) is then

(2.10) $$\lim_{t\to 0} p_{ii}(t) = 1, \qquad i = 1, 2, \cdots.$$

According to Theorem 2.2

$$\lim_{t\to 0} \frac{1 - p_{ii}(t)}{t} = q_i \leq \infty$$

exists for all i. According to Theorem 2.3, if $q_i = \infty$,

$$\lim_{t\to 0} \frac{1 - p_{ij}(t)}{p_{ii}(t)} = \lim_{t\to 0} \frac{p_{ji}(t)}{t} = 0, \qquad j \neq i.$$

On the other hand, if $q_i < \infty$,

$$\lim_{t\to 0} \frac{p_{ij}(t)}{t} = q_{ij} \qquad j \neq i$$

and

$$\lim_{t\to 0} \frac{p_{ji}(t)}{t}$$

exist as finite limits with $\sum_j q_{ij} \leq q_i$. The quantity $\bar{q}(\xi)$ becomes $\sum_j q_{ij}$ (identifying ξ with i). According to Theorem 2.3 (iv), if $\sum_j q_{ij} = q_i$, then Theorem 1.2 is applicable to the infinite dimensional chain. Finally if (2.10) is true uniformly in i, then every q_i is finite and the sequence $\{q_i\}$ is bounded, according to Theorem 2.2, so that every X set is q-bounded. Then $q_i = \sum_j q_{ij}$ for all i. The next theorem shows that in this very special case the sample functions are continuous except for jumps (if the $x(t)$ process is separable), that is, Theorem 1.3 generalizes in this case. However, examples will be given of chains with $\infty > q_i = \sum_j q_{ij}$ for all i and with almost all sample functions having worse discontinuities than jumps. According to Theorem 2.4, $\text{L.U.B.}_i\, q_i = \infty$ in any such example.

It will be useful below to define at this point the statement "$[q(\cdot), q(\cdot, \cdot)]$ is a standard pair of q-functions." This statement is to mean the following:

(a) There is a linear Borel set X; $q(\cdot)$ is defined on X and $q(\cdot, \cdot)$ is defined for $\xi \in X$ and A a Borel subset of $X - \{\xi\}$. (A Borel subset of X on which $q(\cdot)$ is bounded will be called *q-bounded*.)

(b) $q(\cdot)$ is a Baire function; $q(\cdot, \cdot)$ is a Baire function for fixed A, and completely additive in A for fixed ξ. Both functions are finite-valued and non-negative.

(c) $q(\xi) = q(\xi, X - \{\xi\})$. Note that, since q is finite-valued, X is the union of a sequence of q-bounded sets, so that $q(\xi) = \underset{A}{\text{L.U.B.}}\, q(\xi, A)$, for q-bounded $A \subset X - \{\xi\}$.

THEOREM 2.4 *Let $p(\cdot, \cdot, \cdot)$ be a stationary Markov transition function satisfying the continuity condition* (2.1) *uniformly in ξ. Then the pair $q(\cdot)$, $q(\cdot, \cdot)$ defined by Theorems* 2.2 *and* 2.3 *is a standard pair of q-functions and $q(\cdot)$ is bounded. If $\{x(t), 0 \le t < \infty\}$ is a separable Markov process with $p(\cdot, \cdot, \cdot)$ as transition function, almost all the sample functions are step functions.*

Conversely, if $[q(\cdot), q(\cdot, \cdot)]$ is a standard pair of q-functions, and if $q(\cdot)$ is bounded, there is a unique corresponding stationary Markov transition function satisfying the continuity condition (2.1), *and determining $q(\cdot)$, $q(\cdot, \cdot)$ in accordance with Theorems* 2.2 *and* 2.3. *The continuity condition* (2.1) *will then be satisfied uniformly in ξ.*

If a stationary Markov transition function satisfies the continuity condition (2.1) uniformly in ξ, we have seen in Theorem 2.2 that the function $q(\cdot)$ defined in that theorem is bounded. Then $q(\cdot)$ and the $q(\cdot, \cdot)$ defined in Theorem 2.3 constitute a standard pair of q-functions. The proof of the remainder of the theorem follows that of Theorem 1.4, and the subsequent discussion, but will be sketched because the ideas will be generalized below. Let $\{x(t), 0 \le t < \infty\}$ be a separable process determined by the given $p(\cdot, \cdot, \cdot)$ together with an initial probability distribution. If $x(0, \omega) = z_1(\omega)$, and if $q[z_1(\omega)] = 0$, then $x(t, \omega) \equiv z_1(\omega)$, whereas if $q[z_1(\omega)] > 0$ we have seen that there is a $\tau_1(\omega)$, the first discontinuity of the sample function determined by ω, and a $z_2(\omega) \ne z_1(\omega)$, such that $x(t, \omega) = z_1(\omega)$ for $0 \le t < \tau_1(\omega)$, and $x[\tau_1(\omega) +, \omega] = z_2(\omega)$. If $q[z_1(\omega)] = 0$, set $\tau_1(\omega) = \tau_2(\omega) = \cdots = \infty$. If $q[z_2(\omega)] = 0$, set $\tau_2(\omega) = \tau_3(\omega) = \cdots = \infty$; if $q[z_2(\omega)] > 0$, there is a $\tau_2(\omega)$ and a $z_3(\omega) \ne z_2(\omega)$ such that $x(t, \omega) = z_2(\omega)$ for $\tau_1(\omega) < t < \tau_2(\omega)$, $x[\tau_2(\omega)+, \omega] = z_3(\omega)$, and so on. Here

$$\mathbf{P}\{\tau_{n+1}(\omega) - \tau_n(\omega) > \alpha \mid \tau_1, \cdots, \tau_n, z_1, \cdots, z_{n+1}\} = e^{-q(z_{n+1})\alpha}, \qquad \alpha > 0,$$

and

$$\mathbf{P}\{z_{n+1}(\omega) \in A \mid \tau_1, \cdots, \tau_n, z_1, \cdots, z_n\} = \frac{q(z_n, A)}{q(z_n)}$$

with probability 1, with the obvious modifications to take care of the zeroes of $q(\cdot)$. This argument would be correct even if we had supposed only that $[q(\cdot), q(\cdot, \cdot)]$ was a standard pair of q-functions, and gives (almost all) the sample functions of the process for $t \leq \lim_{n\to\infty} \tau_n(\omega)$, except at the jump points. The additional information we have, that $q(\cdot)$ is bounded, is now used just as in the proof of Theorem 1.4 to prove that $\lim_{n\to\infty} \tau_n = \infty$ with probability 1, and it follows that almost all sample functions of the process are step functions. Conversely, if $[q(\cdot), q(\cdot, \cdot)]$ is any standard pair of q-functions, and if z_1 is any random variable with values in X, $z_1, \tau_1, z_2, \tau_2, \cdots$ are defined to have the distributions written above and $x(t)$ is defined by

$$\begin{aligned} x(t, \omega) &= z_1(\omega), \qquad 0 \leq t < \tau_1(\omega) \\ &= z_2(\omega), \qquad \tau_1(\omega) \leq t < \tau_2(\omega) \\ &\cdots\cdots \end{aligned}$$

This definition is effective for $0 \leq t < \lim_{n\to\infty} \tau_n(\omega)$ and again, if $q(\cdot)$ is bounded, as we suppose in this theorem, $\lim_{n\to\infty} \tau_n = \infty$ with probability 1, so that $x(t)$ is defined for all t, with probability 1. The same simple argument used in §1 in the chain case shows that the $x(t)$ process is one with the desired stationary Markov transition function.

THEOREM 2.5 *Suppose that* $[q(\cdot), q(\cdot, \cdot)]$ *is a standard pair of q-functions. There is then at least one corresponding stationary Markov transition function satisfying the continuity condition* (2.1) *and related to* $q(\cdot)$, $q(\cdot, \cdot)$ *by Theorems* 2.2 *and* 2.3. *There is either only one such transition function, and in that case the separable Markov processes with this transition function have sample functions which are almost all step functions, or there are infinitely many such transition functions, and in that case to every transition function corresponds some separable process whose sample functions are step functions with probability* < 1.

To prove this theorem suppose that $[q(\cdot), q(\cdot, \cdot)]$ is a standard pair of q-functions and define $x(t)$ as in the proof of the converse half of Theorem 2.4 for $t < \lim_{n\to\infty} \tau_n(\omega) = \tau_{\underline{\omega}}(\omega)$. We must complete the definition, in case $\tau_{\underline{\omega}}$ is finite with positive probability. In this case one way (but not necessarily the only way) to complete the definition is the following. Let $\pi(\cdot)$ be any probability distribution of Borel subsets of X, and choose $z_{\underline{\omega}+1}$, a random variable independent of the z_j's and τ_j's, with the $\pi(\cdot)$ distribution. Choose $\tau_{\underline{\omega}+1} > \tau_{\underline{\omega}}$ with the distribution determined by

$$\mathbf{P}\{\tau_{\underline{\omega}+1}(\omega) - \tau_{\underline{\omega}}(\omega) > \alpha \mid z_1, \tau_1, z_2, \tau_2, \cdots\} = e^{-q(z_{\underline{\omega}+1})\alpha}, \qquad \alpha > 0,$$

and define $x(t, \omega) = z_{\underline{\omega}+1}(\omega)$ for $\tau_{\underline{\omega}}(\omega) \leq t < \tau_{\underline{\omega}+1}(\omega)$. We then continue as before, letting $x(t, \omega)$ go through transitions determined by the

$q(\cdot, \cdot)/q(\cdot)$ distributions, and determining how to go on at any point like $\tau_{\underline{\omega}}(\omega)$ which is a limit point of jumps by starting off afresh using the $\pi(\cdot)$ distribution. An elementary ordinal number argument shows that, for any t, $x(t)$ is then defined with probability 1. The $x(t)$ process defined in this way is a Markov process with a stationary Markov transition function satisfying the prescribed conditions. In fact, the argument used in the case of §1 is applicable even to this general case. If $\lim_{n\to\infty} \tau_n = \infty$ with probability 1 for every distribution of $x(0)$, only one Markov process can be obtained in this way, aside from the choice of initial distribution, and therefore only one stationary Markov transition function can be obtained. In this case almost all sample functions of any separable process with this transition function are step functions. On the other hand, if for some choice of the initial distribution

$$\mathbf{P}\{\lim_{n\to\infty} \tau_n(\omega) = \infty\} = p < 1,$$

the process finally obtained will depend on the choice of $\pi(\cdot)$, and for each choice the probability that the sample functions are step functions is $p < 1$. This finishes the proof of the theorem.

Suppose now that $p(\cdot, \cdot, \cdot)$ is a stationary Markov transition function satisfying (2.1) whose corresponding pair $[q(\cdot), q(\cdot, \cdot)]$ is a standard pair. Then just as in §1 we can compute the probability ${}_np(t, \xi, A)$ that (for a separable process), if $x(t_0, \omega) = \xi$, then $x(t_0 + t, \omega) \in A$ (where A is any Borel subset of X) and that the transition has been effected in n steps, that is, that the sample function has exactly n discontinuities in the interval $(t_0, t_0 + \tau)$, and that in each of the $n + 1$ open intervals determined by these discontinuities the sample function is identically constant.. Using the same argument as in §1, we find

$$\begin{aligned} (2.11) \qquad {}_0p(t, \xi, A) &= 0 \qquad && \xi \notin A \\ &= e^{-q(\xi)t} && \xi \in A \end{aligned}$$

$$ {}_{n+1}p(t, \xi, A) = \int_0^t ds \int_{X-\{\xi\}} e^{-q(\xi)s}\, {}_np(t - s, \eta, A) q(\xi, d\eta),$$

or alternatively

$$\begin{aligned} (2.11') \qquad {}_0p(t, \xi, A) &= 0, \qquad && \xi \notin A \\ &= e^{-q(\xi)t}, && \xi \in A; \end{aligned}$$

$$ {}_{n+1}p(t, \xi, A) = \int_0^t ds \int_X {}_np(s, \xi, d\eta) \int_{A-A\{\eta\}} e^{-q(\zeta)(t-s)}\, q(\eta, d\zeta).$$

Then $\bar{p}(t, \xi, A)$, defined by

$$\bar{p}(t, \xi, A) = \sum_{n=0}^{\infty} {}_n p(t, \xi, A), \tag{2.12}$$

is the probability of a transition from ξ into a point of A in time t, the transition having been effected in finitely many steps. In particular, if it is known that the sample functions are almost all step functions, as is the case according to Theorem 2.4 if (2.1) holds uniformly in ξ, then $\bar{p}(t, \xi, A) \equiv p(t, \xi, A)$ for all t, ξ, and A. Even without this, however, it is clear that $\bar{p}(t, \xi, A) \leq 1$; according to Theorem 2.5 there are cases in which actually $\bar{p}(t, \xi, A) < 1$. Evidently $\bar{p}(t, \xi, X) \equiv 1$ for all t and ξ if and only if $\bar{p}(t, \xi, A) \equiv p(t, \xi, A)$ for all t, ξ, and A.

Conversely, if $[q(\cdot), q(\cdot, \cdot)]$ is any standard pair of q-functions, (2.11) and (2.12) define a non-negative function $\bar{p}(\cdot, \cdot, \cdot)$ which is a Baire function of ξ for fixed t, A, completely additive in A for fixed t, ξ, which satisfies the Chapman-Kolmogorov equation, and for which $\bar{p}(t, \xi, X) \leq 1$. The analytic procedure of (2.11) and (2.12) can be used to define a stationary Markov transition function in terms of a given pair of standard q-functions if and only if the q-functions match a separable process whose sample functions are almost all step functions. The procedure is therefore applicable, for example, if $q(\cdot)$ is bounded, and in this case the procedure is simply one analytic form of the probability proof of the converse half of Theorem 2.4. According to Theorem 2.5, $\bar{p}(\cdot, \cdot, \cdot)$ can always be increased, if necessary, to a Markov transition function, but if an increase is actually necessary it can be done in infinitely many different ways, leading to different Markov transition functions.

Suppose now that $p(\cdot, \cdot, \cdot)$ is a stationary Markov transition function satisfying the continuity condition (2.1) whose corresponding pair $[q(\cdot), q(\cdot, \cdot)]$ is a standard pair of q-functions. Then

$$p(t, \xi, A) = \int_0^t ds \int_{X-\{\xi\}} e^{-q(\xi)s}\, p(t-s, \eta, A)\, q(\xi, d\eta) + e^{-q(\xi)t}\, \delta(\xi, A), \tag{2.13}$$

where

$$\begin{aligned} \delta(\xi, A) &= 1 \qquad \xi \in A \\ &= 0 \qquad \xi \notin A. \end{aligned}$$

This equation is a generalization of (1.16) and is derived in the same way, by considering a separable process with the given transition probabilities, and observing that the transition from ξ to a point of A can be accomplished either by simply remaining at ξ (if $\xi \in A$) or by remaining at ξ through s time units, then jumping to η from which there is then a transition to a point of A in the remaining time. The rigorous justification of this sort of derivation of equations like (2.13) is the same as that of similar derivations discussed in §1. This equation proves that $p(\cdot, \xi, A)$ is a continuous function of t with a continuous derivative given by

$$\frac{\partial p(t, \xi, A)}{\partial t} = -q(\xi)p(t, \xi, A) + \int_{X-\{\xi\}} p(t, \eta, A)q(\xi, d\eta). \tag{2.14}$$

This equation is a generalization of (1.7). The natural complement of (2.13) is the equation for $p(\cdot, \cdot, \cdot)$ obtained by considering the transition from ξ to A to be accomplished either by simply remaining at ξ (if $\xi \in A$) for a time interval of length t or by going to η at time s, the time of the last jump, jumping to a point of A and remaining at that point through some interval of constancy. Since our hypotheses on $q(\cdot)$, $q(\cdot, \cdot)$ do not insure the existence of such an interval of constancy preceded by a last jump, we have described only one way of going from ξ into A, that is,

$$p(t, \xi, A) \geq \int_0^t ds \int_X p(s, \xi, d\eta) \int_{A-A\{\eta\}} e^{-q(\zeta)(t-s)} q(\eta, d\zeta) + e^{-q(\xi, A)} \delta(\xi, A), \tag{2.13'}$$

and in fact the same type of reasoning gives the slightly more general

$$p(t_2, \xi, A) - \int_A e^{-q(\eta)(t_2-t_1)} p(t_1, \xi, d\eta) \geq \int_{t_1}^{t_2} ds \int_X p(s, \xi, d\eta) \int_{A-A\{\eta\}} e^{-q(\zeta)(t_2-s)} q(\eta, d\zeta), \qquad t_1 < t_2.$$

If we divide both sides of this inequality by $t_2 - t_1$ and take the limit when $t_2 \to t$, $t_1 \to t$, we find that, if A is q-bounded,

$$\frac{\partial p(t, \xi, A)}{\partial t} \geq -\int_A q(\eta, X-A)p(t, \xi, d\eta) + \int_{X-A} q(\eta, A)p(t, \xi, d\eta). \tag{2.14'}$$

This is a generalization of (1.7′). There is equality here for all ξ and q-bounded A if the stochastic transition function $p(\cdot, \cdot, \cdot)$ corresponds to separable processes whose sample functions are almost all step functions. It can be shown that this condition is not necessary, however.

It is interesting to observe that it follows at once from the probability significance of $\bar{p}(\cdot, \cdot, \cdot)$ that this function, which, as we have already remarked, satisfies all the conditions of a stationary Markov transition function except that $\bar{p}(t, \xi, X)$ is not necessarily 1, and satisfies the inequality

$$p(t, \xi, A) \geq \bar{p}(t, \xi, A),$$

also satisfies the backward system of integro-differential equations, and the forward system (with equality). The fact that $\bar{p}(\cdot, \cdot, \cdot)$ satisfies these systems of equations also follows from the fact that (2.11) and (2.11′) summed over n yield (2.13) and (2.13′) (with equality) for $\bar{p}$. The latter pair of equations yields the backward and forward systems of integro-differential equations, on differentiation. Thus the backward system has infinitely many solutions with the same initial conditions $p(0, \xi, A) = \delta(\xi, A)$ unless $\bar{p}(t, \xi, A) \equiv p(t, \xi, A)$, and these solutions include some, like $\bar{p}$, which are not stationary Markov transition functions.

Example 1 *Markov chains with infinitely many states* (*continued*) According to our results, if $\{q_i, i \geq 1\}$ and $\{q_{ij}, i, j \geq 1, i \neq j\}$ are arbitrary sequences of non-negative numbers, with $q_i = \sum_j q_{ij}$ for all i (this is the condition for a standard pair of q-functions in the present case), there is a corresponding stationary Markov transition matrix function $[p_{ij}(\cdot)]$ which satisfies the backward system of differential equations (1.7) (without the restriction $i, j \leq N$ of course). The forward system (1.7′) must be replaced by a system of inequalities obtained by replacing "$=$" in the system (1.7′) by "$\geq$." On the other hand, the matrix function $[\bar{p}_{ij}(\cdot)]$, where $\bar{p}_{ij}(t)$ is the probability of a transition from i to j in time t in finitely many steps, satisfies both backward and forward systems, with equality in each, and with the same initial conditions

$$p_{ij}(0) = \bar{p}_{ij}(0) = \delta_{ij}.$$

Moreover,

$$p_{ij}(t) \geq \bar{p}_{ij}(t).$$

The following very simple case illustrates these results. Suppose that

$$q_{ii+1} = q_i$$

$$q_{ij} = 0, \qquad j \neq i, i + 1.$$

Set $x(0) = 1$ with probability 1, and construct the process corresponding to these q_i's as in Theorem 2.5. In this case the differences $\tau_2 - \tau_1$, $\tau_3 - \tau_2, \cdots$ are mutually independent, and $\tau_{n+1} - \tau_n$ is a positive random variable with density function $q_n e^{-q_n\lambda}$, $\lambda > 0$. Then $\mathbf{E}\{\tau_{n+1} - \tau_n\} = 1/q_n$ so that, if no q_n vanishes and if $\sum_n \dfrac{1}{q_n} < \infty$, it follows that $\lim_{n\to\infty} \tau_n < \infty$ with probability 1. (A more careful examination of the partial sums of the series, say by means of characteristic functions, shows at once that the sufficient condition of convergence we have obtained here is also necessary, but we shall not use this fact.) Suppose that the q_n's are chosen so that this series converges. One simple way to define the $\pi(\cdot)$ distribution of Theorem 2.5 in the present case, that is, to define the distribution of $x(\tau_{\underline{\omega}}+)$, is to set $x(\tau_{\underline{\omega}}+) = 1$ with probability 1, so that after any limit point of jumps the sample functions return to the value 1, but of course there are infinitely many $\pi(\cdot)$ distributions, giving different types of sample functions and different transition probabilities to the resulting process. The backward system of differential equations is

$$p_{ik}'(t) = -q_i p_{ik}(t) + q_i p_{i+1k}(t) \qquad i, k \geq 1,$$

and the forward system of differential inequalities is

$$\begin{aligned} p_{i1}'(t) &\geq -p_{i1}(t)q_1 && i \geq 1 \\ p_{ik}'(t) &\geq -p_{ik}(t)q_k + p_{ik-1}(t)q_k + p_{ik-1}(t)q_{k-1} && i \geq 1 \\ & && k > 1. \end{aligned}$$

According to the forward system applied to $\bar{p}_{ij}(t)$, in which case there is equality, $\bar{p}_{21}(t) \equiv 0$, and it is of course true that no transition from state 2 to state 1 is possible in a finite number of jumps as we have defined the q_{ij}'s. However, if $\sum_n \frac{1}{q_n} < \infty$, such a transition is possible in infinitely many jumps as we chose the $\pi(\cdot)$ distribution above. Other choices exist (for example concentrating this distribution at the value 2 instead of 1) for which such a transition is impossible even in infinitely many jumps.

We have investigated certain types of Markov transition functions in this section, those characterized by the continuity condition (2.1), by methods which reveal the probability significance of each step of the reasoning. The results not explicitly involving sample functions can of course be obtained without reference to the theory of probability. For example, let $p(\cdot, \cdot, \cdot)$ be any stochastic transition function satisfying (2.1) and suppose that the limits $q(\xi)$, $q(\xi, A)$ in (2.2) and (2.6) exist and determine a standard pair of q-functions. Then $p(\cdot, \cdot, \cdot)$ must have the form

$$p(t, \xi, A) = [1 - tq(\xi, A)]\,\delta(\xi, A) + tq(\xi, A - \{\xi\}) + o(t),$$

for each ξ and A. Equation (2.14) is then easily derived by an elementary manipulation of the Chapman-Kolmogorov equation, and then the study of (2.14) and (2.14′) becomes the study of the solutions of these equations under the initial conditions $p(0, \xi, A) = \delta(\xi, A)$.

If $p(\cdot, \cdot; \cdot, \cdot)$ is a Markov transition function (non-stationary case) we sketch a derivation of the generalizations of (2.14) and (2.14′). Suppose that

$$\begin{aligned} p(s, \xi;\ t, A) &= 1 - q(t, \xi)(t - s) + o(t - s) && \xi \in A \\ &= q(t, \xi, A)(t - s) + o(t - s) && \xi \notin A \end{aligned}$$

where

$$q(t, \xi) \geq 0, \qquad q(t, \xi, A) \geq 0, \qquad q(t, \xi, X - \{\xi\}) = q(t, \xi)$$

and $q(t, \xi, \cdot)$ is completely additive in $A \subset X - \{\xi\}$. Then

$$\left.\frac{\partial p(s, \xi; t, A)}{\partial s}\right|_{s=t} = q(t, \xi) \qquad \xi \in A$$

$$= -q(t, \xi, A) \qquad \xi \notin A$$

$$\left.\frac{\partial p(s, \xi; t, A)}{\partial t}\right|_{t=s} = -q(s, \xi) \qquad \xi \in A$$

$$= q(s, \xi, A) \qquad \xi \notin A,$$

and under conditions we shall not discuss here (2.14) and (2.14′) generalize to

$$\frac{\partial p(s, \xi; t, A)}{\partial s} = q(s, \xi)p(s, \xi; t, A) - \int_{X-\{\xi\}} p(s, \eta; t, A)q(s, \xi, d\eta)$$

$$\frac{\partial p(s, \xi; t, A)}{\partial t} \geq -\int_A q(t, \eta, X - A)p(s, \xi; t, d\eta)$$

$$+ \int_{X-A} q(t, \eta, A)p(s, \xi; t, d\eta).$$

These equations are generalizations of (1.17) and (1.17′) and are called the backward and forward equations respectively, as are their specializations (2.14) and (2.14′).

3. The diffusion equations and the corresponding Markov processes

This section is devoted to (real) Markov continuous parameter processes which are of the following type: $x(t_2) - x(t_1)$, the increment between times t_2 and t_1, is a sum of small increments $dx(t)$, each of which is Gaussian with mean $m\,dt$ and variance $\sigma^2\, dt$. These two quantities are of order dt, and m and σ are functions of t and $x(t)$. This is, of course, a rough statement which is only intended to suggest the motivation for the discussion to be given. We write

$$dx(t) = m[t, x(t)]\, dt + \sigma[t, x(t)]\, dy(t). \tag{3.1}$$

Here the $y(t)$ process is the Brownian motion process with variance parameter 1 (see II §9 and VIII §2), that is, it is a real Gaussian process with independent increments and

$$\mathbf{E}\{y(t_2) - y(t_1)\} = 0, \qquad \mathbf{E}\{[y(t_2) - y(t_1)]^2\} = |t_2 - t_1|.$$

The sample functions of a separable Brownian motion process are almost all continuous functions, but almost none are of bounded variation in any finite interval (see VIII §2). Equation (3.1) is only to be considered suggestive for the moment. It will be given a precise interpretation below.

The material in VII §3, IX §2, and IX §5 will be accepted as known throughout this section.

If $\sigma \equiv 0$ in (3.1) it is natural to interpret the equation as the non-probabilistic differential equation

$$\frac{dx}{dt} = m(t, x).$$

In this case probabilistic concepts can enter only by way of the initial conditions.

If $\sigma = \sigma(\cdot)$ depends on t but not on x in (3.1), and, if $m = 0$, it is natural to interpret the equation as a symbolic version of

$$x(t) - x(t_0) = \int_{t_0}^{t} \sigma(s)\, dy(s).$$

(See IX §2 for a discussion of this stochastic integral.) The $x(t)$ process is in this case essentially the Brownian motion process with a change of variable in the time parameter. The transition probability distribution function is given by

$$p(s, \xi;\ t, \eta) = \mathbf{P}\{x(t, \omega) \le \eta \mid x(s, \omega) = \xi\} = \frac{1}{\sqrt{2\pi\Delta}} \int_{-\infty}^{\eta-\xi} e^{-\lambda^2/2\Delta}\, d\lambda,\ s < t,$$

where

$$\Delta = \int_{s}^{t} \sigma(\tau)^2\, d\tau.$$

It follows that

$$\frac{\partial p(s, \xi;\ t, \eta)}{\partial s} = -\frac{\sigma(s)^2}{2}\frac{\partial^2 p(s, \xi;\ t, \eta)}{\partial \xi^2} \tag{3.2}$$

and

$$\frac{\partial p(s, \xi;\ t, \eta)}{\partial t} = \frac{\sigma(t)^2}{2}\frac{\partial^2 p(s, \xi, t, \eta)}{\partial \eta^2}. \tag{3.2'}$$

The first equation is called the *backward equation* because it involves differentiation with respect to the initial time; the second equation is called the *forward equation* because it involves differentiation with respect to the final time.

We shall not investigate in detail the generalization of the backward and forward differential equations (3.2) and (3.2′) for a general m and σ, since we are interested in the processes rather than in the differential equations. We therefore restrict ourselves to the following remarks. Under any reasonable interpretation of (3.1) it can be concluded that

$$\begin{aligned} &\lim_{h\downarrow 0} \mathbf{E}\left\{\frac{x(t+h)-x(t)}{h} \mid x(t,\omega)=\xi\right\} = m[t,\xi], \\ (3.3)\qquad & \\ &\lim_{h\downarrow 0} \mathbf{E}\left\{\frac{[x(t+h)-x(t)]^2}{h} \mid x(t,\omega)=\xi\right\} = \sigma[t,\xi]^2. \end{aligned}$$

Consider now the class of Markov processes for which there are functions $m(\cdot,\cdot)$, $\sigma(\cdot,\cdot)$ satisfying (3.3). Under various regularity assumptions discussed below it is shown that the transition probability function $p(\cdot,\cdot;\cdot,\cdot)$ satisfies the backward diffusion equation

$$\frac{\partial p(s,\xi;\,t,\eta)}{\partial s} + m(s,\xi)\frac{\partial p}{\partial \xi} + \frac{\sigma(s,\xi)^2}{2}\frac{\partial^2 p}{\partial \xi^2} = 0 \tag{3.4}$$

and the forward diffusion equation

$$\frac{\partial p'(s,\xi;\,t,\eta)}{\partial t} + \frac{\partial}{\partial \eta}[m(t,\eta)p'] - \frac{1}{2}\frac{\partial^2}{\partial \eta^2}[\sigma(t,\eta)^2 p'] = 0, \qquad p' = \frac{\partial p}{\partial \eta}. \tag{3.4'}$$

The forward equation is called the Fokker-Planck equation and is usually the more natural equation to consider in physical problems. Kolmogorov derived both equations in the first systematic treatment of this type of Markov process.

Note that the backward equation is a parabolic partial differential equation in s and ξ for $s \leq t$; t and η enter only by way of the initial condition

$$\begin{aligned} p(s,\xi;\,t,\eta)\,\big|_{s=t} &= 1 \qquad \xi < \eta \\ &= 0 \qquad \xi > \eta. \end{aligned}$$

The forward equation is a parabolic partial differential equation in t and η for $t \geq s$; s and ξ enter by way of the initial condition

$$\begin{aligned} p(s,\xi;\,t,\eta)\,\big|_{t=s} &= 1 \qquad \eta > \xi \\ &= 0 \qquad \eta < \xi. \end{aligned}$$

The hypotheses usually imposed to derive these differential equations are the following (described only qualitatively)

F_1 It is supposed that $p(\cdot,\cdot;\cdot,\cdot)$ has appropriate regularity properties (differentiability and so on).

F_2 It is supposed that the limits in (3.3) exist, and define functions $m(\cdot,\cdot)$ and $\sigma(\cdot,\cdot)$ with appropriate regularity properties.

F_3 It is supposed that, for every $\varepsilon > 0$,

$$\int_{|\eta-\xi|>\varepsilon} d_\eta p(s, \xi;\ t, \eta) = \mathbf{P}\{|x(t, \omega) - x(s, \omega)| > \varepsilon \mid x(s, \omega) = \xi\}$$
$$= o(t - s) \qquad s < t.$$

[It is also possible to rewrite (3.3) using truncated variables to avoid the hypothesis that the first and second moments of $x(t) - x(s)$ exist.]

The condition F_3 is *not* satisfied for the processes discussed in §1, for which the probability

$$\mathbf{P}\{x(t, \omega) \neq x(s, \omega) \mid x(s)\}$$

is in general of the order of $t - s$. The sample functions of the separable processes of §1 are typically step functions, progressing by jumps (although this is not true in all cases). The sample functions of the processes under discussion here are typically continuous, but there are exceptions to this, and as in §1 it is to be expected that in a large class of these exceptional cases the backward equation holds but the forward equation must be replaced by an inequality. The theory is still incomplete on this point as on many others relating to these processes.

In the present section we shall adopt an interpretation of (3.1) which makes it possible to solve this equation and thus find a separable Markov process satisfying (3.3), almost all of whose sample functions are continuous. It will then be shown that conversely, if $m(\cdot, \cdot)$ and $\sigma(\cdot, \cdot)$ are given functions, any Markov process satisfying (3.3), which has continuous sample functions with probability 1, can be obtained as a solution of (3.1).

It has been shown by Feller that under suitable restrictions on $m(\cdot, \cdot)$ and $\sigma(\cdot, \cdot)$ the diffusion equations with the initial conditions stated above can be solved to give the stochastic transition function of a Markov process satisfying (3.3), and that this solution is unique. Moreover, Fortet has shown that under Feller's conditions (at least if $\sigma = 1$) almost all sample functions of the corresponding separable processes are continuous.

According to the remarks in the preceding paragraphs the processes obtained by Feller must be exactly those found by solving (3.1) [where it is supposed that Feller's conditions on $m(\cdot, \cdot)$, $\sigma(\cdot, \cdot)$ are strengthened if necessary to agree with those which must be imposed in the discussion of the preceding paragraph].

The results discussed in the preceding paragraph imply that if $m(\cdot, \cdot)$ and $\sigma(\cdot, \cdot)$ are sufficiently well-behaved the stochastic transition functions obtained by solving (3.1) must be the same as those obtained by solving the Kolmogorov-Fokker-Planck diffusion equations, and therefore these stochastic transition functions must have various partial derivatives. No direct proof of this fact has yet been given.

Finally we remark that the processes discussed in this section and the preceding one are special cases of a type which includes them both, whose stochastic transition functions satisfy integrodifferential equations obtained by combining those obtained in the preceding section with the Kolmogorov-Fokker-Planck equations. We omit any further discussion of this general class.

We now proceed to a discussion of (3.1). Let the range of t be the finite interval $[a, b]$. The natural interpretation of (3.1) is

$$x(t) - x(a) = \int_a^t m[s, x(s)]\, ds + \int_a^t \sigma[s, x(s)]\, dy(s). \tag{3.1$'$}$$

This equation must be solved for an $x(t)$ process for which the two integrals on the right are meaningful. Now for any sample function of any $x(t)$ process the first integrand becomes an ordinary function of s, and the usual criteria of integrability are applicable. The second integral has been defined (see IX §5) if, for example, $\sigma[t_0, x(t_0)]$ is for each t_0 a random variable independent of the aggregate of differences $\{y(b) - y(s), s > t_0\}$. Fortunately the intuitive picture of the $x(t)$ process which makes $x(t_0)$ the sum of $x(a)$ and suitably transposed and scaled $y(s)$ increments $dy(s)$ with $a \leq s \leq t_0$ matches this restriction. Thus the interpretation of (3.1) as (3.1′) becomes a practical possibility. We proceed to carry it out in detail, following Ito.

We make the following hypotheses:

H_1 $m(\cdot, \cdot)$ and $\sigma(\cdot, \cdot)$ are Baire functions of the pair (t, ξ) for $a \leq t \leq b$, $-\infty < \xi < \infty$;

H_2 There is a constant K for which

$$|m(t, \xi)| \leq K(1 + \xi^2)^{1/2}$$
$$0 \leq \sigma(t, \xi) \leq K(1 + \xi^2)^{1/2}.$$

H_3 $m(\cdot, \cdot)$ and $\sigma(\cdot, \cdot)$ satisfy a uniform Lipschitz condition in ξ,

$$|m(t, \xi_2) - m(t, \xi_1) \leq K|\xi_2 - \xi_1|$$
$$|\sigma(t, \xi_2) - \sigma(t, \xi_1)| \leq K|\xi_2 - \xi_1|$$

where K is independent of t and ξ. It is no restriction to suppose this K to be the same as the one in H_2.

Assuming hypotheses H_1, H_2, and H_3, we shall find an $x(t)$ process with the following properties:

P_1 The $x(t)$ sample functions are almost all continuous in $[a, b]$.

P_2
$$\int_a^b \mathbf{E}\{x(t)^2\}\, dt < \infty.$$

P_3 For each $t_0 \in (a, b)$, $x(t_0) - x(a)$ is independent of the aggregate of differences $\{y(b) - y(s), s > t_0\}$.

P_4 For each $t \in [a, b]$, (3.1′) is true with probability 1.

It will be shown that the $x(t)$ process is essentially uniquely determined by these properties, and even satisfies

$$P_2' \qquad \mathbf{E}\{\max_{a \leq t \leq b} x(t, \omega)^2\} < \infty.$$

LEMMA 3.1 *If an $x(t)$ process has properties* P_1, P_2, P_3, *and if* $m(\cdot, \cdot)$, $\sigma(\cdot, \cdot)$ *satisfy conditions* H_1, H_2, H_3, *then any $\hat{x}(t)$ process defined by*

$$\hat{x}(t) = \int_a^t m[s, x(s)]\, ds + \int_a^t \sigma[s, x(s)]\, dy(s) \tag{3.5}$$

has properties P_2 *and* P_3. *The second integral in* (3.5) *can be defined for each t in such a way that the $x(t)$ process also has property* P_1, *and the $\hat{x}(t)$ process will then have property* P_2'.

According to H_1, H_3, and P_1 the first integrand in (3.5) is for almost all sample functions a bounded Baire function of t. Hence the first integral defines a continuous function of t, with probability 1. The second integral is a special case of the stochastic integral of IX §5, because the qualitative conditions imposed there are satisfied, and because

$$\int_a^t \mathbf{E}\{\sigma[s, x(s)]^2\}\, ds \leq K^2 \int_a^b [1 + \mathbf{E}\{x(s)^2\}] < \infty.$$

The second integral in (3.5) is thus well-defined, and is uniquely defined for each t, neglecting values on sets of probability 0. The $x(t)$ process obviously satisfies P_3 for any choice of these integrals. Property P_2 is easily verified by a direct calculation, but it will follow from the fact to be proved below that even P_2' is true if these integrals are chosen properly, so the calculation will be omitted. Now, it is shown in IX §5 that the second integral in (3.5) defines a martingale as t varies, and moreover that it can be defined for each t to get continuous sample functions, with probability 1. With this definition the $\hat{x}(t)$ process has property P_1. It will then have property P_2' because

$$\begin{aligned}\mathbf{E}\{\max_{a \leq t \leq b} \left| \int_a^t m[s, x(s)]\, ds \right|^2\} &\leq \mathbf{E}\{[\int_a^b |m[s, x(s)]|\, ds]^2\} \\ &\leq (b - a)K^2 \int_a^b \mathbf{E}\{1 + x(s)^2\}\, ds < \infty,\end{aligned}$$

and by the continuous parameter version of VII, Theorem 2.4, applied to the absolute value of the last term in (3.5), which determines a semi-martingale,

$$\mathbf{E}\{\operatorname*{Max}_{a\leq t\leq b}|\int_a^t \sigma[s,x(s)]\,dy(s)|^2\}\leq 4\mathbf{E}\{|\int_a^b \sigma[s,x(s)]\,dy(s)|^2\}$$
$$= 4\int_a^b \mathbf{E}\{\sigma[s,x(s)]^2\}\,ds$$
$$\leq 4K^2\int_a^b [1+\mathbf{E}\{x(s)^2\}]\,ds<\infty.$$

Equation (3.1′) is solved by successive approximation. Let $x(a)$ be any random variable with $\mathbf{E}\{x(a)^2\}<\infty$, independent of the aggregate of differences $\{y(t_2)-y(t_1),\ t_1,\ t_2\in[a,b]\}$. Let the $x_0(t)$ process be any process with properties P_1, P_2, P_3. [For example, take $x_0(t)=0$.] According to the lemma it is then possible to define $x_n(t)$ for $n>1$ inductively by

$$x_n(t)=x(a)+\int_a^t m[s,x_{n-1}(s)]\,ds+\int_a^t \sigma[s,x_{n-1}(s)]\,dy(s) \tag{3.6}$$

in such a way that every $x_n(t)$ process has properties P_1, P_2', P_3. We shall prove that with this definition

$$\lim_{n\to\infty} x_n(t)=x(t),\qquad a\leq t\leq b \tag{3.7}$$

uniformly in t with probability 1, defining an $x(t)$ process with properties P_1, P_2', P_3 and that

$$\begin{aligned}
&\lim_{n\to\infty}\int_a^t m[s,x_n(s)]\,ds=\int_a^t m[s,x(s)]\,ds \qquad a\leq t\leq b\\
&\lim_{n\to\infty}\int_a^t \sigma[s,x_n(s)]\,dy(s)=\int_a^t \sigma[s,x(s)]\,dy(s)
\end{aligned} \tag{3.8}$$

uniformly in t, with probability 1. The $x(t)$ process will then be a solution of (3.1′). To prove these facts we make the definitions

$$\Delta_n x(t)=x_n(t)-x_{n-1}(t)$$
$$\Delta_n m(t)=m[t,x_n(t)]-m[t,x_{n-1}(t)]$$
$$\Delta_n\sigma(t)=\sigma[t,x_n(t)]-\sigma[t,x_{n-1}(t)],$$

so that, by H_3,

$$|\Delta_n m(t)|\leq K|\Delta_n x(t)|,\qquad |\Delta_n\sigma(t)|\leq K|\Delta_n x(t)|.$$

Then

$$\mathbf{E}\{[\Delta_n x(t)]^2\} \leq 2\mathbf{E}\{|\int_a^t \Delta_{n-1} m(s)\,ds|^2\} + 2\mathbf{E}\{|\int_a^t \Delta_{n-1}\sigma(s)\,dy(s)|^2\}$$

$$\leq 2K^2(b-a+1)\int_a^t \mathbf{E}\{[\Delta_{n-1}x(s)]^2\}\,ds, \qquad n > 1.$$

Hence

$$(3.9) \quad \mathbf{E}\{[\Delta_n x(t)]^2\} \leq [2K^2(b-a+1)]^{n-1}\int_a^t \frac{(t-s)^{n-2}}{(n-2)!}\mathbf{E}\{[\Delta_1 x(s)]^2\}\,ds$$

$$\leq \frac{c^n}{n!}, \qquad a \leq t \leq b,$$

for some constant c. Using this inequality,

$$\mathbf{P}\{\underset{a\leq t\leq b}{\text{Max}} \,|\int_a^t \Delta_n m(s)\,ds\,| \geq 2^{-n}\} \leq \mathbf{P}\{\int_a^b K|\Delta_n x(s)|\,ds \geq 2^{-n}\}$$

$$\leq 4^n\mathbf{E}\{[\int_a^b K|\Delta_n x(s)|\,ds]^2\}$$

$$\leq \frac{4^n(b-a)K^2c^n}{n!}.$$

Since the last term is the general term of a convergent series,

$$(3.10) \qquad \underset{a\leq t\leq b}{\text{Max}} \,|\int_a^t \Delta_n m(s)\,ds| < 2^{-n}$$

for sufficiently large n, with probability 1 (Borel-Cantelli lemma, III, Theorem 1.2). According to IX §5, the process

$$\{\int_a^t \Delta_n\sigma(s)\,dy(s),\ a \leq t \leq b\}$$

is a martingale. Then the family of squares of these random variables is a semi-martingale, to which the continuous parameter version of VII, Theorem 3.2, is applicable. In view of (3.9), we obtain

$$\mathbf{P}\{\underset{a\leq t\leq b}{\text{Max}} \,|\int_a^t \Delta_n\sigma(s)\,dy(s)| \geq 2^{-n}\} \leq 4^n\mathbf{E}\{[\int_a^t \Delta_n\sigma(s)\,dy(s)]^2\}$$

$$= 4^n\int_a^t \mathbf{E}\{[\Delta_n\sigma(s)]^2\}\,ds$$

$$\leq 4^nK^2\int_a^b \mathbf{E}\{[\Delta_n x(s)]^2\}\,ds \leq \frac{4^nK^2c^n}{n!}.$$

Since the last term is the general term of a convergent series,

$$\text{(3.11)} \qquad \operatorname*{Max}_{a \le t \le b} \left| \int_a^t \Delta_n \sigma(s)\, dy(s) \right| < 2^{-n}$$

for sufficiently large n, with probability 1. According to (3.10) and (3.11) the integrals on the right in (3.6) converge uniformly in t, when $n \to \infty$, with probability 1. Hence the limit in (3.7) exists uniformly in t, with probability 1. The $x(t)$ process defined in this way obviously has properties P_1 and P_3. For each t, if $n > m$,

$$\text{(3.12)} \qquad \begin{aligned} \mathbf{E}\{[x_n(t) - x_m(t)]^2\} &= \mathbf{E}\{[\sum_{j=m+1}^{n} \Delta_j x(t)]^2\} \\ &\le \sum_{m+1}^{n} 2^{-m} \sum_{m+1}^{n} 2^j \mathbf{E}\{[\Delta_j x(t)]^2\} \\ &\le 2^{-m} \sum_{1}^{\infty} \frac{(2c)^j}{j!} \to 0 \qquad (m \to \infty). \end{aligned}$$

Then, for each t, $\operatorname*{l.i.m.}_{n\to\infty} x_n(t)$ exists and the mean limit must be $x(t)$ since mean limits and probability 1 limits must coincide (with probability 1). Thus when $n \to \infty$ in (3.12) we find that

$$\mathbf{E}\{[x(t) - x_m(t)]^2\} \le 2^{-m} \sum_{1}^{\infty} \frac{(2c)^j}{j!}$$

so that the $x(t)$ process has property P_2. Finally (3.8) is true uniformly in t, with probability 1 [so that (3.1′) is true] because there is uniform convergence of the integrands with probability 1 in the first limit equation and because, applying the continuous parameter version of VII, Theorem 3.2, as above,

$$\begin{aligned} \mathbf{P}\left\{ \operatorname*{Max}_{a \le t \le b} \left| \int_a^t \left(\sigma[s, x(s)] - \sigma[s, x_n(s)] \right) dy(s) \right| \ge \frac{1}{n} \right\} \\ \le n^2 \int_a^b \mathbf{E}\{(\sigma[s, x(s)] - \sigma[s, x_n(s)])^2\}\, ds \\ \le K^2 n^2 \int_a^b \mathbf{E}\{[x(s) - x_n(s)]^2\}\, ds \\ \le K^2 n^2 2^{-n} \sum_{1}^{\infty} \frac{(2c)^j}{j!}. \end{aligned}$$

Hence, according to the Borel-Cantelli lemma, the maximum involved here will be $< 1/n$ for sufficiently large n, with probability 1, so that the second limit equation in (3.8) is true uniformly in t, with probability 1.

The proof of the existence of a solution to (3.1′) is now complete. The solution (satisfying P_1, P_2, P_3) is essentially uniquely determined by $x(a)$. In fact, if $\Delta(t)$ is the difference of two solutions with the same $x(a)$, the argument leading to (3.9) shows that

$$\mathbf{E}\{\Delta(t)^2\} \leq \frac{c^n}{n!}, \qquad a \leq t \leq b,$$

so that $\Delta(t) = 0$, with probability 1, for each t, and in view of P_1 we then have

$$\mathbf{P}\{\Delta(t, \omega) = 0, a \leq t \leq b\} = 1.$$

Any solution of (3.1′) satisfies

$$x(t) - x(\tau) = \int_\tau^t m[s, x(s)]\, ds + \int_\tau^t \sigma[s, x(s)]\, dy(s). \tag{3.13}$$

We shall always suppose, as we have above, that $x(a)$ is a random variable independent of the aggregate of differences $\{y(t_2) - y(t_1),\ t_1,\ t_2 \in [a, b]\}$. This property reproduces itself in the sense that $x(\tau)$ is then a random variable independent of the aggregate of differences $\{y(t_2) - y(t_1),\ t_1,\ t_2 \in [\tau, b]\}$. According to (3.13) $x(t)$ depends only on $x(\tau)$ and the y-differences for arguments between τ and t. The latter differences are independent of $x(\tau)$, $x(a)$, and (if $s < \tau$) the y-differences for arguments between a and τ, upon which $x(s)$ depends. It follows that the conditional distribution of $x(t)$ for $\{x(s),\ s \leq \tau\}$ given is a function of $x(\tau)$ alone. In other words, the $x(t)$ process is a Markov process, and the conditional distribution of $x(t)$ for $x(\tau, \omega) = \xi$ is the distribution of the solution of (3.1′) with $a = \tau$ and $\mathbf{P}\{x(a, \omega) = \xi\} = 1$. It is not a priori obvious that this uniquely defined conditional distribution satisfies the Chapman-Kolmogorov equation identically, that is, without the necessity of the exclusion of sets of probability 0. The following argument shows that the Chapman-Kolmogorov equation is satisfied identically in the present case: We shall distinguish the probability distribution obtained in (3.1′) when $a = \tau$ and $\mathbf{P}\{x(a, \omega) = \xi\} = 1$ by the subscripts τ, ξ. Suppose that $\tau < s < t$. Then

$$\begin{aligned}\mathbf{P}_{\tau, \xi}\{x(t, \omega) < \lambda\} &= \mathbf{E}_{\tau, \xi}\{\mathbf{P}_{\tau, \xi}\{x(t, \omega) < \lambda \mid x(s, \omega) = \eta\}\} \\ &= \mathbf{E}_{\tau, \xi}\{\mathbf{P}_{s, \eta}\{x(t, \omega) < \lambda\}\},\end{aligned}$$

where we have used the fact that the process is a Markov process with the stated transition probabilities for any initial conditions (independent of later dy's). This equation is precisely the Chapman-Kolmogorov equation.

It will be useful below to have an evaluation of $\mathbf{E}\{\operatorname*{Max}_{a \leq s \leq t} |x(s) - x(a)|^2\}$. Although this can be obtained from the preceding work, it is more instructive to obtain it directly. In the following, K_1, K_2, K_3 will denote constants whose choice will depend only on K and $(b - a)$. We have, using the inequalities obtained in the proof of Lemma 3.1,

$$\mathbf{E}\{\operatorname*{Max}_{a \leq s \leq t} |x(s) - x(a)|^2\} \leq 2\mathbf{E}\{\operatorname*{Max}_{a \leq s \leq t} |\int_a^t m[s, x(s)]\, ds|^2\}$$

$$+ 2\mathbf{E}\{\operatorname*{Max}_{a \leq s \leq t} |\int_a^t \sigma[s, x(s)]\, dy(s)|^2\}$$

$$\leq [K^2(b - a) + 4K^2] \int_a^t [1 + \mathbf{E}\{x(s)^2\}]\, ds$$

$$\leq K_1(t - a)[1 + \mathbf{E}\{x(a)^2\}] + K_1 \int_a^t \mathbf{E}\{|x(s) - x(a)|^2\}\, ds.$$

If now the left side of this inequality is replaced by $\mathbf{E}\{|x(t) - x(a)|^2\}$, the inequality can be integrated to obtain

$$\int_a^t \mathbf{E}\{|x(s) - x(a)|^2\}\, ds \leq K_2(t - a)^2[1 + \mathbf{E}\{x(a)^2\}]$$

and the original inequality then becomes

$$\mathbf{E}\{\operatorname*{Max}_{a \leq s \leq t} |x(s) - x(a)|^2\} \leq K_3(t - a)[1 + \mathbf{E}\{x(a)^2\}]. \tag{3.14}$$

We remark that the expectation may be replaced by a conditional expectation for $x(a, \omega) = \xi$ in this inequality, and of course a may be replaced by any other point of the interval $[a, b]$. This replacement will be made as required without further comment.

The structure of the $x(t)$ process in the small is now easily found. We write

$$x(t + h) - x(t) = \int_t^{t+h} [m[s, x(s)] - m(s, \xi)]\, ds \tag{3.15}$$

$$+ \int_t^{t+h} [\sigma[s, x(s)] - \sigma(s, \xi)]\, dy(s) + \int_t^{t+h} m(s, \xi)\, ds$$

$$+ \int_t^{t+h} \sigma(s, \xi)\, dy(s),$$

and consider the distribution of $x(t+h)-x(t)$ for $x(t,\omega)=\xi$. The sum of the last two terms is Gaussian, with mean and variance

$$\int_t^{t+h} m(s,\xi)\,ds, \qquad \int_t^{t+h} \sigma(s,\xi)^2\,ds$$

respectively. Moreover, this Gaussian random variable is independent of the past, that is, of the aggregate of random variables $\{x(\tau),\ \tau\le t\}$. If the functions m, σ are continuous, the above mean and variance are $hm(t,\xi)$, $h\sigma(t,\xi)^2$ aside from $o(h)$ error terms. Even without the continuity hypothesis the statement is true for each ξ for all values of s except those of a set of Lebesgue measure 0. On the other hand, the first two terms on the right in (3.15) are of order $o(h)$ in the sense that using (3.14), as adjusted to the present situation,

$$\begin{aligned}
(3.16)\qquad \mathbf{E}_{t,\xi}\{\operatorname*{Max}_{t\le\tau\le t+h}\Big|\int_t^{\tau}[m[s,x(s)]-m(s,\xi)]\,ds\Big|^2\}
&\le \mathbf{E}_{t,\xi}\{[\int_t^{t+h}|m[s,x(s)]-m(s,\xi)|\,ds]^2\}\\
&\le K^2h\int_t^{t+h}\mathbf{E}_{t,\xi}\{|x(s)-\xi|^2\}\,ds\\
&\le K^2K_3h^3(1+\xi^2),
\end{aligned}$$

and using the continuous parameter version of VII, Theorem 3.4,

$$\begin{aligned}
(3.17)\quad \mathbf{E}_{t,\xi}\{\operatorname*{Max}_{t\le\tau\le t+h}\Big|\int_t^{\tau}[\sigma[s,x(s)]-\sigma(s,\xi)]\,dy(s)\Big|^2\}
&\le 4\mathbf{E}_{t,\xi}\{\Big|\int_t^{t+h}[\sigma[s,x(s)]-\sigma(s,\xi)]\,dy(s)\Big|^2\}\\
&= 4\int_t^{t+h}\mathbf{E}_{t,\xi}\{|\sigma[s,x(s)]-\sigma(s,\xi)|^2\}\,ds\\
&\le 4K^2\int_t^{t+h}\mathbf{E}_{t,\xi}\{|x(s)-\xi|^2\}\,ds\\
&\le 4K^2K_3h^2(1+\xi^2).
\end{aligned}$$

The above results imply that, for each ξ,

$$
(3.18)\qquad
\begin{aligned}
\mathbf{E}\{x(t+h)-x(t)\mid x(t,\omega)=\xi\} &= \int_t^{t+h} m(s,\xi)\,ds + (1+\xi^2)O(h^{3/2})\\
\mathbf{E}\{[x(t+h)-x(t)]^2\mid x(t,\omega)=\xi\} &= \int_t^{t+h} \sigma(s,\xi)^2\,ds + (1+\xi^2)O(h^2),
\end{aligned}
$$

where the $O(\ )$ terms are uniform in t and ξ. These equations make (3.3) precise for the solutions of the stochastic differential equation (3.1). We observe again that for each ξ the first terms on the right are $h[m(t,\xi)+o(1)]$ and $h[\sigma(t,\xi)^2+o(1)]$ if m and σ are continuous in t, and without the continuity hypothesis the statement is still true (for each ξ) for all values of t except possibly those in a set of Lebesgue measure 0.

Finally we remark that, if $\varepsilon > 0$ and if h is so small that

$$
\left|\int_t^{t+h} m(s,\xi)\,ds\right| < \frac{\varepsilon}{4},
$$

(3.15) yields, in view of (3.16) and (3.17),

$$
(3.19)\qquad \mathbf{P}\{|x(t+h,\omega)-x(t,\omega)| > \varepsilon \mid x(t,\omega)=\xi\} \le \frac{16h^3K^2K_3(1+\xi^2)}{\varepsilon^2}
$$
$$
+\frac{64K^2K_3h^2(1+\xi^2)}{\varepsilon^2} + \mathbf{P}_{t,\xi}\left\{\left|\int_t^{t+h}\sigma(s,\xi)\,dy(s)\right| > \frac{\varepsilon}{4}\right\}.
$$

The last term is

$$
\sqrt{\frac{2}{\pi}}\int_\lambda^\infty e^{-\mu^2/2}\,d\mu,\quad \lambda = \frac{\varepsilon}{4}\left[\int_t^{t+h}\sigma(s,\xi)^2\,ds\right]^{-1/2} \ge \frac{\varepsilon}{4}\left[K^2(1+\xi^2)h\right]^{-1/2}.
$$

According to VIII (2.2) this integral is at most

$$
\sqrt{\frac{2}{\pi}}\frac{e^{-\lambda^2/2}}{\lambda} \le \sqrt{\frac{2}{\pi}}\frac{2}{\lambda^3}.
$$

Hence we have proved

$$
(3.20)\qquad \mathbf{P}\{|x(t+h)-x(t)| > \varepsilon \mid x(t,\omega)=\xi\} \le (1+\xi^2)^{3/2}O(h^{3/2}),
$$

where $O(h^{3/2})$ is uniform in ξ and t. We even have the stronger inequality

$$
(3.20')\qquad \mathbf{P}\{\operatorname*{Max}_{t\le\tau\le t+h}|x(\tau)-x(t)| > \varepsilon \mid x(t,\omega)=\xi\} \le (1+\xi^2)^{3/2}O(h^{3/2}).
$$

In fact, the evaluations we have given prove this inequality also, if (using VIII, Theorem 2.1) the majorant of the last term in (3.19) is doubled.

We have thus verified that the solutions of the stochastic differential equation (3.1) satisfy conditions of the same character as F_2 and F_3 discussed at the beginning of this section. We have already remarked at the beginning of this section that conditions F_1, F_2, and F_3 imply, at least in a wide class of cases, that the sample functions of corresponding separable processes are almost all continuous. Thus it appears that we are discussing here the same general class of processes as that discussed in the theory of diffusion. This fact explains the title of the present section.

We observe that if (3.1′) were replaced by

$$x(t) = \tilde{x}(t) + \int_a^t m[s, x(s)]\, ds + \int_a^t \sigma[s, x(s)]\, dy(s), \tag{3.1″}$$

where the $\tilde{x}(t)$ process is a given process with continuous sample functions, with $\int_a^b \mathbf{E}\{\tilde{x}(t)^2\}\, dt < \infty$, the existence and uniqueness proofs, would require no change. The only difference would be that P_2' would no longer be satisfied unless $\mathbf{E}\{\operatorname*{Max}_{a \le t \le b} \tilde{x}(t)^2\} < \infty$. The $x(t)$ process is a Markov process if for each t the aggregates $\{\tilde{x}(s), a \le s \le t\}$, $\{y(b) - y(s), t \le s \le b\}$ are mutually independent. In particular, if $\tilde{x}(t)$ does not really depend on t, say $\tilde{x}(t) \equiv \tilde{x}$, we have $\tilde{x} = x(a)$ and the special case (3.1′). If $\sigma \equiv 0$ in (3.1″) and if, for each t, $\tilde{x}(t)$ is identically a constant, the probability element of this study disappears. Our work is still applicable, however: it proves the well-known fact that for every continuous t-function $\tilde{x}(\cdot)$ there is one and only one continuous t-function $x(\cdot)$ satisfying

$$x(t) = \tilde{x}(t) + \int_a^t m[s, x(s)]\, ds. \tag{3.1‴}$$

The work of this section has been devoted to the solution of the stochastic differential equation (3.1). We now consider the converse problem: What $x(t)$ processes can be written as solutions of this equation? This problem will be treated in several stages.

THEOREM 3.2 *Let $\{x(t), a \le t \le b\}$ be a separable stochastic process with the following properties:*

(a) *the process is measurable*;

(b) $$\mathbf{E}\{|x(t)|\} < \infty, a \le t \le b; \int_a^b \mathbf{E}\{|x(t)|\}\, dt < \infty;$$

(c) *there is a Baire function* $m(\cdot, \cdot)$ *with*

$$|m(t, \xi)| \leq K(1 + \xi^2)^{1/2},$$

for some constant K, such that, if $a \leq t_1 < t_2 \leq b$,

$$\mathbf{E}\{x(t_2) - x(t_1) \mid x(t), t \leq t_1\} = \mathbf{E}\{\int_{t_1}^{t_2} m[s, x(s)]\, ds \mid x(t), t \leq t_1\}$$

with probability 1.

Then the $\tilde{x}(t)$ *process defined by*

$$\tilde{x}(t) = x(t) - \int_a^t m[s, x(s)]\, ds, \qquad a \leq t \leq b$$

is a separable martingale, and in fact, if $t_1 < t_2$,

$$\mathbf{E}\{\tilde{x}(t_2) \mid x(t), t \leq t_1\} = \tilde{x}(t_1) \tag{3.21}$$

with probability 1.

According to our hypotheses

$$\mathbf{E}\{|x(t_2) - x(t_1)|\} < \infty, \qquad \mathbf{E}\{\int_{t_1}^{t_2} m[s, x(s)]\, ds\} \leq K\mathbf{E}\{\int_{t_1}^{t_2} [1 + |x(s)|]\, ds\}$$

so that the two sides of the equation in (c) are well defined, and $\mathbf{E}\{|\tilde{x}(t)|\} < \infty$ for $a \leq t \leq b$. Moreover,

$$\begin{aligned}
&\mathbf{E}\{\tilde{x}(t_2) - \tilde{x}(t_1) \mid x(t), t \leq t_1\} \\
&\quad = \mathbf{E}\{x(t_2) - x(t_1) \mid x(t), t \leq t_1\} - \mathbf{E}\{\int_{t_1}^{t_2} m[s, x(s)]\, ds \mid x(t), t \leq t_1\} \\
&\quad = 0
\end{aligned}$$

with probability 1, so that (3.21) is true. This equation implies the martingale property for the $\tilde{x}(t)$ process. Since the $x(t)$ process is separable, and since the integral in the definition of $\tilde{x}(t)$ is a continuous function of t for almost all sample functions of the $x(t)$ process, the $\tilde{x}(t)$ process is also separable. We observe that this theorem implies that the sample functions of the $x(t)$ process are (almost all) continuous except for jumps, since this fact is true for the $\tilde{x}(t)$ process sample functions.

THEOREM 3.3 *Let* $\{x(t), a \leq t \leq b\}$ *be a real stochastic process. Let* $\mathscr{F}_t$ *be the smallest Borel field of* ω *sets with respect to which the* $x(s)$*'s with* $s \leq t$ *are measurable. Let* $m(\cdot, \cdot)$, $\sigma(\cdot, \cdot)$ *be functions of* t, ξ. *The following hypotheses are made.*

(a) *Almost all sample functions of the process are continuous in* $[a, b]$.

(b) $\mathbf{E}\{x(t)^2\} < \infty$ *for* $t \in [a, b]$, *and, if* $s < t$,

$$\mathbf{E}\{x(t)^2 \mid \mathscr{F}_s\} \leq z_s$$

with probability 1, *where, for each* s, z_s *is a random variable measurable with respect to the field* $\mathscr{F}_s$, *does not depend on* t, *and* $\mathbf{E}\{z_s\} < \infty$.

(c) *There is a monotone non-decreasing function* $f(\cdot)$, *with* $\lim_{h\to 0} f(h) = 0$ *such that for each* t *and* h *with* $a \leq t < t + h \leq b$

$$\left| \mathbf{E}\{x(t+h) - x(t) \mid \mathscr{F}_t\} - \int_t^{t+h} m[s, x(t)]\, ds \right| \leq [1 + x(t)^2]hf(h)$$

$$\left| \mathbf{E}\{[x(t+h) - x(t)]^2 \mid \mathscr{F}_t\} - \int_t^{t+h} \sigma[s, x(t)]^2\, ds \right| \leq [1 + x(t)^2]hf(h)$$

with probability 1.

(d) $m(\cdot, \cdot)$, $\sigma(\cdot, \cdot)$ *are Baire functions, continuous in their second variables, and there is a constant* K *for which*

$$|m(t, \xi)| \leq K(1 + \xi^2)^{1/2}$$

$$0 \leq \sigma(t, \xi) \leq K(1 + \xi^2)^{1/2}.$$

With these hypotheses it follows that the $x(t) - x(a)$ *process is a Markov process. If* $\sigma(t, \xi)$ *vanishes for no* t, ξ, *there is a Brownian motion process* $\{y(t), a \leq t \leq b\}$ *such that* $x(t)$ *is a solution of the stochastic differential equation* (3.1). *If* $\sigma(t, \xi)$ *may vanish, the statement remains true if a Brownian motion process is adjoined to* ω *space.*

See II §2 for the significance of the adjunction of a process to the given ω space.

Note that, if in addition to the conditions imposed in this theorem $m(\cdot, \cdot)$ and $\sigma(\cdot, \cdot)$ satisfy uniform Lipschitz conditions in ξ, then it follows from our discussion of the stochastic differential equation (3.1) that this equation can be solved to get an $x(t)$ process uniquely determined by $x(a)$ and a given Brownian motion $y(t)$ process, for which the hypotheses of the present theorem are true with

$$z_s = 2K_3(b - a)(1 + x_s^2) + 2x_s^2$$

[see (3.14)] and $f(h) = \text{const. } h^{1/2}$. Thus the conditions under which a given process can be written as a solution of (3.1) are less stringent than the conditions under which we have proved that (3.1) can be solved.

We show first that the hypotheses of Theorem 3.2 are satisfied. Only the third one, the one involving $m(t, \xi)$, is not obvious. To prove this one let $t_1 = s_0 < s_1 < \cdots < s_n = t_2$, $\delta = \operatorname{Max}_j (s_{j+1} - s_j)$. Then

$$\left| \mathbf{E}\{x(t_2) - x(t_1) \mid \mathscr{F}_{t_1}\} - \mathbf{E}\{\int_{t_1}^{t_2} m[s, x(s)]\, ds \mid \mathscr{F}_{t_1}\}\right|$$

$$= \left| \mathbf{E}\left\{ \sum_{j=0}^{n-1} \left[[x(s_{j+1}) - x(s_j)] - \int_{s_j}^{s_{j+1}} m[s, x(s)]\, ds \right] \middle| \mathscr{F}_{t_1} \right\} \right|$$

$$\leq \left| \mathbf{E}\left\{ \sum_{j=0}^{n-1} \int_{s_j}^{s_{j+1}} \left[m[s, x(s)] - m[s, x(s_j)] \right] ds \middle| \mathscr{F}_{t_1} \right\} \right|$$

$$+ \sum_{j=0}^{n-1} (s_{j+1} - s_j) f(s_{j+1} - s_j)\, [1 + \mathbf{E}\{x(s_j)^2 \mid \mathscr{F}_{t_1}\}]$$

with probability 1. It is sufficient to prove that the last two terms in this inequality can be made arbitrarily small by choosing the s_j's properly. The last term is dominated by

$$(b - a) f(\delta)(1 + z_{t_1}),$$

with probability 1, and this goes to 0 with δ. The preceding term has the form $\mathbf{E}\{\varphi \mid \mathscr{F}_{t_1}\}$. It is, therefore, sufficient to prove:

(i) $\varphi \to 0$ when $\delta \to 0$ if the s_j's are chosen properly;

(ii) $\mathbf{E}\{\varphi^2 \mid \mathscr{F}_{t_1}\}$ is bounded independently of δ by some finite ω function.

The first statement is true (with probability 1) because $m(t, \cdot)$ is continuous and almost all sample functions of the $x(t)$ process are continuous. As to (ii), we need only remark that

$$|\varphi| \leq K \sum_{j=0}^{n-1} \int_{s_j}^{s_{j+1}} \left[[1 + x(s)^2]^{1/2} + [1 + x(s_j)^2]^{1/2} \right] ds$$

so that

$$\varphi^2 \leq 2K^2(t_2 - t_1) \left\{ \int_{t_1}^{t_2} [1 + x(s)^2]\, ds + \sum_{j=0}^{n-1} [1 + x(s_j)^2](s_{j+1} - s_j) \right\}$$

and hence

$$\mathbf{E}\{\varphi^2 \mid \mathscr{F}_{t_1}\} \leq 4K^2(t_2 - t_1)^2(1 + z_{t_1}),$$

with probability 1, so that we have obtained the desired bound and thereby shown that Theorem 3.2 is applicable. Applying this theorem, we define an $\tilde{x}(t)$ process by

$$\tilde{x}(t) = x(t) - \int_a^t m[s, x(s)]\, ds,$$

so that the $\tilde{x}(t)$ process is a martingale whose sample functions are almost all continuous.

Define c by

$$c = \mathbf{E}\{[\int_a^b |m[s, x(s)]|\, ds]^2\} \leq (b-a)K^2\mathbf{E}\{\int_a^b [1 + x(s)^2]\, ds\}$$

$$\leq (b-a)^2K^2[1 + \mathbf{E}\{z_a\}] < \infty.$$

Then

$$\mathbf{E}\{\tilde{x}(t)^2\} \leq 2\mathbf{E}\{x(t)^2\} + 2c < \infty$$

so that, using the continuous parameter version of VII, Theorem 3.4, we obtain

$$\mathbf{E}\{\underset{a \leq t \leq b}{\text{Max}}\ x(t)^2\} \leq 2\mathbf{E}\{\underset{a \leq t \leq b}{\text{Max}}\ \tilde{x}(t)^2\} + 2c$$

$$\leq 8\mathbf{E}\{\tilde{x}(b)^2\} + 2c < \infty.$$

If $t_1 = s_0 < s_1 < \cdots < s_n = t_2$ we have, from (3.21),

$$\mathbf{E}\{\tilde{x}(s_k) \mid \mathscr{F}_{s_j}\} = \tilde{x}(s_j) \qquad j < k$$

with probability 1. Hence

$$\mathbf{E}\{\tilde{x}(s_k)\tilde{x}(s_j) \mid \mathscr{F}_{s_j}\} = \tilde{x}(s_j)^2 \qquad j < k$$

with probability 1. Using this fact,

$$\left| \mathbf{E}\{[\tilde{x}(t_2) - \tilde{x}(t_1)]^2 \mid \mathscr{F}_{t_1}\} - \mathbf{E}\{\int_{t_1}^{t_2} \sigma[s, x(s)]^2\, ds \mid \mathscr{F}_{t_1}\} \right|$$

$$= \left| \mathbf{E}\left\{ \sum_{j=0}^{n-1} [\tilde{x}(s_{j+1}) - \tilde{x}(s_j)]^2 - \sum_{j=0}^{n-1} \int_{s_j}^{s_{j+1}} \sigma[s, x(s)]^2\, ds \,\middle|\, \mathscr{F}_{t_1} \right\} \right|$$

$$\leq \left| \mathbf{E}\left\{ \sum_{j=0}^{n-1} \int_{s_j}^{s_{j+1}} [\sigma[s, x(s)]^2 - \sigma[s, x(s_j)]^2]\, ds \,\middle|\, \mathscr{F}_{t_1} \right\} \right|$$

$$+ \sum_{j=0}^{n-1} (s_{j+1} - s_j) f(s_{j+1} - s_j)[1 + \mathbf{E}\{x(s_j)^2 \mid \mathscr{F}\}_{t_1}]$$

with probability 1. We prove that

$$\mathbf{E}\{[\tilde{x}(t_2) - \tilde{x}(t_1)]^2 \mid \mathscr{F}_{t_1}\} = \mathbf{E}\{\int_{t_1}^{t_2} \sigma[s, x(s)]^2\, ds \mid \mathscr{F}_{t_1}\} \tag{3.22}$$

with probability 1 by proving that the last two terms in the preceding inequality can be made arbitrarily small by choosing $\delta = \text{Max}\,(s_{j+1} - s_j)$ small. We have already treated the last term. The next to the last has the form $\mathbf{E}\{\psi \mid \mathscr{F}_{t_1}\}$, where $\psi \to 0$ when $\delta \to 0$, for almost all ω. It will, therefore, according to I §8, CE_5, be sufficient to show that $|\psi| \leq \psi_1$, where ψ_1 is a function independent of δ, with $\mathbf{E}\{\psi_1\} < \infty$, with probability 1. Now

$$\psi \leq \int_{t_1}^{t_2} \sigma[s, x(s)]^2 \, ds + \sum_{j=0}^{n-1} \sigma[s_j, x(s_j)]^2(s_{j+1} - s_j)$$

$$\leq K^2 \int_{t_1}^{t_2} [1 + x(s)^2] \, ds + K^2 \sum_{j=0}^{n} [1 + x(s_j)^2](s_{j+1} - s_j)$$

$$\leq 2K^2(t_2 - t_1) \underset{a \leq t \leq b}{\operatorname{Max}} [1 + x(t)^2] = \psi_1,$$

and we have seen above that $\mathbf{E}\{\psi_1\} < \infty$. Now suppose that $\sigma(t, \xi)$ never vanishes. Then since (3.21) is true there is, according to IX, Theorem 5.3, a Brownian motion process $\{y(t), a \leq t \leq b\}$ such that

$$\tilde{x}(t) - \tilde{x}(a) = \int_a^t \sigma[s, x(s)] \, dy(s),$$

so that

$$x(t) - x(a) = \int_a^t m[s, x(s)] \, ds + \int_a^t \sigma[s, x(s)] \, dy(s).$$

If $\sigma(t, \xi)$ may vanish, the same theorem states that $\tilde{x}(t)$ can be put in this form after a Brownian motion process is adjoined to ω space. In either case it follows that the $x(t) - x(a)$ process is a Markov process, and the theorem is now completely proved.

CHAPTER VII

Martingales

1. Definitions; martingales and semi-martingales

A martingale was defined in II §7 as a (real or complex) stochastic process $\{x_t,\ t \in T\}$ for which $\mathbf{E}\{|x_t|\} < \infty$, $t \in T$, and

$$x_{t_n} = \mathbf{E}\{x_{t_{n+1}} \mid x_{t_1}, \cdots, x_{t_n}\} \tag{1.1}$$

with probability 1, whenever $t_1 < \cdots < t_{n+1}$. Here n is an arbitrary positive integer. It follows that, *if a process* $\{x_t,\ t \in T\}$ *is a martingale, every process* $\{x_t,\ t \in T_1\}$ *with* $T_1 \subset T$ *is also a martingale, and conversely, if the latter process is a martingale for every finite set* $T_1 \subset T$, *then the process* $\{x_t,\ t \in T\}$ *is a martingale.*

In the following, if $\{x_t,\ t \in T\}$ is any stochastic process, and if Λ is an ω set measurable on the sample space of the x_t's (see II §2), we shall attempt to increase the intuitive content of the discussion by describing Λ as a set *determined by conditions on the* x_t's. Thus, if $x_1, \cdots, x_n$ are random variables, an ω set is a set determined by conditions on these random variables if and only if it is an ω set of the form $\{[x_1(\omega), \cdots, x_n(\omega)] \in B\}$, where B is a Borel set in n dimensions (real or complex as the case may be), or if it differs from such a set by one of probability 0.

Going back to the definition of conditional expectation, and changing the notation, (1.1) is equivalent to

$$\int_\Lambda x_s \, d\mathbf{P} = \int_\Lambda x_t \, d\mathbf{P}, \qquad s < t \tag{1.2}$$

for every ω set Λ determined by conditions on a finite number of x_r's with $r \leq s$. This equation is then also true for every ω set determined by conditions on the x_r's with $r \leq s$, since the latter can be approximated arbitrarily closely by the former ones (or apply Supplement, Theorem 2.1). Equation (1.2) is equivalent to

$$x_s = \mathbf{E}\{x_t \mid x_r,\ r \leq s\}, \qquad s < t, \tag{1.1'}$$

where this equation is to hold with probability 1 for each pair s, t. The equality (1.1′) is sometimes used as the defining property of a martingale instead of (1.1). The equality (1.2) will be called the *martingale equality*. If the x_t's are complex, and if (1.2) is true, it is true for the real and imaginary parts separately. Then *if an x_t process is a martingale, the processes defined by the real and imaginary parts of the x_t's are also martingales.* This fact will be used below to reduce theorems on complex martingales to theorems on real ones.

Example 1 (See also Example 1 of II §7.) Let z be a random variable with $\mathbf{E}\{|z|\} < \infty$. Suppose that T is a linear set, and that to each $t \in T$ corresponds a Borel field $\mathscr{F}_t$ of measurable ω sets, with $\mathscr{F}_s \subset \mathscr{F}_t$ when $s < t$. Define

$$x_t = \mathbf{E}\{z \mid \mathscr{F}_t\}.$$

The process $\{x_t, t \in T\}$ is then a martingale, and even the process obtained by adding another parameter value to T, to the right of the given parameter values, and making z correspond to this parameter value, is a martingale. We can prove this statement by using the rules of combination of conditional expectations. Since this method has already been used in the particular case of Example 1 of II §7, however, we prove the result this time by proving the martingale equality directly; that is, we prove that, if $s < t$, and if Λ is an ω set determined by conditions on a finite number of x_r's, with $r \leq s$, then

$$\int_\Lambda x_s \, d\mathbf{P} = \int_\Lambda x_t \, d\mathbf{P},$$

and that moreover this equation remains true for all $s \in T$ if x_t is replaced on the right by z. The following proof is applicable with or without this replacement. The ω set Λ is either itself in $\mathscr{F}_s$ or at least differs from such a set by a set of probability 0. Hence by the definition of conditional expectation (or the identification of x_t with z) both sides of the above equation are equal to

$$\int_\Lambda z \, d\mathbf{P}.$$

Hence the desired equation is true.

As a particular case of Example 1 we obtain Example 1 of II §7, as follows. Let z be as above, and let $w_1, w_2, \cdots$ be any random variables. Then, if x_n is defined by

$$x_n = \mathbf{E}\{z \mid w_1, \cdots, w_n\},$$

the random variables $x_1, x_2, \cdots, z$ constitute a martingale. In fact if $\mathscr{F}_n$ is defined as the Borel field of ω sets determined by conditions on $w_1, \cdots, w_n$, the definition of x_n becomes precisely that given above. In practice $\mathscr{F}_t$ is usually determined as in this particular case; it is the Borel field of ω sets determined by conditions on the random variables of a given collection which depends on t, and increases with t. The general principle we shall use repeatedly is that as more and more conditions are imposed in a conditional expectation, the ordered family of random variables obtained is a martingale.

A *semi-martingale* is a real process $\{x_t, t \in T\}$, defined in the same way as a martingale except that the critical equality is replaced by an inequality; "$=$" is replaced by "$\leq$" in (1.1), (1.2), (1.1′). The inequality

$$\int_\Lambda x_s \, d\mathbf{P} \leq \int_\Lambda x_t \, d\mathbf{P}, \qquad s < t, \tag{1.2s}$$

will be called the semi-martingale inequality. The semi-martingale versions of (1.1), (1.2), (1.1′) are equivalent, and any one can be used as the defining property of a semi-martingale. It will sometimes be convenient although illogical to refer to processes with (1.2s) true with the inequality reversed as *lower semi-martingales.*

The partial sums of any series of non-negative random variables with finite expectations constitute a semi-martingale. More interesting examples will be given below.

We now specialize slightly the martingale and semi-martingale definitions. Let $\{x_t, t \in T\}$ be a stochastic process, with $\mathbf{E}\{|x_t|\} < \infty$, $t \in T$, and suppose that to each $t \in T$ corresponds a Borel field $\mathscr{F}_t$ of measurable ω sets such that

(i) $$\mathscr{F}_s \subset \mathscr{F}_t \qquad s < t;$$

(ii) x_t *is either measurable with respect to the field* $\mathscr{F}_t$ *or is equal for almost all* ω *to a function that is;*
either

(iii) $$x_s = \mathbf{E}\{x_t \mid \mathscr{F}_s\}$$

with probability 1, *whenever* $s < t$, *or*

(iiis) *the process is real and*

$$x_s \leq \mathbf{E}\{x_t \mid \mathscr{F}_s\}$$

with probability 1, *whenever* $s < t$.

Conditions (iii) and (iiis) imply the martingale equality (1.2) and semi-martingale inequality (1.2s) respectively, if $\Lambda \in \mathscr{F}_s$, or if Λ differs by at most an ω set of probability 0 from a set in $\mathscr{F}_s$, in particular if Λ is defined by conditions on the x_r's for $r \leq s$. Then the x_t process is a martingale if (iii) holds, a semi-martingale if (iiis) holds. We shall denote such a martingale or semi-martingale by $\{x_t, \mathscr{F}_t, t \in T\}$ to call attention to the $\mathscr{F}_t$'s, and describe the process as a *martingale or semi-martingale relative to the* $\mathscr{F}_t$*'s.* According to this definition, if $\{y_t,\ t \in T\}$ is any martingale [semi-martingale] and if $\mathscr{G}_t$ is the Borel field of ω sets measurable on the sample space of the y_s's with $s \leq t$, then $\{y_t,\ \mathscr{G}_t,\ t \in T\}$ is a martingale [semi-martingale]. Thus every martingale or semi-martingale is one relative to certain Borel fields of sets, and when the fields are not specified we can always take those defined like the $\mathscr{G}_t$'s in terms of the given random variables.

If $\{x_t,\ \mathscr{F}_t,\ t \in T\}$ is a martingale or semi-martingale, and if $\mathscr{F}_t'$ is the Borel field generated by the sets of $\mathscr{F}_t$ and the sets of probability 0, so that $\mathscr{F}_t'$ consists of the sets of $\mathscr{F}_t$ and of the sets which differ from $\mathscr{F}_t$ sets by sets of probability 0, then $\{x_t,\ \mathscr{F}_t',\ t \in T\}$ is also a martingale or semi-martingale. In other words, it is no restriction on generality to suppose that the $\mathscr{F}_t$'s contain all sets of probability 0 in the first place. If this is true, the alternative in (ii) above is unnecessary since x_t is itself measurable with respect to $\mathscr{F}_t$ if it is equal almost everywhere to a function which is.

Note that, if $\{x_t',\mathscr{F}_t',\ t \in T\}$ and $\{x_t''(\omega),\mathscr{F}_t'',\ t \in T\}$ are both martingales [semi-martingales], defined on the same ω space, then if $\mathscr{F}_t' = \mathscr{F}_t''$, for $t \in T$, the process

$$\{x_t' + x_t'',\mathscr{F}_t',\ t \in T\}$$

is also a martingale [semi-martingale]. The process $\{x_t' + x_t'',\ t \in T\}$ is not necessarily a martingale [semi-martingale] without this identity of the fields involved, however. If $\{x_t,\ \mathscr{F}_t,\ t \in T\}$ is a martingale, and if $x_t = u_t + iv_t$, where u_t and v_t are real, the processes

$$\{u_t, \mathscr{F}_t,\ t \in T\}, \qquad \{v_t. \mathscr{F}_t,\ t \in T\}$$

are martingales.

THEOREM 1.1 (i) *If* $\{x_t,\ \mathscr{F}_t,\ t \in T\}$ *is a semi-martingale, and if* Φ *is a real function of the real variable* λ*, which is monotone non-decreasing and convex, with* $\mathbf{E}\{|\Phi(x_{t_0})|\} < \infty$ *for some* $t_0 \in T$*, then* $\{\Phi(x_t),\ \mathscr{F}_t,\ t \in T,\ t \leq t_0\}$ *is a semi-martingale.*

(ii) *If* $\{x_t,\ \mathscr{F}_t,\ t \in T\}$ *is a martingale, then* $\{|x_t|,\ \mathscr{F}_t,\ t \in T\}$ *is a semi-martingale.*

(iii) *If* $\{x_t,\ \mathscr{F}_t,\ t \in T\}$ *is a real martingale, and if* Φ *is a real function of the real variable* λ *which is continuous and convex, with* $\mathbf{E}\{|\Phi(x_{t_0})|\} < \infty$ *for some* $t_0 \in T$*, then* $\{\Phi(x_t),\ \mathscr{F}_t,\ t \in T,\ t \leq t_0\}$ *is a semi-martingale.*

The proofs of these three statements are the same in principle, so only the proof of (i) will be given. In that case, if $t < t_0$, and if $t \in T$, then we have, using Jensen's inequality (see I §9),

$$\Phi(x_t) \leq \Phi[\mathbf{E}\{x_{t_0} \mid \mathscr{F}_t\}] \leq \mathbf{E}\{\Phi(x_{t_0}) \mid \mathscr{F}_t\} \tag{1.3}$$

and (convexity of Φ) there is a positive constant c such that

$$c\lambda \leq \Phi(\lambda)$$

for sufficiently negative λ. Then $\Phi(x_t)$ is at most equal to the integrable function on the right in (1.3) when $\Phi(x_t)$ is large, and is at least equal to the integrable function $-cx_t$ when $-\Phi(x_t)$ is large. Consequently $\Phi(x_t)$ is itself integrable. Finally, the inequality (1.3) between the extremes must now hold for any pair of parameter values s, s_0 (instead of t, t_0), if $s < s_0 \leq t_0$, and this is one condition necessary and sufficient that $\{\Phi(x_t), \mathscr{F}_t, t \in T, t \leq t_0\}$ be a semi-martingale.

The following are the most important applications of this theorem. (We define $\log^+ \lambda$ as usual as 0 if $\lambda < 1$, $\log \lambda$ if $\lambda \geq 1$.)

(a) *If the x_t process is a semi-martingale, the processes with random variables*

$$\{x_t \log^+ x_t\}, \qquad \left\{\frac{|x_t| + x_t}{2}\right\}$$

are semi-martingales if the relevant expectations exist.

(b) *If the x_t process is a martingale, the processes with random variables*

$$\{|x_t| \log^+ |x_t|\}, \qquad \{|x_t|^\alpha\} \qquad (\alpha \geq 1)$$

are semi-martingales if the relevant expectations exist.

Now consider any series $\sum_j y_j$ of random variables whose expectations exist. The sequence of partial sums is a martingale if and only if

$$\mathbf{E}\{y_{n+1} \mid y_1, \cdots, y_n\} = 0, \qquad n \geq 1 \tag{1.4}$$

with probability 1, according to II §7. The y_j's are easily modified by subtracting the proper expectations to obtain a new series

$$\sum_j [y_j - \mathbf{E}\{y_j \mid y_1, \cdots, y_{j-1}\}]$$

whose summands have the property (1.4), so that the new sequence of partial sums is a martingale. Thus, if in any particular case it is possible to make suitable estimates of the subtracted expectations, a general series can be reduced profitably to one satisfying (1.4). We shall find it more convenient to perform this operation on the partial sums. In terms of these the argument runs as follows. Let $x_1, x_2, \cdots$ be any sequence of random variables whose expectations exist, and define x_n', Δ_n by

$$\begin{gathered} \Delta_1 = 0 \\ x_n = x_n' + \sum_{j=1}^{n} \Delta_j, \qquad \Delta_j = \mathbf{E}\{x_j \mid x_1, \cdots, x_{j-1}\} - x_{j-1}, \qquad j > 1. \end{gathered} \tag{1.5}$$

Then the x_n' process is a martingale (because, if $m < n$, $\mathbf{E}\{x_n' \mid x_1, \cdots, x_m\} = x_m'$). In any particular application enough must be known of the Δ_j's to make the reduction of x_n properties to x_n' properties useful. In particular, if the x_n process is a semi-martingale, the Δ_j's are non-negative. Conversely, if x_n can be put in the form $x_n' + \sum_{j=1}^{n} \Delta_j$, where the x_n' process is a martingale and where more particularly

$$\mathbf{E}\{x_n' \mid x_1, \cdots, x_m\} = x_m', \qquad m < n$$

with probability 1, and where the Δ_j's are non-negative, then the x_n process is a semi-martingale. This characterization of a semi-martingale in terms of a martingale will play an important role below.

Example 2 Let $y_1, y_2, \cdots$ be mutually independent random variables whose expectations exist, and let $x_n = \sum_1^n y_j$. Then the x_n process is a martingale if and only if $\mathbf{E}\{y_j\} = 0$ for $j > 1$, a semi-martingale if and only if the y_j's are real and $\mathbf{E}\{y_j\} \geq 0$ for $j > 1$. In the latter case the representation

$$x_1 = y_1$$
$$x_n = y_1 + \sum_2^n [y_j - \mathbf{E}\{y_j\}] + \sum_2^n \mathbf{E}\{y_j\}$$

is a simple special case of (1.5) in which the Δ_j's are non-negative constants.

If $\{x_n, \mathscr{F}_n, n \geq 1\}$ is a semi-martingale, (1.5) can be generalized slightly to become

$$\Delta_1 = 0$$

(1.5′)

$$x_n = x_n' + \sum_1^n \Delta_j, \qquad \Delta_j = \mathbf{E}\{x_j \mid \mathscr{F}_{j-1}\} - x_{j-1}, \qquad j > 1.$$

In this representation $\{x_n', \mathscr{F}_n, n \geq 1\}$ is a martingale, $\Delta_j \geq 0$, and Δ_j is measurable with respect to $\mathscr{F}_{j-1}$.

In the following we shall say that a process $\{x_t, t \in T\}$ is dominated by a semi-martingale $\{x_t^+, t \in T\}$ if

$$\mathbf{P}\{|x_t(\omega)| \leq x_t^+(\omega)\} = 1, \qquad t \in T.$$

THEOREM 1.2 *Let* $\{x_n, \mathscr{F}_n, n \geq 1\}$ *be a semi-martingale, and consider the representation* (1.5′).

(i) *If* $\text{L.U.B.}_n\ \mathbf{E}\{x_n\} < \infty$, *then* $\sum_1^\infty E\{\Delta_j\} < \infty$, *and* $\sum_1^\infty \Delta_j < \infty$ *with probability* 1.

(ii) *If* $\text{L.U.B.}_n\ \mathbf{E}\{|x_n|\} < \infty$, *then* [*in addition to the conclusions of* (i)], $\text{L.U.B.}_n\ \mathbf{E}\{|x_n'|\} < \infty$.

(iii) *If the* x_n*'s are uniformly integrable, then* [*in addition to the conclusions of* (i) *and* (ii)] *the* x_n'*'s are uniformly integrable.*

(iv) *The semi-martingale* $\{x_n, \mathscr{F}_n, n \geq 1\}$ *is dominated by a semi-martingale, and in fact we can set*

$$x_n^+ = |x_n'| + \sum_1^n \Delta_j. \tag{1.6}$$

Proof of (i) Using the martingale equality, we find

$$\mathbf{E}\{x_n'\} = \mathbf{E}\{x_1'\} = \mathbf{E}\{x_1\},$$

so that

$$\mathbf{E}\{x_n\} = \mathbf{E}\{x_1\} + \sum_1^n \mathbf{E}\{\Delta_j\}. \tag{1.7}$$

Then under the hypothesis of (i) the partial sums on the right are bounded when $n \to \infty$, so that the conclusion of (i) follows at once. We observe that according to the semi-martingale inequality [or (1.7)] the left side of (1.7) is monotone in n, so that the hypothesis of (i) is simply that

$$\lim_{n\to\infty} \mathbf{E}\{x_n\} < \infty.$$

Proof of (ii) If the hypothesis of (i) is strengthened to that of (ii), the left side of the inequality

$$\mathbf{E}\{|x_n'|\} \leq \mathbf{E}\{|x_n|\} + \sum_1^\infty \mathbf{E}\{\Delta_j\}$$

must be bounded in n.

Proof of (iii) If the hypothesis of (ii) is strengthened to that of (iii), the left side of the inequality

$$|x_n'| \leq |x_n| + \sum_1^\infty \Delta_j$$

must be uniformly integrable.

The truth of (iv) is obvious.

In this chapter the parameter values of the processes will not necessarily range over intervals or over sequences of consecutive integers. The properties (1.1′) and the corresponding inequality for semi-martingales are obviously meaningful as long as the range of values of t is an ordered set. For example, we shall sometimes discuss an ordered family of random variables of the type

$$x_1, x_2, \cdots, x_\infty, \quad \text{or} \quad x_{-\infty}, \cdots, x_{-1}, x_0.$$

We shall let the parameter set T be any set of the infinite line $[-\infty, \infty]$, closed at the ends by the addition of the points $\pm\infty$, and topological concepts will be used accordingly. For example, the two parameter sets listed above are closed sets in this topology, but the set of all integers is not closed, because $\pm\infty$ are limit points which are not in the set.

2. Application to games of chance

Suppose a gambler with fortune x_1 plays some game of chance once, and that his fortune after the play is x_2. Then x_2 is a random variable, and the game is usually considered "fair" if

$$\mathbf{E}\{x_2\} = x_1.$$

This definition of fair is of course somewhat arbitrary, although hallowed by tradition. If the gambler then plays the same or some other game, the preceding criterion of fairness becomes

$$\mathbf{E}\{x_3 \mid x_2\} = x_2$$

(with probability 1), where x_3 is his fortune after the second play, and where his choice of the second game may depend on $x_2(\omega)$. Continuing in this way, it becomes clear that one natural definition of fairness is the martingale condition

$$\mathbf{E}\{x_{n+1} \mid x_1, \cdots, x_n\} = x_n, \qquad n = 1, 2, \cdots$$

(with probability 1), where x_n is the gambler's fortune after the $(n-1)$th play. (Since $x_1 =$ const. it makes no difference whether x_1 is present among or absent from the conditioning variables. This irrelevancy of x_1 will not be a fact below, however.)

A closer analysis of the ideas involved in this concept of fairness suggests that the first play should only be considered fair if

$$\mathbf{E}\{x_2 \mid y_1\} = x_1$$

with probability 1, where y_1 indicates one or more random variables representing past history, as known to the gambler, up to the time of the play. At the next play the criterion of fairness becomes

$$\mathbf{E}\{x_3 \mid y_2\} = x_2$$

with probability 1, where y_2 represents the past history known to the gambler, including, for example, the value of x_2, up to the time of the second play, and so on. These fairness conditions are slightly stronger than the first ones given, and suggest the following *mathematical model of a fair game: a fair game is a martingale* $\{x_n, \mathscr{F}_n, n \geq 1\}$ *relative to some stated Borel fields.* The Borel field $\mathscr{F}_n$ represents the influence of the past up to and including time n. In the same spirit we define: *a favorable game is a semi-martingale* $\{x_n, \mathscr{F}_n, n \geq 1\}$ *relative to some stated Borel fields.* We do not make the hypothesis that x_1 is identically constant, with probability 1, although this is suggested by the interpretation we have given.

If the concepts of fairness and of advantageousness are to be self-consistent, fair and advantageous games must remain so when the gambler (or house) adopts certain acceptable practices which have the effect of changing the given process. This fact suggests various theorems which are the subject of the present section, and which have important theoretical applications.

Suppose, for example, that the gambler decides to leave, instead of playing indefinitely, either because he thinks that he has won (or lost) enough, or because he is discouraged by the way the game has been going, or for any other reason. Then the game is still fair (or advantageous) if it was so originally, unless the gambler has quit because he can foresee the future, and knows, for example, that the next plays will go against him (or because he knows that the next plays will go in his favor, and he chivalrously refuses to take advantage of his extrasensory powers). The mathematical formulation of this conduct of the gambler is the following. Let m be a random variable which may take on the value $+\infty$ with positive probability but whose finite values are non-negative integers. (The gambler stops after making m plays.) Define the random variables $\breve{x}_1, \breve{x}_2, \cdots$ by

$$\begin{aligned} \breve{x}_j(\omega) &= x_j(\omega), \qquad & j \leq m(\omega) \\ &= x_{m(\omega)}(\omega), \qquad & j > m(\omega). \end{aligned}$$

It is supposed that the condition $m(\omega) = \mu$ is a condition only on the past up to the μth play; more precisely, it is supposed that

$$\{m(\omega) = \mu\} \in \mathscr{F}_\mu.$$

The transformation from $\{x_n, \mathscr{F}_n, n \geq 1\}$ to $\{\breve{x}_n, n \geq 1\}$ will be said to be a *transformation under a system of optional stopping*. It is natural to expect that under this transformation a fair game remains fair and an advantageous one remains advantageous, that is, that a martingale goes into a martingale and a semi-martingale into a semi-martingale. These invariance properties are the subject of Theorem 2.1. The extensive use of this theorem in §4 to prove convergence properties of martingales shows that it is more profitable to gamble on the convergence of a sequence than on the color of a card. Although Theorem 2.1 is a very special case of Theorem 2.2, we prove it separately here to simplify the reading of §4.

THEOREM 2.1 *Suppose that the semi-martingale* [*martingale*] $\{x_n, \mathscr{F}_n, n \geq 1\}$ *is transformed into the process* $\{\breve{x}_n, n \geq 1\}$ *under optional stopping. Then the* $\breve{x}_n$ *process is a semi-martingale* [*martingale*]. *In the semi-martingale case,*

$$\mathbf{E}\{x_1\} \leq \mathbf{E}\{\breve{x}_n\} \leq \mathbf{E}\{x_n\}, \qquad n \geq 1;$$

in the martingale case,

$$\mathbf{E}\{\breve{x}_n\} = \mathbf{E}\{x_1\}, \qquad n \geq 1.$$

Since $\breve{x}_n$ is equal (in pieces) to $x_1, \cdots, x_n$, it follows that

$$\mathbf{E}\{|\breve{x}_n|\} \leq \sum_1^n \mathbf{E}\{|x_j|\} < \infty.$$

Suppose that the x_n process is a semi-martingale. To prove that the $\breve{x}_n$ process is a semi-martingale, we must prove that, if Λ is any ω set determined by conditions on $\breve{x}_1, \cdots, \breve{x}_n$, then

$$\int_\Lambda \breve{x}_{n+1}\, d\mathbf{P} \geq \int_\Lambda \breve{x}_n\, d\mathbf{P}. \tag{2.1}$$

If m is the function defining the optional stopping, as explained in the definition of this concept, $\breve{x}_j(\omega) = \breve{x}_{m(\omega)}(\omega)$ if $j \geq m(\omega)$. Hence

$$\int_{\Lambda\{m(\omega)\leq n\}} \breve{x}_{n+1}\, d\mathbf{P} = \int_{\Lambda\{m(\omega)\leq n\}} \breve{x}_n\, d\mathbf{P}, \tag{2.2}$$

because the integrands are identical. Now Λ is defined in terms of conditions on $\breve{x}_1, \cdots, \breve{x}_n$, and the latter random variables are defined in terms of $x_1, \cdots, x_n$ in such a way that it is trivial to verify that Λ is also defined by conditions on $x_1, \cdots, x_n$. The ω set $\{m(\omega) \leq n\}$, and therefore its complement $\{m(\omega) > n\}$, are also defined by conditions on $x_1, \cdots, x_n$, and finally this means that $\Lambda\{m(\omega) > n\}$ is so defined. But then, using the semi-martingale property,

$$\int_{\Lambda\{m(\omega)>n\}} x_{n+1}\, d\mathbf{P} \geq \int_{\Lambda\{m(\omega)>n\}} x_n\, d\mathbf{P}, \tag{2.3}$$

and we can replace x_{n+1} by $\breve{x}_{n+1}$ and x_n by $\breve{x}_n$ in the integrands, since

$$x_{n+1}(\omega) = \breve{x}_{n+1}(\omega), \qquad x_n(\omega) = \breve{x}_n(\omega), \qquad \text{if} \quad m(\omega) > n.$$

Adding (2.2) and (2.3), we find that (2.1) is true, that is, the $\breve{x}_n$ process is a semi-martingale. If the x_n process is a martingale, there is equality in (2.3) and therefore in (2.1), so that the $\breve{x}_n$ process is also a martingale. In both cases, $x_1(\omega) \equiv \breve{x}_1(\omega)$. Hence in the martingale case

$$\mathbf{E}\{\breve{x}_n\} = \mathbf{E}\{\breve{x}_1\} = \mathbf{E}\{x_1\}, \qquad n \geq 1.$$

In the semi-martingale case we find, using the semi-martingale property,

$$\mathbf{E}\{x_1\} = \mathbf{E}\{\breve{x}_1\} \leq \mathbf{E}\{\breve{x}_n\} = \sum_{j=1}^{n-1} \int_{\{m(\omega)=j\}} x_j\, d\mathbf{P} + \int_{\{m(\omega)\geq n\}} x_n\, d\mathbf{P} \leq \mathbf{E}\{x_n\}.$$

We now generalize the concept of optional stopping to that of *optional sampling*. Optional sampling transforms the process $\{x_n, \mathscr{F}_n, n \geq 1\}$ into a process $\{\check{x}_n, n \geq 1\}$ as follows. Let $m_1, m_2, \cdots$ be a finite or infinite sequence of integral-valued random variables with the properties

$$1 \leq m_1 \leq m_2 \leq \cdots < \infty$$

$$\{m_j(\omega) = \mu\} \in \mathscr{F}_\mu,$$

the inequality to hold with probability 1. Define

$$\check{x}_j(\omega) = x_{m_j(\omega)}(\omega), \qquad j \geq 1.$$

In gambling terminology, the change from x_n to $\check{x}_n$ corresponds to having the gambler sample his fortune only at certain times dependent on the past and present. In particular, if m is an integral-valued random variable determining optional stopping, as explained above, and if m_j is defined by

$$m_j(\omega) = \text{Min}\,[m(\omega), j],$$

the m_j's satisfy the conditions imposed on an optional sampling m_j sequence, and the optional sampling determined by these m_j's yields the same $\check{x}_n$ process as the optional stopping determined by m.

THEOREM 2.2 *Suppose that the semi-martingale $\{x_n, \mathscr{F}_n, n \geq 1\}$ is transformed into the process $\{\check{x}_n, n \geq 1\}$ by optional sampling. Then, if*

$$\mathbf{E}\{|\check{x}_n|\} < \infty, \qquad n \geq 1, \tag{2.4}$$

and if

$$\liminf_{N\to\infty} \int_{\left\{\substack{m_n(\omega) > N \\ x_N(\omega) > 0}\right\}} x_N \, d\mathbf{P} = 0, \qquad n \geq 1, \tag{2.5}$$

it follows that the $\check{x}_n$ process is also a semi-martingale, with

$$\mathbf{E}\{x_1\} \leq \mathbf{E}\{\check{x}_n\}, \qquad n \geq 1. \tag{2.6}$$

Condition (2.4) *is always satisfied if* $\text{L.U.B.}_n\ \mathbf{E}\{|x_n|\} < \infty$.

Under (2.4) *and*

$$\liminf_{N\to\infty} \int_{\{m_n(\omega) > N\}} |x_N| \, d\mathbf{P} = 0, \qquad n \geq 1, \tag{2.5$'$}$$

(2.6) *can be strengthened to*

$$\mathbf{E}\{x_1\} \leq \mathbf{E}\{\check{x}_n\} \leq \text{L.U.B.}_j\ \mathbf{E}\{x_j\}, \qquad n \geq 1, \tag{2.6$'$}$$

and if the x_n process is a martingale, the $\check{x}_n$ process is also a martingale, and there is then equality in (2.6$'$).

Each of the following conditions C_1–C_4 *implies the validity of* (2.4) *and* (2.5′).

C_1 *The* x_n*'s are uniformly integrable.*

C_2 *Each* m_n *is a bounded random variable* (*with probability* 1). (*This condition is always satisfied in the case of optional stopping.*)

C_3 *There is a constant K such that, for each j,*

$$\mathbf{E}\{m_j\} < \infty,$$

$$\mathbf{E}\{|x_{n+1} - x_n| \mid \mathscr{F}_n\} \leq K, \qquad \text{for} \quad n < m_j(\omega),$$

with probability 1.

C_4 *The* x_n *process is a martingale,* (2.4) *is true, and there is a random variable* $z \geq 0$, *with* $\mathbf{E}\{z\} < \infty$, *and a sequence* $j_1 < j_2 < \cdots$ *of integers, such that*

$$|x_k| \geq |x_{j_i}| - z, \qquad k \geq j_i, i \geq 1,$$

with probability 1. *If* $|x_{j_i}|$ *is replaced by* x_{j_i} *here, the thus weakened condition implies the validity of* (2.4) *and* (2.5) *even if the* x_n *process is only a semi-martingale.*

The proof will be carried through in several steps.

(a) *Proof of* (2.4) *if* $\underset{n}{\text{L.U.B.}}\ \mathbf{E}\{|x_n|\} < \infty$ Fix n, and suppose that the optional stopping transformation defined by m_n takes $\{x_k, k \geq 1\}$ into $\{z_k, k \geq 1\}$, so that

$$\begin{aligned} z_k(\omega) &= x_k(\omega), \qquad && k \leq m_n(\omega) \\ &= x_{m_n(\omega)}(\omega), && k > m_n(\omega). \end{aligned}$$

Then $\lim_{k\to\infty} z_k = x_{m_n} = \breve{x}_n$ with probability 1, and according to Theorem 2.1 the process $\{z_k, k \geq 1\}$ is a semi-martingale. In view of Fatou's theorem, it now suffices to prove that $\underset{k}{\text{L.U.B.}}\ \mathbf{E}\{|z_k|\} < \infty$. Now

$$\begin{aligned} \mathbf{E}\{|z_k|\} &= 2 \int_{\{z_k(\omega)>0\}} z_k \, d\mathbf{P} - \mathbf{E}\{z_k\} \\ &\leq 2 \int_{\left\{\substack{x_k(\omega)>0 \\ m_n(\omega)\geq k}\right\}} x_k \, d\mathbf{P} + 2 \sum_{j=1}^{k-1} \int_{\left\{\substack{x_j(\omega)>0 \\ m_n(\omega)=j}\right\}} x_j \, d\mathbf{P} - \mathbf{E}\{z_1\}, \end{aligned}$$

and, using the facts that $z_1 = x_1$, and that $\{x_j, j \geq 1\}$ is a semi-martingale,

$$\begin{aligned} \mathbf{E}\{|z_k|\} &\leq 2 \int_{\left\{\substack{x_k(\omega)>0 \\ m_n(\omega)\geq k}\right\}} x_k \, d\mathbf{P} + 2 \sum_{j=1}^{k-1} \int_{\left\{\substack{x_k(\omega)>0 \\ m_n(\omega)=j}\right\}} x_k \, d\mathbf{P} - \mathbf{E}\{x_1\} \\ &\leq 2\mathbf{E}\{|x_k|\} - \mathbf{E}\{x_1\}. \end{aligned}$$

(b) *Proof under* (2.4) *and* (2.5) *of the semi-martingale inequality for the* $\breve{x}_n$ *process* We prove that, if Λ is an ω set determined by conditions on $\breve{x}_1, \cdots, \breve{x}_n$, if the x_n process is a semi-martingale, and if (2.4) and (2.5) are true, then

$$\int_\Lambda \breve{x}_n \, d\mathbf{P} \leq \int_\Lambda \breve{x}_{n+1} \, d\mathbf{P}, \tag{2.7}$$

and in fact we shall prove the stronger inequality that, if

$$\Lambda_j = \Lambda\{m_n(\omega) = j\},$$

then

$$\int_{\Lambda_j} \breve{x}_n \, d\mathbf{P} \leq \int_{\Lambda_j} \breve{x}_{n+1} \, d\mathbf{P}. \tag{2.8}$$

Define

$$\begin{aligned} \Lambda_{jk} &= \Lambda\{m_n(\omega) = j, m_{n+1}(\omega) = k\} \qquad k \geq j \\ \Lambda_{jk}' &= \Lambda\{m_n(\omega) = j, m_{n+1}(\omega) > k\} \qquad k \geq j, \\ &= \Lambda_j - \bigcup_{r=j}^{k} \Lambda_{jr}. \end{aligned}$$

Then

$$\Lambda_j \in \mathscr{F}_j, \qquad \Lambda_{jk} \in \mathscr{F}_k, \qquad \Lambda_{jk}' \in \mathscr{F}_k,$$

$$\Lambda_j = \bigcup_{k=j}^{\infty} \Lambda_{jk}.$$

Hence, using the semi-martingale inequality,

$$\begin{aligned} \int_{\Lambda_j} \breve{x}_n \, d\mathbf{P} &= \int_{\Lambda_j} x_j \, d\mathbf{P} = \int_{\Lambda_{jj}} x_j \, d\mathbf{P} + \int_{\Lambda_{jj}'} x_j \, d\mathbf{P} \\ &\leq \int_{\Lambda_{jj}} x_j \, d\mathbf{P} + \int_{\Lambda_{jj}'} x_{j+1} \, d\mathbf{P} \\ &= \int_{\Lambda_{jj}} x_j \, d\mathbf{P} + \int_{\Lambda_{jj+1}} x_{j+1} \, d\mathbf{P} + \int_{\Lambda_{jj+1}'} x_{j+1} \, d\mathbf{P} \\ &\leq \int_{\Lambda_{jj}} x_j \, d\mathbf{P} + \int_{\Lambda_{jj+1}} x_{j+1} \, d\mathbf{P} + \int_{\Lambda_{jj+1}'} x_{j+2} \, d\mathbf{P} \\ &\cdots\cdots\cdots\cdots\cdots \\ &\leq \sum_{k=j}^{N} \int_{\Lambda_{jk}} x_k \, d\mathbf{P} + \int_{\Lambda_{jN}'} x_N \, d\mathbf{P} \\ &= \int_{\bigcup_{k=j}^{N} \Lambda_{jk}} \breve{x}_{n+1} \, d\mathbf{P} + \int_{\Lambda_{jN}'} x_N \, d\mathbf{P}. \end{aligned} \tag{2.9}$$

When $N \to \infty$, the first term in the last line becomes

$$\int_{\Lambda_j} \breve{x}_{n+1}\, d\mathbf{P}.$$

Hence, in view of (2.5),

$$\begin{aligned}\int_{\Lambda_j} \breve{x}_n\, d\mathbf{P} &\leq \int_{\Lambda_j} \breve{x}_{n+1}\, d\mathbf{P} + \liminf_{N\to\infty} \int_{\Lambda_{jN}'} x_N\, d\mathbf{P}\\ &\leq \int_{\Lambda_j} \breve{x}_{n+1}\, d\mathbf{P},\end{aligned}$$

and we have thus derived the desired inequality (2.8). In particular, if the x_n process is a martingale, (2.9) becomes an equality,

$$\int_{\Lambda_j} \breve{x}_n\, d\mathbf{P} = \int_{\bigcup_{k=j}^{N} \Lambda_{jk}} \breve{x}_{n+1}\, d\mathbf{P} + \int_{\Lambda_{jN}'} x_N\, d\mathbf{P}.$$

When $N \to \infty$ in this equation, under the assumption that (2.5′) is true, we obtain (2.8), and therefore also (2.7), with equality. Hence in this case the $\breve{x}_n$ process is a martingale.

(c) *Proof of* (2.6) *under* (2.4) *and* (2.5) Suppose first that $m_1(\omega) \equiv 1$. Then, if the x_n process is a semi-martingale, and if the semi-martingale property is invariant under a given system of optional sampling, the $\breve{x}_n$ process will also be a semi-martingale, so that

$$\mathbf{E}\{\breve{x}_1\} \leq \mathbf{E}\{\breve{x}_n\}, \qquad n \geq 1,$$

and since $x_1(\omega) \equiv \breve{x}_1(\omega)$ we have obtained (2.6). In the martingale case this reasoning yields equality rather than inequality. The proof has supposed that the first random variables of the x_n and $\breve{x}_n$ processes are identical. The following argument shows that this supposition is actually no restriction. In fact, define

$$x_0(\omega) \equiv \mathbf{E}\{x_1\}, \qquad \mathscr{F}_0 = \mathscr{F}_1, \qquad m_0(\omega) \equiv 0.$$

The augmented process $\{x_n, \mathscr{F}_n, n \geq 0\}$ is a semi-martingale or martingale if and only if the given process is; the augmented system of optional sampling takes x_0 into $\breve{x}_0$ but is otherwise the same as before; (2.4) and (2.5) or (2.5′) are valid for the augmented process and optional sampling if they were valid before; finally $\mathbf{E}\{x_0\} = \mathbf{E}\{x_1\}$. Hence the truth of (2.6) for the augmented process and optional sampling, already proved, imply its truth for the original process and optional sampling. This reasoning shows that, if the x_n process is a martingale, and if (2.4) and (2.5′) are true, (2.6) becomes an equality. Finally we prove the right-hand half of (2.6′)

under (2.4) and (2.5′). The inequality is the same as the left-hand half, and there is equality throughout, if the x_n process is a martingale. In the semi-martingale case, we note that, using the semi-martingale inequality,

$$\begin{aligned}\mathbf{E}\{\breve{x}_n\} &= \int_{\{m_n(\omega)\leq N\}} \breve{x}_n\, d\mathbf{P} + \int_{\{m_n(\omega)>N\}} \breve{x}_n\, d\mathbf{P}\\ &= \sum_{j=1}^{N} \int_{\{m_n(\omega)=j\}} x_j\, d\mathbf{P} + \int_{\{m_n(\omega)>N\}} \breve{x}_n\, d\mathbf{P}\\ &\leq \int_{\{m_n(\omega)\leq N\}} x_N\, d\mathbf{P} + \int_{\{m_n(\omega)>N\}} \breve{x}_n\, d\mathbf{P}\\ &\leq \mathbf{E}\{x_N\} + \int_{\{m_n(\omega)>N\}} |x_N|\, d\mathbf{P} + \int_{\{m_n(\omega)>N\}} \breve{x}_n\, d\mathbf{P}.\end{aligned}$$

When $N\to\infty$ this inequality yields the desired inequality, in view of (2.4) and (2.5′).

(d) *Each one of the conditions* C_1–C_4 *implies* (2.4). The hypothesis of C_1 implies that $\text{L.U.B.}_{n}\, \mathbf{E}\{|x_n|\} < \infty$, and thereby, according to part (a) of this proof, that (2.4) is true. Alternatively the truth of (2.4) can also easily be derived from the fact that, according to Theorem 4.1s (ii), there is a semi-martingale $\{x_n^{+}, \mathscr{F}_n, 1\leq n\leq\infty\}$ which dominates the x_n process. (It is easy to show that under C_1 the $\breve{x}_n$'s are even uniformly integrable.) Under C_2, if m_n is bounded by the integer N_n with probability 1,

$$\mathbf{E}\{|\breve{x}_n|\}\leq \sum_{1}^{N_n} \mathbf{E}\{|x_j|\} < \infty,$$

since $\breve{x}_n$ is equal to $x_1, \cdots, x_{N_n}$ in pieces. Under C_3 define

$$y_1 = |x_1|$$

$$y_j = |x_j - x_{j-1}|, \qquad j > 1.$$

Then the y_j's are non-negative random variables, and

$$\mathbf{E}\{y_k \mid \mathscr{F}_{k-1}\}\leq K, \qquad k\leq m_j(\omega),\quad j\geq 1,$$

$$|x_n(\omega)|\leq y_1(\omega) + \cdots + y_n(\omega) \qquad n\leq m_j(\omega),\quad j\geq 1,$$

with probability 1. In the following proof we shall use only the existence of a set of y_n's with these properties, and C_3 could therefore have been stated slightly more generally. Define

$$z_n = \sum_{1}^{n} y_j, \qquad \breve{z}_n = \sum_{1}^{m_n} y_j.$$

Then the $\check{z}_n$ process dominates the x_n process, and

$$\mathbf{E}\{|\check{x}_n|\} = \sum_{k=1}^{\infty} \int_{\{m_n(\omega)=k\}} |\check{x}_n| \, d\mathbf{P} \le \mathbf{E}\{\check{z}_n\} = \sum_{k=1}^{\infty} \int_{\{m_n(\omega)=k\}} z_k \, d\mathbf{P}$$

$$= \sum_{k=1}^{\infty} \int_{\{m_n(\omega)\ge k\}} y_k \, d\mathbf{P}.$$

Since the ω set $\{m_n(\omega) \ge k\}$, as the complement of $\{m_n(\omega) < k\}$, is an ω set in the field $\mathscr{F}_{k-1}$, the last sum can be written in the form

$$\sum_{k=1}^{\infty} \int_{\{m_n(\omega)\ge k\}} \mathbf{E}\{y_k \mid \mathscr{F}_{k-1}\} \, d\mathbf{P} \le K \sum_{k=1}^{\infty} \mathbf{P}\{m_n(\omega) \ge k\} = K\mathbf{E}\{m_n\}.$$

Hence $\mathbf{E}\{\check{x}_n\}$ and $\mathbf{E}\{\check{z}_n\}$ exist, and in fact we have obtained the inequality

$$\mathbf{E}\{|\check{x}_n|\} \le \mathbf{E}\{\check{z}_n\} \le K\mathbf{E}\{m_n\}.$$

Finally, under C_4, (2.4) is true by hypothesis.

(e) *Each one of the conditions* C_1–C_4 *implies* (2.5′) Since

$$\lim_{N\to\infty} \mathbf{P}\{m_n(\omega) > N\} = 0,$$

(2.5′) is true under C_1 (uniform integrability of the x_n's). Under C_2, $\mathbf{P}\{m_n(\omega) > N\} = 0$ for sufficiently large N, so the integral in (2.5′) vanishes for sufficiently large N. Under C_3, using the notation introduced in (d),

$$\int_{\{m_n(\omega)>N\}} |x_N| \, d\mathbf{P} \le \int_{\{m_n(\omega)>N\}} \check{z}_n \, d\mathbf{P},$$

and the integral on the right goes to 0 when $N \to \infty$ because, as we have just proved, $\mathbf{E}\{\check{z}_n\} < \infty$. Under the weak form of C_4, if $j_i > N$, and if the x_n process is a semi-martingale,

$$\int_{\left\{\substack{m_n(\omega)>N \\ x_N(\omega)>0}\right\}} x_N \, dP \le \int_{\left\{\substack{m_n(\omega)>N \\ x_N(\omega)>0}\right\}} x_{N+1} \, d\mathbf{P} \le \int_{\left\{\substack{m_n(\omega)=N+1 \\ x_N(\omega)>0}\right\}} x_{N+1} \, d\mathbf{P} + \int_{\left\{\substack{m_n(\omega)>N+1 \\ x_{N+1}(\omega)>0}\right\}} x_{N+1} \, d\mathbf{P}$$

$$\cdot \; \cdot \; \cdot \; \cdot \; \cdot \; \cdot \; \cdot \; \cdot \; \cdot \; \cdot$$

$$\le \sum_{k=N+1}^{j_i} \int_{\left\{\substack{m_n(\omega)=k \\ x_k(\omega)>0}\right\}} x_k \, d\mathbf{P} + \int_{\left\{\substack{m_n(\omega)>j_i \\ x_{j_i}(\omega)>0}\right\}} x_{j_i} \, d\mathbf{P}$$

$$\le \sum_{k=N+1}^{j_i} \int_{\{m_n(\omega)=k\}} |x_k| \, d\mathbf{P} + \sum_{k=j_i+1}^{\infty} \int_{\{m_n(\omega)=k\}} (|x_k| + z) \, d\mathbf{P}$$

$$\le \int_{\{m_n(\omega)>N\}} (|\check{x}_n| + z) \, d\mathbf{P}.$$

Since the last term goes to 0 when $N \to \infty$, (2.5) is true. If the x_n process is a martingale, then the validity of C_4 for this process implies that of the weak form of C_4 for the $|x_n|$ process, which is a semi-martingale. Hence, by what we have just proved, (2.5) is true for the $|x_n|$ process, that is, (2.5′) is true for the x_n process.

The following example, which has interesting implications in the study of random walks, illustrates the possibility of applying Theorem 2.2 even when there is only a single m_j.

Example 1 Let $\{x_n, n \geq 1\}$ be a real martingale, and suppose that

$$\mathbf{P}\{\underset{n\geq 1}{\text{L.U.B.}}\, x_n(\omega) > \mathbf{E}\{x_1\}\} = 1.$$

Define $m_1(\omega)$ as the first subscript j for which $x_j(\omega) > \mathbf{E}\{x_1\}$. Then, if Theorem 2.2 is applicable, the $\breve{x}_n$ process, where n only takes on the value 1, and $\breve{x}_1 = x_{m_1}$, is a martingale, with

$$\mathbf{E}\{x_1\} = \mathbf{E}\{\breve{x}_1\}.$$

In the present case the martingale property of the $\breve{x}_n$ process is vacuously satisfied but the preceding equality between expectations is obviously false. Hence the hypotheses of Theorem 2.2 cannot be fulfilled, and their failure gives rise to various theorems. For example (see condition C_3), suppose that $y_1, y_2, \cdots$ are mutually independent random variables, with a common distribution function, and that

$$\mathbf{E}\{y_n\} = 0, \qquad \mathbf{E}\{|\, y_n \,|\} > 0,$$

$$x_n = \sum_1^n y_j.$$

Then the x_n process is a martingale, and

$$\mathbf{E}\{x_n\} = 0, \qquad \mathbf{E}\{|x_{n+1} - x_n| \mid x_1, \cdots, x_n\} = \mathbf{E}\{|y_1|\},$$

with probability 1. Thus, if we take $\mathscr{F}_n$ as the field of the ω sets which are determined by conditions on $x_1, \cdots, x_n$, that is, by conditions on $y_1, \cdots, y_n$, the second half of condition C_3 is satisfied, with $K = \mathbf{E}\{|y_1|\}$, so that the first half cannot be satisfied. It follows that, under the stated hypotheses on the y_j's, the condition

$$\mathbf{P}\{\underset{n\geq 1}{\text{L.U.B.}}\, x_n(\omega) > 0\} = 1,$$

which is known always to be satisfied (see Chung and Fuchs [1, 1951]), implies that

$$\mathbf{E}\{m_1\} = \infty.$$

In other words, one is certain to reach the positive half of the coordinate axis in a random walk described by the x_n process, but the expected time to reach it is infinite.

In order to discuss another type of transformation which leaves the martingale and semi-martingale properties invariant we generalize the system theorem of III §5, following Halmos. Consider a fair game, that is, from our point of view, a martingale $\{x_n, \mathscr{F}_n, n \geq 1\}$, and define

$$y_1 = x_1$$

$$y_n = x_n - x_{n-1}, \qquad n > 1,$$

so that y_n is the gain on a play, and the fairness of the game, that is, the martingale equality, is expressed in terms of the y_n's by the condition that

$$\mathbf{E}\{y_{n+1} \mid \mathscr{F}_n\} = 0, \qquad n \geq 1,$$

with probability 1. The Borel field $\mathscr{F}_n$ represents as usual the influence of the past to time n. If a game is fair, the gambler should find that it still seems fair if he decides to skip some games, basing his decision on playing or not playing in terms of the past. This means that, if $\varepsilon_n(\omega) = 1$ or 0 according as he plays or does not play the game with gain $y_n(\omega)$, the condition $\varepsilon_n(\omega) = 1$ must be based on the past before time n, that is,

$$\{\varepsilon_n(\omega) = 1\} \in \mathscr{F}_{n-1}.$$

Let $m_n(\omega)$ be the nth integer j with $\varepsilon_j(\omega) = 1$, so that the gambler now has gains $y_{m_1}, y_{m_2}, \cdots$ instead of $y_1, y_2, \cdots$. Then we expect that the process

$$\{\sum_1^n y_{m_j}, n \geq 1\}$$

is still a martingale. To avoid confusion we restate the hypotheses, omitting references to gambling and expressing everything in terms of the y_j process rather than the x_j process, to facilitate comparison with the system theorem of III §5. We shall include semi-martingales along with martingales, that is, we shall allow favorable games as well as fair games.

Let $\{y_n, n \geq 1\}$ be a stochastic process, and let $\mathscr{F}_1 \subset \mathscr{F}_2 \subset \cdots$ be Borel fields of measurable ω sets with the following properties:

(i) $$\mathbf{E}\{|y_n|\} < \infty, \qquad n \geq 1.$$

(ii) *y_n is either measurable with respect to $\mathscr{F}_n$ or is equal for almost all ω to a function which is. For all $n \geq 1$ either*

(iii) $$\mathbf{E}\{y_{n+1} \mid \mathscr{F}_n\} = 0$$

with probability 1, *or the process is real and*

(iiis) $$\mathbf{E}\{y_{n+1} \mid \mathscr{F}_n\} \geq 0$$

with probability 1.

We shall write $\{y_n, \mathscr{F}_n, n \geq 1\}$ to stress the $\mathscr{F}_n$'s. Let $m_1, m_2, \cdots$ be random variables taking on integral values, and having the following properties:

(iv) $$1 < m_1 < m_2 < \cdots < \infty,$$

(v) $$\{m_j(\omega) = k\} \in \mathscr{F}_{k-1}, \qquad k \geq j,$$

neglecting ω sets of probability 0. Define $\breve{y}_j$ by

$$\breve{y}_j = y_{m_j}, \qquad j \geq 1.$$

The $\breve{y}_j$ process will be said to be obtained from the y_j process by *optional skipping*. If the y_n's are mutually independent, with a common distribution function, and if $\mathscr{F}_n$ is the Borel field of the ω sets determined by conditions on $y_1, \cdots, y_n$, the $\breve{y}_n$'s must also be mutually independent, with the same common distribution function, according to III, Theorem 5.2. The fact that the martingale and semi-martingale properties are also preserved is the content of the following theorem.

THEOREM 2.3 *Suppose that a process $\{y_n, \mathscr{F}_n, n \geq 1\}$ with the properties* (i), (ii), (iiis) *of the preceding paragraph is transformed into the process $\{\breve{y}_n, n \geq 1\}$ by optional skipping, and suppose that*

(2.10) $$\mathbf{E}\{|\breve{y}_n|\} < \infty, \qquad n \geq 1.$$

Then the $\breve{y}_n$ process satisfies

(2.11) $$\mathbf{E}\{\breve{y}_{n+1} \mid \breve{y}_1, \cdots, \breve{y}_n\} \geq 0, \qquad n \geq 1,$$

with probability 1. *If* (iiis) *is replaced by* (iii), *there is equality in* (2.11) *with probability* 1. *The condition* (2.10) *is satisfied if either of the following conditions is satisfied.*

C_1 *Each m_j defining the skipping is a bounded random variable, with probability* 1.

C_2 *There is a number K such that, for each j,*

$$\mathbf{E}\{|y_{n+1}| \mid \mathscr{F}_n\} \leq K, \qquad n \leq m_j(\omega),$$

with probability 1.

To prove (2.11) we prove that, if M is any ω set determined by conditions on $\breve{y}_1, \cdots, \breve{y}_n$, then

$$\int_{\mathrm{M}} \breve{y}_{n+1}\, d\mathbf{P} \geq 0,$$

and in fact we shall prove the stronger inequality that, if

$$\mathrm{M}_j = \mathrm{M}\{m_{n+1}(\omega) = j\}, \qquad j \geq n + 1$$

then

$$\int_{M_j} \breve{y}_{n+1}\, d\mathbf{P} \geq 0, \qquad j \geq n+1,$$

that is,

$$\int_{M_j} y_j\, d\mathbf{P} \geq 0, \qquad j \geq n+1.$$

This inequality follows from (iiis), since $M_j \in \mathscr{F}_{j-1}$ if $j \geq 2$. The condition (2.10) is certainly fulfilled if each m_j is bounded with probability 1, because, if $m_n \leq N$ with probability 1, it follows that

$$\mathbf{E}\{|\breve{y}_n|\} \leq \sum_1^N \mathbf{E}\{|y_k|\} < \infty.$$

The condition (2.10) is also fulfilled if Condition C_2 of the theorem is fulfilled, because in that case

$$\mathbf{E}\{|\breve{y}_n|\} = \sum_{j=1}^{\infty} \int_{\{m_n(\omega)=j\}} |y_j|\, d\mathbf{P} \leq K \sum_{j=1}^{\infty} \mathbf{P}\{m_n(\omega) = j\} = K.$$

More generally, if $\alpha \geq 0$, we could have supposed only that

$$\mathbf{E}\{m_j^{\alpha}\} < \infty, \qquad \mathbf{E}\{|y_{n+1}| \mid \mathscr{F}_n\} \leq Kn^{\alpha}, \qquad n \leq m_j(\omega),$$

with probability 1. This condition reduces to C_2 if $\alpha = 0$.

3. Fundamental inequalities

In the following, if we write $\mathbf{E}\{|x|\} \leq \mathbf{E}\{|y|\}$, this inequality is to be significant, making the obvious conventions, even if one or both sides are $+\infty$.

THEOREM 3.1 *Let $\{x_t, t \in T\}$ be a semi-martingale.*

(i) $\mathbf{E}\{x_t\}$ *is monotone non-decreasing, and is identically constant if and only if the x_t process is a martingale.*

(ii) *If $t_0, t_1 \in T$, with $t_0 \leq t_1$, then*

$$\mathbf{E}\{|x_t|\} \leq -\mathbf{E}\{x_{t_0}\} + 2\mathbf{E}\{|x_{t_1}|\}, \qquad t_0 \leq t \leq t_1.$$

(iii) *If the x_t's are non-negative, and if $t_1 \in T$, the x_t's with $t \leq t_1$ are uniformly integrable.*

(iv) *Suppose that $s_1 \geq s_2 \geq \cdots$, $s_n \in T$. Then the x_{s_n}'s are uniformly integrable if and only if*

$$\lim_{n\to\infty} \mathbf{E}\{x_{s_n}\} > -\infty.$$

Proof of (i) If $s < t$, then

$$x_s \leq \mathbf{E}\{x_t \mid x_s\}$$

with probability 1. Taking expectations of the two sides of this inequality, we find that

$$\mathbf{E}\{x_s\} \leq \mathbf{E}\{x_t\},$$

and there is equality if and only if there is equality with probability 1 in the preceding inequality, that is, there is equality for all pairs s, t if and only if the x_t process is a martingale.

Proof of (ii) If $t_0 \leq t \leq t_1$, then, using the semi-martingale inequality and (i),

$$\begin{aligned} \mathbf{E}\{|x_t|\} &= -\mathbf{E}\{x_t\} + 2 \int_{\{x_t(\omega)>0\}} x_t \, d\mathbf{P} \\ &\leq -\mathbf{E}\{x_{t_0}\} + 2 \int_{\{x_t(\omega)>0\}} x_{t_1} \, d\mathbf{P} \\ &\leq -\mathbf{E}\{x_{t_0}\} + 2\mathbf{E}\{|x_{t_1}|\}. \end{aligned}$$

Proof of (iii) If $t \leq t_1$,

$$\int_{\{x_t(\omega)\geq\lambda\}} x_t \, d\mathbf{P} \leq \int_{\{x_t(\omega)\geq\lambda\}} x_{t_1} \, d\mathbf{P}.$$

Since the integrands are non-negative, uniform integrability (for $t \leq t_1$) is the uniform (for $t \leq t_1$) convergence of the left side of this inequality to 0 when $\lambda \to \infty$. It is therefore sufficient to prove that

$$\lim_{\lambda\to\infty} \mathbf{P}\{x_t(\omega) \geq \lambda\} = 0$$

uniformly for $t \leq t_1$. This is implied by the inequality

$$\lambda \mathbf{P}\{x_t(\omega) \geq \lambda\} \leq \mathbf{E}\{x_t\} \leq \mathbf{E}\{x_{t_1}\}.$$

Proof of (iv) If the x_{s_n}'s are uniformly integrable, $\text{L.U.B.}_n \, \mathbf{E}\{|x_{s_n}|\} < \infty$; this implication has nothing to do with martingale theory. Conversely, if $\lim_{n\to\infty} \mathbf{E}\{x_{s_n}\} > -\infty$, we prove that

$$\lim_{\lambda\to\infty} \int_{\{|x_{s_n}(\omega)|>\lambda\}} |x_{s_n}| \, d\mathbf{P} = 0 \tag{3.1}$$

uniformly in n. Let

$$K = \lim_{n\to\infty} \mathbf{E}\{x_{s_n}\}.$$

Then, using (ii),

$$\lambda \mathbf{P}\{|x_{s_n}(\omega)| > \lambda\} \leq \mathbf{E}\{|x_{s_n}|\} \leq -K + 2\mathbf{E}\{|x_{s_1}|\},$$

so the measure of the domain of integration in (3.1) goes to 0 uniformly in n when $\lambda \to \infty$. Now

$$(3.2) \quad \int_{\{|x_{s_n}(\omega)|>\lambda\}} |x_{s_n}| \, d\mathbf{P} = \int_{\{x_{s_n}(\omega)>\lambda\}} x_{s_n} \, d\mathbf{P} + \int_{\{x_{s_n}(\omega)\geq-\lambda\}} x_{s_n} \, d\mathbf{P} - \mathbf{E}\{x_{s_n}\}$$

$$\leq \int_{\{x_{s_n}(\omega)>\lambda\}} x_{s_N} \, d\mathbf{P} + \int_{\{x_{s_n}(\omega)\geq-\lambda\}} x_{s_N} \, d\mathbf{P} - K \qquad n \geq N$$

$$= \int_{\{|x_{s_n}(\omega)|>\lambda\}} |x_{s_N}| \, d\mathbf{P} + \mathbf{E}\{x_{s_N}\} - K.$$

If $\varepsilon > 0$, we choose N so large that

$$\mathbf{E}\{x_{s_N}\} - K < \frac{\varepsilon}{2},$$

and then choose λ_1 so large that $\mathbf{P}\{|x_{s_n}(\omega)| > \lambda_1\}$ is (for all n) so small that

$$\int_{\{|x_{s_n}(\omega)|>\lambda_1\}} |x_{s_N}| \, d\mathbf{P} < \frac{\varepsilon}{2},$$

and finally choose λ_2 so large that $\lambda_2 \geq \lambda_1$ and that

$$(3.3) \quad \int_{\{|x_{s_n}(\omega)|>\lambda_2\}} |x_{s_n}| \, d\mathbf{P} < \varepsilon, \qquad n < N.$$

With this choice of λ_2, (3.2) and (3.3) imply

$$\int_{\{|x_{s_n}(\omega)|>\lambda_2\}} |x_{s_n}| \, d\mathbf{P} < \varepsilon, \qquad n \geq 1,$$

so that (3.1) is true uniformly in n, as was to be proved.

Examples will be given which will exhibit the fact that (iv) is no longer true if the sequence $\{s_n\}$ is monotone increasing instead of monotone decreasing. In fact, a semi-martingale $\{x_n, 1 \leq n \leq \infty\}$ will be exhibited below with

$$x_n \leq 0, \qquad \mathbf{E}\{x_n\} = -1, \qquad n < \infty, \qquad \lim_{n\to\infty} x_n = 0 = x_\infty$$

(the limit holding with probability 1). The x_n's cannot be uniformly integrable, since $\mathbf{E}\{x_n\}$ does not converge to $\mathbf{E}\{x_\infty\}$.

Many fairly obvious applications of Theorem 3.1 can be obtained by using Theorem 1.1. For example, if $\{x_t, t \in T\}$ is a real or complex martingale, and if, for some t_0 and $\alpha \geq 1$, $\mathbf{E}\{|x_{t_0}|^\alpha\} < \infty$, then the $|x_t|^\alpha$ process for $t \leq t_0$ is a semi-martingale, $\mathbf{E}\{|x_t|^\alpha\}$ is monotone non-decreasing, and the $|x_t|^\alpha$'s are uniformly integrable for $t \leq t_0$. This result is always significant if $\alpha = 1$, but, if $\alpha > 1$, $\mathbf{E}\{|x_t|^\alpha\}$ may never be finite.

According to Theorem 3.1 (i), if the x_t process is a semi-martingale when the x_t's are ordered in the direction of decreasing t as well as of increasing t, it is a real martingale in both orders. We now show that any process which is a martingale in both orders has the property that for every pair of parameter values s, t, $x_s = x_t$ with probability 1. To prove this it is sufficient to prove that if

$$\int_{\Lambda} (y - x)\, d\mathbf{P} = 0$$

for every set $\Lambda = \{y(\omega) \in B\}$ and every set $\Lambda = \{x(\omega) \in B\}$, where B is a Borel set, then $x = y$ with probability 1. We may assume that x and y are real. Then, for every real c,

$$\begin{aligned}\int_{\left\{\substack{y(\omega) > c \\ x(\omega) < c}\right\}} (y - x)\, d\mathbf{P} &= \int_{\{y(\omega) > c\}} (y - x)\, d\mathbf{P} - \int_{\left\{\substack{y(\omega) > c \\ x(\omega) \geq c}\right\}} (y - x)\, d\mathbf{P} \\ &= -\int_{\left\{\substack{y(\omega) > c \\ x(\omega) \geq c}\right\}} (y - x)\, d\mathbf{P} \\ &= -\int_{\{x(\omega) \geq c\}} (y - x)\, d\mathbf{P} + \int_{\left\{\substack{y(\omega) \leq c \\ x(\omega) \geq c}\right\}} (y - x)\, d\mathbf{P} \\ &= \int_{\left\{\substack{y(\omega) \leq c \\ x(\omega) \geq c}\right\}} (y - x) d\mathbf{P}.\end{aligned}$$

Since the first integrand is > 0, and the last is ≤ 0, the first and last integrals must vanish, so that

$$\mathbf{P}\{y(\omega) > c > x(\omega)\} = \mathbf{P}\{y(\omega) < c < x(\omega)\} = 0.$$

Then, restricting c to be rational,

$$\mathbf{P}\{y(\omega) \neq x(\omega)\} \leq \sum_c \mathbf{P}\{y(\omega) > c > x(\omega)\} + \sum_c \mathbf{P}\{y(\omega) < c < x(\omega)\} = 0,$$

as was to be proved.

THEOREM 3.2 *Let $\{x_j,\ 1 \leq j \leq n\}$ be a semi-martingale, and let λ be any real number. Then*

$$\lambda \mathbf{P}\{\operatorname*{Max}_j x_j(\omega) \geq \lambda\} \leq \int_{\{\operatorname*{Max}_j x_j(\omega) \geq \lambda\}} x_n\, d\mathbf{P} \leq \mathbf{E}\{|x_n|\} \tag{3.4$'$}$$

$$\begin{aligned}\lambda \mathbf{P}\{\operatorname*{Min}_j x_j(\omega) \leq \lambda\} &\geq \int_{\{\operatorname*{Min}_j x_j(\omega) \leq \lambda\}} x_n\, d\mathbf{P} - \mathbf{E}\{x_n - x_1\} \\ &\geq \mathbf{E}\{x_1\} - \mathbf{E}\{|x_n|\}.\end{aligned} \tag{3.4$''$}$$

We prove the two parts of this theorem in different ways, to illustrate the possibilities. Inequality (3.4′) is proved by direct calculation as follows. Let $\Lambda = \{\operatorname{Max}_j x_j(\omega) \geq \lambda\}$, and define the ω set Λ_k as the ω set for which $x_k(\omega)$ is the first $x_j(\omega)$ with $x_j(\omega) \geq \lambda$:

$$\Lambda_1 = \{x_1(\omega) \geq \lambda\},$$

$$\Lambda_k = \{x_j(\omega) < \lambda,\ 1 \leq j < k;\ x_k(\omega) \geq \lambda\}, \qquad k > 1.$$

Then Λ_k is determined by conditions on the x_j's for $j \leq k$, the Λ_k's are disjunct, and $\bigcup_k \Lambda_k = \Lambda$. Using the semi-martingale inequality and the fact that $x_k(\omega) \geq \lambda$ on Λ_k, we find that

$$\int_\Lambda x_n \, d\mathbf{P} = \sum_k \int_{\Lambda_k} x_n \, d\mathbf{P} \geq \sum_k \int_{\Lambda_k} x_k \, d\mathbf{P}$$

$$\geq \lambda \sum_k \mathbf{P}\{\Lambda_k\} = \lambda \mathbf{P}\{\Lambda\},$$

which proves (3.4′).

We prove (3.4″) as an application of a game theorem. Let $\mathrm{M} = \{\operatorname{Min}_j x_j(\omega) \leq \lambda\}$, let $m(\omega)$ be the first value of j for which $x_j(\omega) \leq \lambda$, if $\omega \in \mathrm{M}$, and let $m(\omega) = n$ otherwise. Then the condition $m(\omega) = k$ is a condition on $x_1, \cdots, x_k$. We now apply optional stopping, determined by the random variable m just defined, as described in §2. Then, in the notation of Theorem 2.1, $\breve{x}_n = x_m$, and according to that theorem

$$\mathbf{E}\{x_1\} \leq \mathbf{E}\{\breve{x}_n\},$$

so that

$$\mathbf{E}\{x_1\} \leq \mathbf{E}\{\breve{x}_n\} = \int_{\mathrm{M}} \breve{x}_n \, d\mathbf{P} + \int_{\Omega - \mathrm{M}} x_n \, d\mathbf{P}$$

$$\leq \lambda \mathbf{P}\{\mathrm{M}\} + \mathbf{E}\{x_n\} - \int_{\mathrm{M}} x_n \, d\mathbf{P},$$

which implies (3.4″).

This theorem contains III, Theorem 2.1 (essentially Kolmogorov's generalization of the Chebyshev inequality), because, if $\{x_j,\ 1 \leq j \leq n\}$ is a martingale, and if $\mathbf{E}\{|x_n|^2\} < \infty$, then $\{|x_j|^2,\ 1 \leq j \leq n\}$ is a semi-martingale, so that, if $\varepsilon > 0$ we have, from (3.4′),

$$\mathbf{P}\{\operatorname{Max}_j |x_j(\omega)|^2 \geq \varepsilon^2\} \leq \frac{1}{\varepsilon^2} \mathbf{E}\{|x_n|^2\}.$$

This inequality is precisely III (2.1′).

Let $\xi_1, \cdots, \xi_n$ be any real numbers, and let r_1, r_2 be real numbers with $r_1 < r_2$. The *number of upcrossings of the interval* $[r_1, r_2]$ by $\xi_1, \cdots, \xi_n$ is defined as the number of times the sequence $\xi_1, \cdots, \xi_n$ passes from below r_1 to above r_2. More precisely, let ξ_{ν_1} be the first ξ_i (if any) for which $\xi_i \leq r_1$, and in general let ξ_{ν_j} be the first ξ_i (if any) after $\xi_{\nu_{j-1}}$ for which

$$\xi_i \geq r_2 \qquad (j \text{ even})$$

$$\xi_i \leq r_1 \qquad (j \text{ odd}),$$

so that

$$\xi_{\nu_1} \leq r_1, \qquad \xi_{\nu_2} \geq r_2, \qquad \xi_{\nu_3} \leq r_1, \cdots.$$

Then the number of upcrossings is β, where 2β is the largest even integer j for which ξ_{ν_j} is defined, and $\beta = 0$ if ξ_{ν_2} is not defined.

THEOREM 3.3 *Let $\{x_j, 1 \leq j \leq n\}$ be a semi-martingale, and let $\beta(\omega)$ be the number of upcrossings of $[r_1, r_2]$ by a sample sequence $[x_1(\omega), \cdots, x_n(\omega)]$. Then*

$$\mathbf{E}\{\beta\} \leq \frac{1}{r_2 - r_1} \int_{\{x_n(\omega) \geq r_1\}} (x_n - r_1)\, d\mathbf{P} \leq \frac{\mathbf{E}\{|x_n|\} + |r_1|}{r_2 - r_1}. \tag{3.5}$$

To prove the theorem assume first that the x_j's are non-negative, and that $r_1 = 0$. Define the ω functions $\nu_1, \cdots, \nu_n$ in terms of $\xi_1 = x_1(\omega), \cdots, \xi_n = x_n(\omega)$ as described above, defining $\nu_j(\omega) = n + 1$ if the above definitions leave $\nu_j(\omega)$ undefined. The ν_j's and β are now random variables. Define the random variables $u_2, \cdots, u_n, x$ by

$$\begin{aligned} u_j(\omega) &= 1, \qquad j \leq \nu_1(\omega) \\ &= 1, \qquad \nu_i(\omega) < j \leq \nu_{i+1}(\omega) \qquad (i \text{ even}) \\ &= 0, \qquad \nu_i(\omega) < j \leq \nu_{i+1}(\omega) \qquad (i \text{ odd}) \end{aligned}$$

$$x = x_1 + \sum_2^n u_j(x_j - x_{j-1}).$$

Then

$$\{u_j(\omega) = 1\} = \bigcup_i [\{\nu_i(\omega) < j\} - \{\nu_{i+1}(\omega) < j\}] \qquad (i \text{ even}),$$

so that the ω set on the left is determined by conditions on $x_1, \cdots, x_{j-1}$. Hence, using the semi-martingale property,

$$\int_{\{u_j(\omega)=1\}} (x_j - x_{j-1})\, d\mathbf{P} \geq 0.$$

It follows that

$$\mathbf{E}\{x\} = \mathbf{E}\{x_1\} + \sum_{j=2}^n \int_{\{u_j(\omega)=1\}} (x_j - x_{j-1})\, d\mathbf{P} \geq 0. \tag{3.6}$$

On the other hand,

$$x(\omega) \leq x_n(\omega) - r_2\beta(\omega),$$

so that

$$0 \leq \mathbf{E}\{x\} \leq \mathbf{E}\{x_n\} - r_2\mathbf{E}\{\beta\},$$

as was to be proved. In the general case, without the restrictions that the x_j's be non-negative, and that $r_1 = 0$, define

$$x_j'(\omega) = \text{Max}\,[x_j(\omega) - r_1, 0].$$

Then the x_j' process is a semi-martingale, by Theorem 1.1 (i). The number of upcrossings of $[r_1, r_2]$ by the $x_j(\omega)$'s is the number of upcrossings of $[0, r_2 - r_1]$ by the x_j''s. Hence we can apply the special case of the theorem already derived to obtain

$$\mathbf{E}\{\beta\} \leq \frac{\mathbf{E}\{x_n'\}}{r_2 - r_1} = \frac{1}{r_2 - r_1} \int_{\{x_n(\omega) \geq r_1\}} (x_n - r_1)\, d\mathbf{P},$$

which was to be proved.

The following theorem sharpens Theorem 3.2. The important point about both is that the number of random variables n in the semi-martingale is not involved, so that the theorem is applicable to martingales with infinitely many random variables, as long as there is a last one.

THEOREM 3.4 *If $\{x_j, 1 \leq j \leq n\}$ is a semi-martingale, and if the x_j's are non-negative, then*

$$\begin{aligned} (3.7) \qquad \mathbf{E}\{\text{Max}_j\, x_j^\alpha\} &\leq \frac{e}{e-1} + \frac{e}{e-1}\mathbf{E}\{x_n \log^+ x_n\}, && \alpha = 1, \\ &\leq \left(\frac{\alpha}{\alpha - 1}\right)^\alpha \mathbf{E}\{x_n^\alpha\}, && \alpha > 1. \end{aligned}$$

Note that the coefficient in the second inequality is a decreasing function of α, decreasing to e when $\alpha \to \infty$. In view of Theorem 3.2, (3.7) follows at once from the following theorem, which is stated separately for later reference, and because it has independent interest.

THEOREM 3.4′ *If x and y are non-negative random variables satisfying the inequality*

$$(3.8) \qquad \mathbf{P}\{y(\omega) \geq \lambda\} \leq \frac{1}{\lambda} \int_{\{y(\omega) \geq \lambda\}} x\, d\mathbf{P}$$

for all $\lambda > 0$, then

$$\begin{aligned} (3.7') \qquad \mathbf{E}\{y^\alpha\} &\leq \frac{e}{e-1} + \frac{e}{e-1}\mathbf{E}\{x \log^+ x\}, && \alpha = 1, \\ &\leq \left(\frac{\alpha}{\alpha - 1}\right)^\alpha \mathbf{E}\{x^\alpha\}, && \alpha > 1. \end{aligned}$$

To prove this theorem, let Ψ be a monotone non-decreasing function of λ, $\lambda > 0$, with $\Psi(0) = 0$. Then

$$\begin{aligned}(3.9)\qquad \mathbf{E}\{\Psi(y)\} &= -\int_0^\infty \Psi(\lambda)\, d\mathbf{P}\{y(\omega) \geq \lambda\} \\ &\leq \int_0^\infty \mathbf{P}\{y(\omega) \geq \lambda\}\, d\Psi(\lambda) \\ &\leq \int_0^\infty \frac{d\Psi(\lambda)}{\lambda} \int_{\{y(\omega)\geq\lambda\}} x\, d\mathbf{P} \\ &= \int_\Omega x\, d\mathbf{P} \int_0^{y(\omega)} \frac{d\Psi(\lambda)}{\lambda}.\end{aligned}$$

To prove the first inequality of (3.7′), define Ψ by

$$\begin{aligned}\Psi(\lambda) &= \lambda, \qquad \lambda \geq 1, \\ &= 0, \qquad \lambda < 1.\end{aligned}$$

Then, using (3.9),

$$\mathbf{E}\{y - 1\} \leq \mathbf{E}\{\Psi(y)\} \leq \int_{\{y(\omega)\geq 1\}} x \log y\, d\mathbf{P}.$$

The first inequality of (3.7′) now follows at once, in view of the inequality

$$a \log b \leq a \log^+ a + \frac{b}{e} \qquad (a \geq 0, b > 0).$$

To prove the second inequality of (3.7′), define Ψ by

$$\Psi(\lambda) = \lambda^\alpha.$$

Then, using (3.9),

$$\mathbf{E}\{y^\alpha\} \leq \frac{\alpha}{\alpha - 1}\, \mathbf{E}\{xy^{\alpha-1}\},$$

so that, applying Hölder's inequality,

$$\mathbf{E}\{y^\alpha\} \leq \frac{\alpha}{\alpha - 1}\, \mathbf{E}^{\frac{1}{\alpha}}\{x^\alpha\}\, \mathbf{E}^{1-(1/\alpha)}\{y^\alpha\},$$

which yields the desired inequality in (3.7′).

4. Convergence theorems

We shall prove a succession of convergence theorems for martingales and semi-martingales in this section. The theorems are not much weaker for semi-martingales than for martingales, but to clarify the discussion the martingale theorems and semi-martingale theorems will be stated separately, at the expense of some duplication.

If $x_1, x_2, \cdots$ are the random variables of a martingale, we have seen that, if $\mathbf{E}\{|x_n|^2\} < \infty$, we can write the x_n's in the form

$$x_n = \sum_{j=1}^{n} y_j,$$

where the y_j's are mutually orthogonal. Then

$$\mathbf{E}\{|x_n|^2\} = \sum_{j=1}^{n} \mathbf{E}\{|y_j|^2\},$$

and this exhibits the fact that $\mathbf{E}\{|x_n|^2\}$ is monotone non-decreasing in n, as it should be according to Theorem 3.1 (i). According to IV, Theorem 4.1, $\operatorname*{l.i.m.}_{n\to\infty} x_n$ exists if and only if $\sum_j \mathbf{E}\{|y_j|^2\} < \infty$, that is, if and only if $\lim_{n\to\infty} \mathbf{E}\{|x_n|^2\} < \infty$. The following theorem sharpens this rather superficial result.

THEOREM 4.1 *Let $\{x_n, \mathscr{F}_n, n \geq 1\}$ be a martingale. Then, by Theorem* 3.1 (i),

$$\mathbf{E}\{|x_1|\} \leq \mathbf{E}\{|x_2|\} \leq \cdots.$$

(i) *If* $\lim_{n\to\infty} \mathbf{E}\{|x_n|\} = K < \infty$, *then* $\lim_{n\to\infty} x_n = x_\infty$ *exists with probability* 1 *and* $\mathbf{E}\{|x_\infty|\} \leq K$. *In particular,* $K < \infty$ *if the x_n's are all real and* ≥ 0 *or all real and* ≤ 0.

(ii) *The following conditions are equivalent:*

(a) $K < \infty$, *and the random variables* $x_1, x_2, \cdots, x_\infty$ *constitute a martingale.*

(b) *The random variables* $x_1, x_2, \cdots$ *are uniformly integrable.*

(c) $K < \infty$ *and* $\mathbf{E}\{|x_\infty|\} = K$.

(d) $K < \infty$ *and* $\lim_{n\to\infty} \mathbf{E}\{|x_\infty - x_n|\} = 0$.

If these conditions are satisfied, and if $\mathscr{F}_\infty$ is the smallest Borel field of ω sets with $\mathscr{F}_\infty \supset \bigcup_n \mathscr{F}_n$, then $\{x_n, \mathscr{F}_n, 1 \leq n \leq \infty\}$ is a martingale.

(iii) *If, for some* $\alpha > 1$, $\lim_{n\to\infty} \mathbf{E}\{|x_n|^\alpha\} < \infty$, *then the conditions of* (ii) *are satisfied,* $\mathbf{E}\{|x_\infty|^\alpha\} < \infty$, *and*

$$\lim_{n\to\infty} \mathbf{E}\{|x_\infty - x_n|^\alpha\} = 0.$$

Conversely, if the conditions of (ii) *are satisfied and if* $\mathbf{E}\{|x_\infty|^\alpha\} < \infty$ *for some* $\alpha > 1$, *then*

$$\mathbf{E}\{|x_n|^\alpha\} \leq \mathbf{E}\{|x_\infty|^\alpha\}, \qquad n \geq 1.$$

(iv) *If the* x_n*'s are real, and if*

$$\mathbf{E}\{\underset{n \geq 0}{\text{L.U.B.}}\,(x_{n+1} - x_n)\} < \infty \qquad (x_0 = 0), \tag{4.1}$$

then $\lim_{n\to\infty} x_n(\omega)$ *exists and is finite for almost all* ω *for which* $\limsup_{n\to\infty} x_n(\omega) < \infty$.

(v) *If*

$$\mathbf{E}\{\underset{n \geq 0}{\text{L.U.B.}}\,|x_{n+1} - x_n|^2\} < \infty \qquad (x_0 = 0), \tag{4.2}$$

then $\lim_{n\to\infty} x_n(\omega)$ *exists and is finite for almost all* ω *for which* $\sum_1^\infty \mathbf{E}\{|x_{n+1} - x_n|^2 \mid \mathscr{F}_n\} < \infty$, *and conversely.*

This is the strict sense version of IV, Theorem 7.2.

Proof of (i) It is sufficient to prove this part in the real case and we accordingly assume that the x_n's are real. Suppose that $K < \infty$ and let x_*, x^* be respectively the inferior and superior limits of the sequence $\{x_n\}$. Then

$$\{x^*(\omega) - x_*(\omega) \neq 0\} = \bigcup_{r_1, r_2} \{x^*(\omega) > r_2 > r_1 > x_*(\omega)\} \; (r_i \text{ rational}). \tag{4.3}$$

Fix r_1 and $r_2 > r_1$, and let $\beta_n(\omega)$ be the number of upcrossings of the interval $[r_1, r_2]$ by $x_1(\omega), \cdots, x_n(\omega)$. Then $\beta_n \to \infty$ monotonely whenever $x^*(\omega) > r_2 > r_1 > x_*(\omega)$, although according to Theorem 3.3

$$\mathbf{E}\{\beta_n\} \leq \frac{K + |r_1|}{r_2 - r_1}.$$

These two facts imply that each summand in (4.3) has probability 0, so that $x^* = x_*$ with probability 1, that is, $\lim_{n\to\infty} x_n = x_\infty$ exists as a finite or infinite limit with probability 1. By Fatou's lemma, $|x_\infty| < \infty$ with probability 1, and

$$\mathbf{E}\{|x_\infty|\} \leq \lim_{n\to\infty} \mathbf{E}\{|x_n|\} = K.$$

Moreover, there is equality if and only if $x_1, x_2, \cdots$ are uniformly integrable. In particular, if the x_n's are all ≥ 0 or all ≤ 0,

$$\mathbf{E}\{|x_n|\} \equiv \mathbf{E}\{x_n\} \equiv \text{const.} \qquad \text{or} \qquad \mathbf{E}\{|x_n|\} \equiv -\mathbf{E}\{x_n\} \equiv \text{const.},$$

as the case may be, using Theorem 3.1 (i), and hence $K < \infty$. More generally, if we write $x_n = x_n^+ - x_n^-$, where x_n^+ and x_n^- are non-negative, then K is finite if $\mathbf{E}\{x_n^+\}$ or $\mathbf{E}\{x_n^-\}$ is bounded in n. For example, the first assertion follows from the equality

$$\begin{aligned}\mathbf{E}\{|x_n|\} &= -\mathbf{E}\{x_n\} + 2\mathbf{E}\{x_n^+\}\\ &= -\mathbf{E}\{x_1\} + 2\mathbf{E}\{x_n^+\}.\end{aligned}$$

Thus K is finite if the x_n's are uniformly bounded from above or below by a random variable whose expectation exists.

Proof of (ii) If (b) is true, $K < \infty$. Thus statements (a)–(d) of (ii) all either imply or suppose that $K < \infty$, so that x_∞ is defined in each case. Conditions (c) and (d) are simply necessary and sufficient conditions for uniform integrability of an almost everywhere convergent sequence of functions, and have nothing specific to do with martingales. Condition (a) implies (b), uniform integrability, by Theorem 3.1 (iii) applied to the $|x_n|$ process. Conversely, if there is uniform integrability, we prove that condition (a) is satisfied, that is, that $\{x_n, \mathscr{F}_n, 1 \leq n \leq \infty\}$ is a martingale, by proving that

$$\mathbf{E}\{x_\infty \mid \mathscr{F}_n\} = x_n$$

with probability 1, for every (finite) n. That is, we prove that

$$\int_\Lambda x_\infty \, d\mathbf{P} = \int_\Lambda x_n \, d\mathbf{P}, \qquad \Lambda \in \mathscr{F}_n. \tag{4.4}$$

Since the original sequence is a martingale, if $m > n$,

$$\int_\Lambda x_m \, d\mathbf{P} = \int_\Lambda \mathbf{E}\{x_m \mid \mathscr{F}_n\} \, d\mathbf{P} = \int_\Lambda x_n \, d\mathbf{P}.$$

Now, when $m \to \infty$, $x_m \to x_\infty$ with probability 1, and since there is uniform integrability we can integrate to the limit as $m \to \infty$ on the left to obtain (4.4).

Proof of (iii) If, for some $\alpha > 1$, $\lim_{n\to\infty} \mathbf{E}\{|x_n|^\alpha\} < \infty$, the boundedness of this convergent sequence of expectations implies uniform integrability of the x_n's so that the conditions of (ii) are satisfied. By Fatou's lemma,

$$\mathbf{E}\{|x_\infty|^\alpha\} \leq \lim_{n\to\infty} \mathbf{E}\{|x_n|^\alpha\} < \infty.$$

Then by Theorem 3.1 (iii) or the deeper Theorem 3.4 the sequence $\{|x_n|^\alpha\}$ (and therefore also the sequence $\{|x_\infty - x_n|^\alpha\}$) is uniformly integrable, so that $x_n \to x_\infty$ with probability 1 implies that $\lim_{n\to\infty} \mathbf{E}\{|x_\infty - x_n|^\alpha\} = 0$. Conversely, if the conditions of (ii) are satisfied, $\mathbf{E}\{|x_n|^\alpha\} \leq \mathbf{E}\{|x_\infty|^\alpha\}$ by Theorem 3.1 (i).

Proof of (iv) Suppose that the condition (4.1) is satisfied. Let N be any positive number, and let $m(\omega)$ be the first integer j for which $x_j(\omega) > N$, or let $m(\omega) = +\infty$ if there is no such integer. Define $x_n^{(N)}$ by

$$x_n^{(N)}(\omega) = x_n(\omega), \qquad \text{if} \quad n < m(\omega),$$
$$= x_{m(\omega)}(\omega), \qquad \text{if} \quad n \geq m(\omega), \tag{4.5}$$

and let $w = \underset{n \geq 0}{\text{L.U.B.}} (x_{n+1} - x_n)$. The condition $m(\omega) = k$ is a condition on $x_1, \cdots, x_k$, and the $x_n^{(N)}$ process has been obtained from the x_n process by optional stopping based on the stopping variable m. Then according to Theorem 2.1 the process $\{x_n^{(N)}, n \geq 1\}$ is a martingale. Moreover, $x_n^{(N)} \leq N + w$. According to the remark made at the end of the proof of (i), the fact that $x_n^{(N)}$ is bounded from above by a random variable whose expectation exists implies that $\mathbf{E}\{|x_n^{(N)}|\}$ is bounded in n. It now follows from (i) that $\lim_{n \to \infty} x_n^{(N)}$ exists and is finite with probability 1. Since

$$x_n^{(N)}(\omega) = x_n(\omega) \qquad \text{if} \quad \underset{j}{\text{L.U.B.}}\, x_j(\omega) \leq N,$$

it follows that $\lim_{n \to \infty} x_n(\omega)$ exists and is finite for almost all ω with $\underset{n}{\text{L.U.B.}}\, x_n(\omega) \leq N$, that is, with $\limsup_{n \to \infty} x_n(\omega) < \infty$, since N is arbitrary.

Proof of (v) Let N be any positive number, and let $m(\omega)$ be the first integer j for which $|x_j(\omega)| > N$, or let $m(\omega) = \infty$ if there is no such integer. Define $x_n^{(N)}$ by (4.5) and W^2 as the L.U.B. in (4.2). Then, as in (iv), the $x_n^{(N)}$ process has been obtained from the x_n process by optional stopping. According to Theorem 2.1 the process $\{x_n^{(N)}, n \geq 1\}$ is a martingale, and

$$|x_n^{(N)}| \leq N + W, \qquad \mathbf{E}\{|x_n^{(N)}|^2\} \leq 2N^2 + 2\mathbf{E}\{W^2\} < \infty.$$

Then the conditions of (iii) are fulfilled, and we conclude that the sequence $\{x_n^{(N)}, n \geq 1\}$ converges with probability 1 and in the mean. Since the series $\sum_1^\infty (x_{n+1}^{(N)} - x_n^{(N)})$ is a series of mutually orthogonal functions which converges in the mean, $\sum_1^\infty \mathbf{E}\{|x_{n+1}^{(N)} - x_n^{(N)}|^2\} < \infty$, by IV, Theorem 4.1. Hence (with probability 1)

$$\sum_1^\infty \mathbf{E}\{|x_{n+1}^{(N)} - x_n^{(N)}|^2 \mid \mathscr{F}_n\} < \infty,$$

because the operation $\mathbf{E}\{-\}$ performed on the terms of this series yields a convergent series. Moreover,

$$\mathbf{E}\{|x_{n+1}^{(N)} - x_n^{(N)}|^2 \mid \mathscr{F}_n\} = \mathbf{E}\{|x_{n+1} - x_n|^2 \mid \mathscr{F}_n\}, \qquad n < m(\omega),$$
$$= 0, \qquad n \geq m(\omega),$$

so that

$$\sum_{1}^{\infty} \mathbf{E}\{|x_{n+1} - x_n|^2 \mid \mathscr{F}_n\} < \infty \tag{4.6}$$

almost everywhere where $m(\omega) = \infty$, and therefore almost everywhere where $\lim_{n\to\infty} x_n$ exists and is finite, since N is arbitrary. This proves half the statement of (v). To go in the other direction define $m(\omega)$ as the first integer j for which

$$\sum_{i=1}^{j} \mathbf{E}\{|x_{i+1} - x_i|^2 \mid \mathscr{F}_i\} > N$$

and then define $x_n^{(N)}$ by (4.5). Then, as before, $\{x_n^{(N)}, n \geq 1\}$ is a martingale, and

$$\sum_{1}^{\infty} \mathbf{E}\{|x_{n+1}^{(N)} - x_n^{(N)}|^2 \mid \mathscr{F}_n\} = \sum_{n=1}^{m-1} \mathbf{E}\{|x_{n+1} - x_n|^2 \mid \mathscr{F}_n\} \leq N.$$

Hence, taking expectations,

$$\sum_{1}^{\infty} \mathbf{E}\{|x_{n+1}^{(N)} - x_n^{(N)}|^2\} = \lim_{n\to\infty} \mathbf{E}\{|x_n^{(N)} - x_1^{(N)}|^2\} \leq N,$$

so that $\mathbf{E}\{|x_n^{(N)}|^2\}$ is bounded when $n \to \infty$. It follows from (iii) that $\lim_{n\to\infty} x_n^{(N)}$ exists and is finite, with probability 1, and therefore that $\lim_{n\to\infty} x_n(\omega)$ exists almost everywhere where $m(\omega) = \infty$, that is, almost everywhere where (4.6) is true, since N is arbitrary.

If the sequence $x_1, x_2, \cdots$ is a martingale, and if $\lim_{n\to\infty} \mathbf{E}\{|x_n|\} < \infty$, so that $\lim_{n\to\infty} x_n$ exists, it is not necessarily true that the sequence $x_1, x_2, \cdots, x_\infty$ is a martingale. In other words, Theorem 4.1 (i) describes a situation more general than that of Theorem 4.1 (ii). Although simple examples of this are easily given, we omit them here because such examples appear in a natural way in §8. We shall see in §5 that the situations in Theorem 4.1 (i) and (ii) are identical if the differences $\{x_{n+1} - x_n\}$ are mutually independent, that is, if the x_n's are the partial sums of a series of mutually independent random variables.

The following corollaries exhibit the power of Theorem 4.1. There will be many other applications of the theorem in later sections.

COROLLARY 1 *Let* $y_1, y_2, \cdots$ *be a sequence of uniformly bounded non-negative random variables, and let* $p_j = \mathbf{E}\{y_j \mid y_1, \cdots, y_{j-1}\}$. *Then the series* $\sum_{1}^{\infty} y_j(\omega)$ *converges for almost all* ω *for which* $\sum_{1}^{\infty} p_j(\omega)$ *converges, and conversely.*

If $x_n = \sum_1^n (y_j - p_j)$, the process $\{x_n, n \geq 1\}$ is a martingale. By Theorem 4.1 (iv), applied to this process and to $\{-x_n, n \geq 1\}$, $\lim_{n\to\infty} x_n(\omega)$ exists almost everywhere, and is finite where

$$\limsup_{n\to\infty} x_n(\omega) < \infty \qquad \text{or} \qquad \liminf_{n\to\infty} x_n(\omega) > -\infty.$$

Then

$$\mathbf{P}\{\lim_{n\to\infty} x_n(\omega) = \infty\} = \mathbf{P}\{\lim_{n\to\infty} x_n(\omega) = -\infty\} = 0.$$

It follows that the series in the corollary must converge and diverge together with probability 1, as was to be proved.

COROLLARY 2 *Let* $M_1, M_2, \cdots$ *be measurable* ω *sets, and let* p_n *be the conditional probability of* M_n *relative to* $M_1, \cdots, M_{n-1}$, *that is, relative to the field generated by the latter sets. Then the set of points in infinitely many* M_j*'s, and the set of points of divergence of the series* $\sum_1^\infty p_j(\omega)$, *differ by at most a set of probability* 0.

This corollary is sometimes phrased in a more intuitive fashion as follows. *Neglecting zero probabilities, infinitely many events of a given sequence* $E_1, E_2, \cdots$ *occur if and only if the series of conditional probabilities of the* E_j*'s relative to their predecessors diverges.* (Note that these conditional probabilities are random variables, not necessarily constants.)

This corollary is a generalization of the Borel-Cantelli lemma (III, Theorem 1.2), obtained as a special case of Corollary 1 by setting $y_n(\omega) = 1$ or 0 according as ω is or is not in M_n.

The following theorem is the semi-martingale analogue of Theorem 4.1.

THEOREM 4.1s *Let* $\{x_n, \mathscr{F}_n, n \geq 1\}$ *be a semi-martingale, and let* $\mathscr{F}_\infty$ *be the smallest Borel field of* ω *sets with* $\mathscr{F}_\infty \supset \bigcup_1^\infty \mathscr{F}_n$. *Then, by Theorem* 3.1 (i),

$$\mathbf{E}\{x_1\} \leq \mathbf{E}\{x_2\} \leq \cdots.$$

(i) *If* $\text{L.U.B.}_n\, \mathbf{E}\{|x_n|\} < \infty$, *then* $\lim_{n\to\infty} x_n = x_\infty$ *exists with probability* 1, *and* $\mathbf{E}\{|x_\infty|\} < \infty$. *In particular, if the* x_n*'s are non-positive,* $\mathbf{E}\{|x_n|\}$ *is monotone non-increasing, so that this condition is always satisfied; if the* x_n*'s are non-negative, this condition reduces to* $\lim_{n\to\infty} \mathbf{E}\{x_n\} = K < \infty$. *In the latter case* $\mathbf{E}\{x_\infty\} \leq K$.

(ii) (a) *If the* x_n*'s are uniformly integrable, then*

$$\text{L.U.B.}_n\, \mathbf{E}\{x_n\} < \infty, \qquad \lim_{n\to\infty} \mathbf{E}\{|x_\infty - x_n|\} = 0,$$

and the process $\{x_n, \mathscr{F}_n, 1 \leq n \leq \infty\}$ *is a semi-martingale which is dominated by a semi-martingale relative to the same fields.*

(b) *If* $\text{L.U.B.}_n\, \mathbf{E}\{|x_n|\} < \infty$, *so that* x_∞ *exists, and if the process* $\{x_n, 1 \le n \le \infty\}$ *is a semi-martingale, then*

$$\lim_{n\to\infty} \mathbf{E}\{x_n\} \le \mathbf{E}\{x_\infty\}, \tag{4.7}$$

and there is equality if and only if the x_n*'s are uniformly integrable. In particular, there is always equality if the* x_n*'s are non-negative.*

(iii) *Suppose that the* x_n*'s are non-negative. Then*

$$\mathbf{E}\{x_1^\alpha\} \le \mathbf{E}\{x_2^\alpha\} \le \cdots, \qquad \alpha \ge 1.$$

If, for some $\alpha > 1$, $\lim_{n\to\infty} \mathbf{E}\{x_n^\alpha\} < \infty$, *then the* x_n^α*'s are uniformly integrable, and*

$$\mathbf{E}\{x_\infty^\alpha\} < \infty, \qquad \lim_{n\to\infty} \mathbf{E}\{|x_\infty - x_n|^\alpha\} = 0.$$

Conversely, if the x_n*'s are uniformly integrable, and if* $\mathbf{E}\{x_\infty^\alpha\} < \infty$ *for some* $\alpha > 1$, *then*

$$\lim_{n\to\infty} \mathbf{E}\{x_n^\alpha\} = \mathbf{E}\{x_\infty^\alpha\}.$$

(iv) *If*

$$\mathbf{E}\{\text{L.U.B.}_{n\ge 1}\, [x_{n+1} - \mathbf{E}\{x_{n+1} \mid \mathscr{F}_n\}]\} < \infty, \tag{4.8}$$

then $\lim_{n\to\infty} x_n(\omega)$ *exists and is finite for almost all* ω *for which* $\limsup_{n\to\infty} x_n(\omega) < \infty$.

(v) *If*

$$\mathbf{E}\{\text{L.U.B.}_n\, [x_{n+1} - \mathbf{E}\{x_{n+1} \mid \mathscr{F}_n\}]^2\} < \infty, \tag{4.9}$$

then $\lim_{n\to\infty} x_n(\omega)$ *exists and is finite for almost all* ω *for which*

$$\sum_1^\infty [\mathbf{E}\{x_{n+1}^2 \mid \mathscr{F}_n\} - \mathbf{E}^2\{x_{n+1} \mid \mathscr{F}_n\}] < \infty,$$

$$\sum_1^\infty [\mathbf{E}\{x_{n+1} \mid \mathscr{F}_n\} - x_n] < \infty,$$

and conversely.

Proof of (i) The method of proof of Theorem 4.1 (i) [which is a special case of Theorem 4.1s (i)] is applicable without change to the proof of the existence of the limit x_∞ in the present semi-martingale case. However, we give an instructive alternative proof which reduces the desired result to that in the martingale case. According to Theorem 1.2 (ii), the hypothesis of (i) implies that we can write x_n in the form

$$x_n = x_n' + \sum_1^n \Delta_j, \tag{4.10}$$

where $\{x_n', \mathscr{F}_n, n \geq 1\}$ is a martingale, the Δ_j's are non-negative, $\mathbf{E}\{|x_n'|\}$ is bounded in n, $\sum_1^\infty \Delta_j < \infty$ with probability 1, and $\sum_1^\infty \mathbf{E}\{\Delta_j\} < \infty$. Then, according to Theorem 4.1 (i), $\lim_{n\to\infty} x_n' = x_\infty'$ exists and is finite with probability 1, so that

$$\lim_{n\to\infty} x_n = x_\infty' + \sum_1^\infty \Delta_j = x_\infty$$

with probability 1. The statements of (i) are trivial consequences of what we have now obtained.

Proof of (ii) If the hypothesis of (i) is strengthened to the hypothesis of (ii) (a) that the x_n's are uniformly integrable, then we show that the process $\{x_n, \mathscr{F}_n, 1 \leq n \leq \infty\}$ is a semi-martingale by showing that

$$x_n \leq \mathbf{E}\{x_\infty \mid \mathscr{F}_n\}, \qquad n \geq 1,$$

with probability 1. This can be shown directly, as in the corresponding treatment of the martingale case. There is some interest in the following alternative method, however. According to Theorem 1.2 (iii) the x_n''s are uniformly integrable in this case, and it then follows from Theorem 4.1 (ii) that $\{x_n', \mathscr{F}_n, 1 \leq n \leq \infty\}$ is a martingale, so that

$$\begin{aligned}\mathbf{E}\{x_\infty \mid \mathscr{F}_n\} &\geq \mathbf{E}\{x_\infty' + \sum_1^n \Delta_j \mid \mathscr{F}_n\} \\ &= x_n' + \sum_1^n \Delta_j \\ &= x_n\end{aligned}$$

with probability 1, as was to be proved. Moreover, if we define x_n^+ as the right side of the inequality

$$|x_n| \leq |x_n'| + \sum_1^n \Delta_j, \qquad 1 \leq n \leq \infty,$$

the x_n process is dominated by the semi-martingale $\{x_n^+, \mathscr{F}_n, 1 \leq n \leq \infty\}$. This result completes that of Theorem 1.2 (iv). Going in the other direction, assume the hypotheses of (ii) (b). Then (4.7) is implied by the semi-martingale inequality. Moreover, since the process of double the positive parts of the x_n's, $\{|x_n| + x_n, 1 \leq n \leq \infty\}$, is a semi-martingale, its random variables are uniformly integrable, by Theorem 3.1 (iii). Hence

$$\lim_{n\to\infty} \mathbf{E}\{|x_n| + x_n\} = \mathbf{E}\{|x_\infty| + x_\infty\}.$$

(In particular, there is equality in (4.7) if the x_n's are non-negative.) On the other hand, by Fatou's lemma,

$$\lim_{n\to\infty} \mathbf{E}\{|x_n| - x_n\} \geq \mathbf{E}\{|x_\infty| - x_\infty\}.$$

Combining these two relations, we find again that (4.7) is true, and that there is equality in (4.7) if and only if there is equality in the application of Fatou's lemma above, that is, if and only if the negative parts of the x_n's are uniformly integrable. Since, as we have already remarked, the positive parts of the x_n's are always uniformly integrable, under the present hypotheses, it follows that there is equality in (4.7) if and only if the x_n's are uniformly integrable.

Proof of (iii) If the x_n's are non-negative, the x_n's constitute a semi-martingale for any $\alpha > 1$ for which these random variables have finite expectations. Hence $\mathbf{E}\{x_n^\alpha\}$ is non-decreasing in n. If

$$\lim_{n\to\infty} \mathbf{E}\{x_n^\alpha\} < \infty,$$

for some $\alpha > 1$, the x_n's are uniformly integrable; this inference has nothing to do with martingale theory. Then, by (ii), x_∞ exists and $\{x_n, \mathscr{F}_n, 1 \leq n \leq \infty\}$ is a semi-martingale. By Fatou's lemma, $\mathbf{E}\{x_\infty^\alpha\} < \infty$. Hence, by Theorem 1.1 (i), $\{x_n^\alpha, 1 \leq n \leq \infty\}$ is a semi-martingale also. Since the random variables of this semi-martingale are non-negative, and since there is a last one, the random variables are uniformly integrable, by Theorem 3.1 (iii). Then integration to the limit yields

$$\underset{n}{\text{L.U.B.}}\ \mathbf{E}\{x_n^\alpha\} = \lim_{n\to\infty} \mathbf{E}\{x_n^\alpha\} = \mathbf{E}\{x_\infty^\alpha\} < \infty$$

$$\lim_{n\to\infty} \mathbf{E}\{|x_\infty - x_n|^\alpha\} = 0.$$

To prove the converse half of (iii) we need only remark that, if $\{x_n, 1 \leq n \leq \infty\}$ is a semi-martingale with non-negative random variables, and if $\mathbf{E}\{x_\infty^\alpha\} < \infty$, then the process $\{x_n^\alpha, 1 \leq n \leq \infty\}$ is also a semi-martingale, by Theorem 1.1 (i), so that, according to the semi-martingale inequality,

$$\mathbf{E}\{x_n^\alpha\} \leq \mathbf{E}\{x_\infty^\alpha\}, \qquad n \geq 1.$$

We have already seen that there is equality in the limit when $n \to \infty$ since the x_n's are non-negative.

Proofs of (iv) *and* (v) In the notation of the representation (4.10),

$$x_{n+1} - \mathbf{E}\{x_{n+1} \mid \mathscr{F}_n\} = x_{n+1}' - x_n'$$

$$\mathbf{E}\{x_{n+1}^2 \mid \mathscr{F}_n\} - \mathbf{E}^2\{x_{n+1} \mid \mathscr{F}_n\} = \mathbf{E}\{[x_{n+1}' - x_n']^2 \mid \mathscr{F}_n\}$$

$$\mathbf{E}\{x_{n+1} \mid \mathscr{F}_n\} - x_n = \Delta_{n+1}.$$

With the help of these relations, (iv) and (v) are reduced to the corresponding martingale statements of Theorem 4.1 (iv) and (v), as applied to the x_n' process. Note that, if $\mathbf{E}\{\underset{n\geq 0}{\text{L.U.B.}}\ (x_{n+1} - x_n)\} < \infty$, then condition (4.8) is satisfied.

We remark that both in this theorem and in Theorem 4.1 the αth power,

$$\Phi(s) = s^\alpha,$$

has been used ($\alpha > 1$), but this was only in view of certain applications, and any function Φ which is convex and monotone non-decreasing for $s \geq 0$, with

$$\lim_{s\to\infty} \frac{\Phi(s)}{s} = \infty$$

would serve as well.

THEOREM 4.2 *Let $\{x_n, \mathscr{F}_n, n \leq -1\}$ be a martingale, and define*

$$\mathscr{F}_{-\infty} = \bigcap_{-\infty}^{-1} \mathscr{F}_n.$$

Then $\lim_{n\to-\infty} x_n = x_{-\infty}$ exists and is finite with probability 1, and $\{x_n, \mathscr{F}_n, -\infty \leq n \leq -1\}$ is a martingale. The x_n's are uniformly integrable, and

$$\text{(4.11)} \quad \mathbf{E}\{|x_{-\infty}|\} = \lim_{n\to-\infty} \mathbf{E}\{|x_n|\} \leq \cdots \leq \mathbf{E}\{|x_{-2}|\} \leq \mathbf{E}\{|x_{-1}|\}.$$

If, for some $\alpha \geq 1$, $\mathbf{E}\{|x_{-1}|^\alpha\} < \infty$, then

$$\lim_{n\to-\infty} \mathbf{E}\{|x_{-\infty} - x_n|^\alpha\} = 0.$$

This is the strict sense version of IV, Theorem 7.3. It is sufficient to prove the existence of the limit $x_{-\infty}$ in the real case, and we accordingly assume for the moment that the x_n's are real. Let x_*, x^* be respectively the inferior and superior limits of the sequence $\{x_n\}$. Fix r_1 and r_2 and let $\beta_m(\omega)$ be the number of upcrossings of the interval $[r_1, r_2]$ by $x_m(\omega)$, $\cdots$, $x_{-1}(\omega)$. Then $\beta_m(\omega) \to \infty$ monotonely when $m \to -\infty$, on every ω for which $x^*(\omega) > r_2 > r_1 > x_*(\omega)$, although according to Theorem 3.3

$$\mathbf{E}\{\beta_m\} \leq \frac{\mathbf{E}\{|x_{-1}|\} + |r_1|}{r_2 - r_1}.$$

Then

$$\mathbf{P}\{x^*(\omega) > r_2 > r_1 > x_*(\omega)\} = 0,$$

for every pair r_2, r_1, and as in the proof of Theorem 4.1 (i) it follows that $x_* = x^*$ with probability 1, that is, $\lim_{n\to-\infty} x_n = x_{-\infty}$ exists as a finite or infinite limit. The process $\{|x_n|^\alpha, -\infty < n \leq -1\}$ is a semi-martingale, for any $\alpha \geq 1$ for which

$$\mathbf{E}\{|x_{-1}|^\alpha\} < \infty.$$

Since this semi-martingale has non-negative random variables, these random variables are uniformly integrable, according to Theorem 3.1 (iii). It follows ($\alpha = 1$) that $x_{-\infty}$ is finite-valued with probability 1, that

$$\lim_{n\to-\infty} \mathbf{E}\{|x_n|\} = \mathbf{E}\{|x_{-\infty}|\}, \qquad \lim_{n\to-\infty} \mathbf{E}\{|x_{-\infty} - x_n|\} = 0,$$

and that the same is true for the exponent $\alpha > 1$ if the relevant expectations exist. To identify $x_{-\infty}$ with $\mathbf{E}\{x_{-1} \mid \mathscr{F}_{-\infty}\}$ we must prove that, if $\Lambda \in \mathscr{F}_{-\infty}$, then

$$\int_\Lambda x_{-\infty}\, d\mathbf{P} = \int_\Lambda x_{-1}\, d\mathbf{P}.$$

Now, because of the martingale equality and the fact that $\Lambda \in \mathscr{F}_n$ for every n,

$$\int_\Lambda x_n\, d\mathbf{P} = \int_\Lambda x_{-1}\, d\mathbf{P}, \qquad n \leq -1,$$

and when $n \to -\infty$ this yields the preceding equality (because of the uniform integrability of the x_n's).

THEOREM 4.2s *Let* $\{x_n, \mathscr{F}_n, n \leq -1\}$ *be a semi-martingale. Then* $\lim_{n\to-\infty} x_n = x_{-\infty}$ *exists with probability* 1, *and* $-\infty \leq x_{-\infty} < \infty$ *with probability* 1. *Define*

$$\mathscr{F}_{-\infty} = \bigcap_{-\infty}^{-1} \mathscr{F}_n.$$

Then

$$\mathbf{E}\{x_{-1}\} \geq \mathbf{E}\{x_{-2}\} \geq \cdots,$$

by Theorem 3.1 (ii). *If*

$$\lim_{n\to-\infty} \mathbf{E}\{x_n\} = K > -\infty$$

then $x_{-\infty}$ *is finite with probability* 1, *and*

$$\lim_{n\to-\infty} \mathbf{E}\{x_n\} = K = \mathbf{E}\{x_{-\infty}\}, \qquad \lim_{n\to-\infty} \mathbf{E}\{|x_{-\infty} - x_n|\} = 0.$$

Moreover, $\{x_n, \mathscr{F}_n, -\infty \leq n \leq -1\}$ *is a semi-martingale whose random variables are uniformly integrable. If the* x_n*'s are non-negative, and if, for some* $\alpha > 1$, $\mathbf{E}\{x_{-1}^{\alpha}\} < \infty$, *then*

$$\lim_{n\to-\infty} \mathbf{E}\{|x_{-\infty} - x_n|^{\alpha}\} = 0.$$

The method of proof of Theorem 4.2 is applicable without essential change to the present more general theorem. If $x_n = n$, $K = -\infty$ and $x_{-\infty} = -\infty$ with probability 1. However, Theorem 3.2 is easily applied to deduce the fact that $\text{L.U.B.}_n\, x_n < \infty$ with probability 1, so that the same is true of $x_{-\infty}$. The following proof for the case $K > -\infty$ is illuminating in that it illustrates the close connection between martingales and semi-martingales.

If $\lim_{n\to-\infty} \mathbf{E}\{x_n\}$ is finite, the series of non-negative terms

$$\sum_{-\infty}^{-1} [\mathbf{E}\{x_j \mid \mathscr{F}_{j-1}\} - x_{j-1}]$$

converges with probability 1, because the expectations of the partial sums are dominated by

$$\mathbf{E}\{x_{-1}\} - \lim_{n\to-\infty} \mathbf{E}\{x_n\}.$$

Then we define x_n' by

$$x_n = x_n' + \sum_{-\infty}^{n} [\mathbf{E}\{x_j \mid \mathscr{F}_{j-1}\} - x_{j-1}], \tag{4.12}$$

following (1.5′). The process $\{x_n', \mathscr{F}_n, n \leq -1\}$ is a martingale, and therefore $\lim_{n\to-\infty} x_n' = x_{-\infty}'$ exists and is finite with probability 1, and the process $\{x_n', \mathscr{F}_n, -\infty \leq n \leq -1\}$ is a martingale, by Theorem 4.2. Then

$$\lim_{n\to-\infty} x_n = x_{-\infty} = x_{-\infty}'$$

with probability 1. To prove that the process $\{x_n, \mathscr{F}_n, -\infty \leq n \leq -1\}$ is a semi-martingale we have only to prove that

$$x_{-\infty} \leq \mathbf{E}\{x_n \mid \mathscr{F}_n\}, \qquad n \leq -1,$$

with probability 1, and we have from (4.12), since the summands are non-negative,

$$\mathbf{E}\{x_n \mid \mathscr{F}_{-\infty}\} \geq \mathbf{E}\{x_n' \mid \mathscr{F}_{-\infty}\} = x_{-\infty}.$$

The x_n's are uniformly integrable because the x_n''s are [by Theorem 3.1 (iii)], and the sums in (4.12) are because they are dominated by the infinite sum. [Alternatively the x_n's are uniformly integrable by Theorem 3.1 (iv).] Finally, if the x_n's are non-negative, and if

$$\mathbf{E}\{x_{-1}^{\alpha}\} < \infty$$

for some $\alpha \geq 1$, then the x_n^{α} process is a semi-martingale, with non-negative random variables and a last random variable, so that the x_n^{α}'s are uniformly integrable, by Theorem 3.1 (iii), and (integration to the limit)

$$\lim_{n\to-\infty} \mathbf{E}\{|x_{-\infty} - x_n|^{\alpha}\} = 0.$$

THEOREM 4.3 *Let z be a random variable with $\mathbf{E}\{|z|\} < \infty$, and let $\cdots \subset \mathcal{F}_1 \subset \mathcal{F}_2 \subset \cdots$ be Borel fields of measurable ω sets. Let $\mathcal{F}_{-\infty} = \bigcap_n \mathcal{F}_n$ and let $\mathcal{F}_\infty$ be the smallest Borel field of sets with $\mathcal{F}_\infty \supset \bigcup_n \mathcal{F}_n$. Then*

$$\lim_{n\to-\infty} \mathbf{E}\{z \mid \mathcal{F}_n\} = \mathbf{E}\{z \mid \mathcal{F}_{-\infty}\}$$

(4.13)

$$\lim_{n\to\infty} \mathbf{E}\{z \mid \mathcal{F}_n\} = \mathbf{E}\{z \mid \mathcal{F}_\infty\}$$

with probability 1.

Define

$$x_n = \mathbf{E}\{z \mid \mathcal{F}_n\}, \qquad -\infty \le n \le \infty.$$

The process $\{x_n, \mathcal{F}_n, -\infty \le n \le \infty\}$ is a martingale because $\mathcal{F}_n$ is non-decreasing in n (see Example 1, §1), so the first equation in (4.13) is simply an application of Theorem 4.2. The process $\{|x_n|, -\infty \le n \le \infty\}$ is a semi-martingale with non-negative random variables and a last random variable, and its random variables are therefore uniformly integrable, by Theorem 3.1 (iii). Then, by Theorem 4.1 (ii), $\lim_{n\to\infty} x_n = y$ exists with probability 1, and the process $\{x_n, \mathcal{F}_n, 1 \le n \le \infty\}$ is a martingale if x_∞ is replaced by y. To identify y with x_∞ and thus prove the second equation of (4.13) we note that x_∞ is a conditional expectation, characterized (neglecting values on an ω set of probability 0) by two conditions: it is equal almost everywhere to a random variable measurable with respect to $\mathcal{F}_\infty$, and it has the same integral as z on every $\mathcal{F}_\infty$ set. Now, if $\Lambda \in \mathcal{F}_n$, the martingale equality applied twice yields

$$\int_\Lambda y\, d\mathbf{P} = \int_\Lambda x_n\, d\mathbf{P} = \int_\Lambda z\, d\mathbf{P}.$$

Since the extreme terms are equal for $\Lambda \in \mathcal{F}_n$, and therefore for $\Lambda \in \bigcup_n \mathcal{F}_n$, they are equal for $\Lambda \in \mathcal{F}_\infty$. In fact, $\mathcal{F}_\infty$ is by definition the Borel field generated by the field $\bigcup_n \mathcal{F}_n$, and the extreme terms of the above equality define completely additive functions of $\mathcal{F}_\infty$ sets which are identical on the field $\bigcup_n \mathcal{F}_n$ and therefore are identical on $\mathcal{F}_\infty$ (see Supplement, Theorem 2.1). But then y satisfies the conditions characterizing x_∞, so they are equal with probability 1, as was to be proved.

We remark that, if $\mathcal{F}_n$ is only defined for sufficiently large or sufficiently small n, the relevant half of Theorem 4.3 remains applicable.

The following corollary covers the most important case of Theorem 4.3.

COROLLARY 1 *Let z be any random variable with $\mathbf{E}\{|z|\} < \infty$, and let $y_1, y_2, \cdots$ be any random variables. Then, if $\mathscr{G}_n$ is the Borel field of the ω sets determined by conditions on the y_j's with $j \geq n$,*

$$\lim_{n\to\infty} \mathbf{E}\{z \mid y_n, y_{n+1}, \cdots\} = \mathbf{E}\{z \mid \bigcap_1^\infty \mathscr{G}_n\}$$

$$\lim_{n\to\infty} \mathbf{E}\{z \mid y_1, \cdots, y_n\} = \mathbf{E}\{z \mid y_1, y_2, \cdots\} \tag{4.13'}$$

with probability 1. *In particular, if z is a random variable on the sample space of the y_j's, the second limit can be replaced by z.*

To reduce the first of these equations to the first in (4.13) identify $\mathscr{G}_n$ with $\mathscr{F}_{-n}$. To reduce the second to the second in (4.13) identify with $\mathscr{F}_n$ the Borel field of ω sets determined by conditions on the y_j's for $j \leq n$. In particular, if z is a random variable on the sample space of the y_j's, the second limit is z itself, with probability 1, by definition of conditional expectation.

The following theorem is an immediate consequence of Theorem 4.3, but its form makes it more useful in studying certain problems.

THEOREM 4.4 (i) *Suppose that the random variables $w, \cdots, x_{-2}, x_{-1}$ constitute a martingale relative to the respective fields $\mathscr{F}_w, \cdots, \mathscr{F}_{-2}, \mathscr{F}_{-1}$, and define $\mathscr{F}_{-\infty} = \bigcap_{-\infty}^{-1} \mathscr{F}_n$. Then*

$$\lim_{n\to-\infty} x_n = \mathbf{E}\{z \mid \mathscr{F}_{-\infty}\}$$

with probability 1, *and the random variables*

$$w, \mathbf{E}\{z \mid \mathscr{F}_{-\infty}\}, \cdots, x_{-2}, x_{-1}$$

constitute a martingale relative to the respective Borel fields

$$\mathscr{F}_w, \mathscr{F}_{-\infty}, \cdots, \mathscr{F}_{-2}, \mathscr{F}_{-1}.$$

(ii) *Suppose that the random variables $x_1, x_2, \cdots, z$ constitute a martingale relative to the respective Borel fields $\mathscr{F}_1, \mathscr{F}_2, \cdots, \mathscr{F}_z$, and let $\mathscr{F}_\infty$ be the smallest Borel field of ω sets with $\mathscr{F}_\infty \supset \bigcup_1^\infty \mathscr{F}_n$. Then*

$$\lim_{n\to\infty} x_n = \mathbf{E}\{z \mid F_\infty\}$$

with probability 1, *and the random variables*

$$x_1, x_2, \cdots, \mathbf{E}\{z \mid \mathscr{F}_\infty\}, z$$

constitute a martingale relative to the respective Borel fields

$$\mathscr{F}_1, \mathscr{F}_2, \cdots, \mathscr{F}_\infty, \mathscr{F}_z.$$

This theorem shows how to enlarge the parameter set of certain types of martingales; the discussion of this problem will be taken up again in a later section. The semi-martingale version of the theorem is the following. The details of the proof will be given in the semi-martingale case, since some are not obvious.

THEOREM 4.4s (i) *Suppose that the random variables* $w, \cdots, x_{-2}, x_{-1}$ *constitute a semi-martingale relative to the respective Borel fields* $\mathscr{F}_w, \cdots, \mathscr{F}_{-2}, \mathscr{F}_{-1}$, *and define* $\mathscr{F}_{-\infty} = \bigcap_{-\infty}^{-1} \mathscr{F}_n$. *Then*

$$\lim_{n\to-\infty} x_n = x_{-\infty}$$

exists and is finite with probability 1, *and the random variables* $w, x_{-\infty}, \cdots, x_{-2}, x_{-1}$ *constitute a semi-martingale relative to the respective Borel fields* $\mathscr{F}_w, \mathscr{F}_{-\infty}, \cdots, \mathscr{F}_{-2}, \mathscr{F}_{-1}$.

(ii) *Suppose that the random variables* $x_1, x_2, \cdots, z$ *constitute a semi-martingale relative to the respective Borel fields* $\mathscr{F}_1, \mathscr{F}_2, \cdots, \mathscr{F}_z$, *and let* $\mathscr{F}_\infty$ *be the smallest Borel field of* ω *sets with* $\mathscr{F}_\infty \supset \bigcup_1^\infty \mathscr{F}_n$. *Then* $\lim_{n\to\infty} x_n = x_\infty$ *exists and is finite with probability* 1, *and*

$$\lim_{n\to\infty} \mathbf{E}\{x_n\} \le \mathbf{E}\{x_\infty\} \le \mathbf{E}\{z\}. \tag{4.14}$$

The random variables $x_1, x_2, \cdots, x_\infty, z$ *constitute a semi-martingale (and, if so, necessarily one relative to the respective Borel fields* $\mathscr{F}_1, \mathscr{F}_2, \cdots, \mathscr{F}_\infty, \mathscr{F}_z$) *if and only if the first two members of this continued inequality are equal, or equivalently if and only if the* x_n's *are uniformly integrable. The stated conditions will be satisfied, for example, if the* x_n's *are non-negative, or more generally if the* $x_1, x_2, \cdots, z$ *process is dominated by a semi-martingale.*

In connection with the last point we remark that, if the x_n's are non-negative, we can assume that z is non-negative also, since the positive parts of the $x_1, x_2, \cdots, z$ process also constitute a semi-martingale.

Proof of (i) According to Theorem 4.1s the limit $x_{-\infty}$ exists as stated in (i), and the random variables $x_{-\infty}, \cdots, x_{-2}, x_{-1}$ constitute a semi-martingale relative to the indicated fields. There only remains the proof that

$$w \le \mathbf{E}\{x_{-\infty} \mid \mathscr{F}_w\}$$

with probability 1, that is,

$$\int_\Lambda w\, d\mathbf{P} \le \int_\Lambda x_{-\infty}\, d\mathbf{P}, \qquad \Lambda \in \mathscr{F}_w.$$

Now this semi-martingale inequality is true by hypothesis if $x_{-\infty}$ is replaced by x_n, and when $n \to -\infty$ we obtain the desired inequality, because

$$\lim_{n\to-\infty} \mathbf{E}\{|x_{-\infty} - x_n|\} = 0$$

by Theorem 4.1s.

Proof of (ii) Under the hypotheses of (ii),

$$\text{L.U.B.}_{n}\ \mathbf{E}\{|x_n|\} < \infty,$$

according to Theorem 3.1 (ii). It follows that x_∞ exists. The rest of (ii) then follows easily from Theorem 4.1s (ii), except for the inequality relating the last two members of (4.14). The proof of the inequality is given as follows. Let K be any real number, and define

$$\Phi(s) = \text{Max}\,[0, s].$$

Then by Theorem 1.1 (i) the random variables

$$\Phi(x_1 - K), \Phi(x_2 - K), \cdots, \Phi(z - K)$$

constitute a semi-martingale. Hence

$$\mathbf{E}\{\Phi(x_n - K)\} \le \mathbf{E}\{\Phi(z - K)\},$$

so that when $n \to \infty$ we obtain, using Fatou's lemma,

$$\mathbf{E}\{\Phi(x_\infty - K)\} \le \mathbf{E}\{\Phi(z - K)\},$$

that is,

$$\int_{\{x_\infty(\omega) > K\}} (x_\infty - K)\, d\mathbf{P} \le \int_{\{z(\omega) > K\}} (z - K)\, d\mathbf{P}.$$

Then, if $K < 0$,

$$\begin{aligned}\int_{\{x_\infty(\omega) > K\}} x_\infty\, d\mathbf{P} &\le \int_{\{z(\omega) > K\}} z\, d\mathbf{P} + K\mathbf{P}\{x_\infty(\omega) > K\} - K\mathbf{P}\{z(\omega) > K\} \\ &\le \int_{\{z(\omega) > K\}} z\, d\mathbf{P} - K\mathbf{P}\{x_\infty(\omega) \le K\} \\ &\le \int_{\{z(\omega) > K\}} z\, d\mathbf{P} + \int_{\{x_\infty(\omega) \le K\}} |x_\infty|\, d\mathbf{P},\end{aligned}$$

and this inequality becomes the desired one when $K \to -\infty$.

5. Application to the theory of sums of independent random variables

Although the zero-one law is easily proved directly (see III, Theorem 1.1), it is instructive to derive it from martingale theory. Suppose then that $x_1, x_2, \cdots$ are mutually independent random variables. The zero-one law states that, if x is for every n a random variable on the sample space of $x_n, x_{n+1}, \cdots$, then $x = \text{const.}$ with probability 1. To derive this result from martingale theory, assume first that $\mathbf{E}\{|x|\} < \infty$. Then, since for every n the random variable x is independent of the random variables $x_1, \cdots, x_n$, it follows that

$$\mathbf{E}\{x \mid x_1, \cdots, x_n\} = \mathbf{E}\{x\}$$

with probability 1. When $n \to \infty$ this becomes, according to Theorem 4.3, Corollary 1,

$$x = \mathbf{E}\{x\}$$

with probability 1, and this is the desired result. If $\mathbf{E}\{x\}$ does not exist, define $y_N(\omega)$ as $x(\omega)$ if $|x(\omega)| \le N$ and as 0 otherwise. Then the result just obtained implies that $y_N = \mathbf{E}\{y_N\}$ with probability 1. This is impossible for all N unless $x = \text{const.}$ with probability 1.

Martingale theory can be used to lay the basis for the study of the convergence of a series $\sum_1^\infty y_j$ of mutually independent random variables. Rather than showing how this can be done in detail we prove a fundamental theorem on convergence, by martingale methods, and then go on to apply Theorem 4.1 to the partial sums of the series $\sum_1^\infty y_j$. In the following we shall suppose for simplicity that the y_j's are real.

Let y_j have characteristic function Φ_j, and suppose that for some λ no $\Phi_j(\lambda)$ vanishes. Then it is easily verified that, using this λ, the $\tilde{x}_n$ process defined by

$$\tilde{x}_n = \frac{e^{i\lambda \sum_1^n y_j}}{\prod_1^n \Phi_j(\lambda)}, \qquad n = 1, 2, \cdots$$

is a martingale. It was shown in III §2 that, if the infinite product $\prod_1^\infty \Phi_j$ is convergent on a λ set of positive Lebesgue measure, then the series $\sum_1^\infty y_j$ converges with probability 1. This result, which implies that convergence in measure or in the mean of the series $\sum_1^\infty y_j$ implies convergence with probability 1 (see Corollary 2 to III, Theorem 2.7), will now be derived by an analysis of the $\tilde{x}_n$ process. Dropping some of the first terms of the series $\sum_1^\infty y_j$, if necessary, we can suppose that

$\prod_1^\infty |\Phi_j(\lambda)| > \frac{1}{2}$ on a λ set A of positive Lebesgue measure. Then, if $\lambda \in A$, $|\tilde{x}_n(\omega)| < 2$, and therefore $\lim_{n\to\infty} \tilde{x}_n$ exists with probability 1, by Theorem 4.1. Consequently

$$\lim_{n\to\infty} e^{i\lambda \sum_1^\infty y_j}$$

exists with probability 1 for each λ in A. Then, except for an ω set of probability 0,

$$\lim_{n\to\infty} e^{i\lambda \sum_1^\infty y_j(\omega_0)} = f(\lambda, \omega_0)$$

exists for almost all λ in A (Fubini's theorem). To finish the proof we show that, if ω_0 is not in the exceptional set, $\sum_1^\infty y_j(\omega_0)$ converges. To show this let A_1 be any Lebesgue measurable subset of A of finite positive measure. Then

$$\lim_{n\to\infty} \int_{A_1} e^{i\lambda \sum_1^n y_j(\omega_0)}\, d\lambda = \int_{A_1} f(\lambda, \omega_0)\, d\lambda.$$

It follows that, if $\limsup_{n\to\infty} \left| \sum_1^n y_j(\omega_0)\right| = \infty$, then the right side must vanish (for every A_1) by the classical Lebesgue theorem on trigonometric integrals. But then $f(\lambda, \omega_0) = 0$ for almost all $\lambda \in A$, and this contradicts the obvious fact that $|f| = 1$. Then $\limsup_{n\to\infty} \left| \sum_1^n y_j(\omega_0)\right| < \infty$. If s_1 and s_2 are unequal limiting values of the partial sums of the series $\sum_1^\infty y_j(\omega_0)$ it follows that s_1 and s_2 are finite and

$$e^{i\lambda s_1} = e^{i\lambda s_2}$$

for almost all $\lambda \in A$. But this equality is impossible for two values of λ whose ratio is irrational. Hence there is only a single limiting value of the partial sums of the series $\sum_1^\infty y_j(\omega_0)$, that is, the series converges, as was to be proved.

Next we apply Theorem 4.1 to the x_n process defined by $x_n = \sum_1^n y_j$, assuming that $\mathbf{E}\{y_j\} = 0$ for $j > 1$, so that the x_n process is a martingale. By Theorem 4.1 (i),

$$\mathbf{E}\{|x_1|\} \le \mathbf{E}\{|x_2|\} \le \cdots,$$

$\sum_1^\infty y_j$ converges with probability 1 if $\lim_{n\to\infty} \mathbf{E}\{|x_n|\} = K < \infty$, and in that case $\mathbf{E}\{|x_\infty|\} \leq K$. The following theorem provides a converse to this result which is not true for all martingales.

THEOREM 5.1 *Let $y_1, y_2, \cdots$ be mutually independent real random variables with $\mathbf{E}\{y_j\} = 0$ for $j > 1$, and suppose that $\sum_1^\infty y_j$ converges with probability 1 to a sum x_∞ with $\mathbf{E}\{|x_\infty|\} < \infty$. Then $\mathbf{E}\{x_\infty\} = \mathbf{E}\{y_1\}$ and, if $x_n = \sum_1^n y_j$,*

$$\mathbf{E}\{\underset{n}{\text{L.U.B.}}\, |x_n|^\alpha\} \leq 8\mathbf{E}\{|x_\infty|^\alpha\}, \qquad \alpha \geq 1. \tag{5.1}$$

The right side may of course be infinite if $\alpha > 1$. If $\alpha > 1$, Theorem 3.4 is applicable, and implies (5.1) for $\alpha > 1.3$. We therefore need only prove (5.1) for $\alpha \leq 1.3$, but this restriction on α will not be made until it is stated explicitly. First suppose that the y_j's are symmetrically distributed. Then by a trivial extension of III, Theorem 2.2,

$$\mathbf{P}\{\underset{n}{\text{L.U.B.}}\, x_n(\omega) \geq \lambda\} \leq 2\mathbf{P}\{x_\infty(\omega) \geq \lambda\} = \mathbf{P}\{|x_\infty(\omega)| \geq \lambda\}, \qquad \lambda \geq 0,$$

and it follows that

$$\mathbf{E}\{\underset{n}{\text{L.U.B.}}\, |x_n|^\alpha\} = \int_0^\infty \mathbf{P}\{\underset{n}{\text{L.U.B.}}\, |x_n(\omega)|^\alpha \geq \lambda\}\, d\lambda \leq 2 \int_0^\infty \mathbf{P}\{|x_\infty(\omega)|^\alpha \geq \lambda\}\, d\lambda$$

$$= 2\mathbf{E}\{|x_\infty|^\alpha\}.$$

In the unsymmetric case let $y_1^*, y_2^*, \cdots$ be random variables independent of each other and of the y_j's, and let y_n^* have the same distribution as y_n, for every n. Let $x_n^* = \sum_1^n y_j^*$, $x_\infty^* = \sum_1^\infty y_j^*$. Since the $(y_j - y_j^*)$'s are symmetrically distributed, we have

$$\mathbf{E}\{\underset{n}{\text{L.U.B.}}\, |x_n - x_n^*|^\alpha\} \leq 2\mathbf{E}\{|x_\infty - x_\infty^*|^\alpha\} \leq 2^{\alpha+1}\mathbf{E}\{|x_\infty|^\alpha\}. \tag{5.2}$$

Next we show that

$$\mathbf{E}\{x_\infty \mid x_1, \cdots, x_n\} = x_n. \tag{5.3}$$

In fact,

$$\begin{aligned} \mathbf{E}\{x_\infty \mid x_1, \cdots, x_n\} &= \mathbf{E}\{x_\infty - x_n \mid x_1, \cdots, x_n\} + x_n \\ &= \mathbf{E}\{x_\infty - x_n\} + x_n = \mathbf{E}\{x_\infty\} - \mathbf{E}\{y_1\} + x_n, \end{aligned} \tag{5.4}$$

and $\mathbf{E}\{x_\infty\} = \mathbf{E}\{y_1\}$ because, when $n \to \infty$, x_n and the left side of the equation go to x_∞ (by Theorem 4.3, Corollary 1). Thus (5.3) is true and

(Example 1 of §1) the process $\{x_n, 1 \le n \le \infty\}$ is a martingale. If $\alpha \ge 1$ we have, from Theorem 4.1 (iii),

$$\mathbf{E}\{|x_1|^\alpha\} \le \mathbf{E}\{|x_2|^\alpha\} \le \cdots, \qquad \lim_{n\to\infty} \mathbf{E}\{|x_n|^\alpha\} = \mathbf{E}\{|x_\infty|^\alpha\}.$$

Now, if $\alpha \ge 1$,

$$|x_n|^\alpha \le 2^{\alpha-1} \operatorname*{L.U.B.}_m |x_m - x_m{}^*|^\alpha + 2^{\alpha-1}|x_n{}^*|^\alpha.$$

Hence

$$|x_n|^\alpha \le 2^{\alpha-1} \mathbf{E}\{\operatorname*{L.U.B.}_m |x_m - x_m{}^*|^\alpha \mid x_1, x_2, \cdots\} + 2^{\alpha-1} \mathbf{E}\{|x_n{}^*|^\alpha\},$$

so that, using (5.2),

(5.5)

$$\begin{aligned}\mathbf{E}\{\operatorname*{L.U.B.}_n |x_n|^\alpha\} &\le 2^{\alpha-1} \mathbf{E}\{\operatorname*{L.U.B.}_m |x_m - x_m{}^*|^\alpha\} + 2^{\alpha-1} \operatorname*{L.U.B.}_m \mathbf{E}\{|x_m|^\alpha\} \\ &\le 2^{2\alpha} \mathbf{E}\{|x_\infty|^\alpha\} + 2^{\alpha-1} \mathbf{E}\{|x_\infty|^\alpha\} \\ &\le 8\mathbf{E}\{|x_\infty|^\alpha\}, \qquad 1 \le \alpha \le 1.3.\end{aligned}$$

Theorem 4.1 gave general convergence criteria for martingales $\{x_n, n \ge 1\}$. In particular, suppose that $x_n = \sum_1^n y_j$, where the y_j's are mutually independent, and $\mathbf{E}\{y_j\} = 0, j > 1$. Then we have two theorems which allow the strengthening of Theorem 4.1: The zero-one law (III, Theorem 1.1) implies that the sequence $\{x_n\}$ converges to a finite limit with probability 0 or 1; Theorem 5.1 states that in case there is convergence with probability 1, to the limit x_∞, then, if $\mathbf{E}\{|x_\infty|\} < \infty$, it follows that $\mathbf{E}\{\text{L.U.B.}\ |x_n|\} < \infty$. This means among other things that parts (i) and (ii) of Theorem 4.1 coalesce in this special case. The strengthened version of Theorem 4.1 in this special case is the following:

Let $y_1, y_2, \cdots$ be mutually independent random variables, with

$$\mathbf{E}\{|y_j|\} < \infty, \qquad j \ge 1$$
$$\mathbf{E}\{y_j\} = 0, \qquad j > 1.$$

Then, if $x_n = \sum_1^n y_j$,

$$\mathbf{E}\{|x_1|^\alpha\} \le \mathbf{E}\{|x_2|^\alpha\} \le \cdots, \qquad \alpha \ge 1.$$

(i) *If* $\lim_{n\to\infty} \mathbf{E}\{|x_n|^\alpha\} < \infty$ *for some* $\alpha \ge 1$, *then* $\lim_{n\to\infty} x_n = x_\infty$ *exists with probability* 1, $\mathbf{E}\{\operatorname*{L.U.B.}_n |x_n|^\alpha\} < \infty$, $\lim_{n\to\infty} \mathbf{E}\{|x_n - x_\infty|^\alpha\} = 0$, *and the sequence* $\{x_n, n \le \infty\}$ *is a martingale. Conversely, if* $\lim x_n = x_\infty$ *exists with probability* 1, *and if, for some* $\alpha \ge 1$, $\mathbf{E}\{|x_\infty|^\alpha\} < \infty$, *then* $\lim_{n\to\infty} \mathbf{E}\{|x_n|^\alpha\} < \infty$.

(ii) *If the y_j's are real, and if*

$$\mathbf{E}\{\text{L.U.B.}_n\, y_n\} < \infty,$$

then $\lim_{n\to\infty} x_n$ *exists and is finite with probability* 1 *if* $\limsup_{n\to\infty} x_n < \infty$ *with positive probability.*

(iii) *If*

$$\mathbf{E}\{\text{L.U.B.}_n\, |y_n|^2\} < \infty, \tag{5.6}$$

then $\lim_{n\to\infty} x_n$ *exists and is finite with probability* 1 *if and only if* $\sum_1^\infty \mathbf{E}\{|y_n|^2\} < \infty$.

Parts (i) and (ii) need no further comment. Part (iii) states that if (5.6) is true then $\sum_1^\infty y_j$ converges with probability 1 if and only if there is convergence in the mean. The "if" half of this statement is not interesting, because according to the Corollary to III, Theorem 2.7, convergence in the mean of any series of mutually independent random variables implies convergence with probability 1. The "only if" half generalizes III, Theorem 2.4, which replaces (5.6) by the stronger hypothesis that the y_j's are uniformly bounded.

Before continuing, we remark that, if z_1 and z_2 are mutually independent random variables, then, if $\mathbf{E}\{|z_1 + z_2|\} < \infty$, it follows that $\mathbf{E}\{|z_1|\}$ and $\mathbf{E}\{|z_2|\}$ are also finite. It is sufficient to prove that $\mathbf{E}\{|z_1|\} < \infty$, and this follows from the inequality

$$\mathbf{E}\{|z_1 + z_2|\} \geq \int_{\{|z_2(\omega)| \leq a\}} [|z_1| - a]\, d\mathbf{P} = \mathbf{P}\{|z_2(\omega)| \leq a\}[\mathbf{E}\{|z_1|\} - a].$$

THEOREM 5.2 *Let $y_1, y_2, \cdots$ be mutually independent random variables, and suppose that $\sum_1^\infty y_j$ converges with probability* 1 *to a sum x_∞ with $\mathbf{E}\{|x_\infty|^\alpha\} < \infty$ for some $\alpha \geq 1$. Then $\mathbf{E}\{|y_j|^\alpha\} < \infty$ for all j,*

$$\mathbf{E}\{x_\infty\} = \sum_1^\infty \mathbf{E}\{y_j\},$$

$$\mathbf{E}\{\text{L.U.B.}_n\, |\sum_1^n y_j|^\alpha\} \leq 8\cdot 2^{2\alpha-1}\, \mathbf{E}\{|x_\infty|^\alpha\} + 2^{\alpha-1}\, \text{L.U.B.}_n\, |\sum_1^n \mathbf{E}\{y_j\}|^\alpha, \tag{5.7}$$

and

$$\lim_{n\to\infty} \mathbf{E}\{|x_\infty - \sum_1^n y_j|^\alpha\} = 0. \tag{5.8}$$

The last statement, that the partial sums converge in the mean with index α, follows from the fact that according to (5.7) the quantity in the brace in (5.8) is dominated by a function with a finite expectation, and (5.8) will accordingly not be mentioned further. Since the y_j's are mutually independent, and since $x_\infty = y_n + \sum_{j \neq n} y_j$, with $\mathbf{E}\{|x_\infty|\} < \infty$, it follows that $\mathbf{E}\{|y_n|\} < \infty$ also. Let $x_n = \sum_1^n y_j$. Then (5.4) becomes

$$\mathbf{E}\{x_\infty \mid x_1, \cdots, x_n\} = \mathbf{E}\{x_\infty\} - \mathbf{E}\{x_n\} + x_n.$$

We now find when $n \to \infty$ that

$$\mathbf{E}\{x_\infty\} = \lim_{n \to \infty} \mathbf{E}\{x_n\}.$$

It follows that

$$\sum_1^\infty [y_j - \mathbf{E}\{y_j\}] = x_\infty - \mathbf{E}\{x_\infty\}.$$

Then, by Theorem 5.1,

$$\mathbf{E}\{\text{L.U.B.}_n \mid \sum_1^n [y_j - \mathbf{E}\{y_j\}]|^\alpha\} \leq 8\mathbf{E}\{|x_\infty - \mathbf{E}\{x_\infty\}|^\alpha\},$$

so that

$$\begin{aligned}\mathbf{E}\{\text{L.U.B.}_n \mid \sum_1^n y_j|^\alpha\} &\leq 8 \cdot 2^{\alpha-1}\, \mathbf{E}\{|x_\infty - \mathbf{E}\{x_\infty\}|^\alpha\} \\ &\quad + 2^{\alpha-1}\, \text{L.U.B.}_n \mid \sum_1^n \mathbf{E}\{y_j\}|^\alpha \\ &\leq 8 \cdot 2^{2\alpha-1}\, \mathbf{E}\{|x_\infty|^\alpha\} + 2^{\alpha-1}\, \text{L.U.B.}_n \mid \sum_1^n \mathbf{E}\{y_j\}|^\alpha.\end{aligned}$$

This completes our discussion of the application of martingale theory to the study of series of mutually independent random variables. It is significant that, even though the general martingale convergence theorems can be strengthened when martingales defined by the partial sums of mutually independent random variables are studied, martingale theory is useful in this very strengthening.

The main stress in this section has been on the application of Theorem 4.1. Many of the results could have been obtained, however, by applying Theorem 4.2 to the martingale with random variables $\cdots, z_{-2}, z_{-1}$, where

$$z_{-n} = \mathbf{E}\{\sum_1^\infty y_j \mid y_n, y_{n+1}, \cdots\} = \sum_n^\infty y_j + \mathbf{E}\{\sum_1^{n-1} y_j\}.$$

(Here we are assuming that $\sum_1^\infty y_j$ converges with probability 1 and that the expectation of the sum exists.)

6. Application to the strong law of large numbers

Let $y_1', y_2', \cdots$ be mutually independent random variables with a common distribution function, and suppose that $\mathbf{E}\{|y_1'|\} < \infty$. Then

$$\lim_{n\to\infty} \frac{y_1' + \cdots + y_n'}{n} = \mathbf{E}\{y_1'\} \tag{6.1}$$

with probability 1. This theorem was proved in III (Theorem 5.1) as an application of a theorem on the convergence of an infinite series of mutually independent random variables. It can also be deduced as a special case of the strong law of large numbers for strictly stationary processes (X, Theorem 2.1). The wide sense version of (6.1) is the following. Let $Y_1, Y_2, \cdots$ be an orthonormal sequence of random variables. Then

$$\operatorname*{l.i.m.}_{n\to\infty} \frac{Y_1 + \cdots + Y_n}{n} = 0. \tag{6.1'}$$

The wide sense version is trivial because

$$\mathbf{E}\{|\frac{Y_1 + \cdots + Y_n}{n}|^2\} = \frac{1}{n} \to 0 \qquad (n \to \infty).$$

In IV §7 a proof of (6.1′) was given as an application of the convergence theorems for wide sense martingales. We now show how this proof can be translated into a proof of (6.1) by replacing the wide sense concepts by strict sense concepts.

According to Theorem 4.3, Corollary 1, if $y_n = y_1' + \cdots + y_n'$,

$$\lim_{n\to\infty} \mathbf{E}\{y_1' \mid y_n, y_{n+1}, \cdots\} = x_{-\infty}$$

exists with probability 1. Now it is clear that

$$\mathbf{E}\{y_1' \mid y_n, y_{n+1}, \cdots\} = \mathbf{E}\{y_1' \mid y_n, y_{n+1}', y_{n+2}', \cdots\}$$

and, since $y_{n+1}', y_{n+2}', \cdots$ are independent of the pair y_1', y_n,

$$\mathbf{E}\{y_1' \mid y_n, y_{n+1}', \cdots\} = \mathbf{E}\{y_1' \mid y_n\},$$

with probability 1. Thus

$$\lim_{n\to\infty} \mathbf{E}\{y_1' \mid y_1' + \cdots + y_n'\} = x_{-\infty}$$

with probability 1.

Finally, by symmetry,

$$\begin{aligned}
\mathbf{E}\{y_1' \mid y_1' + \cdots + y_n'\} &= \mathbf{E}\{y_j' \mid y_1' + \cdots + y_n'\} \qquad j \leq n \\
&= \frac{1}{n} \sum_{j=1}^{n} \mathbf{E}\{y_j' \mid y_1' + \cdots + y_n'\} \\
&= \mathbf{E}\{\frac{1}{n} \sum_{j=1}^{n} y_j' \mid y_1' + \cdots + y_n'\} \\
&= \frac{y_1' + \cdots + y_n'}{n},
\end{aligned}$$

so that (6.1) is proved except for the identification of the limit. To identify the limit we note that in the first place $x_{-\infty}$ is unaffected by changes in any finite number of y_j's and therefore, by the zero-one law (III, Theorem 1.1), $x_{-\infty} = \text{const.} = \mathbf{E}\{x_{-\infty}\}$ with probability 1. In the second place according to Theorem 4.4 the sequence

$$x_{-\infty}, \cdots, \mathbf{E}\{y_1' \mid y_2, y_3, \cdots\}, \mathbf{E}\{y_1' \mid y_1, y_2, \cdots\}, y_1'$$

is a martingale and therefore $\mathbf{E}\{x_{-\infty}\} = \mathbf{E}\{y_1'\}$.

It would be interesting to generalize the above proof to obtain a martingale proof of the general case of the strong law of large numbers for strictly stationary processes (X, Theorem 2.1, ergodic theorem), but no such proof has ever been given. Wiener has pointed out that the ergodic theorem is really an integration theorem, closely related to the fundamental theorem of the calculus. Martingale theorems are also essentially integration theorems, although in a somewhat different context. In fact, the equation

$$\mathbf{E}\{x_n \mid x_1, \cdots, x_{n-1}\} = x_{n-1},$$

whose validity with probability 1 is the defining property of a martingale $\{x_n, n \geq 1\}$, shows that x_{n-1} is obtained from x_n by integrating out one variable. More insight in this direction will be given in the following section.

7. Application to integration in infinitely many dimensions

The significance of the martingale convergence theorems as integration theorems is exhibited very clearly in the following examples, first discussed (from another point of view) by Jessen. Let the basic ω space be the space of infinite sequences $(\eta_1, \eta_2, \cdots)$, $0 \leq \eta_j \leq 1$, and let the given probability measure be infinite dimensional Lebesgue measure. Then, if y_j is the jth coordinate variable, $y_1, y_2, \cdots$ are mutually independent random variables, each uniformly distributed over the interval [0, 1]. Let z be a random variable with $\mathbf{E}\{|z|\} < \infty$ and consider the integrals

$$
\begin{aligned}
x_{-n}(\omega) &= \int_0^1 \cdots \int_0^1 z(\eta_1, \eta_2, \cdots)\, d\eta_1 \cdots d\eta_{n-1} = \mathbf{E}\{z \mid y_n, y_{n+1}, \cdots\} \\
x_n(\omega) &= \int_0^1 \cdots \int_0^1 z(\eta_1, \eta_2, \cdots)\, d\eta_{n+1}\, d\eta_{n+2} \cdots = \mathbf{E}\{z \mid y_1, \cdots, y_n\}.
\end{aligned}
\tag{7.1}
$$

The integral forms make it intuitively reasonable that

$$
\begin{aligned}
&\lim_{n \to -\infty} x_n = \mathbf{E}\{z\} \\
&\lim_{n \to \infty} x_n = z
\end{aligned}
\tag{7.2}
$$

with probability 1. Actually the expressions for the x_n's as conditional expectations show that the sequences $\cdots, x_{-2}, x_{-1}$ and $x_1, x_2, \cdots$ are martingales. Hence (Theorem 4.2) $\lim_{n \to -\infty} x_n = x_{-\infty}$ with probability 1 and (Theorem 4.3, Corollary 1) $\lim_{n \to \infty} x_n = z$ with probability 1. It remains to identify $x_{-\infty}$ with $\mathbf{E}\{z\}$. Since $x_{-\infty}$ is independent of $y_1, \cdots, y_n$ for every n, $x_{-\infty} = \mathbf{E}\{x_{-\infty}\}$, with probability 1 (zero-one law), and by Theorem 4.2 $\mathbf{E}\{x_{-\infty}\} = \mathbf{E}\{z\}$, so that $x_{-\infty} = \mathbf{E}\{z\}$ with probability 1.

Note that the results (and proofs) are valid whenever the y_j's are mutually independent; the hypothesis of a common distribution function (with constant density) only served to simplify the integration notation.

8. Application to the theory of derivatives

Let Ω be an abstract space, and let $\mathbf{P}\{\cdot\}$ be a probability measure of ω sets, as usual. For each n let $M_0^{(n)}, M_1^{(n)}, \cdots$ be finitely or denumerably infinitely many disjunct measurable ω sets, with union Ω. Let $\mathscr{F}_n$ be the Borel field of ω sets which are unions of $M_j^{(n)}$'s (fixed n). The class $\bigcup_n \mathscr{F}_n$ is a field. Let $\mathscr{F}_\infty$ be the Borel field generated by this field. We suppose that each $M_j^{(n+1)}$ is a subset of some $M_k^{(n)}$. Then

$$\mathscr{F}_1 \subset \mathscr{F}_2 \subset \cdots.$$

Let φ be any function of sets of the field $\bigcup_n \mathscr{F}_n$, completely additive on $\mathscr{F}_n$ for each n, and define the ω function x_n by

$$
\begin{aligned}
x_n(\omega) &= \frac{\varphi(M_j^{(n)})}{\mathbf{P}\{M_j^{(n)}\}}, \qquad \omega \in M_j^{(n)}, \qquad \text{if } \mathbf{P}\{M_j^{(n)}\} > 0, \\
&= 0, \qquad \omega \in M_j^{(n)}, \qquad \text{if } \mathbf{P}\{M_j^{(n)}\} = 0.
\end{aligned}
\tag{8.1}
$$

Then, if $m < n$, it is trivial to verify that

$$\mathbf{E}\{x_n \mid \mathscr{F}_m\} = x_m$$

with probability 1, so that $\{x_n, \mathscr{F}_n, n \geq 1\}$ is a martingale. The present section is devoted to various applications of this fact. Note that the results are not really limited to cases where the basic measure is a probability measure, since any finite-valued measure (not identically 0) can be reduced to a probability measure by the introduction of a suitable proportionality factor.

In the following we shall say that we are *in the Lebesgue case* if Ω is the linear interval [0, 1], if the measurable ω sets are the Lebesgue measurable subsets of [0, 1], and if $\mathbf{P}$ measure is Lebesgue measure.

Example 1 (absolutely continuous φ) Suppose that φ is absolutely continuous, that is, that there is an $\mathscr{F}_\infty$ measurable ω function x, with $\mathbf{E}\{|x|\} < \infty$, such that

$$\varphi(\Lambda) = \int_\Lambda x \, d\mathbf{P}, \qquad \Lambda \in \mathscr{F}_\infty.$$

Then it is trivial to verify that

$$\mathbf{E}\{x \mid \mathscr{F}_n\} = x_n$$

with probability 1. Then in this case the random variables $x_1, x_2, \cdots, x$ form a martingale, and, according to Theorem 4.3,

$$\lim_{n\to\infty} x_n = \mathbf{E}\{x \mid \mathscr{F}_\infty\} = x \tag{8.2}$$

with probability 1. In this case, according to Theorem 3.1 applied to the semi-martingale $|x_1|, |x_2|, \cdots, |x|$, the x_n's are uniformly integrable. Conversely, if the x_n's are uniformly integrable, then $\lim_{n\to\infty} x_n = x_\infty$ exists with probability 1, $\lim_{n\to\infty} \mathbf{E}\{|x_\infty - x_n|\} = 0$, and the process $\{x_n, 1 \leq n \leq \infty\}$ is a martingale, by Theorem 4.1 (ii). But then, if we define φ_1 by

$$\varphi_1(\Lambda) = \int_\Lambda x_\infty \, d\mathbf{P},$$

we have (martingale property)

$$\varphi_1(\Lambda) = \int_\Lambda x_n \, d\mathbf{P}, \qquad \Lambda \in \bigcup_m \mathscr{F}_m$$

for n so large that $\Lambda \in \mathscr{F}_n$, so that when $n \to \infty$ we find that

$$\varphi(\Lambda) = \varphi_1(\Lambda), \qquad \Lambda \in \bigcup_m \mathscr{F}_m.$$

This equality is therefore true for $\Lambda \in \mathscr{F}_\infty$ (see Supplement, Theorem 2.1). We have thus shown that the x_n's are uniformly integrable if and only if

φ is absolutely continuous (relative to $\mathscr{F}_\infty$). If φ is defined on the field $\mathscr{F}$ of all measurable ω sets, and is absolutely continuous relative to $\mathscr{F}$, with density function $X[x]$ relative to $\mathscr{F}[\mathscr{F}_\infty]$, then

$$\mathbf{E}\{X \mid \mathscr{F}_\infty\} = x, \qquad \mathbf{E}\{X \mid \mathscr{F}_n\} = \mathbf{E}\{x \mid \mathscr{F}_n\} = x_n,$$

with probability 1, and the above argument remains valid in every detail. Finally, $x = X$ with probability 1 if and only if X is equal almost everywhere to a function measurable with respect to $\mathscr{F}_\infty$. For example, in the Lebesgue case if the $M_j^{(n)}$'s are intervals, and if δ_n is the least upper bound of the lengths of $M_1^{(n)}, M_2^{(n)}, \cdots$, then, if $\lim_{n\to\infty} \delta_n = 0$, it is clear that $\mathscr{F}_\infty$ contains every subinterval of [0, 1], and therefore every Borel subset of this interval. It follows that in this case $x = X$ with probability 1.

Example 2 (singular φ) Suppose that φ is singular, that is, we suppose that φ is defined on the sets of $\mathscr{F}_\infty$, that φ is completely additive, and that there is a $M \in \mathscr{F}_\infty$, the singular set of φ, such that $\varphi(\Lambda) \neq 0$ for some $\Lambda \subset M$, $\Lambda \in \mathscr{F}_\infty$, and that

$$\mathbf{P}\{M\} = 0$$

$$\varphi(\Lambda) = 0, \qquad \Lambda \subset \Omega - M \qquad (\Lambda \in \mathscr{F}_\infty).$$

Suppose first that φ is non-negative. Then the x_n's are non-negative; so, by Theorem 4.1 (i), $\lim_{n\to\infty} x_n = x_\infty$ exists and is finite with probability 1. By Fatou's lemma,

$$\int_\Lambda x_\infty \, d\mathbf{P} \le \liminf_{n\to\infty} \int_\Lambda x_n \, d\mathbf{P},$$

if Λ is measurable. On the other hand,

$$\varphi(\Lambda) = \int_\Lambda x_n \, d\mathbf{P}, \qquad \Lambda \in \mathscr{F}_m, \qquad m \le n.$$

Then combining these two relations we find that

$$\int_\Lambda x_\infty \, d\mathbf{P} \le \varphi(\Lambda), \qquad \Lambda \in \bigcup_n \mathscr{F}_n.$$

Since $\mathscr{F}_\infty$ is the Borel field generated by the field $\bigcup_n \mathscr{F}_n$, this inequality must be true for $\Lambda \in \mathscr{F}_\infty$, since it is true for $\Lambda \in \bigcup_n \mathscr{F}_n$ (see Supplement, Theorem 2.1). But then, if M is the singular set of φ,

$$\int_{\Omega - M} x_\infty \, d\mathbf{P} \le \varphi(\Omega - M) = 0,$$

so that (since $x_\infty \geq 0$) $x_\infty = 0$ with probability 1. If φ is not necessarily non-negative, it can be expressed as the difference between two non-negative singular functions so that again we deduce that $\lim_{n\to\infty} x_n = 0$ with probability 1.

Example 3 (φ of bounded variation) Suppose that φ is defined on the $\mathscr{F}_\infty$ sets and is completely additive. We assume, as we always do in this book in considering set functions, that φ is finite-valued. Then according to a standard theorem of the theory of set functions φ is of bounded variation. That is there is a K such that, for any disjunct $\mathscr{F}_\infty$ sets $\Lambda_1, \Lambda_2, \cdots$,

$$\sum_j |\varphi(\Lambda_j)| \leq K.$$

Then

$$\mathbf{E}\{|x_n|\} = \sum_j |\varphi(\mathrm{M}_j^{(n)})| \leq K,$$

so that, by Theorem 4.1 (i), $\lim_{n\to\infty} x_n = x_\infty$ exists and is finite with probability 1. The function x_∞ is called the derivative of φ with respect to the given probability measure, relative to the net of the $\mathrm{M}_j^{(n)}$'s, and we have thus proved that every completely additive set function has a derivative with respect to a given measure, relative to a given net. Since φ is the sum of an absolutely continuous and a singular function, relative to $\mathscr{F}_\infty$, the results of the preceding examples identify the derivative x_∞ as the density of the absolutely continuous component of φ relative to $\mathscr{F}_\infty$. If φ is defined on $\mathscr{F}$, this derivative is the density of the absolutely continuous component of φ relative to $\mathscr{F}$ if and only if the net is fine enough, more precisely if and only if that density is equal almost everywhere to a function measurable with respect to $\mathscr{F}_\infty$ and if the singular component of φ relative to $\mathscr{F}$ has an $\mathscr{F}_\infty$ set as singular set.

Example 4 (Lebesgue case) In the Lebesgue case, for each n let $0 = \xi_0^{(n)} < \cdots < \xi_{a_n}^{(n)} = 1$, where each $\xi_j^{(n+1)}$ is a $\xi_k^{(n)}$, and define $\mathrm{M}_j^{(n)}$ by

$$\begin{aligned} \mathrm{M}_j^{(n)} &= [0, \xi_1^{(n)}], \qquad & j = 0 \\ &= (\xi_j^{(n)}, \xi_{j+1}^{(n)}], \qquad & j > 0. \end{aligned}$$

Let F be a numerically valued real function on [0, 1], and define

$$\varphi(\bigcup_{j \in J} \mathrm{M}_j^{(n)}) = \sum_{j \in J} [F(\xi_{j+1}^{(n)}) - F(\xi_j^{(n)})].$$

Then the definition of x_n becomes

$$x_n(\xi) = \frac{F(\xi_{j+1}^{(n)}) - F(\xi_j^{(n)})}{\xi_{j+1}^{(n)} - \xi_j^{(n)}}, \qquad \xi \in \mathrm{M}_j^{(n)}.$$

We suppose in the following that

$$\lim_{n\to\infty} \operatorname{Max}_{j} (\xi_{j+1}^{(n)} - \xi_j^{(n)}) = 0,$$

so that $\mathscr{F}_\infty$ is the class of Borel subsets of [0, 1]. Suppose that F is a function of bounded variation, and assume as known the fact that the derivative F' exists almost everywhere, that the absolutely continuous component F_1 of F is the integral of F',

$$F = F_1 + F_2, \qquad F_1(\xi) = \int_0^\xi F'(\eta)\, d\eta,$$

and that the singular component F_2 of F' has derivative 0 almost everywhere. Then, according to Example 3,

$$\lim_{n\to\infty} x_n = F'$$

with probability 1. On the other hand, the results of Example 3 can also be used to derive the above-stated facts about the existence and significance of F'. We omit a discussion of this point. Now let R be the set of $\xi_j^{(n)}$'s, $j, n \geq 0$. According to Example 1, if φ is absolutely continuous (which is equivalent to the condition that F coincides on R with an absolutely continuous function defined on [0, 1]), the x_n's are uniformly integrable, and conversely, if the x_n's are uniformly integrable, F coincides on R with an absolutely continuous function defined on [0, 1]. Interesting pathological examples of martingales can be obtained by choosing F to be a singular function. The following is such an example, in which F is constant except for a jump at a single point. Let F be defined by

$$\begin{aligned} F(\xi) &= 0, \qquad 0 \leq \xi < \tfrac{1}{2} \\ &= 1, \qquad \tfrac{1}{2} \leq \xi \leq 1, \end{aligned}$$

and set

$$\xi_j^{(n)} = \frac{j}{2^{n+1}}, \qquad j = 0, \cdots, 2^{n+1}.$$

Then

$$\begin{aligned} x_n(\xi) &= 0, \qquad 0 \leq \xi \leq \frac{1}{2} - \frac{1}{2^{n+1}} \\ &= 2^{n+1}, \qquad \frac{1}{2} - \frac{1}{2^{n+1}} < \xi \leq \frac{1}{2} \\ &= 0, \qquad \frac{1}{2} < \xi \leq 1. \end{aligned}$$

The process $\{x_n, n \geq 1\}$ is a martingale, the x_n's are non-negative, $\mathbf{E}\{x_n\} = 1$, and $\lim_{n\to\infty} x_n = x_\infty = 0$ with probability 1;. in fact, $\xi = \frac{1}{2}$ is the only point where there is not convergence to 0. The process $\{x_n, 1 \leq n \leq \infty\}$ is not a martingale, since $\mathbf{E}\{x_1\} \neq \mathbf{E}\{x_\infty\}$. This is the first example we have given of a martingale $x_1, x_2, \cdots$ with limit x_∞, for which $x_1, x_2, \cdots, x_\infty$ is not also a martingale.

9. Application to likelihood ratios in statistics

We now consider II §7, Example 3: $y_1, y_2, \cdots$ are random variables; the distribution of $y_1, \cdots, y_n$ is determined by a Baire density function $p_n(\cdot, \ldots, \cdot)$, or alternatively by $q_n(\cdot, \ldots, \cdot)$. Define x_n by

$$x_n = \frac{q_n(y_1, \cdots, y_n)}{p_n(y_1, \cdots, y_n)}.$$

Then, if $y_1, y_2, \cdots$ are random variables with distributions determined by the p_j's, x_n is a random variable, defined with probability 1. We have seen in II §7 that the sequence $x_1, x_2, \cdots$ is a martingale if $q_n(\xi, \cdots, \xi_n) = 0$ whenever $p_n(\xi_1, \cdots, \xi_n) = 0$. We do not make this hypothesis here; without it the discussion of II §7 proves that the sequence is a lower semi-martingale, that is, that $-x_1, -x_2, \cdots$ is a semi-martingale. Since the $-x_j$'s are non-positive, Theorem 4.1s states that

$$\lim_{n\to\infty} x_n = x_\infty \tag{9.1}$$

exists with probability 1 and

$$1 \geq \mathbf{E}\{x_1\} \geq \mathbf{E}\{x_2\} \geq \cdots, \qquad \mathbf{E}\{x_\infty\} \leq \lim_{n\to\infty} \mathbf{E}\{x_n\}. \tag{9.2}$$

These two statements can be considered a generalized maximum likelihood statement. In fact, the general idea of the principle of maximum likelihood in statistics is that, if $y_1, \cdots, y_n$ probabilities are determined by the density $p_n(\cdot, \ldots, \cdot)$, then $x_n \leq 1$, that is,

$$q_n(y_1, \cdots, y_n) \leq p_n(y_1, \cdots, y_n) \tag{9.3}$$

in some average sense. Besides (9.2) the inequality

$$\mathbf{E}\{\log x_n\} \leq \log \mathbf{E}\{x_n\} = \log \mathbf{E}\{q_n/p_n\} \leq \log 1 = 0 \tag{9.4}$$

is an expression of this fact, and this inequality is fundamental in the study of the consistency of maximum likelihood estimates in statistics. The idea of (9.3) is sometimes expressed by the statement that, if $[y_1(w), \cdots, y_n(\omega)]$ is a set of sample values, these actually obtained values are more probable when calculations are made in terms of the correct density than when calculations are made in terms of any other.

If proper hypotheses are imposed on the p_n's and q_n's, x_n for large n will exhibit this tendency to be ≤ 1 in a striking way. For example, suppose that the p_n's and q_n's correspond to independent y_j's with common distributions, that is,

$$p_n(y_1, \cdots, y_n) = \prod_1^n p_1(y_j)$$

$$x_n = \prod_1^n \frac{q_1(y_j)}{p_1(y_j)}.$$

$$q_n(y_1, \cdots, y_n) = \prod_1^n q_1(y_j)$$

We prove that in this case

$$\lim_{n\to\infty} x_n = x_\infty = 0$$

with probability 1, unless $p_1(\xi) = q_1(\xi)$ for almost all values of ξ (Lebesgue measure), that is, unless the p_1 and q_1 distributions are identical. This is a result of precisely the type desired.

(a) *We prove first that* $x_\infty = 0$ *with probability* 0 *or* 1. In fact, from the form of the infinite product defining x_∞ it is clear that, if $x_\infty{}^*$ has the same distribution as x_∞ and is independent of x_∞, then $x_\infty x_\infty{}^*$ also has the distribution of x_∞. Then

$$\eta = \mathbf{P}\{x_\infty(\omega)x_\infty{}^*(\omega) = 0\} = 1 - (1 - \eta)^2$$

so that

$$\eta = 0 \qquad \text{or} \qquad \eta = 1.$$

(b) *If* $x_\infty = 0$ *with probability* 0, *we shall show that* $x_\infty = 1$ *with probability* 1. In fact, in this case $\log (x_\infty x_\infty{}^*) = \log x_\infty + \log x_\infty{}^*$ has the same distribution as $\log x$ and $\log x^*$. Then, if Φ is their common characteristic function, $\Phi^2 = \Phi$, which implies that $\Phi(t) \equiv 1$ [since $\Phi(0) = 1$ and Φ is continuous], so that $\log x_\infty = 0$ with probability 1. It follows that $x_\infty = 1$ with probability 1.

(c) *Finally it is clear that if* $x_\infty = 1$ *with probability* 1,

$$\prod_2^\infty \frac{q_1(y_j)}{p_1(y_j)} = 1$$

with probability 1 *also. Hence* $q_1(y_1)/p_1(y_1) = 1$ *with probability* 1. This means that $q_1(\xi) = p_1(\xi)$ almost everywhere where $p_1(\xi) > 0$, that is, almost everywhere on the ξ-axis since both functions have the integral 1 over the whole axis.

We have thus verified the general principle underlying the method of maximum likelihood for mutually independent random variables with a common distribution function: unless the two distributions are identical, the likelihood ratio x_n converges to zero with probability 1 when $n \to \infty$ (computing probability by means of the p_1 distribution). A statistician can use this principle as follows. He chooses a positive constant a and decides that, if a given sample $y_1(\omega), \cdots, y_n(\omega)$ makes $x_n(\omega) \geq a$, he will act as if the p distribution were the correct one; otherwise he will act as if the q distribution were the correct one. We have just proved that if n is large the statistician will act incorrectly with small probability, and in fact that in a sequence of trials there is probability 1 that the statistician will finally stop making mistakes. (In statistical language, his procedure is "consistent.") Ordinarily there is a whole family of distributions to choose from, $q(\cdot) = q(\theta, \cdot)$ depends on the parameter θ, one value of which is known to give the correct distribution $q(\theta_0, \cdot) = p(\cdot)$. The method of maximum likelihood, given a sample $y_1(\omega), \cdots, y_n(\omega)$, chooses $\theta_n[y_1(\omega), \cdots, y_n(\omega)]$ as that value θ maximizing $\prod_1^n q[\theta, y_j(\omega)]$, if there is one. According to the theorem just proved, as long as θ is chosen so that

$$\prod_1^n q[\theta, y_j(\omega)] \geq a \prod_1^n q[\theta_0, y_j(\omega)],$$

where a is a fixed positive constant, we cannot keep on picking the same incorrect value of θ. Further hypotheses must be imposed, however, to insure that $\theta_n \to \theta_0$ with probability 1.

10. Application to sequential analysis

Let $y_1, y_2, \cdots$ be mutually independent real random variables, with a common distribution function, and suppose that $\mathbf{E}\{y_j\} = a$ exists. Let m be an integral-valued random variable, with the property that for each k the condition $m(\omega) = k$ is a condition on only the first k y_j's, that is, the ω set $\{m(\omega) = k\}$ is determined by conditions on $y_1, \cdots, y_k$. Define x' by

$$x' = y_1 + \cdots + y_m.$$

Then it is important for certain problems in sequential analysis to find conditions under which

$$\mathbf{E}\{x'\} = a\mathbf{E}\{m\}. \tag{10.1}$$

This is easily solved by martingale theory. In fact, if x_n is defined by

$$x_n = \sum_1^n (y_j - a) = \sum_1^n y_j - na,$$

we have seen that the sequence $x_1, x_2, \cdots$ is a martingale. According to Theorem 2.2, if $\mathbf{E}\{m\} < \infty$, and if for some constant K,

$$\mathbf{E}\{|x_{n+1} - x_n| \mid x_1, \cdots, x_n\} = \mathbf{E}\{|y_{n+1} - a|\} \le K, \qquad n \le m(\omega), \tag{10.2}$$

with probability 1, then the "sequence" x_1, x_m, obtained by optional sampling, is a martingale, with $\mathbf{E}\{x_1\} = \mathbf{E}\{x_m\}$. In the present case (10.2) is certainly satisfied with $K = \mathbf{E}\{|y_1 - a|\}$. The equality $\mathbf{E}\{x_1\} = \mathbf{E}\{x_m\}$ means that

$$0 = \mathbf{E}\{x_1\} = \mathbf{E}\{x' - ma\} = \mathbf{E}\{x'\} - a\mathbf{E}\{m\},$$

the desired equation. The variance σ'^2 of x' is found as follows. The sequence

$$\{(\sum_1^n y_j - na)^2 - n\sigma^2, n = 1, 2, \cdots\}, \qquad \sigma^2 = \mathbf{E}\{(y_j - a)^2\},$$

is easily seen to be a martingale, if $\mathbf{E}\{|y_1|^2\} < \infty$, and in fact

$$\mathbf{E}\{(\sum_1^\nu y_j - \nu a)^2 - \nu\sigma^2 \mid y_1, \cdots, y_n\} = (\sum_1^n y_j - na)^2 - n\sigma^2, \qquad \nu > n,$$

with probability 1. Let $\mathscr{F}_n$ be the Borel field of the ω sets determined by conditions on $y_1, \cdots, y_n$. Then we have proved that

$$\{(\sum_1^n y_j - na)^2 - n\sigma^2, \mathscr{F}_n, n \ge 1\}$$

is a martingale. The condition C_3 of Theorem 2.2 becomes, as applied to this process,

$$\mathbf{E}\{|(y_{n+1} - a)[y_{n+1} - a + 2\sum_1^n (y_j - a)] - \sigma^2| \mid \mathscr{F}_n\} \le K, \qquad n \le m(\omega).$$

This will surely be true for properly chosen K if $\mathbf{E}\{y_1^2\} < \infty$ and if

$$|x_n| = |\sum_1^n (y_j - a)| \le \text{const.}, \qquad n \le m(\omega).$$

The condition is satisfied, for example, if $m(\omega)$ is defined as the first integer j with $|x_j(\omega)| \ge K_1$ and if $|y_1| \le K_2$, where K_1 and K_2 are given constants.

The above type of argument, already applied to find the expectation and variance of x', can be used to find all the semi-invariants of the x' distribution. Rather than doing this, however, we shall show how the argument also yields the fundamental theorem of sequential analysis. Define $\Phi(z)$ by

$$\Phi(z) = \mathbf{E}\{e^{zy_j}\}$$

for z complex. Then the sequence $u_1, u_2, \cdots$, where

$$u_n = \frac{e^{z\left(\sum_1^n y_j\right)}}{\Phi(z)^n},$$

is a martingale, for any value of z for which $\Phi(z)$ exists. In fact, if $n < \nu$,

$$\mathbf{E}\{u_\nu \mid \mathscr{F}_n\} = \mathbf{E}\{u_\nu \mid y_1, \cdots, y_n\} = \frac{e^{z\sum_1^n y_j}}{\Phi(z)^n} \frac{\mathbf{E}\{e^{z\sum_{n+1}^\nu y_j}\}}{\Phi(z)^{\nu-n}} = u_n,$$

with probability 1. Hence $\{u_n, \mathscr{F}_n, n \geq 1\}$ is a martingale. Then, if Theorem 2.2 can be applied,

$$1 = \mathbf{E}\{u_n\} = \mathbf{E}\{u_m\} = \mathbf{E}\left\{\frac{e^{z\sum_1^m y_j}}{\Phi(z)^m}\right\},$$

and this is the fundamental theorem of sequential analysis. The condition C_3 of Theorem 2.2 becomes here: there is a constant K such that

$$\mathbf{E}\left\{\left|\frac{e^{z\sum_1^{n+1} y_j}}{\Phi(z)} - e^{z\sum_1^n y_j}\right| \mid y_1, \cdots, y_n\right\} \leq K\,|\Phi(z)|^n, \qquad n \leq m(\omega),$$

with probability 1, that is,

$$|e^{z\sum_1^n y_j}|\,\mathbf{E}\left\{\left|\frac{e^{zy_1}}{\Phi(z)} - 1\right|\right\} \leq K\,|\Phi(z)|^n, \qquad n \leq m(\omega),$$

with probability 1. If there is a value of z for which $\Phi(z)$ is defined, with $|\Phi(z)| \geq 1$, and if the real part of zx_n is $\leq K_1$ for some K_1 when $n \leq m(\omega)$, then this condition is satisfied.

Throughout the above discussion it was supposed that the y_j's have a common distribution function, but the methods are obviously applicable in the general case.

11. Continuous parameter martingales

In the previous sections we have for the most part restricted our attention to discrete parameter martingales and semi-martingales, although the definitions of these types of families of random variables only require that the parameter range T be an ordered set. In the present section we shall, as always, suppose that T is a linear point set which may include the points $\pm\infty$, but shall usually impose no further restriction. The theorems discussed will include the theorems on sequences of random variables proved in previous sections as special cases, but the point of the section is the application to the continuous parameter case, in particular to the case in which T is an interval, and this explains the section title.

The extensions of the discrete parameter theorems of the previous sections will be identified only as the *general parameter set* versions of the corresponding discrete parameter theorems. Some earlier theorems are not given general parameter set extensions, because either the extensions are uninteresting, unnecessary (because the earlier theorems impose no restrictions on the parameter sets, for example Theorems 1.1 and 3.1), or unknown. On the latter possibility we remark that it is not known under what circumstances the representation (1.5′) of the general random variable of a semi-martingale as the sum of the general random variable of a martingale and a partial sum of a series of non-negative random variables is valid when the parameter set is an interval.

We defer temporarily the general parameter set versions of the game theorems of §2.

(THEOREM 3.2) *If $\{x_t, t \in T\}$ is a separable semi-martingale, if λ is real, and if T has a maximum value b, then*

$$\lambda \mathbf{P}\{\underset{t \in T}{\text{L.U.B.}}\, x_t(\omega) \geq \lambda\} \leq \int_{\{\underset{t\in T}{\text{L.U.B.}}\, x_t(\omega) \geq \lambda\}} x_b \, d\mathbf{P} \leq \mathbf{E}\{|x_b|\},$$

$$\lambda \mathbf{P}\{\underset{t \in T}{\text{G.L.B.}}\, x_t(\omega) \leq \lambda\} \geq \int_{\{\underset{t\in T}{\text{G.L.B.}}\, x_t(\omega) \leq \lambda\}} x_b \, d\mathbf{P} - \mathbf{E}\{x_b\} + \underset{t \in T}{\text{G.L.B.}}\, \mathbf{E}\{x_t\}$$

$$\geq \underset{t \in T}{\text{G.L.B.}}\, \mathbf{E}\{x_t\} - \mathbf{E}\{|x_b|\}.$$

To prove the theorem we remark first that by Theorem 3.2 these inequalities are true if T is replaced by a finite subset of itself, including the point b. Using the obvious limiting procedure, we then see that the inequalities are true if T is replaced by an enumerable subset of itself including the point b. Finally, the theorem is true as stated, because by the definition of a separable stochastic process (see II §2) there is an enumerable subset S of T with the property that (neglecting an ω set of probability 0) a sample function has the same bounds on T as on S. Since the theorem is true when T is replaced by S (which we can assume contains the point b), it is true as stated. Note that the second inequality will be trivially true if $\underset{t \in T}{\text{G.L.B.}}\, \mathbf{E}\{x_t\} = -\infty$. If T has a minimum value, a, we have

$$\underset{t \in T}{\text{G.L.B.}}\, \mathbf{E}\{x_t\} = \mathbf{E}\{x_a\};$$

if T has no minimum value, we have

$$\underset{t \in T}{\text{G.L.B.}}\, \mathbf{E}\{x_t\} = \lim_{t \to a} \mathbf{E}\{x_t\}, \qquad a = \underset{t \in T}{\text{G.L.B.}}\, t.$$

(THEOREM 3.4) This theorem is generalized exactly as was Theorem 3.2, and its statement will be omitted.

(THEOREM 4.1) Part (v) does not seem to have an interesting general parameter set extension. Part (iv) will be extended later. Parts (i), (ii), (iii) are easily extended, and we give the extension of (i) to exhibit the new form of the statement. *Let $\{x_t, t \in T\}$ be a martingale. Then $\mathbf{E}\{|x_t|\}$ is monotone non-decreasing in t. Suppose that $b = \underset{t \in T}{\text{L.U.B.}}\, t \notin T$.*

(i) *If*

$$\lim_{t \to b} \mathbf{E}\{|x_t|\} = K < \infty,$$

then there is a random variable x_b, with $\mathbf{E}\{|x_b|\} \leq K$, such that, if $\{s_n\}$ is a sequence in T, $\lim_{n \to \infty} s_n = b$ implies that $\lim_{n \to \infty} x_{s_n} = x_b$ with probability 1. *If the x_t process is separable, this limit relation can be strengthened to the relation $\lim_{s \to b} x_s = x_b$ with probability* 1. *In particular, $K < \infty$ if the x_t's are all real and ≥ 0 or all real and ≤ 0.*

In this statement b may be finite or infinite. To prove the existence of x_b we note first that, if $K < \infty$, and if $\{s_n\}$ is a sequence of parameter values converging monotonely to b, then the process $\{x_{s_n}, n \geq 1\}$ is a martingale, with $\mathbf{E}\{|x_{s_n}|\} \leq K$, so that $\lim_{n \to \infty} x_{s_n}$ exists and is finite with probability 1 by Theorem 4.1 (i). The limiting random variable must be independent of the monotone sequence $\{s_n\}$, neglecting values on sets of probability 0, because any two sequences $\{s_n\}$ can be combined into a single one which corresponds to a sequence of random variables convergent with probability 1. Moreover, the limit must also exist with probability 1 if the sequence $\{s_n\}$ is convergent to b, but is not necessarily monotone, because such a sequence can be reordered to be monotone. There thus exists an x_b, defined in terms of sequential approach to b, as stated. If the x_t process is separable, we have proved (II, Theorem 2.3) that sequential approach can be replaced by ordinary approach. The remaining statements in the general parameter set version of Theorem 4.1 (i) are proved in exactly the same way as in the discrete parameter case.

(THEOREM 4.1s) This theorem is generalized just as Theorem 4.1 is, and the proof is again carried through by reduction to the discrete parameter case.

(THEOREMS 4.2, 4.2s) The generalizations are carried out in the same way as those of Theorems 4.1 and 4.1s.

(THEOREM 4.3) *Let z be a random variable, with $\mathbf{E}\{|z|\} < \infty$, and let $\mathscr{F}_t$ be a Borel field of measurable ω sets for each t in the linear set T, with $\mathscr{F}_s \subset \mathscr{F}_t$ for $s < t$. Define*

$$a = \underset{t \in T}{\text{G.L.B.}}\, t, \qquad b = \underset{t \in T}{\text{L.U.B.}}\, t$$

$$\mathscr{F}_{a+} = \bigcap_{t \in T} \mathscr{F}_t$$

and let $\mathscr{F}_{b-}$ be the smallest Borel field of ω sets with $\mathscr{F}_{b-} \supset \bigcup_{t \in T} \mathscr{F}_t$. Then the conditional expectation $\mathbf{E}\{z \mid \mathscr{F}_t\}$ can be defined for each $t \in T$ in such a way that

$$\lim_{t \to a} \mathbf{E}\{z \mid \mathscr{F}_t\} = \mathbf{E}\{z \mid \mathscr{F}_{a+}\}$$

$$\lim_{t \to b} \mathbf{E}\{z \mid \mathscr{F}_t\} = \mathbf{E}\{z \mid F_{b-}\}$$

with probability 1.

To prove this theorem we remark that the limit equations are true, by Theorem 4.3, if t goes to its limit along a sequence of values. Moreover, the stochastic process $\{\mathbf{E}\{z \mid \mathscr{F}_t\}, t \in T\}$ can be made separable by a proper choice of the conditional expectations, since each can be changed arbitrarily on an ω set of zero probability. If a choice making the process separable has been made, the sequential approach of t to its limit is no longer necessary, by II, Theorem 2.3.

The fact that the stochastic process of this theorem can be defined to be separable means that the sample function properties of Theorems 11.2 and 11.5 are true of this process when so defined.

Example 1 Let $\{x_n, n \geq 0\}$ be a martingale, and define a process $\{x_t, 0 \leq t < \infty\}$ by

$$x_t = x_n, \qquad n \leq t < n + 1 \qquad (n \geq 0).$$

Then the x_t process is a separable martingale. The sample functions have no discontinuities except at integral values of t, but will be discontinuous at such a value $t = n$ unless $x_n(\omega) = x_{n-1}(\omega)$.

Example 2 Let $\{x_t, 0 \leq t < \infty\}$ be a process with independent increments, with

$$\mathbf{E}\{x_t - x_0\} = m(t).$$

Then the process

$$\{x_t - x_0 - m(t), 0 \leq t < \infty\}$$

is a martingale. This is the type of continuous parameter martingale which corresponds to the discrete parameter type defined by the partial sums of a series of mutually independent random variables with zero expectations. In the particular case of a Poisson process (see II §9 and VIII §4)

$$m(t) = \text{const. } t.$$

Now in this case it is shown in VIII §4 that, if $x_0(\omega) \equiv 0$, x_t process sample functions can be used to represent the number of events of a certain type that have occurred between times 0 and t: almost all sample functions are monotone non-decreasing and increase in unit jumps, if the process is separable. Moreover, for each t,

$$\lim_{s \to t} x_s = x_t$$

with probability 1, that is, the jump points vary from sample function to sample function in such a way that, although the probability of a jump at any particular value of t is 0, the probability of a jump somewhere in an interval is positive.

It will be shown below that, as far as continuity of sample functions is concerned, the two preceding examples are characteristic of the general case. Almost all sample functions of a separable martingale are continuous except for discontinuities at which both left- and right-hand (finite) limits exist, and there are at most enumerably many parameter values where the probability of a discontinuity is positive. The Brownian motion process (II §9 and VIII §2) is a non-trivial example of a separable martingale (a special case of Example 2) almost all of whose sample functions are continuous.

THEOREM 11.1 *Let $\{x_t, t \in T\}$ be a stochastic process, and let T_1 be a set of limit points of T. Suppose that, if $t \in T_1$, at least one of the stochastic limits*

$$p \lim_{s \uparrow t} x_s = x_{t-}, \qquad p \lim_{s \downarrow t} x_s = x_{t+}$$

exists. There is then an at most enumerable subset T_0 of T_1 such that, if $t \in T_1 - T_0$, then both stochastic limits x_{t-} and x_{t+} are defined, and

$$\begin{aligned} x_{t-} &= x_{t+} \\ &= x_t \qquad (\textit{if } t \in T) \end{aligned}$$

with probability 1.

In the most important applications of this theorem T is an interval, and the stochastic limits x_{t-}, x_{t+} are known to exist at every point of T. The theorem then asserts that, for each t not in some exceptional set which is at most enumerable,

$$x_t = x_{t-} = x_{t+}$$

with probability 1.

To prove the theorem define the distance between any two random variables x, y as the greatest lower bound of values of ε for which

$$\mathbf{P}\{|x(\omega) - y(\omega)| \geq \varepsilon\} \leq \varepsilon.$$

With this definition of distance $d(x, y)$, $\lim_{n\to\infty} d(x, x_n) = 0$ if and only if $p \lim_{n\to\infty} x_n = x$. Then, for each $t \in T$, the random variable x_t is a point of a complete metric space, so that the random variables of the x_t process define a t function f, $t \in T$, with values in this metric space. By hypothesis $f(t-)$ or $f(t+)$ exists for every $t \in T_1$. To prove the theorem it must be proved that the T_1 set whose points are not limit points of T from both sides, or are limit points from both sides but are points where the oscillation of f is positive, is an at most enumerable set. It is sufficient to prove that the T_1 set $T_1(n)$, the points of which are not limit points of T from both sides, or are limit points from both sides but are points where the oscillation of $f(t)$ is $\geq 1/n$, is at most enumerable. If $t \in T_1(n)$ and if t is not a limit point of T from both sides, it cannot be a limit point of T_1 from both sides. Hence there is an interval with t as one endpoint, containing t but no other point of T_1. If $t \in T_1(n)$ and if $f(t-)$ [$f(t+)$] exists, t is the right [left] endpoint of an interval containing t but no other point of $T_1(n)$. We thus obtain a set of intervals which we can suppose chosen to be disjunct, each containing a single point of $T_1(n)$, and all points of this set are contained in the intervals. The set $T_1(n)$ is therefore at most enumerable, because a set of disjunct intervals is at most enumerable.

In the following, a point $t_0 \in T$ will be called a *fixed point of discontinuity of a stochastic process* $\{x_t, t \in T\}$ if it is false that whenever $s_n \to t_0$,

$$\lim_{n\to\infty} x_{s_n} = x_{t_0}$$

with probability 1. If the process is separable, it follows that t_0 is a fixed point of discontinuity if and only if it is false that

$$\lim_{s\to t} x_s = x_{t_0}$$

with probability 1, that is, if and only if there is positive probability for a sample function discontinuity at t_0. Any point of discontinuity of a sample function of a separable process, not a fixed point of discontinuity, will be called a moving point of discontinuity. Even if there are no fixed points of discontinuity, it does not follow that the sample functions of a separable stochastic process are almost all (that is, with probability 1) continuous functions, because there may be moving points of discontinuity, as in the case of the Poisson process (see Example 2 above).

THEOREM 11.2 *Let $\{x_t,\ t \in T\}$ be a semi-martingale, and let a, b be respectively the minimum and maximum values of the closure of T. Define T' as the set of limit points of T, except that b is to be excluded from T' unless $b \in T$, and a is to be excluded from T' unless*

$$\operatorname*{G.L.B.}_{s \in T} \mathbf{E}\{x_t\} > -\infty.$$

(i) *To each point $t \in T'$ which is a limit point of T from the left [right] there corresponds a random variable x_{t-} [x_{t+}] such that, if $s_n \to t$ with $s_n < t$ [$s_n > t$] and $s_n \in T$, then*

$$\lim_{n\to\infty} x_{s_n} = x_{t-} \qquad [\lim_{n\to\infty} x_{s_n} = x_{t+}]$$

with probability 1. *If the x_t process is separable, these sequential limits can be replaced by ordinary limits*

$$\lim_{s\uparrow t} x_s = x_{t-} \qquad [\lim_{s\downarrow t} x_s = x_{t+}].$$

(ii) *Except possibly for the points of an at most enumerable t set, for each $t \in T'$ the following equation holds with probability* 1 *between as many of the three members as are defined:*

$$x_{t-} = x_t = x_{t+}.$$

In particular, at most enumerably many parameter points are fixed points of discontinuity.

Let $t \in T'$ be a limit point of T from the left. The existence of x_{t-} as described follows from an application of the general parameter set version of Theorem 4.1s to the semi-martingale $\{x_s,\ s \in T,\ s < t\}$ if we can show that, for some $t_0 < t$,

$$\operatorname*{L.U.B.}_{t_0 < s < t} \ \mathbf{E}\{|x_s|\} < \infty.$$

To show this let t_1 be a point of T with $t_1 \geq t$. There is such a point since $t \in T'$. Choose t_0 any point of T with $t_0 < t$. Then, by Theorem 3.1 (ii),

$$\mathbf{E}\{|x_s|\} \leq -\mathbf{E}\{x_{t_0}\} + 2\mathbf{E}\{|x_{t_1}|\}, \qquad t_0 \leq s \leq t_1.$$

The general parameter set version of Theorem 4.1s is thus applicable, and asserts the existence of x_{t-}. The existence of x_{t+} is proved by applying Theorem 4.2s. Finally (ii) follows from Theorem 11.1. We observe that the point b could have been allowed in T' also under the condition that $\mathbf{E}\{|x_s|\}$ is bounded near b.

Let $\{x_t,\ t \in T\}$ be a semi-martingale. In §1, §2, and §3 we discussed various conditions under which the x_t's are uniformly integrable. The following theorem gives necessary and sufficient conditions.

THEOREM 11.3 *In Theorem* 11.2,

$$\mathbf{E}\{x_s\} \leq \mathbf{E}\{x_t\}, \qquad s < t,$$

$$\lim_{s \uparrow t} \mathbf{E}\{x_s\} \leq \mathbf{E}\{x_{t-}\} \leq \mathbf{E}\{x_t\} \leq \mathbf{E}\{x_{t+}\} = \lim_{s \downarrow t} \mathbf{E}\{x_s\}.$$

If $t_1 \in T$, the x_t's with $t \leq t_1$ are uniformly integrable if and only if

$$\underset{s \in T}{\text{G.L.B.}}\ \mathbf{E}\{x_s\} > -\infty, \qquad \lim_{s \uparrow t} \mathbf{E}\{x_s\} = \mathbf{E}\{x_{t-}\}, \qquad t \leq t_1$$

whenever t is a limit point of T from the left.

The inequalities relating the expectations may be vacuously true at some values of t, since the random variables in question are not necessarily defined for all t, but the continued inequality of the theorem holds for all t in the sense that, whenever two of the random variables involved do exist, the stated inequality between their expectations is true. The inequalities have been proved in the discrete parameter case, and no further discussion of them is needed here. The uniform integrability statement is deduced from discrete parameter special cases as follows. The indicated x_t's are uniformly integrable if and only if the x_t's of every sequence $\{x_{s_n}\}$ are uniformly integrable, and we may even restrict our attention to monotone sequences $\{s_n\}$. Now, if $\{s_n\}$ is monotone decreasing, we have proved [Theorem 3.1 (iv)] that the x_{s_n}'s are uniformly integrable if and only if

$$\lim_{n \to \infty} \mathbf{E}\{x_{s_n}\} > -\infty;$$

if $\{s_n\}$ is monotone increasing, with $s_n \to t$, we have proved [Theorem 4.1s (ii)] that the x_{s_n}'s are uniformly integrable if and only if

$$\lim_{n \to \infty} \mathbf{E}\{x_{s_n}\} = \mathbf{E}\{x_{t-}\}.$$

These two are the stated conditions of the present theorem.

According to Theorem 11.3, the x_t's of a semi-martingale, with $t \leq t_1$ and t_1 in the parameter set, will be uniformly integrable, for example, if $\mathbf{E}\{x_t\}$ is bounded in t and skips no values when t increases, that is, if $\mathbf{E}\{x_t\}$ runs through the values of an interval (which may degenerate into a single point). This is true, in particular, if the x_t process is a martingale, when $\mathbf{E}\{x_t\}$ is identically constant. However, nothing new is obtained in this case, since, according to Theorem 3.1 (iii) (applied to the absolute values of the martingale variables), if the x_t process is a martingale, the x_t's with $t \leq t_1$ are uniformly integrable.

Theorems 4.4 and 4.4s showed how to extend the parameter sets of certain simple types of martingales and semi-martingales. The following theorem is the general parameter set version of these theorems.

THEOREM 11.4 *Let $\{x_t, \mathscr{F}_t, t \in T\}$ be a semi-martingale [martingale], and let I be the closed interval with endpoints the minimum and maximum values of the closure of T, except that the right-hand endpoint is to be excluded from I unless this endpoint is in T, and the left-hand endpoint is to be excluded in the semi-martingale case unless* $\underset{t \in T}{\text{G.L.B.}}\, \mathbf{E}\{x_t\} > -\infty$. *Then it is possible to define x_t and $\mathscr{F}_t$ for $t \in I - T$ in such a way that the process $\{x_t, \mathscr{F}_t, t \in I\}$ is a semi-martingale [martingale].*

Suppose that the x_t process is a semi-martingale [martingale]. If $t \in I - T$ and if t is a limit point of T from the right, define

$$x_t = x_{t+}, \qquad \mathscr{F}_t = \bigcap_{s>t} \mathscr{F}_s.$$

The process with the thus enlarged parameter set is easily seen to be a semi-martingale [martingale] using the general parameter set version of Theorem 4.1s [Theorem 4.1]. (See also Theorem 4.4.) Let $[c, d]$ be a closed interval whose endpoints but no other points lie in the closure of T. Then x_d and $\mathscr{F}_d$ are already defined, and we define

$$x_t = x_d, \qquad \mathscr{F}_t = \mathscr{F}_d, \qquad (c < t < d),$$

also defining

$$x_c = x_d, \qquad \mathscr{F}_c = \mathscr{F}_d$$

if x_c and $\mathscr{F}_c$ are not already defined. The resulting process $\{x_t, \mathscr{F}_t, t \in I\}$ is then a semi-martingale [martingale]. Finally we remark that the extension we have given is not the only one that can be given in some cases.

Theorem 11.4 shows that in considering martingales and semi-martingales we can assume, if desirable, that the parameter set is an interval.

Example 3 In §8 an example was given of a martingale $\{x_n, n \geq 1\}$ satisfying the conditions

$$x_n \geq 0, \qquad \mathbf{E}\{x_n\} = 1, \qquad n \geq 1;$$

$$\lim_{n \to \infty} x_n = 0$$

with probability 1. Define

$$\hat{x}_n = -x_n, \qquad \hat{x}_\infty = 1.$$

The process $\{\hat{x}_n, 1 \leq n \leq \infty\}$ is then a semi-martingale whose random variables are not uniformly integrable. The set I of Theorem 11.4 becomes the interval $[1, \infty]$, and $\hat{x}_{\infty-} = 0$ with probability 1. The definition of $\hat{x}_t$ for t not an integer, as given in the proof of Theorem 11.4, becomes

$$\hat{x}_t = \hat{x}_n, \qquad n - 1 < t \leq n, \qquad n = 2, 3, \cdots.$$

In this example $\hat{x}_\infty$ could have been taken as any non-negative random variable with a finite expectation, but no choice would make the process a martingale.

In the following we use the concept of jump of a function at a point, as defined in V §1, a discontinuity at which the function has one-sided limits, and at which the value of the function lies between these limits. If the function is complex-valued we shall say it has a jump if its real and imaginary parts do, or if one does and the other is continuous.

THEOREM 11.5 *Except possibly for a set of sample functions of probability* 0, *the sample functions of a separable semi-martingale* $\{x_t, t \in T\}$ *have the following properties:*

(i) *They are bounded on every t set of the form* $[a_1, b_1]T$ *with* $a_1, b_1 \in T$, *or simply* $b_1 \in T$ *if the process is a martingale.*

(ii) *They have finite left- [right-] hand limits at every* $t \in T$ *which is a limit point of* T *from the left [right].*

(iii) *Their discontinuities are jumps, except perhaps at the fixed points of discontinuity.*

Since a function which has finite left- and right-hand limits at all values of the argument where such limits are definable has at most enumerably many points of discontinuity, almost all sample functions of a separable martingale are this regular. The theorem is true for complex martingales because it is true for the corresponding real and imaginary parts.

The generalized parameter set version of Theorem 3.2 implies (i) trivially. In proving (ii) and (iii) we shall suppose that T is infinite. Otherwise there would be nothing to prove. According to (i), there is an ω set Λ_1 of probability 0, such that every sample function of the x_t process corresponding to an $w \notin \Lambda_1$ is bounded in every interval with endpoints in T. By definition of separability of a stochastic process, there is a sequence $\{t_n\} \subset T$, dense in T, and an ω set Λ_2 of probability 0, such that every sample function corresponding to a point $\omega \notin \Lambda_2$ has the same lower and upper bounds on open intervals as on the t_n's in the intervals, that is,

$$\underset{t \in IT}{\text{G.L.B.}}\, x_t(\omega) = \underset{t_j \in I}{\text{G.L.B.}}\, x_{t_j}(\omega), \qquad \underset{t \in IT}{\text{L.U.B.}}\, x_t(\omega) = \underset{t_j \in I}{\text{L.U.B.}}\, x_{t_j}(\omega), \qquad \omega \notin \Lambda_2$$

Now let a_1, b_1 be points of T, with $a_1 < b_1$, chosen so that $(a_1, b_1)T$ is not empty. Let $t_1^{(n)}, t_2^{(n)}, \cdots$ be a_1, b_1, and those of the first n t_j's in (a_1, b_1), ordering the $t_j^{(n)}$'s so that $t_1^{(n)} < t_2^{(n)} < \cdots$. Let r_1, r_2 be real numbers with $r_1 < r_2$, and let $\beta_n(\omega)$ be the number of upcrossings of $[r_1, r_2]$ by $x_{t_1^{(n)}}(\omega), x_{t_2^{(n)}}(\omega), \cdots$. According to Theorem 3.3,

$$\mathbf{E}\{\beta_n\} \le \frac{\mathbf{E}\{|x_{b_1}|\} + |r_1|}{r_2 - r_1}.$$

Then, if $M_{nk} = \{\beta_n(\omega) \geq k\}$,

$$\mathbf{P}\{M_{nk}\} \leq \frac{\mathbf{E}\{|x_{b_1}|\} + |r_1|}{k(r_2 - r_1)}, \qquad k \geq 1,$$

so that

$$\mathbf{P}\{\bigcup_n M_{nk}\} \leq \frac{\mathbf{E}\{|x_{b_1}|\} + |r_1|}{k(r_2 - r_1)}. \tag{11.1}$$

Now suppose that a sample function $g(\cdot)$ corresponding to a point $\omega \notin \Lambda_2$ has an oscillatory discontinuity at some point $s \in [a_1, b_1]T$, with either

$$\limsup_{t \uparrow s} g(t) > r_2 > r_1 > \liminf_{t \uparrow s} g(t), \qquad t \in [a_1, b_1]T,$$

or the same inequality with $t \downarrow s$. Then this same inequality is true if t approaches s remaining in $\{t_j\}$, so that the number of upcrossings of $[r_1, r_2]$ by $g(t_1^{(n)}), g(t_2^{(n)}), \cdots$ becomes infinite when $n \to \infty$. Thus, if M is the ω set corresponding to the sample functions g of this type,

$$M \subset \bigcup_n M_{nk}, \qquad k = 1, 2, \cdots. \tag{11.2}$$

According to (11.1) the intersection in k of the ω sets on the right in (11.2) has probability 0. Let $\Lambda(r_1, r_2, a_1, b_1)$ be this intersection, and define

$$\Lambda_3 = \bigcup_{r_1, r_2, a_1, b_1} \Lambda(r_1, r_2, a_1, b_1),$$

where r_1, r_2 vary over all rational numbers with $r_1 < r_2$ and a_1 [b_1] is either the minimum [maximum] value of T if there is one, or in the contrary case a_1 [b_1] varies over some sequence of values in T converging to $\underset{t \in T}{\text{G.L.B.}}\, t$ [$\underset{t \in T}{\text{L.U.B.}}\, t$]. Then

$$\mathbf{P}\{\Lambda_3\} = 0,$$

and any sample function of the process corresponding to a point ω not in $\Lambda_1 \cup \Lambda_2 \cup \Lambda_3$ has finite left- and right-hand limits at each point of discontinuity. Moreover, because of the defining properties of the t_j's, any discontinuity of such a sample function at a point other than a t_j must be a jump. Now, if a t_j is not a fixed point of discontinuity, the sample functions are continuous at t_j with probability 1. Hence, if a further ω set Λ_4 of probability 0 is excluded, we can say that any discontinuity of a sample function not corresponding to an ω in one of the Λ_j's is a jump, unless the point of discontinuity is simultaneously a t_j and a fixed point of discontinuity. This completes the proof of the theorem.

Theorem 11.5, as well as some of the earlier results of this section, have related to separable semi-martingales, not semi-martingales in general. However, if the stochastic process $\{x_t, t \in T\}$ is a semi-martingale, it follows from II §2 (see II, Theorem 2.4) that there is a semi-martingale $\{\tilde{x}_t, t \in T\}$ such that, for each t,

$$\mathbf{P}\{\tilde{x}_t(\omega) = x_t(\omega)\} = 1$$

and that the $\tilde{x}_t$ process is separable. The separable process results are then applicable to the $\tilde{x}_t$ process. The results can also be expressed without the use of the concept of separability. To show how this can be done we rephrase Theorem 11.5 (ii) in this way:

[THEOREM 11.5 (ii)] *Let $\{x_t, t \in T\}$ be a semi-martingale, and let T_1 be a finite or enumerable subset of T. Then the sample functions of the x_t process almost all have the following property: they coincide on T_1 with functions defined on T which have finite left- [right-] hand limits at every $t \in T$ which is a limit point of T from the left [right].*

The truth of the original version of Theorem 11.5 (ii) for the $\tilde{x}_t$ process defined above implies the truth of this alternative version, because

$$\mathbf{P}\{\tilde{x}_t(\omega) = x_t(\omega), t \in T_1\} = 1.$$

Moreover, the alternative version implies the original one if we choose T_1 as the finite or enumerable set involved in the definition of separability.

We now continue our discussion of the general parameter set versions of the discrete parameter theorems of previous sections.

(§5, Zero-one law) *Let $\{x_t, t \in T\}$ be a process with independent increments, and let z be a random variable which for each $b_1 < b = \underset{t \in T}{\text{L.U.B.}}\, t$ is a random variable on the sample space of differences $x_t - x_s$ with*

$$b_1 < s,t < b.$$

Then $z(\omega) \equiv$ const. with probability 1. The proof can be given directly or derived from martingale theory as in the discrete parameter case.

In the following we shall be interested in separable stochastic processes $\{x_t, t \in T\}$, and shall be particularly interested in the differences $x_t - x_a$. We shall use without further reference the obvious fact that, if the x_t process is separable, and if z is any random variable, then the $x_t - z$ process is also separable.

(THEOREM 5.1) *Let $\{x_t, t \in T\}$ be a separable stochastic process with independent increments, and suppose that T has initial point a and last point b. Then, if* $\mathbf{E}\{x_t - x_a\} \equiv 0$,

$$\mathbf{E}\{\underset{t \in T}{\text{L.U.B.}}\, |x_t - x_a|^\alpha\} \leq 8\mathbf{E}\{|x_b - x_a|^\alpha\}, \alpha \geq 1.$$

(This result could be stated slightly more generally to follow Theorem 5.1 more closely, but the generalization is an immediate consequence of Theorem 5.1 and has no independent interest.) This theorem can be proved by translating the proof of Theorem 5.1 into the continuous parameter case, or it can be reduced to Theorem 5.1 by remarking that according to Theorem 5.1 the desired inequality is true if the left side is replaced by

$$\mathbf{E}\{\text{L.U.B.}_j\, |x_{t_j} - x_a|^\alpha\}$$

if $\{t_j\}$ is a finite subset of T, and therefore if $\{t_j\}$ is an enumerably infinite subset of T. Since the x_t process is separable, the sequence $\{t_j\}$ can be chosen to make

$$\text{L.U.B.}_j\, |x_{t_j} - x_a|^\alpha = \text{L.U.B.}_{t \in T}\, |x_t - x_a|^\alpha$$

with probability 1, and the theorem is now completely proved.

The extension of Theorem 5.2 to more general parameter sets is obvious, and we omit an explicit formulation.

(§6, Strong law of large numbers for mutually independent random variables with a common distribution function) *Let* $\{x_t, 0 \le t < \infty\}$ *be a separable stochastic process with stationary independent increments, and suppose that* $\mathbf{E}\{x_t - x_0\} \equiv 0$. *Then*

$$\lim_{t\to\infty} \frac{x_t}{t} = 0 \tag{11.3}$$

with probability 1.

To prove this theorem we observe first that, according to the discrete parameter version of §6,

$$\lim_{n\to\infty} \frac{x_n - x_0}{n} = \lim_{n\to\infty} \frac{1}{n} \sum_{j=1}^{n} (x_j - x_{j-1}) = \mathbf{E}\{x_1 - x_0\} = 0, \tag{11.4}$$

and

$$\lim_{n\to\infty} \frac{1}{n} \sum_{j=1}^{n} \text{L.U.B.}_{j-1\le t\le j}\, |x_t - x_j| = \mathbf{E}\{\text{L.U.B.}_{0\le t\le 1}\, |x_t - x_0|\}$$

with probability 1. (According to the general parameter version of Theorem 5.1 the last expectation is finite.) Then

$$\lim_{n\to\infty} \frac{1}{n-1} \sum_{j=1}^{n-1} \text{L.U.B.}_{j-1\le t\le j}\, |x_t - x_j| = \mathbf{E}\{\text{L.U.B.}_{0\le t\le 1}\, |x_t - x_0|\},$$

and we can obviously replace $1/(n-1)$ by $1/n$ here. Subtracting the resulting equation from the preceding limit equation, we find that

$$\lim_{n\to\infty} \frac{1}{n} \text{L.U.B.}_{n-1\le t\le n}\, |x_t - x_0| = 0$$

with probability 1, and this result combined with (11.4) yields

$$\lim_{t\to\infty} \frac{x_t - x_0}{t} = \lim_{t\to\infty} \frac{x_t}{t} = 0$$

with probability 1. This theorem could also have been proved by showing that, if $t \geq 1$,

$$\mathbf{E}\{x_1 - x_0 \mid x_s - x_0, s \geq t\} = \mathbf{E}\{x_1 - x_0 \mid x_t - x_0\} = \frac{x_t - x_0}{t},$$

and then using the continuous parameter version of Theorem 4.2 or Theorem 4.3 to show that the terms of the equality converge to a limit when $t \to \infty$. The limit is a constant by the zero-one law, and the constant is 0 because the limit is a random variable in the same martingale as $x_1 - x_0$, and hence has the same expectation.

The following preliminary discussion will lead to the general parameter set versions of Theorems 2.1 and 2.2. The treatment will be more unified here than in §2 because Theorem 2.1 was given special treatment in §2 in view of the references to it in §4.

The point of the general parameter set versions of Theorems 2.1 and 2.2 is that they presuppose a semi-martingale or martingale $\{x_t, t \in T(x)\}$, and assert that a new process $\{\breve{x}_\alpha, \alpha \in T(\breve{x})\}$, obtained from the x_t process by a certain type of sampling, is also a semi-martingale or martingale. In Theorems 2.1 and 2.2 both $T(x)$ and $T(\breve{x})$ are sets of integers. In the more general case to be discussed now, each of these two parameter sets will be an arbitrary linear set. We make the following hypotheses, which reduce in the discrete parameter case to those made in §2.

OS_1 *$\{x_t, t \in T(x)\}$ is a stochastic process.*

OS_2 *For each $t \in T(x)$, $\mathscr{F}_t$ is a Borel field of measurable ω sets, with the following properties:*

(a) $$\mathscr{F}_s \subset \mathscr{F}_t, \qquad s < t;$$

(b) *x_t is either measurable with respect to $\mathscr{F}_t$ or is equal for almost all ω to a function that is.*

OS_3 *Almost all sample functions of the x_t process have finite limits from the right,*

$$x_{t+} = \lim_{s \downarrow t} x_s$$

for all $t \in T(x)$.

OS_4 *$\{\tau_\alpha, \alpha \in T(\tau)\}$ is a stochastic process defined on the same ω space as the x_t process, and has the following properties:*

(a) *for each $\alpha \in T(\tau)$, the values taken on by τ_α lie in the set $T(x)$;*

(b) *$\tau_\alpha(\omega)$ is monotone non-decreasing in α for fixed ω;*

(c) *if $\alpha \in T(\tau)$, either*

$$\{\tau_\alpha(\omega) \leq s\} \in \mathscr{F}_s, \qquad s \in T(x),$$

or the ω set on the left differs by at most a set of probability 0 from an $\mathscr{F}_s$ set.

If each $\mathscr{F}_t$ contains all ω sets of probability 0, the weaker alternatives in OS_2 (b) and OS_4 (c) need not be given. Moreover, it is no restriction to assume that each $\mathscr{F}_t$ contains the ω sets of probability 0, since, if this is not true initially, $\mathscr{F}_t$ can be replaced by the Borel field generated by the sets of $\mathscr{F}_t$ and those of probability 0.

We now define the process $\{\breve{x}_\alpha, \alpha \in T(\tau)\}$ obtained from the x_t process by *optional sampling*. For each α let S_α be the (at most enumerable) set of values which τ_α takes on with positive probability. Define $\breve{x}_\alpha(\omega)$ by

$$(11.5) \qquad \begin{aligned} \breve{x}_\alpha(\omega) &= x_{\tau_\alpha(\omega)}(\omega), && \text{if} \quad \tau_\alpha(\omega) \in S_\alpha, \\ &= x_{\tau_\alpha(\omega)+}(\omega), && \text{if} \quad \tau_\alpha(\omega) \notin S_\alpha. \end{aligned}$$

This definition is meaningful on the ω set of probability 1 corresponding to the x_t process sample functions which have limits from the right at all values of the argument, except that $\breve{x}_\alpha$ is not defined when τ_α takes on with probability 0 one of the at most enumerably many values in $T(x)$ which are not limit points of $T(x)$ from the right. In other words, we have defined $\breve{x}_\alpha$ for each $\alpha \in T(\tau)$, with probability 1. We shall prove that with this definition $\breve{x}_\alpha$ is a random variable, that is, a measurable ω function. We expect that just as in the discrete parameter case martingale and semi-martingale properties of the x_t process go over into similar properties of the $\breve{x}_\alpha$ process. Note that OS_3 is satisfied whenever the x_t process is a separable semi-martingale or martingale, by Theorem 11.5.

The most important type of optional sampling is *optional stopping*, defined as follows. Let OS_1, OS_2, OS_3 be satisfied, and let τ be a random variable, $-\infty \leq \tau(\omega) \leq \infty$, satisfying

OS_4' (a) The values $\neq \operatorname{L.U.B.}_{t \in T(x)} t$ taken on by τ lie in the set $T(x)$.

(b) *Either*

$$\{\tau(\omega) \leq s\} \in \mathscr{F}_s, \qquad s \in T(x)$$

or the set on the left differs by at most a set of probability 0 *from an* $\mathscr{F}_s$ *set.*

Then, if we define $T(\tau) = T(x)$, and

$$\tau_t(\omega) = \operatorname{Min}\,[t, \tau(\omega)], \qquad t \in T(x),$$

the τ_t's satisfy OS_4 and the optional sampling determined in this way will be called optional stopping. According to the definition,

$$\begin{aligned} \breve{x}_t(\omega) &= x_{t+}(\omega), && \text{if} \quad t \leq \tau(\omega) \quad \text{and} \quad \mathbf{P}\{\tau(\omega) \geq t\} = 0, \\ &= x_t(\omega), && \text{if} \quad t \leq \tau(\omega) \quad \text{and} \quad \mathbf{P}\{\tau(\omega) \geq t\} > 0 \\ &= x_{\tau(\omega)+}(\omega), && \text{if} \quad t > \tau(\omega), \end{aligned}$$

except that, if τ takes on any value s with positive probability,

$$\breve{x}_t(\omega) = x_s(\omega) \qquad \text{if} \quad t > s \quad \text{and} \quad \tau(\omega) = s.$$

The following approximation procedure will be used. Suppose that OS_1–OS_4 are satisfied. For each positive integer q choose finitely many points

$$a_1^{(q)} < a_2^{(q)} < \cdots \qquad a_j^{(q)} \in T(\tau)$$

in such a way that the first q points of S_α enumerated in some order are $a_j^{(q)}$'s and that every point of $T(x)$ in the interval $[-q, q]$ is within distance $1/q$ of some $a_j^{(q)}$, and that the infinite points of $T(x)$, if any, are $a_j^{(q)}$'s. Define

$$\begin{aligned} \breve{x}_\alpha^{(q)}(\omega) &= x_{a_1^{(q)}}(\omega), &&\text{if } \tau_\alpha(\omega) \le a_1^{(q)}, \\ &= x_{a_j^{(q)}}(\omega), &&\text{if } a_{j-1}^{(q)} < \tau_\alpha(\omega) \le a_j^{(q)}, \quad j > 1 \\ &= 0, &&\text{if } \operatorname*{Max}_j a_j^{(q)} < \tau_\alpha(\omega). \end{aligned}$$

Then $\breve{x}_\alpha^{(q)}$ is a measurable ω function, that is, a random variable, and

$$\lim_{q\to\infty} \breve{x}_\alpha^{(q)} = \breve{x}_\alpha$$

with probability 1. Hence $\breve{x}_\alpha$ is a random variable. We shall find it helpful to specify more closely a Borel field $\breve{\mathscr{F}}_\alpha$ with respect to which $\breve{x}_\alpha$ is measurable. The important point will be that this Borel field will be defined in terms of the $\mathscr{F}_t$'s and τ_t's, and will thus not depend on the x_t's. Fix $\alpha \in T(\tau)$. For each positive integer q let $\mathscr{F}_\alpha^q$ be the Borel field generated by the ω sets of probability 0 and by those of the form

$$\Lambda\{a < \tau_\alpha(\omega) \le b\}, \qquad \Lambda \in \bigcup_{c\le b} \mathscr{F}_c,$$

where a, b, c are not necessarily finite, no one of the first q points of S_α is an interior point of an interval $(a, b]$, and $\arctan b - \arctan a < 1/q$. It is no restriction to take $b \in T(x)$ and $c = b$ here. If $b = -\infty \in T(x)$, we understand by the above

$$\Lambda\{\tau_\alpha(\omega) = -\infty\}, \qquad \Lambda \in \mathscr{F}_{-\infty}.$$

Define

$$\breve{\mathscr{F}}_\alpha = \bigcap_q \mathscr{F}_\alpha^q.$$

Then the approximant $\breve{x}_\alpha^{(q)}$ defined above is measurable with respect to $\mathscr{F}_\alpha^r$ for large q, and therefore $\breve{x}_\alpha$ is measurable with respect to $\mathscr{F}_\alpha^r$ for every r, so that $\breve{x}_\alpha$ is measurable with respect to $\breve{\mathscr{F}}_\alpha$. Moreover, if $\alpha < \beta$, it follows that $\breve{\mathscr{F}}_\alpha \subset \breve{\mathscr{F}}_\beta$. To prove this we prove that $\mathscr{F}_\alpha^q \subset \mathscr{F}_\beta^q$ for all q, by proving that each set of the class generating $\mathscr{F}_\alpha^q$ is an $\mathscr{F}_\beta^q$ set. It is sufficient to consider $\breve{\mathscr{F}}_\alpha$ sets of the form

$$\Lambda_1 = \Lambda\{a < \tau_\alpha(\omega) \le b\}, \qquad \Lambda \in \mathscr{F}_b,\ b \in T(x),$$

with b finite. Trivial changes in the argument take care of non-finite b. Note that $\Lambda_1 \in \mathscr{F}_b$. Since $\tau_\alpha \leq \tau_\beta$,

$$\Lambda_1 = \Lambda_1\{a < \tau_\beta(\omega) \leq b\} \bigcup_{j=1}^{n} \Lambda_1\{b + (j-1)\delta < \tau_\beta(\omega) \leq b + j\delta\} \\ \cup \Lambda_1\{b + n\delta < \tau_\beta(\omega) \leq \infty\},$$

and this union exhibits Λ_1 as an $\mathscr{F}_\beta{}^q$ set, if $\delta > 0$, if arctan $\delta < 1/q$, and if $\pi/2 -$ arctan $(b + n\delta) < 1/q$.

The point of the following discussion is to prove that under suitable regularity conditions a semi-martingale [martingale] $\{x_t, \mathscr{F}_t, t \in T(x)\}$ goes into a semi-martingale [martingale] $\{\check{x}_\alpha, \check{\mathscr{F}}_\alpha, t \in T(\tau)\}$ under optional sampling. Note that, if Φ is a Baire t, λ function, and if φ_t is defined by

$$\varphi_t(\omega) = \Phi[t, x_t(\omega)],$$

then, if the x_t process satisfies OS_1 and OS_2, the φ_t process satisfies the same conditions with the same family of Borel fields. Moreover, if φ_t satisfies OS_3 and if under optional sampling determined by τ_α's satisfying OS_4 the φ_t process goes into a $\check{\varphi}_\alpha$ process, then $\{\check{\varphi}_\alpha, \check{\mathscr{F}}_\alpha, \alpha \in T(\tau)\}$, if specified regularity hypotheses are satisfied, will also be a semi-martingale [martingale]. Here $\check{\mathscr{F}}_\alpha$, as defined above, does not depend on the choice of Φ. In particular, if Φ is continuous,

$$\check{\varphi}_\alpha(\omega) = \Phi[\tau_\alpha(\omega), \check{x}_\alpha(\omega)],$$

and the φ_t process will satisfy OS_3 if the x_t process satisfies OS_3. This is the most important special case.

We now discuss the general parameter set version of Theorem 2.2. The results will be divided into a succession of lemmas and theorems for greater clarity. These results will contain Theorem 2.2 as a special case, and will even sharpen this case in some directions. The special case of optional stopping, corresponding to Theorem 2.1, will not be stated separately. We shall assume throughout the proofs that the processes are real, omitting the trivial remarks necessary to extend the discussion to the complex case. Throughout the discussion we assume that $\{x_t, \mathscr{F}_t, t \in T(x)\}$ is a semi-martingale or martingale satisfying OS_1, OS_2, OS_3, and that it is transformed into the process $\{\check{x}_\alpha, \check{\mathscr{F}}_\alpha, \alpha \in T(\tau)\}$ by τ_α's satisfying OS_4.

LEMMA 11.1 *If $\alpha \in T(\tau)$, $s \in T(x)$, $\Lambda \in \check{\mathscr{F}}_\alpha$, then*

$$\Lambda\{\tau_\alpha(\omega) \leq s\} \in \mathscr{F}_s. \tag{11.6}$$

To prove this relation suppose first that $s \in S_\alpha$, say that s is the q_0th point of S_α according to the enumeration of S_α used in the definition of the $\mathscr{F}_\alpha{}^q$'s. Then, if $q \geq q_0$, we shall strengthen the lemma by proving that (11.6) is true even for $\Lambda \in \mathscr{F}_\alpha{}^q$. It is sufficient only to consider sets Λ of the form

$$\Lambda = M\{c_1 < \tau_\alpha(\omega) \le c_2\}, \qquad M \in \mathscr{F}_c, \quad c \in T(x), \quad c \le c_2, \quad s \notin (c_1, c_2),$$

since the sets of this type generate $\mathscr{F}_\alpha{}^q$. For such a set,

$$\begin{aligned} \Lambda\{\tau_\alpha(\omega) \le s\} &= \Lambda \in \mathscr{F}_s, \qquad &\text{if } s \ge c_2 \\ &= \text{null set} \in \mathscr{F}_s, \qquad &\text{if } s \le c_1, \end{aligned}$$

as was to be proved. As a second case, suppose that s is not a limit point of $T(x)$ on the right. (The point may or may not be in S_α.) Choose q so large that

$$\arctan s_1 - \arctan s > \frac{1}{q} \qquad \text{if } s_1 > s, \quad s_1 \in T(x).$$

Then again we strengthen the lemma, proving that (11.6) is true even if $\Lambda \in \mathscr{F}_\alpha{}^q$. It is sufficient only to consider sets Λ of the same form as above, except that s may now be in (c_1, c_2). However, according to the hypothesis on q, if $s \in (c_1, c_2)$, there is no other point of $T(x)$ in this interval to the right of s. Hence

$$\begin{aligned} \Lambda\{\tau_\alpha(\omega) \le s\} &= \Lambda \in \mathscr{F}_s, \qquad &\text{if } s \ge c_2, \\ &= \text{null set} \in \mathscr{F}_s, \qquad &\text{if } s \le c_1, \\ &\in \mathscr{F}_s, \qquad &\text{if } c_1 < s < c_2. \end{aligned}$$

Finally suppose that $s \notin S_\alpha$, and that s is a limit point of $T(x)$ on the right. We prove first that, if $s_1 > s$ and $s_1 \in T(x)$, then

$$(11.6') \qquad M\{\tau_\alpha(\omega) \le s\} \in \mathscr{F}_{s_1}$$

if

$$M \in \mathscr{F}_\alpha{}^q, \qquad \frac{1}{q} < \arctan s_1 - \arctan s.$$

It is sufficient only to consider $\mathscr{F}_\alpha{}^q$ sets M of the form

$$M = M_1\{c_1 < \tau_\alpha(\omega) \le c_2\}, \qquad M_1 \in \mathscr{F}_c, \quad c \in T(x), \quad c \le c_2.$$

For such a set, our hypothesis on q implies that $s_1 > c_2$ if $s > c_1$, so that

$$\begin{aligned} M\{\tau_\beta(\omega) < s\} &= M \in \mathscr{F}_{s_1}, \qquad &\text{if } s \ge c_2, \\ &= \text{null set} \in \mathscr{F}_{s_1}, \qquad &\text{if } s \le a, \\ &\in \mathscr{F}_{s_1}, \qquad &\text{if } c_1 < s < c_2. \end{aligned}$$

We have thus proved (11.6′). This relation implies

$$M\{\tau_\alpha(\omega) < s\} \in \mathscr{F}_s,$$

and thereby implies the desired relation (11.6), since, according to the hypothesis that $s \notin S_\alpha$,

$$\mathbf{P}\{\tau_\alpha(\omega) = s\} = 0,$$

so that

$$\mathbf{M}\{\tau_\alpha(\omega) = s\} \in \mathscr{F}_s.$$

LEMMA 11.2 *Suppose that the x_t process is dominated by a semi-martingale. Then if $\alpha \in T(\tau)$ and if $s \in T(x)$, the random variable $\breve{x}_\alpha$ is integrable on the ω set $\{\tau_\alpha(\omega) \leq s\}$, and the integrability is uniform in α. If s is an $a_j^{(q)}$ for every q, the random variables of the sequence $\{\breve{x}_\alpha^{(q)}, q \geq 1\}$ are uniformly integrable on the ω set $\{\tau_\alpha(\omega) \leq s\}$, and the degree of uniformity does not depend on α or on the choice of the $a_j^{(q)}$'s.*

We recall that the semi-martingale $\{x_t^+, \mathscr{F}_t, t \in T(x)\}$ is said to dominate $\{x_t, \mathscr{F}_t, t \in T(x)\}$ if

$$\mathbf{P}\{|x_t(\omega)| \leq x_t^+(\omega)\} = 1, \qquad t \in T(x).$$

If the x_t process is a martingale, we can take $x_t^+ = |x_t|$. We can of course take $x_t^+ = x_t$ if the x_t process is a semi-martingale with non-negative random variables. More generally, since the positive part of the x_t process, $\{(|x_t| + x_t)/2, \mathscr{F}_t, t \in T(x)\}$, is a semi-martingale, the condition that an x_t^+ process exist is a condition on the order of magnitude of the negative part of x_t, $-(|x_t| - x_t)/2$. If there is a random variable $z \geq 0$ for which

$$\mathbf{P}\{x_t(\omega) \geq -z(\omega)\} = 1, \qquad t \in T(x)$$

$$\mathbf{E}\{z\} < \infty,$$

then an x_t^+ semi-martingale which dominates the x_t process can be defined by

$$x_t^+ = \mathbf{E}\{z \mid \mathscr{F}_t\} + (|x_t| + x_t)/2.$$

According to Theorem 1.2 (iv), every semi-martingale whose parameter set is the set of positive integers is dominated by a semi-martingale, so that in this case an x_t^+ process always exists. This fact explains the lack of mention of an x_t^+ process in Theorems 2.1 and 2.2. We remark, however, that even without the existence of the x_t^+ process it is possible to prove that $|\breve{x}_\alpha|$ is integrable on the ω set $\{\tau_\alpha(\omega) \leq s\}$, $s \in T(\tau)$, under the relatively weak assumption that

$$\underset{t \in T(x)}{\text{G.L.B.}}\ \mathbf{E}\{x_t\} > -\infty.$$

To prove the lemma, define

$$\mathbf{M}_1^q = \{\tau_\alpha(\omega) \leq a_1^{(q)}\},$$

$$\mathbf{M}_j^q = \{a_{j-1}^{(q)} < \tau_\alpha(\omega) \leq a_j^{(q)}\}, \qquad j > 1.$$

Then, using the semi-martingale property of the x_t^+ process,

$$(11.7)\qquad \int_{\left\{\substack{|\breve{x}_\alpha^{(q)}(\omega)|>\lambda\\ \tau_\alpha(\omega)\le s}\right\}} |\breve{x}_\alpha^{(q)}|\,d\mathbf{P} = \sum_{a_j^{(q)}\le s} \int_{\mathrm{M}_j^q\{|x_{a_j^{(q)}}(\omega)|>\lambda\}} |x_{a_j^{(q)}}|\,d\mathbf{P}$$

$$\le \sum_{a_j^{(q)}\le s} \int_{\mathrm{M}_j^q\{|x_{a_j^{(q)}}(\omega)|>\lambda\}} x_{a_j^{(q)}}^+\,d\mathbf{P}$$

$$\le \sum_{a_j^{(q)}\le s} \int_{\mathrm{M}_j^q\{|x_{a_j^{(q)}}(\omega)|>\lambda\}} x_s^+\,d\mathbf{P}$$

$$\le \int_{\{\operatorname*{Max}_{t\in S} x_t^+(\omega)>\lambda\}} x_s^+\,d\mathbf{P},$$

where S is a finite $T(x)$ set in the interval $(-\infty, s]$, namely the $a_j^{(q)}$'s in that interval. Now, according to Theorem 2.2,

$$(11.8)\qquad \mathbf{P}\{\operatorname*{Max}_{t\in S}\ x_t^+(\omega)>\lambda\}\le\frac{1}{\lambda}\,\mathbf{E}\{x_s^+\}.$$

Hence

$$\lim_{\lambda\to\infty} \mathbf{P}\{\operatorname*{Max}_{t\in S} x_t^+(\omega)>\lambda\}=0$$

uniformly in S. Hence the left side of (11.7) goes to 0 when $\lambda\to\infty$, uniformly in α, q, and in the choices of the $a_j^{(q)}$'s. When $q\to\infty$, (11.7) yields, in view of the uniformity of the integrability,

$$(11.9)\qquad \int_{\left\{\substack{|\breve{x}_\alpha(\omega)|>\lambda\\ \tau_\alpha(\omega)\le s}\right\}} |\breve{x}_\alpha|\,d\mathbf{P}\le \int_{\{\operatorname*{L.U.B.}_{t\in U} x_t^+(\omega)>\lambda\}} x_s^+\,d\mathbf{P},$$

where U is a certain finite or enumerably infinite $T(x)$ set in the interval $(-\infty, s]$. Since (11.8) holds for finite sets, it must hold for enumerably infinite sets. Hence the probability of the domain of integration on the right in (11.9) goes to 0, when $\lambda\to\infty$, uniformly in U, and this implies the truth of the lemma.

LEMMA 11.3 *Suppose that the x_t process is dominated by a semi-martingale. Let $\alpha, \beta\in T(\tau)$, $\alpha\le\beta$, $s\in T(x)$, $\Lambda\in\mathscr{F}_\alpha$. Then, if the x_t process is a semi-martingale,*

$$(11.10)\qquad \int_{\Lambda\{\tau_\alpha(\omega)\le s\}} \breve{x}_\alpha\,d\mathbf{P}\le \int_{\Lambda\{\tau_\beta(\omega)\le s\}} \breve{x}_\beta\,d\mathbf{P}+\int_{\Lambda\left\{\substack{\tau_\beta(\omega)>s\\ \tau_\alpha(\omega)\le s}\right\}} x_s\,d\mathbf{P}$$

$$(11.11)\qquad \int_{\Lambda\{\tau_\alpha(\omega)\le s\}} \breve{x}_\alpha\,d\mathbf{P}\le \int_{\Lambda\{\tau_\alpha(\omega)\le s\}} x_s\,d\mathbf{P}.$$

If the x_t process is a martingale, these inequalities become equalities.

Suppose that the x_t process is a semi-martingale, and choose $a_j^{(q)}$'s to match both τ_α and τ_β, so that both $\breve{x}_\alpha^{(q)}$ and $\breve{x}_\beta^{(q)}$ are now defined, and

$$\lim_{q\to\infty} \breve{x}_\alpha^{(q)} = \breve{x}_\alpha, \qquad \lim_{q\to\infty} \breve{x}_\beta^{(q)} = \breve{x}_\beta$$

with probability 1. Define

$$\Lambda_j^q = \Lambda\{a_{j-1}^{(q)} < \tau_\alpha(\omega) \leq a_j^{(q)}\},$$

$$\Lambda_{jk}^q = \Lambda_j^q\{a_{k-1}^{(q)} < \tau_\beta(\omega) \leq a_k^{(q)}\}, \qquad k \geq j,$$

$$\mathrm{M}_{jk}^q = \Lambda_j^q\{\tau_\beta(\omega) > a_k^{(q)}\} = \Lambda_j^q - \bigcup_{r\leq k} \Lambda_{jr}^q, \qquad k \geq j.$$

Then, by Lemma 11.1,

$$\Lambda_j^q \in \mathscr{F}_{a_j^{(q)}}, \qquad \Lambda_{jk}^q \in \mathscr{F}_{a_k^{(q)}}, \qquad \mathrm{M}_{jk}^q \in \mathscr{F}_{a_k^{(q)}},$$

so that, using the semi-martingale property of the x_t process,

$$\begin{aligned}
(11.12) \qquad \int_{\Lambda_j^q} \breve{x}_\alpha^{(q)}\, d\mathbf{P} &= \int_{\Lambda_j^q} x_{a_j^{(q)}}\, d\mathbf{P} \\
&= \int_{\Lambda_{jj}^q} x_{a_j^{(q)}}\, d\mathbf{P} + \int_{\mathrm{M}_{jj}^q} x_{a_j^{(q)}}\, d\mathbf{P} \\
&\leq \int_{\Lambda_{jj}^q} x_{a_j^{(q)}}\, d\mathbf{P} + \int_{\mathrm{M}_{jj}^q} x_{a_{j+1}^{(q)}}\, d\mathbf{P} \\
&= \int_{\Lambda_{jj}^q} x_{a_j^{(q)}}\, d\mathbf{P} + \int_{\Lambda_{jj+1}^q} x_{a_{j+1}^{(q)}}\, d\mathbf{P} + \int_{\mathrm{M}_{jj+1}^q} x_{a_{j+1}^{(q)}}\, d\mathbf{P} \\
&\qquad \cdots\cdots \\
&\leq \sum_{k=j}^{N} \int_{\Lambda_{jk}^q} x_{a_k^{(q)}}\, d\mathbf{P} + \int_{\mathrm{M}_{jN}^q} x_{a_N^{(q)}}\, d\mathbf{P}, \qquad N \geq j \\
&= \int_{\Lambda_j^q\{\tau_\beta(\omega)\leq a_N^{(q)}\}} \breve{x}_\beta^{(q)}\, d\mathbf{P} + \int_{\mathrm{M}_{jN}^{(q)}} x_{a_N^{(q)}}\, d\mathbf{P}.
\end{aligned}$$

Now suppose that, for each q, s is some $a_j^{(q)}$. Then, choosing N above so that $a_N^{(q)} = s$, and summing over $j \leq N$, we find

$$\int_{\Lambda\{\tau_\alpha(\omega)\leq s\}} \breve{x}_\alpha^{(q)}\, d\mathbf{P} \leq \int_{\Lambda\{\tau_\beta(\omega)\leq s\}} \breve{x}_\beta^{(q)}\, d\mathbf{P} + \int_{\Lambda\{\tau_\alpha(\omega)\leq s,\ \tau_\beta(\omega)>s\}} x_s\, d\mathbf{P}.$$

According to Lemma 11.2 the integrands are uniformly (in q) integrable over the indicated integration sets. Hence when $q \to \infty$ we obtain

(11.10). To obtain (11.11), apply the semi-martingale inequality to the first line of (11.12) to obtain

$$\int_{\Lambda_j^q} \breve{x}_\alpha^{(q)}\, d\mathbf{P} \leq \int_{\Lambda_j^q} x_s\, d\mathbf{P}, \qquad j \leq N.$$

Summing over $j \leq N$, and letting $q \to \infty$ yields (11.11). In the martingale case, all the above inequalities become equalities, so that there is then equality in (11.10) and (11.11).

THEOREM 11.6 *Suppose that the x_t process is dominated by a semi-martingale, and that*

$$\underset{t \in T(x)}{\text{L.U.B.}}\ t = b \in T(x).$$

Then, if the process $\{x_t, \mathscr{F}_t, t \in T(x)\}$ is a semi-martingale [martingale], the process $\{\breve{x}_\alpha, \breve{\mathscr{F}}_\alpha, \alpha \in T(\tau)\}$ is also a semi-martingale [martingale], with

$$\underset{t \in T(x)}{\text{G.L.B.}}\ \mathbf{E}\{x_t\} \leq \mathbf{E}\{\breve{x}_\alpha\} \leq \mathbf{E}\{x_b\}. \tag{11.13}$$

In the martingale case, this inequality becomes an equality.

By Lemma 11.2 with $s = b$, $\mathbf{E}\{|\breve{x}_\alpha|\} < \infty$. If $s = b$ in (11.10), we obtain

$$\int_{\Lambda} \breve{x}_\alpha\, d\mathbf{P} \leq \int_{\Lambda} \breve{x}_\beta\, d\mathbf{P}, \qquad \Lambda \in \breve{\mathscr{F}}_\alpha, \quad \alpha < \beta.$$

This is the semi-martingale inequality for the $\breve{x}_\alpha$ process, and according to Lemma 11.3 there is equality if the x_t process is a martingale, so that in this case the $\breve{x}_\alpha$ process is also a martingale. If $\Lambda = \Omega$, and $s = b$ in (11.11), we obtain the right-hand half of (11.13), with equality in the martingale case. Since the first and third terms of (11.13) are equal in the martingale case, there is equality throughout in that case. There remains the proof of the left half of (11.13) in the semi-martingale case. If $T(x)$ has a minimum value a, the proof of the left half of (11.13) follows that of the corresponding inequality in the discrete parameter case (in which $a = 1$) given in the proof of Theorem 2.2, and will be omitted. We now prove that we can always assume that $T(x)$ has a minimum value, because, if there is no such value initially, one can be adjoined, and the truth of the inequalities in question for the extended processes will imply their truth for the original ones. We can assume that $T(x)$ is bounded, making a bounded monotone transformation of this set, if necessary, to insure the boundedness. Suppose then that $T(x)$ has no minimum value, and define $a = \underset{t \in T(x)}{\text{G.L.B.}}\ t$. By hypothesis the x_t process is dominated by a (non-negative) semi-martingale, whose random variables are uniformly integrable for $t \leq t_1 \in T(x)$ (t_1 fixed) according to Theorem 3.1 (iii). Then the x_t's are also uniformly integrable for $t \leq t_1$, so that

$$\lim_{t \to a} \mathbf{E}\{x_t\} > -\infty.$$

There are then, according to Theorem 11.2 (i), random variables which we shall denote by x_{a-1}, x_{a-1}^+, such that

$$\lim_{t \to a} x_t = x_{a-1}, \qquad \lim_{t \to a} x_t^+ = x_{a-1}^+$$

with probability 1, if $t \to a$ along a parameter sequence. We adjoin the point $a - 1$ to $T(x)$, and define

$$\mathscr{F}_{a-1} = \bigcap_{t > a} \mathscr{F}_t,$$

thus obtaining the desired new parameter set with a minimum value. We chose $a - 1$ here instead of a to insure that OS_3 would be verified (vacuously) at the new parameter value.

THEOREM 11.7 *Suppose that the x_t process is dominated by a semi-martingale. Then*

$$\mathbf{E}\{|\breve{x}_\alpha|\} \leq 3 \underset{t \in T(x)}{\text{L.U.B.}} \mathbf{E}\{|x_t|\}, \tag{11.14}$$

and the factor 3 can be replaced by 1 if the x_t's are non-negative.

We can assume that $T(x)$ is bounded in proving this theorem, making a bounded monotone transformation of this set to insure boundedness, if necessary. Let L be the L.U.B. on the right in (11.14). Let $t \in T(x)$, and define τ_s' by

$$\begin{aligned} \tau_s'(\omega) &= \operatorname{Min}\,[s, \tau_\alpha(\omega)], \qquad s \leq t, \qquad s \in T(x), \\ &= t, \qquad\qquad\qquad\qquad s = t + 1. \end{aligned}$$

Let

$$T(x)' = T(x) \cap [-\infty, t],$$

and let $T(\tau)'$ consist of the points of $T(x)'$ and in addition the point $t + 1$. Then the family $\{\tau_s', \mathscr{F}_s, s \in T(\tau)'\}$ satisfies OS_4. Let the process $\{x_s, s \in T(x)'\}$ go into the process $\{x_s', s \in T(\tau)'\}$ under the optional sampling determined by the τ_s' family. Since $T(x)'$ has a last element t, the x_s' process is a semi-martingale, according to Theorem 11.6. Hence (11.13) is satisfied for this optional sampling transformation, yielding

$$\mathbf{E}\{x_s'\} \geq \underset{s \in T(x)}{\text{G.L.B.}} \mathbf{E}\{x_s\} \geq -L.$$

According to Theorem 3.1 (ii),

$$\begin{aligned}\mathbf{E}\{|x_s'|\} &\leq -\mathbf{E}\{x_s'\} + 2\mathbf{E}\{|x_{t+1}'|\} \\ &= -\mathbf{E}\{x_s'\} + 2\mathbf{E}\{|x_t|\} \leq 3L.\end{aligned}$$

In view of the definition of τ_s', this inequality for $s = t$ is equivalent to

$$\int_{\{\tau_\alpha(\omega) \leq t\}} |\breve{x}_\alpha| \, d\mathbf{P} \leq 3L$$

which yields (11.14) when t increases. As the proof shows, the bound $3L$ can be replaced by

$$2L - \underset{s \in T(x)}{\text{G.L.B.}} \ \mathbf{E}\{x_s\}.$$

In particular, if the x_t's are non-negative, (11.13) yields

$$\int_{\{\tau_\alpha(\omega) \leq t\}} \breve{x}_\alpha \, d\mathbf{P} = \mathbf{E}\{x_t'\} \leq \mathbf{E}\{x_t\} \leq L.$$

This inequality yields (11.14) with L on the right instead of $3L$, when t increases. The device used here could have been used to deduce the right-hand half of (11.13) above.

LEMMA 11.4 *Suppose that the x_t process is dominated by a semi-martingale. Then if $\alpha \in T(\tau)$, $s_1, s_2 \in T(x)$, $s_1 < s_2$, $\Lambda \in \mathscr{F}_{s_1}$,*

$$\int_{\Lambda\{\tau_\alpha(\omega) > s_1\}} x_{s_1} \, d\mathbf{P} \leq \int_{\Lambda\{s_1 < \tau_\alpha \leq s_2\}} \breve{x}_\alpha \, d\mathbf{P} + \int_{\Lambda\{\tau_\alpha(\omega) > s_2\}} x_{s_2} \, d\mathbf{P}.$$

To prove this lemma define the random variables τ', τ'' by

$$\begin{aligned}\tau'(\omega) \equiv s_1; \qquad \tau''(\omega) &= s_1, &&\text{if } \tau_\alpha(\omega) \leq s_1, \\ &= \tau_\alpha(\omega), &&\text{if } s_1 < \tau_\alpha(\omega) \leq s_2, \\ &= s_2, &&\text{if } \tau_\alpha(\omega) > s_2.\end{aligned}$$

The family of random variables consisting of the two random variables τ', τ'' satisfies OS_4. Let the process $\{x_t, \mathscr{F}_t, s \in [s_1, s_2]T(x)\}$ go into the process consisting of the two random variables x_{s_1}, $\breve{x}$ under the optional sampling determined by τ', τ''. Since the parameter set of the original process has a last element s_2, the process $\{x_{s_1}, \breve{x}\}$ is a semi-martingale, according to Theorem 11.6, and we obtain, applying the semi-martingale inequality,

$$\begin{aligned}\int_{\Lambda\{\tau''(\omega) > s_1\}} x_{s_1} \, d\mathbf{P} &\leq \int_{\Lambda\{\tau''(\omega) > s_1\}} \breve{x} \, d\mathbf{P} \\ &\leq \int_{\Lambda\{s_1 < \tau_\alpha(\omega) \leq s_2\}} \breve{x}_\alpha \, d\mathbf{P} + \int_{\Lambda\{\tau_\alpha(\omega) > s_2\}} x_{s_2} \, d\mathbf{P},\end{aligned}$$

as was to be proved.

THEOREM 11.8 *Suppose that the x_t process is dominated by a semi-martingale $\{x_t^+, \mathscr{F}_t, t \in T(x)\}$ and that*

$$b = \underset{t \in T(x)}{\text{L.U.B.}}\ t \notin T(x).$$

(i) *If the x_t process is a semi-martingale, if*

$$\mathbf{E}\{|\breve{x}_\alpha|\} < \infty, \qquad \alpha \in T(\tau), \tag{11.15}$$

and if

$$\liminf_{s \to b} \int_{\substack{\{\tau_\alpha(\omega) > s\} \\ \{x_s(\omega) > 0\}}} x_s \, d\mathbf{P} = 0, \qquad \alpha \in T(\tau), \tag{11.16}$$

it follows that the $\breve{x}_\alpha$ process is also a semi-martingale, with

$$\underset{t \in T(x)}{\text{G.L.B.}}\ \mathbf{E}\{x_t\} \le \mathbf{E}\{\breve{x}_\alpha\}, \qquad \alpha \in T(\tau). \tag{11.17}$$

(ii) *Under* (11.15) *and*

$$\liminf_{s \to b} \int_{\{\tau_\alpha(\omega) > s\}} |x_s| \, d\mathbf{P} = 0, \qquad \alpha \in T(\tau), \tag{11.16'}$$

(11.17) *can be strengthened to*

$$\underset{t \in T(x)}{\text{G.L.B.}}\ \mathbf{E}\{x_t\} \le \mathbf{E}\{\breve{x}_\alpha\} \le \underset{t \in T(x)}{\text{L.U.B.}}\ \mathbf{E}\{x_t\}, \qquad \alpha \in T(\tau), \tag{11.17'}$$

and it then follows that, if the x_t process is a martingale, the $\breve{x}_\alpha$ process is also a martingale, with equality in (11.17′).

(iii) *Each of the following conditions implies the validity of* (11.15) *and* (11.16′).

C_1 *The x_t's are uniformly integrable.*

C_2 *Each τ_α is bounded from above* (*with probability* 1) *by a value in $T(x)$.*

C_3 *There is a constant $K > 0$ with the following properties. The parameter set $T(x)$ contains the integers $\ge K$ (but not $b = \infty$). For each integer $n \ge K$, and each $\alpha \in T(\tau)$,*

$$\mathbf{E}\{x_{n+1}^+ - x_n^+ \mid \mathscr{F}_n\} \le K \quad \textit{for} \quad n < \tau_\alpha(\omega) \tag{11.18}$$

with probability 1. *Moreover,*

$$\mathbf{E}\{|\tau_\alpha| + \tau_\alpha\} < \infty, \qquad \alpha \in T(\tau). \tag{11.19}$$

(iv) *The following condition C_4 [C_4'] implies the validity of* (11.15) *and* (11.16) [(11.16′)].

C_4 (11.15) *is true and there are a random variable $z \ge 0$, with $\mathbf{E}\{z\} < \infty$, and a sequence $t_1 < t_2 < \cdots$, $t_n \in T(x)$, $t_n \to b$, such that*

$$\mathbf{P}\{|x_t(\omega)| \ge x_{t_i}(\omega) - z(\omega)\} = 1, \qquad t \ge t_i,\ i \ge 1. \tag{11.20}$$

C_4' *The x_t process is a martingale,* (11.15) *is true, and there are a random variable $z \geq 0$, with* $\mathbf{E}\{z\} < \infty$, *and a sequence $t_1 < t_2 < \cdots$, $t_n \in T(x)$, $t_n \to b$, such that*

$$\mathbf{P}\{|x_t(\omega)| \geq |x_{t_i}(\omega)| - z(\omega)\} = 1, \qquad t \geq t_i,\ i \geq 1. \tag{11.20'}$$

Proofs of (i), (ii) Although we could use the result of Theorem 11.6 to prove (i), by means of the device used in the proof of Lemma 11.3, we do not do so, because we have already set up the necessary inequalities. In fact, according to Lemma 11.3, if $s \in T(x)$, α, $\beta \in T(\tau)$, $\alpha < \beta$, $\Lambda \in \mathscr{F}_\alpha$, and if the x_t process is a semi-martingale,

$$\begin{aligned} \int_{\Lambda\{\tau_\alpha(\omega) \leq s\}} \breve{x}_\alpha \, d\mathbf{P} &\leq \int_{\Lambda\{\tau_\beta(\omega) \leq s\}} \breve{x}_\beta \, d\mathbf{P} + \int_{\Lambda\left\{\substack{\tau_\beta(\omega) > s \\ \tau_\alpha(\omega) \leq s}\right\}} x_s \, d\mathbf{P} \\ &\leq \int_{\Lambda\{\tau_\beta(\omega) \leq s\}} \breve{x}_\beta \, d\mathbf{P} + \int_{\left\{\substack{\tau_\beta(\omega) > s \\ x_s(\omega) > 0}\right\}} x_s \, d\mathbf{P}. \end{aligned} \tag{11.21}$$

When $s \to b$ we obtain, under (11.15) and (11.16),

$$\int_\Lambda \breve{x}_\alpha \, d\mathbf{P} \leq \int_\Lambda \breve{x}_\beta \, d\mathbf{P},$$

so that the $\breve{x}_\alpha$ process is a semi-martingale. Inequality (11.17) is proved in precisely the same way as the left half of (11.13). Under (11.15) and (11.16′) the right-hand half of (11.17′) follows from (11.11) with $\Lambda = \Omega$, when $s \to b$. This finishes the proof of (i), (ii) in the semi-martingale case. In the martingale case the first line of (11.21) is an equality, and leads, when $s \to b$, to the martingale equality for the $\breve{x}_\alpha$ process. In the martingale case the extreme terms of (11.17′) are equal.

Proof of (iii) If the x_t's are uniformly integrable, that is, if C_1 is satisfied, then $\mathbf{E}\{|x_t|\}$ is bounded in t, and then $\mathbf{E}\{|\breve{x}_\alpha|\} < \infty$ by Theorem 11.7. Thus (11.15) is satisfied. The limit relation (11.16′) is also true in this case, since as $s \to b$ the domain of integration in (11.16′) decreases to a set of probability 0. If each τ_α is bounded from above with probability 1 by a value in $T(x)$, that is, if C_2 is satisfied, (11.15) is true according to Lemma 11.2, and (11.16′) is also true because in this case the limit inferior in (11.16′) is actually attained for s near b. (The whole treatment of C_2 can trivially be reduced to the case discussed in Theorem 11.6.)

If C_3 is satisfied, set

$$\begin{aligned} y_K &= x_K^+, \\ y_j &= x_j^+ - x_{j-1}^+, \qquad j > K. \end{aligned}$$

Here we have supposed that K is an integer. If it is not, it can be replaced by any larger number that is. Define

$$\begin{aligned} w(\omega) &= x_j^+(\omega), \qquad \text{if } j-1 < \tau_\alpha(\omega) \le j, \quad j > K, \\ &= 0, \qquad \text{if } \tau_\alpha(\omega) \le K. \end{aligned}$$

Then

$$\begin{aligned} \mathbf{E}\{w\} &= \sum_{K+1}^{\infty} \int_{\{j-1<\tau_\alpha(\omega)\le j\}} x_j^+ \, d\mathbf{P} \\ &= \lim_{n\to\infty} \sum_{K+1}^{n} \int_{\{j-1<\tau_\alpha(\omega)\le j\}} (x_K^+ + y_{K+1} + \cdots + y_j) \, d\mathbf{P} \\ &= \int_{\{K<\tau_\alpha(\omega)\le n\}} x_K^+ \, d\mathbf{P} + \lim_{n\to\infty} \Big[\sum_{K+1}^{n} \int_{\{\tau_\alpha(\omega)>j-1\}} y_j \, d\mathbf{P} \\ &\qquad - \int_{\{\tau_\alpha(\omega)>n\}} (x_n^+ - x_K^+) \, d\mathbf{P} \Big] \\ &\le \mathbf{E}\{x_K^+\} + \limsup_{n\to\infty} \sum_{K+1}^{n} \int_{\{\tau_\alpha(\omega)>j-1\}} y_j \, d\mathbf{P}. \end{aligned}$$

Now the ω set $\{\tau_\alpha(\omega) > j-1\}$ is an $\mathscr{F}_{j-1}$ set by hypothesis. Hence we can use condition C_3 to continue this inequality, obtaining

$$\begin{aligned} \mathbf{E}\{w\} &\le \mathbf{E}\{x_K^+\} + \sum_{K+1}^{\infty} K\mathbf{P}\{\tau_\alpha(\omega) > j-1\} \\ &\le \mathbf{E}\{x_K^+\} + K\mathbf{E}\{|\tau_\alpha| + \tau_\alpha\} < \infty. \end{aligned}$$

Now suppose that $j-1$ and j are $a_k^{(q)}$'s. Then

$$\begin{aligned} \int_{\{j-1<\tau_\alpha(\omega)\le j\}} |\breve{x}_\alpha^{(q)}| \, d\mathbf{P} &\le \sum_{j-1<a_k^{(q)}\le j} \int_{\{a_{j-1}^{(q)}<\tau_\alpha(\omega)\le a_j^{(q)}\}} x_{a_k^{(q)}}^+ \, d\mathbf{P} \\ &\le \int_{\{j-1<\tau_\alpha(\omega)\le j\}} x_j^+ \, d\mathbf{P}. \end{aligned}$$

When $q \to \infty$, this inequality becomes, in view of the uniformity (in q) of the integrability of the integrand on the left,

$$\int_{\{j-1<\tau_\alpha(\omega)\le j\}} |\breve{x}_\alpha| \, d\mathbf{P} \le \int_{\{j-1<\tau_\alpha(\omega)\le j\}} x_j^+ \, d\mathbf{P}.$$

Thus we obtain, if s is an integer and $s \ge K$,

$$\begin{aligned} \int_{\{\tau_\alpha(\omega)>s\}} |\breve{x}_\alpha| \, d\mathbf{P} &= \sum_{s+1}^{\infty} \int_{\{j-1<\tau_\alpha(\omega)\le j\}} |\breve{x}_\alpha| \, d\mathbf{P} \\ &\le \sum_{s+1}^{\infty} \int_{\{j-1<\tau_\alpha(\omega)\le j\}} x_j^+ \, d\mathbf{P} \\ &= \int_{\{\tau_\alpha(\omega)>s\}} w \, d\mathbf{P}. \end{aligned}$$

Since $|\check{x}_\alpha|$ is already known to have a finite integral on $\{\tau_\alpha(\omega) \le s\}$, from Lemma 11.2, we have now proved that (11.15) is true under condition C_3. Since the last integral in the preceding inequality goes to 0 when $s \to \infty$, (11.16′) is also true.

Proof of (iv) Under C_4, (11.15) is satisfied by hypothesis. To prove (11.16) we note that, under C_4,

$$|\check{x}_\alpha(\omega)| \ge x_{t_i}(\omega) - z(\omega) \qquad \text{if} \quad \tau_\alpha(\omega) \ge t_i,$$

with probability 1. It then follows that, if i is fixed and $s > t_i$, $s \in T(x)$,

$$(11.22) \qquad \int\limits_{\substack{\tau_\alpha(\omega)>s \\ x_s(\omega)>0}} x_s \, d\mathbf{P} \le \int\limits_{\substack{s<\tau_\alpha(\omega)\le t_i \\ x_s(\omega)>0}} |\check{x}_\alpha| \, d\mathbf{P} + \int\limits_{\substack{\tau_\alpha(\omega)>t_i \\ x_s(\omega>0)}} x_{t_i} \, d\mathbf{P}$$

$$\le \int\limits_{\{s<\tau_\alpha(\omega)\le t_i\}} |\check{x}_\alpha| \, d\mathbf{P} + \int\limits_{\{\tau_\alpha(\omega)>t_i\}} (|\check{x}_\alpha| + z) \, d\mathbf{P}$$

$$= \int\limits_{\{\tau_\alpha(\omega)>s\}} (|\check{x}_\alpha| + z) \, d\mathbf{P} \to 0, \qquad (s \to b).$$

Thus (11.16) is satisfied. Finally under C_4' the argument just given can be applied to $|x_t|$ instead of x_t to give the desired conclusion.

The proof of the theorem has now been completed. In the applications the field $\mathscr{F}_t$ is ordinarily taken as the field of ω sets determined by conditions on the x_s's for $s \le t$, and this will be understood below unless other definitions are given.

Note added in proof. In Theorems 11.6, 11.7, 11.8, the condition A that the x_t process be dominated by a semi-martingale is not a restriction on the x_t process in the martingale case, but is apparently one if this process is a semi-martingale. This restriction can be eliminated as follows. If the x_t process is a semi-martingale, and if $u_{nt} = \max(x_t, n)$, where n is a negative integer, the u_{nt} process is also a semi-martingale, and

$$|u_{nt}| \le -n + \max(x_t, 0).$$

Thus the u_{nt} process is dominated by the semi-martingale defined by the right side of this inequality. Theorems 11.6, 11.7, 11.8 can now be applied to the u_{nt} process, and yield corresponding theorems for the x_t process when $n \to -\infty$. In this way it is found that Theorem 11.7 is true without A, and that Theorem 11.6 is true if A is replaced by the weaker condition A_1 that $\underset{t \in T}{\text{G.L.B.}}\ \mathbf{E}\{x_t\} > -\infty$. If A is dropped in the statement of Theorem 11.8, the following other changes should be made: condition A_1 should be made a part of (iii) C_2; in (iii) C_3, x_j^+ should be interpreted as $\max(x_j, 0)$, for $j = n, n+1$, A_1 should be made a part of the condition, and the thus modified condition then implies the validity of (11.15) and (11.16) [not of (11.16′)]. Note that, in the discussion of (iv) C_4, it is now necessary to set $s = t_i$.

The definition we have given of the optional sampling transformation is the useful one for most purposes. However, it can be modified in various ways. For example, each S_α could be arbitrarily augmented by an at most enumerable $T(x)$ set, thus modifying the definition of $\check{x}_\alpha$ in (11.5). This modification would require no change in either the results we have obtained or their proofs.

As an application of optional sampling, we discuss the extension of a discrete parameter sequential analysis result obtained in §10. Let $\{y_t, 0 \leq t \leq \infty\}$ be a separable stochastic process with stationary independent increments, and suppose that $y_0 = 0$. The following regularity hypotheses are made. We shall see in VIII that they are largely implied by the qualitative description of the process just given.

SA_1 *Almost all sample functions of the y_t process are continuous on the right at all parameter values.*

SA_2 $\mathbf{E}\{|y_t|\} < \infty$ *for all t, and there is a constant a such that*

$$\mathbf{E}\{y_t\} = at, \qquad t \geq 0.$$

Let τ be a random variable satisfying the condition OS_3' for a function which determines an optional stopping procedure. We wish to evaluate $\mathbf{E}\{y_\tau\}$, and it is natural to conjecture that

$$\mathbf{E}\{y_\tau\} = a\mathbf{E}\{\tau\}.$$

From now on we therefore impose the additional restriction that $\mathbf{E}\{\tau\}$ exists. To derive the desired equation we proceed just as in the discrete parameter case of §10. Let $x_t = y_t - at$. Then the x_t process is a martingale with independent increments. We shall apply Theorem 11.8 to this process. The set $T(\tau)$ in the present application consists of only a single point, and the corresponding τ_α is simply τ. Then condition C_3 is satisfied, with

$$K = \mathbf{E}\{|x_1|\}.$$

According to Theorem 11.8,

$$\mathbf{E}\{x_0\} = \mathbf{E}\{\breve{x}_\alpha\},$$

that is, in the present case,

$$0 = \mathbf{E}\{x_\tau\} = \mathbf{E}\{y_\tau - a\tau\} = \mathbf{E}\{y_\tau\} - a\mathbf{E}\{\tau\},$$

and this equation gives the desired evaluation of $\mathbf{E}\{y_\tau\}$.

Example 4 Let $\{x_n, n \geq 0\}$ be a martingale, and define τ_t as the largest integer $\leq t$, for $0 \leq t < \infty$. Then the family of random variables $\{\tau_t, 0 \leq t < \infty\}$ (all the random variables are of course constants) determines a system of optional sampling, and the $\breve{x}_t$ process obtained by the sampling is given by

$$\breve{x}_t = x_n, \qquad n \leq t < n+1,$$

so that the new martingale becomes that of Example 1. In this case $T(x)$ is the set of non-negative integers and $T(\tau)$ the set of non-negative real numbers.

Example 5 Let $\{x_t, 0 \leq t < \infty\}$ be a separable Brownian motion process (see Example 2) with $x_0 = 0$. It will be proved in VIII that almost all sample functions of this process are everywhere continuous.

For each ω corresponding to a continuous sample function let $\tau(\omega)$ be the first parameter value at which the sample function takes on the value d. Here d is a non-zero constant, to be held fast in the following discussion. The random variable τ satisfies the condition OS_4' and can therefore be used to define a system of optional stopping, yielding an $\breve{x}_t$ process with

$$\begin{aligned} \breve{x}_t &= x_t(\omega), & \tau(\omega) &\geq t, \\ &= x_{\tau(\omega)}(\omega), & \tau(\omega) &< t. \end{aligned}$$

According to Theorem 11.8 the $\breve{x}_t$ process is a martingale, with

$$\mathbf{E}\{\breve{x}_t\} = 0,$$

even though we have now of course lost the symmetry of the x_t distribution. Continuing the discussion of this example, we now show that $\mathbf{E}\{\tau\} = \infty$. To do this, define a system of optional sampling, with $T(\tau)$ containing only a single point, and let the corresponding τ_α be τ. Then the hypotheses of Theorem 11.8 would not be satisfied, or we would have

$$0 = \mathbf{E}\{x_t\} = \mathbf{E}\{\breve{x}_\alpha\} = \mathbf{E}\{x_\tau\} = d \neq 0,$$

because $x_\tau = d$ by definition of τ. In particular, Condition C_3 of Theorem 11.8 cannot be satisfied. However, (11.18) is satisfied with

$$K = \mathbf{E}\{|x_1|\}.$$

Hence (11.19), the last part of this condition, cannot be satisfied, that is, $\mathbf{E}\{\tau\} = \infty$.

Example 6 Let $\{y_t, 0 \leq t < \infty\}$ be a separable Poisson process, with

$$y_0 = 0, \qquad \mathbf{E}\{y_t\} = ct.$$

It is shown in VIII that the sample functions of the y_t process are almost all continuous except for unit jumps. For each ω corresponding to such a sample function let $\tau(\omega)$ be the parameter value where the sample function has its first jump. Then τ has density of distribution ce^{-cs}, $s \geq 0$, and

$$\mathbf{E}\{\tau\} = \frac{1}{c}.$$

Define

$$x_t = y_t - ct.$$

Then the x_t process is a martingale. We shall illustrate Theorem 11.8 by an example in which we can make explicit calculations. Define a system of optional sampling in which $T(\tau)$ contains only a single point, with corresponding $\tau_\alpha = \tau$, so that $\breve{x}_\alpha = x_{\tau+}$. Then

$$\breve{x}_\alpha = x_{\tau+} = 1 - c\tau.$$

Theorem 11.8 is applicable (using Condition C_3), and we find that

$$0 = \mathbf{E}\{x_t\} = \mathbf{E}\{x_{\tau+}\} = \mathbf{E}\{1 - c\tau\}.$$

This equation checks our previous evaluation of $\mathbf{E}\{\tau\}$. The example also illustrates the continuous parameter version of a sequential analysis result discussed above.

[Theorem 3.1s (iv)] Let $\{x_t, a \leq t < b\}$ *be a separable semi-martingale, and suppose that almost all sample functions of the process are continuous. Then* $\lim_{t\to b} x_t$ *exists and is finite almost everywhere where* $\limsup_{t\to b} x_t < \infty$.

Before proving this, we remark that by Theorem 11.5 the ω sets

$$\{\text{L.U.B.}_t\, x_t(\omega) < \infty\}, \qquad \{\limsup_{t\to b} x_t(\omega) < \infty\}$$

differ by at most a set of probability 0. Let $\tau(\omega)$ be the first parameter value (if any) at which $x_t(\omega) \geq d$, where d is a constant, and define a system of optional stopping based on τ. Then the $\breve{x}_t - d$ process has non-positive random variables, and is a semi-martingale. Hence, according to the general parameter set version of Theorem 4.1s (i), $\lim_{t\to b} \breve{x}_t$ exists and is finite with probability 1, that is, $\lim_{t\to b} x_t(\omega)$ exists and is finite almost everywhere that $\text{L.U.B.}_t\, x_t(\omega) \leq d$, that is, almost everywhere that $\text{L.U.B.}_t\, x_t(\omega) < \infty$, since d is arbitrary. This completes the proof. Essentially the same reasoning would derive the same conclusion if the hypothesis of continuity of the sample functions were replaced by the hypothesis

$$\mathbf{E}\{\text{L.U.B.}_t\, (x_{t+} - x_{t-})\} < \infty.$$

(The measurability of the indicated L.U.B. would also have to be assumed.)

We now make a slight digression into discrete parameter martingale theory. Let $y_1, \cdots, y_n$ be real random variables, with

$$x_k = \sum_1^k y_j,$$

satisfying the conditions

$$(11.23) \qquad \mathbf{E}\{y_1\} = 0, \qquad \mathbf{E}\{y_j \mid y_1, \cdots, y_{j-1}\} = 0, \qquad j > 1$$

and

$$(11.24) \qquad \mathbf{E}\{y_1^2\} = \sigma_1^2, \qquad \mathbf{E}\{y_j^2 \mid y_1, \cdots, y_{j-1}\} = \sigma_j^2, \qquad j > 1$$

with probability 1. Here $\sigma_1, \cdots, \sigma_n$ are non-negative constants. Define

$$\hat{\sigma}_k^2 = \sum_1^k \sigma_j^2.$$

We have already remarked that (11.23) simply states that the process $\{x_j, j \leq n\}$ is a martingale, with

$$\mathbf{E}\{x_j\} = 0, \qquad j \geq 1.$$

Let $\mathscr{F}_j$ be the Borel field of the ω sets determined by conditions on $x_1, \cdots, x_j$, that is, on $y_1, \cdots, y_j$. Then (11.23) and (11.24) imply

$$\mathbf{E}\{x_j^2 - \hat{\sigma}_j^2 \mid \mathscr{F}_{j-1}\} = x_{j-1}^2 - \hat{\sigma}_{j-1}^2, \qquad j > 1,$$

with probability 1. In other words, if (11.23) and (11.24) are true, the process $\{x_j^2 - \hat{\sigma}_j^2, \mathscr{F}_j, j \leq n\}$ is also a martingale. (One way of looking at this fact is to observe that the x_j^2 process is a semi-martingale, since the x_j process is a martingale, and in the representation (1.5′) of this semi-martingale we have the case in which the Δ_j's reduce to constants.) Conversely, if (11.23) is true, and if $\{x_j^2 - \hat{\sigma}_j^2, \mathscr{F}_j, j \leq n\}$ is a martingale, then (11.24) is true. In particular, (11.24) is true if the y_j's are mutually independent, with zero expectations and finite variances. Now in this case the central limit theorem is applicable (see III §4), and shows, under appropriate further restrictions, essentially that $\operatorname{Max}_j |y_j|$ is small, that x_n has an approximately Gaussian distribution. The characteristic function proof is simply that, if Φ_j is the characteristic function of y_j and Φ that of x_n,

$$\log \Phi_j(\lambda) = -\frac{\sigma_j^2}{2}\lambda^2$$

(approximately) so that

$$\log \Phi(\lambda) = \sum_1^n \log \Phi_j(\lambda) = -\frac{\hat{\sigma}_n^2}{2}\lambda^2$$

(approximately). Now the reasoning used here is applicable in the general case, when the y_j's only satisfy (11.23) and (11.24), as follows. Neglecting error terms,

$$\Phi_j(\lambda) = \mathbf{E}\{e^{i\lambda x_j}\} = \mathbf{E}\left\{e^{i\lambda x_{j-1}} \middle| \mathbf{E}\{e^{i\lambda y_j} \mid \mathscr{F}_{j-1}\}\right\}$$

$$= \mathbf{E}\{e^{i\lambda x_{j-1}}[1 - \frac{\sigma_j^2}{2}\lambda^2]\}$$

$$= \mathbf{E}\{e^{i\lambda x_{j-1}}\}\left[1 - \frac{\sigma_j^2\lambda^2}{2}\right] \qquad j > 1$$

so that

$$\log \Phi_j(\lambda) - \log \Phi_{j-1}(\lambda) = -\frac{\hat{\sigma}_j^2\lambda^2}{2}, \qquad j > 1,$$

approximately. Adding these equations, we get

$$\log \Phi(\lambda) = -\frac{\hat{\sigma}_n^2\lambda^2}{2}$$

approximately, as in the case of sums of mutually independent y_j's. Thus it is clear that the central limit theorem is applicable to martingales much as it is to sums of mutually independent random variables. This fact was first observed by Lévy. We shall not go into details in the discrete parameter case, but we shall need one continuous parameter result in this direction. We shall discuss real martingales $\{x_t, \mathscr{F}_t, a \leq t \leq b\}$ with $\mathbf{E}\{x_t^2\} < \infty$. The $x_t - x_a$ process is then also a martingale, with respect to the same fields, so the $(x_t - x_a)^2$ process is a semi-martingale with respect to these fields. Hence the function F defined by

$$F(t) = \mathbf{E}\{(x_t - x_a)^2\}$$

is monotone non-decreasing, and as a matter of fact it is trivial to verify that the x_t process has orthogonal increments, so that

$$F(t) - F(s) = \mathbf{E}\{(x_t - x_s)^2\}, \qquad s < t.$$

The fixed points of discontinuity of the x_t process are the points of discontinuity of F. In analogy with (11.24) we impose the condition

$$\mathbf{E}\{(x_t - x_s)^2 \mid \mathscr{F}_s\} = F(t) - F(s), \tag{11.24'}$$

with probability 1, for each pair s, t with $s < t$. This is equivalent to the condition that the process

$$\{x_t^2 - F(t), \mathscr{F}_t, a \leq t \leq b\}$$

be a martingale. In particular, if F is a continuous function, that is, if the x_t process has no fixed point of discontinuity, the change of parameter from t to t', where $t = F(t')$, reduces the situation to the special case in which $F(t) \equiv t - a$, and this is the case we now treat.

THEOREM 11.9 *Let $\{x_t, \mathscr{F}_t, a \leq t \leq b\}$ be a real martingale, and suppose that almost all sample functions of the process are continuous. Suppose that*

$$\mathbf{E}\{x_t^2\} < \infty, \qquad a \leq t \leq b,$$

and that, for each pair s, t with $s < t$,

$$\mathbf{E}\{(x_t - x_s)^2 \mid \mathscr{F}_s\} = t - s \tag{11.25}$$

with probability 1, *that is, we suppose that $\{x_t^2 - t, \mathscr{F}_t, a \leq t \leq b\}$ is a martingale. Then it follows that the x_t process has independent increments, and is in fact a Brownian motion process.*

We shall see that the main point of the proof is that $x_b - x_a$ is Gaussian. This character of the distribution is intuitively clear, because, for large n, $x_b - x_a$ can be expressed as a sum of small random variables

$$x_b - x_a = \sum_1^n y_j, \qquad y_j = x_{a+cj/n} - x_{a+c(j-1)/n}, \qquad c = b - a.$$

The y_j's satisfy (11.24) with $\sigma_j^2 = c/n$, so that one naturally expects the central limit theorem to be applicable to give the desired result. The following formal proof of this theorem exhibits the simplifications that can be effected by a proper use of Theorem 11.8. To simplify the notation we take $a = 0$, $b = 1$. Moreover, we shall suppose, as we have shown we can, that each $\mathscr{F}_t$ contains all ω sets of probability 0. Replacing x_t by $x_t - x_0$ if necessary, we can suppose that $x_0 = 0$. Let ε be a positive number, let n be a positive integer, and let $\tau(n, \varepsilon, \omega)$ be the first value of t for which

$$\operatorname*{Max}_{\substack{s_1, s_2 \leq t \\ |s_2 - s_1| \leq 1/n}} |x_{s_2}(\omega) - x_{s_1}(\omega)| = \varepsilon,$$

or 1 if there is no such value of t. Here we ignore the discontinuous sample functions. Then $0 < \tau \leq 1$ and the condition $\tau(\omega) > s$ involves restrictions on continuous sample functions only at parameter values $\leq s$, so that $\{\tau(\omega) > s\} \in \mathscr{F}_s$. In fact, considering only continuous sample functions, if $0 < s < 1$, and if r_1, r_2 are restricted to be rational,

$$\{\tau(\omega) > s\} = \bigcup_{m=1}^{\infty} \bigcap_{\substack{0 \leq r_1, r_2 \leq s \\ |r - r_2| \leq 1/n}} \{|x_{r_1}(\omega) - x_{r_2}(\omega)| \leq \varepsilon - 1/m\} \in \mathscr{F}_s.$$

Then $\{\tau(\omega) \leq s\} \in \mathscr{F}_s$, and we use τ to define a system of optional stopping, so that $\breve{x}_t$ is given by

$$\begin{aligned} \breve{x}_t(\omega) &= x_t(\omega), \qquad & t \leq \tau(\omega), \\ &= x_{\tau(\omega)}(\omega), \qquad & \tau(\omega) < t. \end{aligned}$$

Since both the x_t and $x_t^2 - t$ processes are martingales, Theorem 11.8 (i), (iii), states that the processes

$$\{\breve{x}_t, \breve{\mathscr{F}}_t, 0 \leq t \leq 1\}, \qquad \{\breve{x}_t^2 - \operatorname{Min}[t, \tau], \breve{\mathscr{F}}_t, 0 \leq t \leq 1\}$$

are martingales, so that

$$\int_\Lambda \{\breve{x}_s^2 - \operatorname{Min}[s, \tau]\}\, d\mathbf{P} = \int_\Lambda \{\breve{x}_t^2 - \operatorname{Min}[t, \tau]\}\, d\mathbf{P}, \qquad \Lambda \in \breve{\mathscr{F}}_s, \quad s < t.$$

Then

$$\begin{aligned} \int_\Lambda \mathbf{E}\{\breve{x}_t^2 - \breve{x}_s^2 \mid \breve{\mathscr{F}}_s\}\, d\mathbf{P} &= \int_\Lambda (\breve{x}_t^2 - \breve{x}_s^2)\, d\mathbf{P} \\ &= \int_\Lambda (\operatorname{Min}[t, \tau] - \operatorname{Min}[s, \tau])\, d\mathbf{P} \\ &\leq \int_\Lambda (t - s)\, d\mathbf{P}. \end{aligned}$$

The integrand on the left is measurable with respect to the field $\breve{\mathscr{F}}_s$, and this inequality holds for every $\Lambda \in \breve{\mathscr{F}}_s$. Hence

$$\mathbf{E}\{\breve{x}_t^2 - \breve{x}_s^2 \mid \breve{\mathscr{F}}_s\} \leq t - s \tag{11.26}$$

with probability 1. The $\breve{x}_t$ process thus has almost the same three basic properties as the x_t process: the process $\{\breve{x}_t, \breve{\mathscr{F}}_t, 0 \le t \le 1\}$ is a martingale; the $\breve{x}_t$ process sample functions are almost all continuous; the equality (11.25) is weakened to (11.26) (which is equivalent to the statement that the process $\{\breve{x}_t^2 - t, \breve{\mathscr{F}}_t, 0 \le t \le 1\}$ is a lower semi-martingale). Moreover, the $\breve{x}_t$ process is simple in that small t increments are uniformly bounded. In fact, if

$$y_j = \breve{x}_{j/n} - \breve{x}_{(j-1)/n} \qquad j = 1, \cdots, n,$$

then

$$|y_j| \le \varepsilon. \tag{11.27}$$

The martingale property of the $\breve{x}_t$ process and (11.26) imply

(11.28)

$$\begin{aligned} &\mathbf{E}\{y_1\} = 0, &&\mathbf{E}\{y_j \mid y_1, \cdots, y_{j-1}\} = 0 \\ &\sigma_1^2 = \mathbf{E}\{y_1^2\} \le \frac{1}{n}, &&\sigma_j^2 = \mathbf{E}\{y_j^2 \mid y_1, \cdots, y_{j-1}\} \le \frac{1}{n} \end{aligned} \qquad j = 1, \cdots, n$$

with probability 1. We shall prove that $\breve{x}_1$ is nearly normally distributed, using (11.27) and (11.28). We shall choose $n = n(\varepsilon)$ as described below, so that $n \to \infty$ when $\varepsilon \to 0$.

If $j \ge 1$,

$$\begin{aligned} \mathbf{E}\{e^{i\lambda \breve{x}_{j/n}}\} &= \mathbf{E}\{e^{i\lambda \breve{x}_{(j-1)/n}}\, \mathbf{E}\{e^{i\lambda y_j} \mid y_1, \cdots, y_{j-1}\}\} \\ &= \mathbf{E}\left\{e^{i\lambda \breve{x}_{(j-1)/n}} \left[1 - \sigma_j^2 \frac{\lambda^2}{2}(1 + o(1))\right]\right\} \\ &= \mathbf{E}\left\{e^{i\lambda \breve{x}_{(j-1)/n} - \sigma_j^2 \frac{\lambda^2}{2}(1+o(1))}\right\} \end{aligned}$$

where $o(1)$ here and in the following represents any expression which goes to 0 with ε, uniformly in any other variables involved as long as λ is restricted to a finite interval, whereas $O(1)$ represents any expression which remains bounded under the same conditions. It follows that, if $j \ge 1$,

$$\begin{aligned} &\left|\mathbf{E}\{e^{i\lambda \breve{x}_{j/n}}\} - \mathbf{E}\{e^{i\lambda \breve{x}_{(j-1)/n} - \frac{\lambda^2}{2n}}\}\right| \\ &\quad = \left|\mathbf{E}\left\{e^{i\lambda \breve{x}_{(j-1)/n} - \frac{\lambda^2}{2n}}\left[e^{-\frac{\lambda^2}{2}\left[\sigma_j^2(1+o(1)) - \frac{1}{n}\right]} - 1\right]\right\}\right| \\ &\quad \le \mathbf{E}\{|e^{\frac{\lambda^2}{2}\left(\frac{1}{n} - \sigma_j^2\right) + \sigma_j^2 o(1)} - 1|\} \\ &\quad \le O(1)\, \mathbf{E}\left\{\left(\frac{1}{n} - \sigma_j^2\right) + \frac{o(1)}{n}\right\} \\ &\quad \le O(1)\left[\frac{1}{n} - \mathbf{E}\{y_j^2\}\right] + \frac{o(1)}{n}. \end{aligned}$$

Then

$$| \mathbf{E}\{e^{i\lambda\breve{x}_{j/n}}\}e^{\frac{j}{2n}\lambda^2} - \mathbf{E}\{e^{i\lambda\breve{x}_{(j-1)/n}}\}e^{\frac{j-1}{2n}\lambda^2}|$$

$$\leq O(1)\left[\frac{1}{n} - \mathbf{E}\{y_j^2\}\right] + \frac{o(1)}{n}, \qquad j \geq 1.$$

Adding these inequalities, we find

$$| \mathbf{E}\{e^{i\lambda\breve{x}_1}\}e^{\lambda^2/2} - 1| \leq O(1)[1 - \sum_{j=1}^{n} \mathbf{E}\{y_j^2\}] + o(1)$$

$$= O(1)[1 - \mathbf{E}\{\breve{x}_1^2\}] + o(1),$$

so that

$$|\mathbf{E}\{e^{i\lambda\breve{x}_1}\} - e^{-\lambda^2/2}| \leq O(1)[1 - \mathbf{E}\{\breve{x}_1^2\}] + o(1).$$

Now, for a given ε, $\mathbf{P}\{\breve{x}_1(\omega) = x_1(\omega)\}$ can be made arbitrarily near 1 by choosing n large. Hence the characteristic functions of $\breve{x}_1$ and x_1 can be made arbitrarily close to each other, uniformly in any given finite interval, by choosing n large. That is, $n = n(\varepsilon)$ can be chosen so large that

$$|\mathbf{E}\{e^{i\lambda x_1}\} - e^{-\lambda^2/2}| \leq O(1)[1 - \mathbf{E}\{\breve{x}_1^2\}] + o(1).$$

Moreover, since $\sigma_j^2 \leq 1/n$, we know that $\mathbf{E}\{\breve{x}_1^2\} \leq 1$, so that

$$0 \leq 1 - \mathbf{E}\{\breve{x}_1^2\} = \mathbf{E}\{x_1^2 - \breve{x}_1^2\} = \int_{\{\tau(\omega)<1\}} (x_1^2 - x_\tau^2)\, d\mathbf{P}$$

$$\leq \int_{\{\tau(\omega)<1\}} x_1^2\, d\mathbf{P}.$$

Now we have chosen n large to make $\mathbf{P}\{\tau(\omega) = 1\}$ nearly 1 so that $\mathbf{P}\{x_1(\omega) = \breve{x}_1(\omega)\}$ will be near 1. Hence the last integral in the above inequality is $o(1)$, and this proves that

$$\mathbf{E}\{e^{i\lambda x_1}\} = e^{-\lambda^2/2} + o(1).$$

Since only the last term $o(1)$ can depend on ε, this term must actually be 0, so that x_1 must have a Gaussian distribution with expectation 0 and variance 1. A trivial modification of this discussion shows that, if $0 \leq t_0 < \cdots < t_k \leq 1$,

$$\mathbf{E}\{e^{i\sum_{j=1}^{k} \lambda_j(x_{t_j} - x_{t_{j-1}})}\} = \mathbf{E}\{e^{i\sum_{j=1}^{k-1} \lambda_j(x_{t_j} - x_{t_{j-1}})}\}e^{-(\lambda_k^2/2)(t_k - t_{k-1})}.$$

Then

$$\mathbf{E}\{e^{i\sum_{j=1}^{k} \lambda_j(x_{t_j} - x_{t_{j-1}})}\} = \prod_{j=1}^{k} e^{-(\lambda_j^2/2)(t_j - t_{j-1})},$$

that is, the x_t family is Gaussian and has independent increments, as stated in the theorem.

12. Applications of martingale theory to sample function continuity of various types of processes

(a) *Application to Markov processes* Let $\{x_t, a \le t \le b\}$ be a Markov process with a specified transition probability distribution function. That is, we suppose that there is a given function p of ξ, s, η, t, defining a Baire function of ξ for fixed s, η, t, and defining a distribution function in η for fixed ξ, s, t. For each ξ, s, t, with $s < t$, it is supposed that

$$p(x_s, s;\ \eta, t) = \mathbf{P}\{x_t(\omega) \le \eta \mid x_s\},$$

with probability 1, and it is supposed that the Chapman-Kolmogorov equation is satisfied identically in ξ, s, η, t for $s < t$,

$$p(\xi, s;\ \eta, t) = \int_{-\infty}^{\infty} p(\zeta, u;\ \eta, t)\, d_\zeta p(\xi, s;\ \zeta, u), \qquad s < u < t.$$

Then the stochastic process $\{x_t, a \le t \le b\}$ defined by

$$\tilde{x}_t = p(x_t, t;\ \eta, b) = \mathbf{P}\{x_b(\omega) \le \eta \mid x_s, s \le t\}$$

is a martingale, since as t increases more and more conditions are introduced in the conditional probability on the right. The known continuity properties of martingale sample functions imply corresponding properties of the x_t process sample functions. For example, if the Markov process is a chain with stationary transition probabilities (see VI §1), we define $\tilde{x}_t$ (modifying the above definition slightly) as

$$\tilde{x}_t = p_{x_t j}(b - t),$$

where

$$p_{ij}(t) = \mathbf{P}\{x_{s+t} = j \mid x_s(\omega) = i\},$$

and the continuity properties of the $\tilde{x}_t$ process sample functions can be used to derive the fact (see VI, Theorem 1.4) that the sample functions of a separable chain with a finite number of states are almost all step functions.

(b) *Application to processes with independent increments* Let $\{x_t, a \le t \le b\}$ be a real separable process with independent increments. Suppose that for each $t \in [a, b]$ the limits

$$\lim_{s \uparrow t} x_s = x_{t-}, \qquad \lim_{s \downarrow t} x_s = x_{t+}$$

exist and are finite with probability 1. It is proved in VIII, Theorem 6.3, that then almost all sample functions of the process are bounded. (Theorem 6.3 assumes that the x_t process is centered, which besides the above conditions imposes the condition that a difference $x_t - x_{t-}$ or $x_{t+} - x_t$ is not identically constant with probability 1 unless the constant is 0, but this condition is not used in the proof.) We now apply martingale theory to the study of the continuity properties of the x_t process, and we prove Lévy's result (VIII, Theorem 7.2) that *almost all sample functions of the process are continuous except for discontinuities at which both left- and right-hand finite limits exist.* To prove this let Φ_t be the characteristic function of $x_t - x_a$,

$$\Phi_t(\mu) = \mathbf{E}\{e^{i\mu(x_t - x_a)}\}$$

and define

$$\tilde{x}_t = \frac{e^{i\mu(x_t - x_a)}}{\Phi_t(\mu)}.$$

This definition assumes that μ is chosen so that $\Phi_t(\mu) \neq 0$. If $\delta > 0$ is chosen so small that

$$\Phi_b(\mu) \neq 0, \qquad |\mu| \leq \delta,$$

the equation

$$\Phi_b(\mu) = \Phi_t(\mu)\mathbf{E}\{e^{i\mu(x_b - x_t)}\}$$

shows that

$$\Phi_t(\mu) \neq 0, \qquad |\mu| \leq \delta, \qquad a \leq t \leq b.$$

Thus, for each μ with $|\mu| \leq \delta$, there is an $\tilde{x}_t$ process. Let $\mathscr{F}_t$ be the Borel field of the ω sets determined by conditions on $x_s - x_a$ for $s \leq t$. Then the process $\{\tilde{x}_t, \mathscr{F}_t, a \leq t \leq b\}$ is a martingale, because, if $s < t$,

$$\mathbf{E}\{\tilde{x}_t \mid \mathscr{F}_s\} = \frac{\mathbf{E}\{e^{i\mu(x_t - x_s)}\}e^{i\mu(x_s - x_a)}}{\Phi_t(\mu)} = \tilde{x}_s$$

with probability 1. In the following we shall choose a set R in the interval $[a, b]$. A function g defined on $[a, b]$ will be said to have the property C_R if it coincides on R with a function defined on $[a, b]$ which has left- and right-hand limits at all points of $[a, b]$. In other words, g has the property C_R if it can be redefined at the points of $[a, b] - R$ in such a way that the resulting function has left- and right-hand limits at all points of $[a, b]$. Let R be any denumerable subset of $[a, b]$. Then, even though the $\tilde{x}_t$ process (with μ fixed) is not necessarily a separable process, we have seen that Theorem 11.5 is applicable to give the result that almost all sample functions of the $\tilde{x}_t$ process have the property C_R. Now by our hypotheses $\Phi_t(\mu)$ has for each μ left- and right-hand limits in t at all points of $[a, b]$. Hence for each μ almost every sample function of the x_t process has the property that the t function

$$e^{i\mu[x_\cdot(\omega) - x_a(\omega)]}$$

has the property C_R. Then almost every sample function of the x_t process has the property that simultaneously for all rational μ in the interval $[-\delta, \delta]$ the corresponding t function above, and hence its imaginary part, has the property C_R. Now if a real function $f(\cdot)$ is bounded, and if $|\mu f| \leq \pi/2$, then $\sin \mu f$ has the property C_R if and only if f has this property. It follows that almost all sample functions of the x_t process have this property. By hypothesis, the x_t process is separable. Let R be the parameter set involved in the definition of separability. Then the fact that almost all x_t process sample functions have the property C_R means that almost all have finite left- and right-hand limits at all points of $[a, b]$, as was to be proved.

(c) *Application to Gaussian processes* It is instructive to apply martingale theory to the study of the conditional expectations involved in Gaussian processes. Suppose first (discrete parameter case) that $\{x_n, n \geq 0\}$ is a Gaussian process. Then $\mathbf{E}\{x_0 \mid x_1, \cdots, x_n\}$ is a linear combination of $x_1, \cdots, x_n$ and, since (Theorem 4.3, Corollary 1)

$$\mathbf{E}\{x_0 \mid x_1, x_2, \cdots\} = \lim_{n\to\infty} \mathbf{E}\{x_0 \mid x_1, \cdots, x_n\}$$

with probability 1, the conditional expectation on the left, as a limit of Gaussian variables, is itself Gaussian. More generally, it is clear that any conditional expectations of x_m's, for finitely or infinitely many given x_j's, have a joint Gaussian distribution since these conditional expectations are the limits of linear combinations of x_j's. (The Gaussian distributions may of course be degenerate.) In the general case if $\{x_t, t \in T\}$ is a Gaussian process, consider a conditional expectation of the form

$$\mathbf{E}\{x_t \mid x_s, s \in T_1\}.$$

There is a finite or enumerable sequence $\{s_n\}$ in T_1 such that

$$\mathbf{E}\{x_t \mid x_s, s \in T_1\} = \mathbf{E}\{x_t \mid x_{s_j}, j \geq 1\}$$

with probability 1 (I, Theorem 8.2), so that according to the discrete parameter analysis just made the conditional expectation under consideration must be a Gaussian variable, and more generally any set of conditional expectations of this form is itself a Gaussian process. As an example, suppose that $\tilde{x}_t$ is defined by

$$\tilde{x}_t = \mathbf{E}\{x_b \mid x_s, s \leq t\}.$$

Then the $\tilde{x}_t$ family is a Gaussian martingale. (It is easily seen that a Gaussian martingale which has no fixed points of discontinuity must be a Brownian motion aside from an additive constant and a change of the time parameter, because the family has uncorrelated increments.) We observe that $\tilde{x}_t$ can be considered a prediction of x_b, knowing the "past" of x_s up to time t (see XII).

CHAPTER VIII

Processes with Independent Increments

1. General remarks

Processes with independent increments were defined in II §9. The integral parameter case has already been discussed; in fact we have remarked in II §9 that, if the process with random variables $x_1, x_2, \cdots$ has independent increments, x_n is the nth partial sum of a series of mutually independent random variables, and such sums have been discussed in III. In practice the nomenclature "independent increments" is ordinarily used only in the continuous parameter case. The nomenclatures "differential process," "additive process," and "integral with independent random elements" have also been used. The latter nomenclature is suggested by the formal expression

$$x_t - x_a = \int_a^t d_s x_s,$$

in which the differential elements are mutually independent.

Processes with independent increments have wide sense versions, processes with uncorrelated or orthogonal increments. The latter are treated in IV (discrete parameter case) and IX (continuous parameter case).

It will be convenient in several of the following sections to use the notation $x(t)$ rather than x_t for the general random variable of the process under consideration.

If $\{x_t, t \in T\}$ is a process with independent increments, and if f is a function defined on T, then $\{x_t - f(t), t \in T\}$ is a process which also has independent increments. It will be convenient below to replace x_t by $x_t - f(t)$, where f is chosen to obtain simple continuity properties for the new sample functions. Moreover, if T has a minimum value a, it is usually convenient to replace x_t by $x_t - x_a$, to obtain a new process, also with independent increments, whose sample functions all vanish at a.

2. Brownian movement process

This process was defined in II §9. It is a real process, $\{x(t),\ t \in T\}$, where T is usually taken to be an interval, in fact usually either $(-\infty, \infty)$ or $[0, \infty)$, with independent Gaussian increments satisfying

$$\text{(2.1)} \qquad \mathbf{E}\{x(t) - x(s)\} = 0, \qquad \mathbf{E}\{[x(t) - x(s)]^2\} = \sigma^2|t - s|,$$

where σ is a positive constant. The process is of great importance because of its central role in the theory of stationary Gaussian processes (see X and XI) and because of its numerous physical applications.

The inequality

$$\text{(2.2)} \qquad \int_\lambda^\infty e^{-\xi^2/2}\, d\xi \le \frac{1}{\lambda}\int_\lambda^\infty \xi e^{-\xi^2/2}\, d\xi = \frac{e^{-\lambda^2/2}}{\lambda} \qquad (\lambda > 0)$$

will be useful below.

THEOREM 2.1 (*Separable Brownian movement process*)

$$\text{(2.3)} \qquad \mathbf{P}\{\underset{0\le t\le T}{\text{L.U.B.}}\ [x(t, \omega) - x(0, \omega)] \ge \lambda\} = 2\mathbf{P}\{x(T, \omega) - x(0, \omega) \ge \lambda\}$$
$$\le \frac{\sigma}{\lambda}\sqrt{\frac{2T}{\pi}}\, e^{-\lambda^2/(2\sigma^2 T)}.$$

Suppose $0 = t_0 < \cdots < t_n = T$. Then, according to III, Theorem 2.2,

$$\text{(2.4)} \qquad \mathbf{P}\{\underset{j\le n}{\text{Max}}\ [x(t_j, \omega) - x(0, \omega)] \ge \lambda\} \le 2\mathbf{P}\{x(T, \omega) - x(0, \omega) \ge \lambda\}.$$

Now the left side of (2.3) is obtained by choosing a denser and denser sequence $t_1, \cdots, t_n$ and going to the limit (II, Theorem 2.2). Hence

$$\text{(2.3')} \qquad \mathbf{P}\{\underset{0\le t\le T}{\text{L.U.B.}}\ [x(t, \omega) - x(0, \omega)] \ge \lambda\} \le 2\mathbf{P}\{x(T, \omega) - x(0, \omega) \ge \lambda\}.$$

To prove the reverse inequality set $t_j = jT/n$ and apply the other half of III, Theorem 2.2, obtaining, for each $\varepsilon > 0$,

$$\mathbf{P}\{\underset{0\le t\le T}{\text{L.U.B.}}\ [x(t, \omega) - x(0, \omega)] \ge \lambda\}$$
$$\ge \mathbf{P}\{\underset{j\le n}{\text{Max}}\ [x(t_j, \omega) - x(0, \omega)] \ge \lambda\}$$
$$\ge 2\mathbf{P}\{x(t, \omega) - x(0, \omega) \ge \lambda + 2\varepsilon\} - 2n\mathbf{P}\left\{x\left(\frac{T}{n}, \omega\right) - x(0, \omega) \ge \varepsilon\right\}$$
$$\ge 2\mathbf{P}\{x(T, \omega) - x(0, \omega) \ge \lambda + 2\varepsilon\} - \frac{2n\, e^{-\varepsilon^2 n/2\sigma^2 T}\, \sigma\sqrt{T}}{\sqrt{2\pi}\, \varepsilon\sqrt{n}}$$
$$\to 2\mathbf{P}\{x(T, \omega) - x(0, \omega) \ge \lambda\} \qquad (n \to \infty, \quad \varepsilon \to 0).$$

THEOREM 2.2 *Almost all sample functions of a separable Brownian movement process are continuous.*

Suppose for definiteness that the range of the parameter is the interval $[0, \infty)$. We prove that

$$\text{(2.5)} \qquad \left|x(t, \omega) - x\left(\frac{j}{N}, \omega\right)\right| \leq N^{-1/4}, \qquad \text{if} \quad \left|t - \frac{j}{N}\right| \leq \frac{1}{N}, \quad j \leq N^2,$$

for sufficiently large N, except for an ω set of probability 0. Using Theorem 2.1,

$$\mathbf{P}\left\{ \underset{\substack{|t-j/N| \leq 1/N \\ 1 \leq j \leq N^2}}{\text{L.U.B.}} \left|x(t, \omega) - x\left(\frac{j}{N}, \omega\right)\right| \geq N^{-1/4}\right\}$$

$$\leq N^2 \mathbf{P}\left\{ \underset{|t-1/N| \leq 1/N}{\text{L.U.B.}} \left|x(t, \omega) - x\left(\frac{1}{N}, \omega\right)\right| \geq N^{-1/4}\right\}$$

$$\leq \frac{2\sigma\sqrt{2}}{\sqrt{\pi}} N^{7/4} e^{-N^{1/2}/2\sigma^2}.$$

The last term is the Nth term of a convergent series; hence (Borel-Cantelli lemma) the event whose probability is being measured happens only a finite number of times, with probability 1; that is, the indicated L.U.B. is $\leq N^{-1/4}$ for large N, with probability 1. Thus (2.5) is true, and (2.5) states (with specific estimates of the ε's and δ's involved) that almost all sample functions are uniformly continuous in every finite t interval.

The proof of continuity of the sample functions was made dependent on the evaluation of $\mathbf{P}\{ \underset{0 \leq t \leq T}{\text{L.U.B.}} [x(t, \omega) - x(0, \omega)] \geq \lambda\}$ in Theorem 2.1. Actually, all that was needed was an upper bound for this probability, that is, only half of Theorem 2.1 was needed, (2.3′). Suppose then that instead of (2.3) only (2.3′) had been proved. The proof of Theorem 2.2 would need no change. The exact evaluation (2.3) and similar exact evaluations are easily made, once sample function continuity has been proved, using what is known as the reflection principle of Desiré André. To illustrate the method, (2.3) will be derived, assuming only sample function continuity.

Obviously

$$\text{(2.6)} \qquad \mathbf{P}\{ \underset{0 \leq t \leq T}{\text{Max}} [x(t, \omega) - x(0, \omega)] \geq \lambda;\ x(T, \omega) - x(0, \omega) \geq \lambda\}$$

$$= \mathbf{P}\{x(T, \omega) - x(0, \omega) \geq \lambda\},$$

where we can write Max instead of L.U.B. because the sample functions are continuous (with probability 1). On the other hand, consider the continuous sample functions satisfying

$$\max_{0 \le t \le T} [x(t, \omega) - x(0, \omega)] \ge \lambda, \; x(T, \omega) - x(0, \omega) < \lambda. \tag{2.7}$$

If $\tau(\omega)$ *is the first value of* t *for which* $x(t, \omega) - x(0, \omega) = \lambda$, *the changes in* $x(t, \omega)$ *after* $\tau(\omega)$ *are independent of the changes before* $\tau(\omega)$, *and are equally likely to be positive or negative*; *we shall not change the probabilities by reflecting the sample curve for* $t > \tau(\omega)$ *in the line* $x = \lambda$. That is to say,

$$\begin{aligned} \mathbf{P}\{\max_{0 \le t \le T} [x(t, \omega) - x(0, \omega)] \ge \lambda, \; x(T, \omega) - x(0, \omega) < \lambda\} \\ = \mathbf{P}\{\max_{0 \le t \le T} [x(t, \omega) - x(0, \omega)] \ge \lambda, \; x(T, \omega) - x(0, \omega) > \lambda\} \\ = \mathbf{P}\{x(T, \omega) - x(0, \omega) > \lambda\}. \end{aligned} \tag{2.8}$$

Adding (2.6) to (2.8), we obtain (2.3). This proof is, of course, not complete without further elaboration of the italicized statement. The point is, however, that the parallel approximate analysis, for t running through only a discrete set of points $T/n, 2T/n, \cdots, T$ (which we have carried through) gives in the limit $(n \to \infty)$ the exact analysis, and, if desired, the italicized statement can be considered an abbreviation of this limit reasoning even if the rather delicate detailed justification of the italicized statement is omitted.

The sample functions of a separable Brownian movement process, although (almost all) continuous are very irregular in nature. It will be shown below for example that, for each fixed t_0,

$$\limsup_{t \to t_0} \frac{x(t) - x(t_0)}{t - t_0} = \infty \tag{2.9}$$

with probability 1. In other words, at each t_0 the sample functions have infinite upper derivatives with probability 1. This means (by II, Theorem 2.5 this process is measurable) that almost all sample functions have the property that the upper derivates are $+\infty$ for all values of t except for a set of values of t of Lebesgue measure 0. The exceptional t set will vary from sample function to sample function. To prove (2.9) suppose that λ is positive and consider the probability

$$\mathbf{P}\{\underset{t_0 < t \le t_0 + \delta}{\text{L.U.B.}} \frac{x(t, \omega) - x(t_0, \omega)}{t - t_0} \ge \lambda\}.$$

It is sufficient to prove that the probability goes to 1 when $\delta \to 0$, and this follows from

$$\mathbf{P}\left\{\underset{t_0<t\le t_0+\delta}{\text{L.U.B.}} \frac{x(t, \omega) - x(t_0, \omega)}{t - t_0} \ge \lambda\right\} \ge \mathbf{P}\{\underset{t_0\le t\le t_0+\delta}{\text{Max}} [x(t, \omega) - x(t_0, \omega)] \ge \lambda\delta\}$$

$$= 2\mathbf{P}\{x(t_0 + \delta, \omega) - x(t_0, \omega) \ge \lambda\delta\}$$

$$= 2\mathbf{P}\left\{\frac{x(t_0 + \delta, \omega) - x(t_0, \omega)}{\sqrt{\delta}} \ge \lambda \sqrt{\delta}\right\} \to 1 \ (\delta \to 0)$$

since $[x(t_0 + \delta, \omega) - x(t_0, \omega)]/\sqrt{\delta}$ is normally distributed with expectation 0 and variance σ^2 independent of δ.

We can state even more striking facts about the irregular character of the Brownian movement sample functions. Let f be any fixed continuous function of t, and suppose that $0 = t_0 < \cdots < t_n = T$. Then

$$\sum_{j=0}^{n-1} [f(t_{j+1}) - f(t_j)]^2 \le \underset{j\le n-1}{\text{Max}} |f(t_{j+1}) - f(t_j)| \sum_{j=0}^{n} |f(t_{j+1}) - f(t_j)|.$$

Hence, if the function is continuous in the interval $[0, T]$, and if the sum on the right is bounded independently of the choice of n and of the t_j's, that is, if f is of bounded variation (or, geometrically, if the graph has finite length), the sum on the left will go to 0 when $\underset{j\le n-1}{\text{Max}} (t_{j+1} - t_j) \to 0$. The following theorem, according to which this sum approaches 0 for almost no sample function, thus shows that almost no Brownian movement sample function has bounded variation.

THEOREM 2.3 (*Brownian movement process*) *Let* $t_0, t_1, \cdots$ *be everywhere dense in the interval* $[0, T]$; *and let* $t_0^{(n)}, \cdots, t_n^{(n)}$ *be the numbers* $t_0, \cdots, t_n$ *arranged in numerical order,* $t_0^{(n)} < \cdots < t_n^{(n)}$. *Then*

$$\lim_{n\to\infty} \sum_{j=0}^{n-1} [x(t_{j+1}^{(n)}) - x(t_j^{(n)})]^2 = \sigma^2 T$$

with probability 1, *and the limit relation is also true in the sense of convergence in the mean.*

In fact, let S_n be the sum on the left and suppose first that t_0 and t_1 are $0, T$ so that $t_0^{(n+1)}, \cdots, t_{n+1}^{(n+1)}$ is obtained by inserting t_{n+1} *between* two of the $t_j^{(n)}$'s. Then we shall show that the sequence $\cdots, S_2, S_1$ (note the order) constitutes a martingale. It is sufficient to show that, for every pair of positive integers m, n,

$$\mathbf{E}\{S_n \mid S_{n+m}, \cdots, S_{n+1}\} = S_{n+1}$$

with probability 1; that is to say,

$$\mathbf{E}\{S_n - S_{n+1} \mid S_{n+m}, \cdots, S_{n+1}\} = 0 \tag{2.10}$$

with probability 1. The details of the symmetry argument used to prove this equation will be omitted except for the case $m = 2$. (If the equation is true for $m = 2$, it is necessarily also true for $m = 1$.) There are two possibilities: either t_{n+1} and t_{n+2} are both inserted in one of the intervals $(t_j^{(n)}, t_{j+1}^{(n)})$ or they are inserted in different intervals. The two are treated in the same way, and we shall consider only the first. In this case (2.10) is a consequence of the following statement: let x_1, x_2, x_3, s be mutually independent random variables, of which the first three are Gaussian, with zero means. Then

(2.11)

$$\begin{aligned}\mathbf{E}\{(x_1+x_2+x_3)^2+s-[(x_1+x_2)^2+x_3^2+s] \mid x_1^2+x_2^2+x_3^2+s, (x_1+x_2)^2+x_3^2+s\}\\ = \mathbf{E}\{2(x_1 + x_2)x_3 \mid x_1^2 + x_2^2 + x_3^2 + s, (x_1 + x_2)^2 + x_3^2 + s\} = 0,\end{aligned}$$

with probability 1. To prove this relation we observe first that (symmetry)

$$\mathbf{E}\{2(x_1 + x_2)x_3 \mid x_1, x_2, x_3^2\} = 0,$$

with probability 1. Next, since s is independent of the x_j's it follows that

$$\mathbf{E}\{2(x_1 + x_2)x_3 \mid x_1, x_2, x_3^2, s\} = 0 \tag{2.12}$$

with probability 1, and (2.11) is deduced by taking the conditional expectation of both sides of this equation with respect to the conditioning variables in (2.11). [The point is that the conditions in (2.11) are less restrictive than those in (2.12).]

Since the sequence $\cdots, S_2, S_1$, is a martingale, $\lim_{n\to\infty} S_n$ exists with probability 1 (VII, Theorem 4.2). To show that the limit is $\sigma^2 T$ we show that the l.i.m. is $\sigma^2 T$,

$$\begin{aligned}\mathbf{E}\{(S_n - \sigma^2 T)^2\} &= 2\sigma^4 \sum_{j=0}^{n-1} (t_{j+1}^{(n)} - t_j^{(n)})^2\\ &\leq 2\sigma^4 T \operatorname*{Max}_j (t_{j+1}^{(n)} - t_j^{(n)}) \to 0 \qquad (n \to \infty).\end{aligned}$$

This finishes the proof in the special case $t_0 = 0$, $t_1 = T$. In the general case, if we define S_n' by

$$S_n' = [x(t_0^{(n)}) - x(0)]^2 + S_n + [x(T) - x(t_n^{(n)})]^2,$$

it follows from what we have just proved that

$$\lim_{n\to\infty} S_n' = \sigma^2 T$$

with probability 1. The theorem is then true in the general case, because

$$\lim_{n\to\infty} x(t_0^{(n)}) = x(0), \qquad \lim_{n\to\infty} x(t_n^{(n)}) = x(T)$$

with probability 1.

3. Physical applications of the Brownian movement process

Let $x(t)$ be the x coordinate at time t of a particle in some medium. Much of the following analysis is applicable to cases as different as:

(a) The particle is a molecule of a liquid or gas.

(b) The particle is of microscopic size, in a fluid, say a colloidal particle.

(c) The particle is a star; the medium is the stellar universe.

The English biologist Brown observed in 1826 that a particle in case (b) makes irregular apparently spontaneous movements, now known to be caused by the impacts on it of the molecules of the medium. This motion is called Brownian motion, and the Brownian motion (or "movement") process is used to analyze this motion. The essential condition of the following analysis is that there are only negligible bonds between the particle and those of the surrounding medium, except at the times of impact. The analysis is thus applicable, for example, to the motion of a molecule of a gas at low pressure. An "impact" is the proximity to a particle under analysis of the force field of a particle of the medium; it is unnecessary to identify an impact with something like a collision of two billiard balls.

The impacts on a Brownian particle follow each other in an irregular fashion, at a high rate (in the relevant time scale), and the displacement component $x(t+s)-x(t)$ is thus the sum of a large number of small displacement components, if s is large compared to the time between impacts. It is natural to consider the function $x(\cdot)$ as a sample function of a stochastic process, and the problem is to describe this continuous parameter process. It is supposed that the medium is in macroscopic equilibrium, and this hypothesis is reasonably translated to mean that the distribution of $x(t+s)-x(t)$ is symmetric, and does not depend on t. As a first approximation it is supposed that the random variable $x(t+s)-x(t)$ is for $s>0$ independent of the motion at times well before t, and this leads to the still stronger hypothesis that the $x(t)$ process has independent increments. The central limit theorem now suggests that $x(t+s)-x(t)$, as a sum of nearly independent small random variables, has a Gaussian distribution. If these hypotheses are all accepted, the $x(t)$ process must be what we have called the Brownian motion process. In the case of a particle in a fluid the parameter σ^2 of the process can be identified with $2D^2$, where D is the diffusion constant of the fluid, so that

$$\mathbf{E}\{[x(t)-x(0)]^2\} = 2D^2t.$$

This formula is due to Einstein. As noted by him, and as the above "derivation" makes clear, the formula can only be expected to be a crude approximation when t is of the order of magnitude of the time between molecular impacts. In any event, the reasoning used here, undisturbed by any specific particle dynamics, can hardly be considered more than suggestive. Indeed a more sophisticated mathematical approach might have concluded only that the $x(t)$ process should have stationary independent increments, and that the distribution of any increment $x(t+s)-x(t)$ should be infinitely divisible (cf. III, §4). The Brownian movement process is not the only one satisfying these conditions (cf. §7 of this chapter). However, the others could be excluded by examining the exact hypotheses of the relevant form of the central limit theorem and then asserting flatly that these seem to be reasonably well verified experimentally, or by noting that (cf. Theorem 7.1) the others do *not* have continuous sample functions, and discontinuous trajectories are repugnant to practical sensibilities. However much or little such a discussion is convincing, the fact is that it at least suggests the applicability of a separable Brownian movement process, and this mathematical model has in fact been checked empirically, or at any rate a few of its implications have been checked. It is satisfactory that the sample functions of the process are continuous; it is less satisfactory that (cf. §2) these sample functions are not of bounded variation, so that the trajectories have infinite length. Moreover, the velocities do not exist; in fact, we proved in §2 that the upper and lower derivates of a sample function at a given point $t=t_0$ are $+\infty$ and $-\infty$ respectively with probability 1. These properties of the process should not be taken too seriously from a practical point of view since they are *in the small* properties of the sample functions, involving increments $x(t+s)-x(t)$ with s small, and we have already remarked that the fit of theory to practice cannot be expected to be good for properties involving small increments. Closer analyses of Brownian movements have led to a somewhat different mathematical model with an $x(t)$ process which has sample functions with continuous derivatives. As would be expected, $\mathbf{E}\{[x(t+s)-x(t)]^2\} \sim s^2\mathbf{E}\{x'(t)^2\}$ is of the order of s^2 for small s in this process.

4. Poisson process

This type of process was defined in II §9. A Poisson process is a real process with stationary independent increments; the increments are integral-valued, and

$$\mathbf{P}\{x(t,\omega)-x(s,\omega)=m\}=\frac{e^{-c(t-s)}c^m(t-s)^m}{m!}, \qquad t>s, \quad m=0,1,2,\cdots \tag{4.1}$$

where c is a positive constant. We shall need the expectations

$$\begin{aligned} &\mathbf{E}\{x(t)-x(s)\}=c(t-s)\\ &\mathbf{E}\{[x(t)-x(s)-c(t-s)]^2\}=c(t-s)\end{aligned} \qquad t>s. \tag{4.2}$$

According to (4.1), $\mathbf{P}\{x(t, \omega) \geq x(s, \omega)\} = 1$, for fixed s and t. Then, if $t_1, t_2, \cdots$ is any sequence of t values, $x(t) - x(0)$ is monotone non-decreasing and integral-valued, if t is restricted to the t_j's, with probability 1. That is, almost all sample functions (if considered defined only on the t_j's) are monotone non-decreasing, with integral-valued increments. Now (II §2), if the process is separable, the sequence $\{t_j\}$ can be chosen in such a way that almost every sample function $x(t)$ has the same upper and lower bounds on every open interval as on the t_j's in the interval. Hence, excluding a collection of sample functions of probability 0, each sample function has the following properties: it is monotone non-decreasing; it increases only in jumps, of integral-valued magnitudes, that is, if the sample function $f(\cdot)$ has a jump at t_0,

$$f(t_0 -) \leq f(t_0) \leq f(t_0 +) = f(t_0 -) + n,$$

where n is a positive integer; $f(t) - f(0)$ is then integral-valued except perhaps at the jump points. Moreover, we shall now show that the amount of the jump, n above, can be taken as 1, that is, the probability of the class of sample functions which ever have a jump of magnitude greater than 1 is 0. It is obviously sufficient to prove that the probability of a jump of magnitude greater than 1 at some point in any given *finite* interval, say $(0, T)$ is 0. This follows from the limit equation

$$\begin{aligned} \mathbf{P}\left\{\operatorname*{Max}_{0<j<n}\left[x\left(\frac{j+1}{n}T, \omega\right) - x\left(\frac{j-1}{n}T, \omega\right)\right] > 1\right\} \\ \leq \sum_{j=1}^{n-1} \mathbf{P}\left\{x\left(\frac{j+1}{n}T, \omega\right) - x\left(\frac{j-1}{n}T, \omega\right) > 1\right\} \\ = (n-1)\left(1 - e^{-2cT/n} - e^{-2cT/n}\frac{2cT}{n}\right) \to 0 \qquad (n \to \infty), \end{aligned}$$

since, if a sample function has a jump of magnitude greater than 1 in the interval $(0, T)$, the indicated maximum will be greater than 1 for every n.

The sample functions of a separable Poisson process are not continuous functions; in fact, the probability of continuity throughout an interval $(t, t + T)$ is e^{-cT}, which goes to 0 when T becomes infinite. Nevertheless there is continuity at each value t_0 of t with probability 1, since

$$\mathbf{P}\{x(t_0 + \varepsilon, \omega) - x(t_0 - \varepsilon, \omega) > 0\} = 1 - e^{-2c\varepsilon} \to 0 \qquad (\varepsilon \to 0).$$

The point is that the probability of continuity simultaneously at all the points of an interval is less than 1.

It is convenient in some applications to make a semantic transformation here, and to describe each jump of a sample function as an *event*. The number of events occurring *within* the interval (s, t) is then $x(t-)-x(s+)$ (if we ignore completely here, as we shall in the following, all sample functions except those which are monotone and increase only in unit jumps). We observe that we have just shown that $x(t-)-x(s+) = x(t)-x(s)$ with probability 1, so that in counting events in an interval, if zero probabilities can be neglected, it makes no difference whether the endpoints of the interval are considered or not. In the present terminology the first equation of (4.2) states that the expected number of events in an interval of length l is cl. The constant c is thus the (average) rate of occurrence of the events.

If events occur in accordance with the Poisson law as described here, they are sometimes described as "purely random," or in the physical literature sometimes simply as "random." This distribution of events can be described as follows, without the terminology of stochastic processes:

Events are occurring in such a way that the probability that m will occur in the open time interval (s, t) *is given by the right side of* (4.1) (*and this is then also the probability that m will occur in the closed time interval* $[s, t]$); *if* $\mu_1, \cdots, \mu_n$ *are the numbers of events which occur in n intervals of time, disjunct except possibly for endpoints, then* $\mu_1, \cdots, \mu_n$ *are mutually independent random variables. Here n is an arbitrary positive integer.*

We observe that the conditional distribution of events in the interval (s, t), under the hypothesis that m have occurred there is that of m points chosen independently in the interval, each with constant probability density $1/(t-s)$, so that the probability density that the m points will be at $t_1, \cdots, t_m$ is $m!/(t-s)^m$. In fact, the probability that $m_1, \cdots, m_n$ events occur respectively in non-overlapping intervals in (s, t) of lengths $l_1, \cdots, l_n$ with $\sum_1^n l_j = (t-s)$, if $m = \sum_1^n m_j$ events are known to have occurred in (s, t), is

$$\frac{\prod_{j=1}^{n} e^{-cl_j} \frac{(cl_j)^{m_j}}{m_j!}}{e^{-c(t-s)} \frac{[c(t-s)]^m}{m!}} = \left[\prod_{j=1}^{n} \frac{l_j^{m_j}}{t-s}\right] \frac{m!}{m_1! \cdots m_n!},$$

which reduces to

$$\frac{m!}{(t-s)^m} \prod_1^m d\xi_j$$

if m of the m_j's are 1 (and the others 0) and if the corresponding l_j's are replaced by $d\xi_j$'s. Thus the Poisson distribution of events can be obtained in a given finite time interval of length l by first determining the number μ of events to occur in the interval,

$$\mathbf{P}\{\mu(\omega) = m\} = e^{-cl}\frac{(cl)^m}{m!}$$

and then choosing $\mu(\omega)$ points independently in the interval each with density of distribution $1/l$ in the interval. The Poisson distribution of events can be approximated in an infinite time interval, say $(0, \infty)$ in the same spirit, as follows. Choose μ_T points in the interval $(0, T)$, choosing them independently, each with density of distribution $1/T$ in the interval. Here μ_T is for each T a constant which satisfies the equation

$$\lim_{T\to\infty} \frac{\mu_T}{T} = c;$$

for example, μ_T may be the integer closest to cT. Then when $T \to \infty$ the distribution of events approaches the Poisson distribution of events in the sense that, if $I_1, \cdots, I_n$ are any intervals, disjunct except possibly for endpoints, of lengths $l_1, \cdots, l_n$ respectively, and if $\mu_j(T)$ is the number of points chosen in I_j, then $\mu_1(T), \cdots, \mu_n(T)$ are random variables for which

$$\lim_{T\to\infty} \mathbf{P}\{\mu_j(\omega) = m_j, j = 1, \cdots, n\} = \prod_{j=1}^{n} e^{-cl_j}\frac{(cl_j)^{m_j}}{m_j!}.$$

In fact, if the interval $(0, T)$ includes the I_j's and if $l = \sum_{j=1}^{n} l_j$, $m = \sum_{1}^{n} m_j$, the probability on the left is

$$\left(\frac{l_1}{T}\right)^{m_1} \cdots \left(\frac{l_n}{T}\right)^{m_n} \left(\frac{T-l}{T}\right)^{\mu_T - m} \frac{\mu_T!}{m_1! \cdots m_n!(\mu_T - m)!}$$

$$= \left(1 - \frac{l}{T}\right)^{\mu_T - m} \left(\frac{\mu_T}{T}\right) \cdots \left(\frac{\mu_T - m + 1}{T}\right) \frac{l_1^{m_1} \cdots l_n^{m_n}}{m_1! \cdots m_n!}$$

$$\to e^{-cl} c^m \frac{l_1^{m_1} \cdots l_n^{m_n}}{m_1! \cdots m_n!} \qquad (T \to \infty),$$

as was to be proved.

It is important to find simple qualitative conditions under which a distribution of events follows the Poisson law. One such set is the following, written for the parameter interval $[0, \infty)$. Each event is identified with a point on the time axis, that is, at most one event happens at any moment.

(a) *Only a finite number of events occur in any finite time interval.* We write $x(t)$ for the number of events occurring in the interval $0 \leq s < t$. Then $x(t)$ is a random variable for every $t > 0$; we define $x(0) = 0$.

(b) *The $x(t)$ process has independent increments.*

(c) *The distribution of $x(t) - x(s)$ depends only on $t - s$, that is, the $x(t)$ process has stationary increments.*

To show that a process satisfying (a), (b), and (c) is a Poisson process let $\Phi(t)$ be the probability that no events occur in the interval $[0, t)$,

$$\Phi(t) = \mathbf{P}\{x(t, \omega) - x(0, \omega) = 0\}, \qquad t > 0.$$

The function $\Phi(t)$ does not vanish identically or an event would be certain to occur in any time interval, no matter how small, contradicting (a). By (b) and (c), if $s > 0$ and if $t > 0$,

$$\Phi(s + t) = \Phi(s)\Phi(t), \qquad 0 \leq \Phi(t) \leq 1. \tag{4.3}$$

According to (4.3), if $\Phi(t_0) = 0$,

$$0 = \Phi(t_0) = \Phi(t_0/2)^2 = \cdots$$

so that $\Phi(t)$ vanishes arbitrarily near $t = 0$, and therefore vanishes identically since Φ is monotone non-increasing. Since this implication is false, $\Phi(t)$ can never vanish. The only monotone non-vanishing solution of (4.3) has the form

$$\Phi(t) = e^{-ct}$$

for some constant $c \geq 0$. The case $c = 0$ corresponds to a degenerate Poisson distribution where no events ever occur. We finish the proof by proving the Poisson formula (4.1). It is no restriction to take $s = 0$. Since the probability that an event will occur at a given time is 0 (because $\lim_{t \to 0} \Phi(t) = 1$), we have

$$\mathbf{P}\{x(t, \omega) - x(0, \omega) = m\} = \mathbf{P}\,\{\text{at least one event occurs in each of } m \text{ of the intervals } (jt/n, (j + 1)t/n),\ j = 0, \cdots, n - 1, \text{ and no event in the remaining intervals}\} + q_n, \tag{4.4}$$

where

$$|q_n| \leq \mathbf{P}\{\text{two or more events occur in at least one interval } (jt/n, (j + 1)t/n)\}.$$

Let Λ_n be the ω set described in the preceding line. Then, if $\omega \in \Lambda_n$ for infinitely many values of n, it follows that $\omega \in \Lambda_n$ for all large values of n, and corresponding to this point ω there are infinitely many events in $(0, t)$. Thus, by (a), Λ_n converges when $n \to \infty$ to an ω set of zero probability. The first term on the right in (4.4) has the evaluation

$$\frac{n!}{m!(n - m)!} (e^{-ct/n})^{n-m} (1 - e^{-ct/n})^m \to \frac{e^{-ct}(ct)^m}{m!} \qquad (n \to \infty).$$

Thus (4.4) becomes (4.1) (with $s = 0$) when $n \to \infty$.

Note that, if property (c) is dropped, we no longer can write Φ explicitly, but it must nevertheless be monotone non-increasing. If it is supposed that the probability is 0 that an event occur at any preassigned time, Φ must be continuous. The argument just used shows then that, if $\Psi = \log \Phi$,

$$\mathbf{P}\{x(t, \omega) - x(s, \omega) = m\} = e^{-[\Psi(s)-\Psi(t)]} \frac{[\Psi(s) - \Psi(t)]^m}{m!}, \qquad s < t.$$

In other words, the process is the Poisson process after a change of variable on the time axis.

Finally, another important property of the Poisson distribution of events over the interval $[0, \infty)$ is the following: *Let s_1 be the time to the first event, and for $j > 1$ let s_j be the time between the $(j-1)$th and jth event. Then $s_1, s_2, \cdots$ are mutually independent random variables with the common distribution function*

$$\begin{aligned} \mathbf{P}\{s_1(\omega) \le \lambda\} &= 1 - e^{-c\lambda} & \lambda \ge 0 \\ &= 0 & \lambda < 0. \end{aligned}$$

To avoid notational complexities we only consider s_1 and s_2. In the first place

$$\mathbf{P}\{s_1(\omega) \le t_1\} = \mathbf{P}\{x(t_1, \omega) - x(0, \omega) \ge 1\} = 1 - e^{-ct_1}, \qquad t_1 \ge 0,$$

which verifies the statement for s_1. In the second place

$$\begin{aligned} &\mathbf{P}\{s_1(\omega) \le t_1, s_2(\omega) \le t_2\} \\ &\le \sum_{j=0}^{n-1} \mathbf{P}\Big\{x\left(\frac{jt_1}{n}, \omega\right) - x(0, \omega) = 0, x\left(\frac{j+1}{n} t_1, \omega\right) - x\left(\frac{jt_1}{n}, \omega\right) = 1, \\ &\qquad\qquad x\left(\frac{j+1}{n} t_1 + t_2, \omega\right) - x\left(\frac{j+1}{n} t_1, \omega\right) \ge 1\Big\} \\ &+ \sum_{j=0}^{n-1} \mathbf{P}\Big\{x\left(\frac{jt_1}{n}, \omega\right) - x(0, \omega) = 0, x\left(\frac{j+1}{n} t_1, \omega\right) - x\left(\frac{jt_1}{n}, \omega\right) \ge 2\Big\}. \end{aligned}$$

Now the second sum here is $o(1/n)$ as $n \to \infty$ since it is dominated by

$$\sum_{j=0}^{n-1} e^{-cjt_1/n}\left(1 - e^{-ct_1/n} - e^{-ct_1/n} \frac{ct_1}{n}\right) = \frac{1 - e^{-ct_1}}{1 - e^{-ct_1/n}}\, o\left(\frac{1}{n^2}\right) = o\left(\frac{1}{n}\right),$$

and the first sum is

$$\sum_{j=0}^{n-1} e^{-cjt_1/n - ct_1/n}\left(\frac{c}{n} t_1\right)(1 - e^{-ct_2}) \to (1 - e^{-ct_1})(1 - e^{-ct_2}), \qquad n \to \infty.$$

Hence

$$\mathbf{P}\{s_1 \le t_1, s_2 \le t_2\} \le (1 - e^{-ct_1})(1 - e^{-ct_2}). \tag{4.5}$$

To prove the reverse inequality we observe that the left side of (4.5) is at least

$$\sum_{j=0}^{n-1} \mathbf{P}\left\{x\left(\frac{jt_1}{n}, \omega\right) - x(0, \omega) = 0, x\left(\frac{j+1}{n} t_1, \omega\right) - x\left(\frac{jt_1}{n}, \omega\right) = 1,\right.$$

$$\left. x\left(\frac{jt_1}{n} + t_2, \omega\right) - x\left(\frac{j+1}{n} t_1, \omega\right) \geq 1\right\}$$

$$= \sum_{j=0}^{n-1} e^{-\frac{cjt_1}{n}} \frac{ct_1}{n} e^{-\frac{ct_1}{n}} (1 - e^{-c[t_2 - (t_1/n)]})$$

$$\to (1 - e^{-ct_1})(1 - e^{-ct_2}), \qquad n \to \infty.$$

It is sometimes useful to consider a Poisson process with parameter running from $-\infty$ to ∞ instead of merely from 0 to ∞. If this is done one can define s', the total time between the events that immediately precede and follow a given time t_0. Then s' has the distribution of $s_1 + s_2$, so that it has density of distribution $c^2\lambda e^{-c\lambda}$ as compared with density $ce^{-c\lambda}$ for the s_j's. In discussing the "time between events" one must be careful to distinguish between s' and s_j.

5. Application of the Poisson process to molecular and stellar distributions

Consider a set of points on a finite or infinite interval, say the stars of a one-dimensional universe. Let the coordinates of the points be $\cdots$, $x_0, x_1, \cdots$. It is supposed that the points are uniformly distributed, and the question is how they can move to preserve this property. In other words, what kind of motion is compatible with the statistical equilibrium of the system of particles as distinguished from the stationarity of the process governing the individual particles? The following two examples will clarify the possibilities.

(a) Suppose that there are infinitely many particles on an infinite line, in a Brownian motion. If $x(t)$ is the coordinate of one of these particles at time t, we have seen that the $x(t)$ process is the Brownian movement process, for which $\mathbf{E}\{[x(t) - x(0)]^2\} = 2Dt \to \infty$ $(t \to \infty)$. Hence the particle at any given time t will, if t is large, be far out with probability near 1. In other words, the individual particles tend to diffuse outward; the $x(t)$ process is of course not stationary. Nevertheless, it is intuitively reasonable that a system of *infinitely* many particles can present a picture of a stationary system; the particles from far out can change places with those near the origin.

(b) Suppose that there are finitely many particles diffusing in the interval (a, b), being reflected whenever they reach a or b ("reflecting barrier"). In this case it seems reasonable that, defining $x(t)$ as in (a), the $x(t)$ process tends to stationarity as $t \to \infty$, regardless of the initial conditions, and in fact will be a stationary process if the proper initial conditions are imposed. In this case then, if the various particles move independently, the *system* will be stationary, regardless of the number of particles.

We shall consider only the case of an infinite interval. The first thing to do is to define what one means by a distribution of infinitely many particles with density c over the ξ-axis. The definition is sometimes given in terms of a limit idea, as follows: choose μ_l particles in the interval $(-l, l)$, choosing them independently, each with constant density of distribution $1/(2l)$ in the interval. Then let $l \to \infty$ and $\mu_l \to \infty$ so that $\mu_l/2l \to c$. According to §4, this is simply an indirect way of stating that the points are distributed over the infinite axis in accordance with a Poisson process with parameter c. (Note that the parameter is ξ here, representing distance rather than time; the "events" are the x_j's, the particle positions. The events are now distributed over the whole axis instead of the half-axis.)

We can drop this indirect approach, and suppose simply that for each t we have a sequence of random variables $\cdots, x_0(t), x_1(t), \cdots$, where $x_j(t)$ is the position of the jth particle at time t. At $t = 0$ the particles have the Poisson distribution over the ξ-axis, with c particles per unit length, and are numbered so that $\cdots < x_0(0) < x_1(0) < \cdots$. We wish to find conditions on the $x_j(t)$ processes which insure that at all future times the particles will still have the Poisson distribution over the ξ-axis with the same density c. The $x_j(t)$ processes are *not* mutually independent, because of the prescribed inequalities at $t = 0$. *We suppose that the distribution of* $x_j(t) - x_j(0)$ *does not depend on* j, *for* $t > 0$. *Moreover we suppose that, for each* $t > 0$, *the classes of random variables*

$$\{x_j(t) - x_j(0), -\infty < j < \infty\}, \qquad \{x_j(0), -\infty < j < \infty\},$$

are mutually independent, and that the random variables of the first class are mutually independent.

Let $\mu_1(\omega), \cdots, \mu_n(\omega)$ be the number of $x_j(t, \omega)$'s, for some fixed $t > 0$, in n finite intervals $I_1, \cdots, I_n$ of lengths $l_1, \cdots, l_n$. The intervals are supposed disjunct, except possibly for endpoints. We shall prove that the particles are distributed in the same way at time t as at time 0 by proving

$$\mathbf{P}\{\mu_j(\omega) = m_j, j = 1, \cdots, n\} = \prod_{j=1}^{n} \frac{(cl_j)^{m_j}}{m_j!} e^{-cl_j}. \tag{5.1}$$

Note that this does *not* state that any individual particle has a position whose distribution at time t is independent of the time. In fact, if the individual particles are for example subject to the Brownian movement process, then each one will tend to be far from $\xi = 0$ for large t. The statement means, however, that the system itself is in macroscopic equilibrium, *considering only relative positions*. If each particle is moving in the direction of increasing ξ with constant unit velocity, the conditions imposed on the motion are satisfied, and (5.1) is obviously true. To prove (5.1) we note first that, if $a < b$ and if F is a distribution function, then

$$\int_{-\infty}^{\infty} [F(b-t) - F(a-t)]\, dt = \int_{-\infty}^{\infty} dt \int_{a-t+}^{b-t+} dF(s) = \int_{-\infty}^{\infty} dF(s) \int_{a-s}^{b-s} dt = b - a.$$

It will be convenient to write $F(I)$ for $F(b) - F(a)$ if I is the interval $(a, b]$, and we understand by $I - t$ the interval I translated through $-t$ units. The preceding equation can be written in the form

$$\int_{-\infty}^{\infty} F(I - t)\, dt = b - a. \tag{5.2}$$

This equation remains true if I is an interval sum, interpreting $F(I - t)$ in the obvious way, if $b - a$ is replaced by the length of I. Let F be the distribution function of $x_j(t) - x_j(0)$. The probability that m_j of the particles at time 0 are in the interval $|\xi| < \alpha/2$ and go into I_j at time $t, j = 1, \cdots, n$ is the probability (summed over $k \geq \sum_{j=1}^{n} m_j = m$) that k of the particles are initially in the interval $|\xi| < \alpha/2$ and that m_j of these are in I_j at time $t, j = 1, \cdots, n$; this probability is

$$\sum_{k=m}^{\infty} e^{-c\alpha} \frac{(c\alpha)^k}{k!} \frac{k!}{m_1! \cdots m_n!(k-m)!} \left[\prod_{j=1}^{n} \left(\int_{-\alpha/2}^{\alpha/2} F(I_j - \xi) \frac{d\xi}{\alpha} \right)^{m_j} \right] \cdot \left[\int_{-\alpha/2}^{\alpha/2} [1 - F(I - \xi)] \frac{d\xi}{\alpha} \right]^{k-m}$$

where $I = \bigcup_{j=1}^{n} I_j$. This probability can be written in the form

$$e^{-c\alpha} \prod_{j=1}^{n} \frac{\left[c \int_{-\alpha/2}^{\alpha/2} F(I_j - \xi)\, d\xi\right]^{m_j}}{m_j!} \sum_{k=0}^{\infty} \frac{\left[c \int_{-\alpha/2}^{\alpha/2} [1 - F(I - \xi)]\, d\xi\right]^k}{k!}$$

$$= \prod_{j=1}^{n} \frac{\left[c \int_{-\alpha/2}^{\alpha/2} F(I_j - \xi)\, d\xi\right]^{m_j}}{m_j!} e^{-c \int_{-\alpha/2}^{\alpha/2} F(I - \xi)\, d\xi}$$

and, using (5.2), the last term converges to the right side of (5.1), as $\alpha \to \infty$, as was to be proved.

As a particular case, suppose that the particle velocities exist, that is, almost all the sample functions of the $x_j(t)$ processes are absolutely continuous. If $v_j(t, \omega) = dx_j(t, \omega)/dt$, the conditions we have imposed above on the $x_j(t)$ processes, will be satisfied if the following conditions are satisfied. *If n is any positive integer, and if $0 \leq t_1 < \cdot \cdot \cdot < t_n$, the joint distribution of $v_j(t_1), \cdot \cdot \cdot, v_j(t_n)$ does not depend on j. The classes of random variables*

$$\{v_j(t), 0 \leq t < \infty, -\infty < j < \infty\}, \qquad \{x_j(0), -\infty < j < \infty\},$$

are mutually independent, and the classes

$$\{v_j(t), 0 \leq t < \infty\}, \qquad -\infty < j < \infty,$$

are mutually independent. Thus, under these hypotheses, a Poisson distribution of particles at time 0 is reproduced at all later times. Note that although the system is then in macroscopic equilibrium we have not supposed that the $v_j(t)$ process is stationary, or even that the above hypotheses are self-reproducing. For example, $x_j(s)$ may not be independent of the $v_j(t)$'s for $t > s$ except when $s = 0$.

All the above considerations have been one-dimensional. In r dimensions a Poisson distribution of particles is one in which, if $I_1, \cdot \cdot \cdot, I_n$ are r-dimensional intervals, disjunct except perhaps for points on $(r-1)$-dimensional faces, of r-dimensional volumes $l_1, \cdot \cdot \cdot, l_n$, and if μ_j is the number of particles in I_j, then (5.1) is true. The translation of the results of this section into r dimensions is trivial.

6. The centering of the general process with independent increments

If $\{x_t, t \in T\}$ is a process with independent increments, and if f is any function of $t \in T$, the process $\{x_t - f(t), t \in T\}$ also has independent increments. Lévy has shown that f can be chosen in such a way that sample functions of the $x_t - f(t)$ process have simple continuity properties. As a first step in obtaining his results, we prove the following theorem.

THEOREM 6.1 *Let $\{x_t, t \in T\}$ be a process with independent increments. Define T' as the set of limit points of T, except that the minimum and maximum values of the closure of T are to be excluded from T' unless they are in T. There is then a function f, defined for $t \in T$, such that, if $z_t = x_t - f(t)$, the process $\{z_t, t \in T\}$ is a process with independent increments, with the following properties:*

(i) *To each point $t \in T'$ which is a limit point of T from the left* [*right*] *there corresponds a random variable $z_{t-}[z_{t+}]$ such that, if $s_n \to t$ with $s_n < t$ $[s_n > t]$ and $s_n \in T$, then*

$$\lim_{n \to \infty} z_{s_n} = z_{t-} \qquad [\lim_{n \to \infty} z_{s_n} = z_{t+}]$$

with probability 1. *If the z_t process is separable, these sequential limits can be replaced by ordinary limits*

$$\lim_{s\uparrow t} z_s = z_{t-} \qquad [\lim_{s\downarrow t} z_s = z_{t+}] \qquad (s \in T)$$

with probability 1.

(ii) *If any difference $z_t - z_s$, or any such difference with t replaced by $t+$ or $t-$, or s replaced by $s+$ or $s-$, is identically constant with probability* 1, *the constant is* 0.

(iii) *Except possibly for the points of an at most enumerable t set $S \subset TT'$, for each $t \in T$ the following equation holds with probability* 1, *between as many of the members as are defined:*

$$z_{t-} = z_t = z_{t+},$$

that is, at most enumerably many parameter points are fixed points of discontinuity of the z_t process. The set S is independent of the choice of the function f.

Any function f with the properties described in the theorem will be called a *centering function* of the x_t process, and a process for which $f = 0$ is a centering function will be called a *centered process*. For example, the z_t process of the theorem is a centered process. The centering function f is not uniquely determined by the process, since, if g is a function defined and continuous on the closure of T, $f + g$ is also a centering function. It will be sufficient to prove the theorem in the real case, which we shall assume from now on, because in the complex case the real and imaginary parts of the x_t's can be treated separately. Our choice of a centering function is motivated by the following fact. *If $\{w_n, n \geq 1\}$ is a sequence of real random variables, and if there is a sequence $\{c_n\}$ of constants such that*

$$\lim_{n\to\infty} (w_n - c_n) = w$$

exists and is finite with probability 1, *then, if c_n' is for each n, the unique constant satisfying*

$$\mathbf{E}\{\arctan (w_n - c_n')\} = 0,$$

it follows that $\lim_{n\to\infty} (w_n - c_n')$ exists and is finite with probability 1. *In other words, $\lim_{n\to\infty} (c_n' - c_n)$ exists and is finite.* To prove this assertion, let c be a finite limiting value of the sequence $\{c_n' - c_n\}$. Then

$$\lim_{n\to\infty}{}' (w_n - c_n') = w - c$$

with probability 1, where the prime signifies that $n \to \infty$ along some sequence of integers. Hence

$$0 = \lim_{n\to\infty}{}' \mathbf{E}\{\arctan (w_n - c_n')\} = \mathbf{E}\{\arctan (w - c)\}.$$

Thus there is at most one finite limiting value of the sequence $\{c_n' - c_n\}$, since c is uniquely determined by this equation. Since an infinite limiting value must satisfy this same equation, with the obvious conventions, there can be no infinite limiting value. Hence $\lim_{n\to\infty} (c_n' - c_n)$ exists, as was to be proved.

With this fact in mind, we define $f(t)$, for each $t \in T$, to satisfy

$$\mathbf{E}\{\arctan [x_t - f(t)]\} = 0, \qquad t \in T,$$

and define

$$z_t = x_t - f(t).$$

Proof of (i) Suppose that $s_n \to t$, $s_n < t$, $s_n \in T$, and suppose that $s_1 \leq s_2 \leq \cdots$. The series

$$\sum_1^\infty (x_{s_{j+1}} - x_{s_j})$$

is a series of mutually independent random variables, and the equation

$$\sum_1^n (x_{s_{j+1}} - x_{s_j}) + (x_t - x_{s_{n+1}}) = x_t - x_{s_1},$$

in which the parenthetical differences on the left are mutually independent, and in which the right side does not involve n, implies that the series converges with probability 1 when centered (III, Theorem 2.8), that is.

$$\lim_{n\to\infty} (x_{s_n} - c_n)$$

exists and is finite with probability 1, for some choice of centering constants $c_1, c_2, \cdots$. Then, because of the way z_{s_n} was defined, it follows that $\lim_{n\to\infty} z_{s_n}$ exists and is finite with probability 1. The hypothesis that the sequence $\{s_n\}$ is monotone can now be dropped, because, if the sequence is not monotone in the first place, it can be reordered to be monotone. Finally, if

$$s_n' \to t, \qquad s_n' < t, \qquad s_n' \in T$$

$$s_n'' \to t, \qquad s_n'' < t, \qquad s_n'' \in T,$$

then

$$\lim_{n\to\infty} z_{s_n'} = \lim_{n\to\infty} z_{s_n''}$$

with probability 1, because the sequences $\{s_n'\}$, $\{s_n''\}$ can be combined into a single sequence $\{s_n\}$ convergent to t from below, yielding a convergent sequence $\{z_{s_n}\}$. Thus z_{t-} can be defined as the limit of z_s when $s \uparrow t$ sequentially. The definition of z_{t+} is made in the same way, in terms of sequential approach from above. (Use the preceding results, replacing x_t by $x_t' = x_{-t}$ to convert approach from above to approach from below.) In particular, if the z_t process is separable the sequential one-sided limits can, as always in the separable case, be replaced by ordinary limits.

Proof of (ii) Since

$$\mathbf{E}\{\arctan z_t\} = 0, \qquad t \in T,$$

integration to the limit yields

$$\mathbf{E}\{\arctan z_{t-}\} = \mathbf{E}\{\arctan z_{t+}\} = 0, \qquad t \in T'.$$

These equations imply that, if one of the differences described in (ii) is constant with probability 1, then the constant must be 0.

Proof of (iii) The result of (i) implies that

$$p \lim_{s \uparrow t} z_s = z_{t-}, \qquad p \lim_{s \downarrow t} z_s = z_{t+}, \qquad t \in T'$$

whenever the right sides are defined. Then (iii) follows as a direct application of VII, Theorem 11.1. The set S of fixed points of discontinuity is independent of the choice of centering function, because a change of centering function can only change the differences $z_t - z_{t-}$ and $z_{t+} - z_t$ by constants, and such a change cannot reduce one of these differences to 0, by (ii). If $t \in T'T$ is a limit point of T from both sides, it is not a fixed point of discontinuity if and only if

$$z_{t+} - z_{t-} = (z_t - z_{t-}) + (z_{t+} - z_t) = 0$$

with probability 1, because the sum of two mutually independent random variables is constant with probability 1 if and only if each summand is. (This statement follows from the fact that the product of two characteristic functions has modulus identically 1 if and only if the same is true of each factor.)

In the following we shall adopt the convention that, if $\{x_t, t \in T\}$ is a centered process with independent increments and if $t \in T$ is not a limit point of T from the left [right], then x_{t-} [x_{t+}] is defined as x_t.

The proof of Theorem 6.1 has been completed, but it may be instructive to note that the set S of the theorem can be described without the use of a centering function. In fact, we now prove that, *if a is a parameter value, held fast below, then $S(a, \infty]$ is the set of discontinuities of the monotone non-increasing t function defined by*

$$\int_0^1 |\mathbf{E}\{e^{i\mu(x_t - x_a)}\}| \, d\mu, \qquad t > a.$$

To prove this, let Φ_t be the characteristic function of $x_t - x_a$,

$$\Phi_t(\mu) = \mathbf{E}\{e^{i\mu(x_t - x_a)}\}.$$

Then, since

$$\Phi_{t+h}(\mu) = \Phi_t(\mu)\mathbf{E}\{^{i\mu(x_{t+h} - x_t)}\}, \qquad h > 0,$$

it follows that $|\Phi_t(\mu)|$ is monotone non-increasing in t for fixed μ. Let f be any centering function of the process, let $z_t = x_t - f(t)$, and let Ψ_t be the characteristic function of $z_{t+} - z_{t-}$. Then $|\Psi_t| = 1$ when $t \in S$, but $|\Psi_t(\mu)| < 1$ at some points of the interval $[0, 1]$ if $t \in S$, since, for such a value of t, $z_{t+} - z_{t-}$ is not with probability 1 a constant. Moreover, the characteristic function of $z_t - z_a$ has absolute value $|\Phi_t|$, and since the characteristic function of $z_t - z_a$ jumps by Ψ_s at a point $s \in S$ but is continuous at other points of T, it follows that $|\Phi.|$ jumps by $|\Psi_s|$ at a point $s \in S$, but is continuous at other points of T. Then S is the set of points of discontinuity of the t function defined by

$$\int_0^1 |\Phi_t(\mu)| \, d\mu, \qquad t > a,$$

as was to be proved.

Although there is no reason to prefer one centering function to another as far as Theorem 6.1 is concerned, we shall see that certain centering functions have special advantages.

THEOREM 6.2 *Let $\{x_t, t \in T\}$ be a process with independent increments, and let I be the closed interval with endpoints the minimum and maximum values of the closure of T except that the endpoints are themselves included only if they are in T. Then it is possible to define x_t for $t \in I - T$ in such a way that the process $\{x_t, t \in T\}$ has independent increments, and that the latter process is centered if the former is.*

Let f be a centering function of the process, and let $z_t = x_t - f(t)$. If $t \in I - T$ and if t is a limit point of T from the right, define $x_t = z_{t+}$.

The remaining points of I are in semi-open or open intervals $[c, d)$ or (c, d) with d either in T or not in T but a limit point of T from the right. Define $x_t = x_d$ in each such interval. The process $\{x_t, t \in I\}$ as so defined has independent increments. If the x_t process is centered, we can take $f(t) \equiv 0$, and in that case the extended x_t process will also be centered.

THEOREM 6.3 *Let $\{x_t, t \in T\}$ be a centered separable process with independent increments. Then, if $c, d \in T$, almost all sample functions of the process are bounded for $c \leq t \leq d$.*

It is sufficient to prove this in the real case, and we shall accordingly assume that the process is real from now on. Let $m(t)$ be a median of $x_d - x_t$. If $s_n \uparrow t$ $[s_n \downarrow t]$ with $c \leq s_n \leq d$, and $s_n \in T$, then every limiting value of the sequence $\{m(s_n)\}$ must be a median of $x_d - x_{t-}$ $[x_d - x_{t+}]$. It follows that m is a bounded function of t for $c \leq t \leq d$; say $|m(t)| \leq K$. Define

$$z_t = x_t - x_c + m(t), \qquad c \leq t \leq d.$$

Then the z_t process has independent increments, and 0 is a median of $z_d - z_t$. If

$$c = t_0 < \cdots < t_n = d, \qquad t_j \in T,$$

it follows from III, Theorem 2.2, that

$$\mathbf{P}\{\operatorname*{Max}_j z_{t_j}(\omega) \geq \lambda\} \leq 2\mathbf{P}\{z_d(\omega) \geq \lambda\}, \qquad \lambda > 0.$$

Hence

$$\mathbf{P}\{\operatorname*{Max}_j [x_{t_j}(\omega) - x_c(\omega)] \geq \lambda + K\} \leq 2\mathbf{P}\{x_d(\omega) - x_c(\omega) \geq \lambda\}.$$

Since this inequality is true for all finite subsets $\{t_j\}$ of $T[c, d]$, it is also true for enumerably infinite subsets, and therefore (separability of the x_t process)

$$\mathbf{P}\{\operatorname*{L.U.B.}_{t \in T[c, d]} [x_t(\omega) - x_c(\omega)] \geq \lambda + K\} \leq 2\mathbf{P}\{x_d(\omega) - x_c(\omega) \geq \lambda\}.$$

Applying this result to the $-x_t$ process and combining the two inequalities, we obtain

$$\mathbf{P}\{\operatorname*{L.U.B.}_{t \in T[c, d]} |x_t(\omega) - x_c(\omega)| \geq \lambda + K\} \leq 2\mathbf{P}\{|x_d(\omega) - x_c(\omega)| \geq \lambda\}. \tag{6.1}$$

This inequality implies that almost all sample functions of the x_t process are bounded in the interval $[c, d]$.

We now construct an example which will clarify the role of the fixed points of discontinuity of a process with independent increments. Let $t_1, t_2, \cdots$ be a finite or enumerably infinite linear set, and let I be the closed interval with endpoints the minimum and maximum values of the closure of $\{t_j\}$ except that the endpoints are not included in I unless they themselves are t_j's. It is supposed that to each t_j corresponds a pair of real random variables u_j, v_j, with the following properties:

DP$_1$ *The random variables* $u_1, v_1, u_2, v_2, \cdots$ *are mutually independent.*

DP$_2$ $\mathbf{P}\{u_j(\omega)^2 + v_j(\omega)^2 > 0\} > 0, j \geq 1$, *and, if a* u_j *or a* v_j *is identically constant with probability* 1, *the constant is* 0.

DP$_3$ *For every closed finite interval* $J \subset I$ *the series*

$$\sum_{t_j \in J} u_j, \qquad \sum_{t_j \in J} v_j$$

converge with probability 1, *no matter how the summands are ordered* (*that is, in the terminology of* III §2, *both these series, in some order of summation, have absolute centering constants* 0, 0, $\cdots$).

We recall from III §2 that the sums in DP_3 will be independent of the order of summation, neglecting values on ω sets of zero probability.

Let α be any point of I, fixed hereafter, and define the random variable x_t for $t \in I$ by

$$(6.2) \qquad x_t = \sum_{\alpha \le t_j \le t} u_j + \sum_{\alpha \le t_j < t} v_j, \qquad t \ge \alpha,$$

$$= -\sum_{t < t_j < \alpha} u_j - \sum_{t \le t_j < \alpha} v_j, \qquad t < \alpha.$$

Then the process $\{x_t, t \in I\}$ has independent increments, and

$$x_t - x_s = \sum_{s < t_j \le t} u_j + \sum_{s \le t_j < t} v_j, \qquad s < t.$$

Note that, for each t, x_t is defined with probability 1 according to this definition, but x_t may not be defined simultaneously for all t with probability 1, since we have not supposed that the series in DP_3 converge absolutely with probability 1. However, we simply define each x_t arbitrarily where it is not already defined. If $s_n \uparrow t \notin \{t_j\}$, then

$$x_t - x_{s_1} = \sum_1^\infty (x_{s_{n+1}} - x_{s_n})$$

and, if $s_n \downarrow t \notin \{t_j\}$,

$$x_{s_1} - x_t = \sum_1^\infty (x_{s_n} - x_{s_{n+1}})$$

with probability 1, by III, Theorem 2.7, Corollary 1. Hence in each case $\lim_{n\to\infty} x_{s_n} = x_t$ with probability 1. Similarly

$$\lim_{n\to\infty} x_{s_n} = x_{t_j} - u_j \qquad s_n \uparrow t_j$$

$$= x_{t_j} + v_j \qquad s_n \downarrow t_j$$

with probability 1. The x_t process is centered, and

$$u_j = x_{t_j} - x_{t_j-}, \qquad v_j = x_{t_j+} - x_{t_j}.$$

The t_j's are the fixed points of discontinuity of the process.

We summarize these results, and add another, in the following theorem.

THEOREM 6.4 *If an x_t process with independent increments is defined by* (6.2), *under* DP_1, DP_2, DP_3, *then the fixed points of discontinuity are the t_j's with the discontinuities indicated in the preceding equations. Moreover, if the process is defined in such a way that it is separable, then almost all the sample functions are continuous except at the t_j's.*

Only the last statement of the theorem remains to be proved. It is much stronger than the statement that only the t_j's are fixed points of discontinuity. The continuity properties of this process are at the other extreme from those of a separable Poisson process in which (except in degenerate cases) almost all the sample functions are discontinuous, even though there are no fixed points of discontinuity. To prove the theorem we shall reduce it to the case in which the u_j's and v_j's are uniformly bounded and have zero means. It is sufficient to prove that almost all sample functions of the x_t process are continuous, except at the t_j's, in every closed interval $J \subset I$ whose endpoints are t_j's. To avoid complicating the notation we can, therefore, suppose in the first place that DP_3 is true with J replaced by I, where I is closed. We suppose then that the series $\sum_1^\infty u_j$, $\sum_1^\infty v_j$ converge with probability 1, regardless of the order of summation. We now apply the three-series theorem (III, Theorem 2.5). Define

$$\begin{aligned} u_j'(\omega) &= u_j(\omega) \qquad && |u_j(\omega) - m_{uj}| \leq 1 \\ &= m_{uj} && |u_j(\omega) - m_{uj}| > 1, \\ v_j'(\omega) &= v_j(\omega) && |v_j(\omega) - m_{vj}| \leq 1 \\ &= m_{vj} && |v_j(\omega) - m_{vj}| > 1, \end{aligned}$$

where u_j and v_j have median values m_{uj}, m_{vj} respectively, and let σ_j^2 be the variance of $u_j' + v_j'$. Then, according to the three-series theorem, the series

$$\sum_1^\infty \mathbf{E}\{u_j'\}, \qquad \sum_1^\infty \mathbf{E}\{v_j'\}, \qquad \sum_1^\infty \sigma_j^2,$$

$$\sum_1^\infty \mathbf{P}\{u_n(\omega) \neq u_n'(\omega)\}, \qquad \sum_1^\infty \mathbf{P}\{v_n(\omega) \neq v_n'(\omega)\},$$

all converge, and even converge absolutely, since there is convergence for every order of summation. The series $\sum_1^\infty u_j'$, $\sum_1^\infty v_j'$ are series of mutually independent random variables, with absolute centering constants $0, 0, \cdots$ [because $u_j(\omega) = u_j'(\omega)$, $v_j(\omega) = v_j'(\omega)$ for sufficiently large j, with probability 1]. Hence we can define an x_t' process based on the u_j''s and v_j''s just as the x_t process was based of the u_j's and v_j's. By Kolmogorov's generalization of the Chebyshev inequality, III, Theorem 2.1, if a is the initial point of I, and if $a = s_0 < \cdots < s_n$, then

$$\mathbf{P}\{\operatorname*{Max}_j |x_{s_j}'(\omega) - \mathbf{E}\{x_{s_j}'\} - [x_a'(\omega) - \mathbf{E}\{x_a'\}]| \geq \delta\} \leq \frac{\mathbf{E}\{(x_{s_n}' - x_a')^2\}}{\delta^2}$$

$$\leq \frac{\sum_1^\infty \sigma_j^2}{\delta^2}.$$

Now suppose that

$$\sum_{1}^{\infty} [|\mathbf{E}\{u_j'\}| + |\mathbf{E}\{v_j'\}|] \leq \delta,$$

$$\sum_{1}^{\infty} [\mathbf{P}\{u_j(\omega) \neq u_j'(\omega)\} + \mathbf{P}\{v_j(\omega) \neq v_j'(\omega)\}] \leq \delta. \tag{6.3}$$

Then

$$\mathbf{P}\{\operatorname{Max}_j |x_{s_j}'(\omega) - x_a'(\omega)| \geq 2\delta\} \leq \frac{\sum_{1}^{\infty} \sigma_j^2}{\delta^2},$$

and

$$\mathbf{P}\{x_{s_j}(\omega) = x_{s_j}'(\omega), j \leq n\} \geq 1 - \delta.$$

Hence

$$\mathbf{P}\{\operatorname{Max}_j |x_{s_j}(\omega) - x_a(\omega)| \geq 2\delta\} \leq \frac{\sum_{1}^{\infty} \sigma_j^2}{\delta^2} + \delta.$$

Since the x_t process is separable, this inequality implies the inequality

$$\mathbf{P}\{\operatorname{L.U.B.}_t |x_t(\omega) - x_a(\omega)| \geq 2\delta\} \leq \frac{\sum_{1}^{\infty} \sigma_j^2}{\delta^2} + \delta. \tag{6.4}$$

Now suppose that a sample function of the x_t process has a discontinuity, not at a t_j, at which the oscillation is at least 4δ. For such a sample function

$$\operatorname{L.U.B.}_t |x_t(\omega) - x_a(\omega)| \geq 2\delta.$$

In other words, the ω set corresponding to such sample functions is included in an ω set of probability at most the right side of (6.4), that is, the outer measure p_δ of the former ω set satisfies the inequality

$$p_\delta \leq \frac{\sum_{1}^{\infty} \sigma_j^2}{\delta^2} + \delta.$$

Moreover, if a finite number of the t_j's, together with the corresponding u_j's and v_j's are deleted from this development, the sample function discontinuities are not changed except at the deleted t_j's. Hence

$$p_\delta \leq \frac{\sum_{k}^{\infty} \sigma_j^2}{\delta^2} + \delta,$$

for every k for which (6.3) is true with the sums over $j \geq k$ (instead of $j \geq 1$). The inequality for p_δ is therefore true for sufficiently large k, and we find that

$$p_\delta \leq \delta.$$

Since p_δ is monotone non-increasing in δ, this inequality means that $p_\delta = 0$ for every δ, that is, almost all sample functions of the x_t process are continuous except at the t_j's (where their discontinuities have already been analyzed), as was to be proved.

We observe that by application (b) of VII §12 we could have deduced without any calculation whatever the fact that almost all x_t process sample functions have left- and right-hand limits at all their discontinuities (which implies that they have at most enumerably many discontinuities) but some calculation is necessary to show that almost all the sample functions are continuous except at the t_j's.

Now consider any real process $\{x_t, t \in T\}$ which has independent increments. It will be convenient to suppose that T is an interval, and we shall do so. This is no real restriction, according to Theorem 6.2. Let f_1 be a centering function of the x_t process, and let $\{t_j\}$ be the set of fixed points of discontinuity of the centered process. Define $\tilde{u}_j$, $\tilde{v}_j$ as the jumps of the centered process on the left and right at t_j,

$$\tilde{u}_j = x_{t_j} - f_1(t_j) - [x_{\cdot} - f_1(\cdot)]\big|_{t_j-}, \qquad \tilde{v}_j = [x_{\cdot} - f_1(\cdot)]\big|_{t_j+} - [x_{t_j} - f_1(t_j)].$$

It is clear that $\tilde{u}_1, \tilde{u}_2, \cdots, \tilde{v}_1, \tilde{v}_2, \cdots$ are mutually independent random variables. We have seen in III §2 that, if a series of mutually independent random variables converges with probability 1 when centered, there are absolute centering constants (which can always be taken as truncated expectations for example) for which the centered series will converge with probability 1 regardless of the order of summation, and for which every subseries has this same property. Let $\{u_j\}$, $\{v_j\}$ be the sequences $\{\tilde{u}_j\}$, $\{\tilde{v}_j\}$ when centered by subtraction of truncated expectations, as defined in III §2. Then, if $J : [a, b]$ is a closed subinterval of T, if $\{t_{a_n}\}$ is the subset of $\{t_j\}$ in J, and if Δ_n is defined by

$$\sum_1^n u_{a_j} + \Delta_n = x_b - x_a,$$

it follows that the random variables on the left are mutually independent. But then, according to III, Theorem 2.8, the series $\sum_1^\infty u_{a_j}$ converges with probability 1 when centered. The same argument is applicable to the series $\sum_1^\infty v_{a_j}$. Since the centering has already been done, we have proved that, if J is any closed interval of parameter values, the series

$$\sum_{t_j \in J} u_j, \qquad \sum_{t_j \in J} v_j$$

converge with probability 1 regardless of the order of summation. Now define $x_t^{(d)}$ as the right side of (6.2). We have seen (Theorem 6.4) that the $x_t^{(d)}$ process has independent increments, is centered, and has the t_j's as its fixed points of discontinuity, with

$$x_{t_j}^{(d)} - x_{t_j-}^{(d)} = u_j, \qquad x_{t_j+}^{(d)} - x_{t_j}^{(d)} = v_j.$$

The process $\{x_t - f_1(t) - x_t^{(d)},\ t \in T\}$ has independent increments. Let f_2 be a centering function of this process, and define $x_t^{(c)}$ by

$$x_t = f(t) + x_t^{(d)} + x_t^{(c)}, \qquad f = f_1 + f_2. \tag{6.5}$$

Then the $x_t^{(c)}$ process has independent increments, is centered, and has no fixed points of discontinuity. Moreover, the two processes

$$\{x_t^{(d)},\ t \in T\}, \qquad \{x_t^{(c)},\ t \in T\}$$

are mutually independent. Note that f is a centering function of the x_t process. It is not an arbitrary centering function, but one chosen in such a way that the discontinuities at the fixed points of discontinuity have special properties. The $x_t^{(d)}$ process is the type described in Theorem 6.4. The decomposition (6.5) is due to Lévy. In any particular case all components need not be present. The decomposition (6.5) is obviously also applicable to complex processes with independent increments.

7. The character of the distribution functions and the continuity of the sample functions

Let $\{x_t,\ a \leq t \leq b\}$ (a, b finite) be a process with independent increments; suppose that the process is centered and that there are no fixed points of discontinuity. Let $\Phi_{s,t}$ be the characteristic function of $x_t - x_s$ ($s \leq t$),

$$\Phi_{s,t}(\mu) = \mathbf{E}\{e^{i\mu(x_t - x_s)}\}, \qquad s \leq t.$$

Then, if $s_1 < s_2 < s_3$,

$$\Phi_{s_1, s_3} = \Phi_{s_1, s_2} \cdot \Phi_{s_2, s_3} \tag{7.1}$$

because the process has independent increments. Moreover,

$$\lim_{t \to s} \Phi_{s,t}(\mu) = 1$$

uniformly in s, t, μ for μ in any finite interval, because the process is centered and there are no fixed points of discontinuity. This means, by (7.1), that $\Phi_{s,t}(\mu)$ is continuous in s, t, μ with value 1 when $s = t$. Then, if we write $\Phi_{s,t}$ in the form

$$\Phi_{s,t} = \prod_0^{n-1} \Phi_{s_j, s_{j+1}} \qquad s_j = s + j(t-s)/n,$$

the characteristic function $\Phi_{s,t}$ is expressed as the product of characteristic functions which can be made uniformly close to 1 in every finite μ interval. It follows that the distribution of $x_t - x_s$ is infinitely divisible (III §4), so that (III, Theorem 4.1) for each $t > 0$,

$$\log \Phi_{a,t}(\mu) = i\mu\gamma_t + \int_{-\infty}^{\infty} \left(e^{i\mu\lambda} - 1 - \frac{i\mu\lambda}{1+\lambda^2}\right) \frac{1+\lambda^2}{\lambda^2}\, dG(t, \lambda), \tag{7.2}$$

where $G(t, \cdot)$ is monotone non-decreasing continuous on the right and bounded in λ, with $G(t, -\infty) = 0$, and γ_t is a constant. The left side of this equation is continuous in t. It follows that $\gamma.$ is continuous, and

$$\lim_{t_n \to t} G(t_n, \lambda) = G(t, \lambda)$$

at all points of continuity (in λ) of G. [Cf. the discussion in III §4 of the determination of γ and G by the corresponding distribution, and note for use below that according to the reasoning of that section the function G is uniquely determined by the left side of (7.2) even if G is merely supposed of bounded variation in λ rather than monotone.] The above reasoning applied to the interval $[s, t]$ gives the same formula for $\Phi_{s,t}$ with $\gamma_{s,t}$, $G(s, t, \lambda)$ in place of γ_t and $G(t, \lambda)$. On the other hand, according to (7.1),

$$\log \Phi_{s,t}(\mu) = i(\gamma_t - \gamma_s)\mu + \int_{-\infty}^{\infty} \left(e^{i\mu\lambda} - 1 - \frac{i\mu\lambda}{1+\lambda^2}\right) \cdot \frac{1+\lambda^2}{\lambda^2}\, d[G(t, \lambda) - G(s, \lambda)]. \tag{7.3}$$

We deduce that

$$\gamma_{s,t} = \gamma_t - \gamma_s$$

$$G(s, t, \lambda) = G(t, \lambda) - G(s, \lambda),$$

so that (7.3) exhibits $\log \Phi_{s,t}$ in the Lévy-Khintchine expression for an infinitely divisible distribution. Then, if $s < t$, $G(t, \cdot) - G(s, \cdot)$ must be non-negative and monotone non-decreasing in λ. In other words, G is monotone non-decreasing in both λ and t, $d_\lambda G$ and $d_t G$ are monotone in t and λ, and G can be used to define a measure $\iint d_t d_\lambda G$ in t, λ space. This measure will prove important.

Conversely, suppose that for all s, t, with $0 \leq s < t$, $\Phi_{s,t}(\mu)$ is determined by (7.3), where $\gamma.$ and G have the properties described. There is then a centered process with independent increments and no fixed points of discontinuity, obtained by assigning to $x_t - x_s$ the distribution determined by (7.3).

If the distribution of $x_t - x_s$ depends only on $t - s$, and if the process has independent increments, it is said to have *stationary independent increments*. If such a process is centered, it can have no fixed points of discontinuity, because the process has the same stochastic properties at each value of the parameter. In this case (7.3) combined with (7.1) yields, if $a = 0$,

$$\gamma_{s+t} = \gamma_s + \gamma_t \qquad G(s + t, \lambda) = G(s, \lambda) + G(t, \lambda)$$

so that

$$\gamma_t = t\gamma, \qquad G(t, \lambda) = tG(\lambda)$$

for some constant γ and function $G(\cdot)$. Thus (7.3) becomes

$$\log \Phi_{s,t}(\mu) = i(t - s)\gamma\mu + (t - s) \int_{-\infty}^{\infty} \left(e^{i\mu\lambda} - 1 - \frac{i\mu\lambda}{1 + \lambda^2}\right) \frac{1 + \lambda^2}{\lambda^2}\, dG(\lambda). \tag{7.3'}$$

Example 1 If the process is the Brownian movement process of §2, $\gamma = 0$ and $G(\cdot)$ is constant except for a jump of magnitude σ^2 at $\lambda = 0$.

Example 2 If the process is the Poisson process of §4, $\gamma = 1$ and $G(\cdot)$ is constant except for a jump of magnitude $c/2$ at $\lambda = 1$.

Example 3 Consider the following process: Events are to occur in accordance with a Poisson distribution, at average rate c. The random variable x_t is defined as the sum of N independent random variables, each with a given distribution function F, where N is the number of events that have occurred between times 0 and t inclusive. In other words, at each event a drawing is made from the F distribution and x_t is the cumulative sum of the numbers drawn. The x_t process has stationary independent increments; it is centered and has no fixed points of discontinuity. The characteristic function $\Phi_{0,t}$ is easily evaluated:

$$\Phi_{0,t}(\mu) = \sum_0^{\infty} e^{-ct} \frac{(ct)^n}{n!} \left[\int_{-\infty}^{\infty} e^{i\mu\lambda}\, dF(\lambda)\right]^n,$$

so that

$$\log \Phi_{0,t}(\mu) = ct \int_{-\infty}^{\infty} (e^{i\mu\lambda} - 1)\, dF(\lambda).$$

Making the proper identifications in (7.3′), we find that

$$\gamma = c \int_{-\infty}^{\infty} \frac{\lambda\, dF(\lambda)}{1 + \lambda^2}, \qquad G(\lambda) = c \int_{-\infty}^{\lambda} \frac{\mu^2}{1 + \mu^2}\, dF(\mu).$$

Example 4 *Gaussian case* Suppose that $\{x_t, a \leq t \leq b\}$ is a process with independent increments and that $x_b - x_a$ has a Gaussian distribution. According to a theorem of Cramér, if the sum of two mutually independent random variables is Gaussian, each of the random variables is itself Gaussian, and it follows that every difference $x_t - x_s$ is Gaussian. In the case which interests us, when the process is centered and has no fixed points of discontinuity, it is unnecessary to use the Cramér theorem to obtain this result. In fact in this case, using the notation of (7.2), $G(b, \cdot)$ is constant except for a jump at $\lambda = 0$. Since $d_\lambda G(t, \lambda)$ is monotone non-decreasing in t, it follows that $G(t, \cdot)$ must also be constant except for a possible jump when $\lambda = 0$. Then the same is true of $G(t, \cdot) - G(s, \cdot)$, so that $x_t - x_s$ is Gaussian. Now define

$$\sigma(t)^2 = \mathbf{E}\{(x_t - x_a)^2\}.$$

If $\sigma(\cdot)^2$ has the form const. $(t - a)$, the x_t process is the Brownian motion process with parameter interval $[a, b]$. If not, $\sigma(\cdot)^2$ is still continuous, and, if $y_t = x_s$, $t = \sigma^2(s)$, the y_t process is the Brownian motion process in the interval $[0, \sigma(b)^2]$. We conclude that the sample functions of the x_t process are (almost all) continuous functions in $[a, b]$, if the process is separable.

The following theorem strengthens the results obtained in discussing this example, and adds a converse.

THEOREM 7.1 *Let $\{x_t, a \leq t \leq b\}$ be a centered process with independent increments and no fixed points of discontinuity.*

(i) *The distribution of every difference $x_t - x_s$ is infinitely divisible.*
(ii) *The following conditions are equivalent:*

(a) *$x_b - x_a$ is Gaussian.*
(b) *Every difference $x_t - x_s$ is Gaussian.*
(c) *If R is a denumerable subset of $[a, b]$, dense in $[a, b]$, almost all sample functions coincide on R with functions defined and continuous on $[a, b]$ (that is, if the process is separable, almost all sample functions are continuous on $[a, b]$.)*
(d) *If $a = s_0 < \cdots < s_n = b$, and if $\delta = \operatorname{Max}_j (s_{j+1} - s_j)$, then*

$$\lim_{\delta \to 0} \sum_{j=0}^{n-1} \mathbf{P}\{|x_{s_{j+1}}(\omega) - x_{s_j}(\omega)| > \varepsilon\} = 0. \tag{7.4}$$

We have already proved (i), the equivalence of (ii) (a) and (ii) (b), and the fact that (ii) (b) implies (ii) (c). To finish the proof we prove that (ii) (c) implies (ii) (d) and that (ii) (d) implies (ii) (a).

If (ii) (c) is true, and if the process is separable, almost all sample functions are uniformly continuous on $[a, b]$, and this implies that

$$\lim_{\delta \to 0} \mathbf{P}\{\operatorname{Max}_j |x_{s_{j+1}}(\omega) - x_{s_j}(\omega)| > \varepsilon\} = 0 \tag{7.4'}$$

for all $\varepsilon > 0$. We have already remarked (III §1) and used several times the fact that (7·4′) and (7.4) are equivalent, that is, the probability of at least one of a number of mutually independent events and their expected number go to 0 together. Hence (ii) (d) is true. If the process is not separable, we change each x_t on an ω set of probability 0 to make the process separable. This change does not affect the validity of (7.4) or (7.4′). Finally, if (ii) (d) is true, (ii) (a) is also true, by one form of the central limit theorem (see the Corollary to III, Theorem 4.1).

It is convenient, in analyzing the structure of a process with independent increments, to write (7.3) in Lévy's original form,

$$\log \Phi_{s,t}(\mu) = i(\gamma_t - \gamma_s)\mu - \frac{\sigma_t^2 - \sigma_s^2}{2}\mu^2 + \left[\int_{-\infty}^{0-} + \int_{0+}^{\infty}\left(e^{i\mu\lambda} - 1 - \frac{i\mu\lambda}{1+\lambda^2}\right) d_\lambda[F(t,\lambda) - F(s,\lambda)]\right], \tag{7.5}$$

where

$$F(t,\lambda) = \int_{-\infty}^{\lambda} \frac{1+\alpha^2}{\alpha^2}\, d_\alpha G(t,\alpha), \qquad \lambda < 0$$

$$= -\int_{\lambda}^{\infty} \frac{1+\alpha^2}{\alpha^2}\, d_\alpha G(t,\alpha), \quad \lambda > 0$$

$$\sigma_t^2 = G(t, 0+) - G(t, 0-).$$

Then $F(t, \cdot)$ is monotone non-decreasing in λ for $\lambda > 0$ and for $\lambda < 0$ and $\iint d_t d_\lambda F(t,\lambda)$ defines a t, λ measure whose significance will be discussed below. The function $F(t, \cdot)$ can be supposed continuous in λ on the right for $\lambda \neq 0$; it vanishes at $\pm\infty$. Although F may not be bounded near $\lambda = 0$,

$$\left[\int_{0+}^{1} + \int_{-1}^{-0}\right] \lambda^2\, d_\lambda F(t,\lambda) < \infty.$$

If the x_t process is a Poisson process, with average occurrence rate c, then $\gamma_t = ct/2$, $\sigma_t^2 \equiv 0$ and $F(t, \cdot)$ jumps ct units at $\lambda = 1$ but otherwise does not contribute to (7.5). More generally, define x_t by

$$x_t = \sum_{j=1}^{n} \lambda_j x_t^{(j)} + \sigma y_t + \beta t,$$

where the $x_t^{(1)}, \cdots, x_t^{(n)}$, y_t processes are mutually independent: the $x_t^{(j)}$ process is a Poisson process with average occurrence rate c_j; the y_t process is a Brownian motion process with variance parameter 1; $\lambda_1, \cdots, \lambda_n, c_1, \cdots, c_n, \sigma, \beta$ are constants; $\sigma \geq 0$; the λ_j's are distinct and none vanish. Then the x_t process is a centered process with stationary independent increments, with

$$\gamma_t = t\left(\beta + \sum_{j=1}^{n} \frac{c_j \lambda_j}{1 + \lambda_j^2}\right) \qquad \sigma_t^2 = \sigma^2 t$$

and $F(t, \lambda) = tF(\lambda)$, where $F(\cdot)$ increases for $\lambda \neq 0$ only in jumps of magnitude c_j at λ_j. In this way, then, the most general centered process with stationary independent increments can be approximated by the sum of a Brownian motion process, a linear combination of Poisson processes and a linear function, approximated in such a way that the first two terms on the right in (7.5) are exactly reproduced and the integral replaced by a Riemann-Stieltjes sum. In the approximating process the integral

$$\int_s^t \int_\lambda^\mu d_t d_\alpha F(t, \alpha)$$

is simply the expected number of jumps (occurrences in the component Poisson processes) of the sample functions, of magnitude between λ and μ between times s and t. This argument is easily extended to the non-stationary case, and makes it plausible that in the most general case almost all the sample functions of a centered separable process with independent increments and no fixed points of discontinuity are continuous except for jumps, and the above double integral is the expected number of jumps of the sample functions, of magnitude between λ and μ between times s and t. These results of Lévy's will now be proved.

THEOREM 7.2 *Except possibly for a set of sample functions of probability* 0 *the sample functions of a separable centered process with independent increments* $\{x_t, t \in T\}$ *have the following properties:*

(i) *They are bounded on every parameter set of the form* $[c, d]T$, *with* $c, d \in T$.

(ii) *They have finite left-* [*right-*] *hand limits at every* $t \in T$ *which is a limit point of* T *from the left* [*right*].

(iii) *Their discontinuities are jumps, except perhaps at the fixed points of discontinuity.*

This theorem should be compared with the corresponding martingale theorem, VII, Theorem 11.5. Part (i) is simply a restatement of Theorem 6.3, repeated here only for the sake of completeness. Part (ii) was proved in VII §12 as an application of martingale theory [application (b)]. The fact that the parameter set of that application is an interval is irrelevant, by Theorem 6.2. Part (iii) follows from (ii) and the definition of separability (see the reasoning used in the proof of VII, Theorem 11.5). A function which is continuous except for jumps is necessarily continuous except perhaps at the points of a finite or enumerable set. Hence almost all sample functions under consideration have this property.

As three examples we mention the separable Brownian motion process whose sample functions are (almost all) continuous; the separable Poisson process whose sample functions are (almost all) monotone, increasing in unit jumps from the left-hand limit at a discontinuity to the right-hand limit, but whose points of discontinuity vary from function to function; and the processes considered in Theorem 6.4, whose sample functions are almost all continuous except at the fixed points of discontinuity.

We close this section by a discussion of the distribution of the jumps of the sample functions of a centered separable process $\{x_t,\ a \le t \le b\}$ with no fixed points of discontinuity. Let $\nu_{\lambda,t}$ be the number of jumps of magnitude (that is, right-hand limit minus left-hand limit) $\le \lambda$ (< 0) between a and t. Then $\nu_{\lambda,t}$ is a finite-valued integral-valued random variable. The $\nu_{\lambda,t}$ process (fixed λ) obviously has independent increments and it is centered; it has stationary increments if the x_t process has. In the latter case the $\nu_{\lambda,t}$ process is a Poisson process, since it satisfies the qualitative defining characteristics (a), (b), (c) detailed in §4. If the x_t process does not have stationary increments, the $\nu_{\lambda,t}$ process is a Poisson process except for a change of the time variable. We wish to prove

$$\mathbf{E}\{\nu_{\lambda,t}\} = \int_a^t \int_{-\infty}^{\lambda} d_\alpha\, d_s F(s, \alpha) = F(t, \lambda), \tag{7.6}$$

where $F(\cdot, \cdot)$ is the function in (7.5) supposed continuous on the right, and thus explain the significance of the (t, λ) measure defined by $F(\cdot, \cdot)$. [The corresponding identification of $-F(\cdot, \cdot)$ for $\lambda > 0$ with the expected number of jumps of magnitude $> \lambda$ between 0 and t is treated in the same way or reduced to the previous case by replacing x_t by $-x_t$.] To make the desired identification we need only go back to the derivation of (7.5) in III §4, which can be outlined as follows: Write $x_t - x_0$ in the form

$$\begin{aligned} x_t - x_0 &= \sum_{j=0}^{n-1} (x_{s_{j+1}} - x_{s_j}) \qquad s_j = a + \frac{j}{n}(b - a) \\ &= \sum_{j=0}^{n-1} m_j^{(n)} + \sum_{j=0}^{n-1} (x_{s_{j+1}} - x_{s_j} - m_j^{(n)}) \end{aligned}$$

where $m_j^{(n)}$ is a centering constant, obtained as described in III §4; it is the expectation of $x_{s_{j+1}} - x_{s_j}$ after truncation of the latter around a median value. Since the x_t process has no fixed points of discontinuity, the $m_j^{(n)}$'s are uniformly small for n large. The form (7.5) is then obtained, by characteristic functions, with

$$(7.7)\quad F(t, \lambda) - F(s, \lambda) = \lim_{n\to\infty} \sum_{j=0}^{n-1} \mathbf{P}\{x_{s_{j+1}}(\omega) - x_{s_j}(\omega) - m_j^{(n)} \leq \lambda\}, \qquad \lambda < 0,$$

at least at all the λ continuity points of this limit. Now, fixing t, let A be the (at most enumerable) set of values of λ at which $F(t, \cdot)$ has a discontinuity in λ or for which the probability is positive that there is a sample function jump of magnitude λ between 0 and t. Let $\nu_{\lambda, t}^{(n)}$ be the number of values of j for which a sample function satisfies the condition between the braces in (7.7). Then

$$(7.8)\qquad \lim_{n\to\infty} \nu_{\lambda, t}^{(n)} = \nu_{\lambda, t} \qquad \lambda \notin A$$

for almost all sample functions, and (7.7) becomes

$$(7.7')\qquad F(t, \lambda) = \lim_{n\to\infty} \mathbf{E}\{\nu_{\lambda, t}^{(n)}\}, \qquad \lambda \notin A.$$

Moreover, since $\nu_{\lambda, t}^{(n)}$ is the sum of n mutually independent random variables, each of which takes on only the values 1 or 0, its variance is the sum of the variances of the summed variables, which is at most the sum of their second moments, which is in turn the sum of their first moments,

$$\mathbf{E}\{\nu_{\lambda, t}^{(n)2}\} - \mathbf{E}^2\{\nu_{\lambda, t}^{(n)}\} \leq \mathbf{E}\{\nu_{\lambda, t}^{(n)}\} \to F(t, \lambda).$$

Then $\mathbf{E}\{\nu_{\lambda, t}^{(n)2}\}$ remains bounded when $n \to \infty$. Hence integration to the limit in (7.8) is legitimate, and implies, using (7.7′), the desired relation (7.6) for $\lambda \notin A$. Then (7.6) is true for all $\lambda < 0$ by continuity, as was to be proved.

CHAPTER IX

Processes with Orthogonal Increments

1. Continuity properties

Processes with orthogonal increments were defined in II §10. It was proved there that to each such process $\{y_t, t \in T\}$ corresponds a monotone non-decreasing function F, uniquely determined up to an additive constant, satisfying

$$\mathbf{E}\{|y_t - y_s|^2\} = F(t) - F(s), \qquad s < t, \tag{1.1}$$

which we shall write in the symbolic form

$$\mathbf{E}\{|dy_t|^2\} = dF(t).$$

The continuity properties of F determine those of the y_t process, in the following way.

THEOREM 1.1 *Let $\{y_t, t \in T\}$ be a process with orthogonal increments. Define T' as the set of limit points of T except that the minimum and maximum values of the closure of T are to be excluded from T' unless they are in T.*

(i) *To each point $t \in T'$ which is a limit point of T from the left* [*right*] *there corresponds a random variable y_{t-} [y_{t+}] such that*

$$\operatorname*{l.i.m.}_{s \uparrow t} y_s = y_{t-} \ [\operatorname*{l.i.m.}_{s \downarrow t} y_s = y_{t+}].$$

(ii) *Except possibly at the points of an at most enumerable t set, for each t the following equation holds with probability* 1, *between as many of the members as are defined:*

$$y_{t-} = y_t = y_{t+}.$$

This theorem is almost obvious from (1.1). To prove that there is a y_{t-}, for example, we need only remark that, if $t \in T'$ and if t is a limit point of T from the left, then $F(s)$ is bounded for $s < t$ and

$$\lim_{s_1, s_2 \uparrow t} \mathbf{E}\{|y_{s_1} - y_{s_2}|^2\} = \lim_{s_1, s_2 \uparrow t} [F(s_2) - F(s_1)] = 0,$$

so that $\underset{s \uparrow t}{\text{l.i.m.}}\, y_s = t_{t-}$ exists. The exceptional t set of (ii) is the set of jump points of the monotone function F, and obviously

$$F(t) - F(t-) = \mathbf{E}\{|y_t - y_{t-}|^2\}$$
$$F(t+) - F(t) = \mathbf{E}\{|y_{t+} - y_t|^2\}$$
$$F(t+) - F(t-) = \mathbf{E}\{|y_{t+} - y_{t-}|^2\}.$$

Let I be the closed interval whose endpoints are the minimum and maximum values of the closure of T, except that these endpoints are themselves included only if they are in T. Then we can define y_t for $t \in I - T$, in such a way that the resulting process has orthogonal increments. To do this suppose that $t \in I - T$ and that t is a limit point of T from the right. Then define $y_t = y_{t+}$. The remaining points of $I - T$ are in disjunct semi-open or open intervals $[c, d)$ or (c, d) with d either in T or not in T but a limit point of T from the right. Define $y_t = y_d$ in each such interval to complete the definition of the extended y_t process.

Throughout the rest of this chapter the parameter range T will be an interval. The result just obtained shows how little a restriction this is.

2. Stochastic integrals

Let $\{y_t, \ t \in T\}$ be a process with orthogonal increments. In the following, T will always be a finite or infinite interval. Let Φ be a fixed t function (that is, one not depending on ω). We shall define

$$\varphi = \int_A \Phi(t)\, d_t y_t$$

for a large class of functions Φ and sets $A \subset T$. The integral φ will be a random variable. Since the sample functions of the y_t process are not of bounded variation except in special cases, φ cannot be defined as an ordinary Stieltjes integral in the individual sample functions. The integral in question is, however, a generalized Stieltjes integral, and in an attempt to make it look more like one we shall adopt the notation $y(t)$ instead of y_t, writing the integral

$$\varphi = \int_A \Phi(t)\, dy(t).$$

In the following we shall assume that the parameter range T is the infinite line $(-\infty, \infty)$. The modification necessary for T a different interval will be obvious. We first define φ when Φ is a step function of a special type and $A = T$. If $a_1 < \cdots < a_n$, and if

$$\begin{aligned} \Phi(t) &= 0, \ t < a_1 \\ &= c_j, \ a_{j-1} \leq t < a_j, \qquad j \leq n, \\ &= 0, \ t \geq a_n, \end{aligned}$$

then we define

$$(2.1)\qquad \varphi = \int_T \Phi(t)\,dy(t) = \sum_2^n c_j[y(a_j-) - y(a_{j-1}-)].$$

In fact, we shall accept as φ any random variable equal almost everywhere to the sum on the right. The integral will, however, always be understood to be one particular random variable, not a class of equivalent ones.

As defined by (2.1) the integral is determined uniquely by Φ, neglecting φ values on sets of zero probability, linear combinations of Φ's correspond to the same linear combinations of the corresponding φ's, and

$$(2.2)\qquad \mathbf{E}\Big\{\Big[\int_T \Phi(t)\,dy(t)\Big]\overline{\Big[\int_T \Psi(t)\,dy(t)\Big]}\Big\} = \int_T \Phi(t)\overline{\Psi(t)}\,dF(t).$$

Equality (2.2) can be interpreted as follows: we have set up a correspondence between certain functions of t (step functions Φ) and random variables φ. If distance between Φ's and distance between φ's is defined by

$$(2.3)\qquad \begin{aligned} \|\Phi_1 - \Phi_2\| &= \Big[\int_T |\Phi_1(t) - \Phi_2(t)|^2\,dF(t)\Big]^{1/2} \\ \|\varphi_1 - \varphi_2\| &= [\mathbf{E}\{|\varphi_1 - \varphi_2|^2\}]^{1/2}, \end{aligned}$$

equation (2.2) implies that distance is preserved by the correspondence. Now suppose that Φ is a limit (in the sense of the above distance) of a sequence $\{\Phi_n\}$ of step functions of the above type. Then

$$\|\Phi - \Phi_m\|^2 = \int_T |\Phi(t) - \Phi_n(t)|^2\,dF(t) \to 0 \qquad (n \to \infty),$$

that is,

$$\operatorname*{l.i.m.}_{n\to\infty} \Phi_n = \Phi$$

[where in the l.i.m. we are using the weighting $dF(t)$]. It follows, since distance is preserved, that $\operatorname*{l.i.m.}_{n\to\infty} \varphi_n$ also exists, defining a random variable φ. This random variable, as a limit in the mean, is defined uniquely, neglecting values on an ω set of probability 0. Aside from this intrinsic lack of uniqueness, φ is independent of the particular sequence $\{\Phi_n\}$ chosen, since two sequences converging in the mean to Φ can be combined to form a single sequence converging in the mean to Φ, whose corresponding sequence of random variables converges in the mean. We define

$$\int_T \Phi(t)\,dy(t)$$

as the limit φ obtained in this way, or rather as any random variable equal with probability 1 to a limit obtained in this way. The class of functions Φ for which the integral is defined is the class of functions of t which are measurable with respect to the Lebesgue-Stieltjes dF measure

$$\int_A dF(t)$$

and for which

$$\int_T |\Phi(t)|^2 \, dF(t) < \infty.$$

(This definition is simply an application of the principle that a uniformly continuous function defined on a point set of a metric space, and taking on values in a complete metric space, can be defined on the closure of its domain of definition by continuity.) Equation (2.2) is true in the general case since it is true for step function integrands. Finally, if A is a Borel t set or more generally is any t set measurable with respect to the dF measure, and if $\Phi_A(t)$ is defined as $\Phi(t)$ on A and 0 otherwise, we define

$$\int_A \Phi(t) \, dy(t) = \int_T \Phi_A(t) \, dy(t)$$

if the integral on the right exists. With this definition (2.2) is true for T replaced by A. The stochastic integral is obviously linear and homogeneous in the integrand and additive in the domain of integration, neglecting values of the integral on ω sets of probability 0.

In many applications,

$$\mathbf{E}\{y(t) - y(s)\} = 0.$$

If this is true, then

$$\mathbf{E}\{\int_A \Phi(t) \, dy(t)\} = 0$$

because this equation will obviously be true for the step function integrands and $A = T$.

Note that, if $T = (-\infty, \infty)$,

$$\text{(2.4)} \qquad \underset{\substack{a \to -\infty \\ b \to +\infty}}{\text{l.i.m.}} \int_{[a,b]} \Phi(t) \, dy(t) = \int_{-\infty}^{\infty} \Phi(t) \, dy(t)$$

since, if the integral on the right exists,

$$\text{(2.5)} \quad \mathbf{E}\{|\int_{-\infty}^{\infty} \Phi(t) \, dy(t) - \int_{[a,b]} \Phi(t) \, dy(t)|^2\}$$

$$= \int_{\{t \notin [a,b]\}} |\Phi(t)|^2 \, dF(t) \to 0 \qquad (a \to -\infty, b \to +\infty).$$

It is easy to show that, if Φ is continuous in the finite interval $[a, b]$, or even merely Riemann-Stieltjes integrable with respect to F, then

$$\text{l.i.m.}_{\delta \to 0} \sum_{j=0}^{n-1} \Phi(t_j')[y(t_{j+1}) - y(t_j)] = \int_{a-}^{b+} \Phi(t)\, dy(t), \tag{2.6}$$

where

$$a = t_0 < \cdots < t_n = b, \quad t_j \le t_j' \le t_{j+1}, \quad \delta = \operatorname*{Max}_j (t_{j+1} - t_j),$$

and the t_j's are points of continuity of F. Here we interpret the integral as over the closed interval $[a, b]$, and if F is not continuous at a, b we replace $y(t_0)$ by $y(t_0 -)$ and $y(t_n)$ by $y(t_n +)$ on the left. To prove this limit equation let $\Psi(t)$ be defined by

$$\begin{aligned} \Psi(t) &= \Phi(t) - \Phi(t_j'), & t &\in [t_j, t_{j+1}), \\ &= \Phi(t) - \Phi(t_n'), & t &= t_n. \end{aligned}$$

Then the absolute value squared of the difference between sum and integral in (2.6) has expectation

$$\mathbf{E}\{|\sum_j \int_{t_j}^{t_{j+1}} [\Phi(t_j') - \Phi(t_j)]\, dy(t)|^2\} = \int_{a-}^{b+} |\Psi(t)|^2\, dF(t).$$

The function Ψ is bounded, independently of the t_j's and t'_j's, and goes to 0 with δ for almost all t (dF measure) according to the hypotheses imposed on Φ. Then the right side of this equality goes to 0 with δ, as was to be proved. If t_j's are allowed to be at discontinuities of F, the approximating sums will have to correspond to semi-closed intervals, but the proof needs no change.

In the following we shall need a criterion for convergence. *Since*

$$\int_A |\Phi(t) - \Phi_n(t)|^2\, dF(t) = \mathbf{E}\left\{\left| \int_A \Phi(t)\, dy(t) - \int_A \Phi_n(t)\, dy(t) \right|^2\right\},$$

the sequence of stochastic integrals $\{\int_A \Phi_n(t)\, dy(t)\}$ *converges in the mean to* $\int_A \Phi(t)\, dy(t)$ *if and only if the integrands converge in the mean to* $\Phi(t)$ [*weighting* $dF(t)$ *on* A].

Suppose now that the process $\{y(t), t \in T\}$ has orthogonal increments, with $\mathbf{E}\{|dy|^2\} = dF$, where T is an interval as usual. We shall have occasion to use double integrals of the form

$$\iint_A \Phi(s, t)\, ds\, dy(t), \tag{2.7}$$

where A is a two-dimensional Borel set, or more generally is measurable with respect to the two-dimensional Lebesgue-Stieltjes $ds\, dF$ measure

$$\iint_B ds\, dF(t).$$

This integral will be defined in terms of the iterated integrals. The following theorem will be needed.

THEOREM 2.1 *Let $\{y(t),\, t \in T\}$ be a process with orthogonal increments, with $\mathbf{E}\{|dy|^2\} = dF$, let $\Phi(\cdot, \cdot)$ be measurable with respect to the Lebesgue-Stieltjes $ds\, dF$ measure, and suppose that*

$$\int_T |\Phi(s, t)|^2\, dF(t) < \infty$$

for almost all s (Lebesgue measure). Then the stochastic integral

$$z(s) = \int_T \Phi(s, t)\, dy(t)$$

can be defined in such a way that the $z(s)$ process is measurable.

The problem is to take advantage of the lack of uniqueness of the stochastic integral to define $z(s)$ to get a measurable s, ω function. To prove the theorem suppose first that $\Phi(s, t)$ is a finite sum of the form

$$\Phi(s, t) = \sum_j \Phi_{1j}(s)\Phi_{2j}(t)$$

with the first factors Lebesgue measurable, the second factors measurable with respect to dF measure, and

$$\int_T |\Phi_{2j}(t)|^2\, dF(t) < \infty.$$

Then the evaluation

$$\int_T \Phi(s, t)\, dy(t) = \sum_j \Phi_{1j}(s) \int_T \Phi_{2j}(t)\, dy(t)$$

shows that the stochastic integral on the left can be expressed as the sum of products of functions measurable in each variable and therefore measurable in the pair. The general case is then treated by the usual approximation procedure.

We proceed to the definition of the double integral (2.7). It is sufficient to consider only the case when A is the infinite strip $-\infty < s < \infty$, $t \in T$, since the integral can be defined for other sets A by setting the integrand equal to 0 in the complement of A. Suppose then that Φ is a function defined on the infinite strip, measurable with respect to $ds\, dF(t)$ measure, and that either

$$\int_T dF(t) \Big[\int_{-\infty}^{\infty} |\Phi(s, t)|\, ds \Big]^2 < \infty \tag{2.8'}$$

or

$$\int_{-\infty}^{\infty} ds \Big[\int_T |\Phi(s, t)|^2\, dF(t) \Big]^{1/2} < \infty. \tag{2.8''}$$

If (2.8′) is true, the iterated integral

$$z' = \int_T dy(t) \int_{-\infty}^{\infty} \Phi(s, t)\, ds \tag{2.9'}$$

is well defined, with $\mathbf{E}\{|z'|^2\}$ dominated by the left side of (2.8′). If (2.8″) is true, the iterated integral

$$z'' = \int_{-\infty}^{\infty} ds \int_T \Phi(s, t)\, dy(t) \tag{2.9''}$$

is well defined, if the result of the first integration is chosen to be s, ω measurable, as it can be according to Theorem 2.1, and $\mathbf{E}\{|z''|\}$ is dominated by the right side of (2.8″). Moreover,

$$\mathbf{P}\{z'(\omega) = z''(\omega)\} = 1 \tag{2.10}$$

if both (2.8′) and (2.8″) are true, that is, the order of integration is immaterial. To show this suppose that Φ is given by

$$\Phi(s, t) = \sum_{j=1}^{n} \Phi_{1j}(s)\Phi_{2j}(t),$$

where Φ_{1j} is Lebesgue measurable, Φ_{2j} is measurable with respect to dF measure, and

$$\int_{-\infty}^{\infty} |\Phi_{1j}(s)|\, ds < \infty, \Big[\int_T |\Phi_{2j}(t)|^2\, dF(t) \Big]^{1/2} < \infty.$$

Then (2.8′) and (2.8″) are both true, and it is trivial to verify that (2.10) is true. The proof in the general case that when (2.8′) and (2.8″) are both true (2.10) is also necessarily true can be effected by the usual approximation procedure. The double integral (2.7) with A the infinite strip $-\infty < s < \infty$, $t \in T$ is now defined when either (2.8′) or (2.8″) is true as an iterated integral. This integral, according to our definition, is any one of a number of random variables, any two of which are equal with probability 1, and containing any random variable equal to one of them with probability 1.

As an application of this double integral consider the following special case. Let h be an absolutely continuous t function in the interval $[a, b] \subset T$.

Define
$$\begin{aligned} \Phi(s, t) &= h'(s), \qquad s \leq t, \\ &= 0, \qquad s > t. \end{aligned}$$

Then, evaluating the double integral by iterated integration in both orders,

$$\begin{aligned} (2.11) \quad \int_a^b \int_a^b \Phi(s, t)\, ds\, dy(t) &= \int_a^b [h(t) - h(a)]\, dy(t) = \int_a^b [y(b) - y(s)]h'(s)\, ds \\ &= [h(b) - h(a)][y(b) - y(a)] \\ &\quad - \int_a^b [y(t) - y(a)]h'(t)\, dt \end{aligned}$$

with probability 1, if F is continuous at a and b. We have thus obtained the formula for integration by parts. If F need not be continuous at a and b, the formula is modified slightly, the modification depending on whether or not a and b are included in the domain of integration.

In many applications the $y(t)$ process has uncorrelated rather than orthogonal increments. If this is true, the process with variables $\{y(t) - m(t)\}$ has orthogonal increments, if we define $m(t)$ by

$$m(t) = \mathbf{E}\{y(t) - y(0)\}.$$

We then extend the definition of the stochastic integral as follows:

$$(2.12) \qquad \int_A \Phi(t)\, dy(t) = \int_A \Phi(t)\, d[y(t) - m(t)] + \int_A \Phi(t)\, dm(t),$$

where it is supposed that m is so regular that the last integral can be defined in the usual way. With this definition, if

$$\mathbf{E}\{|y(t) - m(t) - [y(s) - m(s)]|^2\} = F(t) - F(s), \qquad s < t,$$

then

$$\begin{aligned} (2.13) \quad & \mathbf{E}\left\{\int_A \Phi(t)\, dy(t)\right\} = \int_A \Phi(t)\, dm(t) \\ & \mathbf{E}\left\{\int_A \Phi(t)\, dy(t) \overline{\int_A \Psi(t)\, dy(t)}\right\} \\ &\qquad = \int_A \Phi(t)\overline{\Psi(t)}\, dF(t) + \int_A \Phi(t)\, dm(t) \overline{\int_A \Psi(t)\, dm(t)}. \end{aligned}$$

It will be useful to have the variance of the stochastic integral,

$$\mathbf{E}\{|\int_A \Phi(t)\,dy(t) - \int_A \Phi(t)\,dm(t)|^2\} = \int_A |\Phi(t)|^2\,dF(t). \tag{2.14}$$

This stochastic integral was first used by Wiener [who supposed the $y(t)$ process to be the Brownian movement process]. We shall see the theoretical applications to stationary processes in later chapters. Two simple practical applications will be given here.

3. Application to Campbell's theorem

Suppose that events are occurring in accordance with a Poisson process, at a rate $c > 0$. Each event has a certain intensity u, and has an effect $u\Phi(t)$ after t time units have passed. Let $\theta(t)$ be the sum of the effects of all events occurring prior to time t.

$$\theta(t) = \sum_j \Phi(t - t_j)u_j, \tag{3.1}$$

where $t_1, t_2, \cdots$ are the times of the events occurring before time t and $u_1, u_2, \cdots$ are the intensities at these times. It is supposed that $u_1, u_2, \cdots$ are mutually independent random variables with a common distribution function. Let $\{y(t), -\infty < t < \infty\}$ be a stochastic process whose sample functions are constant between the events, and increase by the corresponding intensity at each event. Then the $y(t)$ process has stationary independent increments, and

$$\begin{aligned} \mathbf{E}\{y(s+t) - y(s)\} &= c\alpha t \\ \mathbf{E}\{[y(s+t) - y(s) - c\alpha t]^2\} &= c\beta t \end{aligned} \tag{3.2}$$

where

$$\mathbf{E}\{u_1\} = \alpha, \qquad \mathbf{E}\{u_1^2\} = \beta. \tag{3.3}$$

Finally, we can now write $\theta(t)$ in the form

$$\theta(t) = \int_{-\infty}^{t} \Phi(t-s)\,dy(s) = \int_{-\infty}^{\infty} \Phi(t-s)\,dy(s) \tag{3.4}$$

if we set $\Phi(t) = 0$ for $t \le 0$. Thus, $\theta(t)$ defines a strictly stationary stochastic process, a process of moving averages. With the notation

$$\int_0^\infty \Phi(t)\,dt = a, \qquad \int_0^\infty \Phi(t)^2\,dt = b \tag{3.5}$$

we find

$$\mathbf{E}\{\theta(t)\} = c\alpha a$$

(3.6)

$$\mathbf{E}\{[\theta(t) - c\alpha a]^2\} = c\beta \int_0^\infty \Phi(t)^2\,dt = c\beta b.$$

In particular, if the u distribution is concentrated at a single value, $\beta = \alpha^2$. In this case the evaluation of the variance of $\theta(t)$ just obtained is known as Campbell's theorem.

According to Theorem 2.1, $\theta(t)$ can be defined in such a way, for each t, that the $\theta(t)$ process is measurable. It is then possible to identify the expectations in (3.6) with time averages, as follows. The $\theta(t)$ process is strictly stationary and metrically transitive (see XI §1), and it then follows from the strong law of large numbers for strictly stationary processes (ergodic theorem, XI, Theorem 2.1) that

$$\lim_{s\to\infty} \frac{1}{s}\int_0^s \theta(t)\,dt = c\alpha a$$

(3.6′)

$$\lim_{s\to\infty} \frac{1}{s}\int_0^s [\theta(t) - c\alpha a]^2\,dt = c\beta b$$

with probability 1.

In most applications the rate c at which events take place is very large, and if this is so it is easily proved, by characteristic functions for example, that the $y(t)$ increments are nearly Gaussian, and that the $\theta(t)$ process is nearly Gaussian. This means that $\{y(t) - c\alpha t, -\infty < t < \infty\}$ is very nearly a Brownian movement process (see VIII §2). To make this more concrete let $\{y_1(t), -\infty < t < \infty\}$ be a process with independent Gaussian increments satisfying (3.2), so that the process $\{y_1(t) - c\alpha t, -\infty < t < \infty\}$ is a Brownian movement process. Define $\theta_1(t)$ by

$$\theta_1(t) = \int_{-\infty}^{\infty} \Phi(t-s)\,dy_1(s). \tag{3.4′}$$

Then $\theta_1(t)$ defines a strictly stationary Gaussian process to which the $\theta(t)$ process is asymptotic for large c, and it is the θ_1 rather than the θ process which is usually treated in the applications.

4. Fourier transform of a process with orthogonal increments

We shall need the Fourier transform of a process $\{y(t), -\infty < t < \infty\}$ with orthogonal increments, satisfying

$$\mathbf{E}\{|dy(t)|^2\} = \sigma^2\,dt \qquad \sigma > 0. \tag{4.1}$$

We wish to find a second process $\{y^*(t) - \infty < t < \infty\}$, also with orthogonal increments and satisfying (4.1), for which (formally)

$$
\begin{aligned}
y'(t) &= \int_{-\infty}^{\infty} e^{2\pi its} y^{*\prime}(s)\, ds \\
y^{*\prime}(t) &= \int_{-\infty}^{\infty} e^{-2\pi its} y'(s)\, ds.
\end{aligned}
\tag{4.2}
$$

These equations are of course meaningless as they stand, since $y'(t)$ does not necessarily exist. We interpret them by formal integration between λ and μ,

$$
\begin{aligned}
y(\mu) - y(\lambda) &= \int_{-\infty}^{\infty} \frac{e^{2\pi is\mu} - e^{2\pi is\lambda}}{2\pi is}\, dy^*(s) \\
y^*(\mu) - y^*(\lambda) &= \int_{-\infty}^{\infty} \frac{e^{-2\pi is\mu} - e^{-2\pi is\lambda}}{-2\pi is}\, dy(s).
\end{aligned}
\tag{4.3}
$$

We shall show that the second equation in (4.3) defines a y^* process with orthogonal increments satisfying (4.1) and the first equation of (4.3). We shall use the notation $\Phi_{\lambda,\,\mu}$ for the function which is 1 between λ and μ and 0 otherwise; $\Phi_{\lambda,\,\mu}{}^*$ will denote its Fourier transform, the second integrand in (4.3). Then, from the second equation in (4.3),

$$
\begin{aligned}
&\mathbf{E}\{[y^*(\mu_1) - y^*(\lambda_1)]\, \overline{[y^*(\mu_2) - y^*(\lambda_2)]}\} \\
&\quad = \sigma^2 \int_{-\infty}^{\infty} \Phi_{\lambda_1,\,\mu_1}{}^*(s) \overline{\Phi_{\lambda_2,\,\mu_2}{}^*(s)}\, ds = \sigma^2 \int_{-\infty}^{\infty} \Phi_{\lambda_1,\,\mu_1}(s) \overline{\Phi_{\lambda_2,\,\mu_2}(s)}\, ds,
\end{aligned}
\tag{4.4}
$$

using the Parseval identity for Fourier transforms. In particular, if $\lambda_1 < \mu_1 \leq \lambda_2 < \mu_2$, the last integral vanishes. Hence the y^* process [defined by the second equation in (4.3) with $\lambda = 0$ say] has orthogonal increments. If $\lambda_1 = \lambda_2 < \mu_1 = \mu_2$, the last integral in (4.4) becomes $\mu_1 - \lambda_1$. Thus (4.1) is satisfied by the y^* process. In order to prove the first equation of (4.3) we prove the more general

$$
\int_{-\infty}^{\infty} \overline{f(s)}\, dy(s) = \int_{-\infty}^{\infty} \overline{f^*(s)}\, dy^*(s).
\tag{4.5}
$$

Here f and f^* are Fourier transforms of each other,

$$f(t) = \int_{-\infty}^{\infty} e^{2\pi its} f^*(s)\, ds$$
$$(4.6)$$
$$f^*(t) = \int_{-\infty}^{\infty} e^{-2\pi its} f(s)\, ds,$$

and (4.5) will be proved for every Lebesgue-measurable f whose square is integrable, that is, for every Lebesgue-measurable f^* whose square is integrable. For such functions the correspondence between f and f^* is given by the Fourier-Plancherel theorem and in (4.6) the integrals must be interpreted as certain limits in the mean, in accordance with this theorem, so that equations (4.6) are only true for almost all t. It is sufficient to prove (4.5) in order to prove the first equation in (4.3), because, if $f = \Phi_{\lambda,\mu}$, (4.5) reduces to the first equation in (4.3). We already know that (4.5) holds for $f^* = \Phi_{\lambda,\mu}$; this is the second equation in (4.3). The class of functions f^* for which (4.5) holds is a closed linear manifold of functions, defining the distance between f_1^* and f_2^* as

$$\left[\int_{-\infty}^{\infty} |f_1^* - f_2^*|^2\, ds\right]^{1/2}.$$

Since this manifold contains every $\Phi_{\lambda,\mu}$, it contains every f^* of the stated class.

5. A generalization of the stochastic integral of §2

Let $\{y(t),\ t \in T\}$ be a stochastic process and let Φ be a function of $t \in T$ and ω. Stochastic integrals

$$\varphi = \int_A \Phi(t, \omega)\, dy(t)$$

are used in many places in this book. In each case the $y(t)$ process is of some special type, the functions Φ are restricted to some linear class which depends on the given $y(t)$ process, and A is restricted to some specified class of t sets. The general principle involved in the definition of the stochastic integral is the following. It is always supposed that T is an interval, which may be infinite. Under the hypothesis that $A = T$ and that Φ is a step function, the stochastic integral is defined as the obvious Riemann-Stieltjes sum. Keeping $A = T$, the stochastic integral is then defined for the general Φ by a limit procedure. Finally the integral over A is defined by

$$\int_A \Phi(t, \omega)\, dy(t) = \int_T \Phi_A(t, \omega)\, dy(t),$$

where

$$\Phi_A(t, \omega) = \Phi(t, \omega), \qquad t \in A,$$
$$= 0, \qquad t \notin A.$$

This procedure has already been used in §2. In the present section a related case will be treated. The definition will be given for $A = T = (-\infty, \infty)$. The extension to other cases will be obvious. The following hypotheses are made.

I_1 *The process* $\{y(t), \mathscr{F}_t, -\infty < t < \infty\}$ *is a martingale.* (See VII §1.) *There is a monotone non-decreasing function F such that, if* $s < t$,

$$\mathbf{E}\{|y(t) - y(s)|^2\} = \mathbf{E}\{|y(t) - y(s)|^2 \mid \mathscr{F}_s\} = F(t) - F(s)$$

with probability 1.

In particular, if $F(t) \equiv \text{const. } t$, if the martingale is real, and if almost all its sample functions are continuous, the $y(t)$ process is a Brownian motion process, according to VII, Theorem 11.9, and this is the most important special case.

I_2 Φ *is a* (t, ω) *function measurable with respect to* $dt\, d\mathbf{P}$ *measure. For each s,* $\Phi(s, \cdot)$ *is* ω *measurable with respect to the field* $\mathscr{F}_s$. *Finally*

$$\int_{-\infty}^{\infty} \mathbf{E}\{|\Phi(t, \omega)|^2\}\, dF(t) < \infty.$$

The stochastic integral

$$\int_{-\infty}^{\infty} \Phi(t, \omega)\, dy(t) \tag{5.1}$$

will be defined in such a way that, if the integrands Φ and Ψ correspond to the integrals φ and ψ,

$$\begin{aligned} \mathbf{E}\{\varphi\} &= \mathbf{E}\left\{\int_{-\infty}^{\infty} \Phi(t, \omega)\, dy(t)\right\} = 0, \\ \mathbf{E}\{\varphi\bar{\psi}\} &= \int_{-\infty}^{\infty} \mathbf{E}\{\Phi(t, \omega)\, \overline{\Psi(t, \omega)}\} dF(t). \end{aligned} \tag{5.2}$$

These equations generalize (2.13), to which they reduce if Φ and Ψ are functions of t alone. The $y(t)$ process has orthogonal increments because of I_1. The stochastic integral of this section is more general than that discussed in §2 in that the integrand may depend on ω as well as on t, but less general in that the $y(t)$ process is a martingale, instead of merely having orthogonal increments.

We shall use the following fact in the discussion below without further comment. *If* $\Phi(\cdot)$ *is measurable with respect to the field* $\mathscr{F}_{t_0}$, *and if* $\mathbf{E}\{|\Phi(\omega)|\} < \infty$, *then* (assuming I_1), *if* $t_0 \leq t_1 < t_2$,

$$\mathbf{E}\{|\Phi(\omega)|\,|y(t_2) - y(t_1)|^2\} < \infty \tag{5.3}$$

and

$$\begin{aligned} &\mathbf{E}\{\Phi(\omega)[y(t_2) - y(t_1)]\} = 0 \\ &\mathbf{E}\{\Phi(\omega)|y(t_2) - y(t_1)|^2\} = \mathbf{E}\{\Phi(\omega)\}[F(t_2) - F(t_1)]. \end{aligned} \tag{5.4}$$

To prove (5.3) and (5.4) we remark that, if (5.3) is true, then

$$\mathbf{E}\{|\Phi(\omega)|\,|y(t_2) - y(t_1)|\} < \infty,$$

by Schwarz's inequality, and hence

$$\begin{aligned} \mathbf{E}\{\Phi(\omega)[y(t_2) - y(t_1)]\} &= \mathbf{E}\Big\{\Phi(\omega)\mathbf{E}\{y(t_2) - y(t_1) \mid \mathscr{F}_{t_1}\}\Big\} \\ &= \mathbf{E}\{\Phi(\omega)0\} = 0. \end{aligned}$$

Then the first equation of (5.4) is true, and the second is proved in the same way. Thus we have only to prove that, if $\mathbf{E}\{|\Phi(\omega)|\} < \infty$, (5.3) must be true. To prove this define

$$\begin{aligned} \Phi_n(\omega) &= \Phi(\omega), \qquad |\Phi(\omega)| \leq n \\ &= 0, \qquad |\Phi(\omega)| > n. \end{aligned}$$

Then (5.3) is true with Φ replaced by Φ_n, so that from what we have already proved

$$\mathbf{E}\{|\Phi_n(\omega)|\,|y(t_2) - y(t_1)|^2\} = \mathbf{E}\{|\Phi_n(\omega)|\}[F(t_2) - F(t_1)].$$

When $n \to \infty$ this equation implies (5.3).

We now define the stochastic integral (5.1). Assume first that Φ is a function of the form

$$\begin{aligned} \Phi(t, \omega) &= 0, \qquad t < a_1 \\ &= \Phi_j(\omega), \quad a_j \leq t < a_{j+1} \qquad j \leq n - 1 \\ &= 0 \qquad a_n \leq t, \end{aligned}$$

where $a_1 < \cdots < a_n$, Φ_j is measurable with respect to $\mathscr{F}_{a_j}$, and $\mathbf{E}\{|\Phi_j(\omega)|^2\} < \infty$. Any such function will be called a (t, ω) step function.

In this case we define

$$\varphi = \int_{-\infty}^{\infty} \Phi(t, \omega)\, dy(t) = \sum_j \Phi_j(\omega)[y(a_{j+1}-) - y(a_j-)],$$

and in fact we accept as the integral any random variable equal almost everywhere to the sum on the right. With this definition φ is determined uniquely by Φ, neglecting values on ω sets of measure 0, and (5.2) is true. We have thus set up a correspondence between (t, ω) step functions Φ and certain random variables φ in which linear combinations of Φ's yield the corresponding linear combinations of φ's. If distances between pairs of functions Φ satisfying I_2 and between pairs of random variables φ, with $\mathbf{E}\{|\varphi|^2\} < \infty$, are defined respectively by

$$||\Phi_1 - \Phi_2|| = \left[\int_{-\infty}^{\infty} \mathbf{E}\{|\Phi_1(t, \omega) - \Phi_2(t, \omega)|^2\}\, dF(t)\right]^{1/2}$$

$$||\varphi_1 - \varphi_2|| = \mathbf{E}^{1/2}\{|\varphi_1 - \varphi_2|^2\},$$

this correspondence is distance preserving because (5.2) is satisfied.

Then, just as in §2, if Φ is a limit (in the sense of Φ distance) of a sequence of (t, ω) step functions, we define the stochastic integral (5.1) as the limit (in the sense of φ distance, that is, limit in the mean) of the corresponding integrals. With this definition (5.2) will clearly be satisfied in all cases. There remains the proof that the closure of the class of (t, ω) step functions includes all functions satisfying I_2; that is, the proof that, if $\mathfrak{M}$ is the class of functions satisfying I_2 which can be approximated arbitrarily closely (in terms of Φ distance) by (t, ω) step functions, $\mathfrak{M}$ includes all functions satisfying I_2. We proceed to this proof. We can write $F(t)$ in the form

$$F(t) = F_1(t) + F_2(t),$$

where F_1 is a monotone non-decreasing function increasing only in jumps, at $\tau_1, \tau_2, \cdots$, the discontinuities of F, and F_2 is monotone non-decreasing and continuous. We shall suppose in the following that $F_2(t) \equiv t$. Even if this is not so, the change of variable $t' = F_2(t)$, $\tilde{y}(t) = y(t')$, will make it so. (If F_2 is bounded, the integrals will then become integrals between finite limits, but this causes no change in the argument. An obvious convention is to be used if the transformation from t to t' is not one to one.) We prove first that for a given k, if $\Phi(\cdot)$ is an ω function measurable with respect to the field $\mathscr{F}_{\tau_k}$, with $\mathbf{E}\{|\Phi(\omega)|^2\} < \infty$, then the t, ω function defined by

$$\begin{aligned} \Phi(t, \omega) &= \Phi(\omega), \qquad && t = \tau_k, \\ &= 0, && t \neq \tau_k, \end{aligned}$$

is in the class $\mathfrak{M}$. We prove this by exhibiting a sequence of (t, ω) step functions which converges to $\Phi(\cdot, \cdot)$ in the sense of Φ distance. In fact, $\Phi_n(\cdot, \cdot)$ defined by

$$\begin{aligned} \Phi_n(t, \omega) &= \Phi(\omega), \qquad \tau_k \leq t < \tau_k + 1/n, \\ &= 0, \qquad \text{otherwise,} \end{aligned}$$

is a (t, ω) step function and

$$||\Phi(\cdot, \cdot) - \Phi_n(\cdot, \cdot)||^2 = \mathbf{E}\{|\Phi(\omega)|^2\}[F(\tau_k + 1/n -) - F(\tau_k +)] \to 0 \qquad (n \to \infty).$$

The class $\mathfrak{M}$ is a linear manifold. Hence finite sums of functions $\Phi(\cdot, \cdot)$ of the above type are also in $\mathfrak{M}$. Since the class $\mathfrak{M}$ is closed in the sense of Φ distance, a function $\Phi(\cdot, \cdot)$ defined by

$$\begin{aligned} \Phi(t, \omega) &= \Phi_k(\omega), \qquad t = \tau_k, \quad k = 1, 2, \cdots \\ &= 0, \qquad \text{otherwise,} \end{aligned}$$

is also in $\mathfrak{M}$, if each Φ_k is measurable with respect to the field $\mathscr{F}_{\tau_k}$ and if

$$\sum_k \mathbf{E}\{|\Phi_k(\omega)|^2\}[F(\tau_k +) - F(\tau_k -)] < \infty.$$

In fact, the function $\Phi(\cdot, \cdot)$ is then the limit in the sense of Φ distance of finite sums of the functions of the first type considered. The functions which we have now proved are in $\mathfrak{M}$ are precisely the functions $\Phi(\cdot, \cdot)$ which satisfy I_2 and which vanish except when $t \in \{\tau_k\}$. We next suppose that F is continuous, so that $F(t) \equiv F_2(t) \equiv t$, and prove that, if $\Phi(\cdot, \cdot)$ satisfies I_2, then $\Phi(\cdot, \cdot) \in \mathfrak{M}$. It is clearly sufficient to prove this for functions $\Phi(\cdot, \cdot)$ which are bounded and which vanish for t outside some finite interval. Suppose then that $\Phi(\cdot, \cdot)$ has these properties. Define $\alpha_n(t)$ by

$$\alpha_n(t) = \frac{j}{2^n}, \qquad \frac{j}{2^n} \leq t < \frac{j+1}{2^n}, \qquad j = 0, \pm 1, \cdots.$$

Then $\Phi[\alpha_n(t - s) + s, \omega]$ defines a (t, ω) step function, and it is sufficient to prove that s can be chosen in such a way that

$$\lim_{n \to \infty} ||\Phi[\alpha_n(t - s) + s, \omega] - \Phi(t, \omega)|| = 0.$$

To do this we prove first that, if f is a bounded Lebesgue-measurable function of s which vanishes outside some finite interval, then

$$\lim_{h \to 0} \int_{-\infty}^{\infty} |f(s + h) - f(s)|^2 \, ds = 0. \tag{5.5}$$

In fact, for every $\varepsilon > 0$ there is a continuous function f_ε, vanishing outside some finite interval, with

$$\int_{-\infty}^{\infty} |f(s) - f_\varepsilon(s)|^2\, ds \leq \varepsilon^2$$

so that, using Minkowski's inequality,

$$\limsup_{h\to 0}\Big[\int_{-\infty}^{\infty} |f(s+h) - f(s)|^2\, ds\Big]^{1/2} \leq \limsup_{h\to 0}\Big[\int_{-\infty}^{\infty} |f_\varepsilon(s+h) - f_\varepsilon(s)|^2\, ds\Big]^{1/2} + 2\varepsilon = 2\varepsilon.$$

According to (5.5),

$$\lim_{h\to 0} \int_{-\infty}^{\infty} |\Phi(s+h, \omega) - \Phi(s, \omega)|^2\, ds = 0$$

for almost all ω. Then, for each t,

$$\lim_{n\to\infty} \int_{-\infty}^{\infty} |\Phi[\alpha_n(t) + s, \omega] - \Phi(t+s, \omega)|^2\, ds = 0$$

for almost all ω. Hence

$$\lim_{n\to\infty} \int_{\Omega} \int_{-\infty}^{\infty} \int_{-\infty}^{\infty} |\Phi[\alpha_n(t) + s, \omega] - \Phi(t+s, \omega)|^2\, ds\, dt\, d\mathbf{P} = 0,$$

that is, the integrand converges to 0 in (s, t, ω) measure when $n \to \infty$. There is then a sequence of integers $\{n_j\}$ and a value of s such that

$$\lim_{j\to\infty} \int_{\Omega} \int_{-\infty}^{\infty} |\Phi[\alpha_{n_j}(t) + s, \omega] - \Phi(t+s, \omega)|^2\, dt\, d\mathbf{P}$$

$$= \lim_{j\to\infty} \int_{\Omega} \int_{-\infty}^{\infty} |\Phi[\alpha_{n_j}(t-s) + s, \omega] - \Phi(t, \omega)|^2\, dt\, d\mathbf{P} = 0$$

for almost all ω, as was to be proved.

We have shown that the class $\mathfrak{M}$ of functions Φ contains all the functions satisfying I_2 if F is continuous, and we showed earlier that in any case $\mathfrak{M}$ contains all functions satisfying I_2 which vanish except at the discontinuities of F. These two results are combined as follows. Let $\Phi(\cdot, \cdot)$ satisfy I_2, and define

$$\begin{aligned} \Phi_1(t, \omega) &= \Phi(t, \omega), \qquad && t \in \{\tau_k\} \\ &= 0, && t \notin \{\tau_k\} \\ \Phi_2(t, \omega) &= \Phi(t, \omega) - \Phi_1(t, \omega). \end{aligned}$$

Then $\Phi_1(\cdot, \cdot)$ and $\Phi_2(\cdot, \cdot)$ both satisfy I_2, and $\Phi_1(\cdot, \cdot) \in \mathfrak{M}$ since it vanishes except when $t \in \{\tau_k\}$. There remains the proof that $\Phi_2(\cdot, \cdot)$, which satisfies I_2 and vanishes if $t \in \{\tau_k\}$, is also in $\mathfrak{M}$. We have already shown that there is a sequence $\{\Psi_n(\cdot, \cdot)\}$ of functions in $\mathfrak{M}$ [in fact of (t, ω) step functions] such that

$$\lim_{n\to\infty} \int_{-\infty}^{\infty} \mathbf{E}\{|\Phi(t, \omega) - \Psi_n(t, \omega)|^2\}\, dt = 0.$$

If we modify each $\Psi_n(t, \omega)$ by setting it equal to 0 when $t \in \{\tau_k\}$, the modified function still is in $\mathfrak{M}$ and we now have

$$\lim_{n\to\infty} ||\Phi_2 - \Psi_n||^2 = \lim_{n\to\infty} \int_{-\infty}^{\infty} \mathbf{E}\{|\Phi_2(t, \omega) - \Psi_n(t, \omega)|^2\}\, dF(t) = 0.$$

Then $\Phi_2(\cdot, \cdot) \in \mathfrak{M}$, as was to be proved.

The definition of the integral (5.1) is now complete. We defined it first for $\Phi(\cdot, \cdot)$, a (t, ω) step function, and then for all functions which are limits of (t, ω) step functions in the sense of Φ distance. We then proved that the class of integrands obtained in this way includes all the functions satisfying I_2. Actually it is a slightly larger class in general. The stochastic integral over any Borel t-set, or t-set measurable with respect to dF measure, is defined by setting the integrand equal to 0 outside this set. We observe that because of the method of definition the stochastic integral is only uniquely defined neglecting ω sets of probability 0; that is, to any integrand corresponds a whole family of integrals, any two of which are equal for almost all ω.

If $\Phi(\cdot, \cdot)$ is actually a function of t alone, this integral reduces to the one defined in §2, but in this case, as we have seen in §2, the method of proof only requires that the $y(t)$ process have orthogonal increments.

In particular, if the integrand vanishes except at the set $\{\tau_k\}$ of discontinuities of F, the stochastic integral is easily seen to be

$$\int_{-\infty}^{\infty} \Phi(t, \omega)\, dy(t) = \sum_k \Phi(\tau_k, \omega)[y(\tau_k +) - y(\tau_k -)],$$

where the series (if infinite) converges in the mean. In general, for purposes of evaluation we note that if $a = t_0 < \cdots < t_m = b$, and if $\alpha(\cdot)$ is defined by

$$\alpha(t) = t_j, \qquad t_j \leq t < t_{j+1}, \qquad j < m,$$

then, if F is continuous at the t_j's,

$$\text{(5.6)}\qquad \mathbf{E}\left\{\left|\int_{-\infty}^{\infty}\Big[\Phi(t,\omega)-\Phi[\alpha(t),\omega]\Big]\,dy(t)\right|^2\right\}$$
$$=\int_{-\infty}^{\infty}\mathbf{E}\{|\Phi(t,\omega)-\Phi[\alpha(t),\omega]|^2\}\,dF(t)$$
$$=\sum_{j=0}^{m-1}\int_{t_j}^{t_{j+1}}\mathbf{E}\{|\Phi(t,\omega)-\Phi(t_j,\omega)|^2\}\,dF(t).$$

Example Let $\Phi(t,\omega)=y(t,\omega)-y(a,\omega)$. Then we shall evaluate

$$\int_a^b [y(t)-y(a)]\,dy(t)$$

for two different $y(t)$ processes. In each case $dF(t)=dt$, so that the last term in (5.6) becomes

$$\sum_{j=0}^{m-1}\int_{t_j}^{t_{j+1}}(t-t_j)\,dt\le(b-a)\operatorname*{Max}_j\,(t_{j+1}-t_j).$$

Then, when $\operatorname*{Max}_j\,(t_{j+1}-t_j)$ is small, the stochastic integral of $\Phi[\alpha(t),\omega]$ will be nearly that of $\Phi(t,\omega)$. The former integral is easily calculated, since $\Phi[\alpha(t),\omega]$ defines a (t,ω) step function,

$$\int_a^b\Phi[\alpha(t),\omega]\,dy(t)=\sum_{j=0}^{m-1}[y(t_j)-y(a)][y(t_{j+1})-y(t_j)]$$
$$=\tfrac{1}{2}[y(b)-y(a)]^2-\tfrac{1}{2}\sum_{j=0}^{m-1}[y(t_{j+1})-y(t_j)]^2.$$

It follows that, if $\delta=\operatorname{Max}\,(t_{j+1}-t_j)$,

$$\int_a^b[y(t)-y(a)]\,dy(t)=\tfrac{1}{2}[y(b)-y(a)]^2-\tfrac{1}{2}\operatorname*{l.i.m.}_{\delta\to0}\sum_{j=0}^{m-1}[y(t_{j+1})-y(t_j)]^2.$$

We observe that a formal integration would give only the first term on the right. Suppose now that $y(t)=z(t)-t$, where the $z(t)$ process is a Poisson process (see VIII §4) with rate parameter 1, so that

$$\text{(5.7)}\qquad \mathbf{E}\{dy(t)\}=0,\qquad \mathbf{E}\{[dy(t)]^2\}=dt.$$

Then, with probability 1, $z(t_0)\le\cdots\le z(t_m)$ and, with probability which approaches 1 when $\delta \to 0$, two successive values in this finite sequence are either equal or differ by 1. It follows that for almost all ω

$$\lim_{\delta\to 0} \sum_{j=0}^{m-1} [y(t_{j+1}) - y(t_j)]^2 = y(b) - y(a), \tag{5.8}$$

if the t_j's are restricted to be in some enumerable set. Then

$$\int_a^b [y(t) - y(a)]\, dy(t) = \tfrac{1}{2}[y(b) - y(a)]^2 - \tfrac{1}{2}[y(b) - y(a)].$$

As a second case suppose that the $y(t)$ process is a Brownian motion process with variance parameter 1 (see VIII §2). Then (5.7) is true once more, but (5.8) must be replaced, according to VIII, Theorem 2.3, by

$$\operatorname*{l.i.m.}_{\delta\to 0} \sum_{j=0}^{m-1} [y(t_{j+1}) - y(t_j)]^2 = b - a,$$

so that in this case the stochastic integral in question has the evaluation

$$\int_a^b [y(t) - y(a)]\, dy(t) = \tfrac{1}{2}[y(b) - y(a)]^2 - \tfrac{1}{2}(b - a).$$

We now consider the $x(t)$ process defined by

$$x(t) = \int_a^t \Phi(s, \omega)\, dy(s) \qquad t \geq a. \tag{5.9}$$

This process is not uniquely determined, in the sense that, for each t, $x(t)$ is only determined up to values on a set of ω measure 0, and we are therefore free to modify any given choice of $x(t)$, for each t, on an ω set of measure 0. According to II, Theorem 2.4, this leeway makes it possible to define $x(t)$ in such a way that the $x(t)$ process is separable.

THEOREM 5.1 *Every $x(t)$ process defined by* (5.9) *is a martingale.*

Let $b > a$ be some fixed number. It was shown that there is a sequence $\{\Phi_n(\cdot, \cdot)\}$ of (t, ω) step functions such that

$$\lim_{n\to\infty} \int_a^b \mathbf{E}\{|\Phi(t, \omega) - \Phi_n(t, \omega)|^2\}\, dF(t) = 0.$$

The stochastic integral $x_n(t)$ defined by

$$x_n(t) = \int_a^t \Phi_n(s, \omega)\, dy(s)$$

is a certain finite sum, and, for fixed $t \in [a, b]$,

$$x(t) = \operatorname*{l.i.m.}_{n\to\infty} x_n(t)$$

since

$$\begin{aligned} \mathbf{E}\{|x(t) - x_n(t)|^2\} &= \int_a^t \mathbf{E}\{|\Phi(s, \omega) - \Phi_n(s, \omega)|^2\} \, dF(s) \\ &\leq \int_a^b \mathbf{E}\{|\Phi(s, \omega) - \Phi_n(s, \omega)|^2\} \, dF(s) \to 0. \end{aligned}$$

Now, if $t < t_1$,

$$\mathbf{E}\{x_n(t_1) \mid \mathscr{F}_t\} = x_n(t)$$

with probability 1, by inspection of the sums which define $x_n(t)$, that is,

$$\int_\Lambda x_n(t) \, d\mathbf{P} = \int_\Lambda x_n(t_1) \, d\mathbf{P}, \qquad \Lambda \in \mathscr{F}_t.$$

When $n \to \infty$ this equation becomes

$$\int_\Lambda x(t) \, d\mathbf{P} = \int_\Lambda x(t_1) \, d\mathbf{P},$$

so that

$$\mathbf{E}\{x(t_1) \mid \mathscr{F}_t\} = x(t)$$

with probability 1. This equation implies that the process $\{x(t), \mathscr{F}_t, t \in T\}$ is a martingale.

THEOREM 5.2 *The stochastic integral* (5.9) *can be defined for each t in such a way that the $x(t)$ process is a separable martingale. Almost all sample functions of this process will then have one-sided limits at all points. The fixed points of discontinuity of the $x(t)$ process are points of discontinuity of F. If the sample functions of the $y(t)$ process are almost all continuous, those of a separable $x(t)$ process will almost all be continuous also.*

Only the last two statements of the theorem remain to be proved. Since

$$\mathbf{E}\left\{\left|\int_{t_1}^{t_2} \Phi(t, \omega) \, dy(t)\right|^2\right\} = \int_{t_1}^{t_2} \mathbf{E}\{|\Phi(t, \omega)|^2\} \, dF(t),$$

a fixed point of discontinuity of the $x(t)$ process must be a discontinuity of F. If almost all sample functions of the $y(t)$ process are continuous, F must be continuous, since

$$\mathbf{E}\{|y(t_2) - y(t_1)|^2\} = F(t_2) - F(t_1).$$

Then, according to VII, Theorem 11.9, the $y(t)$ process is the Brownian motion process after a change of the time variable. We prove that almost all the (separable) $x(t)$ process sample functions are continuous in this case by first remarking that the statement is obvious if Φ is a (t, ω) step function, and then proving the general statement by approximation, as follows. It is sufficient to prove it for t in a finite interval $[a, b]$. Let Φ_n be a (t, ω) step function with

$$\int_a^b \mathbf{E}\{|\Phi(t, \omega) - \Phi_n(t, \omega)|^2\} \, dF(t) \leq \frac{1}{n^4}$$

and define an $x_n(t)$ process by

$$x_n(t, \omega) = \int_a^t \Phi_n(s, \omega) \, dy(s).$$

Then

$$x(t, \omega) - x_n(t, \omega) = \int_a^t [\Phi(s, \omega) - \Phi_n(s, \omega)] \, dy(s),$$

so that the $x(t) - x_n(t)$ process is a martingale, which we can suppose is made separable by a proper choice of $x_n(t)$ for each t. Hence, by the continuous parameter version of III, Theorem 2.1, or of VII, Theorem 3.2, applied to the $|x(t) - x_n(t)|^2$ process,

$$\mathbf{P}\left\{\underset{a \leq t \leq b}{\text{L.U.B.}} |x(t) - x_n(t)| \geq \frac{1}{n}\right\} \leq \mathbf{E}\{|x(b) - x_n(b)|^2\} n^2 \leq \frac{n^2}{n^4} = \frac{1}{n^2}.$$

It now follows from the Borel-Cantelli lemma (III, Theorem 1.2), since the right side of the preceding inequality is the general term of a convergent series, that

$$|x(t, \omega) - x_n(t, \omega)| < 1/n, \qquad a \leq t \leq b$$

for sufficiently large n, with probability 1. This implies that, with probability 1, the $x_n(t)$ process sample functions converge uniformly to those of the $x(t)$ process, which proves the desired result.

Certain applications in VI make it desirable to generalize the stochastic integral (5.1), in the case where F is absolutely continuous, by relaxing the condition on the $y(t)$ process. Condition I_1 is replaced by:

I_1' *The process* $\{y(t), \mathscr{F}_t, -\infty < t < \infty\}$ *is a martingale, with*

$$\mathbf{E}\{|y(t) - y(s)|^2\} < \infty$$

for all s *and* t; f *is a non-negative* (t, ω) *function defined and measurable* *with respect to dt d***P** *measure, such that, for each s, f(s, ·) is ω measurable with respect to the field* $\mathscr{F}_s$ *and that, if s < t,*

$$\mathbf{E}\left\{\int_s^t f(u, \omega)\, du\right\} < \infty$$

and

$$\mathbf{E}\{|y(t) - y(s)|^2 \mid \mathscr{F}_s\} = \mathbf{E}\left\{\int_s^t f(u, \omega)\, du \mid \mathscr{F}_s\right\},$$

with probability 1. Condition I_2 is replaced by:

I_2' *Φ is a (t, ω) function measurable with respect to dt d***P** *measure. For each s, Φ(s, ·) is ω measurable with respect to the field* $\mathscr{F}_s$. *Finally*

$$\int_{-\infty}^{\infty} \mathbf{E}\{|\Phi(t, \omega)|^2 f(t, \omega)\}\, dt < \infty.$$

In particular, if f is in fact a function of t alone, the hypotheses become a special case of I_1 and I_2 with $F'(t) = f(t)$.

We now outline the definition of the stochastic integral (5.1) in the present case. Few changes are needed. However, (5.2) is replaced by

$$\mathbf{E}\{\varphi\} = 0 \tag{5.2'}$$

$$\mathbf{E}\{\varphi\bar{\psi}\} = \int_{-\infty}^{\infty} \mathbf{E}\{\Phi(t, \omega)\overline{\Psi(t, \omega)} f(t, \omega)\}\, dt,$$

and, in the discussion of the definition, the distance between functions Φ and Ψ is defined by

$$||\Phi - \Psi|| = \left[\int_{-\infty}^{\infty} \mathbf{E}\{|\Phi(t, \omega) - \Psi(t, \omega)|^2 f(t, \omega)\}\, dt\right]^{1/2}.$$

The new (t, ω) step functions are defined like the old ones, except that, in the notation of the old definition, it is supposed not that Φ_j^2 is integrable but that

$$\int_{a_j}^{a_{j+1}} \mathbf{E}\{|\Phi_j(\omega)|^2 f(t, \omega)\}\, dt < \infty.$$

To show that any Φ satisfying I_2' can be approximated arbitrarily closely in terms of the new distance, by new (t, ω) step functions, we can assume that Φ is bounded, $|\Phi| \leq K$, and vanishes outside a finite interval $[a, b]$. Then Φ satisfies I_2, and hence there is a sequence of (old) (t, ω) step functions $\Phi_1, \Phi_2, \cdots$ such that

$$\lim_{n\to\infty} \int_a^b \mathbf{E}\{|\Phi(t, \omega) - \Phi_n(t, \omega)|^2\}\, dt = 0.$$

Moreover our earlier discussion shows that we can assume that $|\Phi_n| \leq K$. Then Φ_n converges to Φ in $dt\,d\mathbf{P}$ measure, so that $|\Phi - \Phi_n|^2 f$ converges to 0 in $dt\,d\mathbf{P}$ measure, and the latter function is dominated by the $dt\,d\mathbf{P}$ integrable function $4K^2 f$. Hence

$$\lim_{n\to\infty} \int_a^b \mathbf{E}\{|\Phi(t, \omega) - \Phi_n(t, \omega)|^2 f(t, \omega)\}\, dt = 0,$$

and this is the desired approximation result. Theorems 5.1 and 5.2 remain true, and their proofs require no change of method. There are now no fixed points of discontinuity in Theorem 5.2.

Finally we remark that, if $x(t)$ is defined by

$$x(t) = \int_a^t \Phi(s, \omega)\, dy(s),$$

the $x(t)$ process is a martingale, according to Theorem 5.1, and

$$\mathbf{E}\{x(t) \mid \mathscr{F}_s\} = x(s)$$

$$\mathbf{E}\{|x(t) - x(s)|^2 \mid \mathscr{F}_s\} = \mathbf{E}\Big\{\int_s^t |\Phi(u, \omega)|^2 f(u, \omega)\, du \mid \mathscr{F}_s\Big\}$$

$$= \mathbf{E}\Big\{\int_s^t f_1(u, \omega)\, du \mid \mathscr{F}_s\Big\}$$

with probability 1, where

$$f_1(t, \omega) = |\Phi(t, \omega)|^2 f(u, \omega).$$

Thus the $x(t)$ process and function f_1 satisfy the same hypotheses as the $y(t)$ process and function. This means that we can consider stochastic integrals of the form

$$\int \Phi_1(t, \omega)\, dx(t).$$

Moreover,

$$\int_a^b \Phi_1(t, \omega)\, dx(t) = \int_a^b \Phi_1(t, \omega)\, \Phi(t, \omega)\, dy(t),$$

because this equality is true if Φ_1 is a (t, ω) step function. In particular, if Φ does not vanish, or at least vanishes at most on a (t, ω) set of $ds\,d\mathbf{P}$ measure 0,

$$y(t) - y(a) = \int_a^t \frac{dx(s)}{\Phi(s, \omega)}.$$

Now consider the following problem: What $x(t)$ process can be represented in the form (5.9), with the $y(t)$ process a Brownian motion process? As we have seen, it is no restriction to assume that the $x(t)$ process defined in this way is separable. If we assume that the Brownian motion process has variance parameter 1, we find that, if $t_1 < t_2$,

$$\mathbf{E}\{|x(t_2) - x(t_1)|^2 \mid \mathscr{F}_{t_1}\} = \mathbf{E}\left\{\int_{t_1}^{t_2} |\Phi(s, \omega)|^2 \, ds \mid \mathscr{F}_{t_1}\right\} \tag{5.10}$$

with probability 1. Moreover, according to Theorem 5.2 the $x(t)$ process sample functions are almost all continuous. The following theorem shows that these two conditions are sufficient.

THEOREM 5.3 *Let* $\{x(t), \mathscr{F}_t, a \leq t \leq b\}$ *be a martingale with the following properties.*

(i) $$\mathbf{E}\{|x(t)|^2\} < \infty, \qquad a \leq t \leq b.$$

(ii) *Almost all sample functions of the process are continuous in* $[a, b]$.

(iii) *There is a non-negative* (t, ω) *function* Φ, *measurable with respect to* $dt\, d\mathbf{P}$ *measure, such that, for each* s, $\Phi(s, \cdot)$ *is* ω *measurable with respect to the field* $\mathscr{F}_s$, *and such that, if* $t_1 < t_2$, (5.10) *is true with probability* 1. *Then, if* Φ *vanishes almost nowhere on* (t, ω) *space, there is a Brownian motion process* $\{y(t), a \leq t \leq b\}$ *such that for each* $t \in [a, b]$,

$$x(t) = x(a) + \int_a^t \Phi(s, \omega) \, dy(s) \tag{5.11}$$

with probability 1. *Even without this additional hypothesis on the vanishing of* Φ, *the statement is true after a Brownian motion process has been adjoined to the* $x(t)$ *process.*

See II §2 for the meaning of the adjunction of a Brownian motion process to the given process. As a trivial example of a case in which this adjunction is necessary, consider an ω space consisting of a single point, and define $x(t, \omega) = 1$ for all t. The hypotheses of the theorem are certainly satisfied in this case, and $\Phi \equiv 0$. Obviously there is no Brownian motion process $\{y(t), a \leq t \leq b\}$ because ω space is not sufficiently complex to carry such a process.

According to the hypotheses of this theorem we can define a $y(t)$ process by

$$y(t) = \int_a^t \frac{dx(s)}{\Phi(s, \omega)}$$

(defining $1/\Phi(s, \omega)$ to be 0 if $\Phi(s, \omega) = 0$). The $y(t)$ process is a martingale and, if $t_1 < t_2$,

$$\mathbf{E}\{|y(t_2) - y(t_1)|^2 \mid \mathscr{F}_{t_1}\} = \mathbf{E}\left\{\int_{t_1}^{t_2} \frac{|\Phi(s, \omega)|^2}{|\Phi(s, \omega)|^2} ds \mid \mathscr{F}_{t_1}\right\}$$

with probability 1. In particular, if Φ never vanishes, or even if it vanishes almost nowhere in (t, ω), it follows that the right side of this equation becomes $t_2 - t_1$. Moreover, according to Theorem 5.2, the $y(t)$ process can be defined to have almost all its sample functions continuous. Hence by VII, Theorem 11.9, the $y(t)$ process is a Brownian motion process, and the inversion of the defining formula gives (5.10). If Φ may vanish, the right side of the preceding equation is no longer $t_2 - t_1$ and the preceding argument is no longer valid. Suppose, however, that there is a Brownian motion process $\{z(t), a \leq t \leq b\}$ for which, if $t_1 < t_2$, $t_1 \leq t_1' < t_2'$,

$$\mathbf{E}\{z(t_2) \mid \mathscr{F}_{t_1}\} = z(t_1)$$

$$\mathbf{E}\{|z(t_2) - z(t_1)|^2 \mid \mathscr{F}_{t_1}\} = (t_2 - t_1)$$

$$\mathbf{E}\{[x(t_2) - x(t_1)]\overline{[z(t_2') - z(t_1')]} \mid \mathscr{F}_{t_1}\} = 0$$

with probability 1. Such a process can always be obtained by the adjunction procedure. Define

$$\begin{aligned} \Phi_0(t, \omega) &= 1, \qquad \text{if } \Phi(t, \omega) = 0, \\ &= 0, \qquad \text{if } \Phi(t, \omega) \neq 0, \end{aligned}$$

$$y(t) = \int_a^t \frac{dx(s)}{\Phi(s, \omega)} + \int_a^t \Phi_0(s, \omega)\, dz(s),$$

choosing the integrals to get a separable $y(t)$ process. Then the $y(t)$ process has continuous sample functions, by Theorem 5.2. Moreover, since $\Phi_0\Phi = 0$,

$$\mathbf{E}\left\{\left|\int_a^t \Phi_0(s, \omega)\, dx(s)\right|^2\right\} = \int_a^t \mathbf{E}\{|\Phi_0\Phi|^2\}\, ds = 0, \tag{5.12}$$

using the rules of combination we have developed for stochastic integrals. The following equations (true with probability 1) are now easy to verify:

$$\mathbf{E}\{|y(t_2) - y(t_1)|^2 \mid \mathscr{F}_{t_1}\} = \mathbf{E}\left\{\int_{t_1}^{t_2} \left|\frac{\Phi(s, \omega)}{\Phi(s, \omega)}\right|^2 ds + \int_{t_1}^{t_2} |\Phi_0(s, \omega)|^2\, ds \,\middle|\, \mathscr{F}_{t_1}\right\} = t_2 - t_1, \tag{5.13}$$

$$\int_a^t \Phi(s, \omega)\, dy(s) = \int_a^t \frac{\Phi\, dx(s)}{\Phi} = \int_a^t (1 - \Phi_0)\, dx(s). \tag{5.14}$$

Now, from (5.14),

$$x(t) = x(a) + \int_a^t \Phi_0(s, \omega)\, dx(s) + \int_a^t \Phi(s, \omega)\, dy(s) \tag{5.15}$$

with probability 1 and, from (5.12), the first integral on the right may be omitted. Finally the $y(t)$ process is a Brownian motion process because (see VII, Theorem 11.9), its sample functions are almost all continuous, and (5.13) is true. Then (5.15) reduces to (5.11).

CHAPTER X

Stationary Processes—Discrete Parameter

1. Generalities; metric transitivity

(a) *Strictly stationary processes* Strictly stationary processes were defined in II §8. The study of these processes is usually carried out by non-probabilists in the language of measure-preserving transformations, but the connection between the two ideas is not always clearly stated in the literature, and we proceed to develop it in some detail.

We adopt our usual basic hypothesis: there is a probability measure defined on a Borel field of sets of some space Ω. A transformation $\mathbf{T}$ taking points of Ω into points of Ω is called a 1–1 *measure-preserving point transformation* if it is 1–1, has domain and range Ω, and if it and its inverse take measurable sets into measurable sets of the same probability. Such a transformation induces a 1–1 transformation of random variables into random variables (which we shall also denote by $\mathbf{T}$), if we define

$$[\mathbf{T}x](\omega) = x(\mathbf{T}^{-1}\omega).$$

Under this transformation, if x is a random variable which is 1 on a measurable ω set Λ and 0 otherwise, $\mathbf{T}x$ is 1 on the image of Λ under $\mathbf{T}$, and 0 otherwise. For any x, $\mathbf{T}x$ has the same distribution as x, and in fact the stochastic process

$$\{x_n, -\infty < n < \infty\}, \qquad x_n = \mathbf{T}^n x,$$

is strictly stationary. Thus any measure-preserving point transformation can be used to generate strictly stationary stochastic processes.

A set transformation $\mathbf{T}$ defined on the Borel field of measurable ω sets, taking measurable ω sets into measurable ω sets, is called a *measure-preserving set transformation* if the following conditions are satisfied:

MP_1 $\mathbf{T}$ *is single valued, modulo sets of probability* 0: if Λ_1 is an image of Λ under $\mathbf{T}$, the class of all images of Λ is the class of all measurable sets differing from Λ_1 by sets of probability 0. In the following the notation $\mathbf{T}\Lambda$ will always mean a particular image of Λ under $\mathbf{T}$.

MP$_2$ $$\mathbf{P}\{\mathbf{T}\Lambda\} = \mathbf{P}\{\Lambda\}.$$

MP$_3$ *Neglecting* ω *sets of probability* 0,

$$\mathbf{T}(\Lambda_1 \cup \Lambda_2) = \mathbf{T}\Lambda_1 \cup \mathbf{T}\Lambda_2$$

$$\mathbf{T}(\bigcup_{n=1}^{\infty} \Lambda_n) = \bigcup_{n=1}^{\infty} \mathbf{T}\Lambda_n$$

$$\mathbf{T}(\Omega - \Lambda) = \Omega - \mathbf{T}\Lambda.$$

Condition MP$_3$ implies that, neglecting sets of probability 0, finite and enumerably infinite intersections go over into the corresponding intersections under $\mathbf{T}$, and that, if $\Lambda_1 \subset \Lambda_2$, then $\mathbf{T}\Lambda_1 \subset \mathbf{T}\Lambda_2$. If every measurable set is the image of some measurable set under $\mathbf{T}$, this transformation must be 1–1 neglecting sets of probability 0 (that is, the transformation is 1–1 modulo the sets of probability 0) and the inverse $\mathbf{T}^{-1}$ is thus defined, and is also a measure-preserving set transformation. We shall describe this situation by stating that $\mathbf{T}$ has an inverse.

A 1–1 measure-preserving point transformation $\mathbf{T}$ induces a measure-preserving set transformation, if we define the family of images of a set Λ as all the ω sets differing by at most a set of probability 0 from the image of Λ under the point transformation. This set transformation evidently has an inverse. However, a set transformation with an inverse cannot always be generated in this way by a point transformation.

We shall see that measure-preserving set transformations can always be avoided in the theory of probability, in favor of the simpler 1–1 measure-preserving point transformations, but this avoidance obscures the significance of the results to some extent.

If $\mathbf{T}$ is a measure-preserving set transformation, there is one and only one transformation $\mathbf{T}_1$ defined for every random variable, taking random variables into random variables, and having the following properties:

RV$_1$ $\mathbf{T}_1$ *is single-valued modulo the random variables which vanish with probability* 1: if x_1 is an image of x under $\mathbf{T}_1$, the class of all images of x is the class of random variables equal to x_1 with probability 1. In the following, the notation $\mathbf{T}_1 x$ will always mean a particular image of x under $\mathbf{T}_1$.

RV$_2$ $\mathbf{T}_1$ *is consistent with* $\mathbf{T}$: if x is a random variable which is 1 on a measurable ω set Λ and 0 otherwise, then $\mathbf{T}_1 x$ is a random variable which is 1 almost everywhere on $\mathbf{T}\Lambda$ and 0 otherwise. Thus $\mathbf{T}_1$ can be considered an extension of $\mathbf{T}$.

RV_3 $\mathbf{T}_1$ *is linear:* if a, b are constants and if x, y are random variables,

$$\mathbf{T}_1(ax + by) = a\mathbf{T}_1x + b\mathbf{T}_1y$$

with probability 1.

RV_4 $\mathbf{T}_1$ *preserves convergence:* if $\lim_{n\to\infty} x_n = x$ with probability 1, then $\lim_{n\to\infty} \mathbf{T}_1x_n = \mathbf{T}_1x$ with probability 1.

There is at most one transformation $\mathbf{T}_1$ with these properties, since two transformations with these properties would be identical, by RV_2 and RV_3, for the random variables taking on only finitely many values, hence by RV_4 for all random variables. We show that there is one such $\mathbf{T}_1$ by defining it explicitly, as follows. It is sufficient to consider only real random variables, since in the complex case the real and imaginary parts can be treated separately. For each rational r define

$$\Lambda_r = \mathbf{T}\{x(\omega) \le r\},$$

choosing the set transforms, as we certainly can, so that

$$\Lambda_{r_1} \subset \Lambda_{r_2}, \qquad r_1 < r_2$$
$$\bigcap_r \Lambda_r = 0, \qquad \bigcup_r \Lambda_r = \Omega.$$

Now define

$$[\mathbf{T}_1x](\omega) = s \qquad \text{if} \quad \omega \in \bigcap_{r>s} \Lambda_r - \bigcup_{r<s} \Lambda_r.$$

The ω function $\mathbf{T}_1x$ is defined for all ω, and

$$\{[\mathbf{T}_1x](\omega) \le s\} = \{\bigcap_{r>s} \Lambda_r\} = \mathbf{T}\{x(\omega) \le s\},$$

neglecting sets of probability 0, so that $\mathbf{T}_1x$ is a random variable, that is, a measurable ω function. If we define the class of images of x under $\mathbf{T}_1$ as the class of all random variables equal with probability 1 to the image just defined, it is easily verified that $\mathbf{T}_1$ has the desired properties. In the following we shall use the same notation $\mathbf{T}$ for both the set and random variable transformation. In particular, if $\mathbf{T}$ is a measure-preserving set transformation derived from a 1–1 measure-preserving point transformation $\mathbf{S}$, it is clear that

$$[\mathbf{T}x](\omega) = x(\mathbf{S}^{-1}\omega)$$

with probability 1.

If $\mathbf{T}$ is a measure-preserving set transformation, and if x is a random variable, the stochastic process

$$\{x_n, n \ge 0\}, \qquad x_n = \mathbf{T}^n x,$$

is strictly stationary, and, if $\mathbf{T}$ has an inverse, the stochastic process

$$\{x_n, -\infty < n < \infty\}, \qquad x_n = \mathbf{T}^n x,$$

is strictly stationary. Thus a measure-preserving set transformation can be used to generate strictly stationary stochastic processes.

We have supposed above that a measure-preserving set transformation **T** has as domain and range space the class of all measurable ω sets. But it is obvious that nothing whatever would be changed in the discussion if domain and range space were any Borel field of measurable sets including all sets of probability 0. This trivial extension of the idea will be convenient below. The induced transformation of random variables will then have as domain and range space the class of random variables measurable with respect to the field in question.

We now consider a strictly stationary stochastic process $\{x_n, n \geq 0\}$ and study the conditions under which it can be generated by a measure-preserving set transformation, as described above. In studying this stochastic process, the only measurable ω sets of interest are those determined by conditions on the x_j's, that is, those measurable on the sample space of the x_j's. We, therefore, wish to determine a measure-preserving set transformation, with domain the class of sets determined by conditions on the x_j's, with the property that $\mathbf{T}^n x_0 = x_n, n \geq 1$, with probability 1. It will be shown that there is such a transformation and that it is uniquely determined. The transformation is called the *shift*, because $\mathbf{T}x_n = x_{n+1}, n \geq 0$, with probability 1. If there is such a **T**, it will take the ω set

$$\Lambda = \{[x_1(\omega), \cdots, x_n(\omega)] \in A\},$$

where A is a Borel set, into the sets differing by at most sets of probability 0 from the set

$$\{[x_2(\omega), \cdots, x_{n+1}(\omega)] \in A\}.$$

Now if two shift transformations exist they will agree on sets of the above type, and therefore on all the sets of the Borel field determined by those of the above type. They will, therefore, agree on all sets determined by conditions on the x_j's. Thus, if there is a shift, there is only one. There is actually a shift since we can define **T** as indicated for the sets of the above particular type, and then extend the definition to all sets determined by conditions on the x_j's. (In fact, if the distance between measurable ω sets Λ_1, Λ_2 is defined as

$$\mathbf{P}\{\Lambda_1(\Omega - \Lambda_2)\} + \mathbf{P}\{(\Omega - \Lambda_1)\Lambda_2\},$$

the space of measurable sets becomes a complete metric space, and the desired shift becomes a single-valued distance-preserving function on a closed subset of the space, the set E corresponding to the sets determined by conditions on the x_j's. The sets Λ defined above form a subset of E dense in E, and **T** as defined on this subset is distance-preserving and is therefore a uniformly continuous function. There is, therefore, one and only one way of extending the definition of $\mathbf{T}$, preserving continuity, to all of E.) Similarly, if $\{x_n, -\infty < n < \infty\}$ is a strictly stationary stochastic process, there is a uniquely determined measure-preserving set transformation, the shift (relative to this process) with domain the Borel field of sets determined by conditions on the x_j's, such that $\mathbf{T}^n x_0 = x_n$ for all integers n.

We have already remarked that the introduction of measure-preserving set (rather than point) transformations can be dispensed with in this subject. We now proceed to justify this statement. Let $\{x_n, 0 \leq n < \infty\}$ be a strictly stationary stochastic process. We shall find a related stochastic process, which for most purposes can be used to replace the given one, and is generated by a 1–1 measure-preserving point transformation. This is done as follows (see I §5 and §6 and II §1). Let $\tilde{\Omega}$ be the coordinate space of all sequences $\tilde{\omega}$ of numbers $\{\cdot \cdot \cdot, \xi_1, \xi_2, \cdot \cdot \cdot\}$; we define a probability measure in this space by assigning a finite dimensional distribution to each finite set of coordinate variables. Let $\tilde{x}_n$ be the nth coordinate variable of $\tilde{\Omega}$, so that $\tilde{x}_n(\tilde{\omega}) = \xi_n$ if $\tilde{\omega}$ is the point $(\cdot \cdot \cdot, \xi_1, \xi_2, \cdot \cdot \cdot)$. The $\tilde{\omega}$ set

$$\{\tilde{x}_{m_j}(\tilde{\omega}) \in A_j, j = 1, \cdot \cdot \cdot, n\}$$

is assigned the probability

$$\mathbf{P}\{x_{m_j+h}(\omega) \in A_j, j = 1, \cdot \cdot \cdot, n\}.$$

Here the A_j's are Borel sets and h is chosen so large that

$$m_j + h \geq 0, \qquad j = 1, \cdot \cdot \cdot, n.$$

According to Kolmogorov's theorem on infinite dimensional measures (see I §5), this assignment of finite dimensional distributions determines a probability measure on a Borel field of $\tilde{\omega}$ sets. The coordinate function $\tilde{x}_n$ is a random variable on $\tilde{\Omega}$. The strictly stationary stochastic process $\{\tilde{x}_n, -\infty < n < \infty\}$ has the property that the multivariate distribution of the random variables $\tilde{x}_0, \cdot \cdot \cdot, \tilde{x}_n$ on $\tilde{\omega}$ space is the same as that of the random variables $x_0, \cdot \cdot \cdot, x_n$ on ω space. Thus, if only questions involving these distributions are to be considered, we can use the $\tilde{x}_n$ process instead of the x_n process. For example, consider the averages

$$\frac{x_0 + \cdot \cdot \cdot + x_n}{n + 1}, \qquad n \geq 0.$$

The law of large numbers makes assertions about the convergence of these averages when $n \to \infty$. It is clear that these averages converge in the mean, stochastically, or with probability 1, if and only if the corresponding $\tilde{x}_n$ averages do so. From the present point of view, the $\tilde{x}_n$

process is more advantageous than the x_n process in that the shift for the x_n process is generated by a 1–1 measure-preserving point transformation, the transformation which simply shifts each coordinate one step.

We have supposed here that the x_n process has as parameter range the set of non-negative integers. If the parameter range is the set of all integers, the $\tilde{x}_n$ process is defined in the same way, and becomes what we have called in I §6 a representation of the x_n process.

We have now shown that in discussing strictly stationary stochastic processes we are really discussing the iterates of a shift operator acting on a given random variable, and that for many purposes we can even assume that the shift is a 1–1 measure-preserving point transformation. In the following we shall use interchangeably the language of stochastic processes and that of measure-preserving set transformations.

A measurable ω set is called *invariant* under a measure-preserving point or set transformation if it differs from its images by sets of probability 0. Every set of probability 0 or 1 is then invariant. The invariant sets form a Borel field. A random variable x is called invariant under a measure-preserving point or set transformation $\mathbf{T}$ if $\mathbf{T}x = x$ with probability 1. Then every function identically constant with probability 1 is invariant. If $\mathbf{T}$ has an inverse, the same measurable ω sets and random variables are invariant with respect to the inverse as with respect to $\mathbf{T}$.

If Λ is a measurable ω set, and if x is a random variable which is 1 on Λ and 0 otherwise, then Λ is an invariant set if and only if x is an invariant random variable. If x is an invariant random variable, the ω set $\{x(\omega) \in A\}$ is an invariant ω set for every Borel set A. Conversely, if the ω set $\{x(\omega) \in A\}$ is an invariant set for every Borel set A, or even (in the real case) for every interval A, then x is an invariant random variable.

A measure-preserving point or set transformation is called *metrically* transitive if the only invariant sets are those which have probability 0 or 1, that is, if the only invariant random variables are those which are identically constant with probability 1.

We have seen that to each strictly stationary process there corresponds a unique measure-preserving set transformation, the shift, defined on the ω sets determined by conditions on the random variables of the process. Sets and random variables invariant relative to this transformation are called invariant sets and random variables of the process, and the process is said to be metrically transitive if the shift is. A process is metrically transitive if and only if the $\tilde{x}_n$ process defined above, the corresponding process on coordinate space, is metrically transitive.

Let $\mathbf{T}$ be a metrically transitive measure-preserving set transformation, and let x be a random variable. Then the process

$$\{x_n, n \geq 0\}, \qquad x_n = \mathbf{T}^n x,$$

is metrically transitive. To see this we note that, if $\mathbf{T}_1$ is the shift corresponding to this stochastic process, $\mathbf{T}$ and $\mathbf{T}_1$ are identical on the ω sets determined by conditions on the x_j's, that is, $\mathbf{T} = \mathbf{T}_1$ on the domain of definition of $\mathbf{T}_1$. It follows that, if an ω set is invariant with respect to $\mathbf{T}_1$, it will be invariant with respect to $\mathbf{T}$, and therefore that $\mathbf{T}_1$ is metrically transitive, because $\mathbf{T}$ is.

An important particular case of the result of the previous paragraph is the following. Let $\{x_n, -\infty < n < \infty\}$ be a strictly stationary stochastic process which is metrically transitive; then any measurable function on the sample space of the x_j's, that is, any random variable measurable with respect to the field of sets determined by conditions on the x_j's, defines a strictly stationary stochastic process if the x_j's are shifted step by step. For example, if the x_n process is metrically transitive, the processes

$$\{x_n^2, -\infty < n < \infty\} \qquad \{x_n + x_{n+2}^3, -\infty < n < \infty\}$$

are also metrically transitive. The corresponding results hold for a one-sided parameter range.

If $\{x_n, -\infty < n < \infty\}$ is a process which is strictly stationary, the process $\{x_{-n}, -\infty < n < \infty\}$ is also strictly stationary, and has the same invariant random variables as the given process, since the inverse of a measure-preserving set transformation with an inverse has the same invariant random variables as the given transformation. The process $\{x_n, n \geq 0\}$ is also strictly stationary, and has the same invariant random variables as the given process. This assertion is not obvious and will be proved in detail. It will follow that the three processes under discussion here are either simultaneously metrically transitive or none is. Obviously any random variable invariant with respect to the process $\{x_n, n \geq 0\}$ is also invariant with respect to the process $\{x_n, -\infty < n < \infty\}$. To prove the converse we only need to prove that a random variable y which is invariant with respect to the second process is measurable on the sample space of the x_n's with $n \geq 0$. To prove this we remark first that, since the random variable y is measurable on the sample space of the x_n's, to every k corresponds a random variable y_k, measurable with respect to the field of sets determined by conditions on a certain finite number of x_n's, for which

$$\mathbf{P}\{|y(\omega) - y_k(\omega)| > 1/k\} < 2^{-k}$$

(see II §1). Moreover, if desirable, we can replace y_k by $\mathbf{T}^j y_k$ here, where $\mathbf{T}$ is the shift corresponding to the process $\{x_n, -\infty < n < \infty\}$, since, using the invariance of y under $\mathbf{T}$,

$$\mathbf{P}\{|y(\omega) - [\mathbf{T}^j y_k](\omega)| > 1/k\} < 2^{-k}.$$

We, therefore, can even suppose that y_k is measurable with respect to the sample space of the x_n's with $n \geq 0$. Since

$$\lim_{n\to\infty} y_k = y$$

with probability 1, we conclude that y is also measurable on the sample space of the x_n's with $n \geq 0$, as was to be proved. A trivial modification of this reasoning yields the following result, which we state for later reference. *If* $\{x_n, -\infty < n < \infty\}$ *or* $\{x_n, n \geq 0\}$ *is a strictly stationary process, and if* y *is an invariant random variable, then, for every* n, y *is measurable on the sample space of* $x_n, x_{n+1}, \cdots$.

Example 1 *Markov chains* Let $[p_{ij}]$ be an N-dimensional stochastic matrix, that is, suppose that

$$p_{ij} \geq 0, \qquad \sum_{j=1}^{N} p_{ij} = 1.$$

Then we have proved in V §2 that there are numbers $p_1, \cdots, p_N$, called stationary absolute probabilities, satisfying

$$\sum_{i=1}^{N} p_i = 1, \qquad p_i \geq 0, \qquad \sum_{i=1}^{N} p_i p_{ij} = p_j.$$

We shall suppose that the stationary absolute probabilities have been chosen so that $p_i > 0$ if i is in an ergodic class. Such a choice is always possible. (For example, in V, Theorem 2.4, take the coefficients of the linear combination positive for $i \notin F$.) Let $\{x_n, -\infty < n < \infty\}$ be a Markov chain with the given absolute and transition probabilities, so that

$$\mathbf{P}\{x_n(\omega) = i\} = p_i$$

$$\mathbf{P}\{x_{n+1}(\omega) = j \mid x_n(\omega) = i\} = p_{ij}$$

(with probability 1). The x_n process is then a strictly stationary process. Let E be an ergodic class of states, and define

$$\Lambda = \{x_0(\omega) \in E\}.$$

Since $x_n(\omega) \in E$ implies that $x_{n+1}(\omega) \in E$, and conversely, neglecting ω sets of probability 0, it follows that Λ is an invariant set. Moreover,

$$\mathbf{P}\{\Lambda\} = \sum_{i \in E} p_i > 0.$$

If there is another ergodic class, a second invariant set can be defined in terms of it, in the same way. This set will also have positive probability, and will have no points in common with Λ. Hence in this case $0 < \mathbf{P}\{\Lambda\} < 1$, and we have proved that *if there is more than one ergodic class the process is not metrically transitive.* Conversely, *if there is only one ergodic class, the process is metrically transitive.* To prove this we need only prove that every invariant set is of the form $\{x_0(\omega) \in E\}$, and this is a special case of the following theorem.

THEOREM 1.1 *If $\{x_n, n \geq 0\}$ is a stationary Markov process, and if z is an invariant random variable, then z is measurable on the sample space of x_0.*

In terms of sets, rather than random variables, this theorem states that any invariant ω set is determined by conditions on x_0 alone. Of course x_0 can be replaced by any x_k. Because of the set version of this theorem, we can assume, if convenient in proving it, that z is bounded. Actually we prove the statement by assuming that z is integrable and proving that $z = \mathbf{E}\{z \mid x_0\}$ with probability 1. We have already remarked above that the invariance of z implies that z is measurable on the sample space of $x_n, x_{n+1}, \cdots$ for every n. Hence, since the x_n process is a Markov process,

$$\mathbf{E}\{z \mid x_0, \cdots, x_n\} = \mathbf{E}\{z \mid x_n\},$$

with probability 1, and, by VII, Theorem 4.3, Corollary 1,

$$\lim_{n\to\infty} \mathbf{E}\{z \mid x_n\} = z$$

with probability 1. Then

$$\lim_{n\to\infty} \mathbf{P}\{|z(\omega) - \mathbf{E}\{z \mid x_n\}| > \varepsilon\} = 0$$

for every $\varepsilon > 0$. Since z is invariant, the probability on the left is actually independent of n, so that, taking $n = 0$, we find that

$$z = \mathbf{E}\{z \mid x_0\}$$

with probability 1, as was to be proved.

Example 2 *Processes with mutually independent random variables.* Suppose that $\{x_n, n \geq 0\}$ is a stochastic process whose random variables are mutually independent. It is then strictly stationary if and only if the x_n's have a common distribution. According to the following theorem such a stationary process is always metrically transitive.

THEOREM 1.2 *If $x_0, x_1, \cdots$ are mutually independent random variables with a common distribution, the strictly stationary process $\{x_n, n \geq 0\}$ is metrically transitive.*

Since an invariant random variable is measurable on the sample space of $x_n, x_{n+1}, \cdots$ for every n, the only invariant random variables are the constants (with probability 1), by the zero-one law (III, Theorem 1.1), as was to be proved. This theorem also follows readily from Theorem 1.1.

Example 3 *Moving averages* Consider the x_n process defined by

$$x_n = \sum_{m=-\infty}^{\infty} c_m y_{m+n},$$

where $\sum_{-\infty}^{\infty} |c_n|^2 < \infty$, and the y_n's are mutually independent random variables with a common distribution function having zero mean and a finite variance. (The series then converges in the mean and also with probability 1, by III, Theorem 2.3.) The y_n process is metrically transitive, according to Theorem 1.2. Let $\mathbf{T}$ be the shift corresponding to this process. Then the x_n process is also metrically transitive, since it is generated by the metrically transitive measure-preserving set transformation $\mathbf{T}$ applied to x_0.

(b) *Wide sense stationary processes* Wide sense stationary processes were defined in II §8. The concepts connected with strictly stationary processes all have their wide sense counterparts. We suppose, as usual, that there is a probability measure defined on a Borel field of sets of some space Ω. In the following we shall use some of the elementary Hilbert space geometry developed in IV §2. Let $\mathfrak{M}$ be a closed linear manifold of random variables whose squares are integrable. A transformation $\mathbf{U}$ operating on the elements of $\mathfrak{M}$, taking them into elements of $\mathfrak{M}$, is called isometric if the following conditions are satisfied.

IS_1 $\mathbf{U}$ *is single-valued, modulo the random variables which vanish with probability* 1: if x_1 is an image of x under $\mathbf{U}$, the class of all images of x is the class of random variables equal to x_1 with probability 1. In the following the notation $\mathbf{U}x$ will always mean a particular image of x under $\mathbf{U}$.

IS_2 $\mathbf{U}$ *is linear:* if a and b are constants, and if x, y are random variables in $\mathfrak{M}$,

$$\mathbf{U}(ax + by) = a\mathbf{U}x + b\mathbf{U}y$$

with probability 1.

IS_3 $\mathbf{U}$ *preserves the root mean square norm:*

$$\mathbf{E}\{|\mathbf{U}x|^2\} = \mathbf{E}\{|x|^2\}, \qquad x \in \mathfrak{M}.$$

It is easily verified that IS_3 is true if and only if

$$\mathbf{E}\{\mathbf{U}x\overline{\mathbf{U}y}\} = \mathbf{E}\{x\bar{y}\}, \qquad x, y \in \mathfrak{M}.$$

If every element of $\mathfrak{M}$ is the image of some x under $\mathbf{U}$, then $\mathbf{U}$ must be 1–1 modulo the random variables which vanish almost everywhere, so that $\mathbf{U}$ has an isometric inverse. In this case $\mathbf{U}$ is called *unitary*.

If $\mathbf{T}$ is a measure-preserving set transformation with domain space a Borel field $\mathscr{F}$, there is a corresponding isometric transformation of random variables. Let $\mathfrak{M}$ be the closed linear manifold of random variables measurable with respect to $\mathscr{F}$, with integrable squares. Then the random variable transformation $\mathbf{T}$ with domain limited to $\mathfrak{M}$ is isometric, and is unitary if $\mathbf{T}$ has an inverse.

If $\mathbf{U}$ is an isometric transformation, and if x is a random variable in its domain of definition, then the stochastic process

$$\{x_n, n \geq 0\}, \qquad x_n = \mathbf{T}^n x,$$

is stationary in the wide sense. If $\mathbf{U}$ is unitary, the corresponding statement with parameter range $-\infty < n < \infty$ is true.

Conversely let $\{x_n, n \geq 0\}$ be a stochastic process which is stationary in the wide sense. There is then an isometric transformation which generates this process in the sense of the preceding paragraph. To see this define $\mathfrak{M}$ as the closed linear manifold of random variables generated by the x_n's. Then it is clear that there is one and only one isometric transformation $\mathbf{U}$ with the property that $\mathbf{U}x_n = x_{n+1}$, $n \geq 0$, with probability 1. This transformation is called the *shift*. Similarly, if the parameter range of the process is $-\infty < n < \infty$, we obtain a uniquely determined unitary transformation taking x_n into x_{n+1}, $-\infty < n < \infty$, also called the shift (relative to the process involved).

We shall now justify considering isometric and unitary transformations the wide sense versions of measure-preserving set transformations, and measure-preserving set transformations with inverses, respectively, and also show how isometric transformations (rather than unitary ones) can be dispensed with in this study.

Let $\{x_n, n \geq 0\}$ or $\{x_n, -\infty < n < \infty\}$ be a stochastic process which is stationary in the wide sense, and define

$$r(m, n) = \mathbf{E}\{x_m \bar{x}_n\}$$

in the second case. In the first case define

$$r(m, n) = \mathbf{E}\{x_{m+k} \bar{x}_{n+k}\},$$

choosing k so large that $m + k \geq 0$ and $n + k \geq 0$. The precise value of k used is irrelevant because of the hypothesis of stationarity. Then obviously

$$r(m, n) = \bar{r}(n, m),$$

and, if $t_1, \cdots, t_N$ are any integers, the matrix $[r(t_i, t_j)]$ is non-negative definite because

$$\sum_{i,j=1}^{N} r(t_i, t_j) a_i \bar{a}_j = \mathbf{E}\{|\sum_{1}^{N} a_j x_{t_j+k}|^2\} \geq 0,$$

where k is chosen so large that the random variables on the right are defined. According to II, Theorem 3.1 or 3.2, there is then a Gaussian process $\{\hat{x}_n, -\infty < n < \infty\{$ for which

$$\mathbf{E}\{\hat{x}_n\} = 0$$

$$\mathbf{E}\{\hat{x}_m \bar{\hat{x}}_n\} = r(m, n)$$

$$\mathbf{E}\{\hat{x}_m \hat{x}_n\} = 0.$$

This Gaussian process was constructed explicitly in the proof of II, Theorem 3.1, with $\hat{\omega}$ space taken as infinite dimensional coordinate space, $\hat{x}_n$ the nth coordinate variable, and with the probability measure the infinite dimensional measure set up by Kolmogorov.

As far as any theorem only involving the covariance function $r(\cdot, \cdot)$ is concerned, the $\hat{x}_n$ process can be used instead of the x_n process. For example, the averages

$$\frac{x_0 + \cdots + x_n}{n+1}, \qquad n \geq 0,$$

converge in the mean if and only if the corresponding $\hat{x}_n$ averages do. However, the $\hat{x}_n$ process is simpler than the x_n process in many ways. In fact, the $\hat{x}_n$ process is stationary in both the strict and wide senses, the shift (considered a measure-preserving set transformation) is induced by a measure-preserving point transformation, the obvious shift of coordinates, and the isometric shift is unitary. Thus in proving many theorems on wide sense stationary processes we can without loss of generality assume that the isometric shift is unitary, specifically if we are only interested in wide sense theorems (see II §3).

If **U** is an isometric transformation, and if x is a random variable on its manifold of definition, x is called *invariant* if $\mathbf{U}x = x$ with probability 1. For example, the random variables vanishing with probability 1 are invariant. If these are the only invariant random variables, the isometric transformation will be called *metrically transitive in the wide sense.* This terminology will not be used outside this section however, since it is unorthodox, and is introduced here only to clarify the relation between strict and wide sense stationarity.

The random variables invariant with respect to the isometric shift of a wide sense stationary process will be called invariant (wide sense) random variables of the process. The process will be called (in this section only) metrically transitive in the wide sense if its isometric shift is metrically transitive in the wide sense.

If $\{x_n, -\infty < n < \infty\}$ is a process which is stationary in the wide sense, the process $\{x_{-n}, -\infty < n < \infty\}$ is also stationary in the wide sense, and has the same invariant random variables as the given process, since the inverse of a unitary transformation has the same invariant random variables as the given transformation. The process $\{x_n, n \geq 0\}$ is also stationary in the wide sense, and has the same invariant random variables as the given process. This assertion is proved by using the same basic ideas as in the strict sense version given in the first half of this section, and the proof will therefore be omitted. It follows that the three processes involved here are either simultaneously metrically transitive in the wide sense or none is.

The following is the wide sense version of Example 2.

Example 4 *Processes with mutually orthogonal random variables.* Suppose that $\{x_n, n \geq 0\}$ is a stochastic process whose random variables are mutually orthogonal. It is then stationary in the wide sense if and only if $\mathbf{E}\{|x_n|^2\}$ is independent of n. According to the following theorem such a stationary (wide sense) process is always metrically transitive in the wide sense.

THEOREM 1.3 *If $x_0, x_1, \cdots$ are mutually orthogonal random variables, with $\mathbf{E}\{|x_n|^2\}$ independent of n, the stationary (wide sense) process $\{x_n, n \geq 0\}$ is metrically transitive in the wide sense.*

According to the hypothesis the random variables of the process satisfy the conditions

$$\begin{aligned} \mathbf{E}\{x_m\bar{x}_n\} &= 0 \qquad m \neq n \\ &= \sigma^2 \qquad m = n \end{aligned}$$

for some $\sigma \geq 0$. Let $\mathfrak{M}$ be the closed linear manifold of random variables generated by the x_n's. If $\sigma = 0$, every random variable in $\mathfrak{M}$ vanishes with probability 1, so the process is metrically transitive. If $\sigma > 0$, any x in $\mathfrak{M}$ can be written as the sum of its Fourier series in the x_n's,

$$x = \sum_0^\infty a_j x_j \qquad a_j = \frac{1}{\sigma^2}\mathbf{E}\{x\bar{x}_j\},$$

where the series converges in the mean. Then, if x is invariant,

$$\sum_0^\infty a_j x_j = \sum_0^\infty a_j x_{j+1},$$

with probability 1. Hence, equating coefficients,

$$0 = a_0 = a_1 = \cdots,$$

so that $x = 0$ with probability 1, as was to be proved.

2. The strong law of large numbers for strictly stationary stochastic processes

The fundamental theorem of measure-preserving transformations is the *ergodic theorem.* This is usually stated as follows: *Let* $\mathbf{S}$ *be a* 1–1 *measure-preserving point transformation, and let x be a measurable and integrable function on the space involved. Then*

$$\lim_{n\to\infty} \frac{x(\omega) + x(\mathbf{S}\omega) + \cdots + x(\mathbf{S}^n\omega)}{n+1}$$

exists and is finite for almost all ω. Since $\mathbf{S}^{-1}$ is also a measure-preserving point transformation, the above result holds also with $\mathbf{S}^{-1}$ instead of $\mathbf{S}$

(and we shall even show below that the limit is the same, neglecting values on an ω set of measure 0). Since measure-preserving set transformations are slightly more general than measure-preserving point transformations, the following version of the ergodic theorem is slightly more general than that stated above: *Let* $\mathbf{T}$ *be a measure-preserving set transformation, and let* x *be a measurable and integrable function on the space involved. Then*

$$\lim_{n\to\infty} \frac{x + \mathbf{T}x + \cdots + \mathbf{T}^n x}{n+1}$$

exists and is finite for almost all ω. The transformation $\mathbf{T}$ of this version of the ergodic theorem corresponds to $\mathbf{S}^{-1}$ in the previous version. This version of the ergodic theorem is the useful one for the purposes of probability theory, in which the theorem is called the *strong law of large numbers for strictly stationary processes.* Although the probability language makes it possible to give an intuitive description of the limit, the discussion of §1 shows that otherwise the following statement of the theorem in probability language differs only verbally from the measure theoretic statement just given. The $\mathbf{T}$ of the measure theoretic statement becomes the shift transformation of the stochastic process.

THEOREM 2.1 *Let* $\{x_n, n \geq 0\}$ *be a strictly stationary stochastic process, with* $\mathbf{E}\{|x_0|\} < \infty$, *and let* $\mathscr{I}$ *be the Borel field of invariant* ω *sets. Then*

$$\lim_{n\to\infty} \frac{x_0 + \cdots + x_n}{n+1} = \mathbf{E}\{x_0 \mid \mathscr{I}\} \tag{2.1}$$

with probability 1. *In particular, if the process is metrically transitive, the right-hand side of* (2.1) *can be replaced by* $\mathbf{E}\{x_0\}$.

This theorem is sometimes stated using superficially different averages. It is clear, for example, that the content of the theorem is not changed if the average on the left in (2.1) is replaced by

$$\frac{x_k + \cdots + x_{k+n}}{n+1}$$

(fixed k) and that the limit is unaltered. If the parameter range of the process is the set of all integers, we can change the time direction, that is, replace x_j by x_{-j}, and find that the limiting average of $x_{-n}, \cdots, x_0$ when $n \to \infty$ exists with probability 1. Since the invariant sets of the inverse of the shift are the same as those of the shift, the limit is the same as that in (2.1). It then follows that the average

$$\frac{x_{-n} + \cdots + x_n}{2n+1}$$

has the same limit (with probability 1), when $n \to \infty$ However,

$$\lim_{n-m\to\infty} \frac{x_m + \cdots + x_n}{n - m + 1}$$

does not exist with probability 1, in general.

The most important special case of the theorem is that of mutually independent x_n's. In this case the stationarity hypothesis becomes the hypothesis that the x_n's have a common distribution function. According to Theorem 1.2 every process of this type is metrically transitive. Hence the limit in (2.1) is $\mathbf{E}\{x_0\}$ in this case. Two proofs of the strong law of large numbers in this special case have already been given, in III (Theorem 5.1) and VII (§6).

Before proving Theorem 2.1 we prove three lemmas.

LEMMA 2.1 *Let $C_1, \cdots, C_n$ be any real numbers, and let E be the class of integers $m < n$ satisfying*

$$C_m < \operatorname{Max}_{j>m} C_j. \tag{2.2}$$

The class E consists of groups of consecutive integers; if α, β are the first and last integers in such a group, then

$$C_j < C_{\beta+1}, \qquad \alpha \leq j \leq \beta. \tag{2.3}$$

In fact, since $\beta + 1 \notin E$, $C_{\beta+1} \geq \operatorname{Max}_{j>\beta+1} C_j$ (or $\beta + 1 = n$), so that, since $\beta \in E$,

$$C_\beta < \operatorname{Max}_{j>\beta} C_j = C_{\beta+1}.$$

If $\alpha < \beta$, then $\beta - 1 \in E$, and it follows that

$$C_{\beta-1} < \operatorname{Max}_{j>\beta-1} C_j = C_{\beta+1}, \tag{2.4}$$

and so on.

LEMMA 2.2 *Let $\{x_n, n \geq 0\}$ be a real strictly stationary stochastic process, with $\mathbf{E}\{|x_0|\} < \infty$. Then, if β is any constant, if M is an invariant ω set of the process, and if $S_n = x_0 + \cdots + x_{n-1}$,*

$$\int_{M\left\{\text{L.U.B.}_{n\geq 1} \frac{S_n(\omega)}{n} > \beta\right\}} x_0 \, d\mathbf{P} \geq \beta \mathbf{P}\left\{\left[\text{L.U.B.}_{n\geq 1} \frac{S_n(\omega)}{n} > \beta\right] M\right\}. \tag{2.5}$$

Replacing the sequence $\{x_n\}$ by $\{x_n - \beta\}$, so that S_n/n is replaced by $(S_n/n) - \beta$, it is clear that it is sufficient to prove the theorem for $\beta = 0$. Define Λ and Λ_j by

$$\Lambda = \{\text{L.U.B.}_{n\geq 1} S_n(\omega) > 0\}, \qquad \Lambda_j = \{\text{L.U.B.}_{1\leq n\leq j} S_n(\omega) > 0\},$$

so that Λ_j increases to Λ when $j \to \infty$. Apply Lemma 2.1 to a sample sequence of $S_1, \cdots, S_m$, and let N_j be the ω set where j is in the class of integers defined by Lemma 2.1,

$$N_j = \{ \operatorname*{Max}_{j<k\leq m-1} [x_j(\omega) + \cdots + x_k(\omega)] > 0\} = T^j\Lambda_{m-j}.$$

Then according to the lemma,

$$\sum_{\omega \in N_j} x_j(\omega) \geq 0,$$

where the notation signifies that the sum is to include $x_j(\omega)$ if $\omega \in N_j$. It follows that

$$\sum_{j=0}^{m-1} \int_{MN_j} x_j \, d\mathbf{P} \geq 0,$$

and therefore, in view of the fact that $\mathbf{T}$ is probability-preserving, we conclude that

$$(2.6) \qquad 0 \leq \sum_{j=0}^{m-1} \int_{MT^j\Lambda_{m-j}} x_j \, d\mathbf{P} = \sum_{j=0}^{m-1} \int_{M\Lambda_{m-j}} x_0 \, d\mathbf{P} = \sum_{j=1}^{m} \int_{M\Lambda_j} x_0 \, d\mathbf{P}.$$

But

$$(2.7) \qquad \lim_{j\to\infty} \int_{M\Lambda_j} x_0 \, d\mathbf{P} = \int_{M\Lambda} x_0 \, d\mathbf{P},$$

so that, dividing (2.6) through by m, this inequality yields (2.5) (with $\beta = 0$) when $m \to \infty$.

LEMMA 2.3 *Under the hypotheses of Theorem* 2.1, *if the x_n process is real, the random variables*

$$\liminf_{n\to\infty} \frac{S_n}{n} = \tilde{x}_1, \qquad \limsup_{n\to\infty} \frac{S_n}{n} = \tilde{x}_2$$

are invariant.

We give the proof only for $\tilde{x}_1$. We must prove that

$$\liminf_{n\to\infty} \frac{x_1 + \cdots + x_n}{n} = \tilde{x}_1$$

with probability 1. Now

$$\liminf_{n\to\infty} \frac{x_1 + \cdots + x_n}{n} = \liminf_{n\to\infty} \frac{x_0 + \cdots + x_n}{n}$$

$$= \liminf_{n\to\infty} \frac{x_0 + \cdots + x_n}{n+1} = \tilde{x}_1,$$

as was to be proved. Note that as yet we have not proved that $\tilde{x}_1$ and $\tilde{x}_2$ are finite-valued.

We now proceed to the proof of Theorem 2.1. It is sufficient to prove it in the real case. If $\alpha < \beta$, and if $M_{\alpha,\beta}$ is the invariant ω set $\{\tilde{x}_1(\omega) < \alpha < \beta < \tilde{x}_2(\omega)\}$, then evidently

$$M_{\alpha,\beta} = M_{\alpha,\beta}\left\{\underset{n\geq 1}{\text{L.U.B.}}\, \frac{S_n(\omega)}{n} > \beta\right\},$$

so that, from Lemma 2.2,

$$\int_{M_{\alpha,\beta}} x_0\, d\mathbf{P} = \int_{M_{\alpha,\beta}\left\{\underset{n\geq 1}{\text{L.U.B.}}\, \frac{S_n(\omega)}{n} > \beta\right\}} x_0\, d\mathbf{P} \geq \beta\mathbf{P}\{M_{\alpha,\beta}\}. \tag{2.8}$$

Applying this result to $\{-x_n\}$, replacing α, β by $-\beta, -\alpha$,

$$\int_{M_{\alpha,\beta}} x_0\, d\mathbf{P} \leq \alpha\mathbf{P}\{M_{\alpha,\beta}\}. \tag{2.9}$$

Combining (2.8) and (2.9), we find that $\mathbf{P}\{M_{\alpha,\beta}\} = 0$. Since

$$\begin{aligned} \mathbf{P}\{\tilde{x}_1(\omega) < \tilde{x}_2(\omega)\} &= \mathbf{P}\{\bigcup_{\alpha<\beta} M_{\alpha,\beta}\}, \qquad (\alpha, \beta \text{ rational}), \\ &\leq \sum_{\alpha,\beta} \mathbf{P}\{M_{\alpha,\beta}\} = 0, \end{aligned} \tag{2.10}$$

it follows that $\tilde{x}_1 = \tilde{x}_2$ with probability 1. There is thus a finite or infinite limit x. An application of Fatou's lemma to the averages

$$\frac{|x_0| + \cdots + |x_n|}{n+1}$$

shows that x is finite-valued with probability 1, and integrable, but this fact will also be a consequence of the following deeper argument. To identify the limit x with $\mathbf{E}\{x_0 \mid \mathscr{I}\}$ we must prove that x is finite-valued with probability 1, is integrable, and satisfies

$$\int_\Lambda x_0\, d\mathbf{P} = \int_\Lambda x\, d\mathbf{P} \tag{2.11}$$

for every invariant Λ. To prove these assertions it is sufficient to show that the averages

$$\frac{x_0 + \cdots + x_n}{n+1}, \qquad n \geq 0,$$

are uniformly integrable, because then their limit x will be finite-valued with probability 1, integrable, and, because uniform integrability implies that integration to the limit is legitimate,

$$\begin{aligned} \int_\Lambda x_0\, d\mathbf{P} = \frac{1}{n+1}\sum_0^n \int_{T^j\Lambda} x_j\, d\mathbf{P} &= \int_\Lambda \frac{1}{n+1}\sum_0^n x_j\, d\mathbf{P} \\ &\to \int_\Lambda x\, d\mathbf{P} \qquad (n \to \infty), \end{aligned}$$

if Λ is invariant. Thus (2.11) is true. We shall show not only that the ergodic theorem averages are uniformly integrable but that their δth powers are uniformly integrable, if $\delta \geq 1$ and if $\mathbf{E}\{|x_0|^\delta\} < \infty$. To show this let ε_1 be a positive number, and choose ε_2 so small that

$$\int_M |x_j|^\delta \, d\mathbf{P} < \varepsilon_1 \quad \text{if} \quad \mathbf{P}\{M\} < \varepsilon_2, \quad j \geqslant 0.$$

Since the x_j's have a common distribution, there is a positive ε_2 with this property. Hence

$$\text{(2.12)} \qquad \int_M \left| \frac{x_0 + \cdots + x_n}{n+1} \right|^\delta d\mathbf{P} \leq \int_M \frac{|x_0|^\delta + \cdots + |x_n|^\delta}{n+1} \, d\mathbf{P} < \varepsilon_1 \quad \text{if} \quad \mathbf{P}\{M\} < \varepsilon_2,$$

so that the integrands yield uniformly small integrals over sets of small probability. This property implies uniform integrability if it is known that the integrands yield uniformly bounded integrals over the whole space. The latter property follows from the fact that, if $M = \Omega$ in (2.12), the integral on the right reduces to $\mathbf{E}\{|x_0|^\delta\}$. If $\delta > 1$, the uniform integrability just proved also follows from the following inequality, obtained by combining Lemma 2.2, with x_n replaced by $|x_n|$, and VII, Theorem 3.4′,

$$\mathbf{E}\{\underset{n \geq 0}{\text{L.U.B.}} \, [\sum_0^n |x_j|/(n+1)]^\delta\} \leq \frac{e}{e-1} + \frac{e}{e-1} \mathbf{E}\{|x_0| \log^+ |x_0|\}, \qquad \delta = 1$$

$$\leq \left(\frac{\delta}{\delta - 1}\right)^\delta \mathbf{E}\{|x_0|^\delta\}, \qquad \delta > 1.$$

Finally we remark that, if $\mathbf{E}\{|x_0|^\delta\} < \infty$ and if $\delta \geq 1$, then

$$\text{(2.13)} \qquad \lim_{n \to \infty} \mathbf{E}\left\{ \left| \frac{x_0 + \cdots + x_n}{n+1} - \mathbf{E}\{x_0 \mid \mathscr{I}\} \right|^\delta \right\} = 0,$$

because going to the limit under the expectation symbol is now justified by the uniform integrability of the averages. Thus there is mean convergence of order δ.

Theorem 2.1 has the following corollary, which is important in the harmonic analysis of the sample sequences of strictly stationary processes.

COROLLARY *If μ is real and $-\frac{1}{2} < \mu \leq \frac{1}{2}$,*

$$\text{(2.14)} \qquad \lim_{n \to \infty} \frac{1}{n+1} \sum_{j=0}^n x_j e^{-2\pi i j \mu} = \tilde{x}(\mu)$$

exists with probability 1. *The random variable* $\tilde{x}(\mu)$ *has finite expectation,*

$$\mathbf{E}\{\tilde{x}(0)\} = \mathbf{E}\{x_0\}$$

$$\mathbf{E}\{\tilde{x}(\mu)\} = 0, \qquad \mu \neq 0$$

and is transformed by the shift into $e^{2\pi i\mu}\tilde{x}(\mu)$. *Unless* μ *is in a certain (at most denumerable) set,* $\mathbf{P}\{\tilde{x}(\mu, \omega) = 0\} = 1$.

If $\mathbf{E}\{|x_0|^2\} < \infty$, *then* $\mathbf{E}\{|\tilde{x}(\mu)|^2\} < \infty$ *also,*

$$\text{l.i.m.}_{n\to\infty} \frac{1}{n+1}\sum_0^n x_j e^{-2\pi ij\mu} = \tilde{x}(\mu) \tag{2.15}$$

and

$$\mathbf{E}\{\tilde{x}(\mu_1)\overline{\tilde{x}(\mu_2)}\} = 0, \qquad \mu_1 \neq \mu_2.$$

Note that, if $\mu = 0$, $\tilde{x}(\mu)$ becomes the limit in Theorem 2.1. As in the theorem, the average $\frac{1}{n+1}\sum_0^n$ can be replaced by $\frac{1}{n+1}\sum_k^{k+n}$, and so on. To prove the corollary, let φ be a random variable which is uniformly distributed in the interval (0, 1), and which is independent of the x_n process. More precisely, adjoin a space to ω space, as described in II §2, to obtain a random variable φ, uniformly distributed in the interval (0, 1) (where now all random variables are functions on the enlarged space) with the property that, if A is a Borel set and if Λ is a measurable ω set, then

$$\mathbf{P}\{\varphi \in A, \omega \in \Lambda\} = \mathbf{P}\{\varphi \in A\}\mathbf{P}\{\omega \in \Lambda\}.$$

For each μ the stochastic process

$$\{x_n e^{-2\pi i(\phi+n\mu)}, n \geq 0\}$$

is strictly stationary, so that, according to Theorem 2.1,

$$\lim_{n\to\infty} \frac{1}{n+1}\sum_{j=0}^n x_j e^{-2\pi i(\phi+j\mu)} = e^{-2\pi i\phi}\lim_{n\to\infty}\frac{1}{n+1}\sum_{j=0}^n x_j e^{-2\pi ij\mu}$$

exists with probability 1. Thus $\tilde{x}(\mu)$ is defined, and it is obvious that under the shift $\tilde{x}(\mu)$ becomes $\tilde{x}(\mu)e^{2\pi i\mu}$. We have seen in the proof of the theorem that, if $\mathbf{E}\{|x_0|^\delta\} < \infty$ for some $\delta \geq 1$, then the random variables

$$\left[\frac{1}{n+1}\sum_0^n |x_j|\right]^\delta, \qquad n \geq 0,$$

are uniformly integrable (applying the proof to the $|x_j|$'s rather than to the x_j's). Then the random variables

$$\left|\frac{1}{n+1}\sum_0^n x_j e^{-2\pi ij\mu}\right|^\delta, \qquad n \geq 0,$$

are uniformly integrable, and it follows that

$$\lim_{n\to\infty} \mathbf{E}\left\{\left|\frac{1}{n+1}\sum_0^n x_j e^{-2\pi i j\mu} - \tilde{x}(\mu)\right|^\delta\right\} = 0,$$

since the uniform integrability justifies going to the limit under the expectation symbol. Thus in this case there is convergence in the mean with index δ. In particular when $\delta = 1$ we find that the averages in (2.14) are uniformly integrable, so that

$$\begin{aligned}\mathbf{E}\{\tilde{x}(\mu)\} &= \lim_{n\to\infty} \mathbf{E}\left\{\frac{1}{n+1}\sum_{j=0}^n x_j e^{-2\pi i j\mu}\right\}\\ &= \mathbf{E}\{x_0\} \lim_{n\to\infty} \frac{1}{n+1}\sum_{j=0}^n e^{-2\pi i j\mu}\\ &= \mathbf{E}\{x_0\}, \qquad \mu = 0\\ &= 0, \qquad \mu \neq 0.\end{aligned}$$

[The evaluation for $\mu = 0$ is of course also implied by the evaluation of $\tilde{x}(0)$ given in Theorem 2.1.]

If $\mathbf{E}\{|x_0|^2\} < \infty$, the x_n process is stationary in the wide sense, and (2.15) and the mutual orthogonality of the $\tilde{x}(\mu)$'s will be proved in §6 by another method. The proof of (2.15) has already been given (in fact, we have proved a generalized version with any exponent $\delta \geq 1$). Since the shift is probability-preserving, it preserves expectations, and we conclude that, if $\mathbf{E}\{|x_0|^2\} < \infty$,

$$\begin{aligned}\mathbf{E}\{\tilde{x}(\mu_1)\overline{\tilde{x}(\mu_2)}\} &= \mathbf{E}\{\tilde{x}(\mu_1)e^{2\pi i\mu_1}\overline{\tilde{x}(\mu_2)e^{2\pi i\mu_2}}\}\\ &= e^{2\pi i(\mu_1-\mu_2)}\,\mathbf{E}\{\tilde{x}(\mu_1)\overline{\tilde{x}(\mu_2)}\} = 0, \qquad \mu_1 \neq \mu_2,\end{aligned}$$

so that the $\tilde{x}(\mu)$'s are mutually orthogonal. Let $\mu_1, \mu_2, \cdots$ be values of μ for which $\mathbf{P}\{\tilde{x}(\mu_j, \omega) = 0\} < 1$. Then the sequence

$$\left\{\tilde{x}(\mu_j)\mathbf{E}^{-1/2}\{|\tilde{x}(\mu_j)|^2\}\right\}, \qquad j = 1, 2, \cdots$$

is an orthonormal sequence of random variables; x_0 can be expanded in a Fourier series relative to this sequence,

$$x_0 \sim \sum_{k=1}^\infty a_k \tilde{x}(\mu_k)/\mathbf{E}^{1/2}\{|\tilde{x}(\mu_k)|^2\},$$

where

$$a_k = \mathbf{E}\{x_0\overline{\tilde{x}(\mu_k)}\}/\mathbf{E}^{1/2}\{|\tilde{x}(\mu_k)|^2\}$$

and (Bessel's inequality)

$$\mathbf{E}\{|x_0|^2\} \geq \sum_{k=1}^\infty |a_k|^2. \tag{2.16}$$

Now, applying the shift,

$$\begin{aligned}\mathbf{E}\{x_0\overline{\tilde{x}(\mu_k)}\} &= \mathbf{E}\{x_1\overline{\tilde{x}(\mu_k)e^{2\pi i\mu_k}}\} = \mathbf{E}\{x_2\overline{\tilde{x}(\mu_k)e^{4\pi i\mu_k}}\} = \cdots \\ &= \mathbf{E}\left\{\frac{1}{n+1}\sum_{j=0}^{n} x_j e^{-2\pi ij\mu_k}\,\overline{\tilde{x}(\mu_k)}\right\} \\ &\to \mathbf{E}\{|\tilde{x}(\mu_k)|^2\} \qquad (n\to\infty).\end{aligned}$$

Hence (2.16) becomes

$$\mathbf{E}\{|x_0|^2\} \geq \sum_{k=1}^{\infty} \mathbf{E}\{|\tilde{x}(\mu_k)|^2\}.$$

Since the μ_k's are arbitrary, there is at most an enumerable μ set for which $\mathbf{E}\{|\tilde{x}(\mu)|^2\} > 0$, that is, for which $\mathbf{P}\{\tilde{x}(\mu, \omega) = 0\} < 1$. This would finish the proof of the corollary, except that the existence of this μ set is stated in the corollary without the hypothesis that $\mathbf{E}\{|x_0|^2\} < \infty$. To remove this restriction we note that, if $x_j^{(N)}(\omega) = x_j(\omega)$ for $|x_j(\omega)| \leq N$ and is otherwise 0, the $x_j^{(N)}$ process (fixed N) is also strictly stationary, with $\mathbf{E}\{|x_j^{(N)}|^2\} < \infty$. Defining $\tilde{x}^{(N)}(\mu)$ using the $x_j^{(N)}$'s instead of the x_j's we find that, except for an enumerable set $F^{(N)}$, $\mathbf{P}\{\tilde{x}^{(N)}(\mu, \omega) = 0\} = 1$; the previous result is applicable because $x_0^{(N)}$ is bounded. Moreover

$$\begin{aligned}&\mathbf{E}\left\{\left|\frac{1}{n+1}\sum_{j=0}^{n} x_j e^{-2\pi ij\mu} - \frac{1}{n+1} x_j^{(N)} e^{-2\pi ij\mu}\right|\right\} \\ &\qquad \leq \mathbf{E}\left\{\frac{1}{n+1}\sum_{j=0}^{n} |x_j - x_j^{(N)}|\right\} = \mathbf{E}\{|x_0 - x_0^{(N)}|\}.\end{aligned}$$

Hence (Fatou's lemma, or use the uniform integrability) when $n \to \infty$ we have

$$\mathbf{E}\{|\tilde{x}(\mu) - \tilde{x}^{(N)}(\mu)|\} \leq \mathbf{E}\{|x_0 - x_0^{(N)}|\}$$

so that, if $F = \bigcup_{N=1}^{\infty} F^{(N)}$,

$$\mathbf{E}\{|\tilde{x}(\mu)|\} \leq \mathbf{E}\{|x_0 - x_0^{(N)}|\}, \qquad \mu \notin F, \quad N = 1, 2, \cdots.$$

Since the right side goes to 0 when $N \to \infty$, the left side must vanish, and this finishes the proof.

Let $\mathbf{T}$ be the shift corresponding to the x_n process, and let z be a random variable on the sample space of the x_n's, with $\mathbf{E}\{|z|\} < \infty$. Then, according to the corollary,

$$\lim_{n\to\infty} \frac{1}{n+1}\sum_{j=0}^{n} \mathbf{T}^j z e^{-2\pi ij\mu} = \tilde{z}(\mu) \tag{2.14'}$$

exists with probability 1, for all μ. A slight extension of the argument of the preceding paragraph can be used to show that there is an at most enumerable μ set G, independent of the choice of z, such that

$$\mathbf{P}\{\tilde{z}(\mu, \omega) = 0\} = 1, \qquad \mu \notin G.$$

A full understanding of these results cannot be obtained without an understanding of the spectral analysis of unitary operators, for which the reader is referred to the literature of this subject.

3. The covariance function of a stationary stochastic process; examples

In this and the following sections we shall consider stationary (wide sense) processes $\{x_n, -\infty < n < \infty\}$, with particular reference to their harmonic analysis. The translation of the results to the case of the parameter range $n \geq 0$ will usually be obvious.

In any harmonic analysis it is easier to deal with complex valued functions, and series

$$\Sigma a_n e^{2\pi i n \lambda}$$

rather than with real functions, and series

$$\Sigma[a_n \cos 2\pi n\lambda + b_n \sin 2\pi n\lambda].$$

For this reason, complex processes will be considered first, and the results for real processes obtained by specialization.

The covariance function defined by

$$R(n) = \mathbf{E}\{x_{m+n}\bar{x}_m\}$$

plays a fundamental role in the study of processes stationary in the wide sense. The following theorem describes the class of covariance functions.

THEOREM 3.1 *The covariance function is positive definite, that is,*

$$R(-n) = \overline{R(n)}, \tag{3.1}$$
$$\sum_{m,n=1}^{N} R(m-n)\alpha_m\bar{\alpha}_n \geq 0, \qquad N = 1, 2, \cdots,$$

for every set of complex numbers $\alpha_1, \cdots, \alpha_N$. *Conversely any function* R *satisfying* (3.1) *is the covariance function of a stationary* (*wide sense*) *process, which can be taken as real if the covariance function is real.*

In II §3 necessary and sufficient conditions were found that a function be the covariance function of a process, $r(s, t) = \mathbf{E}\{x_s\bar{x}_t\}$. In the present case these conditions reduce to (3.1). The following theorem describes the positive definite functions as Fourier transforms.

THEOREM 3.2 *A function R is positive definite if and only if it can be expressed in the form*

$$R(n) = \int_{-1/2}^{1/2} e^{2\pi in\lambda}\, dF(\lambda) \tag{3.2}$$

where F (defined for $|\lambda| \leq \frac{1}{2}$) is monotone non-decreasing. The function F is uniquely determined, if suitably normalized, by the equations

$$\begin{aligned} \frac{F(\lambda_2+)+F(\lambda_2-)}{2} - \frac{F(\lambda_1+)+F(\lambda_1-)}{2} &= R(0)(\lambda_2-\lambda_1) + \lim_{N\to\infty} \sum_{\substack{-N \\ (n\neq 0)}}^{N} R(n) \frac{e^{-2\pi in\lambda_2} - e^{-2\pi in\lambda_1}}{-2\pi in} \\ &\qquad -\tfrac{1}{2} < \lambda_1 < \lambda_2 < \tfrac{1}{2}, \end{aligned} \tag{3.3}$$

$$F(\tfrac{1}{2}) - F(-\tfrac{1}{2}) = R(0).$$

If R is real (3.2) *can be replaced by*

$$R(n) = \int_0^{1/2} \cos 2\pi n\lambda\, dG(\lambda), \tag{3.2'}$$

where G (defined for $0 \leq \lambda \leq \frac{1}{2}$) is monotone non-decreasing and is uniquely determined, if suitably normalized, by the equations

$$\frac{G(\lambda+)+G(\lambda-)}{2} - G(0) = 2R(0)\lambda + \frac{2}{\pi}\sum_1^\infty R(n)\frac{\sin 2\pi n\lambda}{n} \qquad 0 < \lambda < \tfrac{1}{2}, \tag{3.3'}$$

$$G(\tfrac{1}{2}) - G(0) = R(0).$$

If $R(n)$ is defined by (3.2), $R(-n) = \overline{R(n)}$ and

$$\begin{aligned} \sum_{m,n=1}^{N} R(m-n)\alpha_m\bar{\alpha}_n &= \int_{-1/2}^{1/2} \sum_{m,n=1}^{N} \alpha_m\bar{\alpha}_n e^{2\pi i(m-n)\lambda}\, dF(\lambda) \\ &= \int_{-1/2}^{1/2} \Big|\sum_1^N \alpha_n e^{2\pi in\lambda}\Big|^2\, dF(\lambda) \geq 0. \end{aligned}$$

If b_n is the coefficient of $R(n)$ in (3.3),

$$b_n = \int_{\lambda_1}^{\lambda_2} e^{-2\pi in\lambda}\, d\lambda = \int_{-1/2}^{1/2} e^{-2\pi in\lambda}\, b(\lambda)\, d\lambda,$$

where

$$\begin{aligned} b(\lambda) &= 1, & \lambda_1 < \lambda < \lambda_2, \\ &= \tfrac{1}{2}, & \lambda = \lambda_1, \lambda_2, \\ &= 0, & \lambda < \lambda_1 \text{ or } \lambda > \lambda_2, \end{aligned}$$

except that, if $\lambda_1 = -\frac{1}{2}$, $\lambda_2 < \frac{1}{2}$, we take $b(\frac{1}{2}) = \frac{1}{2}$; if $\lambda_1 > -\frac{1}{2}$, $\lambda_2 = \frac{1}{2}$, we take $b(-\frac{1}{2}) = \frac{1}{2}$; and if $\lambda_1 = -\frac{1}{2}$, $\lambda_2 = \frac{1}{2}$, we take $b(\lambda) \equiv 1$.

Thus b_n is the nth Fourier coefficient of the function $b(\cdot)$. The Fourier series for $b(\cdot)$ converges to $b(\cdot)$, and the partial sums are uniformly bounded, if the Nth partial sum is taken to be $\sum_{-N}^{N}$. Using this fact, we find that

$$\sum_{-N}^{N} b_n R(n) = \int_{-1/2}^{1/2} \sum_{-N}^{N} b_n e^{2\pi i n\lambda}\, dF(\lambda) \to \int_{-1/2}^{1/2} b(\lambda)\, dF(\lambda), \qquad N \to \infty,$$

and this yields (3.3). Conversely suppose that R is a positive definite function. Then, if $\alpha_m = e^{-2\pi i m\lambda}$ in (3.1), we find that

$$f_N(\lambda) = \frac{1}{N} \sum_{m,\, n=1}^{N+1} R(m-n) e^{-2\pi i(m-n)\lambda} = \sum_{-N}^{N} R(n)\left(1 - \frac{|n|}{N}\right) e^{-2\pi i n\lambda} \geq 0.$$

The equations express $f_N(\lambda)$ as the sum of a finite Fourier series, and we have the usual coefficient formula,

(3.4)

$$R(n)\left(1 - \frac{|n|}{N}\right) = \int_{-1/2}^{1/2} e^{2\pi i n\lambda} f_N(\lambda)\, d\lambda$$

$$= \int_{-1/2}^{1/2} e^{2\pi i n\lambda}\, dF_N(\lambda), \qquad |n| \leq N$$

$$0 = \int_{-1/2}^{1/2} e^{2\pi i n\lambda}\, dF_N(\lambda), \qquad |n| > N, \quad F_N(\lambda) = \int_{-1/2}^{\lambda} f_N(\mu)\, d\mu.$$

Now F_N is monotone, vanishes at $-\frac{1}{2}$, and has value $R(0)$ at $\lambda = \frac{1}{2}$. Hence by Helly's theorem there is a convergent subsequence of $\{F_n\}$, that is, there is a sequence of values of N, say $N_1, N_2, \cdots$ for which $N_j \to \infty$ and $F_{N_j}(\lambda) \to F(\lambda)$ for all λ. When $N \to \infty$ along this sequence in (3.4) we obtain (3.2). We have thus proved the theorem in the complex case. If F is normalized by adjusting it so that

$$F(-\tfrac{1}{2}) = 0, \qquad F(\lambda +) = F(\lambda), \qquad -\tfrac{1}{2} \leq \lambda < \tfrac{1}{2},$$

transferring the jump at $-\frac{1}{2}$, if there is one, to $\frac{1}{2}$, F is uniquely determined by (3.3). In particular, suppose that R is real. Then R is even and, since replacing $F(\lambda)$ in (3.2) by $-F(-\lambda)$ replaces $R(n)$ by $R(-n)$ $= R(n)$, the fact that F is uniquely determined if normalized implies that

$$F(\lambda_2+)-F(\lambda_1-)=F(-\lambda_1+)-F(-\lambda_2-), \qquad -\tfrac{1}{2}<\lambda_1<\lambda_2<\tfrac{1}{2}.$$

Then, for any version of F, if G is defined by

$$\begin{aligned} G(\lambda) &= 2[F(\lambda)-F(0+)]+F(0+)-F(0-), && 0<\lambda\leq\tfrac{1}{2},\\ &= 0, && \lambda=0, \end{aligned}$$

(3.2′) will be true. Conversely, given a G for which (3.2′) is true, F can be found for which (3.2) is true, say

$$\begin{aligned} F(\lambda) &= \tfrac{1}{2}[G(\lambda)-G(0+)]+G(0+)-G(0), && \lambda\geq 0,\\ &= -\tfrac{1}{2}[G(-\lambda)-G(0+)], && \lambda<0. \end{aligned}$$

Equation (3.3′) then follows from (3.3). If G is normalized, say by adjusting it so that

$$G(0)=0, \qquad G(\lambda+)=G(\lambda), \qquad 0<\lambda<\tfrac{1}{2},$$

G is uniquely determined by (3.3′).

If F is absolutely continuous, that is, if it is the integral of its derivative, the process is said to have a *spectral density* and F' is called the *spectral density function* (*complex form*). If the process is real, F is absolutely continuous if and only if G is, and in that case $G'=2F'$ is called the *spectral density function* (*real form*). In any case, F is called the spectral distribution function of the corresponding process.

If $\sum_{-\infty}^{\infty}|R(n)|$ converges, there is a continuous spectral density function given by

$$F'(\lambda)=\sum_{-\infty}^{\infty}R(n)e^{-2\pi in\lambda} \tag{3.5}$$

which reduces to

$$G'(\lambda)=2R(0)+4\sum_{1}^{\infty}R(n)\cos 2\pi n\lambda \tag{3.5′}$$

in the real case. In fact, this restriction on the covariance function implies that (3.5) and (3.5′) can be integrated to give (3.3) and (3.3′).

The *spectrum of the process* consists of every number λ_0 in whose neighborhood F actually increases, in the sense that

$$F(\lambda_0+\varepsilon)>F(\lambda_0-\varepsilon)$$

for every $\varepsilon>0$. The numbers in the spectrum are the frequencies which enter effectively in the harmonic analysis of R [given by (3.2) and (3.3)] and also, as will be seen below, in the harmonic analysis of the sample functions of the process. In particular, if the spectrum only contains a finite or enumerable sequence of values of λ, F increases only in jumps, one at each point of the spectrum.

Example 1 Suppose that the random variables $\cdots, x_0, x_1, \cdots$ are mutually orthogonal, with $\mathbf{E}\{|x_n|^2\} = \sigma^2$ for all n. Then the x_n process is stationary in the wide sense, with

$$\begin{aligned} R(n) &= 0, \qquad n \neq 0, \\ &= \sigma^2, \qquad n = 0, \\ F'(\lambda) &\equiv \sigma^2; \end{aligned}$$

all frequencies are equally important.

Example 2 Suppose that the x_n process is a Markov process in the wide sense and is also stationary in the wide sense. Then (V §8)

$$R(n) = a^n R(0), n \geq 0, \qquad |a| \leq 1.$$

If $|a| = 1$,

$$\mathbf{E}\{|x_n - a^n x_0|^2\} = R(0) - \bar{a}^n R(n) - a^n \overline{R(n)} + R(0) = 0,$$

so that

$$x_n = a^n x_0 \qquad n = 0, \pm 1, \cdots,$$

with probability 1, and conversely any sequence of random variables satisfying these equations obviously defines a Markov process (wide sense) which is stationary (wide sense). If $|a| < 1$, it is not immediately clear that R is positive definite. However, assuming for the moment that it is, there is a spectral density which can be evaluated by (3.5),

$$\begin{aligned} F'(\lambda) &= [\sum_0^\infty a^n e^{-2\pi i n\lambda} + \sum_{-\infty}^{-1} \bar{a}^{-n} e^{-2\pi i n\lambda}] R(0) \\ &= \frac{1 - |a|^2}{1 + |a|^2 - 2|a| \cos 2\pi(\lambda - \theta)} R(0), \qquad a = |a| e^{2\pi i\theta}. \end{aligned}$$

Now this function of λ is non-negative and integrable, and therefore defines a spectral density. The corresponding covariance function R defined by (3.2) is the given one, which is thereby proved legitimate. A second method of proving legitimacy would be the actual construction of a process with the desired properties. For example, if $\xi_1, \xi_2, \cdots$ is an orthonormal sequence of random variables and if x_n is defined by

$$x_n = \sum_{j=0}^\infty a^j \xi_{n-j}, \qquad (|a| < 1),$$

the x_n process is Markov and stationary, both in the wide sense, with

$$R(n) = a^n R(0), \qquad n \geq 0.$$

Example 3 Let $\xi_1, \cdots, \xi_k$ be mutually orthogonal random variables, with

$$\mathbf{E}\{|\xi_j|^2\} = \sigma_j^2 > 0,$$

and let $\lambda_1, \cdots, \lambda_k$ be real numbers. Define x_n by

$$x_n = \sum_{j=1}^{k} \xi_j e^{2\pi i n \lambda_j}.$$

Then

$$\mathbf{E}\{x_{m+n}\bar{x}_m\} = \sum_{j=1}^{k} \sigma_j^2 e^{2\pi i n \lambda_j} = R(n)$$

is independent of m, so that the x_n process is stationary in the wide sense. Since the λ_j's can be changed by integral amounts without affecting x_n, it is no restriction to assume that $-\frac{1}{2} < \lambda_j \leq \frac{1}{2}$, and we can also assume that the λ_j's are distinct. With these assumptions the spectral distribution function F of the x_n process increases only at the points $\lambda_1, \cdots, \lambda_k$ (that is, the spectrum only contains k points) with the jump σ_j^2 at λ_j. The real process corresponding to this complex process is defined as follows. Let $u_1, \cdots, u_k, v_1, \cdots, v_k$ be mutually orthogonal real random variables, with

$$\mathbf{E}\{u_j^2\} = \mathbf{E}\{v_j^2\} = \sigma_j^2 > 0$$

and let $\lambda_1, \cdots, \lambda_k$ be any real numbers. Define x_n by

$$x_n = \sum_{j=1}^{k} (u_j \cos 2\pi n\lambda_j + v_j \sin 2\pi n\lambda_j).$$

Then

$$\mathbf{E}\{x_{m+n}x_m\} = \sum_{j=1}^{k} \sigma_j^2 \cos 2\pi n\lambda_j = R(n)$$

is independent of m, so that the x_n process is stationary in the wide sense. As in the complex case, it can be assumed that

$$-\tfrac{1}{2} < \lambda_j \leq \tfrac{1}{2},$$

but we can now confine the λ_j's still further, by replacing any negative λ_j by $|\lambda_j|$, changing the sign of v_j to compensate. Thus it is no restriction to assume that

$$0 \leq \lambda_j \leq \tfrac{1}{2}$$

and that the λ_j's are distinct. With these assumptions G increases only at $\lambda_1, \cdots, \lambda_k$, increasing by σ_j^2 at λ_j; the (real form) spectrum has k points in it. The corresponding complex form spectrum may have either $2k$ or $2k - 1$ points in it. If $\lambda_j > 0$, F increases by $\sigma_j^2/2$ at $\pm\lambda_j$; if

$\lambda_j = 0$, F increases by σ_j^2 at 0. By a proper choice of the λ_j's and ξ_j's, the complex version of Example 3 reduces to this real one.

In particular, if the y_j's and z_j's are Gaussian, with

$$\mathbf{E}\{y_j\} = \mathbf{E}\{z_j\} = 0,$$

the x_n process is stationary in the strict sense.

It will be proved below that every stationary (wide sense) process is either as in Example 3 or can be approximated arbitrarily closely by processes of this type. It will then be clear that the harmonic analysis of stationary processes must play an essential role in their study.

Example 4 Let ξ be any real random variable for which $-\frac{1}{2} < \xi \leq \frac{1}{2}$, and let a be a constant. Define x_n by

$$x_n = ae^{2\pi i n \xi}.$$

Then

$$\mathbf{E}\{x_{m+n}\bar{x}_m\} = |a|^2\mathbf{E}\{e^{2\pi i n \xi}\} = \int_{-1/2}^{1/2} e^{2\pi i n \lambda}\, dF(\lambda) = R(n),$$

where $F/|a|^2$ is the distribution function of ξ. The x_n process is stationary in the wide sense, with spectral distribution function (complex form) F. This example exhibits a stationary (wide sense) process with any pre-assigned spectral distribution function. Note that the sample sequences are in every case periodic (thinking of n as continuous) whereas in Example 3 this is not true, if $k > 1$, except in very special cases. Thus the spectral distribution function does not determine the harmonic analysis of the individual sample functions except in some sort of average sense. The real example corresponding to this complex one is the following. Let y, z be real mutually independent random variables, with $0 < y \leq \frac{1}{2}$ and z uniformly distributed in the interval from $-\frac{1}{2}$ to $\frac{1}{2}$; let σ be a real constant. Define x_n by

$$x_n = \sigma \cos 2\pi(ny + z).$$

Then

$$\mathbf{E}\{x_{m+n}x_m\} = \frac{\sigma^2}{2}\mathbf{E}\{\cos 2\pi n y\} = \int_0^{1/2} \cos 2\pi n\lambda\, dG(\lambda) = R(n),$$

where $2G/\sigma^2$ is the distribution function of y. The x_n process is stationary in the wide sense with spectral distribution function (real form) G.

The sample functions in this example are elementary trigonometric functions, and what randomness there is in the process is obtained by picking values of frequency (y) and phase (z) which determine the sample sequence completely. In general, one picks the whole sample sequence out of a hat, that is, one determines values for $\cdots, x_{-1}, x_0, x_1, \cdots$ in accordance with their individual and joint distributions. Although it may seem at first sight that there is something less random about the process under consideration because a choice of only two sample values $y(\omega)$, $z(\omega)$ determines the full x_n sample sequence, this impression is false. In fact in the most general case, picking a sample sequence is equivalent to picking only a single random variable sample value, simply because infinite dimensional space can be mapped on one-dimensional space in such a way that random variables, say the x_j's, become functions of a single random variable x, so that picking x automatically determines values for all the x_j's. This point is worth more discussion since it is a cause of confusion. From our general point of view sample sequence probabilities are probabilities in ω space, so that choosing a sample sequence means choosing a single point in this space, a single choice. This does not alter the fact that after $\cdots, x_{-1}(\omega), x_0(\omega)$ are all chosen in accordance with the relevant probability distributions, $x_1(\omega)$ may or may not be uniquely determined as a function of these chosen values. The following example is instructive. Let ω space be the real interval [0, 1], and let probability measure be Lebesgue measure on this interval. Let x be the coordinate variable of the interval. Then we can write

$$x(\omega) = \omega = .\,\xi_1\xi_2\cdots,$$

the decimal expansion of ω. If we define the random variable x_n by

$$x_n(\omega) = \xi_n,$$

x_n becomes a single-valued function of x, neglecting certain rational values of ω which form a set of probability 0. The x_n's are mutually independent random variables, with

$$\mathbf{P}\{x_n(\omega) = j\} = \tfrac{1}{10}, \qquad j = 0, \cdots, 9.$$

A choice of $x_1(\omega), x_2(\omega), \cdots$ thus requires, from one point of view, infinitely many mutually independent choices (say by throwing a ten-sided die). But from another point of view all the $x_j(\omega)$'s are uniquely determined by a sample value of one random variable x.

Example 5 Suppose that x_n is defined by

$$x_n = \int_{-1/2}^{1/2} e^{2\pi i n\lambda}\, dy(\lambda),$$

where the $y(\lambda)$ process has orthogonal increments, with

$$\mathbf{E}\{|dy(\lambda)|^2\} = dF(\lambda).$$

Then (IX §2)

$$\mathbf{E}\{x_{m+n}\bar{x}_m\} = \int_{-1/2}^{1/2} e^{2\pi in\lambda}\, dF(\lambda) = R(n)$$

is independent of m, so that the x_n process is stationary in the wide sense with covariance function R and spectral distribution function F. It will be shown in §4 that every stationary (wide sense) process can be represented in this way, and that in the real case the representation can also be put in the form

$$x_n = \int_0^{1/2} \cos 2\pi n\lambda\, du(\lambda) + \sin 2\pi n\lambda\, dv(\lambda),$$

where the $u(\lambda)$, $v(\lambda)$ processes are real, with orthogonal increments, and

$$\mathbf{E}\{|du(\lambda)|^2\} = \mathbf{E}\{|dv(\lambda)|^2\} = dG(\lambda)$$

$$\mathbf{E}\{du(\lambda)\, dv(\mu)\} = 0.$$

Here G is the spectral distribution function (real form) and the last equation is a symbolic version of the statement that every $u(\lambda)$ increment is orthogonal to every $v(\lambda)$ increment. Example 3 above is the particular case in which F is constant except for a finite number of jumps, so that the spectrum of the process only contains a finite number of points. This representation of x_n is called the *spectral representation* (real or complex form, as the case may be).

4. The spectral representation of a stationary process

In the following theorem, and in certain later ones involving the spectral representation of a stationary process, the parameter set is always taken as the set of all integers, rather than, for example, the set of positive integers. We omit the extension to the latter case, which would carry us too far afield. The theorems are not always true in the latter case without certain modifications, but the essential character of the sample sequences is the same.

THEOREM 4.1 *Every stationary (wide sense) process* $\{x_n, -\infty < n < \infty\}$ *has the spectral representation*

$$x_n = \int_{-1/2}^{1/2} e^{2\pi in\lambda}\, dy(\lambda), \tag{4.1}$$

where the $y(\lambda)$ *process has orthogonal increments, and*

$$\mathbf{E}\{|dy(\lambda)|^2\} = dF(\lambda).$$

If a $y(\lambda)$ process with orthogonal increments satisfies (4.1), *with a suitable evaluation of the right-hand side, then*

$$\frac{y(\lambda_2+)+y(\lambda_2-)}{2}-\frac{y(\lambda_1+)+y(\lambda_1-)}{2} = x_0(\lambda_2-\lambda_1)+\underset{N\to\infty}{\text{l.i.m.}}\sum_{\substack{-N\\(n\neq 0)}}^{N} x_n \frac{e^{-2\pi in\lambda_2}-e^{-2\pi in\lambda_1}}{-2\pi in} \quad (-\tfrac{1}{2}<\lambda_1<\lambda_2<\tfrac{1}{2}) \tag{4.2}$$

$$y(\tfrac{1}{2})-y(-\tfrac{1}{2})=x_0,$$

and these equations determine $y(\lambda)$ uniquely, neglecting values on an ω set of probability 0, *if the $y(\lambda)$ process is properly normalized.*

If the x_n process is real, (4.1) *can be put in the form*

$$x_n=\int_0^{1/2}\cos 2\pi n\lambda\, du(\lambda)+\sin 2\pi n\lambda\, dv(\lambda), \tag{4.1'}$$

where the (real) $u(\lambda)$ and $v(\lambda)$ processes have orthogonal increments, with

$$\mathbf{E}\{[du(\lambda)]^2\}=\mathbf{E}\{[dv(\lambda)]^2\}=dG(\lambda), \qquad \lambda>0$$

$$\mathbf{E}\{[du(\lambda)]^2\}=dG(\lambda) \qquad \lambda\geq 0$$

$$\mathbf{E}\{du(\lambda)\, dv(\mu)\}=0 \qquad 0\leq\lambda,\ \mu\leq\tfrac{1}{2}.$$

If $u(\lambda)$, $v(\lambda)$ processes with orthogonal increments satisfy these conditions and (4.1′), *with a suitable evaluation of the right-hand side, then*

$$\frac{u(\lambda+)+u(\lambda-)}{2}-u(0)=2x_0\lambda+\underset{N\to\infty}{\text{l.i.m.}}\sum_{\substack{-N\\(n\neq 0)}}^{N} x_n\frac{\sin 2\pi n\lambda}{\pi n} \quad (0<\lambda<\tfrac{1}{2}) \tag{4.2'}$$

$$u(\tfrac{1}{2})-u(0)=x_0$$

$$\frac{v(\lambda_2+)+v(\lambda_2-)}{2}-\frac{v(\lambda_1+)+v(\lambda_1-)}{2} = -\underset{N\to\infty}{\text{l.i.m.}}\sum_{\substack{-N\\(n\neq 0)}}^{N} x_n\frac{\cos 2\pi n\lambda_2-\cos 2\pi n\lambda_1}{\pi n} \quad (0<\lambda_1,\lambda_2<\tfrac{1}{2})$$

and these equations determine $u(\lambda)$, $v(\lambda)$ uniquely, neglecting values on an ω set of probability 0, *if the $u(\lambda)$, $v(\lambda)$ processes are properly normalized.*

We have already seen (§3, Example 5) that any x_n process defined by (4.1) or (4.1′) with the stated conditions on the $y(\lambda)$, $u(\lambda)$, $v(\lambda)$ processes is stationary in the wide sense. Note that (complex form) if $\mathbf{E}\{dy(\lambda)\}=0$, then $\mathbf{E}\{x_n\} = 0$ for all n and [by (4.2)] conversely, and that the x_n process is Gaussian if and only if the $y(\lambda) - y(0)$ process is Gaussian. Similar remarks hold for the real form.

The proof of the theorem is in part the stochastic analogue of that of §3 Theorem 3.2, and the notation of that proof will be used. Since the Fourier series for $b(\cdot)$ converges boundedly to $b(\cdot)$, it also converges in the mean, with any λ weighting. Hence if x_n is defined by (4.1),

$$\underset{N\to\infty}{\text{l.i.m.}} \sum_{-N}^{N} b_n x_n = \underset{N\to\infty}{\text{l.i.m.}} \int_{-1/2}^{1/2} \left[\sum_{-N}^{N} b_n e^{2\pi i n\lambda}\right] dy(\lambda)$$

$$= \int_{-1/2}^{1/2} \underset{N\to\infty}{\text{l.i.m.}} \left[\sum_{-N}^{N} b_n e^{2\pi i n\lambda}\right] dy(\lambda)$$

$$= \int_{-1/2}^{1/2} b(\lambda)\, dy(\lambda)$$

and this yields (4.2). Here the l.i.m. [ ] under the integral sign is a mean limit with λ weighting $dF(\lambda)$, and the justification for putting the l.i.m. under the integral sign was given in IX §2. Conversely suppose that an x_n process is stationary in the wide sense. It is not difficult to show that the mean limit in (4.2) exists, and this leads to a definition of a $y(\lambda)$ process satisfying (4.2). This $y(\lambda)$ process yields the spectral representation (4.1). Rather than giving the details of this approach, which are surprisingly fussy if the spectral distribution function of the x_n process is not continuous at $\pm \frac{1}{2}$, we outline a procedure due to Cramér which is applicable in more general settings. Let $\mathfrak{M}_0$ be the linear manifold of linear combinations of the x_j's. Define the inner product of two random variables x, y as $\mathbf{E}\{x\bar{y}\}$, and the distance between them as

$$[\mathbf{E}\{|x - y|^2\}]^{1/2}.$$

Let $\mathfrak{M}$ be the closure of $\mathfrak{M}_0$ with this distance definition. Let $\mathfrak{M}_0'$ be the linear manifold of linear combinations of the functions $e^{2\pi i n\lambda}$, $-\infty < n < \infty$, on the interval $[-\frac{1}{2}, \frac{1}{2}]$. Define the inner product of two functions Φ, Ψ on this interval and the distance between them as

$$\int_{-1/2}^{1/2} \Phi(\lambda)\overline{\Psi(\lambda)}\, dF(\lambda), \qquad \left[\int_{-1/2}^{1/2} |\Phi(\lambda) - \Psi(\lambda)|^2\, dF(\lambda)\right]^{1/2},$$

respectively. Here F is the spectral distribution function of the x_n process. Let $\mathfrak{M}'$ be the closure of $\mathfrak{M}_0'$ with this distance definition.

Then $\mathfrak{M}'$ is the class of functions Φ on $[-\frac{1}{2}, \frac{1}{2}]$, which are measurable relative to F measure, with

$$\int_{-1/2}^{1/2} |\Phi(\lambda)|^2 \, dF(\lambda) < \infty.$$

In the following, we consider two random variables identical if they are equal with probability 1, and two functions on $[-\frac{1}{2}, \frac{1}{2}]$ identical if they are equal almost everywhere (F measure]. We shall assume that F is continuous on the right, as we can without restricting the problem. We shall define a 1–1 correspondence between the elements of $\mathfrak{M}$ and $\mathfrak{M}'$ which preserves inner products, and therefore also preserves distance. To do this, define x_n as the random variable corresponding to $e^{2\pi i n\lambda}$, and more generally $\sum_n b_n x_n$ is to correspond to $\sum_n b_n e^{2\pi i n\lambda}$ (finite sums). Since

$$\begin{aligned} \mathbf{E}\{\sum_m b_m x_m \overline{\sum_n c_n x_n}\} &= \sum_{m,\, n} b_m \bar{c}_n R(m-n) \\ &= \int_{-1/2}^{1/2} \sum_m b_m e^{2\pi i m\lambda} \overline{\sum_n c_n e^{2\pi i n\lambda}} \, dF(\lambda), \end{aligned}$$

this correspondence, defined as yet only on $\mathfrak{M}_0$ and $\mathfrak{M}_0'$, is 1–1 and preserves inner products and distance. The correspondence is then extended by continuity to $\mathfrak{M}$ and $\mathfrak{M}'$. Let $y(\lambda)$ be the random variable corresponding to the function which is 1 on the interval $[-\frac{1}{2}, \lambda]$, and vanishes otherwise. The process $\{y(\lambda), -\frac{1}{2} \le \lambda \le \frac{1}{2}\}$ will be shown to yield the representation (4.1). The fact that inner products are preserved in the correspondence defined above implies that the $y(\lambda)$ process has orthogonal increments. The fact that distance is preserved implies that $\mathbf{E}\{|\, dy(\lambda)|^2\} = dF(\lambda)$. For each λ, the random variable $y(\lambda)$ lies in $\mathfrak{M}$. Hence any linear combination of the $y(\lambda)$'s lies in $\mathfrak{M}$, and more generally, according to the definition of stochastic integral in IX §2,

$$\int_{-1/2}^{1/2} \Phi(\lambda) \, dy(\lambda) \in \mathfrak{M}, \qquad \text{if} \quad \Phi \in \mathfrak{M}'.$$

We shall prove that this stochastic integral is the element of $\mathfrak{M}$ corresponding to the integrand Φ in $\mathfrak{M}'$. In fact this assertion is true by definition if Φ is the characteristic function of an interval $[-\frac{1}{2}, \lambda]$. Since the class of Φ's for which it is true form a closed linear manifold, containing these characteristic functions, it contains $e^{2\pi i n\lambda}$ for every n, by an elementary approximation, and therefore this class is $\mathfrak{M}'$ itself. Finally, according to the definition of the correspondence, x_n corresponds to

$e^{2\pi in\lambda}$, and this fact is now one way of indicating the spectral representation (4.1).

If the $y(\lambda)$ process satisfies the condition

$$\mathbf{P}\{y(-\tfrac{1}{2}, \omega) = 0\} = 1$$

$$\mathbf{P}\{y(\lambda +, \omega) = y(\lambda, \omega)\} = 1, \qquad -\tfrac{1}{2} \leq \lambda < \tfrac{1}{2}$$

(corresponding to the condition that the spectral distribution function vanish at $-\frac{1}{2}$ and be continuous on the right), $y(\lambda)$ is uniquely determined up to values on an ω set of probability 0 by (4.2). If the $y(\lambda)$ process does not satisfy this condition, $y(\lambda)$ can be replaced by

$$\begin{aligned} y_1(\lambda) &= 0 & \lambda &= -\tfrac{1}{2} \\ &= y(\lambda +) - y(0 +) & -\tfrac{1}{2} &< \lambda < \tfrac{1}{2} \\ &= x_0 & \lambda &= \tfrac{1}{2} \end{aligned}$$

to obtain a process satisfying this condition and for which (4.1) remains true. Because of the ambiguity in the definition of the stochastic integral, there is equality in (4.1) only for one particular version of the stochastic integral; for other versions, there is only equality with probability 1.

If the x_n process is real, the equality $x_n = \bar{x}_n$ combined with (4.1) implies that, whether the $y(\lambda)$ process is normalized or not,

$$\mathbf{P}\{y(\lambda_2 +, \omega) - y(\lambda_1 -, \omega) = \overline{y(-\lambda_1 +, \omega) - y(-\lambda_2 -, \omega)}\} = 1, \qquad -\tfrac{1}{2} < \lambda_1 < \lambda_2 < \tfrac{1}{2},$$

from which the representation (4.1′) is easily derived. The inverse formula (4.2′) is then derived by taking real and imaginary parts in (4.2). Note that the $dv(\lambda)$ integrand in (4.1′) vanishes at 0 and $\frac{1}{2}$ so that, if 0 or $\frac{1}{2}$ is a fixed point of discontinuity of the $v(\lambda)$ process, the discontinuity makes no contribution to the stochastic integral. If the $u(\lambda)$ and $v(\lambda)$ processes satisfy the conditions

$$\mathbf{P}\{u(0, \omega) = v(0, \omega) = 0\} = 1$$

$$\mathbf{P}\{v(\tfrac{1}{2}, \omega) - v(\tfrac{1}{2} -, \omega) = v(0 +, \omega) - v(0, \omega) = 0\}$$

$$\mathbf{P}\{u(\lambda +, \omega) = u(\lambda, \omega), v(\lambda +, \omega) = v(\lambda, \omega)\} = 1, \qquad 0 \leq \lambda < \tfrac{1}{2},$$

then $u(\lambda)$ and $v(\lambda)$ are uniquely determined up to values on an ω set of probability 0 by (4.2′), and this normalization can always be effected by trivial manipulations, as in the complex case.

If the spectrum of the x_n process only contains a finite number of points, say if F is constant except for jumps of magnitude $\sigma_1^2, \cdots, \sigma_k^2$ at $\lambda_1, \cdots, \lambda_k$, the spectral representation reduces to §3 Example 3,

$$x_n = \int_{-1/2}^{1/2} e^{2\pi i n\lambda}\, dy(\lambda) = \sum_{j=1}^{k} e^{2\pi i n\lambda_j} y_j,$$

where

$$y_j = y(\lambda_j +) - y(\lambda_j -).$$

The process of Example 3 approximates the general case arbitrarily closely in the following sense. The integral

$$\int_{-1/2}^{1/2} e^{2\pi i n\lambda}\, dy(\lambda)$$

is the limit in the mean, for each n, of the appropriate Riemann-Stieltjes sums,

$$\sum_j e^{2\pi i n\lambda_j} [y(\lambda_{j+1}) - y(\lambda_j)],$$

when $\text{Max}\,(\lambda_{j+1} - \lambda_j) \to 0$; each sum defines a process of the type of Example 3 [with $y_j = y(\lambda_{j+1}) - y(\lambda_j)$].

In most practical applications $\mathbf{E}\{x_n\}$ is independent of n. In particular we have seen that, if $\mathbf{E}\{x_n\} = 0$ for all n, then $\mathbf{E}\{dy(\lambda)\} = 0$ and the $y(\lambda)$ process has uncorrelated as well as orthogonal increments. In general, if $\mathbf{E}\{x_n\} = m$ is independent of n, the process with random variables $\cdots, x_0 - m, x_1 - m, \cdots$ is also stationary in the wide sense and it is convenient to use the spectral representation of $x_n - m$ rather than that of x_n. There is actually very little difference between the $y(\lambda)$ processes involved; we have

$$x_n = \int_{-1/2}^{1/2} e^{2\pi i n\lambda}\, dy_1(\lambda), \qquad x_n - m = \int_{-1/2}^{1/2} e^{2\pi i n\lambda}\, dy(\lambda),$$

where $y(\lambda)$ is the same as $y_1(\lambda)$ except for a jump of magnitude m when $\lambda = 0$,

$$\begin{aligned} y(\lambda) &= y_1(\lambda), & \lambda &< 0, \\ &= y_1(\lambda) + m, & \lambda &\geq 0. \end{aligned}$$

5. Spectral decompositions

Let $\{x_n, -\infty < n < \infty\}$ be a stationary (wide sense) process, and suppose that $A_1, \cdots, A_\nu$ are ν disjunct sets in the interval $[-\frac{1}{2}, \frac{1}{2}]$, whose union is this interval. It is supposed that these sets are measurable with respect to the F measure $\int_A dF(\lambda)$. Then it is possible to exhibit the x_n

process as a sum of mutually orthogonal stationary (wide sense) processes whose spectral distributions are confined to the sets $A_1, \cdots, A_\nu$. This is done as follows: If x_n has the spectral representation

$$x_n = \int_{-1/2}^{1/2} e^{2\pi n i\lambda}\, dy(\lambda), \qquad \mathbf{E}\{|dy(\lambda)|^2\} = dF(\lambda),$$

define $x_n^{(j)}$ by

$$x_n^{(j)} = \int_{-1/2}^{1/2} e^{2\pi n i\lambda}\, \Phi_j(\lambda)\, dy(\lambda) \qquad j = 1, \cdots,$$

where $\Phi_j(\lambda)$ is 1 on A_j and 0 otherwise. Then the $x_n^{(j)}$ process is stationary (wide sense), and has spectral distribution function given by

$$F_j(\lambda) = \int_{-1/2}^{\lambda} \Phi_j(\mu)\, dF(\mu).$$

The F_j distribution is thus confined to A_j. Moreover,

$$\mathbf{E}\{x_n^{(j)}\overline{x_m^{(k)}}\} = \int_{-1/2}^{1/2} e^{2\pi i(n-m)\lambda}\, \Phi_j(\lambda)\overline{\Phi_k(\lambda)}\, dF(\lambda) = 0, \qquad j \neq k.$$

The above procedure is also applicable if the A_j's are (denumerably) infinite in number. Note that, if $\mathbf{E}\{x_n\} = 0$ for all n, it follows that $\mathbf{E}\{x_n^{(j)}\} = 0$ for all n also, since then $\mathbf{E}\{dy(\lambda)\} = 0$.

The decomposition we have described here can be effected linearly. To show this, we show that, if (fixed j) $a_{-N}, \cdots, a_N$ are chosen properly, the sum $\sum_{k=-N}^{N} a_k x_{n+k}$ will approximate $x_n^{(j)}$ arbitrarily closely in the mean square sense. Since

$$\sum_{k=-N}^{N} a_k x_{n+k} = \int_{-1/2}^{1/2} e^{2\pi i n\lambda} \sum_{k=-N}^{N} a_k e^{2\pi i k\lambda}\, dy(\lambda),$$

it need only be shown (see IX §2) that, with a proper choice of the a_k's the sum $\sum_{k=-N}^{N} a_k e^{2\pi i k\lambda}$ will approximate $\Phi_j(\lambda)$ arbitrarily closely in the mean square sense with λ weighting $dF(\lambda)$,

$$\int_{-1/2}^{1/2} |\Phi_j(\lambda) - \sum_{k=-N}^{N} a_k e^{2\pi i k\lambda}|^2\, dF(\lambda) \sim 0.$$

This is true if the set A_j is an interval whose endpoints are not points of discontinuity of F, because in that case the partial sums of the Fourier series for Φ_j converge boundedly to Φ_j, neglecting the endpoints of A_j which have $F(\lambda)$ measure 0. The class of functions Φ, measurable with respect to F measure, with

$$\int_{-1/2}^{1/2} |\Phi(\lambda)|^2 \, dF(\lambda) < \infty,$$

contains a subclass of functions which can be approximated in the mean square sense, with λ weighting $dF(\lambda)$, by trigonometric sums. This subclass is obviously a closed linear manifold, and we have just seen that it contains the characteristic functions of intervals whose endpoints are not points of discontinuity of F. The subclass is therefore the whole class, and includes in particular the characteristic function of A_j, as was to be proved.

Certain special cases of this decomposition are very important. Let F be the given spectral distribution function. Then we can write F in the form

$$F = F_1 + F_2 + F_3,$$

where F_1 is the jump function of F, which increases only at the jumps of F by the amount of the jumps, F_2 is the absolutely continuous component of F,

$$F_2(\lambda) = \int_0^\lambda F'(\mu) \, d\mu,$$

and the remainder F_3, continuous and monotone non-decreasing, is the continuous singular component of F. The F_1 distribution is confined to the points where F is discontinuous; the F_2 distribution is confined to the continuity points at which F' is finite; the F_3 distribution is confined to the remainder, the set of measure 0 where F is continuous and F' either does not exist or is $+\infty$. Thus the decomposition of F implies a corresponding decomposition of the x_n process into three mutually orthogonal processes. The first reduces to a simple sum: If $\lambda_1, \lambda_2, \cdots$ are the discontinuities of F supposed continuous on the right, the random variables $y(\lambda_j) - y(\lambda_j -)$ form an orthogonal set, and

$$x_n^{(1)} = \sum_j e^{2\pi i n \lambda_j} [y(\lambda_j) - y(\lambda_j -)].$$

The series converges in the mean (Riesz-Fischer theorem). This is simply a somewhat extended version of §3, Example 3.

It will be seen in Chapter XII that the problem of least squares linear prediction involves separating the $x_n^{(2)}$ out of the x_n process.

6. The law of large numbers for stationary (wide sense) processes

Let $\{x_n, n \geq 0\}$ be a stationary (wide sense) process. Then we shall prove that

$$\text{(6.1)} \qquad \underset{n\to\infty}{\text{l.i.m.}} \frac{1}{n+1} \sum_0^n x_j$$

exists; this is the law of large numbers for these processes. Before giving the proof we outline an indirect proof which has educational (although no other) value. According to II §3 there is a Gaussian $\hat{x}_n$ process not necessarily defined on the same measure space, for which

$$\mathbf{E}\{\hat{x}_n\} = 0$$

$$\mathbf{E}\{\hat{x}_m \bar{\hat{x}}_n\} = \mathbf{E}\{x_m \bar{x}_n\} \qquad m, n \geq 0$$

$$\mathbf{E}\{\hat{x}_m \hat{x}_n\} = 0.$$

The $\hat{x}_n$ process is strictly stationary and is therefore subject to the (strong) law of large numbers, Theorem 2.1, according to which

$$\lim_{n\to\infty} \frac{1}{n+1} \sum_0^n \hat{x}_j$$

exists with probability 1. Since the random variables involved are Gaussian, it is easily verified that the averages cannot converge unless they also converge in the mean. But then the l.i.m. in (6.1) also exists because the x_n and $\hat{x}_n$ processes have the same covariance function, and only the covariance function is involved in checking the existence of this l.i.m. This proof is hardly the best from any point of view, but it illustrates the fact that the theorem in question is the wide sense version of the strong law of large numbers for stationary processes and illustrates the close connection between wide sense and strict sense theorems in general.

THEOREM 6.1 *Let $\{x_n, -\infty < n < \infty\}$ be a stationary (wide sense) process with spectral representation*

$$\text{(6.2)} \qquad x_n = \int_{-1/2}^{1/2} e^{2\pi i n\lambda}\, dy(\lambda) \qquad \begin{aligned} &\mathbf{E}\{|dy(\lambda)|^2\} = dF(\lambda) \\ &F(\lambda+) = F(\lambda). \end{aligned}$$

Then

$$\text{(6.3)} \qquad \underset{n-m\to\infty}{\text{l.i.m.}} \frac{1}{n-m+1} \sum_m^n x_j = y(0) - y(0-),$$

and

$$\text{(6.4)} \quad \mathbf{E}\{|y(0) - y(0-)|^2\} = F(0) - F(0-) = \lim_{n-m\to\infty} \frac{1}{n-m+1} \sum_m^n R(j).$$

The random variable $y(0) - y(0-)$ is invariant under the shift and is the projection of x_0 on the closed linear manifold $\mathfrak{M}_0$ of invariant random variables (wide sense).

$$y(0) - y(0-) = \hat{\mathbf{E}}\{x_0 \mid \mathfrak{M}_0\}.$$

Note that the limit will be 0 if, for example, $\lim_{n\to\infty} R(n) = 0$. Except for the fact that there is more freedom in the averaging in (6.3) than in (2.1′), Theorem 6.1 is in all detail the wide sense version of Theorem 2.1. If the x_n's in Theorem 2.1 are mutually independent, corresponding to the wide sense statement that those in Theorem 6.1 are mutually orthogonal, exactly corresponding proofs of the strong and weak versions have been given in VII §6 and IV §7.

The proof of (6.3) is very simple, since the averages can be evaluated explicitly:

$$\begin{aligned}\frac{1}{n-m+1}\sum_{j=m}^{n} x_j &= \int_{-1/2}^{1/2} \frac{1}{n-m+1}\sum_{j=m}^{n} e^{2\pi i j\lambda}\, dy(\lambda)\\ &= \int_{-1/2}^{1/2} \frac{e^{2\pi i m\lambda}}{n-m+1}\, \frac{1-e^{2\pi i(n-m+1)\lambda}}{1-e^{2\pi i\lambda}}\, dy(\lambda),\end{aligned}$$

and similarly

$$\frac{1}{n-m+1}\sum_{j=m}^{n} R(j) = \int_{-1/2}^{1/2} \frac{e^{2\pi i m\lambda}}{n-m+1}\, \frac{1-e^{2\pi i(n-m+1)\lambda}}{1-e^{2\pi i\lambda}}\, dF(\lambda).$$

The integrand is ≤ 1 in modulus and converges to $f(\lambda)$ when $n \to \infty$, where

$$\begin{aligned} f(\lambda) &= 0 \qquad \lambda \neq 0\\ &= 1 \qquad \lambda = 0.\end{aligned}$$

Since the convergence is bounded convergence, it is also convergence in the mean with λ weighting $dF(\lambda)$. Hence (cf. IX §2)

$$\underset{n-m\to\infty}{\text{l.i.m.}}\ \frac{1}{n-m+1}\sum_{j=m}^{n} x_j = \int_{-1/2}^{1/2} f(\lambda)\, dy(\lambda) = y(0) - y(0-).$$

Since the convergence is bounded convergence,

$$\lim_{n\to\infty} \frac{1}{n-m+1}\sum_{j=m}^{n} R(j) = \int_{-1/2}^{1/2} f(\lambda)\, dF(\lambda) = F(0) - F(0-),$$

and this finishes the proof of (6.3) and (6.4).

The following is the wide sense version of the corollary to Theorem 2.1.

COROLLARY *If* $-\frac{1}{2} < \mu \leq \frac{1}{2}$,

$$\underset{n-m\to\infty}{\text{l.i.m.}} \frac{1}{n-m+1} \sum_{j=m}^{n} x_j e^{-2\pi i j\mu} = y(\mu) - y(\mu -);$$

moreover,

$$\mathbf{E}\{|y(\mu) - y(\mu -)|^2\} = F(\mu) - F(\mu -) = \lim_{n-m\to\infty} \frac{1}{n-m+1} \sum_{j=m}^{n} R(j)e^{-2\pi i j\mu}.$$

To prove the corollary we need only apply Theorem 6.1 to the process

$$\{x_n e^{-2\pi i n\mu}, -\infty < n < \infty\}.$$

This process is stationary in the wide sense, with covariance function given by $R(n)e^{-2\pi i n\mu}$ and spectral distribution function given by $F(\lambda + \mu)$ (mod 1 in the argument).

The theorem and its corollary are applicable to stationary (wide sense) processes whose parameter range is the set of non-negative integers, as far as the existence of the indicated limits goes, since this existence can be expressed in terms of the covariance functions of the processes. Moreover, the limits whose existence is asserted in the corollary are mutually orthogonal for different values of μ, since this orthogonality can also be expressed in terms of the covariance functions. However, the evaluation of the limits in terms of the spectral representation must be sacrificed.

The theorem and corollary show how, by means of a linear operation, the discontinuities of F and the corresponding component of the x_n process (the $x_n^{(1)}$ process in the notation of §5) can be identified. In practice, of course, there is no sampling method (based on finite samples) that can do more than indicate vaguely that any particular λ value is a jump point of F. It is important conceptually to note that in the corollary, if $n = 0$, the jumps are evaluated in terms of the *past* of the process, that is, in terms of x_m for $m \leq 0$. Suppose, for example, that the x_n process has a spectral distribution confined to a finite or infinite sequence of λ values (so that in the notation of §5 the $x_n^{(2)}$ and $x_n^{(3)}$ processes are absent). Then according to Theorem 6.1 and its corollary the $y(\lambda)$ process [which is made up by summing the jumps $y(\mu) - y(\mu -)$] is completely determined by the x_n's for $n \leq 0$. It follows from the spectral representation of x_n that x_n itself must be determined uniquely for all n by the x_n's for $n \leq 0$, and determined linearly, that is, each x_n for $n > 0$ can be approximated arbitrarily closely in the mean by linear combinations of x_n's with $n \leq 0$. That this is not true in general is made clear by noting that it is not true, for example, if the x_n's are mutually orthogonal. (These questions will be studied in great detail in XII.) The fact that this is not true in the general case explains why the $y(\lambda)$ process cannot always be expressed linearly in terms of the x_n's for $n \leq 0$ and why therefore the sums $\sum_{-N}^{N}$ in (4.2) cannot be modified to be sums $\sum_{0}^{N}$ or $\sum_{-N}^{0}$.

THEOREM 6.2 *Under the hypotheses of Theorem* 6.1, *if* $-\frac{1}{2} < \mu \leq \frac{1}{2}$, *then*

$$\lim_{n\to\infty} \frac{1}{n+1} \sum_{0}^{n} x_j e^{-2\pi i j\mu} = 0$$

with probability 1, *if there is a positive K and a positive* α *such that the following equal expressions are bounded as indicated:*

$$\begin{aligned}\mathbf{E}\left\{\left|\frac{1}{n+1}\sum_{0}^{n} x_j e^{-2\pi i j\mu}\right|^2\right\} &= \frac{1}{(n+1)^2}\sum_{j,\,k=0}^{n} R(j-k)e^{-2\pi i(j-k)\mu} \\ &= \frac{1}{n+1}\sum_{j=-n}^{n}\left(1-\frac{|j|}{n+1}\right)R(j)e^{-2\pi i j\mu} \\ &= \int_{-1/2}^{1/2} \frac{\sin^2 \pi(n+1)(\lambda-\mu)}{(n+1)^2 \sin^2 \pi(\lambda-\mu)}\, dF(\lambda) \leq \frac{K}{n^\alpha}.\end{aligned}$$

The equality of these expressions is easily checked, and the proof will be omitted. Note that the expressions are bounded [by $R(0)$] for every process; the restrictive hypothesis of the theorem lies in the first place in the assumption that the expressions go to 0 when $n \to \infty$ (which implies that the mean limit in the preceding corollary is 0), and in the second place in the assumption that the approach to 0 is as fast as K/n^α. The condition of the theorem is satisfied for all μ, with $\alpha = 1$, if $\sum_{0}^{\infty} |R(n)| < \infty$; it is also satisfied for all μ if $|R(n)| < \text{const.}/n^\alpha$. In particular, the condition of the theorem is certainly satisfied for all μ, with $\alpha = 1$, if the x_j's are mutually orthogonal. The theorem has already been proved in that special case (IV, Theorem 5.2).

To prove the theorem when $\mu = 0$, choose β so large that $\beta\alpha > 1$. Then, if $n \geq m^\beta$,

$$\mathbf{E}\left\{\left|\frac{1}{n+1}\sum_{0}^{n} x_j\right|^2\right\} \leq \frac{K}{m^{\alpha\beta}},$$

so that, if $\varepsilon > 0$, and if n_m is the smallest integer $\geq m^\beta$,

$$\sum_{m=1}^{\infty} \mathbf{P}\left\{\left|\frac{1}{n_m+1}\sum_{0}^{n_m} x_j(\omega)\right| \geq \varepsilon\right\} \leq \sum_{m=1}^{\infty} \frac{K}{\varepsilon^2 m^{\alpha\beta}} < \infty.$$

Hence

$$\left|\frac{1}{n_m+1}\sum_{0}^{n_m} x_j(\omega)\right| < \varepsilon$$

for sufficiently large m, with probability 1 (Borel-Cantelli lemma), and since ε is arbitrary it follows that

$$\lim_{m\to\infty} \frac{1}{n_m+1} \sum_0^{n_m+1} x_j = 0 \tag{6.5}$$

with probability 1. Moreover,

$$\mathbf{E}\left\{\operatorname*{Max}_{n_m \le n < n_{m+1}} \left|\frac{1}{n+1}\sum_0^n x_j - \frac{1}{n+1}\sum_0^{n_m} x_j\right|^2\right\}$$

$$\le \frac{1}{(n_m+1)^2}\mathbf{E}\left\{\left[\sum_{n_m+1}^{n_{m+1}} |x_j|\right]^2\right\}$$

$$= \frac{1}{(n_m+1)^2}\sum_{n_m+1}^{n_{m+1}} \mathbf{E}\{|x_j x_k|\}$$

$$\le \frac{R(0)}{n_m^{\ 2}}[n_{m+1} - n_m]^2 \le \frac{K_1}{m^2}$$

for suitable K_1. It follows, as in the argument just used, that

$$\lim_{n\to\infty} \left|\frac{1}{n+1}\sum_0^n x_j - \frac{1}{n+1}\sum_0^{n_m} x_j\right| = 0, \qquad n_m \le n < n_{m+1}, \tag{6.6}$$

with probability 1, and, since the second factor $1/(n+1)$ in (6.6) can be replaced by $1/(n_m+1)$ (because their ratio approaches 1 when $n \to \infty$), (6.5) and (6.6) combine to give the conclusion of the theorem when $\mu = 0$. The general case can be reduced to this one by the trick used in the proof of the corollary to Theorem 6.1.

7. The estimation of $R(\nu)$ and $F(\lambda)$ from sample sequences

It is natural to take as an estimate of $R(\nu)$ the average

$$\frac{1}{n+1}\sum_{j=0}^{n} x_{\nu+j}\bar{x}_j. \tag{7.1}$$

The following theorem justifies this estimate, at least for large n. In this section ν is fixed and $X_n = x_{\nu+n}\bar{x}_n$.

THEOREM 7.1 *Suppose that the x_n and X_n processes are both stationary in the wide sense, that is, that*

$$\mathbf{E}\{|x_n|^2\} < \infty, \qquad \mathbf{E}\{|x_{\nu+n}x_n|^2\} < \infty, \qquad n = 0, \pm 1, \cdots$$

and that

$$\mathbf{E}\{x_{n+m}\bar{x}_n\}, \qquad \mathbf{E}\{x_{\nu+n+m}\bar{x}_{n+m}\overline{x_{\nu+n}\bar{x}_n}\}$$

are independent of n, for all $m = 0, \pm 1, \cdots$. *Then*

$$\text{(7.2)} \qquad \underset{n-m\to\infty}{\text{l.i.m.}} \frac{1}{n-m+1} \sum_{n=m}^{n} x_{\nu+j}\bar{x}_j$$

exists. This limit in the mean is $R(\nu)$ *(with probability* 1*) if and only if*

$$\text{(7.3)} \qquad \lim_{n\to\infty} \frac{1}{n+1} \sum_{j=0}^{n} \mathbf{E}\{x_{j+\nu}\bar{x}_j\overline{x_\nu\bar{x}_0}\} = |R(\nu)|^2.$$

If for some positive K and α

$$\text{(7.4)} \qquad \frac{1}{n+1} \sum_{j=-n}^{n} \left(1 - \frac{|j|}{n+1}\right) \mathbf{E}\{x_{j+\nu}\bar{x}_j\overline{x_\nu\bar{x}_0}\} - |R(\nu)|^2 \leq \frac{K}{n^\alpha}, \qquad n = 1, 2, \cdots,$$

then

$$\text{(7.2')} \qquad \lim_{n\to\infty} \frac{1}{n+1} \sum_{j=0}^{n} x_{\nu+j}\bar{x}_j = R(\nu)$$

with probability 1.

If the x_n *process is strictly stationary as well as stationary in the wide sense, the limit in* (7.2) *with* $m = 0$ *also exists as a limit with probability* 1. *In particular, this is true if the* x_n *process is real and Gaussian, with* $\mathbf{E}\{x_n\} = 0$ *for all n, and in this case* (7.2′) *is true with probability* 1 *for all* ν *if and only if*

$$\text{(7.5)} \qquad \lim_{n\to\infty} \frac{1}{n+1} \sum_{j=0}^{n} |R(j)|^2 = 0.$$

This condition is equivalent to the condition that the spectral distribution function of the x_n *process have no discontinuities.*

According to this theorem the estimate (7.1) of $R(\nu)$ [although it always has the correct expectation $R(\nu)$] is only "consistent" in the usual statistical sense, that is, asymptotically equal to $R(\nu)$, if specific restrictions are imposed on the x_n process. The restrictions have the basic effect of decreasing the influence of the past of the x_n process on the future.

To prove the theorem we note that according to its hypotheses

$$\mathbf{E}\{X_n\} = \mathbf{E}\{x_{\nu+n}\bar{x}_n\} = R(\nu)$$

is independent of n. Hence the process with variables $\{X_n - R(\nu)\}$ is stationary in the wide sense and has zero expectations. It now follows from Theorem 6.1 that

$$\underset{n\to\infty}{\text{l.i.m.}} \frac{1}{n-m+1} \sum_{j=m}^{n} [X_j - R(\nu)] = \underset{n\to\infty}{\text{l.i.m.}} \frac{1}{n-m+1} \sum_{j=m}^{n} x_{\nu+j}\bar{x}_j - R(\nu)$$

exists. Moreover, according to Theorem 6.1 [see (6.4)] this limit is 0 with probability 1 if and only if the Césaro limit at ∞ of the covariance function of the $X_n - R(\nu)$ process is 0. This condition is (7.3). Theorem 6.2 applied to the $X_n - R(\nu)$ process gives condition (7.4) as a sufficient condition for (7.2′). If the x_n process is strictly stationary, the X_n process is also strictly stationary; hence, since $\mathbf{E}\{|X_n|\} \leq \mathbf{E}\{|x_0|^2\}$, the strong law of large numbers is applicable to the X_n process. That is, the limit in (7.2′) exists with probability 1. It may or may not be $R(\nu)$; for example, it is $R(\nu)$, according to Theorem 2.1, if the x_n process is metrically transitive. In particular, if the x_n process is real and Gaussian, it is strictly stationary if $\mathbf{E}\{x_n\} = 0$ for all n, and if it is stationary in the wide sense. If these conditions are true, (7.2′) is true if and only if (7.3) is true, and in this case the expectations can be evaluated; (7.3) becomes

$$\lim_{n\to\infty} \frac{1}{n+1} \sum_{j=0}^{n} [R(j)^2 + R(j-\nu)R(j+\nu)] = 0. \tag{7.6}$$

If this condition is true for $\nu = 0$, (7.5) is satisfied. Conversely, if (7.5) is satisfied, (7.6) is also satisfied, since by Schwarz's inequality

$$\begin{aligned}
&\frac{1}{n+1} \sum_{j=0}^{n} [R(j)^2 + R(j-\nu)R(j+\nu)] \\
&\qquad \leq \frac{1}{n+1} \sum_{j=0}^{n} R(j)^2 + \frac{1}{2n+2} \sum_{j=-\nu}^{n-\nu} R(j)^2 + \frac{1}{2n+2} \sum_{j=\nu}^{n+\nu} R(j)^2 \\
&\qquad = \frac{2}{n+1} \sum_{j=0}^{n} R(j)^2 + o(1).
\end{aligned}$$

The above discussion is entirely symmetric in n, so that the average in (7.2′) can be replaced by $\frac{1}{n+1} \sum_{j=-n}^{0}$. The final statement of the theorem is implied by the following theorem:

THEOREM 7.2 *If $\{R(n)\}$ is any covariance sequence, that is, any positive definite sequence,*

$$R(n) = \int_{-1/2}^{1/2} e^{2\pi i n\lambda}\, dF(\lambda), \tag{7.7}$$

then

$$\lim_{n\to\infty} \frac{1}{n+1} \sum_{j=0}^{n} |R(j)|^2 = \sum_{\lambda} [F(\lambda+) - F(\lambda-)]^2. \tag{7.8}$$

In fact,

$$\frac{1}{n+1}\sum_{j=0}^{n}|R(j)|^2 = \frac{1}{n+1}\int_{-1/2}^{1/2}\int_{-1/2}^{1/2}\sum_{j=0}^{n} e^{2\pi ij(\lambda-\mu)}\,dF(\lambda)\,dF(\mu)$$

$$= \int_{-1/2}^{1/2}\int_{-1/2}^{1/2}\frac{1-e^{2\pi i(n+1)(\lambda-\mu)}}{(n+1)[1-e^{2\pi i(\lambda-\mu)}]}\,dF(\lambda)\,dF(\mu).$$

Repeating an argument already used, since the integrand converges boundedly to $f(\lambda, \mu)$, which is 0 except that $f(\lambda, \lambda) \equiv 1$, it follows that

$$\lim_{n\to\infty}\frac{1}{n+1}\sum_{j=0}^{n}|R(j)|^2 = \int_{-1/2}^{1/2}\int_{-1/2}^{1/2} f(\lambda, \mu)\,dF(\lambda)\,dF(\mu)$$

$$= \int_{-1/2}^{1/2}[F(\lambda+)-F(\lambda-)]\,dF(\lambda) = \sum_{\lambda}[F(\lambda+)-F(\lambda-)]^2.$$

The sum on the right is of course effectively an enumerable sum, since $F(\lambda+)-F(\lambda-) > 0$ on at most an enumerable λ set.

Now consider the problem of estimating the spectral distribution of a stationary process from sample sequences. One way is to estimate a part of the covariance function and to use this to evaluate the spectral distribution function. The following theorem shows the possibility of the direct approach.

THEOREM 7.3 *Suppose that the x_n process is stationary (wide sense) and that for every ν*

$$\lim_{n\to\infty}\frac{1}{n+1}\sum_{j=0}^{n} x_{j+\nu}\,\bar{x}_j = R(\nu), \tag{7.9}$$

with probability 1. *Then, if μ_1 and μ_2 are continuity points of the spectral distribution function F,*

$$\lim_{n\to\infty}\int_{\mu_1}^{\mu_2}\frac{1}{n+1}\Big|\sum_{j=0}^{n} x_j e^{-2\pi ij\lambda}\Big|^2\,d\lambda = F(\mu_2)-F(\mu_1), \tag{7.10}$$

with probability 1. *The limit is uniform in any closed interval of continuity of F, with probability* 1. *If the limit in* (7.9) *exists only as a limit in the mean (or even only a limit in probability), the limit equation* (7.10) *will still be true as a limit in probability.*

To prove this define Φ_n by

$$\Phi_n(\lambda) = \int_{-1/2}^{\lambda}\frac{1}{n+1}\left|\sum_{j=0}^{n} x_j e^{-2\pi ij\mu}\right|^2 d\mu.$$

Then, if $|\nu| < n$,

$$\int_{-1/2}^{1/2} e^{2\pi i\nu\lambda}\, d\Phi_n(\lambda) = \frac{1}{n+1}\sum_{j=0}^{n-\nu} x_{j+\nu}\,\bar{x}_j.$$

Hence, from (7.9),

$$\begin{aligned}\lim_{n\to\infty}\int_{-1/2}^{1/2} e^{2\pi i\nu\lambda}\, d\Phi_n(\lambda) &= \lim_{n\to\infty}\frac{1}{n+1}\sum_{j=0}^{n-\nu} x_{j+\nu}\bar{x}_j \\ &= \lim_{n\to\infty}\frac{1}{n-\nu+1}\sum_{j=0}^{n-\nu} x_{j+\nu}\bar{x}_j \\ &= R(\nu) = \int_{-1/2}^{1/2} e^{2\pi i\nu\lambda}\, dF(\lambda),\end{aligned}$$

with probability 1. Thus the Fourier-Stieltjes coefficients of Φ_n converge to those of F. It follows (Lévy continuity theorem for monotone functions defined in closed finite intervals) that $\Phi_n \to F +$ const. at all the continuity points of F, for almost every sample sequence. This proves the first part of the theorem. If only p lim is supposed in (7.9), p lim can be derived in (7.10) by going to convergent subsequences.

Note that it is incorrect to deduce from (7.10) that

$$\Phi_n'(\lambda) = \frac{1}{n+1}\left|\sum_{j=0}^{n} x_j e^{-2\pi ij\lambda}\right|^2$$

can be used to approximate the spectral density $F'(\lambda)$. In fact, as a counterexample suppose that the x_j's are real and Gaussian, and mutually independent, with $\mathbf{E}\{x_j\} = 0$, $\mathbf{E}\{x_j^2\} = 1$. In this case

$$\begin{aligned} &R(0) = 1 \\ &R(n) = 0, \qquad n \neq 0, \end{aligned} \qquad F'(\lambda) \equiv 1.$$

The approximant $\Phi_n'(0)$ becomes the square of a real Gaussian random variable with mean 0 and variance 1 which of course cannot converge to $F'(0) = 1$ when $n \to \infty$. A similar argument takes care of the other values of λ. A somewhat more elegant complex counterexample is the following: Let the x_j's be mutually independent, with 0 means, each with mutually independent Gaussian real and imaginary parts having expectations 0 and variances $\frac{1}{2}$, so that $\mathbf{E}\{|x_j|^2\} = 1$. Then

$$x_0,\ x_1 e^{2\pi i\lambda},\ x_2 e^{4\pi i\lambda},\ \cdots$$

have these same properties, so that $\Phi_n'(\lambda)$ is the square of the absolute value of a complex Gaussian random variable with mean 0, whose real and imaginary parts are mutually independent and have variance $\frac{1}{2}$. When $n \to \infty$, $\Phi_n'(\lambda)$ cannot possibly converge to $F'(\lambda) = 1$, for any λ.

8. Absolutely continuous spectral distributions and moving averages

The spectral representation

$$x_n = \int_{-1/2}^{1/2} e^{2\pi i n\lambda}\, dy(\lambda), \qquad \mathbf{E}\{|dy(\lambda)|^2\} = dF(\lambda), \tag{8.1}$$

can be modified advantageously if F is absolutely continuous. In that case, if f is a Baire function satisfying

$$|f|^2 = F',$$

there is a $\tilde{y}(\lambda)$ process with orthogonal increments which satisfies

$$x_n = \int_{-1/2}^{1/2} e^{2\pi i n\lambda} f(\lambda)\, d\tilde{y}(\lambda), \qquad \mathbf{E}\{|d\tilde{y}(\lambda)|^2\} = d\lambda. \tag{8.2}$$

If $F'(\lambda)$ never vanishes, (8.2) is true with

$$\tilde{y}(\lambda) = \int_{-1/2}^{\lambda} \frac{1}{f(\mu)}\, dy(\mu).$$

If $F'(\lambda)$ may vanish, the proof needs a bit more care. Let $\{y_1(\lambda), \frac{1}{2} \le \lambda \le \frac{1}{2}\}$ be a stochastic process with orthogonal increments, satisfying

$$\mathbf{E}\{|dy_1(\lambda)|^2\} = d\lambda,$$

$$\mathbf{E}\{[y(\mu_1) - y(\lambda_1)]\overline{[y_1(\mu_2) - y_1(\lambda_2)]}\} = 0 \qquad (\text{all } \lambda_i, \mu_i).$$

It may be necessary to enlarge ω space by adjunction, as explained in II §2, to obtain such a $y_1(\lambda)$ process. Then, if $\Phi(\lambda) = 1$ when $F'(\lambda) = 0$, and $\Phi(\lambda) = 0$ otherwise, we can define $y(\lambda)$ to satisfy (8.2) by

$$\tilde{y}(\lambda) = \int_0^{\lambda} \frac{1}{f(\lambda)}\, dy(\lambda) + \int_0^{\lambda} \Phi(\lambda)\, dy_1(\lambda)$$

[where we take $1/f(\lambda)$ as 0 when $f(\lambda) = 0$]. In all cases, if $\mathbf{E}\{x_n\} = 0$ for all n we can choose $\tilde{y}(\lambda)$ in such a way that $\mathbf{E}\{d\tilde{y}(\lambda)\} = 0$.

As an application of this spectral representation we consider processes of moving averages, that is, processes defined by series of the form

$$x_n = \sum_{j=-\infty}^{\infty} c_j \xi_{n+j} \tag{8.3}$$

where $\cdots, \xi_0, \xi_1, \cdots$ are mutually orthogonal random variables, with

$$\mathbf{E}\{\xi_m\bar{\xi}_n\} = \delta_{mn},$$

so that the ξ_n's form an orthonormal set. It is supposed that $\sum_{n=-\infty}^{\infty} |c_n|^2 < \infty$, and this implies that the series in (8.3) converges in the mean (Riesz-Fischer theorem). We have

$$\mathbf{E}\{x_{m+n}\bar{x}_m\} = \sum_{j=-\infty}^{\infty} c_{j-n}\bar{c}_j = R(n).$$

Since the left side is thus independent of m, the x_n process is stationary in the wide sense. If the ξ_n process is strictly stationary, for example if the ξ_n's are mutually independent, with a common distribution function, the x_n process will also be strictly stationary. We shall now prove that a *stationary* (*wide sense*) *process is a process of moving averages if and only if its spectral distribution is absolutely continuous.* In the first place, if the x_n process is a process of moving averages, given by (8.3), define $c(\lambda)$ by

$$c(\lambda) = \sum_{-\infty}^{\infty} c_k e^{2\pi i k\lambda}$$

where the series converges in the mean. Then, using the Parseval identity,

$$R(n) = \sum_{j=-\infty}^{\infty} c_{j-n}\bar{c}_j = \int_{-1/2}^{1/2} [c(\lambda)e^{2\pi i n\lambda}]\overline{c(\lambda)}\, d\lambda$$

$$= \int_{-1/2}^{1/2} |c(\lambda)|^2 e^{2\pi i n\lambda}\, d\lambda,$$

so that the spectral distribution function is absolutely continuous, with

$$F'(\lambda) = |c(\lambda)|^2.$$

Conversely, if F is an absolutely continuous spectral distribution function of some x_n process, we use the spectral representation (8.2) and insert in it the Fourier series of f, which converges in the mean:

$$f(\lambda) = \sum_{-\infty}^{\infty} \gamma_k e^{2\pi i k\lambda},$$

obtaining

$$x_n = \int_{-1/2}^{1/2} e^{2\pi i n\lambda} \sum_{-\infty}^{\infty} \gamma_k e^{2\pi i k\lambda}\, d\tilde{y}(\lambda)$$

$$= \sum_{-\infty}^{\infty} \gamma_k \xi_{n+k}, \qquad \xi_m = \int_{-1/2}^{1/2} e^{2\pi i m\lambda}\, d\tilde{y}(\lambda).$$

Thus x_n has the representation (8.3); the orthonormality of the ξ_m's is verified by inspection.

It is of some interest to evaluate explicitly the $y(\lambda)$ in the spectral representation (8.1) for a process given by (8.3). This is easily done, and we have

$$y(\lambda_2) - y(\lambda_1) = \sum_{-\infty}^{\infty} \left(\int_{\lambda_1}^{\lambda_2} c(\lambda) e^{-2\pi i \nu \lambda} \, d\lambda \right) \xi_\nu,$$

where $c(\lambda)$ was defined above.

9. Linear operations on stationary processes

Let $\{x_n, -\infty < n < \infty\}$ be a stationary (wide sense) process with spectral representation

$$x_n = \int_{-1/2}^{1/2} e^{2\pi i n \lambda} \, dy(\lambda), \qquad \mathbf{E}\{|dy(\lambda)|^2\} = dF(\lambda). \tag{9.1}$$

By a linear operation on the process we mean a transformation taking x_n into $\hat{x}_n$, where $\hat{x}_n$ is either a finite sum of the form

$$\hat{x}_n = \sum_j c_j x_{n+j}$$

or is a limit in the mean of such sums. The $\hat{x}_n$ process is then stationary (wide sense). Since

$$\sum_j c_j x_{n+j} = \int_{-1/2}^{1/2} e^{2\pi i n \lambda} \left(\sum_j c_j e^{2\pi i j \lambda} \right) dy(\lambda)$$

and since convergence in the mean for stochastic integrals corresponds to convergence in the mean of the integrands [λ weighting $dF(\lambda)$], it is clear that the most general $\hat{x}_n$ process is given by

$$\hat{x}_n = \int_{-1/2}^{1/2} e^{2\pi i n \lambda} c(\lambda) \, dy(\lambda), \qquad \int_{-1/2}^{1/2} |c(\lambda)|^2 \, dF(\lambda) < \infty, \tag{9.2}$$

where $c(\cdot)$ is measurable with respect to F measure. The covariance function of the $\hat{x}_n$ process is given by

$$\hat{R}(n) = \int_{-1/2}^{1/2} e^{2\pi i n \lambda} |c(\lambda)|^2 \, dF(\lambda)$$

so that the spectral distribution function is given by

$$\hat{F}(\lambda) = \int_0^{\lambda} |c(\mu)|^2 \, dF(\mu).$$

In other words spectral intensities are multiplied by the factor $|c(\lambda)|^2$.

The processes of moving averages discussed in §8 are a special case; they are obtained by performing linear operations on an x_n process whose variables are orthogonal to each other.

As an example of an application of a linear operation consider the result of smoothing data by the use of moving averages. To be definite suppose that the smoothing is accomplished by averaging three consecutive x_n's,

$$\hat{x}_n = \tfrac{1}{3}(x_{n-1} + x_n + x_{n+1}).$$

Then $c(\lambda) = \frac{1}{3}(e^{-2\pi i\lambda} + 1 + e^{2\pi i\lambda})$ so that spectral intensities are multiplied by the factor

$$\tfrac{1}{9}\left|e^{-2\pi i\lambda} + 1 + e^{2\pi i\lambda}\right|^2 = \frac{(1 + 2\cos 2\pi\lambda)^2}{9}.$$

This means that frequencies near $\lambda = \frac{1}{3}$ become relatively less important. Any such averaging smooths the data at the cost of changing the frequency relations of the process.

10. Rational spectral densities (in $e^{2\pi i\lambda}$)

In this section we shall treat stationary (wide sense) processes with absolutely continuous spectral distributions having spectral densities $F'(\lambda) \not\equiv 0$ which are rational functions of $e^{2\pi i\lambda}$,

$$F'(\lambda) = c' e^{2\pi i\nu\lambda} \frac{\prod_{j=1}^{\alpha'} (e^{2\pi i\lambda} - w_j')}{\prod_{j=1}^{\beta'} (e^{2\pi i\lambda} - z_j')}, \qquad c' \neq 0, \quad w_j', z_j' \neq 0,$$

where no w_j' is a z_k'. This expression for F' can be put in a simpler form as follows. In the first place no z_j' can have modulus 1, since F' is integrable, and any w_j' of modulus 1 must appear an even number of times since $F' \geq 0$. In the second place, since F' is real,

$$c' e^{2\pi i\nu\lambda} \frac{\prod_{j=1}^{\alpha'} (e^{2\pi i\lambda} - w_j')}{\prod_{j=1}^{\beta'} (e^{2\pi i\lambda} - z_j')} \equiv \bar{c}' e^{-2\pi i\nu\lambda} \frac{\prod_{j=1}^{\alpha'} (e^{-2\pi i\lambda} - \bar{w}_j')}{\prod_{j=1}^{\beta'} (e^{-2\pi i\lambda} - \bar{z}_j')}$$

$$\equiv \bar{c}' e^{2\pi i\lambda(\beta' - \alpha' - \nu)} \frac{\prod_{j=1}^{\alpha'} (1/\bar{w}_j' - e^{2\pi i\lambda})\bar{w}_j'}{\prod_{j=1}^{\beta'} (1/\bar{z}_j' - e^{2\pi i\lambda})\bar{z}_j'}$$

Hence to each w_j' corresponds a $w_k' = 1/\bar{w}_j'$, and to each z_j' corresponds a $z_k' \neq 1/\bar{z}_j'$. Since, if $\zeta \neq 0$,

$$\left| e^{2\pi i\lambda} - \frac{1}{\bar{\zeta}} \right| = \frac{1}{|\zeta|} \left| \bar{\zeta} e^{2\pi i\lambda} - 1 \right| = \frac{1}{|\zeta|} \left| e^{2\pi i\lambda} - \zeta \right|,$$

we can write $F'(\lambda)$, which is non-negative, as the absolute value of any of the above expressions and obtain

$$F'(\lambda) = c \left| \frac{\prod_{j=1}^{\alpha} (e^{2\pi i\lambda} - w_j)}{\prod_{j=1}^{\beta} (e^{2\pi i\lambda} - z_j)} \right|^2 \qquad c > 0, \qquad \begin{matrix} 0 < |w_j| \leq 1 \\ 0 < |z_j| < 1 \end{matrix}.$$

Here the w_j's are the w_j''s of modulus < 1 and also those of modulus 1 counted half as many times as they appear in the w_j''s; the z_j's are the z_j''s of modulus < 1. We thus can write, finally,

$$F'(\lambda) = \left| \frac{\sum_0^{\alpha} A_j e^{2\pi ij\lambda}}{\sum_0^{\beta} B_j e^{2\pi ij\lambda}} \right|^2, \qquad A_0 B_0 A_\alpha B_\beta \neq 0, \tag{10.1}$$

where the roots of $\sum_0^{\beta} B_j z^j$ have modulus < 1 and those of $\sum_0^{\alpha} A_j z^j$ have roots of modulus ≤ 1, and the two polynomials have no common factors.

In particular, if the x_n process is real, F' is even. It follows that to every w_j corresponds a $w_k = \bar{w}_j$, and to every z_j corresponds a $z_k = \bar{z}_j$. Hence the z_j's and w_j's are conjugate imaginary in pairs, and the A_j's and B_j's are real, or can be made so by multiplying the numerator and denominator polynomials by a suitable constant, say $|A_0|/A_0$.

According to §8 the spectral representation of a stationary (wide sense) process of the type considered here can be written in the form

$$x_n = \int_{-1/2}^{1/2} e^{2\pi in\lambda} \frac{\sum_0^{\alpha} A_j e^{2\pi ij\lambda}}{\sum_0^{\beta} B_j e^{2\pi ij\lambda}} \, d\tilde{y}(\lambda), \qquad \mathbf{E}\{|d\tilde{y}(\lambda)|^2\} = d\lambda, \tag{10.2}$$

where the $\tilde{y}(\lambda)$ process has orthogonal increments. The covariance function is

$$R(n) = \int_{-1/2}^{1/2} e^{2\pi in\lambda} \left| \frac{\sum_0^{\alpha} A_j e^{2\pi ij\lambda}}{\sum_0^{\beta} B_j e^{2\pi ij\lambda}} \right|^2 d\lambda. \tag{10.3}$$

Since this integral can be written in the form

$$(10.3') \qquad R(n) = \frac{1}{2\pi i} \int_{|z|=1} z^{n+\beta-\alpha-1} \frac{\sum_0^\alpha A_j z^j \sum_0^\alpha \bar{A}_j z^{\alpha-j}}{\sum_0^\beta B_j z^j \sum_0^\beta \bar{B}_j z^{\beta-j}} dz,$$

$R(n)$ is the nth coefficient in the Laurent expansion in powers of z on $|z| = 1$ of a rational function. It follows that $R(n)$ decreases exponentially when $|n| \to \infty$ (because the Laurent series converges in an annulus containing $|z| = 1$). The value of $R(n)$ can be obtained by residues as follows. Suppose, as above, that the roots of $\sum_0^\beta B_j z^j$ have modulus < 1. Then, if these roots are distinct, say $z_1, \cdots, z_\beta$, residue theory gives the evaluation

$$R(n) = \sum_1^\beta c_j z_j^n, \qquad n \geq 0.$$

(The roots of $\sum_0^\beta \bar{B}_j z^{\beta-j}$ will necessarily have modulus > 1; they are $1/\bar{z}_1, \cdots, 1/\bar{z}_\beta$.) If the process is real, each non-real z_j can be paired with a $z_k = \bar{z}_j$, and $c_k = \bar{c}_j$. If some of the z_j's are multiple roots, the coefficients c_j will become polynomials in n. This form of R shows that R satisfies a linear difference equation with constant coefficients, and in fact

$$(10.4) \qquad \sum_{k=0}^\beta B_k R(n+k) = \frac{1}{2\pi i} \int_{|z|=1} z^{n+\beta-\alpha-1} \frac{\sum_0^\alpha A_j z^j \sum_0^\alpha \bar{A}_j z^{\alpha-j}}{\sum_0^\beta \bar{B}_j z^{\beta-j}} dz$$

$$= 0, \qquad n + \beta - \alpha - 1 \geq 0,$$

$$= A_0 \bar{A}_\alpha / \bar{B}_\beta, \qquad n + \beta - \alpha - 1 = -1,$$

since the quotient in the integrand has no poles when $|z| \leq 1$, and has value $A_0 A_\alpha / \bar{B}_\beta$ at $z = 0$. Define $\xi_{n+\beta}$ by

$$(10.5) \qquad \sum_{j=0}^\beta B_j x_{n+j} = \xi_{n+\beta}.$$

Then

$$\xi_{n+\beta} = \int_{-1/2}^{1/2} e^{2\pi i n\lambda} \sum_0^\alpha A_j e^{2\pi i j\lambda} \, d\tilde{y}(\lambda).$$

Using (10.4) and this expression for $\xi_{n+\beta}$, we find

$$(10.6) \qquad \mathbf{E}\{\xi_n \bar{x}_m\} = 0, \qquad n - m \geq \alpha + 1,$$

$$= A_0 \bar{A}_\alpha / \bar{B}_\beta, \qquad n - m = \alpha,$$

(10.7)
$$\begin{aligned} \mathbf{E}\{\xi_n\bar{\xi}_m\} &= 0, \qquad |n-m| \geq \alpha+1, \\ &= \sum_0^\alpha |A_j|^2, \qquad |n-m| = \alpha. \end{aligned}$$

Case 1 $\beta = 0$, $B_0 = 1$. In this case,

$$F'(\lambda) = \Big|\sum_0^\alpha A_j e^{2\pi ij\lambda}\Big|^2, \qquad A_0 A_\alpha \neq 0,$$

and

$$x_n = \int_{-1/2}^{1/2} e^{2\pi in\lambda} \sum_0^\alpha A_j e^{2\pi ij\lambda}\, d\tilde{y}(\lambda) = \sum_{j=0}^\alpha A_j \zeta_{n+j},$$

where

$$\zeta_k = \int_{-1/2}^{1/2} e^{2\pi ik\lambda} d\tilde{y}(\lambda),$$

so that the sequence $\{\zeta_k\}$ is an orthonormal sequence. Thus in this case (see §8) the x_n process is a process of moving averages, involving averages of $\alpha + 1$ successive terms. Note that $R(n) = 0$ for $|n| > \alpha$. Conversely, if this condition on R is satisfied, F is absolutely continuous, with spectral density the square of the absolute value of a polynomial in $e^{2\pi i\lambda}$ of order $\leq \alpha$, by (3.5).

THEOREM 10.1 *If a stationary (wide sense) x_n process satisfies the difference equation* (10.5), *where $B_0, \cdots, B_\beta$ are arbitrary except that $B_0 B_\beta \neq 0$ and $\sum_0^\beta B_j z^j$ has no zero of modulus* 1, *where the first line of* (10.7) *is true, and $\mathbf{E}\{|\xi_n|^2\} > 0$, then the spectral distribution of the x_n process is absolutely continuous, with derivative given by* (10.1), *for some $A_0, \cdots, A_\alpha$. The finite poles of $\sum_0^\alpha A_j z^j \Big/ \sum_0^\beta B_j z^j$ have modulus < 1 if and only if the first line of* (10.6) *is true.*

Under the stated hypotheses, the ξ_n process is stationary in the wide sense, under Case 1 discussed above. If the x_n process has spectral distribution function F, and if (10.5) is true, the spectral distribution function of the process defined by the left side of (10.5) is (cf. §9) given by

$$\int_{-1/2}^{\lambda} \Big|\sum_0^\beta B_j e^{2\pi ij\mu}\Big|^2 dF(\mu).$$

Thus, equating the two evaluations of the ξ_n process spectral distribution that we have obtained,

$$\int_{-1/2}^{\lambda} \Big|\sum_0^\beta B_j e^{2\pi ij\mu}\Big|^2 dF(\mu) = \int_{-1/2}^{\lambda} \Big|\sum_0^\alpha A_j e^{2\pi ij\mu}\Big|^2 d\mu.$$

Hence F is absolutely continuous, and F' is given by (10.1). If the zeros of $\sum_0^\beta B_j z^j$ have modulus < 1, we have proved that (10.6) is true. The same proof shows that both lines of (10.6) are true if the finite poles of the fraction specified in the statement of the theorem have modulus < 1. Conversely, if the first line of (10.6) is true,

$$0 = \mathbf{E}\{\bar{\xi}_n x_m\} = \frac{1}{2\pi i} \int_{|z|=1} z^{m-n+\beta-1} \left[\frac{\sum_0^\alpha A_j z^j}{\sum_0^\beta B_j z^j} \right] dz, \qquad n - m \geq \alpha + 1.$$

Hence the Laurent expansion of the bracketed quotient on $|z| = 1$ contains no powers of z higher than $\alpha - \beta$, in any event only a finite number of positive powers. It follows that the bracketed quotient has no finite poles for $|z| > 1$, as was to be proved.

THEOREM 10.2 *If a stationary (wide sense) x_n process has an absolutely continuous spectral distribution, with density given by* (10.1), *and if the zeros of $\sum_0^\beta B_j z^j$ have modulus < 1, then R satisfies the difference equation*

$$(10.4') \qquad \sum_0^\beta B_k R(n+k) = 0, \qquad n \geq -\beta + \alpha + 1$$
$$= A_0 \bar{A}_\alpha / \bar{B}_\beta \neq 0, \qquad n = -\beta + \alpha.$$

Conversely, if the covariance function of a stationary (wide sense) process satisfies this difference equation, aside from the specification of the value in the second line of (10.4′), *for some constants $B_0, \cdots, B_\beta$ with $B_0 B_\beta \neq 0$, and some integer $\alpha \geq 0$, then, if $\xi_{n+\beta}$ is defined by* (10.5), *the first lines of* (10.6) *and* (10.7) *are true, and*

$$\mathbf{E}\{\xi_n \bar{x}_{n-\alpha}\} \neq 0, \qquad \mathbf{E}\{|\xi_n|^2\} \neq 0.$$

We have already proved the direct half of this theorem. The verification of the converse is trivial. Note that Theorem 10.1 is available to complete the characterization of the x_n process under the hypotheses of the converse.

In the preceding theorems, we have avoided the case when the difference equation (10.5) is satisfied by $\xi_n = 0$. This case is not really relevant, because here the spectrum of the x_n process only contains finitely many points, that is, the spectral distribution function is constant except for a finite number of jumps.

Case 2 $\alpha = 0$, $A_0 = 1$. In this case, the ξ_n sequence defined above is an orthonormal sequence, by (10.7). If $\mathbf{E}\{\xi_n \bar{x}_m\} = 0$ for $n > m$, that is, according to Theorem 10.1, if the roots of $\sum_{0}^{\beta} B_j z^j$ have modulus < 1, the sum on the left in (10.5) is orthogonal to all the random variables $\cdots, x_{n+\beta-2}, x_{n+\beta-1}$. Hence

$$\frac{B_{\beta-1}x_{n+\beta-1} + \cdots + B_0 x_n}{-B_\beta}$$

is the projection of $x_{n+\beta}$ on the closed linear manifold of random variables generated by all previous x_j's. In particular, if $\beta = 1$, the process is a Markov process in the wide sense. Thus Case 2 can be described as the case covering those processes for which the projection of any x_n on its past involves only a finite segment of the past, and in which this projection is not x_n itself. The last proviso is to exclude the case $\xi_n = 0$. If this projection is considered the linear mean square prediction of x_n in terms of the past, the prediction error is ξ_n/B_β, and the mean square prediction error is

$$\mathbf{E}\{|\xi_n/B_\beta|^2\} = 1/|B_\beta|^2.$$

In particular, if the x_n process is real and Gaussian, and if $\mathbf{E}\{x_n\} = 0$ for all n, the process is strictly stationary, the above projections become conditional expectations, and, if $\beta = 1$, the process becomes a Markov process; if $\beta > 1$ the process is a multiple Markov process.

Finally, we remark that, if $\alpha = 0$ in the hypothesis of Theorem 10.1, that is, if $\alpha = 0$ in (10.7), then, as the proof of the theorem shows, the hypothesis that $\sum_{0}^{\beta} B_j z^j$ has no zero of modulus 1 is unnecessary.

CHAPTER XI

Stationary Processes—Continuous Parameter

1. Generalities; metric transitivity

(a) *Strictly stationary processes* We adopt our usual basic hypotheses: there is a probability measure $\mathbf{P}\{\cdot\}$ defined on a Borel field of ω sets. Here ω is a point of some space Ω. A family $\{\mathbf{T}_t, -\infty < t < \infty\}$ [$\{\mathbf{T}_t, 0 \le t < \infty\}$] of transformations taking points of Ω into points of Ω will be called a *translation group* [*semi-group*] *of measure preserving* 1–1 *point transformations* if each $\mathbf{T}_t$ is a 1–1 measure-preserving point transformation as defined in X §1, and if

$$\mathbf{T}_{s+t} = \mathbf{T}_s\mathbf{T}_t \tag{1.1}$$

identically in the indicated parameters. The transformation $\mathbf{T}_0$ will necessarily be the identity. In the group case $\mathbf{T}_{-t}$ will be the inverse of $\mathbf{T}_t$, and the family $\{\mathbf{T}_{-t}, -\infty < t < \infty\}$ will also be a translation group of measure-preserving 1–1 point transformations.

If x is a random variable, and if $\{\mathbf{T}_t, -\infty < t < \infty\}$ is a translation group of measure-preserving 1–1 point transformations, the stochastic process

$$\{\mathbf{T}_t x, -\infty < t < \infty\}$$

is strictly stationary. The corresponding result holds in the semi-group case.

A family $\{\mathbf{T}_t, -\infty < t < \infty\}$ [$\{\mathbf{T}_t, 0 \le t < \infty\}$] of measure-preserving set transformations will be called a *translation group* [*semi-group*] *of measure-preserving set transformations* if (1.1) is true modulo the sets of probability 0, that is, if for every measurable ω set Λ, and every choice of $\mathbf{T}_s\mathbf{T}_t\Lambda$, the latter choice is one of the images of Λ under $\mathbf{T}_{s+t}$. The transformation $\mathbf{T}_0$ will be the identity in the sense that every image of a measurable set Λ under $\mathbf{T}_t$ will differ from Λ by at most a set of probability 0. In the group case $\mathbf{T}_{-t}$ is the inverse of $\mathbf{T}_t$, and the family $\{\mathbf{T}_{-t}, -\infty < t < \infty\}$ is also a translation group of measure-preserving set transformations.

If x is a random variable, and if $\{\mathbf{T}_t, -\infty < t < \infty\}$ is a translation group of measure-preserving set transformations, the stochastic process

$$\{\mathbf{T}_t x, -\infty < t < \infty\}$$

is strictly stationary, no matter which version of $\mathbf{T}_t x$ is taken for each t. The corresponding result holds in the semi-group case.

Each translation group [semi-group] of measure-preserving 1–1 point transformations induces a translation group [semi-group] of measure-preserving set transformations, in the obvious way (see the analogous discussion in X §1), but all the latter groups and semi-groups cannot be induced in this way.

In the following we shall use a (t, ω) "two-dimensional" measure, the usual direct product of Lebesgue t measure and the given ω probability measure.

Let $\{\mathbf{T}_t, -\infty < t < \infty\}$ be a translation group of measure-preserving 1–1 point transformations and let Λ be a measurable ω set. Consider the (t, ω) set of all pairs (t, ω) with $\omega \in \mathbf{T}_t\Lambda$, that is, the set of all pairs (t, ω) with $\mathbf{T}_{-t}\omega \in \Lambda$. If this (t, ω) set is (t, ω) measurable for each Λ, the translation group is said to be measurable. The corresponding definition of measurability is made in the semi-group case. If x is a random variable,

$$(\mathbf{T}_t x)(\omega) = x(\mathbf{T}_{-t}\omega)$$

is measurable in the pair (t, ω) if the translation group or semi-group, as the case may be, is measurable. The group $\{\mathbf{T}_t, -\infty < t < \infty\}$ is measurable if and only if the inverse group $\{\mathbf{T}_{-t}, -\infty < t < \infty\}$ is measurable.

As an example of measurability of a transformation group suppose that Ω is the linear interval $(0, 1]$, and let ω measure be Lebesgue measure. Let $\mathbf{T}_t$ be the translation (mod 1) through the distance t. The point ω is uniquely determined by the value of $e^{2\pi i\omega}$. To prove that this translation group is measurable we must show that, if Λ is a Lebesgue measurable set on the perimeter C of the unit circle in the complex plane, the (t, ω) set defined by

$$e^{2\pi i(\omega - t)} \in \Lambda$$

is Lebesgue measurable. If Λ is a Borel set on C, this is obvious, since the exponential function is continuous. If Λ is a Lebesgue measurable set on C, it can be expressed as the union of a Borel set Λ_1 and a set Λ_2 of Lebesgue measure 0. It is thus sufficient to show that the (t, ω) set defined by

$$e^{2\pi i(\omega - t)} \in \Lambda_2$$

has two-dimensional Lebesgue measure 0. Since there is a Borel set Λ_3 on C of one-dimensional Lebesgue measure 0, with $\Lambda_2 \subset \Lambda_3$, it is sufficient to show that the (t, ω) set defined by

$$e^{2\pi i(\omega - t)} \in \Lambda_3$$

has two-dimensional Lebesgue measure 0. This (t, ω) set has already been seen to be measurable. For fixed t the ω cross section of this set is a rotation of Λ_3, and so has Lebesgue measure 0. Hence (Fubini's theorem) the (t, ω) set has measure 0, as was to be proved.

Let $\{\mathbf{T}_t, -\infty < t < \infty\}$ be a translation group of measure-preserving set transformations, and let Λ be a measurable ω set. Choose some image $\mathbf{T}_t\Lambda$ of Λ under $\mathbf{T}_t$ for each t, and consider the (t, ω) set of all pairs (t, ω) with $\omega \in \mathbf{T}_t\Lambda$. If for each Λ it is possible to choose the images $\{\mathbf{T}_t\Lambda, -\infty < t < \infty\}$ in such a way that the indicated (t, ω) set is (t, ω) measurable, the translation group is said to be measurable. The corresponding definition of measurability is made in the semi-group case. If a translation group [semi-group] of measure-preserving point transformations is measurable, the induced group [semi-group] of measure-preserving set transformations is measurable. A group $\{\mathbf{T}_t, -\infty < t < \infty\}$ of measure-preserving set transformations is measurable if and only if the inverse group $\{\mathbf{T}_{-t}, -\infty < t < \infty\}$ is measurable. If x is a random variable, $(\mathbf{T}_t x)(\omega)$ is measurable in the pair (t, ω) if the group or semi-group, as the case may be, is measurable, and if for each t the image $\mathbf{T}_t x$ is properly chosen.

We have seen that a translation group or semi-group of measure-preserving point or set transformations in conjunction with a random variable defines a strictly stationary stochastic process. Conversely, just as in the discrete parameter case of X §1, if $\{x_t, -\infty < t < \infty\}$ or $\{x_t, 0 \leq t < \infty\}$ is a strictly stationary stochastic process, there is a corresponding translation group or semi-group of measure-preserving set transformations which together with the random variable x_0 induces the stochastic process. Following the reasoning used in the discrete parameter case, if the ω space of the x_t process is function space, and if x_t is the tth coordinate variable, the transformations of the group or semi-group become point transformations, and by going to function space it is always possible to avoid the use of set (rather than point) transformations, and of semi-groups (rather than groups).

If $\{\mathbf{T}_t, -\infty < t < \infty\}$ $[\{\mathbf{T}_t, 0 \leq t < \infty\}]$ is a measurable translation group [semi-group] of measure-preserving 1–1 point transformations, and if x is a random variable, the x_t process defined by setting

$$x_t(\omega) = [\mathbf{T}_t x](\omega) = x(\mathbf{T}_{-t}\omega)$$

is measurable, since, as we have just remarked, $x_t(\omega)$ is (t, ω) measurable. However, if the group [semi-group] is a group [semi-group] of set transformations, the x_t process defined by setting $x_t = \mathbf{T}_t x$ may or may not be measurable, depending on the choice of $\mathbf{T}_t x$, which is not uniquely determined. In the language of II §2 an x_t process obtained in this way may not be measurable, but it will have a measurable standard modification, that is, there is a measurable $\tilde{x}_t$ process with

$$\mathbf{P}\{x_t(\omega) = \tilde{x}_t(\omega)\} = 1$$

for all t.

A measurable ω set is called *invariant* under a translation group or semi-group of measure-preserving point or set transformations if the set differs from its image under $\mathbf{T}_t$ by at most a set which may depend on t but which has probability 0 for each t. The invariant sets form a Borel field of ω sets. A random variable x is called *invariant* under such a group or semi-group if, for each t, $x = \mathbf{T}_t x$ with probability 1. In the group case, the same measurable ω sets and random variables are invariant with respect to the inverse group $\{\mathbf{T}_{-t}, -\infty < t < \infty\}$ as with respect to the given group. A translation group or semi-group of measure-preserving point or set transformations is called metrically transitive if the only invariant sets are those which have probability 0 or 1 (these sets are always invariant), that is, if the only invariant random variables are those which are identically constant with probability 1 (these random variables are always invariant.

We have seen that to each strictly stationary process there corresponds a unique translation group or semi-group of measure-preserving set transformations. The measure-preserving set transformations are defined on the Borel field of sets determined by conditions on the random variables of the process. The transformations are called *shifts*, and the group or semi-group is called the shift group or semi-group. Sets and random variables invariant relative to the shift group or semi-group are called invariant sets and random variables of the process, and the process is called metrically transitive if the shift group or semi-group is metrically transitive. A process is metrically transitive if and only if the corresponding coordinate space process, for which the shift transformations become point transformations, is metrically transitive. (This assumes that the measure in the coordinate space is the Kolmogorov measure determined by that of the finite dimensional sets, without further extensions.) If $\{\mathbf{T}_t, -\infty < t < \infty\}$ or $\{\mathbf{T}_t, 0 \leq t < \infty\}$ is a metrically transitive translation group or semi-group and if x is a random variable, the x_t process determined by setting $x_t = \mathbf{T}_t x$ is metrically transitive. If $\{x_t, -\infty < t < \infty\}$ is a strictly stationary stochastic process, the processes $\{x_{-t}, -\infty < t < \infty\}$ and $\{x_t, 0 \leq t < \infty\}$ are also strictly stationary.

These three processes have the same invariant sets and invariant random variables, so they are all metrically transitive if any one is. (See the discrete parameter argument in X §1.) If the first or second of the three processes is measurable, the other two are; but, if only the third is known to be measurable, all one can say is that the other two at least have measurable standard modifications.

Let $\{x_t, -\infty < t < \infty\}$ be a stochastic process, and let $\mathscr{F}_d$ be the Borel field of ω sets generated by those of the form

$$\{x_t(\omega) - x_s(\omega) \in G\}$$

where G is open, that is, $\mathscr{F}_d$ is the smallest Borel field with respect to which the x_t process differences are measurable. Then the sets of $\mathscr{F}_d$ are called the *difference sets* and $\mathscr{F}_d$ is called the *difference field.* It is sometimes useful to call the x_t process strictly stationary with respect to the difference field if, whenever $t_1 < \cdots < t_n$, the multivariate distribution of the random variables

$$x_{t_2} - x_{t_1}, \cdots, x_{t_n} - x_{t_{n-1}}$$

is independent of translations of the t axis. In particular, every process which is strictly stationary is also strictly stationary with respect to the difference field. The Brownian motion process is an example of a process which is strictly stationary with respect to the difference field but not strictly stationary. Repeating an argument we have used already, it is easily seen that if an x_t process is strictly stationary with respect to the difference field there is a unique measure-preserving set transformation (shift) $\mathbf{T}_t$ for which

$$\mathbf{T}_t(x_{t_2} - x_{t_1}) = x_{t_2+t} - x_{t_1+t}.$$

The shifts, defined on the sets of $\mathscr{F}_d$, form a translation group or semi-group depending on the parameter set of the given process. The concepts of invariant sets and random variables, and of metric transitivity (all relative to the difference field) are now defined as usual in terms of the shift transformations.

Example 1 *Markov chains* The continuous parameter discussion of this example is parallel to that given in the discrete parameter case, and will be omitted. The obvious continuous parameter version of X, Theorem 1.1 is true, and its proof requires only obvious changes.

Example 2 *Processes with independent increments* This example corresponds to the discrete parameter example of processes with mutually independent random variables. Suppose that $\{x_t, 0 \leq t < \infty\}$ is a process with independent increments. It is then strictly stationary relative to the difference fields if and only if the increments are strictly stationary. According to the following theorem such a process is always metrically transitive.

THEOREM 1.1 *If* $\{x_t, 0 \leq t < \infty\}$ *is a process with strictly stationary independent increments, it is metrically transitive relative to the difference field.*

The proof follows that of the discrete parameter case (see X, Theorem 1.2) and will be omitted. It is made to depend on a continuous parameter version of the zero-one law, a version formulated in terms of processes with independent increments.

Example 3 *Moving averages* Let the process $\{y_t, -\infty < t < \infty\}$ be a process with (strictly) stationary independent increments for which

$$\mathbf{E}\{|y_t - y_s|^2\} < \infty$$

$$\mathbf{E}\{y_t - y_s\} = 0.$$

Consider the x_t process defined by

$$x_t = \int_{-\infty}^{\infty} c(s)\, dy(s+t) = \int_{-\infty}^{\infty} c(s-t)\, dy(s).$$

The y_t process is metrically transitive relative to the difference field, by Theorem 1.1. Let $\{\mathbf{T}_t, -\infty < t < \infty\}$ be the corresponding shift group. The x_t process is strictly stationary, and metrically transitive, because it is generated by the metrically transitive group $\{\mathbf{T}_t, -\infty < t < \infty\}$ applied to x_0, $x_t = \mathbf{T}_t x_0$.

(b) *Wide sense stationary processes* We use in the following the concepts of X §1 (b). A family of transformations $\{\mathbf{U}_t, -\infty < t < \infty\}$ $[\{\mathbf{U}_t, 0 \leq t < \infty\}]$ operating on the random variables of a closed linear manifold of random variables will be called a *translation group* [*semi-group*] *of isometric transformations* if the transformations are isometric and if

$$\mathbf{U}_{s+t} = \mathbf{U}_s \mathbf{U}_t$$

for all s, t (modulo the random variables which vanish almost everywhere). The transformation $\mathbf{U}_0$ will necessarily be the identity. In the group case $\mathbf{U}_{-t}$ will be the inverse of $\mathbf{U}_t$, the isometric transformations will be unitary, and $\{\mathbf{U}_{-t}, -\infty < t < \infty\}$ will also be a translation group of unitary transformations, called the *inverse group.*

If $\{\mathbf{U}_t, -\infty < t < \infty\}$ or $\{\mathbf{U}_t, 0 \leq t < \infty\}$ is a translation group or semi-group of isometric transformations, and if x is a random variable in the domain of definition of the transformations, the x_t process defined by $x_t = \mathbf{U}_t x$ is stationary in the wide sense. Conversely, if $\{x_t, -\infty < t < \infty\}$ or $\{x_t, 0 \leq t < \infty\}$ is a process which is stationary in the wide sense, there is a corresponding translation group or semi-group of isometric transformations such that, for each t, $x_t = \mathbf{U}_t x_0$ with probability 1. The isometries are defined on the closed linear manifold generated by the x_t's.

If $\{\mathbf{U}_t, -\infty < t < \infty\}$ or $\{\mathbf{U}_t, 0 \leq t < \infty\}$ is a translation group or semi-group of isometric transformations, and if x is a random variable such that $\mathbf{U}_t x = x$ with probability 1 for each t, the random variable x is called *invariant*. The transformation group or semi-group is called *metrically transitive in the wide sense* if the only invariant random variables are those which vanish with probability 1. If the isometric transformations are shift isometries derived from a process which is stationary in the wide sense, the invariant random variables are said to be invariant random variables of the process, and the process is said to be metrically transitive in the wide sense if the shift group or semi-group of isometries is metrically transitive in the wide sense.

If the stochastic process $\{x_t, -\infty < t < \infty\}$ is stationary in the wide sense, the processes $\{x_{-t}, -\infty < t < \infty\}$ and $\{x_t, 0 \leq t < \infty\}$ are also stationary in the wide sense. These three processes have the same wide sense invariant random variables, and it follows that either all three processes or none of the three are metrically transitive in the wide sense.

If $\{x_t, -\infty < t < \infty\}$ is a stochastic process, with

$$\mathbf{E}\{|x_t - x_s|^2\} < \infty,$$

let $\mathfrak{M}_d$ be the closed linear manifold generated by the differences $x_t - x_s$. The manifold $\mathfrak{M}_d$ will be called the difference manifold of the process, and the random variables in it the difference random variables of the process. If the process is stationary in the wide sense, relative to the difference random variables, that is, if the process has stationary increments in the wide sense, there is a translation group $\{\mathbf{U}_t, -\infty < t < \infty\}$ of unitary transformations defined on $\mathfrak{M}_d$ such that

$$\mathbf{U}_t(x_{t_2} - x_{t_1}) = x_{t_2+t} - x_{t_1+t}$$

with probability 1. The concepts of invariant random variables and metric transitivity in the wide sense relative to the difference manifold are then referred back to the $\mathbf{U}_t$ group. The corresponding remarks for a process $\{x_t, 0 \leq t < \infty\}$ lead to a $\mathbf{U}_t$ semi-group.

The following is the wide sense version of Example 2.

Example 4 *Processes with orthogonal increments* A process $\{x_t, 0 \leq t < \infty\}$ with orthogonal increments is stationary in the wide sense relative to the difference manifold if and only if its increments are stationary in the wide sense, and we have seen in IV that this is true if and only if

$$\mathbf{E}\{|x_t - x_s|^2\} = \text{const. } |t - s|.$$

According to the following theorem such a process is always metrically transitive in the wide sense relative to the difference manifold.

THEOREM 1.2 *If $\{x(t),\ 0 \leq t < \infty\}$ is a process with stationary (wide sense) orthogonal increments, it is metrically transitive (wide sense) relative to the difference manifold.*

Suppose that

$$\mathbf{E}\{|dx(t)|^2\} = a\,dt.$$

If $a = 0$, then $x(t) = x(s)$ with probability 1, for each pair s, t, and the difference manifold $\mathfrak{M}_d$ therefore only contains random variables which vanish with probability 1. In this case, then, the theorem is trivially true. If $a > 0$, consider the class of random variables $\mathfrak{M}$ of the form

$$x = \int_0^\infty c(t)\,dx(t),$$

where $c(\cdot)$ is Lebesgue measurable and

$$\int_0^\infty |c(t)|^2\,dt < \infty.$$

Since

$$x(t_2) - x(t_1) = \int_{t_1}^{t_2} dx(t),$$

$\mathfrak{M}$ includes every difference $x(t_2) - x(t_1)$. Since

$$\mathbf{E}\{|\int_0^\infty c_1(t)\,dx(t) - \int_0^\infty c_2(t)\,dx(t)|^2\} = a\int_0^\infty |c_1(t) - c_2(t)|^2\,dt,$$

root mean square distance between integrands $c(\cdot)$ is equal, apart from a non-zero constant factor, to root mean square distance between the random variables x. Since the class of integrands is the closed linear manifold generated by the integrands which are 1 on finite intervals and vanish otherwise, the class of integrals is the closed linear manifold generated by $x(t)$ differences, that is, $\mathfrak{M} = \mathfrak{M}_d$. The relation between integrand $c(t)$ and $x \in \mathfrak{M}_d$ determined by the integral transformation is one to one, modulo integrands which vanish almost everywhere and random variables which vanish with probability 1. In particular, if x is an invariant random variable in the difference manifold,

$$x = \int_0^\infty c(t)\,dx(t) = \int_0^\infty c(t)\,dx(t+s) = \int_s^\infty c(t-s)\,dx(t)$$

with probability 1, for each $s > 0$. Then, comparing integrands, $c(t) = 0$ for almost all t in the interval $(0, s)$. It follows that $c(t) = 0$ for almost all t, so that $x = 0$ with probability 1, and the $x(t)$ process is metrically transitive (wide sense) with respect to the difference manifold, as was to be proved.

2. The strong law of large numbers for strictly stationary stochastic processes

The ergodic theorem in the continuous parameter case, whose probability name heads this section, is usually stated as follows. *Let* $\{\mathbf{S}_t, -\infty < t < \infty\}$ *be a measurable translation group of measure-preserving* 1–1 *point transformations, and let* x *be a measurable and integrable function on the space involved. Then*

$$\lim_{t\to\infty} \frac{1}{t} \int_0^t x(\mathbf{S}_s\omega)\, ds$$

exists and is finite for almost all ω. Since the inverse group $\{\mathbf{S}_{-t}, -\infty < t < \infty\}$ is a group of the same type, $\mathbf{S}_{-s}$ can be substituted for $\mathbf{S}_s$ in the above integrand. In this form the theorem can be generalized slightly by replacing the group of point transformations by a semi-group of set transformations to obtain the following slightly more general version. *Let* $\{\mathbf{T}_t, 0 \leq t < \infty\}$ *be a measurable translation semi-group of measure-preserving set transformations, and let* x *be a measurable and integrable function on the space involved. Then if for each* t *the random variable* $\mathbf{T}_t x$ *is chosen to make* $(\mathbf{T}_t x)(\omega)$ *measurable in the pair* (t, ω), *it follows that*

$$\lim_{t\to\infty} \frac{1}{t} \int_0^t (\mathbf{T}_s x)(\omega)\, ds$$

exists and is finite for almost all ω. The $\mathbf{T}_t$ group corresponds to the inverse group of the $\mathbf{S}_t$ group in the previous version. According to §1 the following statement of the theorem differs only verbally from the one just given as far as the existence of the limit is concerned. In this form the theorem is also called the strong law of large numbers for strictly stationary stochastic processes (continuous parameter case).

THEOREM 2.1 *Let* $\{x_t, 0 \leq t < \infty\}$ *be a measurable strictly stationary stochastic process, with* $\mathbf{E}\{|x_0|\} < \infty$, *and let* $\mathscr{I}$ *be the Borel field of invariant* ω *sets. Then*

$$\lim_{t\to\infty} \frac{1}{t} \int_0^t x_s(\omega)\, ds = \mathbf{E}\{x_0 \mid \mathscr{I}\} \tag{2.1}$$

with probability 1. *In particular, if the process is metrically transitive, the right-hand side of* (2.1) *can be replaced by* $\mathbf{E}\{x_0\}$.

It is clear that the content of the theorem is not changed if the average on the left in (2.1) is replaced by

$$\frac{1}{t}\int_{u}^{u+t} x_s(\omega)\,ds$$

(fixed u) and that the limit is unaltered. If the parameter range of the process is the whole t line, we can replace x_t by x_{-t}, and find that the corresponding limiting average exists. Since the invariant sets of the inverse shift group are the same as those of the shift group, the limiting average is the same as that in (2.1). It then follows that also

$$\lim_{t\to\infty}\frac{1}{2t}\int_{-t}^{t} x_s(\omega)\,ds = \mathbf{E}\{x_0 \mid \mathscr{I}\}$$

with probability 1. However, the limit

$$\lim_{t_2-t_1\to\infty}\frac{1}{t_2-t_1}\int_{t_1}^{t_2} x_s(\omega)\,ds$$

does not exist with probability 1, in general.

Since it is supposed in the theorem that the process is measurable, the sample functions are almost all measurable. Moreover, since

$$\int_a^b \mathbf{E}\{|x_s(\omega)|\}\,ds = (b-a)\mathbf{E}\{|x_0(\omega)|\} < \infty,$$

it follows from II, Theorem 2.7, that the sample functions are almost all Lebesgue integrable over finite intervals. Thus the integral averages of the theorem are defined with probability 1. The existence of the limiting averages can be proved by giving the continuous parameter version of the discrete parameter proof (X, Theorem 2.1) or can be reduced to the conclusion of that theorem as follows. Define $\hat{x}_m$, $\hat{y}_m$ by

$$\hat{x}_m(\omega) = \int_m^{m+1} x_s(\omega)\,ds, \qquad \hat{y}_m(\omega) = \int_m^{m+1} |x_s(\omega)|\,ds.$$

The $\hat{x}_m$ and $\hat{y}_m$ processes are strictly stationary integral parameter processes. Hence by X, Theorem 2.1,

$$\lim_{n\to\infty}\frac{1}{n}\sum_0^{n-1}\hat{x}_j(\omega) = \lim_{n\to\infty}\frac{1}{n}\int_0^n x_s(\omega)\,ds \tag{2.2}$$

exists and is finite with probability 1, and the same is true for the $\hat{y}_j$ averages. The latter fact implies that

$$(2.3) \qquad \lim_{n\to\infty} \frac{\hat{y}_n(\omega)}{n} = \lim_{n\to\infty} \frac{1}{n} \int_n^{n+1} |x_s(\omega)|\, ds = 0$$

with probability 1. If $[t]$ is the largest integer $\leq t$,

$$\frac{1}{t}\int_0^t x_s(\omega)\, ds = \frac{1}{[t]}\left\{\int_0^{[t]} x_s(\omega)\, ds\right\}\frac{[t]}{t} + \varepsilon_t$$

where, using (2.3),

$$(2.4) \qquad |\varepsilon_t| = \left|\frac{1}{t}\int_{[t]}^t x_s(\omega)\, ds\right| \leq \frac{1}{[t]}\int_{[t]}^{[t]+1} |x_s(\omega)|\, ds \to 0 \qquad (t \to \infty)$$

with probability 1. Hence the existence of the limit in (2.2) with probability 1 combined with (2.4) implies the existence of the limit in (2.1) with probability 1. The rest of the theorem is proved in exactly the same way as the analogous part of X, Theorem 2.1. As in the discrete parameter case, if $\delta \geq 1$ and if $\mathbf{E}\{|x_0|^\delta\} < \infty$,

$$\lim_{t\to\infty} \mathbf{E}\left\{\left|\frac{1}{t}\int_0^t x_s(\omega)\, ds - \mathbf{E}\{x_0 \mid \mathscr{I}\}\right|^\delta\right\} = 0,$$

that is, there is convergence to the limit in the mean of order δ. The corollary to X, Theorem 2.1, becomes here

COROLLARY *If μ is real,*

$$\lim_{t\to\infty} \frac{1}{t}\int_0^t x_s(\omega) e^{-2\pi i s\mu}\, ds = \tilde{x}(\mu)$$

exists with probability 1. *The random variable $\tilde{x}(\mu)$ has finite expectation,*

$$\mathbf{E}\{\tilde{x}(0)\} = \mathbf{E}\{x_0\}$$

$$\mathbf{E}\{\tilde{x}(\mu)\} = 0, \qquad \mu \neq 0,$$

and is transformed by the shift transformation with parameter value t into $e^{2\pi i t\mu}\,\tilde{x}(\mu)$. Unless μ is in an (at most denumerable) exceptional set,

$$\mathbf{P}\{\tilde{x}(\mu) = 0\} = 1;$$

if $\mathbf{E}\{|x_0|^2\} < \infty$, *then* $\mathbf{E}\{|\tilde{x}(\mu)|^2\} < \infty$ *also,*

$$\text{(2.5)} \qquad \underset{t\to\infty}{\text{l.i.m.}} \frac{1}{t}\int_0^t x_s(\omega)e^{-2\pi i s\mu}\, ds = \tilde{x}(\mu),$$

and

$$\mathbf{E}\{\tilde{x}(\mu_1)\overline{\tilde{x}(\mu_2)}\} = 0, \qquad \mu_1 \neq \mu_2.$$

The proof follows the discrete parameter proof and will be omitted. As in the theorem, the average $\frac{1}{t}\int_0^t$ can be replaced by the average $\frac{1}{t}\int_u^{u+t}$, and so on.

3. The covariance function of a stationary process; examples

In this and the following sections we shall consider stationary (wide sense) complex processes $\{x(t),\ t \in T\}$ as defined in II §8, with particular reference to their harmonic analysis. Since most of the theorems are exact analogues of those in the discrete parameter case, the details will be omitted. One new hypothesis will be introduced: it will be assumed that

$$\text{(3.1)} \qquad \lim_{t-s\to 0} \mathbf{E}\{|x(t) - x(s)|^2\} = 0.$$

According to II §2 this continuity hypothesis implies that some standard modification of the $x(t)$ process is separable and measurable. We recall that the transition from the original process to a standard modification does not affect the joint distributions of finite aggregates of $x(t)$'s. Conversely it can be shown that, if a standard modification of the $x(t)$ process is measurable, then (3.1) is satisfied. Thus (3.1) is a minimal continuity hypothesis, and we shall assume that it is satisfied whenever we discuss processes stationary in the wide sense. The parameter set of such a process will always be taken to be either $(-\infty, \infty)$ or $[0, \infty)$.

If a process is measurable, we have seen in II §2 that almost all its sample functions are Lebesgue measurable. Moreover, if the process is stationary in the wide sense as well as measurable, the squares of its sample functions are almost all Lebesgue integrable over every finite interval, using II, Theorem 2.7, because then

$$\int_a^b \mathbf{E}\{|x(t)|^2\}\, dt = (b-a)\mathbf{E}\{|x(0)|^2\} < \infty.$$

The covariance function of a process which is stationary in the wide sense is defined by

$$R(t) = \mathbf{E}\{x(s+t)\overline{x(s)}\}.$$

The following theorem describes the class of covariance functions.

THEOREM 3.1 *If* (3.1) *is satisfied, the covariance function R is positive definite, that is, it is continuous, and*

$$R(-t) = \overline{R(t)}$$

$$\sum_{m,n=1}^{N} R(t_m - t_n)\alpha_m \bar{\alpha}_n \geq 0$$

for every finite set of parameter values $t_1, \cdots, t_N$ *and complex numbers* $\alpha_1, \cdots, \alpha_N$. *Conversely, any function R satisfying these conditions, and continuous when* $t = 0$, *is everywhere continuous, and is the covariance function of a stationary (wide sense) process. If the covariance function is real, the latter process can be taken as real.*

The continuity of the covariance function follows from the inequality

$$\begin{aligned} |R(t) - R(s)| &= |\mathbf{E}\{[x(t) - x(s)]\overline{x(0)}\}| \\ &\leq [\mathbf{E}\{|x(t) - x(s)|^2\}\mathbf{E}\{|x(0)|^2\}]^{1/2} \end{aligned}$$

in which the right side approaches 0 when $t - s \to 0$, by (3.1). The rest of the proof follows that of X, Theorem 3.1, and will be omitted.

The continuous parameter version of X, Theorem 3.2, is

THEOREM 3.2 *A function R is positive definite if and only if it can be expressed in the form*

$$R(t) = \int_{-\infty}^{\infty} e^{2\pi i t\lambda}\, dF(\lambda), \tag{3.2}$$

where F is monotone non-decreasing and bounded. The function F is uniquely determined, if suitably normalized, by the equation

$$\begin{aligned} &\frac{F(\lambda_2 +) + F(\lambda_2 -)}{2} - \frac{F(\lambda_1 +) + F(\lambda_1 -)}{2} \\ &\qquad = \lim_{T\to\infty} \int_{-T}^{T} \frac{e^{-2\pi i\lambda_2 t} - e^{-2\pi i\lambda_1 t}}{-2\pi i t} R(t)\, dt. \end{aligned} \tag{3.3}$$

If R is real, (3.2) *can be replaced by*

$$R(t) = \int_{0}^{\infty} \cos 2\pi t\lambda\, dG(\lambda), \tag{3.2$'$}$$

where G is monotone non-decreasing and bounded, and is uniquely determined, if suitably normalized, by

$$\begin{aligned} &\frac{G(\lambda +) + G(\lambda -)}{2} - G(0) \\ &\qquad = \frac{2}{\pi} \int_{0}^{\infty} \frac{\sin 2\pi\lambda t}{t} R(t)\, dt, \qquad \lambda > 0. \end{aligned} \tag{3.3$'$}$$

The proof is simply a transcription of that of X, Theorem 3.2, to the continuous parameter case, and will therefore only be sketched. It is immediately verified that (3.2) defines a positive definite function whenever F is monotone non-decreasing and bounded. The formula (3.3) is simply Lévy's formula for a distribution function in terms of its characteristic function. From the present point of view the natural proof of Lévy's formula is the following. Suppose $\lambda_1 < \lambda_2$, and define Φ by

$$\begin{aligned}\Phi(\lambda) &= 1, \qquad \lambda_1 < \lambda < \lambda_2,\\ &= \tfrac{1}{2}, \qquad \lambda = \lambda_1, \lambda_2,\\ &= 0, \qquad \lambda < \lambda_1 \text{ or } \lambda > \lambda_2.\end{aligned}$$

Then, if Φ^* is the Fourier transform of Φ,

$$\Phi^*(t) = \int_{-\infty}^{\infty} \Phi(\lambda) e^{-2\pi i t \lambda}\, d\lambda = \int_{\lambda_1}^{\lambda_2} e^{-2\pi i t \lambda}\, d\lambda,$$

the inverse Fourier transform is Φ, in the sense that

$$\Phi(\lambda) = \lim_{T\to\infty} \int_{-T}^{T} \Phi^*(t) e^{2\pi i \lambda t}\, dt$$

and the convergence is bounded convergence. Using this fact, we find that

$$\begin{aligned}\lim_{T\to\infty} \int_{-T}^{T} \Phi^*(t) R(t)\, dt &= \lim_{T\to\infty} \int_{-T}^{T} \Phi^*(t)\, dt \int_{-\infty}^{\infty} e^{2\pi i t \lambda}\, dF(\lambda)\\ &= \lim_{T\to\infty} \int_{0}^{\infty} dF(\lambda) \int_{-T}^{T} \Phi^*(t) e^{2\pi i t \lambda}\, dt\\ &= \int_{-\infty}^{\infty} \Phi(\lambda)\, dF(\lambda),\end{aligned}$$

and this is precisely (3.3).

Conversely, suppose that R is a positive definite function of t. Then the discrete parameter derivation of R as a Fourier-Stieltjes transform can be carried through in the continuous parameter case, when properly modified, or the latter case is reduced to the discrete parameter case as follows. For every $\varepsilon > 0$, $R(n\varepsilon)$ defines a positive definite function of the integer n. Hence (X, Theorem 3.2)

$$R(n\varepsilon) = \int_{-1/2}^{1/2} e^{2\pi i n \lambda}\, d\hat{F}_\varepsilon(\lambda) = \int_{-1/2\varepsilon}^{1/2\varepsilon} e^{2\pi i n \varepsilon \lambda}\, dF_\varepsilon(\lambda),$$

where F_ε is monotone non-decreasing, with total increase $R(0)$. In the following, ε will go to 0 along the values $1/2, 1/2^2, 1/2^3, \cdots$. By Helly's theorem, if we define $F_\varepsilon(\lambda) = 0$ for $\lambda \leq -1/2\varepsilon$ and $F_\varepsilon(\lambda) = R(0)$ for $\lambda > 1/2\varepsilon$, when $\varepsilon \to 0$ there is a subsequence of ε's along which $\lim\limits_{\varepsilon\to 0} F_\varepsilon(\lambda) = F(\lambda)$ exists for all λ. If integration to the limit can be justified, the evaluation of $R(n\varepsilon)$ then becomes (3.2), valid if t has the form $n/2^m$ for some integral n and positive integral m; it will then be valid for all t by continuity. To justify integration to the limit it need only be shown that

$$\lim_{A\to\infty} \int_{-A}^{A} dF_\varepsilon(\lambda) = R(0) \tag{3.4}$$

uniformly in ε, when $\varepsilon \to 0$ along some sequence. Now, if $\varepsilon = 2^{-m}$, if $T = 2^{m-r}\varepsilon = 2^{-r}$, and if $r < m$,

$$\left|2^{-(m-r)} \sum_{0}^{2^{m-r}-1} R(n\varepsilon)\right| = \left|2^{-(m-r)} \int_{-1/2\varepsilon}^{1/2\varepsilon} \frac{1 - e^{2\pi i T\lambda}}{1 - e^{2\pi i\varepsilon\lambda}}\, dF_\varepsilon(\lambda)\right|$$

$$\leq \int_{-A}^{A} dF_\varepsilon(\lambda) + \int_{|\lambda|>A} \frac{2\varepsilon}{T\left|1 - e^{2\pi i\varepsilon\lambda}\right|}\, dF_\varepsilon(\lambda).$$

When $m \to \infty$ this becomes

$$\left|\frac{1}{T}\int_0^T R(t)\, dt\right| \leq \liminf_{m\to\infty} \int_{-A}^{A} dF_\varepsilon(\lambda) + \frac{1}{\pi AT} \limsup_{m\to\infty} \int_{|\lambda|>A} dF(\lambda)$$

$$= \left[\liminf_{m\to\infty} \int_{-A}^{A} dF_\varepsilon(\lambda)\right]\left(1 - \frac{1}{\pi AT}\right) + \frac{R(0)}{\pi AT}.$$

Hence

$$\left|\frac{1}{T}\int_0^T R(t)\, dt\right| \leq \liminf_{A\to\infty} \liminf_{m\to\infty} \int_{-A}^{A} dF_\varepsilon(\lambda).$$

The right side is at most $R(0)$; the left side becomes $R(0)$ when $T \to 0$. It follows that the right side is $R(0)$, and this is simply another way of stating (3.4) (with uniformity). This finishes the proof of the theorem for R complex. If F is normalized by adjusting it so that

$$F(-\infty) = 0, \qquad F(\lambda +) = F(\lambda),$$

F is uniquely determined by (3.3). As in the discrete parameter case, if R is real, $dF(\lambda)$ is even, and (3.3′) is obtained with

$$\begin{aligned} G(\lambda) &= 2[F(\lambda) - F(0+)] + F(0+) - F(0-), \qquad && \lambda > 0, \\ &= 0, && \lambda = 0, \end{aligned}$$

and G is uniquely determined by (3.3′), if it is normalized by adjusting it so that

$$G(0) = 0, \qquad G(\lambda+) = G(\lambda), \qquad \lambda > 0.$$

If F is absolutely continuous, F' is called the *spectral density function* (*complex form*) of the process; in the real case F is absolutely continuous if and only if G is, and in that case $G' = 2F'$ is called the *spectral density function* (*real form*) of the process. When the phrase "spectral density" is used it is always to be understood that the spectral distribution is absolutely continuous.

If $\int_{-\infty}^{\infty} |R(t)|\, dt < \infty$ there is a continuous spectral density function given by

$$F'(\lambda) = \int_{-\infty}^{\infty} R(t)e^{-2\pi i\lambda t}\, dt \tag{3.5}$$

which reduces to

$$G'(\lambda) = 4\int_{0}^{\infty} R(t) \cos 2\pi\lambda t\, dt \tag{3.5′}$$

in the real case. In fact, this restriction on the covariance function implies that (3.5) and (3.5′) can be integrated to give (3.3) and (3.3′).

The spectrum of the process, as in the discrete parameter case, consists of the numbers λ_0 in the neighborhood of which F is actually increasing. These numbers are the frequencies which enter effectively in the harmonic analysis of both the covariance function and the sample functions of the process.

Example 1 Suppose that the random variables $\{x(t)\}$ of a process satisfy

$$\begin{aligned} \mathbf{E}\{x(s)\overline{x(t)}\} &= 0, \qquad && s \neq t, \\ &= \sigma^2 > 0, && s = t. \end{aligned}$$

The process is then stationary in the wide sense, with

$$\begin{aligned} R(t) &= 0, \qquad && t \neq 0, \\ &= \sigma^2, && t = 0, \end{aligned}$$

but we have excluded it from consideration since the continuity condition (3.1) is not satisfied.

Example 2 Suppose that the $x(t)$ process is a Markov process in the wide sense, and is also stationary in the wide sense. Then (V §8)

$$R(t) = e^{-ct}R(0), \qquad t \geq 0 \qquad c = \alpha + i\beta, \qquad \alpha \geq 0.$$

If $\alpha = 0$,

$$R(t) = e^{-2\pi it\beta}R(0) \qquad -\infty < t < \infty$$

and (cf. X §3, Example 2)

$$x(t) = e^{-2\pi it\beta}x(0), \qquad -\infty < t < \infty,$$

with probability 1, for each t, and conversely any family of random variables satisfying this equation obviously defines a Markov process (wide sense) which is stationary (wide sense). If $\alpha > 0$, and assuming for the moment that R is actually a covariance function, that is, positive definite, so that there is a spectral density which can be evaluated by (3.5),

$$F'(\lambda) = R(0)\int_0^\infty e^{-ct-2\pi i\lambda t}\,dt + R(0)\int_{-\infty}^0 e^{\bar{c}t-2\pi i\lambda t}\,dt$$

$$= \frac{2\alpha R(0)}{\alpha^2 + \beta^2 + 4\pi\lambda\beta + 4\pi^2\lambda^2}.$$

Since this function of λ is non-negative and integrable it defines a spectral density. The corresponding covariance function defined by (3.2) is the given one, which is thereby proved legitimate. If $\{\xi(t), -\infty < t < \infty\}$ is a process with orthogonal increments, with

$$\mathbf{E}\{|d\xi(t)|^2\} = dt,$$

if c is a constant with a positive real part, and if $x(t)$ is defined by

$$x(t) = \int_0^\infty e^{-cs}\,d\xi(t - s),$$

the $x(t)$ process has covariance function proportional to e^{-ct} $(t \geq 0)$. It will be shown in §8 that every stationary (wide sense) Markov (wide sense) process can be put in this form, aside from a multiplicative constant, unless it has the form described above when $\alpha = 0$.

In particular, if the process is real, $c = \alpha$ and

$$F'(\lambda) = \frac{2cR(0)}{c^2 + 4\pi^2\lambda^2}.$$

Example 3 Let $\xi_1, \cdots, \xi_k$ be mutually orthogonal random variables, with

$$\mathbf{E}\{|\xi_j|^2\} = \sigma_j^2 > 0,$$

and let $\lambda_1, \cdots, \lambda_k$ be distinct real numbers; define $x(t)$ by

$$x(t) = \sum_{j=1}^{k} \xi_j e^{2\pi i t \lambda_j}.$$

Then

$$\mathbf{E}\{x(s+t)\overline{x(s)}\} = \sum_{j=1}^{k} \sigma_j^2 e^{2\pi i t \lambda_j} = R(t)$$

is independent of s, so that the $x(t)$ process is stationary in the wide sense. The spectrum contains only the points $\lambda_1, \cdots, \lambda_k$, at which the spectral distribution function has the jumps $\sigma_1^2, \cdots, \sigma_k^2$ respectively. Conversely, if a covariance function R is given by this formula, it will be proved below that $x(t)$ must have the stated form. The real process corresponding to this complex process is defined as follows. Let $u_1, \cdots, u_k, v_1, \cdots, v_k$ be mutually orthogonal real random variables, with

$$\mathbf{E}\{u_j^2\} = \mathbf{E}\{v_j^2\} = \sigma_j^2 > 0,$$

and let $\lambda_1, \cdots, \lambda_k$ be any real numbers. Define $x(t)$ by

$$x(t) = \sum_{j=1}^{k} u_j \cos 2\pi t \lambda_j + v_j \sin 2\pi t \lambda_j.$$

Then

$$\mathbf{E}\{x(s+t)\overline{x(s)}\} = \sum_{j=1}^{k} \sigma_j^2 \cos 2\pi t \lambda_j = R(t)$$

is independent of s, so that the $x(t)$ process is stationary in the wide sense. We can replace any negative λ_j by $|\lambda_j|$, changing the sign of v_j to compensate. Thus it is no restriction to assume that $\lambda_j \geq 0$ (and that the λ_j's are distinct). With these assumptions the spectral distribution (real form) increases only at $\lambda_1, \cdots, \lambda_k$, having jump σ_j^2 at λ_j. The spectral distribution function (complex form) increases only at $\pm\lambda_j$, having jump σ_j^2 at λ_j if $\lambda_j = 0$, $\sigma_j^2/2$ at λ_j and $-\lambda_j$ otherwise. With a proper choice of the λ_j's and ξ_j's the complex version of Example 3 reduces to this real one.

In particular, if the u_j's and v_j's are Gaussian, with $\mathbf{E}\{u_j\} = \mathbf{E}\{v_j\} = 0$, the $x(t)$ process is stationary in the strict sense.

It will be proved below that every stationary (wide sense) process is either as in Example 3 or can be approximated arbitrarily closely by processes of this type.

Example 4 Let ξ be any real random variable, and let a be a constant. Define $x(t)$ by

$$x(t) = ae^{2\pi it\xi}.$$

Then

$$\mathbf{E}\{x(s+t)\overline{x(s)}\} = |a|^2 \mathbf{E}\{e^{2\pi it\xi}\} = \int_{-\infty}^{\infty} e^{2\pi it\lambda}\, dF(\lambda) = R(t),$$

where $F/|a|^2$ is the distribution function of ξ. The $x(t)$ process is stationary in the wide sense, with spectral distribution function F. This example exhibits a stationary (wide sense) process with any preassigned spectral distribution. Note that the sample functions are periodic. The choice of a sample function is the choice of the frequency ξ. Since (cf. Example 3) the sample functions of a stationary process need not be periodic, the spectral distribution function does not determine the harmonic analysis of individual sample functions, except in some sort of average sense. This statement is of course not necessarily true if only certain classes of processes are considered. For example, if the processes considered are real and Gaussian, with $\mathbf{E}\{x(t)\} = 0$, the spectral distribution function determines the covariance function and thereby the joint distribution of every finite set of $x(t)$'s.

It is obvious from X §3, Example 4, what the corresponding real process is, and no discussion of this process is necessary. The further remarks made on the randomness of this process in the discrete parameter case are also applicable to the present case.

Example 5 Suppose that $x_h(t)$ is defined by

$$x_h(t) = \frac{y(t+h) - y(t)}{h},$$

where the $y(t)$ process has orthogonal increments, and

$$\mathbf{E}\{|dy(t)|^2\} = \sigma^2\, dt, \qquad \sigma > 0.$$

Then

$$\begin{aligned} \mathbf{E}\{x_h(s+t)\overline{x_h(s)}\} &= 0 && |t| \geq h \\ &= \frac{\sigma^2}{h^2}(h - |t|), && |t| \leq h; \end{aligned}$$

the $x_h(t)$ process is stationary (wide sense) for fixed h, with covariance function R given by the preceding formula, and spectral density F' given by

$$F'(\lambda) = \int_{-h}^{h} e^{-2\pi i\lambda t} \frac{\sigma^2}{h^2}(h - |t|)\, dt = \frac{1 - \cos 2\pi\lambda h}{2\pi^2\lambda^2 h^2}\, \sigma^2.$$

This spectral density is very nearly σ^2 if h is small. In other words, although $x_h(t)$ does not necessarily converge to a limiting random variable $x_0(t)$ when $h \to 0$, its spectral distribution acts as if $x_0(t) = y'(t)$ existed and defined a stationary process with spectral density σ^2. No process with constant spectral density exists, since the spectral density has a finite integral over the range $(-\infty, +\infty)$, but the behavior of $x_h(t)$ for small h makes it plausible that whenever a $y'(t)$ appears symbolically, as in the stochastic integrals $\int_{-\infty}^{\infty} f(t)\, dy(t)$ of IX, the $y'(t)$ (although symbolic) will act, from the point of view of harmonic analysis, like a stationary wide sense process with constant spectral density. Examples of this will appear below.

Example 6 Suppose that $x(t)$ is defined by

$$x(t) = \int_{-\infty}^{\infty} e^{2\pi it\lambda}\, dy(\lambda),$$

where the $y(t)$ process has orthogonal increments with

$$\mathbf{E}\{|dy(\lambda)|^2\} = dF(\lambda) \qquad F(-\infty) = 0, \qquad F(\lambda+) = F(\lambda).$$

Then (IX §2)

$$\mathbf{E}\{x(s+t)\overline{x(s)}\} = \int_{-\infty}^{\infty} e^{2\pi it\lambda}\, dF(\lambda) = R(t)$$

is independent of s, so that the $x(t)$ process is stationary in the wide sense with covariance function R and spectral distribution function F. It will be shown in §4 that every stationary (wide sense) process can be represented in this way, and that in the real case the representation can also be put in the form

$$x(t) = \int_0^{\infty} \cos 2\pi t\lambda\, du(\lambda) + \sin 2\pi t\lambda\, dv(\lambda),$$

where the $u(\lambda)$, $v(\lambda)$ processes are real, with orthogonal increments, and

$$\mathbf{E}\{|du(\lambda)|^2\} = \mathbf{E}\{|dv(\lambda)|^2\} = dG(\lambda)$$

$$\mathbf{E}\{du(\lambda)\, dv(\mu)\} = 0.$$

Example 3 above is the particular case in which F is constant except for a finite number of jumps, so that the spectrum of the process only contains a finite number of points. This representation of $x(t)$ is called the *spectral representation* (real or complex form as the case may be).

4. The spectral representation of a stationary process

THEOREM 4.1 *Every stationary (wide sense) process* $\{x(t), -\infty < t < \infty\}$ *satisfying* (3.1) *has the spectral representation*

$$x(t) = \int_{-\infty}^{\infty} e^{2\pi it\lambda}\, dy(\lambda), \tag{4.1}$$

where the $y(\lambda)$ *process has orthogonal increments, and*

$$\mathbf{E}\{|dy(\lambda)|^2\} = dF(\lambda).$$

If a $y(\lambda)$ *process with orthogonal increments satisfies* (4.1), *with a suitable evaluation of the right-hand side, then*

$$\frac{y(\lambda_2+) + y(\lambda_2-)}{2} - \frac{y(\lambda_1+) + y(\lambda_1-)}{2} = \operatorname*{l.i.m.}_{T\to\infty} \int_{-T}^{T} \frac{e^{-2\pi it\lambda_2} - e^{-2\pi it\lambda_1}}{-2\pi it}\, x(t)\, dt \qquad -\infty < \lambda_1, \lambda_2 < \infty, \tag{4.2}$$

and this equation determines $y(\lambda)$ *uniquely, neglecting values on an* ω *set of probability* 0, *if the* $y(\lambda)$ *process is properly normalized.*

If the $x(t)$ *process is real,* (4.1) *can be put in the form*

$$x(t) = \int_{0}^{\infty} \cos 2\pi t\lambda\, du(\lambda) + \sin 2\pi t\lambda\, dv(\lambda), \tag{4.1'}$$

where the (real) $u(\lambda)$, $v(\lambda)$ *processes have orthogonal increments, with*

$$\begin{aligned} \mathbf{E}\{[du(\lambda)]^2\} &= \mathbf{E}\{[dv(\lambda)]^2\} = dG(\lambda), && \lambda > 0 \\ \mathbf{E}\{[du(\lambda)]^2\} &= dG(\lambda) && \lambda \geq 0 \\ \mathbf{E}\{du(\lambda)\, dv(\mu)\} &= 0 && 0 \leq \lambda, \mu < \infty. \end{aligned}$$

If $u(\lambda)$, $v(\lambda)$ *processes with orthogonal increments satisfy these conditions and* (4.1′), *with a suitable evaluation of the right-hand side, then*

$$\frac{u(\lambda+) + u(\lambda-)}{2} - u(0) = \operatorname*{l.i.m.}_{T\to\infty} \int_{-T}^{T} \frac{\sin 2\pi t\lambda}{\pi t}\, x(t)\, dt \qquad 0 < \lambda < \infty$$

$$\frac{v(\lambda_2+) + v(\lambda_2-)}{2} - \frac{v(\lambda_1+) + v(\lambda_1-)}{2} = \operatorname*{l.i.m.}_{T\to\infty} \int_{-T}^{T} \frac{\cos 2\pi t\lambda_2 - \cos 2\pi t\lambda_1}{-\pi t}\, x(t)\, dt \qquad 0 < \lambda_1, \lambda_2 < \infty, \tag{4.2'}$$

and these equations determine $u(\lambda)$, $v(\lambda)$ *uniquely, neglecting values on an* ω *set of probability* 0, *if the* $u(\lambda)$, $v(\lambda)$ *processes are properly normalized.*

The proof of this theorem is identical with that of X, Theorem 4.1 (except that Fourier series are replaced by Fourier integrals), and will be omitted. We remark that the normalized $v(\lambda)$ process would have

$$\mathbf{P}\{v(0+,\omega) = v(0,\omega)\} = 1,$$

since a discontinuity at 0 would make no contribution to the stochastic integral in (4.1), and can therefore be subtracted out. The integrals in (4.2′) can be interpreted as Lebesgue integrals involving sample functions of the $x(t)$ process, if this process is measurable (see §3), and otherwise as Lebesgue integrals involving sample functions of a measurable standard modification of the $x(t)$ process, as explained in II §2.

The general discussion of X §4 is applicable to the continuous parameter case, and will not be repeated, except for the discussion of the significance of Example 3. As in the discrete parameter case, if the spectrum of the $x(t)$ process contains only a finite number of points, the spectral representation reduces to a finite sum of the type discussed in §3, Example 3,

$$x(t) = \sum_j e^{2\pi it\lambda_j}\,\xi_j = \sum_j e^{2\pi it\lambda_j}\,[y(\lambda_j+) - y(\lambda_j-)],$$

where the ξ_j's are mutually orthogonal. Moreover, in the general case each Riemann-Stieltjes sum approximating $x(t)$ is of this same type,

$$\sum_j e^{2\pi it\lambda_j}\,\xi_j = \sum_j e^{2\pi it\lambda_j}\,[y(\lambda_{j+1}) - y(\lambda_j)].$$

Thus Example 3 can be used to approximate the general case in the sense that to every $\varepsilon > 0$ corresponds a stationary wide sense process of the type in Example 3, with variables $\{x_\varepsilon(t)\}$, satisfying

$$\mathbf{E}\{|x(t) - x_\varepsilon(t)|^2\} < \varepsilon \qquad -\infty < t < \infty;$$

we need only take as $x_\varepsilon(t)$ an appropriate Riemann-Stieltjes sum of the spectral representation of $x(t)$. This approximation is the justification for the following procedure commonly used by engineers and physicists in examining stationary (wide sense) processes. They write a series as in Example 3 to define $x(t)$ and then increase the number of λ_j's and adjust the corresponding σ_j^2's to get the function

$$\sum_{\lambda_j \le \lambda} \sigma_j^2$$

to approximate the desired spectral distribution. This asymptotic procedure is correct in that it approximates the spectral representation stochastic integral by sums, but there is frequently no reason to use the approximating sums rather than the integral in this connection any more than there is to replace integrals by approximating sums in other parts of mathematics.

5. Spectral decompositions

Let $A_1, \cdots, A_\nu$ be disjunct λ sets, measurable with respect to F measure, whose union is the whole λ axis. Then, as in the discrete parameter case, a stationary (wide sense) process $\{x(t), -\infty < t < \infty\}$ with spectral distribution function F can be exhibited as the sum of mutually orthogonal processes of the same type, whose spectral distributions are confined to the respective sets, $A_1, \cdots, A_\nu$. One example is obtained (as in X §5) by the standard decomposition of the spectral distribution function F,

$$F = F_1 + F_2 + F_3,$$

where F_1 is the jump function of F, F_2 is the absolutely continuous component, and F_3 is the continuous singular component. These three monotone functions increase on disjunct sets (see X §5), and thus correspond to a decomposition of the process,

$$x(t) = x^{(1)}(t) + x^{(2)}(t) + x^{(3)}(t).$$

If $\lambda_1, \lambda_2, \cdots$ are the points of discontinuity of the spectral distribution function F, if F is right continuous, and if $\mathbf{E}\{[dF(\lambda)]^2\} = dF(\lambda)$,

$$x^{(1)}(t) = \sum_j e^{2\pi it\lambda}[y(\lambda_j) - y(\lambda_j -)],$$

where the bracketed random variables form an orthogonal set. The series converges in the mean for each fixed t. The covariance function is uniformly almost periodic,

$$R^{(1)}(t) = \int_{-\infty}^{\infty} e^{2\pi it\lambda}\, dF_1(\lambda) = \sum_j e^{2\pi it\lambda_j}[F(\lambda_j) - F(\lambda_j -)].$$

6. The law of large numbers for stationary (wide sense) processes

The theorems of X §6 go over without any difficulty into the continuous parameter case; it is only necessary to replace sums by integrals in the statements of the theorems and in their proofs. The proofs will therefore be omitted.

THEOREM 6.1 *Let the $x(t)$ process be stationary (wide sense) with spectral representation*

$$x(t) = \int_{-\infty}^{\infty} e^{2\pi it\lambda}\, dy(\lambda), \qquad \mathbf{E}\{|dy(\lambda)|^2\} = dF(\lambda), \quad F(\lambda +) = F(\lambda).$$

Then

$$\underset{T\to\infty}{\text{l.i.m.}}\ \frac{1}{T}\int_0^T x(t)\,dt = y(0) - y(0-),$$

and

$$\mathbf{E}\{|y(0) - y(0-)|^2\} = F(0) - F(0-) = \lim_{T\to\infty} \frac{1}{T}\int_0^T R(t)\,dt.$$

The limit will then be 0 if, for example, $\lim_{t\to\infty} R(t) = 0$ or if $\int_0^\infty |R(t)|\,dt < \infty$.

COROLLARY *If μ is any real number,*

$$\underset{T\to\infty}{\text{l.i.m.}}\ \frac{1}{T}\int_0^T x(t)e^{-2\pi it\mu}\,dt = y(\mu) - y(\mu-)$$

and

$$\mathbf{E}\{|y(\mu) - y(\mu-)|^2\} = F(\mu) - F(\mu-) = \lim_{T\to\infty} \frac{1}{T}\int_0^T R(t)e^{-2\pi it\mu}\,dt.$$

As in the discrete parameter case, the one-sided averages above can be replaced by two-sided averages: $\frac{1}{T}\int_0^T$ can be replaced by $\frac{1}{2T}\int_{-T}^T$ or even by $\frac{1}{T'-T}\int_T^{T'}$ where $T' - T \to \infty$. The fact that one-sided averages are admissible is, however, of fundamental importance (cf. X §6).

THEOREM 6.2 *The limit in the corollary to Theorem* 6.1 (*in particular when $\mu = 0$, the limit in the theorem*) *exists with probability* 1, *and is* 0, *if there is a positive K and a positive α such that the following equal expressions are bounded as indicated.*

$$\begin{aligned}
\mathbf{E}\left\{\left|\frac{1}{T}\int_0^T x(t)e^{-2\pi i\mu t}\,dt\right|^2\right\} &= \frac{1}{T^2}\int_0^T\int_0^T R(t-s)e^{-2\pi i\mu(t-s)}\,ds\,dt \\
&= \frac{1}{T}\int_0^T \left(1 - \frac{|t|}{T}\right) R(t)e^{-2\pi i\mu t}\,dt \\
&= \int_{-\infty}^{\infty} \frac{\sin^2 \pi T(\mu-\lambda)}{\pi^2T^2(\mu-\lambda)^2}\,dF(\lambda) \\
&\le \frac{K}{T^\alpha}.
\end{aligned}$$

The condition of the theorem is satisfied for all μ, with $\alpha = 1$, if $\int_0^\infty |R(t)|\,dt < \infty$; it is also satisfied for all μ if $R(t) \leq \text{const.}\ |t|^{-\alpha}$ for some $\alpha > 0$.

7. The estimation of $R(t)$ and $F(\lambda)$ from sample functions

Theorems 7.1 and 7.2 of X go directly over to the continuous parameter case (replacing averages by integral averages), and the continuous parameter statements will be omitted.

It is important to note that $R(t)$ and $F(\lambda)$ cannot in general be determined from a knowledge of the sample functions in a finite interval. In fact, suppose that the stochastic process were known completely in the interval $|t| \leq T$; this of course is more than could be known from sampling. Suppose even that the process is known to be Gaussian, with $\mathbf{E}\{x(t)\} = 0$. Then $R(t)$ would be known for $|t| \leq 2T$, but $R(t)$ could not be determined for all t, in general, and therefore the spectral distribution could not be determined (in both cases we mean with complete accuracy of course) because examples of pairs of positive definite functions, that is, covariance functions, have been given which are identical in an interval containing $t = 0$.

Theorem 7.3 of X goes over into the continuous parameter case with no difficulty, and shows how to estimate the spectral distribution in practice:

$$\lim_{t\to\infty} \int_{\mu_1}^{\mu_2} \frac{1}{t} \left| \int_0^t e^{-2\pi i s\lambda}\, x(s)\, ds \right|^2 d\lambda = F(\mu_2) - F(\mu_1), \tag{7.1}$$

if μ_1 and μ_2 are continuity points of the spectral distribution function F, but again the differentiated version

$$\lim_{t\to\infty} \frac{1}{t} \left| \int_0^t e^{-2\pi i s\lambda}\, x(s)\, ds \right|^2 = F'(\lambda)$$

is incorrect even if F is absolutely continuous. The limit in (7.1) is a limit in probability for each pair μ_1, μ_2, and is even a limit with probability 1 if for each t,

$$\lim_{T\to\infty} \frac{1}{T} \int_0^T x(s+t)\overline{x(s)}\, ds = R(t),$$

with probability 1.

8. Absolutely continuous spectral distributions and moving averages

The spectral representation

$$(8.1)\qquad x(t) = \int_{-\infty}^{\infty} e^{2\pi it\lambda}\, dy(\lambda), \qquad \mathbf{E}\{|dy(\lambda)|^2\} = dF(\lambda)$$

can be replaced by

$$(8.2)\qquad x(t) = \int_{-\infty}^{\infty} e^{2\pi it\lambda} f(\lambda)\, d\tilde{y}(\lambda), \qquad \mathbf{E}\{|d\tilde{y}(\lambda)|^2\} = d\lambda,$$

if F is absolutely continuous and if $|f|^2 = F'$, just as in the discrete parameter case, and the proof will be omitted. If $\mathbf{E}\{x(t)\} = 0$, $\mathbf{E}\{dy(\lambda)\} = 0$, and $\tilde{y}(\lambda)$ can be chosen in such a way that $\mathbf{E}\{d\tilde{y}(\lambda)\} = 0$.

In the continuous parameter case a process of moving averages is defined as a process given by an expression of the form

$$x(t) = \int_{-\infty}^{\infty} C^*(\lambda)\, d\xi(\lambda + t) = \int_{-\infty}^{\infty} C^*(\lambda - t)\, d\xi(\lambda),$$

where C^* is a Lebesgue measurable function,

$$\int_{-\infty}^{\infty} |C^*(\lambda)|^2\, d\lambda < \infty,$$

and the $\xi(\lambda)$ process has orthogonal increments, with $\mathbf{E}\{|d\xi(\lambda)|^2\} = d\lambda$. With this definition,

$$\mathbf{E}\{x(s+t)\overline{x(s)}\} = \int_{-\infty}^{\infty} C^*(\lambda - t)\overline{C^*(\lambda)}\, d\lambda$$

$$= \int_{-\infty}^{\infty} |C(\lambda)|^2\, e^{2\pi it\lambda}\, d\lambda,$$

where C is the Fourier transform of C^*,

$$C(\lambda) = \int_{-\infty}^{\infty} e^{2\pi i\lambda\mu} C^*(\mu)\, d\mu$$

$$\left(= \underset{A\to\infty}{\text{l.i.m.}} \int_{-A}^{A} e^{2\pi i\lambda\mu} C^*(\mu)\, d\mu \right).$$

Hence the $x(t)$ process is stationary in the wide sense, with

$$R(t) = \int_{-\infty}^{\infty} |C(\lambda)|^2 e^{2\pi it\lambda}\, d\lambda;$$

the spectral distribution is absolutely continuous, with density $|C|^2$. Conversely suppose that an $x(t)$ process has an absolutely continuous spectral distribution. We can then write $x(t)$ in the form (8.2), with $\int_{-\infty}^{\infty} |f(\lambda)|^2 \, d\lambda < \infty$; it will be helpful to change the notation, and write instead

$$x(t) = \int_{-\infty}^{\infty} e^{2\pi i t\lambda} f(\lambda) \, d\xi^*(\lambda), \qquad \mathbf{E}\{|d\xi^*(\lambda)|^2\} = dF(\lambda). \tag{8.2'}$$

We can write f as a Fourier transform,

$$f(\lambda) = \int_{-\infty}^{\infty} e^{2\pi i \lambda\mu} f^*(\mu) \, d\mu.$$

If (8.2′) is written symbolically as

$$x(t) = \int_{-\infty}^{\infty} e^{2\pi i t\lambda} f(\lambda) \xi^{*\prime}(\lambda) \, d\lambda \tag{8.2''}$$

and if a $\xi(\lambda)$ process is defined as the Fourier transform of the $\xi^*(\lambda)$ process, so that, symbolically (cf. IX §4)

$$\xi'(\lambda) = \int_{-\infty}^{\infty} e^{2\pi i \lambda\mu} \xi^{*\prime}(\mu) \, d\mu,$$

Parseval's identity applied formally to (8.2″) yields

$$x(t) = \int_{-\infty}^{\infty} f^*(\mu) \, \xi'(t + \mu) \, d\mu = \int_{-\infty}^{\infty} f^*(\mu - t) \, d\xi(\mu).$$

This equation is correct (ignoring the middle term) because the Fourier transform ξ process was defined in IX §4 precisely to make these formal operations correct. Thus we have proved that *a stationary (wide sense) process is a process of moving averages if and only if its spectral distribution is absolutely continuous.*

As an example consider the stationary (wide sense) Markov (wide sense) processes, discussed in §3, Example 2. The spectral distribution of such a process is absolutely continuous, with

$$F'(\lambda) = \frac{2\alpha R(0)}{|c + 2\pi i\lambda|^2}, \qquad R(t) = e^{-ct} R(0), \qquad (t \geq 0),$$

where c has positive real part α (we exclude the degenerate case $\alpha = 0$). In this case it is easily verified that we can take

$$\begin{aligned} C^*(t) &= [2\alpha R(0)]^{1/2} e^{ct}, \qquad & t \leq 0 \\ &= 0 & t > 0 \end{aligned}$$

$$C(\lambda) = \frac{[2\alpha R(0)]^{1/2}}{c + 2\pi i\lambda}$$

so that $x(t)$ can be put in the form

$$x(t) = [2\alpha R(0)]^{1/2} \int_0^\infty e^{-c\lambda}\, d\xi(t-\lambda).$$

It is easy to derive the (wide sense) Markov property from this representation of $x(t)$ in terms of the past of a process with orthogonal increments.

9. Linear operations on stationary processes

Let $\{x(t)\}$ be the variables of a stationary (wide sense) process with spectral representation

$$x(t) = \int_{-\infty}^{\infty} e^{2\pi i t\lambda}\, dy(\lambda) \qquad \mathbf{E}\{|dy(\lambda)|^2\} = dF(\lambda). \tag{9.1}$$

By a linear operation on the $x(t)$ process we mean a transformation taking the $x(t)$ process into an $\hat{x}(t)$ process, where $\hat{x}(t)$ is either a finite sum of the form

$$\hat{x}(t) = \sum_j C_j x(t+t_j) = \int_{-\infty}^{\infty} e^{2\pi i t\lambda} [\sum_j C_j e^{2\pi i t_j \lambda}]\, dy(\lambda)$$

or a limit in the mean of such finite sums. Since mean limits on the left correspond to mean limits of the bracket [with λ weighting $dF(\lambda)$], the most general $\hat{x}(t)$ process is given by

$$\hat{x}(t) = \int_{-\infty}^{\infty} e^{2\pi i t\lambda}\, C(\lambda)\, dy(\lambda).$$

Here $C(\lambda)$ is any limit in the mean of finite sums $\sum_j C_j e^{2\pi i t_j \lambda}$ and as such may be any function which is measurable with respect to F and for which

$$\int_{-\infty}^{\infty} |C(\lambda)|^2\, dF(\lambda) < \infty.$$

The function C is called the *gain* of the operation. Thus every linear operation has a gain, and every gain determines a linear operation which defines a new stationary (wide sense) process. The new process satisfies the continuity condition (3.1) because

$$\mathbf{E}\{|\hat{x}(t) - \hat{x}(s)|^2\} = \int_{-\infty}^{\infty} |e^{2\pi i t\lambda} - e^{2\pi i s\lambda}|^2 |C(\lambda)|^2\, dF(\lambda)$$

$$\to 0, \qquad t - s \to 0.$$

The new covariance function is given by

$$\hat{R}(t) = \int_{-\infty}^{\infty} e^{2\pi it\lambda} |C(\lambda)|^2 \, dF(\lambda),$$

so that the new spectral distribution function is given by

$$\hat{F}(\lambda) = \int_{-\infty}^{\lambda} |C(\mu)|^2 \, dF(\mu);$$

spectral intensities are multiplied by the factor $|C(\lambda)|^2$. If the gain is identically 1, the linear operation is the identity.

If a linear operation with gain C_1 takes an $x(t)$ process into an $x_1(t)$ process, and if one with gain C_2 takes the $x_1(t)$ process into an $x_2(t)$ process, the one with gain C_1C_2 takes the $x(t)$ process into the $x_2(t)$ process, and, if $x(t)$ is given by (9.1),

$$x_2(t) = \int_{-\infty}^{\infty} C_1(\lambda)C_2(\lambda)e^{2\pi it\lambda} \, dy(\lambda).$$

The only conditions (besides measurability) on C_1 and C_2 are that

$$\int_{-\infty}^{\infty} |C_1(\lambda)|^2 \, dF(\lambda) < \infty, \qquad \int_{-\infty}^{\infty} |C_1(\lambda)C_2(\lambda)|^2 \, dF(\lambda) < \infty.$$

If in addition

$$\int_{-\infty}^{\infty} |C_2(\lambda)|^2 \, dF(\lambda) < \infty,$$

the linear operations with gains C_2 and C_1 can be performed successively on the $x(t)$ process and on the resulting process, obtaining again the $x_2(t)$ process. Thus when several linear operations are performed successively the result is a linear operation with gain the product of the individual gains, and the operations are commutative in the sense that if the operations can be performed in different orders the resulting process is unaffected by changes in the order.

The sum of two linear operations is defined as the one with gain the sum of the gains of the operations. It yields an $x(t)$ which is the sum of the processes resulting from the operations.

Example 1 *Differentiation* Suppose that $\int_{-\infty}^{\infty} \lambda^2 \, dF(\lambda) < \infty$, and consider the gain $2\pi i\lambda$,

$$\hat{x}(t) = \int_{-\infty}^{\infty} 2\pi i\lambda e^{2\pi it\lambda} \, dy(\lambda).$$

Since

$$\frac{x(t+h)-x(t)}{h}=\int_{-\infty}^{\infty}\frac{e^{2\pi i(t+h)\lambda}-e^{2\pi it\lambda}}{h}\,dy(\lambda),$$

the fact that [with λ weighting $dF(\lambda)$]

$$\underset{h\to 0}{\text{l.i.m.}}\,\frac{e^{2\pi i(t+h)\lambda}-e^{2\pi it\lambda}}{h}=2\pi i\lambda e^{2\pi it\lambda}$$

implies that

$$\underset{h\to 0}{\text{l.i.m.}}\,\frac{x(t+h)-x(t)}{h}=\hat{x}(t)=\int_{-\infty}^{\infty}2\pi i\lambda e^{2\pi it\lambda}\,dy(\lambda),$$

so that $\hat{x}(t)$ is a mean square derivative; in this extended sense the $x(t)$ process sample functions have derivatives. If now as usual $x(t, \omega)$ is the value assumed by the random variable $x(t)$ at the point ω, we shall strengthen the foregoing result by proving that, if the $x(t)$ process is separable, almost all sample functions of the process are absolutely continuous, and in fact that, if $x'(\cdot, \omega)$ denotes the derived function of the sample function $x(\cdot, \omega)$, then, for each t,

$$x'(t, \cdot)=\hat{x}(t)$$

with probability 1. Without the separability hypothesis we can only prove the (equivalent) result that, if R is any denumerable parameter set, almost all sample functions coincide on R with functions defined for all t, which are absolutely continuous, and whose derived functions satisfy the above relation. For given t the random variable $\hat{x}(t)$ is uniquely defined only up to values on ω sets of probability 0. We have seen in §3 that we can therefore suppose that this random variable is defined in such a way that the $\hat{x}(t)$ process is measurable, and that almost all sample functions of the $\hat{x}(t)$ process are Lebesgue integrable over finite intervals. Then

$$\int_0^t \hat{x}(s, \omega)\,ds$$

defines an absolutely continuous function of t, for almost all ω, and each such t function has derivative $\hat{x}(t, \omega)$ for almost all t. Moreover, dropping ω from the notation, as usual,

$$\begin{aligned}\int_0^t \hat{x}(s)\,ds&=\int_0^t ds\int_{-\infty}^{\infty}2\pi i\lambda e^{2\pi is\lambda}\,dy(\lambda)=\int_{-\infty}^{\infty}(e^{2\pi it\lambda}-1)\,dy(\lambda)\\&=x(t)-x(0),\end{aligned}$$

with probability 1, since (IX §2) the order of integration can be reversed in the iterated integral. This equation is to be interpreted as follows: for each t the left and right sides are equal, with probability 1. Then the two sides are equal with probability 1 simultaneously for all values of t in any prescribed denumerable set S. If S is chosen, as can be done since the $x(t)$ process is separable by hypothesis, so that the upper and lower limits of almost all $x(t)$ process sample functions coincide on open intervals with these limits for t restricted to S in these intervals, it follows that almost every $x(t)$ process sample function $x(\cdot, \omega)$ is absolutely continuous, with derived function $\hat{x}(\cdot, \omega)$, as was to be proved.

We have shown that the operation with gain $2\pi i\lambda$ corresponds to mean square and ordinary differentiation. Conversely, suppose that mean square differentiation is possible; that is, we suppose that

$$\underset{h\to 0}{\text{l.i.m.}}\, [x(t+h) - x(t)]/h$$

exists; in λ language this means that we suppose that [using λ weighting $dF(\lambda)$]

$$\underset{h\to 0}{\text{l.i.m.}}\, \frac{e^{2\pi i(t+h)\lambda} - e^{2\pi it\lambda}}{h}$$

exists. The limit must be $2\pi i\lambda e^{2\pi it\lambda}$ since where both mean and ordinary limits exist they must agree, and it follows that

$$\int_{-\infty}^{\infty} \left|2\pi i\lambda e^{2\pi it\lambda}\right|^2 dF(\lambda) = 4\pi^2 \int_{-\infty}^{\infty} \lambda^2\, dF(\lambda) < \infty.$$

Thus we have shown that the operation of mean square (which implies ordinary) differentiation is possible if and only if $\int_{-\infty}^{\infty} \lambda^2\, dF(\lambda) < \infty$, that is, if and only if high frequencies are not too prevalent. Note that the existence of ordinary derivatives does *not* necessarily imply that of mean square derivatives. To see this consider the following example. Let the $y(t)$ process be a separable Poisson process with parameter $c > 0$ and let $x(t) = y(t+1) - y(t) - c$. The $x(t)$ process is strictly stationary, as well as stationary in the wide sense; the spectral distribution is absolutely continuous with density

$$\frac{1 - \cos 2\pi\lambda}{2\pi^2\lambda^2}\, c.$$

Since

$$\int_{-\infty}^{\infty} \lambda^2 \frac{1-\cos 2\pi\lambda}{2\pi^2\lambda^2}\, c\, d\lambda = +\infty,$$

the mean square derivative does not exist. On the other hand, $x'(t, \omega)$ exists and is 0 for each ω, except at a denumerable t set, considering only those sample functions, whose aggregate has probability 1, which change only in unit jumps.

Example 2 *Integration* Suppose the gain C can be expressed as the Fourier transform of an integrable function,

$$C(\lambda) = \int_{-\infty}^{\infty} e^{2\pi i\lambda\mu}\, C^*(\mu)\, d\mu, \quad \int_{-\infty}^{\infty} |C^*(\mu)|\, d\mu < \infty.$$

The corresponding linear operation can always be performed, since C is bounded and continuous. This operation can be identified with an integral averaging,

$$\begin{aligned}\hat{x}(t) &= \int_{-\infty}^{\infty} e^{2\pi it\lambda} C(\lambda)\, dy(\lambda) = \int_{-\infty}^{\infty} e^{2\pi it\lambda}\, dy(\lambda) \int_{-\infty}^{\infty} e^{2\pi i\lambda\mu} C^*(\mu)\, d\mu \\ &= \int_{-\infty}^{\infty} C^*(\mu)\, d\mu \int_{-\infty}^{\infty} e^{2\pi i(t+\mu)\lambda}\, dy(\lambda) \\ &= \int_{-\infty}^{\infty} C^*(\mu)\, x(t+\mu)\, d\mu.\end{aligned}$$

(The interchange of order of integration is justified by the discussion in IX §2.) Conversely suppose that one begins with an integral average

$$\int_{-\infty}^{\infty} C^*(\mu) x(t+\mu)\, d\mu,$$

with as yet unspecified conditions on the averaging function C^*. The natural condition to impose is the absolute convergence of the double integral

$$\mathbf{E}\left\{\int_{-\infty}^{\infty} C^*(\mu) x(t+\mu)\, d\mu\right\}$$

and this condition leads back to the preceding condition that $\int_{-\infty}^{\infty} |C^*(\mu)|\, d\mu < \infty$ by way of the inequality

$$\begin{aligned}\mathbf{E}\left\{\int_{-\infty}^{\infty} |C^*(\mu) x(t+\mu)|\, d\mu\right\} &= \int_{-\infty}^{\infty} |C^*(\mu)| \mathbf{E}\{|x(t+\mu)|\}\, d\mu \\ &\le \int_{-\infty}^{\infty} |C^*(\mu)| \mathbf{E}^{1/2}\{|x(t+\mu)|^2\}\, d\mu \\ &= \mathbf{E}^{1/2}\{|x(0)|^2\} \int_{-\infty}^{\infty} |C^*(\mu)|\, d\mu.\end{aligned}$$

In the following therefore, in performing averaging, we shall always assume that the averaging function C^* is absolutely integrable over $(-\infty, \infty)$. The gain will then be its Fourier transform C, so that spectral intensities will be multiplied by $|C|^2$. It is instructive to consider the following degenerate case: let a $\xi(t)$ process have orthogonal increments with $\mathbf{E}\{|d\xi(t)|^2\} = \sigma^2\, dt$. Then *formally* $\xi'(t)$ is, as we have seen, a process with constant spectral density σ^2. According to this, the integral average

$$x(t) = \int_{-\infty}^{\infty} C^*(\mu)\xi'(t+\mu)\, d\mu = \int_{-\infty}^{\infty} C^*(\mu)\, d\xi(t+\mu)$$

should have spectral density $|C|^2\sigma^2$; as a matter of fact, the $x(t)$ process is a process of moving averages with this spectral density (cf. §8); in this case the appropriate condition on C^* is that

$$\int_{-\infty}^{\infty} |C^*(\mu)|^2\, d\mu < \infty.$$

If several operations of integral averaging are performed successively, the result is a linear operation which is also an integral average, and the averaging functions combine by convolution. We prove this for two operations. Suppose, then, that C_1^*, C_2^* are two averaging functions; we must prove that, if C^* is defined by

$$C^*(\mu) = \int_{-\infty}^{\infty} C_1^*(\mu - \lambda)C_2^*(\lambda)\, d\lambda,$$

then C^* is the averaging function corresponding to the repeated operation, that is, $\int_{-\infty}^{\infty} |C^*(\mu)|\, d\mu < \infty$ and C^* is an averaging function with gain (Fourier transform) C_1C_2. On the first point,

$$\int_{-\infty}^{\infty} |C^*(\mu)|\, d\mu \le \int_{-\infty}^{\infty}\int_{-\infty}^{\infty} |C_1^*(\mu-\lambda)C_2^*(\lambda)|\, d\lambda\, d\mu$$
$$= \int_{-\infty}^{\infty} |C_1^*(\mu)|\, d\mu \int_{-\infty}^{\infty} |C_2^*(\lambda)|\, d\lambda < \infty.$$

On the second,

$$\int_{-\infty}^{\infty} e^{2\pi i t\mu} C^*(\mu)\, d\mu = \int_{-\infty}^{\infty} e^{2\pi i t\mu}\, d\mu \int_{-\infty}^{\infty} C_1^*(\mu-\lambda)C_2^*(\lambda)\, d\lambda$$
$$= \int_{-\infty}^{\infty} e^{2\pi i t\lambda} C_2^*(\lambda)\, d\lambda \int_{-\infty}^{\infty} e^{2\pi i t(\mu-\lambda)} C_1^*(\mu-\lambda)\, d(\mu-\lambda)$$
$$= C_1(\mu)C_2(\mu).$$

We have of course here simply verified the well-known fact that convolution of functions corresponds to multiplication of their Fourier transforms. The various interchanges of orders of integration were all justified by the absolute convergence of the integrals involved.

We now consider gains of the form $2\pi i\lambda C$, where C is, as above, the Fourier transform of an integrable C^*, and (the necessary condition for a gain)

$$4\pi^2 \int_{-\infty}^{\infty} \lambda^2 |C(\lambda)|^2 \, dF(\lambda) < \infty \tag{9.2}$$

is satisfied. Under these conditions the linear operation can be considered the result of two successive operations: integral averaging with averaging function C^* followed by differentiation. The condition (9.2) legitimizes the differentiation. Since we have not supposed that $\int_{-\infty}^{\infty} \lambda^2 \, dF(\lambda) < \infty$, it may not be possible to reverse the order of the two operations. However, we shall show that the result of the successive operations can always be written in the form

$$\hat{x}(t) = \int_{-\infty}^{\infty} C^*(\mu) \, dx(t + \mu), \tag{9.3}$$

which reduces to the result of differentiation and integral averaging,

$$\hat{x}(t) = \int_{-\infty}^{\infty} C^*(\mu) x'(t + \mu) \, d\mu,$$

if $\int_{-\infty}^{\infty} \lambda^2 \, dF(\lambda) < \infty$, so that the $x'(t)$ process exists. We put this problem in a more general setting by discussing in detail stochastic integrals of the form

$$\int_{-\infty}^{\infty} f^*(t) \, dx(t)$$

for $x(t)$ processes which are stationary in the wide sense. If the spectral representation of $x(t)$ is

$$x(t) = \int_{-\infty}^{\infty} e^{2\pi i t\lambda} \, dy(\lambda), \qquad \mathbf{E}\{|dy(\lambda)|^2\} = dF(\lambda),$$

and, if we observe our usual convention relating starred and unstarred functions,

$$f(\lambda) = \int_{-\infty}^{\infty} e^{2\pi i\lambda t} f^*(t) \, dt, \tag{9.4}$$

then formally

$$\text{(9.5)} \qquad \int_{-\infty}^{\infty} f^*(t)\, dx(t) = 2\pi i \int_{-\infty}^{\infty} \int_{-\infty}^{\infty} \lambda e^{2\pi i t\lambda} f^*(t)\, dt\, dy(\lambda)$$

$$= 2\pi i \int_{-\infty}^{\infty} \lambda f(\lambda)\, dy(\lambda).$$

Now integrals like the last one have been defined in IX §2. The condition imposed in that section becomes

$$\text{(9.6)} \qquad \int_{-\infty}^{\infty} \lambda^2 |f(\lambda)|^2\, dF(\lambda) < \infty$$

here. We *define* the left side of (9.5) as the last integral in (9.5) for every function f^* whose corresponding Fourier transform f satisfies (9.6). The method of defining f must be unique up to a λ set over which $\int \lambda^2\, dF(\lambda)$ vanishes (or the stochastic integral would not be uniquely defined), and it should be linear in the sense that $af^* + bg^*$ should correspond to $af + bg$ (or the stochastic integral would not be linear in the integrand). Finally f should be the usual Fourier transform if f^* is, say, absolutely integrable over $(-\infty, \infty)$. It will be sufficient below if we restrict ourselves to functions f^* which are integrable over all finite intervals, for which

$$\lim_{A\to\infty} \int_{-A}^{A} e^{2\pi i\lambda t} f^*(t)\, dt$$

exists for all λ, defining the transform f which satisfies (9.6). This class is certainly admissible in accordance with the principles just stated. According to the definition (9.5)

$$\text{(9.7)} \qquad \mathbf{E}\left\{ \int_{-\infty}^{\infty} f^*(t)\, dx(t) \overline{\int_{-\infty}^{\infty} g^*(t)\, dx(t)} \right\} = 4\pi^2 \int_{-\infty}^{\infty} \lambda^2 f(\lambda)\overline{g(\lambda)}\, dF(\lambda).$$

In particular, if $\int_{-\infty}^{\infty} \lambda^2\, dF(\lambda) < \infty$, the derived $x'(t)$ process exists,

$$x'(t) = 2\pi i \int_{-\infty}^{\infty} \lambda e^{2\pi i t\lambda}\, dy(\lambda),$$

and (combining the present definition with the discussion of Example 2)

$$\int_{-\infty}^{\infty} f^*(t)\, dx(t) = \int_{-\infty}^{\infty} f^*(t) x'(t)\, dt,$$

if

$$\int_{-\infty}^{\infty} |f^*(t)|\, dt < \infty.$$

Using the preceding results, we find that, as stated above, if C^* is absolutely integrable over $(-\infty, \infty)$ and if C satisfies (9.6),

$$\int_{-\infty}^{\infty} C^*(\mu)\, dx(t+\mu) = \int_{-\infty}^{\infty} C^*(\mu - t)\, dx(\mu) = 2\pi i \int_{-\infty}^{\infty} \lambda C(\lambda) e^{2\pi i t\lambda}\, dy(\lambda),$$

that is, the operation with gain $2\pi i\lambda C$ can be written in the form (9.3).

10. Rational spectral densities

In this section we shall treat stationary (wide sense) processes with absolutely continuous spectral distributions having rational spectral densities $F'(\lambda) \not\equiv 0$,

$$F'(\lambda) = c \frac{\prod_j (\lambda - w_j')}{\prod_j (\lambda - z_j')}, \qquad c \neq 0.$$

The numerator and denominator can be supposed to have no common roots. The reality of F' means that

$$c \frac{\prod_j (\lambda - w_j')}{\prod_j (\lambda - z_j')} = \bar{c} \frac{\prod_j (\lambda - \bar{w}_j')}{\prod_j (\lambda - \bar{z}_j')},$$

so that the imaginary z_j''s and the w_j''s must be conjugate in pairs. Moreover, since F' is integrable, no z_j can be real and, since $F' \geq 0$, every real w_j must be a root of even multiplicity and $c = \bar{c} > 0$. Hence, using the fact that $|\lambda - \xi| = |\lambda - \bar{\xi}|$, we can write F' in the form

$$F'(\lambda) = c \frac{\left| \sum_{j=1}^{\alpha} (\lambda - w_j) \right|^2}{\left| \sum_{j=1}^{\beta} (\lambda - z_j) \right|^2} \tag{10.1}$$

$$= \left| \frac{\sum_{j=0}^{\alpha} A_j \lambda^j}{\sum_{j=0}^{\beta} B_j \lambda^j} \right|, \qquad A_\alpha B_0 B_\beta \neq 0, \quad \beta > \alpha.$$

Here $\beta > \alpha$ because F' is integrable, and we can suppose, if convenient, that the roots of the denominator and the imaginary roots of the numerator have positive imaginary part. We shall always suppose that numerator and denominator have no common root. If the roots are chosen as described, the A_j's and B_j's are uniquely determined up to a proportionality factor.

In particular, if the $x(t)$ process is real, F' is even, and we can write (10.1) in the form

$$F'(\lambda) = \left| \frac{\sum_0^\alpha A_j'(i\lambda)^j}{\sum_0^\beta B_j'(i\lambda)^j} \right|^2 \qquad \begin{array}{l} A_j = A_j' i^j \\ B_j = B_j' i^j \end{array} \tag{10.1'}$$

where the A_j''s and B_j''s are real.

According to §8 the spectral representation of the process can be written in the form

$$x(t) = \int_{-\infty}^{\infty} e^{2\pi i t\lambda} \frac{\sum_0^\alpha A_j\lambda^j}{\sum_0^\beta B_j\lambda^j} dz(\lambda), \qquad \mathbf{E}\{|dz(\lambda)|^2\} = d\lambda, \tag{10.2}$$

where the $z(\lambda)$ process has orthogonal increments. The covariance function is given by

$$\begin{aligned} R(t) &= \int_{-\infty}^{\infty} e^{2\pi i t\lambda} \left| \frac{\sum_0^\alpha A_j\lambda^j}{\sum_0^\beta B_j\lambda^j} \right|^2 d\lambda \\ &= \int_{-\infty}^{\infty} e^{2\pi i t\lambda} \frac{\sum_0^\alpha A_j\lambda^j \sum_0^\alpha \bar{A}_j\lambda^j}{\sum_0^\beta B_j\lambda^j \sum_0^\beta \bar{B}_j\lambda^j} d\lambda. \end{aligned} \tag{10.3}$$

According to a standard residue argument, $R(t)/(2\pi i)$ is the sum of the residues of the last integrand in the upper half-plane, if $t \geq 0$. Moreover, if L is a simple closed rectifiable curve in the upper half-plane, which contains in its interior the zeros of the denominator in the upper half-plane, then $R^{(k)}(t)$ is given by

$$R^{(k)}(t) = \int_L e^{2\pi i t\lambda} \frac{\sum_0^\alpha A_j\lambda^j \sum_0^\alpha \bar{A}_j\lambda^j}{\sum_0^\beta B_j\lambda^j \sum_0^\beta \bar{B}_j\lambda^j} (2\pi i\lambda)^k \, d\lambda, \qquad \begin{array}{l} k \geq 0, \\ t \geq 0. \end{array} \tag{10.4}$$

Here $R^{(k)}(0)$ is to be interpreted as the one-sided derivative $R^{(k)}(0+)$. Hence $R^{(k)}(t)/(2\pi i)$ is the sum of the residues of the integrand in (10.4) in the upper half-plane. The function R is thus indefinitely differentiable for $t \geq 0$, and also for $t \leq 0$, since $R(-t) = \overline{R(t)}$. The residue evaluation of $R(t)$ yields, if we suppose that the zeros of $\sum_0^\beta B_j z^j$ have positive imaginary parts,

$$R(t) = \sum_j C_j e^{2\pi i t z_j}, \qquad t \geq 0, \qquad [R(-t) = \overline{R(t)}], \tag{10.5}$$

where C_j is a polynomial in t, and the z_j's are zeros of $\sum_0^\beta B_j z^j$. If R is real, we can write

$$R(t) = \sum_j (C_j' \cos 2\pi a_j t + C_j'' \sin 2\pi a_j t) e^{-2\pi t b_j},$$

where C_j', C_j'' are real polynomials in t, and a_j, b_j are respectively the real and imaginary parts of z_j.

Thus R has one-sided derivatives of all orders at 0, and $R^{(k)}(0+)/(2\pi i)$ is the sum of the residues of the integrand in (10.4) (with $t = 0$) in the upper half-plane. Similarly, $-R^{(k)}(0-)/(2\pi i)$ is the sum of the residues of the same function in the lower half-plane. Hence the quantity

$$\frac{R^{(k)}(0+) - R^{(k)}(0-)}{2\pi i}$$

is the sum of all the residues of this function in the finite plane, that is, the coefficient of $1/\lambda$ in the power series expansion of the function in a neighborhood of ∞. Hence

$$\begin{aligned} R^{(k)}(0+) - R^{(k)}(0-) &= 0, & k &\leq 2\beta - 2\alpha - 2, \\ &= \frac{|A_\alpha|^2}{|B_\beta|^2}(2\pi i)^{2\beta-2\alpha}, & k &= 2\beta - 2\alpha - 1. \end{aligned} \tag{10.6}$$

The first line here is also an obvious consequence of (10.3), since differentiation under the sign of integration in (10.3) is legitimate for the values of k involved. Note that, according to our evaluation of $R(t)$, $R(t) \to 0$ exponentially when $|t| \to \infty$.

We can deduce immediately, from either (10.4) or (10.5), that R satisfies the differential equations

$$\begin{aligned} \sum_0^\beta \frac{B_j}{(2\pi i)^j} R^{(j)}(t) &= 0, & t &\geq 0+, \\ \sum_0^\beta \frac{\bar{B}_j}{(2\pi i)^j} R^{(j)}(t) &= 0, & t &\leq 0-, \end{aligned} \tag{10.7}$$

where the qualifications on the right mean that the equations hold on the closed half-lines, using the appropriate one-sided derivatives when $t = 0$. Then

$$\sum_{j,k=0}^\beta \frac{B_j \bar{B}_k}{(2\pi i)^{j+k}} R^{(j+k)}(t) = 0, \qquad t \neq 0. \tag{10.8}$$

According to (10.6), this differential equation cannot be satisfied when $t = 0$, because the first 2β derivatives of R do not exist at that point. Even without (10.6), the non-existence of these derivatives is readily deduced from the fact that R is bounded, whereas no solution of (10.8) for all t is bounded if $\sum_0^\beta B_j z^j$ has no real zero.

THEOREM 10.1 *If the stationary (wide sense) process* $\{x(t), -\infty < t < \infty\}$ *has an absolutely continuous spectral distribution, with density given by* (10.1), *and if the zeros of* $\sum_0^\beta B_j z^j$ *are in the upper half-plane, then R satisfies the differential equations* (10.7), *and the boundary conditions* (10.6). *Conversely, if the covariance function of a stationary (wide sense) process satisfies* (10.7), *for some constants* $B_0, \cdots, B_\beta$, *if* $B_0 B_\beta \neq 0$, *and if no zero of* $\sum_0^\beta B_j z^j$ *is real, then, if m is the smallest value of k for which* $R^{(k)}(0+) \neq R^{(k)}(0-)$, *it follows that the process has an absolutely continuous spectral distribution, with density* (10.1), *for some constants* $A_0, \cdots, A_\alpha$, *where* $2\alpha = 2\beta - m - 1$.

We have already proved the direct half of this theorem. To prove the converse, we note that, if R satisfies (10.7), it must be indefinitely differentiable for $t \geq 0+$. Moreover, as a solution of (10.7), R must be given by (10.5), where the z_j's are zeros of $\sum_0^\beta B_j z^j$ and the C_j's are polynomials in t. Since R is bounded, and since there are no real z_j's, by hypothesis, only z_j's with positive imaginary part can actually appear in the expression for $R(t)$. Then $R(t)$ vanishes exponentially at $\pm\infty$. Hence [cf. 3.5)], F is absolutely continuous, with

$$F'(\lambda) = \int_{-\infty}^{\infty} e^{-2\pi i\lambda t} R(t)\, dt,$$

and, integrating by parts,

$$\begin{aligned} F'(\lambda) &= \int_{-\infty}^{\infty} \frac{e^{-2\pi i\lambda t}}{2\pi i\lambda} R^{(1)}(t)\, dt \\ &= \sum_{j=2}^{n} \frac{R^{(j-1)}(0+) - R^{(j-1)}(0-)}{(2\pi i\lambda)^j} + \int_{-\infty}^{\infty} \frac{e^{-2\pi i\lambda t}}{(2\pi i\lambda)^n} R^{(n)}(t)\, dt, \qquad n \geq 2. \end{aligned}$$

Then

$$\lambda^n F'(\lambda) = p_{n-m-1} + \frac{1}{(2\pi i)^n} \int_{-\infty}^{\infty} e^{-2\pi i\lambda t} R^{(n)}(t)\, dt, \qquad n \geq 0,$$

where p_k is a polynomial of degree k, if $k \geq 0$, and vanishes identically if $k < 0$. It follows that

$$\begin{aligned} (10.9) \qquad |\sum_0^\beta B_j \lambda^j|^2 F'(\lambda) &= \sum_{j,\,k=0}^{\beta} B_j \bar{B}_k \lambda^{j+k} F'(\lambda) \\ &= A(\lambda) + \int_{-\infty}^{\infty} e^{-2\pi i\lambda t} \sum_{j,k=0}^{\beta} \frac{B_j \bar{B}_k}{(2\pi i)^{j+k}} R^{(j+k)}(t)\, dt, \end{aligned}$$

where $A(\cdot)$ is a polynomial of degree $2\beta - m - 1$. Since (10.7) implies (10.8), the integral vanishes, and (10.9) reduces to

$$|\sum_0^\beta B_j\lambda^j|^2 F'(\lambda) = A(\lambda),$$

and we have now proved the converse half of the theorem. Note that, in the discussion of the converse, the expression (10.5) for $R(t)$ can contain only z_k's with positive imaginary part, since R is bounded. If $\sum_0^\beta B_j z^j$ has zeros with negative imaginary parts, they do not appear in (10.5). It then follows that R satisfies a differential equation of order $< \beta$, so that, in the representation of F' just obtained, the numerator and denominator polynomials have common roots.

We proceed to analyze in more detail the case when $\alpha = 0$, $A_0 = 1$ in (10.1), corresponding to Case 2 in the discrete parameter discussion of X §10. In this case

$$F'(\lambda) = \frac{1}{|\sum_0^\beta B_j\lambda^j|^2} \qquad \beta > 0, \quad B_0 B_\beta \neq 0,$$

and it is convenient to write the spectral representation in the form

$$x(t) = \int_{-\infty}^{\infty} \frac{e^{2\pi i t\lambda}}{\sum_{j=0}^\beta B_j\lambda^j}\, dz^*(\lambda), \qquad \mathbf{E}\{|dz^*(\lambda)|^2\} = d\lambda.$$

Then

$$\int_{-\infty}^{\infty} \lambda^{2(\beta-1)}\, dF(\lambda) < \infty,$$

so that the first $\beta - 1$ derived processes exist; if the process is separable, the sample functions have $\beta - 1$ derivatives with

$$x^{(k)}(t) = \int_{-\infty}^{\infty} \frac{(2\pi i\lambda)^k e^{2\pi i t\lambda}}{\sum_{j=0}^\beta B_j\lambda^j}\, dz^*(\lambda), \qquad k \leq \beta - 1.$$

Hence *formally*

$$(10.10) \quad B_0 x(t) + \frac{B_1}{2\pi i} x'(t) + \cdots + \frac{B_\beta}{(2\pi i)^\beta} x^{(\beta)}(t) = \int_{-\infty}^{\infty} e^{2\pi i t\lambda}\, dz^*(\lambda)$$

$$= z'(t),$$

where (cf. IX §4) the $z(t)$ process is the Fourier transform of the $z^*(\lambda)$ process. That is to say, formally the $x(t)$ sample functions satisfy a stochastic version of (10.5), a differential equation of order β, linear with constant coefficients (which are real if the process is real) whose right-hand side is the (fictitious) derivative of a process with orthogonal stationary (wide sense) increments. With some stretching of the mathematics the fictitious $z'(t)$ here can be said to have the property that

$$\mathbf{E}\{z'(s)\overline{z'(t)}\} = 0, \qquad s \neq t,$$

corresponding to the orthogonality of the $z(t)$ increments. Thus the differential equation is the analogue of the difference equation X (10.5) in the discrete parameter discussion, X §10, Case 2. In order to justify the symbolism of (10.10) we prove that, for a large class of functions f^*,

$$\begin{aligned}(10.10') \quad B_0 \int_{-\infty}^{\infty} f^*(t)x(t)\,dt + \cdots + \frac{B_{\beta-1}}{(2\pi i)^{\beta-1}} \int_{-\infty}^{\infty} f^*(t)x^{(\beta-1)}(t)\,dt \\ + \frac{B_\beta}{(2\pi i)^\beta} \int_{-\infty}^{\infty} f^*(t)\,dx^{(\beta-1)}(t) = \int_{-\infty}^{\infty} f^*(t)\,dz(t).\end{aligned}$$

In this equation the terms on the left are ordinary Lebesgue integrals of sample functions, except for the last one. The last one was defined in §9 for functions f^* satisfying

$$\int_{-\infty}^{\infty} |f^*(t)|\,dt < \infty, \qquad \int_{-\infty}^{\infty} \lambda^{2\beta}|f(\lambda)|^2\,dF(\lambda) < \infty,$$

where

$$f(\lambda) = \int_{-\infty}^{\infty} e^{2\pi i\lambda t} f^*(t)\,dt.$$

The integral on the right was defined in IX §2 for functions f^* satisfying

$$\int_{-\infty}^{\infty} |f^*(t)|^2\,dt < \infty.$$

We shall prove that (10.10′) holds with probability 1 for any Baire function f^* satisfying all these conditions. For example, if f^* is continuous with a continuous derivative, in some finite closed interval, and vanishes outside the interval, these conditions will certainly be satisfied. To prove (10.10′) we need only remark that in accordance with the integration rules already derived [in particular see (9.5)], the left side of (10.10′) can be expressed in the form

$$\int_{-\infty}^{\infty} dz^*(\lambda) \int_{-\infty}^{\infty} \frac{\sum_0^\beta B_j\lambda^j}{\sum_0^\beta B_j\lambda^j} f^*(t)e^{2\pi i\lambda t}\, dt$$

$$= \int_{-\infty}^{\infty} f(\lambda)\, dz^*(\lambda)$$

$$= \int_{-\infty}^{\infty} f^*(t)\, dz(t)$$

since the $z(\lambda)$ process is the Fourier transform of the $z^*(\lambda)$ process.

Conversely suppose that the first $\beta - 1$ derived processes of a stationary (wide sense) $x(t)$ process exist and that there is a $z(t)$ process with orthogonal increments, with $\mathbf{E}\{|dz(t)|^2\} = dt$, for which (10.10) holds in the sense that (10.10′) holds. Then we prove that F is absolutely continuous, with $F'(\lambda) = |\sum_0^\beta B_j\lambda^j|^{-2}$. The principle of the proof is that the (fictitious) $z'(t)$ process has constant spectral density 1 and that the left side of (10.10), as the result of a linear operation on the $x(t)$ process with gain $\sum_0^\beta B_j\lambda^j$, has spectral density $|\sum_0^\beta B_j\lambda^j|^2F'(\lambda)$. Equating the two spectral densities gives the required result. More rigorously, we observe that, by (10.10′),

$$\int_{-\infty}^{\infty} f^*(-s+t)\left[\sum_{j=0}^{\beta-1} \frac{B_j}{(2\pi i)^j} x^{(j)}(t)\, dt + \frac{B_\beta}{(2\pi i)^\beta} dx^{(\beta-1)}(t)\right]$$

$$= \int_{-\infty}^{\infty} f^*(-s+t)\, dz(t)$$

with probability 1, if f^* satisfies the conditions imposed above. Now the left side of this equation is the result of a linear operation on the $x(t)$ process with gain $f(\lambda)\sum_0^\beta B_j\lambda^j$, so that the process on the left has spectral distribution function given by

$$\int_{-\infty}^{\lambda} |f(\mu)\sum_0^\beta B_j\mu^j|^2\, dF(\mu).$$

The right side of this equation defines a process of moving averages, with spectral distribution function given by

$$\int_{-\infty}^{\lambda} |f(\mu)|^2\, d\mu.$$

Equating these two distribution functions, we find that any discontinuity of F must be a zero of $\sum_0^\beta B_j\lambda^j$, and that between discontinuities F is absolutely continuous, with

$$F'(\lambda) = \frac{1}{|\sum_0^\beta B_j\lambda^j|^2}.$$

Since F' is integrable, the denominator does not vanish. Then F is continuous, and therefore has the stated form, as was to be proved.

From the point of view of prediction theory it is important to know when the $z'(t)$ in (10.10) is orthogonal to the past of $x(t)$, that is, when

$$\mathbf{E}\{[z(t_2) - z(t_1)]\overline{x(s)}\} = 0, \qquad s \leq t \leq t_1 < t_2.$$

This will now be shown to be true if and only if the roots of $\sum_0^\beta B_j z^j$ all have positive imaginary parts. In fact, if f^* in (10.10′) is defined by

$$\begin{aligned} f^*(t) &= 1 \qquad t_1 \leq t \leq t_2 \\ &= 0 \qquad \text{otherwise,} \end{aligned}$$

we find that

$$\begin{aligned} \sum_0^{\beta-1} \frac{B_j}{(2\pi i)^j} \int_{t_1}^{t_2} \mathbf{E}\{x^{(j)}(t)\overline{x(s)}\}\, dt + \frac{B_\beta}{(2\pi i)^\beta} \mathbf{E}\{[x^{(\beta-1)}(t_2) - x^{(\beta-1)}(t_1)]\, \overline{x(s)}\} \\ = \mathbf{E}\{[z(t_2) - z(t_1)]\overline{x(s)}\}. \end{aligned}$$

The left side of this equation can be put in the form

$$\begin{aligned} \sum^{\beta-1} \int_{t_1}^{t_2} dt \int_{-\infty}^{\infty} \frac{B_j\lambda^j e^{2\pi i\lambda(t-s)}}{|\sum_0^\beta B_j\lambda^j|^2}\, d\lambda + \int_{-\infty}^{\infty} \frac{B_\beta\lambda^\beta[e^{2\pi i\lambda(t_2-s)} - e^{2\pi i\lambda(t_1-s)}]}{2\pi i\lambda|\sum_0^\beta B_j\lambda^j|^2}\, d\lambda \\ = \int_{-\infty}^{\infty} \frac{e^{2\pi i\lambda(t_2-s)} - e^{2\pi i\lambda(t_1-s)}}{2\pi i\lambda \sum_0^\beta \bar{B}_j\lambda^j}\, d\lambda. \end{aligned}$$

If the roots of $\sum_0^\beta B_j z^j$ have positive imaginary parts, those of $\sum_0^\beta \bar{B}_j z^j$ have negative imaginary parts and the last integral above must vanish, by a residue argument. The same argument shows that the integral does not vanish for all s, t_1, t_2 obeying the stated inequalities unless the roots of $\sum_0^\beta B_j z^j$ all have positive imaginary parts.

Thus, if $F'(\lambda) = |\sum_0^\beta B_j \lambda^j|^{-2}$, the sample functions satisfy a stochastic differential equation (10.10) of a very simple type. Since $|\lambda - \xi| = |\lambda - \bar{\xi}|$ for real λ, and since we have *not* supposed in discussing this differential equation that the roots of $\sum_0^\beta B_j z^j$ have positive imaginary parts, the B_j's, and thereby the differential equation, can be changed without affecting the process simply by replacing roots by their conjugates. Thus the sample functions of the given process satisfy many differential equations (10.10), but there is only one for which the $z'(t)$ is orthogonal to the past of the $x(t)$ process in the sense described above, that for which the roots of $\sum_0^\beta B_j z^j$ all lie in the upper half-plane.

The differential equation (10.10) is of a standard type whose solution is known,

$$(10.11)\qquad x(t) = \sum_{j=0}^{\beta-1} a_j(t - t_0) x^{(j)}(t_0) + \int_{t_0} g(t - s)\, dz(s), \qquad t > t_0$$

where $a_0, \cdots, a_{\beta-1}, g$ are functions which can be written down explicitly in terms of the given coefficients $B_0, \cdots, B_\beta$. This solution is of course usually used for a fixed function $z'(t)$ on the right in (10.10), but the right side of (10.11) is defined in the present circumstances, and, since the stochastic integral involved obeys the usual rules, $x(t)$ as defined by (10.11) must actually satisfy (10.10), that is, (10.10'). If the roots of $\sum_0^\beta B_j z^j$ have positive imaginary parts, we have seen that the z increments involved in (10.11) are orthogonal to each random variable $x(t)$ for $t \leq t_0$. Then the integral in (10.11) is orthogonal to $x(t)$ for $t \leq t_0$, so that

$$\sum_{j=0}^{\beta-1} a_j(t - t_0) x^{(j)}(t_0)$$

is the projection of $x(t)$ on the past up to time t_0. In particular, if $\beta = 1$ the process is a Markov process in the wide sense. In that case (10.11) becomes

$$(10.11')\quad x(t) = e^{-b(t-t_0)} x(t_0) + \frac{2\pi i}{B_1} \int_{t_0}^{t} e^{-b(t-s)}\, dz(s), \qquad b = \frac{B_0}{B_1} 2\pi i$$

where $B_0 + B_1 z = 0$ defines a root $-b/2\pi i$ with positive imaginary part, so that b has a positive real part. When $t_0 \to -\infty$, (10.11′) becomes

$$x(t) = \frac{2\pi i}{B_1} \int_{-\infty}^{t} e^{-b(t-s)} \, dz(s) \tag{10.12′}$$

and in the same way (10.11) becomes

$$x(t) = \int_{-\infty}^{t} g(t-s) \, dz(s). \tag{10.12}$$

This expression displays the $x(t)$ process as a process of moving averages of a special type; $x(t)$ depends only on the past of the z process. The form (10.12′) has already been written down in §3 (Example 2).

11. Processes with stationary (wide sense) increments

Suppose that $\cdots, x_0, x_1, \cdots$ are random variables for which

$$\mathbf{E}\{|x_n - x_m|^2\} < \infty$$

for all m, n and for which

$$r(m_1, n_1;\ m_2, n_2) = \mathbf{E}\{(x_{n_1} - x_{m_1})\overline{(x_{n_2} - x_{m_2})}\}$$

is independent of translations of the parameter axis, that is,

$$r(m_1 + h, n_1 + h;\ m_2 + h, n_2 + h) = r(m_1, n_1, m_2, n_2)$$

for all integral h. Then the x_n process will be said to have stationary (wide sense) increments. The continuous parameter case is obtained by allowing the parameter values to be any real numbers, and we shall always also make the following continuity hypothesis [writing in this case $x(t)$ instead of x_t]

$$\lim_{h\to 0} \mathbf{E}\{|x(t+h) - x(t)|^2\} = \lim_{h\to 0} r(0, h;\ 0, h) = 0.$$

In the discrete parameter case the x_n process has stationary (wide sense) increments if and only if the differences $\cdots, x_1 - x_0, x_2 - x_1, \cdots$ constitute a stationary (wide sense) process, in which event we can write

$$x_{j+1} - x_j = \int_{-1/2}^{1/2} e^{2\pi i j\lambda} \, dy(\lambda), \qquad \mathbf{E}\{|dy(\lambda)|^2\} = dF(\lambda),$$

where the $y(\lambda)$ process has orthogonal increments. Then

$$x_n - x_m = \sum_{j=m}^{n-1} (x_{j+1} - x_j) = \int_{-1/2}^{1/2} \frac{e^{2\pi i n\lambda} - e^{2\pi i m\lambda}}{e^{2\pi i\lambda} - 1} \, dy(\lambda) \tag{11.1}$$

and

$$r(m_1, n_1;\ m_2, n_2) = \int_{-1/2}^{1/2} \frac{(e^{2\pi i n_1\lambda} - e^{2\pi i m_1\lambda})\overline{(e^{2\pi i n_2\lambda} - e^{2\pi i m_2\lambda})}}{|e^{2\pi i\lambda} - 1|^2} \, dF(\lambda). \tag{11.2}$$

The principal purpose of the following analysis is to derive the analogues of these formulas in the continuous parameter case, and only this case will be considered in the remainder of this section.

Example 1 If an $x_1(t)$ process is stationary in the wide sense, $x(t) = \int_0^t x_1(s)\, ds$ defines a process whose increments are stationary in the wide sense.

Example 2 If an $x_1(t)$ process is stationary in the wide sense and if x_0 is an arbitrary random variable, $x(t) = x_0 + x_1(t)$ defines a process whose increments are stationary in the wide sense. If the spectral representation of the $x_1(t)$ process is

$$x_1(t) = \int_{-\infty}^{\infty} e^{2\pi it\lambda}\, dy_1(\lambda), \qquad \mathbf{E}\{|dy_1(\lambda)|^2\} = dF_1(\lambda)$$

we have

$$x(t) - x(s) = \int_{-\infty}^{\infty} (e^{2\pi it\lambda} - e^{2\pi is\lambda})\, dy_1(\lambda)$$

$$r(s_1, t_1;\ s_2, t_2) = \int_{-\infty}^{\infty} (e^{2\pi it_1\lambda} - e^{2\pi is_1\lambda})\overline{(e^{2\pi it_2\lambda} - e^{2\pi is_2\lambda})}\, dF_1(\lambda).$$

We shall prove that Example 2 approximates the general case in the sense that every process with stationary (wide sense) increments has the spectral representation

$$x(t) - x(s) = \int_{-\infty}^{\infty} \frac{e^{2\pi it\lambda} - e^{2\pi is\lambda}}{2\pi i\lambda}(1 + \lambda^2)^{1/2}\, dy(\lambda), \tag{11.1'}$$

where the $y(\lambda)$ process has orthogonal increments with

$$\mathbf{E}\{|dy(\lambda)|^2\} = dH(\lambda)$$

and H is a bounded monotone function. It is clear that (11.1′) always defines such a process, with

$$\begin{aligned} r(s_1, t_1;\ s_2, t_2) &= \mathbf{E}\{x(t_1) - x(s_1)[\overline{x(t_2) - x(s_2)}]\} \\ &= \int_{-\infty}^{\infty} \frac{(e^{2\pi it_1\lambda} - e^{2\pi is_1\lambda})\overline{(e^{2\pi it_2\lambda} - e^{2\pi is_2\lambda})}}{4\pi^2\lambda^2}(1 + \lambda^2)\, dH(\lambda). \end{aligned} \tag{11.2'}$$

[In (11.1′) and (11.2′) the integrands are defined by continuity when $\lambda = 0$; the first is $t - s$, the second $(t_1 - s_1)(t_2 - s_2)$.] Formulas (11.1′) and (11.2′) are the continuous parameter versions of (11.1) and (11.2).

If $\int_{-1}^{1} \frac{dH(\lambda)}{\lambda^2} < \infty$ (and only if this is true), (11.1′) can be written in the form of Example 2,

$$x(t) = \int_{-\infty}^{\infty} e^{2\pi it\lambda}\, dy_1(\lambda) + x_0,$$

$$\mathbf{E}\{|dy_1(\lambda)|^2\} = \frac{1+\lambda^2}{4\pi^2\lambda^2}\, dH(\lambda)$$

$$x_0 = x(0) - y_1(\infty) + y_1(-\infty)$$

$$y_1(\lambda) = \int_{-\infty}^{\lambda} \frac{(1+\mu^2)^{1/2}}{2\pi i\mu}\, dy(\mu).$$

In the most general case, according to (11.1′), $x(t)$ is very nearly a sum of the form

$$x(t) = x(0) + ty_0 + \sum_{j=1}^{n} e^{2\pi it\lambda_j}\, y_j,$$

where $y_0, y_1, \cdots$ are mutually orthogonal.

To derive the representation (11.1′) we observe first that since the $x(t)$ process has stationary (wide sense) increments the same is true of the discrete parameter process (m fixed) $\{x[(n+1)/m)] - x(n/m), n = 0, \pm 1, \cdots\}$. The spectral representation of this process has the form (after a change of the integration variable)

$$x\left(\frac{(n+1)}{m}\right) - x\left(\frac{n}{m}\right) = \int_{-m/2}^{m/2} e^{2\pi in\lambda/m}\, dy_m(\lambda),$$

where the $y_m(\lambda)$ process has orthogonal increments with

$$\mathbf{E}\{|dy_m(\lambda)|^2\} = dF_m(\lambda).$$

Then, if t is a multiple of $1/m$,

$$x(t) - x(0) = \int_{-m/2}^{m/2} \frac{e^{2\pi it\lambda} - 1}{e^{2\pi it\lambda/m} - 1}\, dy_m(\lambda),$$

so that

$$\mathbf{E}\{|x(t) - x(0)|^2\} = \int_{-m/2}^{m/2} \left|\frac{e^{2\pi it\lambda} - 1}{e^{2\pi i\lambda/m} - 1}\right|^2 dF_m(\lambda),$$

and, more generally, if s_1, t_1, s_2, t_2 are multiples of $1/m$,

$$\mathbf{E}\{[x(t_1) - x(s_1)][\overline{x(t_2) - x(s_2)}]\} = \int_{-m/2}^{m/2} \frac{(e^{2\pi it_1\lambda} - e^{2\pi is_1\lambda})(\overline{e^{2\pi it_2\lambda} - e^{2\pi is_2\lambda}})}{|e^{2\pi i\lambda/m} - 1|^2}\, dF_m(\lambda).$$

Then, if H_m is defined by

$$H_m(\lambda) = \int_{-m/2}^{\lambda} \frac{4\pi^2\mu^2}{(1+\mu^2)|e^{2\pi i\mu/m}-1|^2}\,dF_m(\mu),$$

we have

$$\mathbf{E}\{[x(t_1)-x(s_1)]\overline{[x(t_2)-x(s_2)]}\}$$
$$= \int_{-m/2}^{m/2} \frac{(e^{2\pi it_1\lambda}-e^{2\pi is_1\lambda})\overline{(e^{2\pi it_2\lambda}-e^{2\pi is_2\lambda})}}{4\pi^2\lambda^2}(1+\lambda^2)\,dH_m(\lambda).$$

It will be convenient to define $H_m(\lambda)$ for $\lambda > m/2$ as $H_m(m/2)$ and for $\lambda < -m/2$ as $H_m(-m/2) = 0$. It will be proved below that the sequence $\{H_m\}$ is bounded, and that

$$\int_{|\lambda|>\alpha} dH_m(\lambda)$$

is uniformly small if α is large. By Helly's theorem there is a convergent subsequence $\{H_{a_n}\}$. Let H be the limit function. If s_1, t_1, s_2, t_2 are arbitrary, if in the preceding equation these numbers are replaced by the nearest multiple of $1/m$, and if then $m \to \infty$ along the sequence $\{a_n\}$, the equation becomes (11.2′). We must still prove the stated properties of the sequence $\{H_m\}$ however. In the first place, if t is a multiple of $1/m$ we have

$$\mathbf{E}\{|x(t)-x(0)|^2\} = \int_{-m/2}^{m/2} |e^{2\pi it\lambda}-1|^2\frac{1+\lambda^2}{4\pi^2\lambda^2}\,dH_m(\lambda) \tag{11.3}$$
$$\geq \int_{-\alpha}^{\alpha} \frac{\sin^2\pi t\lambda}{\pi^2\lambda^2}\,dH_m(\lambda)$$
$$\geq \frac{4t^2}{\pi^2}\int_{-\alpha}^{\alpha} dH_m(\lambda)$$

for any α, if t is so small that $t\alpha \leq \frac{1}{2}$. In the second place, if N is any positive integer,

$$\frac{1}{N}\sum_{j=1}^{N}\mathbf{E}\left\{\left|x\left(\frac{j}{m}\right)-x(0)\right|^2\right\}$$
$$= \frac{1}{N}\int_{-m/2}^{m/2}\sum_{j=1}^{N}|e^{2\pi ij\lambda/m}-1|^2\frac{1+\lambda^2}{4\pi^2\lambda^2}\,dH_m(\lambda)$$
$$= 2\int_{-m/2}^{m/2}\left[1-\frac{-\sin\pi\lambda/m+\sin[(N+1/2)2\pi\lambda/m]}{2N\sin\pi\lambda/m}\right]\frac{1+\lambda^2}{4\pi^2\lambda^2}\,dH_m(\lambda),$$

so that, if $N/m = t \geq 1/\alpha$,

$$(11.4) \quad \frac{1}{N}\sum_{j=1}^{N} \mathbf{E}\left\{\left|x\left(\frac{j}{m}\right) - x(0)\right|^2\right\} \geq \int_{|\lambda|>\alpha} 2\left[1 - \frac{1}{2} - \frac{1}{4t\alpha}\right]\frac{1}{4\pi^2}\, dH_m(\lambda)$$

$$\geq \frac{1}{8\pi^2}\int_{|\lambda|>\alpha} dH_m(\lambda),$$

Since by our continuity hypothesis the left side of this inequality approaches 0 when $t \to 0$, the right side must go to 0 (uniformly in m) when $\alpha \to \infty$. Moreover, (11.3) and (11.4) taken together imply that the sequence $\{H_m\}$ is bounded. This completes the derivation of the stated properties of the sequence $\{H_m\}$ and thereby of (11.2′).

Before deriving (11.1′) we discuss integrals of the form

$$\int_{-\infty}^{\infty} f^*(t)\, dx(t),$$

where the $x(t)$ process has stationary (wide sense) increments with (11.2′) valid, and f^* is a fixed function. In accordance with the usual procedure we define the integral first in the obvious way for simple step functions f^*, that is, those which vanish for large t and take on only a finite number of values, each on an interval or on finitely many intervals. For these functions, if we continue our previous convention relating starred and unstarred functions, so that

$$f(\lambda) = \int_{-\infty}^{\infty} e^{2\pi i\lambda t} f^*(t)\, dt,$$

we have, from (11.2′),

$$(11.5) \qquad \mathbf{E}\left\{\left|\int_{-\infty}^{\infty} f^*(t)\, dx(t)\right|^2\right\} = \int_{-\infty}^{\infty} |f(\lambda)|^2(1 + \lambda^2)\, dH(\lambda),$$

and, more generally,

$$(11.5') \quad \mathbf{E}\left\{\int_{-\infty}^{\infty} f^*(t)\, dx(t)\,\overline{\int_{-\infty}^{\infty} g^*(t)\, dx(t)}\right\} = \int_{-\infty}^{\infty} f(\lambda)\overline{g(\lambda)}\,(1 + \lambda^2)\, dH(\lambda).$$

We extend the definition to a larger class of functions in such a way that (11.5′) remains true. The functions will be Baire functions f^*, integrable over all finite intervals, for which the limit

$$f(\lambda) = \lim_{A\to\infty}\int_{-A}^{A} e^{2\pi i\lambda t} f^*(t)\, dt$$

exists for all λ and satisfies

$$(11.6) \qquad \int_{-\infty}^{\infty} |f(\lambda)|^2\,(1 + \lambda^2)\, dH(\lambda) < \infty.$$

If distance between random variables is defined as root mean square distance,

$$\mathbf{E}^{1/2}\{|x_1 - x_2|^2\},$$

and that between functions f^* is also defined as root mean square distance, with λ weighting $(1 + \lambda)\,dH(\lambda^2)$ between the corresponding functions f,

$$\Big[\int_{-\infty}^{\infty} |f_1(\lambda) - f_2(\lambda)|^2 (1 + \lambda^2)\,dH(\lambda)\Big]^{1/2}$$

the stochastic integral sets up a correspondence between certain functions of t (simple step functions) and random variables, which preserves linearity and distance [because of (11.5)]. Then the stochastic integral is defined by continuity (cf. a similar discussion in IX §2) for all f satisfying (11.6) and therefore for all f^* of the class considered. Equation (11.5′) remains true since it is true for simple step functions. In particular, if the $x(t)$ process is stationary in the wide sense this stochastic integral reduces to that defined in §9.

We can now derive (11.1′). If such a formula holds, and if

$$\hat{y}(\lambda) = \int_0^{\lambda} (1 + \mu^2)^{1/2}\,dy(\mu),$$

(11.1′) becomes, after a formal differentiation,

$$x'(t) = \int_{-\infty}^{\infty} e^{2\pi i t\lambda}\hat{y}'(\lambda)\,d\lambda,$$

so that we expect a reciprocal formula which is an integrated version of

$$\hat{y}'(\lambda) = \int_{-\infty}^{\infty} e^{-2\pi i\lambda t}x'(t)\,dt.$$

Actually we shall prove that, if $\hat{y}(\lambda)$ is defined by

$$\hat{y}(\lambda) = \int_{-\infty}^{\infty} \frac{e^{-2\pi i\lambda t} - 1}{-2\pi i t}\,dx(t)$$

[with possible normalizations to make $\hat{y}(\lambda +) = \hat{y}(\lambda)$], then the $\hat{y}(\lambda)$ process has orthogonal increments, and if finally $y(\lambda)$ is defined by

$$y(\lambda) = \int_{-\infty}^{\lambda} \frac{d\hat{y}(\mu)}{(1 + \mu^2)^{1/2}}, \tag{11.7}$$

the $y(\lambda)$ process has orthogonal increments, with $\mathbf{E}\{|dy(\lambda)|^2\} = dH(\lambda)$ [assuming that $H(\lambda +) = H(\lambda)$], and (11.1′) is true. The above formula for $\hat{y}(\lambda)$ leads to

$$\hat{y}(\lambda_2) - \hat{y}(\lambda_1) = \int_{-\infty}^{\infty} \frac{e^{-2\pi i\lambda_2 t} - e^{-2\pi i\lambda_1 t}}{-2\pi i t}\, dx(t),$$

and this stochastic integral, like the one defining $\hat{y}(\lambda)$, has already been discussed. We have, from (11.5) and (11.5′),

$$\mathbf{E}\{|\hat{y}(\lambda_2) - \hat{y}(\lambda_1)|^2\} = \int_{\lambda_1}^{\lambda_2} (1 + \lambda^2)\, dH(\lambda), \qquad \lambda_1 < \lambda_2$$

$$\mathbf{E}\{[y(\lambda_2) - y(\lambda_1)]\,\overline{[y(\mu_2) - y(\mu_1)]}\} = 0, \qquad \lambda_1 < \lambda_2 \leq \mu_1 < \mu_2$$

if $\lambda_1, \lambda_2, \mu_1, \mu_2$ are points of continuity of $H(\lambda)$. The $\hat{y}(\lambda)$ process thus has orthogonal increments for λ restricted to continuity points of H, and we define it at the discontinuities to make $\hat{y}(\lambda +) = \hat{y}(\lambda)$. Then, if also $H(\lambda +) = H(\lambda)$, we have $\mathbf{E}\{|d\hat{y}(\lambda)|^2\} = (1 + \lambda^2)\, dH(\lambda)$. Finally $y(\lambda)$ is defined by (11.7), so that $\mathbf{E}\{|dy(\lambda)|^2\} = dH(\lambda)$. The $\hat{y}(\lambda)$ and $y(\lambda)$ processes as defined satisfy

$$\int_{-\infty}^{\infty} f(\lambda)\, d\hat{y}(\lambda) = \int_{-\infty}^{\infty} f(\lambda)(1 + \lambda^2)^{1/2}\, dy(\lambda) = \int_{-\infty}^{\infty} f^*(t)\, dx(t) \tag{11.8}$$

if $f(\lambda)$ is 1 on a finite interval whose endpoints are points of continuity of H, and 0 otherwise. In fact, this is precisely the definition of the $\hat{y}(\lambda)$ process. This equation is therefore true (with probability 1) by the usual extension argument, for all f, f^* for which the integral on the right is defined. In particular, if f^* is 1 on a finite interval and 0 otherwise, (11.8) reduces to the desired equation (11.1′).

Once (11.1′) has been proved, it follows that the $y(\lambda)$ in this formula is essentially uniquely determined. In fact, if in the following we assume that (11.1′) is true, and that $y(\lambda +) = y(\lambda)$, $H(\lambda +) = H(\lambda)$, then (11.8) is true for $f^*(t)$ defined as 1 on a finite interval and 0 otherwise. It is then true for all functions f, f^* for which the integral on the right is defined. In particular, if $f(\lambda)$ is 1 on a finite interval, $\frac{1}{2}$ at the endpoints, and 0 otherwise, (11.8) becomes

$$\frac{\hat{y}(\lambda_2) + \hat{y}(\lambda_2 -)}{2} - \frac{\hat{y}(\lambda_1) + \hat{y}(\lambda_1 -)}{2} = \int_{-\infty}^{\infty} \frac{e^{-2\pi i\lambda_2 t} - e^{-2\pi i\lambda_1 t}}{-2\pi i t}\, dx(t). \tag{11.9}$$

Taking the expectation of the square of the modulus of both sides of (11.9), we find an expression determining H,

$$\int_{\lambda_1}^{\lambda_2} (1 + \mu^2)\, dH(\mu) = \mathbf{E}\left\{\left| \int_{-\infty}^{\infty} \frac{e^{-2\pi i t\lambda_2} - e^{-2\pi i t\lambda_1}}{2\pi t}\, dx(t)\right|^2\right\} \tag{11.10}$$

if λ_1 and λ_2 are points of continuity of H. The right-hand side can also be expressed in terms of the covariance function $r(s_1, t_1; s_2, t_2)$.

Example 1 (Continued) If an $x_1(t)$ process is stationary in the wide sense, its spectral representation leads to

$$\int_0^t x_1(s)\,ds = \int_{-\infty}^{\infty} \frac{e^{2\pi it\lambda} - 1}{2\pi i\lambda}\,dy_1(\lambda), \qquad \mathbf{E}\{|dy_1(\lambda)|^2\} = dF_1(\lambda).$$

Thus in this case the $x(t)$ process defined by the integral on the left has stationary (wide sense) increments, and in the spectral representation (11.1′)

$$(1 + \lambda^2)^{1/2}\,dy(\lambda) = dy_1(\lambda)$$
$$(1 + \lambda^2)\,dH(\lambda) = dF_1(\lambda).$$

Thus in this case $\int_{-\infty}^{\infty} \lambda^2\,dH(\lambda) < \infty$. Conversely, if an $x(t)$ process has stationary (wide sense) increments, and if $\int_{-\infty}^{\infty} \lambda^2\,dH(\lambda) < \infty$, (11.1′) can be put in the form

$$x(t) - x(s) = \int_{-\infty}^{\infty} \frac{e^{2\pi it\lambda} - e^{2\pi is\lambda}}{2\pi i\lambda}\,dy_1(\lambda), \qquad y_1(\lambda) = \int_{-\infty}^{\lambda} (1 + \mu^2)\,dy(\mu).$$

Since

$$\underset{s\to t}{\text{l.i.m.}}\ \frac{e^{2\pi it\lambda} - e^{2\pi is\lambda}}{2\pi i\lambda(t - s)} = e^{2\pi it\lambda},$$

where l.i.m. refers to λ weighting $(1 + \lambda^2)\,dH(\lambda)$ on the λ axis, it follows that

$$\underset{s\to t}{\text{l.i.m.}}\ \frac{x(t) - x(s)}{t - s} = x_1(t) = \int_{-\infty}^{\infty} e^{2\pi it\lambda}\,dy_1(\lambda).$$

The $x_1(t)$ process so defined is stationary in the wide sense and will be called the derived process of the $x(t)$ process. The $x(t)$ process sample functions are almost all absolutely continuous, if the process is separable, and $x_1(t) = x'(t)$. (Cf. the discussion of Example 1, §9.) We have thus proved that a process with stationary (wide sense) increments has the form of Example 1, that is, the process has a derived process, if and only if

$$\int_{-\infty}^{\infty} \lambda^2\,dH(\lambda) < \infty.$$

We have seen in this chapter that an $x(t)$ process with stationary (wide sense) orthogonal increments satisfies $\mathbf{E}\{|dx(t)|^2\} = \sigma^2\,dt$ and acts in many ways as if the derived $x'(t)$ process existed as a stationary (wide sense) process with spectral density σ^2. If the orthogonality hypothesis is dropped we have the processes under discussion in this section, and the formal version of (11.1′),

$$x'(t) = \int_{-\infty}^{\infty} e^{2\pi i t\lambda}(1 + \lambda^2)^{1/2}\, dy(\lambda),$$

suggests that the most general process with stationary (wide sense) increments acts as if the derived $x'(t)$ process existed as a stationary (wide sense) process with spectral distribution function given by $\int_{-\infty}^{\lambda} (1 + \mu^2)\, dH(\mu)$. [This function of λ is not in general a proper spectral distribution function since $\int_{-\infty}^{\infty} (1 + \mu^2)\, dH(\mu)$ may be infinite.] This somewhat vague statement has exactly the same kind of verification as in the special case of orthogonal increments, and we omit the details. We remark, however, that the statement is true (with no fictions) if the derived process actually exists, and remark also that the formal differential equation

$$\sum_{j=0}^{\beta} \frac{B_j}{(2\pi i)^j} X^{(j)}(t) = x'(t)$$

can be interpreted for the general $x(t)$ process with stationary (wide sense) increments in exactly the same way as in §10 where the increments were orthogonal. There is a stationary (wide sense) $X(t)$ process which is a solution of this equation, with spectral distribution function given by

$$\int_{-\infty}^{\lambda} \frac{(1 + \mu^2)\, dH(\mu)}{\left|\sum_{0}^{\beta} B_j \mu^j\right|^2}.$$

In particular, if the $x(t)$ process has orthogonal increments the numerator in the integrand becomes const. $d\mu$, as was worked out in detail in §10.

CHAPTER XII

Linear Least Squares Prediction—Stationary (Wide Sense) Processes

1. General principles (discrete parameter)

Let $\{x_n, -\infty < n < \infty\}$ be a stationary (wide sense) process with spectral distribution function F. In the following we shall describe as "F measure" the measure $\int_E dF(\lambda)$ of λ sets E. This is Lebesgue-Stieltjes measure on the λ axis, and the corresponding measurable sets and measurable functions will be called the sets and functions measurable with respect to F. (They include the Borel sets and the Baire functions.) It will be useful to introduce a systematic notation for certain closed linear manifolds (see IV §1 and §2) of random variables and of functions measurable with respect to F. In the latter case we use λ weighting $dF(\lambda)$, that is, F measure is used in the integration defining the root mean square distance between functions which determines the closure property. In the list below $\mathfrak{M}\{\cdot\ \cdot\ \cdot\}$ means the closed linear manifold generated by the elements or sets of elements in the braces.

$$\mathfrak{M}_n = \mathfrak{M}\{x_j, j \le n\} \qquad {}_n\mathfrak{M} = \mathfrak{M}\{e^{2\pi i j\lambda}, j \le n\}$$

$$\mathfrak{M}_{-\infty} = \bigcap_{-\infty}^{\infty} \mathfrak{M}_n \qquad {}_{-\infty}\mathfrak{M} = \bigcap_{-\infty}^{\infty} {}_n\mathfrak{M}$$

$$\mathfrak{M}_\infty = \mathfrak{M}\{x_j, -\infty < j < \infty\} \qquad {}_\infty\mathfrak{M} = \mathfrak{M}\{e^{2\pi i j\lambda}, -\infty < j < \infty\}$$

$$= \mathfrak{M}\{\bigcup_{-\infty}^{\infty} \mathfrak{M}_n\} \qquad = \mathfrak{M}\{\bigcup_{-\infty}^{\infty} {}_n\mathfrak{M}\}.$$

It is clear that the functions in ${}_n\mathfrak{M}$ consist of those in ${}_m\mathfrak{M}$ multiplied by $e^{2\pi i(n-m)\lambda}$. The manifold ${}_\infty\mathfrak{M}$ consists of the functions Φ which are measurable with respect to F and for which $\int_{-1/2}^{1/2} |\Phi(\lambda)|^2\, dF(\lambda) < \infty$. In fact, these functions Φ certainly define a closed linear manifold $\mathfrak{M}$, with ${}_\infty\mathfrak{M} \subset \mathfrak{M}$ since $e^{2\pi in\lambda} \in \mathfrak{M}$ for all n. Conversely, an elementary Fourier series argument shows that ${}_{\infty}\mathfrak{M}$ includes every function which takes on only a finite number of values, each on an interval, and since these functions are dense in $\mathfrak{M}$ the closure property implies $\mathfrak{M} \subset {}_{\infty}\mathfrak{M}$.

Consider the problem of approximating $x_{n+\nu}$ by linear combinations $\sum_{j=0}^{N-1} b_j x_{n-j}$, minimizing the mean square error

$$\mathbf{E}\{|x_{n+\nu} - \sum_{j=0}^{N-1} b_j x_{n-j}|^2\}.$$

We have proved in IV §3 that if N is fixed there is a minimizing linear combination, which we denoted by $\hat{\mathbf{E}}\{x_{n+\nu} \mid x_{n-N+1}, \cdots, x_n\}$, the projection of $x_{n+\nu}$ on the linear manifold generated by the random variables $x_{n-N+1}, \cdots, x_n$. If N is unrestricted we rephrased the problem as follows: it is desired to minimize $\mathbf{E}\{|x_{n+\nu} - \varphi|^2\}$ for $\varphi \in \mathfrak{M}_n$. In this case also there is a solution, $\hat{\mathbf{E}}\{x_{n+\nu} \mid \mathfrak{M}_n\}$, the projection of $x_{n+\nu}$ on $\mathfrak{M}_n$. In both cases the solution is unique, neglecting zero probabilities (see IV §3).

In particular, if the process is real and Gaussian, and if $\mathbf{E}\{x_j\} = 0$ for all j, the solutions to these minimum problems are respectively the conditional expectations of $x_{n+\nu}$ relative to the sets of random variables $x_{n-N+1}, \cdots, x_n$ and $\cdots, x_{n-1}, x_n$. It will be remembered that projections have also been called wide sense conditional expectations.

The random variables $\hat{\mathbf{E}}\{x_{n+\nu} \mid x_{n-N+1}, \cdots, x_n\}$ and $\hat{\mathbf{E}}\{x_{n+\nu} \mid \cdots, x_{n-1}, x_n\}$ will be called respectively the linear least squares prediction of $x_{n+\nu}$ in terms of $x_{n-N+1}, \cdots, x_n$ and in terms of the complete past to x_n.

We shall not discuss non-linear least squares prediction except to remark on the definition and existence of a solution. The problem of non-linear least squares prediction is that of approximating $x_{n+\nu}$ by functions $f(x_n, \cdots, x_{n-N+1})$, minimizing

$$\mathbf{E}\{|x_{n+\nu} - f(x_n, \cdots, x_{n-N+1})|^2\}.$$

Here the function f is to be a random variable measurable on the sample space of $x_{n-N+1}, \cdots, x_n$, and it is supposed that $\mathbf{E}\{|f|^2\} < \infty$. The admissible random variables (for n, N fixed) constitute a closed linear manifold, and there is a unique minimizing f, the projection g of $x_{n+\nu}$ on this linear manifold. This solution is characterized by the fact that it is in the manifold described and that $x_{n+\nu} - g$ is orthogonal to the manifold. These two properties characterize the function $\mathbf{E}\{x_{n+\nu} \mid x_{n-N+1}, \cdots, x_n\}$. Hence g is simply the conditional expectation. Thus the random variables

$$\hat{\mathbf{E}}\{x_{n+\nu} \mid x_{n-N+1}, \cdots, x_n\}, \qquad \mathbf{E}\{x_{n+\nu} \mid x_{n-N+1}, \cdots, x_n\}$$

are respectively the best linear and best (unrestricted) least squares prediction. The extension to predictions based on the full past to x_n is obvious. In particular, if the process is real and Gaussian with $\mathbf{E}\{x_j\} = 0$

for all j, linear and unrestricted prediction coincide. In the language of II §3, linear prediction is general prediction in the wide sense.

Note that according to IV, Theorem 7.4, and its strict sense version VII, Theorem 4.3,

$$\underset{N\to\infty}{\text{l.i.m.}}\ \hat{\mathbf{E}}\{x_{n+\nu} \mid x_{n-N+1}, \cdots, x_n\} = \hat{\mathbf{E}}\{x_{n+\nu} \mid \mathfrak{M}_n\}$$

$$\lim_{N\to\infty} \mathbf{E}\{x_{n+\nu} \mid x_{n-N+1}, \cdots, x_n\} = \mathbf{E}\{x_{n+\nu} \mid \cdots, x_{n-1}, x_n\}$$

(the second limit holding with probability 1). These limit equations justify the use of predictions based on part of the past to approximate those based on the full past.

Throughout the rest of this chapter we shall consider almost exclusively predictions based on the full past. We shall denote by $\varphi_{n,\nu}$ the linear least squares prediction of $x_{n+\nu}$ in terms of the past up to x_n,

$$\varphi_{n,\nu} = \hat{\mathbf{E}}\{x_{n+\nu} \mid \mathfrak{M}_n\}.$$

2. Linear least squares prediction as polynomial approximation

The spectral representation of x_n,

$$x_n = \int_{-1/2}^{1/2} e^{2\pi i n\lambda}\, dy(\lambda), \qquad \mathbf{E}\{|dy(\lambda)|^2\} = dF(\lambda), \tag{2.1}$$

where the $y(\lambda)$ process has orthogonal increments, will play an essential role in this section. We have seen in IX §2 that, if $\Phi \in {}_{\infty}\mathfrak{M}$, the stochastic integral

$$\varphi = \int_{-1/2}^{1/2} \Phi(\lambda)\, dy(\lambda) \tag{2.2}$$

defines a random variable with $\mathbf{E}\{|\varphi|^2\} < \infty$. For example, if $\Phi = e^{2\pi i n\lambda}$, $\varphi = x_n$. It follows at once from the discussion in IX §2 that $\varphi \in \mathfrak{M}_\infty$, and conversely that to any $\varphi \in \mathfrak{M}_\infty$ corresponds a $\Phi \in {}_{\infty}\mathfrak{M}$ such that (2.2) is satisfied. The random variable φ is determined uniquely by Φ, neglecting zero probabilities, and the function Φ is determined uniquely by φ, neglecting sets of zero F measure. The manifold $\mathfrak{M}_n$ corresponds to ${}_n\mathfrak{M}$ and $\mathfrak{M}_{-\infty}$ corresponds to ${}_{-\infty}\mathfrak{M}$. Any finite sum $\sum_n \gamma_n x_n$ corresponds to $\sum_n \gamma_n e^{2\pi i n\lambda}$, with

$$\mathbf{E}\{|\gamma_n x_n|^2\} = \int_{-1/2}^{1/2} |\sum_n \gamma_n e^{2\pi i n\lambda}|^2\, dF(\lambda).$$

Using this correspondence, the problem of finding $\varphi_{0,\nu}$ can be put in λ language as follows. The function $\Phi_\nu \in {}_0\mathfrak{M}$ is to be found which minimizes

$$\int_{-1/2}^{1/2} |e^{2\pi i\nu\lambda} - \Phi(\lambda)|^2\, dF(\lambda), \qquad \Phi \in {}_0\mathfrak{M}.$$

In other words, we are to approximate $e^{2\pi i\nu\lambda}$ by linear combinations of the functions $\{e^{2\pi in\lambda}, n \leq 0\}$. The equality

$$\mathbf{E}\{|x_\nu - \sum_{j=0}^{N-1} b_j x_{-j}|^2\} = \int_{-1/2}^{1/2} |e^{2\pi i\nu\lambda} - \sum_{j=0}^{N-1} b_j e^{-2\pi ij\lambda}|^2 \, dF(\lambda)$$

exhibits the complete equivalence of the formulations. The solution Φ_ν is the projection of $e^{2\pi i\nu\lambda}$ on ${}_0\mathfrak{M}$, and is uniquely determined, neglecting sets of F measure 0. In the correspondence (2.2) we have

$$\varphi_{0,\nu} = \int_{-1/2}^{1/2} \Phi_\nu(\lambda) \, dy(\lambda).$$

More generally, it is clear that we have

$$\varphi_{n,\nu} = \int_{-1/2}^{1/2} e^{2\pi in\lambda}\Phi_\nu(\lambda) \, dy(\lambda). \tag{2.3}$$

From one point of view the approximation problems discussed here, one in terms of random variables, the other in terms of the corresponding λ functions, are not only equivalent but even identical. To see this suppose for simplicity that $F(\infty) = 1$. Define probability as the measure defined by the function F on the λ axis. Then the sequence $\{e^{2\pi in\lambda}, -\infty < n < \infty\}$ is a sequence of random variables which is stationary in the wide sense, with spectral distribution function F. The prediction problem for the stochastic process defined in this way is precisely the prediction problem for the x_n process as described above, when translated into λ language.

The prediction problem in λ language is closely related to classical polynomial approximation problems. In fact, consider the problem of minimizing $|P(z) - z^{-\nu}|$ for fixed $\nu > 0$ when P runs through all polynomials of degree $N - 1$. If "minimizing" is taken to mean "in the mean square sense on $|z| = 1$ with arbitrary weighting," the problem is to minimize

$$\int_{-1/2}^{1/2} |P(e^{2\pi i\lambda}) - e^{-2\pi i\nu\lambda}|^2 \, dF(\lambda)$$

for given ν and F; if $P(z) = \sum_0^{N-1} \bar{b}_j z^j$, this integral becomes

$$\int_{-1/2}^{1/2} |e^{2\pi i\nu\lambda} - \sum_0^{N-1} b_j e^{-2\pi ij\lambda}|^2 \, dF(\lambda) = \mathbf{E}\{|x_{n+\nu} - \sum_{j=0}^{N-1} b_j x_{n-j}|^2\}.$$

Thus this polynomial approximation problem is identical with what we have called above the prediction problem using only a finite part of the past.

The function Φ_ν will be called the prediction function for lag ν. If σ_ν^2 is the mean square error of the prediction,

$$(2.4) \qquad \sigma_\nu^2 = \mathbf{E}\{|x_{n+\nu} - \varphi_{n,\nu}|^2\} = \int_{-1/2}^{1/2} |e^{2\pi i\nu\lambda} - \Phi_\nu(\lambda)|^2 \, dF(\lambda).$$

Since σ_ν^2 is the lower limit of the error in approximating $x_{n+\nu}$ by linear combinations of the random variables $\cdots, x_{n-1}, x_n$, whereas $\sigma_{\nu+1}^2$ is the lower limit of the error in approximating $x_{n+\nu}$ by linear combinations of the random variables $\cdots, x_{n-2}, x_{n-1}$, we must have $\sigma_\nu^2 \leq \sigma_{\nu+1}^2$. Hence

$$0 \leq \sigma_1^2 \leq \sigma_2^2 \leq \cdots.$$

The process is called *regular* if $\sigma_1^2 > 0$, *non-regular* or *deterministic* if $\sigma_1^2 = 0$. In the regular case, x_{n+1} does not lie in $\mathfrak{M}_n$, so that the sequence of linear manifolds $\cdots, \mathfrak{M}_0, \mathfrak{M}_1, \cdots$ is (strictly) increasing. In the non-regular case x_{n+1} lies in $\mathfrak{M}_n$, so that these linear manifolds are all the same; each is $\mathfrak{M}_\infty$, and all the σ_j's vanish. The σ_j^2's may not be strictly increasing in the regular case. In fact, if the x_n's form an orthonormal set, $1 = \sigma_1^2 = \sigma_2^2 = \cdots$ (see §3). The past completely determines the future in the non-regular case. This may also be true even in the regular case, but in the regular case the future cannot be determined completely by a *linear operation on the past.*

We shall solve the prediction problem by finding $\varphi_{n,\nu}$ and Φ_ν explicitly. We observe that formally, if Φ_ν is given by

$$\Phi_\nu(\lambda) = \sum_{j=0}^{\infty} b_j e^{-2\pi i j\lambda},$$

$\varphi_{n,\nu}$ is given by

$$\varphi_{n,\nu} = \sum_{j=0}^{\infty} b_j x_{n-j},$$

which exhibits $\varphi_{n,\nu}$ as a linear operation on the past. It may not be possible to express Φ_ν as the sum of a nicely convergent series of this form, however, and $\varphi_{n,\nu}$ may have to be expressed in some more complicated way also.

3. Solution of the prediction problem in simple cases (discrete parameter)

As a trivial example of the prediction problem we are studying in this chapter, suppose that the x_n's are mutually orthogonal, so that

$$dF(\lambda) = \mathbf{E}\{|x_0|^2\} \, d\lambda.$$

Then $x_{n+\nu}$ is orthogonal to $\mathfrak{M}_n$, so that $\varphi_{n,\nu}(\omega) = 0$ and $\Phi_\nu(\lambda) = 0$. Note that we have not assumed that $\mathbf{E}\{x_j\} = 0$ in this discussion.

Next suppose that the spectral distribution function F is absolutely continuous, with

$$F'(\lambda) = \frac{1}{|\sum_0^\beta B_j e^{2\pi i j\lambda}|^2}, \qquad B_0 B_\beta \neq 0. \tag{3.1}$$

The polynomial $\sum_0^\beta B_j z^j$ cannot vanish when $|z| = 1$, since F' is integrable. At the expense of changing the B_j's, without changing the value of the polynomial on $|z| = 1$, we can suppose that all its roots have modulus < 1 (see X §10). Then (see X §10, Case 2) x_{n+1} can be written in the form

$$x_{n+1} = -\frac{1}{B_\beta}(B_{\beta-1}x_n + \cdots + B_0 x_{n-\beta+1}) + \frac{\xi_{n+1}}{B_\beta}, \tag{3.2}$$

where the ξ_j's form an orthonormal set, and each ξ_{n+1} is orthogonal to the random variables $\cdots, x_{n-1}, x_n$, that is, ξ_{n+1} is orthogonal to $\mathfrak{M}_n$. The function

$$\varphi_{n,1} = -\frac{1}{B_\beta}(B_{\beta-1}x_n + \cdots + B_0 x_{n-\beta+1}) \tag{3.3}$$

is then the prediction of x_{n+1} for lag 1, since as so defined $\varphi_{n,1} \in \mathfrak{M}_n$ and $x_{n+1} - \varphi_{n,1} = \xi_{n+1}/B_\beta$ is orthogonal to $\mathfrak{M}_n$. The prediction function Φ_1 is given by

$$\Phi_1(\lambda) = -\frac{1}{B_\beta}(B_{\beta-1} + \cdots + B_0 e^{-2\pi i(\beta-1)\lambda}).$$

The mean square prediction error (lag 1) is

$$\sigma_1^2 = \mathbf{E}\left\{\left|\frac{\xi_{n+1}}{B_\beta}\right|^2\right\} = \frac{1}{|B_\beta|^2}.$$

To find the prediction for lag greater than 1, (3.2) can be iterated. For example,

$$\begin{aligned} x_{n+2} &= -\frac{1}{B_\beta}[B_{\beta-1}x_{n+1} + \cdots + B_0 x_{n-\beta+2}] + \frac{\xi_{n+2}}{B_\beta} \\ &= \sum_{=0}^{\beta-2} \frac{B_{\beta-1}B_{\beta-j-1} - B_\beta B_{\beta-j-2}}{B_\beta^2} x_{n-j} + \frac{B_{\beta-1}B_0}{B_\beta^2} x_{n-\beta+1} \\ &\qquad + \frac{\xi_{n+2}}{B_\beta} - \frac{B_{\beta-1}}{B_\beta^2}\xi_{n+1}. \end{aligned}$$

This equation shows that $\varphi_{n,2}$ is given by the terms on the right in $x_{n-\beta+1}, \cdots, x_n$ and

$$\sigma_2^2 = \mathbf{E}\left\{\left|\frac{\xi_{n+2}}{B_\beta} - \frac{B_{\beta-1}\xi_{n+1}}{B_\beta^2}\right|^2\right\} = \frac{1}{|B_\beta|^2} + \frac{|B_{\beta-1}|^2}{|B_\beta|^4}.$$

The case $\beta = 1$ (Markov process in the wide sense) has special interest. In this case (see X §3, Example 2) the formulas are particularly simple:

$$F'(\lambda) = \frac{1}{|B_1 e^{2\pi i\lambda} + B_0|^2}$$

$$R(n) = \frac{c^n}{(1 - |c|^2)|B_1|^2} \qquad n \geq 0 \qquad c = -\frac{B_0}{B_1}$$

(3.4)

$$\varphi_{n,\nu} = c^\nu x_n \qquad \Phi_\nu(\lambda) = c^\nu$$

$$\sigma_\nu^2 = \frac{1}{|B_1|^2}(1 + |c|^2 + \cdots + |c|^{2\nu-2}).$$

Note that $\varphi_{n-\nu,\nu}$, the prediction of x_n with lag ν, goes to 0 when $\nu \to \infty$. In other words, the predicted value of x_n based on the remote past is 0, and the corresponding mean square error is $R(0) = \mathbf{E}\{|x_n|^2\}$.

The processes with spectral density (3.1) are regular and have the property that the prediction $\varphi_{n,1}$ depends only on a finite number of the past x_j's. Conversely, a regular process with this property must satisfy a difference equation of the form (3.2), where $B_0 B_\beta \neq 0$, $\mathbf{E}\{|\xi_{n+1}|^2\} = 1$, and ξ_{n+1} is orthogonal to $\mathfrak{M}_n$. According to X §10 the x_n process must then have an absolutely continuous spectral distribution with density (3.1), and the roots of $\sum_0^\beta B_j z^j$ must have modulus < 1.

Obviously a process is deterministic and has the property that $\varphi_{n,1}$ depends only on a finite number of past x_j's if and only if it satisfies a homogeneous difference equation

$$B_\beta x_{n+1} + \cdots + B_0 x_{n-\beta+1} = 0, \qquad B_0 B_\beta \neq 0 \tag{3.5}$$

where we have supposed that $\varphi_{n,1}$ depends on β past x_j's. A necessary and sufficient condition for (3.5) is that F be constant except for a finite number of jumps. In fact,

(3.6)

$$\mathbf{E}\{|B_\beta x_{n+1} + \cdots + B_0 x_{n-\beta+1}|^2\} = \int_{-1/2}^{1/2} |B_\beta e^{2\pi i\beta\lambda} + \cdots + B_0|^2 \, dF(\lambda).$$

If (3.5) is true, the integral in (3.6) vanishes, and this is impossible unless $F(\lambda)$ remains constant except at the $(\leq \beta)$ zeros of the integrand. Conversely, if F is constant except for β jumps, $B_0, \cdots, B_\beta$ can be chosen to make the integrand in (3.6) vanish at each jump. Then the integral will vanish and (3.5) will be true.

The following example exhibits the result of combining a regular and a deterministic process. Let $\{u_n, -\infty < n < \infty\}$ be a stationary (wide sense) Markov (wide sense) process with spectral density given by

$$F_u'(\lambda) = \frac{1}{|e^{2\pi i\lambda} - c|^2}, \qquad |c| < 1,$$

so that, by (3.4) with $B_1 = 1$,

$$R_u(n) = \frac{c^n}{1 - |c|^2}, \qquad n \geq 0$$

$$\varphi_{u,n,\nu} = c^\nu x_n$$

$$\sigma_{u,\nu}^{\ 2} = 1 + |c|^2 + \cdots + |c|^{2\nu-2}$$

Let v be orthogonal to all the u_n's, with $\mathbf{E}\{|v|^2\} = 1$, and define a v_n process by setting $v_n = v$ for all n. The spectral distribution of the v_n process is concentrated at the origin,

$$\begin{aligned} F_v(\lambda) &= 0 \qquad \lambda < 0 \\ &= 1 \qquad \lambda \geq 0, \end{aligned}$$

and the prediction of v_n with any lag is of course v_n. Finally set $x_n = u_n + v_n$ so that the spectral distribution function F of the x_n process is $F_u + F_v$. To evaluate the prediction $\varphi_{n,1}$ of x_{n+1} with lag 1 note that

$$\underset{m\to\infty}{\text{l.i.m.}} \frac{u_n + \cdots + u_{n-m}}{m+1} = 0.$$

This can be derived either from the law of large numbers for stationary (wide sense) processes (X Theorem 6.1) or by an explicit evaluation of the expectation of the square of the modulus of the above ratio. The random variable

$$\varphi = cx_n + (1 - c)\,\underset{m\to\infty}{\text{l.i.m.}} \frac{x_n + \cdots + x_{n-m}}{m+1} = cu_n + v_n \tag{3.7}$$

lies in $\mathfrak{M}_n$, and $x_{n+1} - \varphi = u_{n+1} - cu_n$ is orthogonal to $\mathfrak{M}_n$. Hence $\varphi = \varphi_{n,1}$, and we have proved that the prediction $\varphi_{n,1}$ is the sum of the individual predictions cu_n and v_n relative to the u_n and v_n processes. If the x_j's on the left in (3.7) are replaced by u_j's, the right side becomes cu_n, the u prediction of u_{n+1}; if the x_j's are replaced by v_j's, the right side becomes $v = v_{n+1}$, the prediction of v_{n+1}. Hence the calculation of the prediction can be made in the same way for the x_n, u_n, and v_n processes.

In other words, the prediction function can be chosen to be the same for all three; explicitly

$$\Phi_1(\lambda) = c + (1 - c) \operatorname*{l.i.m.}_{m\to\infty} \frac{1 + \cdots + e^{-2\pi i m\lambda}}{m + 1},$$

where the λ weighting for l.i.m. is $dF(\lambda)$, and therefore

$$\begin{aligned} \Phi_1(\lambda) &= c \qquad \lambda \neq 0 \\ &= 1 \qquad \lambda = 0. \end{aligned}$$

We have already observed in (3.4) that $\Phi_1(\lambda)$ for the u_n process should be identically c. There is no contradiction here since, in terms of $dF_u(\lambda)$ weighting, a change of Φ_1 at one point makes no difference (that is, each point has F_u measure 0). Of course, the linear operation (3.7) is needlessly complicated for the prediction of the u_n process, corresponding to the fact that Φ_1 is needlessly complicated for the prediction function of the u_n process. The point is that, as defined, Φ_1 is the prediction function for the x_n, u_n, and v_n processes, and it will be seen in §4 that this is characteristic of the general case. Finally note that the prediction error with lag ν of the x_n process is

$$\sigma_\nu^2 = 1 + |c|^2 + \cdots + |c|^{2\nu-2},$$

the same as that of the u_n process, but when $\nu \to \infty$ this approaches

$$\frac{1}{1 - |c|^2} < 1 + \frac{1}{1 - |c|^2} = \mathbf{E}\{|x_0|^2\}.$$

As a last example suppose that F is an absolutely continuous spectral distribution function with density F' of the form

$$F'(\lambda) = \left| \sum_0^\infty c_j e^{-2\pi i j\lambda} \right|^2, \qquad c_0 \neq 0,$$

where the series converges uniformly, and suppose that $\sum_0^\infty c_j z^{-j}$ converges uniformly for $|z| \geq 1 - \varepsilon$ for some $\varepsilon > 0$ and does not vanish there. For example, we have seen in X §10 that, if F' is a rational function of $e^{2\pi i\lambda}$, it can be written in the form

$$F'(\lambda) = \left| \frac{\sum_0^\alpha A_j z^j}{\sum_0^\beta B_j z^j} \right|^2, \qquad A_0 B_0 A_\alpha B_\beta \neq 0$$

where $z = e^{2\pi i\lambda}$ and where the denominator polynomial does not vanish for $|z| \geq 1$. Then the above hypotheses are satisfied in this case. Under the above hypotheses we can define the prediction function Φ_ν by

$$\Phi_\nu(\lambda) = e^{2\pi i\nu\lambda} \frac{\sum_{j=\nu}^{\infty} c_j e^{-2\pi i j\lambda}}{\sum_{j=0}^{\infty} c_j e^{-2\pi i j\lambda}}.$$

In fact, as so defined, $\Phi_\nu \in {}_0\mathfrak{M}$ since we can put $z = e^{2\pi i\lambda}$ in

$$\frac{z^\nu \sum_{j=\nu}^{\infty} c_j z^{-j}}{\sum_{j=0}^{\infty} c_j z^{-j}} = \frac{c_\nu}{c_0} + \sum_{j=1}^{\infty} d_j z^{-j}, \qquad |z| \geq 1 - \frac{\varepsilon}{2},$$

where the series converges uniformly, and $e^{2\pi i\nu\lambda} - \Phi_\nu$ is orthogonal to ${}_0\mathfrak{M}$ [λ weighting $dF(\lambda)$] because

$$\int_{-1/2}^{1/2} [e^{2\pi i\nu\lambda} - \Phi_\nu(\lambda)] e^{2\pi i m\lambda}\, dF(\lambda)$$
$$= \int_{-1/2}^{1/2} e^{2\pi i\nu\lambda} \sum_0^{\nu-1} c_j e^{-2\pi i j\lambda} \sum_0^{\infty} \bar{c}_j e^{2\pi i j\lambda} e^{2\pi i m\lambda}\, d\lambda = 0$$

if $m \geq 0$.

Thus the prediction problem is solved once F' can be put in the form of the preceding paragraph. Much of the next section will be devoted to an examination of the implications of such a representation of F' with weaker convergence hypotheses.

4. General solution of the prediction problem (discrete parameter)

THEOREM 4.1 *If $\{\xi_n, -\infty < n < \infty\}$ is an orthonormal sequence of random variables, if $\sum_0^\infty |c_n|^2 < \infty$, $c_0 \neq 0$, and if x_n is defined by*

$$x_n = \sum_0^\infty c_j \xi_{n-j}, \tag{4.1}$$

then the x_n process is stationary in the wide sense and has spectral density $|\sum_0^\infty c_n e^{-2\pi i n\lambda}|^2$. The process is regular and the mean square prediction error σ_ν^2 for lag ν satisfies the inequality

$$\sigma_\nu^2 \geq |c_0|^2 + \cdots + |c_{\nu-1}|^2. \tag{4.2}$$

More generally, (4.2) *remains true if instead of supposing that* x_n *is given by* (4.1) *it is only supposed that the* x_n *process has spectral distribution function* F *(not necessarily absolutely continuous) with*

$$F'(\lambda) = \left| \sum_{n=0}^{\infty} c_n e^{-2\pi i n \lambda} \right|^2, \qquad c_0 \neq 0, \qquad \sum_0^{\infty} |c_n|^2 < \infty, \tag{4.3}$$

for almost all λ *(Lebesgue measure). In either case there cannot be equality in* (4.2) *for any* ν *if* $\sum_0^{\infty} c_n z^n$ *has any zeros in* $|z| < 1$.

We observe that the series for x_n in (4.1) converges in the mean, by IV, Theorem 4.1. According to X §8 an x_n process can be written in the form (4.1) with the stated conditions on the c_n's if and only if the spectral distribution is absolutely continuous and F' is given by (4.3).

We prove (4.2) for $\nu = 1$; the proof in the general case is similar. The mean square prediction error of any finite prediction sum for lag 1 is easily calculated, and we find

$$\begin{aligned}
\mathbf{E}\{|x_n - \sum_{j=1}^{N} b_j x_{n-j}|^2\} &= \int_{-1/2}^{1/2} |e^{2\pi i n \lambda} - \sum_{j=1}^{N} b_j e^{2\pi i (n-j)\lambda}|^2 \, dF(\lambda) \\
&\geq \int_{-1/2}^{1/2} \left| \left[1 - \sum_{j=1}^{N} b_j e^{-2\pi i j \lambda}\right] \left[\sum_0^{\infty} c_j e^{-2\pi i j \lambda}\right] \right|^2 d\lambda \\
&= \int_{-1/2}^{1/2} |c_0 + \text{const. } e^{-2\pi i \lambda} + \text{const. } e^{-4\pi i \lambda} + \cdots|^2 \, d\lambda \\
&= |c_0|^2 + \cdots \geq |c_0|^2.
\end{aligned}$$

Thus $\sigma_1^2 \geq |c_0|^2$. Now suppose $\sum_0^{\infty} c_j z_0^{\,j} = 0$, $|z_0| < 1$. Then we shall prove that $\sigma_1^2 > |c_0|^2$. In the first place, since $\sum_0^{\infty} \bar{c}_j \bar{z}_0^{\,j} = 0$,

$$\sum_0^{\infty} \bar{c}_n z^n = (z - \bar{z}_0) \sum_0^{\infty} \bar{c}_n' z^n, \qquad c_0' = -\frac{c_0}{z_0}$$

so that, going to $|z| = 1$,

$$\sum_0^{\infty} \bar{c}_j e^{2\pi i j \lambda} = (e^{2\pi i \lambda} - \bar{z}_0) \sum_0^{\infty} \bar{c}_j' e^{2\pi i j \lambda},$$

where both series converge in the mean. Moreover,

$$\left|\sum_0^\infty c_j e^{-2\pi i j\lambda}\right| = \left|\sum_0^\infty \bar{c}_j e^{2\pi i j\lambda}\right| = \left|\frac{1-e^{2\pi i\lambda}z_0}{e^{2\pi i\lambda}-\bar{z}_0}\right|\left|\sum_0^\infty \bar{c}_j e^{2\pi i j\lambda}\right|$$

$$= |1-e^{2\pi i\lambda}z_0|\left|\sum_0^\infty \bar{c}_j' e^{2\pi i j\lambda}\right| = \left|\sum_0^\infty \bar{c}_j'' e^{2\pi i j\lambda}\right|$$

$$= \left|\sum_0^\infty c_j'' e^{-2\pi i j\lambda}\right|, \qquad c_0'' = c_0' = -\frac{c_0}{z_0},$$

where the last series converges in the mean, so that $\sum_0^\infty |c_j''|^2 < \infty$. Now

$$\tilde{x}_n = \sum_{j=0}^\infty c_j'' \xi_{n-j}$$

defines a process with the same spectral density

$$\left|\sum_0^\infty c_j e^{-2\pi i j\lambda}\right|^2 = \left|\sum_0^\infty c_j'' e^{-2\pi i j\lambda}\right|^2$$

as the x_n process. Hence it has the same prediction error, and by the inequality already proved

$$\sigma_1^2 \geq |c_0''|^2 = \frac{|c_0|^2}{|z_0|^2} > |c_0|^2.$$

This finishes the proof of the lemma.

We now show that any regular process can be decomposed into a regular and a non-regular process of canonical types. Let $\cdots, x_0, x_1, \cdots$ be the variables of a regular process, and define ξ_n by

$$x_n - \varphi_{n-1,1} = \sigma_1 \xi_n \qquad \sigma_1 = \mathbf{E}^{1/2}\{|x_n - \varphi_{n-1,1}|^2\} > 0. \tag{4.4}$$

Since $x_n - \varphi_{n-1,1}$ is orthogonal to $\mathfrak{M}_{n-1}$, it is orthogonal to every $x_m - \varphi_{m-1,1}$ with $m < n$. Thus the ξ_j's form an orthonormal set, and we can write x_n as the sum of a Fourier series in the ξ_j's and a remainder v_n,

$$x_n = u_n + v_n, \qquad u_n = \sum_{j=0}^\infty c_j \xi_{n-j}, \qquad \sum_0^\infty |c_j|^2 < \infty,$$
$$c_j = \mathbf{E}\{x_n \bar{\xi}_{n-j}\}, \qquad (c_0 = \sigma_1). \tag{4.5}$$

Thus u_n is the projection of x_n on the closed linear manifold generated by the ξ_j's and v_n is orthogonal to this manifold. From the definition of ξ_n, $\xi_n \in \mathfrak{M}_n$; hence u_n and $v_n = x_n - u_n$ are in $\mathfrak{M}_n$. To recapitulate,

$$\mathbf{E}\{\xi_m \bar{\xi}_n\} = \delta_{m,n},$$
$$\mathbf{E}\{\xi_m \bar{v}_n\} = \mathbf{E}\{u_m \bar{v}_n\} = 0, \qquad \text{all } m, n,$$
$$\mathbf{E}\{\xi_n \bar{x}_m\} = 0, \qquad m < n,$$
$$u_n, v_n, \xi_n \in \mathfrak{M}_n.$$

The geometrical significance of (4.5) can be seen as follows. The orthogonality and inclusion relations just noted imply that ξ_n is orthogonal to $\mathfrak{M}_{n-1}$. Then, if $w \in \mathfrak{M}_{-\infty} = \bigcap_{\infty}^{\infty} \mathfrak{M}_n$, w is orthogonal to every ξ_m. Conversely, suppose that, for some n, $w \in \mathfrak{M}_n$ and that w is orthogonal to every ξ_m. Then w is in the closed linear manifold generated by ξ_n and the random variables in $\mathfrak{M}_{n-1}$ (this is one way of describing $\mathfrak{M}_n$), and, since ξ_n is orthogonal to both w and $\mathfrak{M}_{n-1}$, it follows that $w \in \mathfrak{M}_{n-1}$. This argument is, in symbols,

$$w = \hat{\mathbf{E}}\{w \mid \mathfrak{M}_n\} = \hat{\mathbf{E}}\{w \mid \mathfrak{M}_{n-1}, \xi_n\} = \hat{\mathbf{E}}\{w \mid \mathfrak{M}_{n-1}\} + \hat{\mathbf{E}}\{w \mid \xi_n\}$$
$$= \hat{\mathbf{E}}\{w \mid \mathfrak{M}_{n-1}\} \in \mathfrak{M}_{n-1}.$$

Repeating this argument, it follows that $w \in \mathfrak{M}_j$ for all j, that is, $w \in \mathfrak{M}_{-\infty}$. Thus, for each n, $\mathfrak{M}_{-\infty}$ consists of all the random variables in $\mathfrak{M}_n$ orthogonal to $\xi_n, \xi_{n-1}, \cdots$; that is to say, the manifold $\mathfrak{M}_{-\infty}$ together with the one-dimensional manifolds of the multiples of $\xi_n, \xi_{n-1}, \cdots$ are mutually orthogonal and generate $\mathfrak{M}_n$. Equation (4.5) expresses x_n as the sum of its projections on these orthogonal manifolds. The term v_n, the projection on $\mathfrak{M}_{-\infty}$, is the contribution to x_n from the remote past.

Let $\mathfrak{M}_{v,n} \subset \mathfrak{M}_{-\infty}$ be the closed linear manifold generated by $v_n, v_{n-1}, \cdots$. We prove that $\mathfrak{M}_{v,n} = \mathfrak{M}_{-\infty}$ for all n by proving that, if $w \in \mathfrak{M}_{-\infty}$, w is its own projection on $\mathfrak{M}_{v,n}$. In fact, since the ξ_j's are orthogonal to the v_k's,

$$\hat{\mathbf{E}}\{w \mid v_n, v_{n-1}, \cdots\} = \hat{\mathbf{E}}\{w \mid v_n, v_{n-1}, \cdots, \xi_n, \xi_{n-1}, \cdots\}, \tag{4.6}$$

with probability 1. According to (4.5) the closed linear manifold generated by the v_j's and ξ_j's for $j \le n$ is $\mathfrak{M}_n$. Hence the right side of (4.6) is $\hat{\mathbf{E}}\{w \mid \mathfrak{M}_n\}$, and this is w because $w \in \mathfrak{M}_{-\infty} \subset \mathfrak{M}_n$.

A similar argument shows that ξ_n lies in the closed linear manifold generated by $u_n, u_{n-1}, \cdots$. We give the argument in geometrical terms this time. The random variable ξ_n lies in $\mathfrak{M}_n$, which is the closed linear manifold generated by $u_n, u_{n-1}, \cdots, v_n, v_{n-1}, \cdots$, by (4.5). Since the v_j's are orthogonal to ξ_n and to $u_n, u_{n-1}, \cdots$, they are superfluous here. Thus the closed linear manifold generated by $u_n, u_{n-1}, \cdots$ is that generated by $\xi_n, \xi_{n-1}, \cdots$.

The u_n and v_n processes are stationary in the wide sense. Since $\mathfrak{M}_{v,n} = \mathfrak{M}_{v,n-1}$ $(= \mathfrak{M}_{-\infty})$, v_n is in the closed linear manifold generated by $v_{n-1}, v_{n-2}, \cdots$; in other words, the v_n process is non-regular. Let F_u and F_v be the spectral distribution function of the u_n and v_n processes respectively. Since $x_n = u_n + v_n$, and since the u_j's are orthogonal to the v_k's,

$$F = F_u + F_v.$$

This orthogonality also means that any approximating sum $\sum_{j=1}^{N} b_j x_{n-j}$ used in prediction theory splits into orthogonal sums,

$$\sum_{j=1}^{N} b_j x_{n-j} = \sum_{j=1}^{N} b_j u_{n-j} + \sum_{j=1}^{N} b_j v_{n-j},$$

one sum involving only the u_n process, the other only the v_n process. We shall show that a single function Φ_ν will serve as the prediction function of the x_n, the u_n, and the v_n process, so that a linear combination as above which is very nearly the best approximation to x_n splits into the sum of the same linear combination of u_k's giving nearly the best approximation to u_n and the same linear combination of v_k's giving nearly the best approximation to v_n.

The prediction $\varphi_{n-\nu,\,\nu}$ can now be written down explicitly,

$$\varphi_{n-\nu,\,\nu} = \hat{\mathbf{E}}\{x_n \mid \mathfrak{M}_{n-\nu}\} = \sum_{j=\nu}^{\infty} c_j \xi_{n-j} + v_n. \tag{4.7}$$

In fact, the right-hand side is a random variable in $\mathfrak{M}_{n-\nu}$, because

$$\xi_{n-j} \in \mathfrak{M}_{n-j} \subset \mathfrak{M}_{n-\nu}, \qquad j \geq \nu$$

$$v_n \in \mathfrak{M}_{-\infty} \subset \mathfrak{M}_{n-\nu},$$

and, using this definition of $\varphi_{n-\nu,\,\nu}$,

$$x_n - \varphi_{n-\nu,\,\nu} = \sum_{j=0}^{\nu-1} c_j \xi_{n-j},$$

which is orthogonal to $\mathfrak{M}_{n-\nu}$. The same reasoning shows that the infinite sum on the right in (4.7) is the linear least squares prediction of u_n in terms of $u_{n-\nu}, u_{n-\nu-1}, \cdots$. The mean square error in predicting u_n is the same as that in predicting x_n,

$$\sigma_\nu^2 = \mathbf{E}\{|\sum_{j=0}^{\nu-1} c_j \xi_{n-j}|^2\} = |c_0|^2 + \cdots + |c_{\nu-1}|^2. \tag{4.8}$$

If Φ_ν is the prediction function of the x_n process, for lag ν,

$$\begin{aligned} \sigma_\nu^2 &= \int_{-1/2}^{1/2} |e^{2\pi i\nu\lambda} - \Phi_\nu(\lambda)|^2 \, dF(\lambda) \\ &= \int_{-1/2}^{1/2} |e^{2\pi i\nu\lambda} - \Phi_\nu(\lambda)|^2 \, dF_u(\lambda) + \int_{-1/2}^{1/2} |e^{2\pi i\nu\lambda} - \Phi_\nu(\lambda)|^2 \, dF_v(\lambda). \end{aligned} \tag{4.9}$$

Now, since Φ_ν is by definition in the closed linear manifold ${}_0\mathfrak{M}$ generated by the functions $\{e^{2\pi ij\lambda}, j \leq 0\}$ with λ weighting $dF(\lambda)$, Φ_ν must be in the closed linear manifolds generated by the same functions but using the smaller weighting $dF_u(\lambda)$ and $dF_v(\lambda)$. Hence the two integrals in the second line of (4.9) are at least equal respectively to the mean square prediction errors for lag ν of the u_n and v_n processes, namely σ_ν^2 and 0. It follows that the first integral is σ_ν^2 and the second one 0. In other words, Φ_ν is simultaneously the prediction function for lag ν of the x_n the u_n and the v_n processes.

The ξ_n process is stationary in the wide sense with spectral density identically 1. If we write ξ_n in its spectral representation,

$$\xi_n = \int_{-1/2}^{1/2} e^{2\pi in\lambda}\, dy_\xi(\lambda), \qquad \mathbf{E}\{|dy_\xi(\lambda)|^2\} = d\lambda,$$

where the $y_\xi(\lambda)$ process is a process with orthogonal increments, we find

$$u_n = \sum_{j=0}^{\infty} c_j \xi_{n-j} = \int_{-1/2}^{1/2} e^{2\pi in\lambda} \left(\sum_{j=0}^{\infty} c_j e^{-2\pi ij\lambda}\right) dy_\xi(\lambda). \tag{4.10}$$

If we write v_n in its spectral representation,

$$v_n = \int_{-1/2}^{1/2} e^{2\pi in\lambda}\, dy_v(\lambda), \qquad \mathbf{E}\{|dy_v(\lambda)|^2\} = dF_v(\lambda),$$

the y_v increments are orthogonal to the y_ξ increments. The spectral representation of x_n can now be put in the form

$$x_n = u_n + v_n = \int_{-1/2}^{1/2} e^{2\pi in\lambda} [\sum_{j=0}^{\infty} c_j e^{-2\pi ij\lambda}\, dy_\xi(\lambda) + dy_v(\lambda)]$$

so that the $y(\lambda)$ of the spectral representation (2.1) of x_n is given by

$$y(\lambda) = \int_{-1/2}^{\lambda} [\sum_{j=0}^{\infty} c_j e^{-2\pi ij\mu}\, dy_\xi(\mu) + dy_v(\mu)].$$

We now can find the λ functions corresponding to ξ_n, u_n, v_n, and $\varphi_{n,\nu}$ under the correspondence (2.2), and incidentally evaluate F_v. If ξ_n corresponds to Φ,

$$\xi_n = \int_{-1/2}^{1/2} \Phi(\lambda) [\sum_{j=0}^{\infty} c_j e^{-2\pi ij\lambda}\, dy_\xi(\lambda) + dy_v(\lambda)]$$

$$= \int_{-1/2}^{1/2} e^{2\pi in\lambda}\, dy_\xi(\lambda),$$

we find, on comparing integrands, that

$$\Phi(\lambda) \sum_{j=0}^{\infty} c_j e^{-2\pi ij\lambda} = e^{2\pi in\lambda}$$

for almost all λ (Lebesgue measure) and

$$\Phi(\lambda) = 0$$

for almost all λ (F_v measure). These two conditions are incompatible unless there is a λ set S of Lebesgue measure 0 such that

$$\int_S dF_v(\lambda) = F_v(\tfrac{1}{2}) - F_v(-\tfrac{1}{2}),$$

that is, unless F_v is a singular monotone function. In the following we shall call S the *set of increase* of F_v. Since the expression for u_n shows that the u_n process has an absolutely continuous spectral distribution with density (see X §8), given by

$$F_u'(\lambda) = |\sum_0^\infty c_j e^{-2\pi ij\lambda}|^2,$$

we have thus proved that the decomposition of x_n into a regular u_n process and a deterministic v_n process corresponds in terms of spectral distributions to the representation of F as the sum of its absolutely continuous and singular components F_u and F_v respectively. The function Φ corresponding to ξ_n is uniquely determined, neglecting values on a set of F measure 0, by the above equations. We can take

$$\Phi(\lambda) = \frac{e^{2\pi in\lambda}}{\sum_0^\infty c_j e^{-2\pi ij\lambda}}, \qquad \lambda \notin S$$
$$= 0 \qquad \lambda \in S.$$

It follows that we can take as the λ functions corresponding to u_n, v_n, $\varphi_{n-\nu,\nu}$

$$u_n: \begin{cases} e^{2\pi in\lambda}, & \lambda \notin S, \\ 0, & \lambda \in S, \end{cases} \qquad v_n: \begin{cases} 0, & \lambda \notin S \\ e^{2\pi in\lambda}, & \lambda \in S, \end{cases}$$

$$\varphi_{n-\nu,\nu}: \begin{cases} \dfrac{\sum_{j=\nu}^\infty c_j e^{2\pi i(n-j)\lambda}}{\sum_{=0}^\infty c_j e^{-2\pi ij\lambda}}, & \lambda \notin S \\ e^{2\pi in\lambda}, & \lambda \in S, \end{cases}$$

and, since Φ_ν, the prediction function for lag ν, corresponds to $\varphi_{0,\nu}$,

$$\Phi_\nu(\lambda) = e^{2\pi i\nu\lambda} \frac{\sum_{j=\nu}^\infty c_j e^{-2\pi ij\lambda}}{\sum_{j=0}^\infty c_j e^{-2\pi ij\lambda}}, \qquad \lambda \notin S, \tag{4.11}$$
$$= e^{2\pi i\nu\lambda}, \qquad \lambda \in S.$$

Thus u_n and v_n can be expressed in the form

$$u_n = \int_{S'} e^{2\pi i n\lambda}\, dy(\lambda), \qquad v_n = \int_{S} e^{2\pi i n\lambda}\, dy(\lambda),$$

where S' is the complement of S in $[-\frac{1}{2}, \frac{1}{2}]$. In X §5 we have discussed the decompositions of a process into processes with disjunct spectra and shown how they can be accomplished by linear operations. We have now shown that in the regular case the decomposition separating out the singular component of the spectral distribution can be accomplished by a linear operation on the past and corresponds to separating out the deterministic component of the process.

We observe that $\varphi_{n-\nu,\nu}$, the prediction of x_n in terms of $x_{n-\nu}, x_{n-\nu-1}, \cdots$, becomes v_n when $\nu \to \infty$. Only the deterministic component can be predicted from the remote past. As would be expected,

$$\lim_{\nu\to\infty} \sigma_\nu^2 = \mathbf{E}\{|u_n|^2\}.$$

In the decomposition of a regular process the regular component must be present, but of course the deterministic component, the v_n process, may not appear at all, that is, we may have $x_n = u_n$.

The decomposition can be summarized as follows:

THEOREM 4.2 *Let $\{x_n, -\infty < n < \infty\}$ be a regular process. Then x_n can be written in the form*

$$x_n = \sum_{j=0}^{\infty} c_j \xi_{n-j} + v_n = u_n + v_n,$$

where

$$\sum_0^\infty |c_j|^2 < \infty, \qquad c_0 > 0,$$

$$\mathbf{E}\{\xi_m \bar{\xi}_n\} = \delta_{m,n}, \qquad \mathbf{E}\{\xi_m \bar{v}_n\} = 0 \qquad \textit{all } m, n,$$

$$\xi_n \in \mathfrak{M}_n, \qquad v_n \in \mathfrak{M}_{-\infty}.$$

There is only one sequence of constants $\{c_n\}$ and only one sequence of random variables $\{\xi_n\}$ satisfying these conditions.

In this representation the u_n process is regular and the v_n process is deterministic. The prediction $\varphi_{n-\nu,\nu}$ is given by (4.7), and the prediction error is $\sigma_\nu^2 = \sum_0^{\nu-1} |c_j|^2$, which is the same as that for the u_n process. The spectral distributions of the u_n and v_n processes are respectively the absolutely continuous and singular components of the spectral distribution of the x_n process.

All that remains to be proved is the uniqueness of the c_j's and ξ_j's under the indicated conditions. These conditions imply that $c_j = \mathbf{E}\{x_n \bar{\xi}_{n-j}\}$. Moreover, $\varphi_{n-1,1}$ must be given by

$$\varphi_{n-1,1} = \sum_{j=1}^{\infty} c_j \xi_{n-j} + v_n,$$

since as so defined $\varphi_{n-1,1} \in \mathfrak{M}_{n-1}$ and $x_n - \varphi_{n-1,1}$ is orthogonal to $\mathfrak{M}_{n-1}$. Then $x_n - \varphi_{n-1,1} = c_0 \xi_n$ is uniquely determined by the process. Since the process is regular, $|c_0|^2 = \sigma_1^2 > 0$; c_0 is then uniquely determined by the hypothesis that it is positive, and ξ_n is thereby also uniquely determined, that is, neglecting values on a set of probability 0. Finally c_j for $j > 0$ is now uniquely determined by the Fourier coefficient formula already written.

The theorems of IV §6 make it possible to put the preceding results in an analytic form.

THEOREM 4.3 *A process is regular if and only if $F'(\lambda) > 0$ for almost all λ (Lebesgue measure) and*

$$\int_{-1/2}^{1/2} \log F'(\lambda)\, d\lambda > -\infty. \tag{4.12}$$

In the regular case the constants $\{c_n\}$ of Theorem 4.2 satisfy and are uniquely determined by the conditions

$$c_0 = e^{\frac{1}{2}\int_{-1/2}^{1/2} \log F'(\lambda)\, d\lambda},$$

$$\sum_0^{\infty} |c_n|^2 < \infty, \qquad \sum_0^{\infty} c_n z^n \neq 0 \qquad (|z| < 1) \tag{4.13}$$

$$F'(\lambda) = \Big|\sum_0^{\infty} c_n e^{-2\pi i n\lambda}\Big|^2,$$

the last to hold for almost all λ (Lebesgue measure). These constants can be calculated by using

$$\tfrac{1}{2} \log F'(\lambda) \sim \sum_{-\infty}^{\infty} a_n e^{2\pi i n\lambda} \ \Big(\textit{that is, } a_n = \tfrac{1}{2}\int_{-1/2}^{1/2} \log F'(\lambda) e^{-2\pi i n\lambda}\, d\lambda\Big)$$

(4.14)

$$\sum_0^{\infty} \bar{c}_n z^n = e^{a_0 + 2\sum_1^{\infty} a_n z^n}, \qquad |z| < 1.$$

Note that $\sigma_1^2 = c_0^2$ has now been evaluated explicitly. This theorem now requires little proof. We have already seen that in the regular case F_u is the absolutely continuous component of F. Then

$$F'(\lambda) = F_u'(\lambda) = \Big|\sum_0^{\infty} c_j e^{-2\pi i j\lambda}\Big|^2 \tag{4.15}$$

for almost all λ (Lebesgue measure). Hence by IV, Theorem 6.2, $F'(\lambda) > 0$ for almost all λ and (4.12) is true. Conversely, if $F'(\lambda) > 0$ for almost all λ and if (4.12) is true, F' can be written as the square of the modulus of a series of the type in (4.13) according to IV, Theorem 6.2, and the regularity follows from Theorem 4.1. Thus (4.12) is necessary and sufficient for regularity.

According to Theorem 4.1 the constants $\{c_n\}$ of Theorem 4.2 must have the property that $\sum_0^\infty c_n z^n \neq 0$ for $|z| < 1$, since the mean square error of the x_n and u_n processes satisfies (4.2) with equality. To derive the evaluation of c_0 in (4.13) we note that if (4.12) is satisfied we can, according to IV, Theorem 6.2, write F' in the form

$$F'(\lambda) = |\sum_0^\infty \bar{d}_j e^{2\pi i j\lambda}|^2, \qquad \sum_0^\infty |d_j|^2 < \infty,$$

with

$$d_0 = e^{\frac{1}{2}\int_{-1/2}^{1/2} \log F'(\lambda)\, d\lambda}$$

But then, according to Theorem 4.1, the prediction error $\sigma_1^2 = c_0^2$ for lag 1 of the u_n process, a process with spectral density F', is at least d_0^2. Hence $c_0 \geq d_0$. On the other hand, according to IV, Theorem 6.2 [see IV, (6.7')], the reverse inequality is necessarily true, so that $c_0 = d_0$, as was to be proved. Thus the conditions (4.13) are satisfied. Conversely, according to IV, Theorem 6.2, the conditions (4.13) uniquely determine the c_n's and they can be calculated explicitly by (4.14).

Note that, if x_n can be written in the form

$$(4.16) \qquad x_n = \sum_{j=0}^\infty c_j \xi_{n-j}, \qquad \sum_{j=0}^\infty |c_j|^2 < \infty, \qquad \mathbf{E}\{\xi_j \bar{\xi}_k\} = \delta_{jk},$$

with not all c_j's vanishing, it is no restriction, shifting subscripts on the ξ_j's if necessary, to suppose that $c_0 \neq 0$. The process is regular by Theorem 4.1. Since F has no singular component in this case, the process has no deterministic component. The c_j's in (4.16) may not be the uniquely determined c_j's of Theorems 4.2 and 4.3, however, even if the ξ_j's are chosen, as can always be done and as we shall suppose has been done, so that c_0 is real and positive. In fact, if (4.13) is not satisfied there will be a uniquely determined *different* sequence $\{c_n'\}$ for which it is, and there will be a *different* orthonormal sequence $\{\xi_n'\}$ such that

$$(4.16') \qquad x_n = \sum_{j=0}^\infty c_j' \xi_{n-j}', \qquad c_0' > c_0.$$

If x_n can be written in the form

$$x_n = \sum_{j=-\infty}^{\infty} c_j \xi_{n-j}, \qquad \sum_{-\infty}^{\infty} |c_j|^2 < \infty, \qquad \mathbf{E}\{\xi_j \bar{\xi}_k\} = \delta_{jk}, \tag{4.17}$$

the process has an absolutely continuous spectral distribution function F, with

$$F'(\lambda) = \left| \sum_{-\infty}^{\infty} \bar{c}_j e^{2\pi i j \lambda} \right|^2. \tag{4.18}$$

Since any integrable non-negative function can be put in the form (4.18) (the sum is simply the Fourier series of any square root of F'), the only restriction implicit in (4.17) is that F be absolutely continuous. The x_n process may be regular, and then will have no deterministic component, or it may be non-regular.

At the other extreme, if the x_n process has a distribution function which is singular, then the process cannot be regular, because $F'(\lambda) = 0$ except on a λ set of Lebesgue measure 0. It is interesting to observe that, although the deterministic component of a regular process is always of this type, this is *not* the general non-regular case. In fact, by Theorem 4.3 the spectral distribution of a non-regular process may be absolutely continuous, as long as (4.12) is false.

It will be useful to have an explicit representation of the linear manifolds $\mathfrak{M}_n$ and $\mathfrak{M}_{-\infty}$ in λ language. The manifold $\mathfrak{M}_n$ has corresponding λ manifold ${}_n\mathfrak{M}$, the closed linear manifold generated by $\{e^{2\pi i m \lambda},\ m \leq n\}$ [λ weighting $dF(\lambda)$] and ${}_{-\infty}\mathfrak{M} = \bigcap_{-\infty}^{\infty} {}_n\mathfrak{M}$ corresponds to $\mathfrak{M}_{-\infty}$. We observe that the definitions of ${}_n\mathfrak{M}$ and ${}_{-\infty}\mathfrak{M}$ depend on the choice of F but do not involve any probability concepts, and we shall phrase their description without involving these concepts.

THEOREM 4.4 *Let F be a monotone non-decreasing function for $|\lambda| \leq \frac{1}{2}$, with $F(\lambda +) = F(\lambda)$, $F(-\frac{1}{2}) = 0$. Let F_v be its singular component, and let S, of Lebesgue measure* 0, *be the set of increase of F_v.*

(i) *If $F'(\lambda) = 0$ at most on a set of Lebesgue measure* 0, *and if*

$$\int_{-1/2}^{1/2} \log F'(\lambda)\, d\lambda > -\infty, \tag{4.19}$$

there is one and only one sequence $\{c_j\}$ satisfying (4.13). *The c_j's are determined, for example, by* (4.14). *The manifold ${}_{-\infty}\mathfrak{M}$ consists of all functions α which are measurable with respect to F, vanish for almost all λ (Lebesgue measure), and for which*

$$\int_{-1/2}^{1/2} |\alpha(\lambda)|^2\, dF_v(\lambda) < \infty.$$

The manifold ${}_n\mathfrak{M}$ *consists of all functions of the form*

$$\begin{array}{ll} \dfrac{e^{2\pi in\lambda}\sum_0^\infty \gamma_j e^{-2\pi ij\lambda}}{\sum_0^\infty c_j e^{-2\pi ij\lambda}} & \lambda \notin S \\ \alpha(\lambda) & \lambda \in S \end{array}$$

where α *is as just described and* $\sum_0^\infty |\gamma_j|^2 < \infty$.

(ii) *If the hypothesis of* (i) *is not satisfied,* ${}_{-\infty}\mathfrak{M} = {}_n\mathfrak{M}$ *for all n and this manifold consists of all functions* β *which are measurable with respect to F and for which*

$$\int_{-1/2}^{1/2} |\beta(\lambda)|^2\, dF(\lambda) < \infty.$$

(i′) *If there are constants* $\{c_j\}$ *with* $\sum_0^\infty |c_j|^2 < \infty$ *such that* (4.18) *is true, then, if* ${}_n\mathfrak{M}$ *is as described in* (i), *the* c_j*'s are the uniquely determined constants described in* (i), *aside from a multiplicative constant of modulus* 1.

(ii′) *If* ${}_\infty\mathfrak{M}$ *is as described in* (ii), (4.19) *is false.*

The theorem is obviously true [with Case (ii)] if $F(\frac{1}{2}) = 0$, and in the following it will be convenient to exclude this possibility. If λ is considered a random variable, with distribution function $F/F(\frac{1}{2})$, the x_n process defined by

$$x_n = F(\tfrac{1}{2})^{1/2} e^{2\pi in\lambda}$$

is stationary in the wide sense with spectral distribution function F. The theorem thus becomes a theorem on stationary (wide sense) processes, and the previous theorems can be used to deduce it. Only the description of the manifolds ${}_n\mathfrak{M}$ and ${}_{-\infty}\mathfrak{M}$ requires any comment. In the regular case we have seen that $\mathfrak{M}_{-\infty}$ is the closed linear manifold generated by the v_n's, that is, ${}_{-\infty}\mathfrak{M}$ is the closed linear manifold [λ weighting $dF(\lambda)$] generated by the functions

$$\begin{array}{ll} 0 & \lambda \notin S \\ e^{2\pi in\lambda} & \lambda \in S, \qquad n = 0, \pm 1, \cdots, \end{array}$$

and this is the manifold described in (i). The manifold $\mathfrak{M}_n$ is the closed linear manifold generated by $\mathfrak{M}_{-\infty}$, and the ξ_j's for $j \leq n$, that is, ${}_n\mathfrak{M}$ is the closed linear manifold [λ weighting $dF(\lambda)$] generated by ${}_{-\infty}\mathfrak{M}$ and the functions

$$\begin{array}{ll} \dfrac{e^{2\pi ik\lambda}}{\sum_0^\infty c_j e^{-2\pi ij\lambda}} & \lambda \notin S, \qquad k \leq n \\ 0 & \lambda \in S. \end{array}$$

Then considering the functions only on the complement of S [where $dF(\lambda)$ weighting means $|\sum_0^\infty c_j e^{-2\pi i j\lambda}|^2\, d\lambda$ weighting], ${}_n\mathfrak{M}$ consists of all functions of the form

$$\frac{\gamma(\lambda)}{\sum_0^\infty c_j e^{-2\pi i j\lambda}},$$

where γ is a function in the closed linear manifold [λ weighting $d\lambda$] generated by $\{e^{2\pi i j\lambda}, j \leq n\}$, that is, γ consists of all functions

$$e^{2\pi i n\lambda} \sum_0^\infty \gamma_j e^{-2\pi i j\lambda}, \qquad \sum_0^\infty |\gamma_j|^2 < \infty.$$

In the non-regular case ${}_n\mathfrak{M} = {}_\infty\mathfrak{M}$ consists of all the functions in the closed linear manifold generated by $\{e^{2\pi i j\lambda}, j = 0, \pm 1, \cdots\}$, and this is the manifold described in (ii).

5. General solution of the prediction problem (continuous parameter)

Let $\{x(t), -\infty < t < \infty\}$ be a continuous parameter stationary (wide sense) process. We suppose throughout that the continuity condition

$$\lim_{t\to 0} \mathbf{E}\{|x(t) - x(0)|^2\} = 0$$

is satisfied. Let G be the spectral distribution function of the process. As in the discrete parameter case, we introduce a systematic notation for certain closed linear manifolds of random variables and μ functions [using μ weighting $dG(\mu)$ in the latter case].

$$\mathfrak{N}(r, s) = \mathfrak{M}\{x(t), r < t \leq s\} \qquad (r, s)\mathfrak{N} = \mathfrak{M}\{e^{2\pi i t\mu}, r < t \leq s\}$$

$$\mathfrak{N}_t = \mathfrak{N}(-\infty, t) \qquad {}_t\mathfrak{N} = (-\infty, t)\mathfrak{N}$$

$$\mathfrak{N}_{-\infty} = \bigcap_{-\infty}^{\infty} \mathfrak{N}_t \qquad {}_{-\infty}\mathfrak{N} = \bigcap_{-\infty}^{\infty} {}_t\mathfrak{N}$$

$$\mathfrak{N}_\infty = \mathfrak{M}\{x(t), -\infty < t < \infty\} \qquad {}_\infty\mathfrak{N} = \mathfrak{M}\{e^{2\pi i t\mu}, -\infty < t < \infty\}$$

The functions in ${}_t\mathfrak{N}$ consist of those in ${}_s\mathfrak{N}$ multiplied by $e^{2\pi i(t-s)\mu}$. The manifold ${}_\infty\mathfrak{N}$ consists of the functions Ψ measurable with respect to G for which $\int_{-\infty}^{\infty} |\Psi(\mu)|^2\, dG(\mu) < \infty$. (See the corresponding descriptions in §1.)

The problem of predicting $x(t)$ on the basis of a finite segment of the past or of the full past using linear least squares prediction is that of minimizing

$$\mathbf{E}\{|x(t) - \psi|^2\}, \qquad \psi \in \mathfrak{N}(r, s),$$

where r, s are fixed and $r < s < t$. The solution is $\hat{\mathbf{E}}\{x(t) \mid \mathfrak{M}(r, s)\}$, and as in the discrete parameter case

$$\underset{r \to -\infty}{\text{l.i.m.}}\ \hat{\mathbf{E}}\{x(t) \mid \mathfrak{M}(r, s)\} = \hat{\mathbf{E}}\{x(t) \mid \mathfrak{M}_s\},$$

so that prediction on the basis of a finite segment of the past becomes that based on the full past as the segment increases. Because of the continuity condition imposed on the process, $\mathfrak{M}(r, s)$ is the closed linear manifold generated by the random variables $x(t_1)$, $x(t_2)$, $\cdots$, where $\{t_n\}$ is any sequence in the interval (s, t), dense in that interval. Thus it is possible to avoid explicit use of non-denumerably many random variables, but there is no reason to do so.

The remarks made in §1 on non-linear prediction go over to the continuous parameter case without modification.

We shall evaluate $\psi(s, t)$, the prediction of $x(s + t)$ with lag t based on the full past,

$$\psi(s, t) = \hat{\mathbf{E}}\{x(s + t) \mid \mathfrak{M}_s\}.$$

This prediction problem will be solved by reducing it to discrete parameter prediction. To avoid confusion we shall always use F and G respectively to denote discrete and continuous parameter process spectral distribution functions.

The spectral representation

$$x(t) = \int_{-\infty}^{\infty} e^{2\pi i t\mu}\, dy(\mu), \qquad \mathbf{E}\{|dy(\mu)|^2\} = dG(\mu), \tag{5.1}$$

induces a correspondence between random variables and μ functions (see §2) in which $x(t_0)$ corresponds to $e^{2\pi i t_0\mu}$, and more generally, if ψ corresponds to $\Psi(\mu)$,

$$\psi = \int_{-\infty}^{\infty} \Psi(\mu)\, dy(\mu). \tag{5.2}$$

In this correspondence $\psi \in \mathfrak{M}_\infty$ is uniquely determined by $\Psi \in {}_\infty\mathfrak{M}$, neglecting zero probabilities, and Ψ is uniquely determined by ψ, neglecting μ sets of G measure 0. In particular, $\mathfrak{M}_t$ corresponds to ${}_t\mathfrak{M}$ for $-\infty \leq t \leq \infty$.

Instead of stating the prediction problem in the language of random variables it can be stated in μ language: the prediction function Ψ_t is to be found, a function in ${}_0\mathfrak{M}$ which minimizes

$$\int_{-\infty}^{\infty} |e^{2\pi i t\mu} - \Psi(\mu)|^2\, dG(\mu)$$

for functions Ψ in ${}_0\mathfrak{M}$. We have

$$\psi(s, t) = \int_{-\infty}^{\infty} e^{2\pi i s\mu}\, \Psi_t(\mu)\, dy(\mu). \tag{5.3}$$

The function Ψ_t is the projection of $e^{2\pi it\mu}$ on ${}_0\mathfrak{N}$. If σ_t^2 is the mean square prediction error,

$$\sigma_t^2 = \mathbf{E}\{|x(s+t) - \psi(s,t)|^2\} = \int_{-\infty}^{\infty} |e^{2\pi it\mu} - \Psi_t(\mu)|^2 \, dF(\mu). \tag{5.4}$$

As in the discrete parameter case (see §2), σ_t^2 is monotone non-decreasing in t, and either $\sigma_t^2 \equiv 0$ or $\sigma_t^2 > 0$ for all $t > 0$. The first case will be called the non-regular or deterministic case, the second the regular case. To each continuous parameter process with spectral distribution function G we make correspond a discrete parameter process with spectral distribution function F, where

$$F(\lambda) = G(\mu), \qquad \lambda = \frac{1}{\pi}\arctan \mu.$$

Here

$$e^{2\pi i\lambda} = \frac{1+i\mu}{1-i\mu} \qquad \mu = \tan \pi\lambda = -\,i\,\frac{e^{2\pi i\lambda}-1}{e^{2\pi i\lambda}+1}.$$

The manifolds ${}_\infty\mathfrak{M}$ and ${}_\infty\mathfrak{N}$ correspond to each other under this change of variables. The critical closed linear manifold for prediction purposes in the discrete parameter case is ${}_0\mathfrak{M}$, defined in §4 as the closed linear manifold of λ functions, $|\lambda| \le \frac{1}{2}$, generated by the sequence $\{e^{2\pi in\lambda}, n \le 0\}$, λ weighting $dF(\lambda)$. In the continuous parameter case the critical prediction manifold is ${}_0\mathfrak{N}$. The manifold of λ functions ${}_0\mathfrak{M}$ goes into a manifold of μ functions $\mathfrak{N}$, the closed linear manifold [μ weighting $dG(\mu)$] generated by the sequence

$$\left\{\left(\frac{1+i\mu}{1-i\mu}\right)^n, n \le 0\right\}.$$

We prove that $\mathfrak{N} = {}_0\mathfrak{N}$. In the first place the representation

$$\frac{1-i\mu}{1+i\mu} = -1 + \frac{2}{1+i\mu} = -1 + 4\pi\int_{-\infty}^{0} e^{2\pi it\mu}e^{2\pi t}\,dt$$

shows that $\left(\dfrac{1+i\mu}{1-i\mu}\right)^{-1}$, and therefore $\left(\dfrac{1+i\mu}{1-i\mu}\right)^{n}$ for all $n < 0$, can be approximated boundedly and uniformly in every finite μ interval by linear combinations of the $e^{2\pi it\mu}$'s involving only values of $t \le 0$. Then $\mathfrak{N} \subset {}_0\mathfrak{N}$. In the second place, if $t < 0$ the function

$$e^{2\pi t\frac{z-1}{z+1}}$$

is regular for $|z| > 1$, with modulus < 1. Hence it has an expansion in non-positive powers of $z = |z|e^{2\pi i\lambda}$, valid for $|z| > 1$. For each $t < 0$

and $|z| > 1$ this function is therefore a function of λ in ${}_0\mathfrak{M}$ which becomes a function of μ in ${}_0\mathfrak{N}$. When $|z| \to 1$ with λ fixed, this function becomes

$$e^{2\pi it \tan \pi\lambda} = e^{2\pi it\mu},$$

if $\lambda \neq \pm \frac{1}{2}$. Hence, if $t < 0$, $e^{2\pi it\mu}$ can be approximated boundedly by functions in ${}_0\mathfrak{N}$. Then $e^{2\pi it\mu} \in \mathfrak{N}$ if $t \leq 0$, so that ${}_0\mathfrak{N} \subset \mathfrak{N}$, and, combining this with the previously obtained reverse relation, we find that $\mathfrak{N} = {}_0\mathfrak{N}$.

Now a discrete parameter process is deterministic if and only if ${}_n\mathfrak{M} = {}_\infty\mathfrak{M}$ for all n, and it is sufficient if this holds for a single value of n, say $n = 0$. Similarly, a continuous parameter process is deterministic if and only if ${}_0\mathfrak{N} = {}_\infty\mathfrak{N}$. Since ${}_0\mathfrak{M}$ and ${}_\infty\mathfrak{M}$ go into ${}_0\mathfrak{N}$ and ${}_\infty\mathfrak{N}$, we have obtained the result that *a continuous parameter process is deterministic if and only if its corresponding discrete parameter process is deterministic.*

This leads at once to an analytic condition for regularity corresponding to (4.12) in the discrete parameter case.

THEOREM 5.1 *A process is regular if and only if the derivative G' of its spectral distribution function vanishes at most on a set of Lebesgue measure* 0 *and*

$$\int_{-\infty}^{\infty} \frac{\log G'(\mu)}{1 + \mu^2}\, d\mu > -\infty.$$

This theorem follows from Theorem 4.3, since

$$\log F'(\lambda)\, d\lambda = \frac{\log [G'(\mu)\pi(1 + \mu^2)]}{\pi(1 + \mu^2)}\, d\mu$$

so that the integrals

$$\int_{-1/2}^{1/2} \log F'(\lambda)\, d\lambda, \qquad \int_{-\infty}^{\infty} \frac{\log G'(\mu)}{1 + \mu^2}\, d\mu$$

always are finite and infinite together.

Before proceeding further we find the μ counterpart of the λ Fourier series

$$\gamma(\lambda) = \sum_{j=0}^{\infty} \gamma_j e^{-2\pi ij\lambda}, \qquad \sum_{0}^{\infty} |\gamma_j|^2 < \infty.$$

Using the relation between λ and μ, we find that

$$\frac{\gamma(\lambda)}{\sqrt{\pi}(1 + i\mu)} = \frac{1}{\sqrt{\pi}} \sum_{j=0}^{\infty} \gamma_j \frac{(1 - i\mu)^j}{(1 + i\mu)^{j+1}},$$

and since, for properly chosen $A_0, \cdots, A_j$,

$$\frac{(1 - i\mu)^j}{\sqrt{\pi}(1 + i\mu)^{j+1}} = \sum_{r=0}^{j} \frac{A_r}{(1 + i\mu)^{r+1}} \tag{5.5}$$

$$= \sum_{r=0}^{j} \frac{(-1)^r(2\pi)^{r+1}A_r}{r!} \int_{-\infty}^{0} e^{2\pi i\mu t} t^r e^{2\pi t}\, dt$$

$$= \int_{-\infty}^{\infty} e^{2\pi i\mu t} f_j(t)\, dt,$$

where

$$f_j(t) = e^{2\pi t} \sum_{r=0}^{j} \frac{(-1)^r(2\pi)^{r+1}A_r}{r!} t^r \qquad t < 0$$

$$= 0 \qquad t \geq 0,$$

we have finally

$$\frac{\gamma(\lambda)}{\sqrt{\pi}(1 + i\mu)} = \sum_{0}^{\infty} \gamma_j \int_{-\infty}^{\infty} e^{2\pi i\mu t} f_j(t)\, dt = \int_{-\infty}^{\infty} e^{2\pi i\mu t} \sum_{0}^{\infty} \gamma_j f_j(t)\, dt.$$

Since the λ functions $\{e^{2\pi ij\lambda}\}$ form an orthonormal sequence in $[-\frac{1}{2}, \frac{1}{2}]$ (weighting $d\lambda$), the μ functions

$$\left\{\frac{1}{\sqrt{\pi}} \frac{(1 - i\mu)^j}{(1 + i\mu)^{j+1}}\right\}$$

form an orthonormal sequence in $(-\infty, \infty)$ (weighting $d\mu$). The f_j's are thus Fourier transforms of the functions of an orthonormal sequence and hence (Parseval identity) themselves form an orthonormal sequence in $(-\infty, \infty)$ (weighting dt). The series $\sum_{1}^{\infty} \gamma_j f_j$ therefore converges in the mean, and the above operations were all legitimate. The fact that the Fourier series for γ involves no $e^{2\pi ij\lambda}$ with $j > 0$ corresponds to the fact that $\gamma(\lambda)/(1 + i\mu)$ as a function of μ is the Fourier transform of a function vanishing for positive values of the argument.

In the discrete parameter case a condition of the form $\sum_{0}^{\infty} c_j z^j \neq 0$ for $|z| < 1$ was important. The c_j's were constants with $\sum_{0}^{\infty} |c_j|^2 < \infty$. If we make the transformation

$$z = \frac{1 - iw}{1 + iw},$$

this condition becomes

$$\sum_{0}^{\infty} c_j \left(\frac{1 - iw}{1 + iw}\right)^j \neq 0, \qquad \mathfrak{I}w < 0.$$

Now define

$$c^*(t) = \sum_0^\infty c_j f_j(t), \qquad c(\mu) = \int_{-\infty}^{\infty} e^{2\pi i \mu t} c^*(t)\, dt.$$

The function f_j was defined above so that

$$\frac{(1 - iw)^j}{\sqrt{\pi}(1 + iw)^{j+1}} = \int_{-\infty}^{\infty} e^{2\pi i w t} f_j(t)\, dt$$

for w real. This relation then obviously holds for $\mathfrak{I}(w) < 0$ also so that the condition $\sum_0^\infty c_j z^j \neq 0$, $|z| < 1$, becomes

$$\int_{-\infty}^{\infty} e^{2\pi i w t} c^*(t)\, dt \neq 0, \qquad \mathfrak{I}(w) < 0.$$

[Since $c^*(t) = 0$ for $t \geq 0$, the integrand is exponentially small when $|t|$ is large.]

We can now obtain the desired prediction theorems in the continuous parameter case. It will be convenient to derive them in an order different from that of the corresponding theorems in §4. Corresponding to Theorem 4.4 we have:

THEOREM 5.2 *Let G be a bounded monotone non-decreasing function, on $(-\infty, \infty)$, with $G(\mu +) = G(\mu)$, $G(-\infty) = 0$. Let G_v be its singular component, and let S, of Lebesgue measure 0, be the set of increase of G_v.*

(i) *If $G'(\mu) = 0$ at most on a set of Lebesgue measure 0, and if*

$$\int_{-\infty}^{\infty} \frac{\log G'(\mu)}{1 + \mu^2}\, d\mu > -\infty, \tag{5.6}$$

there is a Lebesgue measurable function c^* such that

$$c^*(t) = 0, \qquad t \geq 0, \qquad \int_{-\infty}^{\infty} |c^*(t)|^2\, dt < \infty,$$

$$\int_{-\infty}^{\infty} c^*(t) e^{2\pi i w t}\, dt \neq 0, \qquad \mathfrak{I}w < 0, \tag{5.7}$$

$$\log \left[\int_{-\infty}^{\infty} c^*(t) e^{2\pi t}\, dt \right] = \frac{1}{2\pi} \int_{-\infty}^{\infty} \frac{\log G'(\mu)}{1 + \mu^2}\, d\mu$$

(where the integral on the left is real and positive), and that, if c is defined by

$$c(\mu) = \int_{-\infty}^{\infty} e^{2\pi i\mu t}c^*(t)\,dt,$$

then

$$G(\mu) = \int_{-\infty}^{\mu} |c(\mu')|^2\,d\mu' + G_v(\mu). \tag{5.8}$$

The function c^* is uniquely determined, neglecting sets of Lebesgue measure 0, by these conditions.

The manifold ${}_{-\infty}\mathfrak{N}$ consists of all functions β which are measurable with respect to G vanish for almost all μ (Lebesgue measure) and for which

$$\int_{-\infty}^{\infty} |\beta(\mu)|^2\,dG_v(\mu) < \infty.$$

The manifold ${}_t\mathfrak{N}$ consists of all functions of the form

$$\begin{aligned} &e^{2\pi it\mu}\frac{\int_0^{\infty} \gamma(s)e^{-2\pi i\mu s}\,ds}{c(\mu)}, && \mu \notin S \\ &\beta(\mu), && \mu \in S, \end{aligned}$$

where $\beta \in {}_{-\infty}\mathfrak{N}$ and $\int_0^{\infty} |\gamma(s)|^2\,ds < \infty$.

(ii) *If the hypothesis of* (i) *is not satisfied, ${}_{-\infty}\mathfrak{N} = {}_t\mathfrak{N}$ for all t and this manifold consists of all functions β which are measurable with respect to G and for which*

$$\int_{-\infty}^{\infty} |\beta(\mu)|^2\,dG(\mu) < \infty.$$

(i′) *If there is a Lebesgue measurable c^*, with $\int_{-\infty}^{\infty} |c^*(t)|^2\,dt < \infty$ and $c^*(t) = 0$ for $t > 0$, if* (5.8) *is true, where c is the Fourier transform of c^*, and if ${}_t\mathfrak{N}$ is as described in* (i), *then c^* is the uniquely determined function described in* (i), *aside from a proportionality constant of modulus* 1.

(ii′) *If ${}_{-\infty}\mathfrak{N}$ is as described in* (ii), (5.6) *is false.*

The representation of G in (i) is the exact counterpart of the representation of F in Theorem 4.4. We have already observed that ${}_0\mathfrak{M}$ corresponds in μ language to ${}_0\mathfrak{N}$. Then by Theorem 4.4 the description of ${}_t\mathfrak{N}$ in the regular case is at least correct for $t = 0$. It is correct for all t since ${}_t\mathfrak{N}$ consists of the functions in ${}_0\mathfrak{N}$ multiplied by $e^{2\pi it\mu}$. The rest of the theorem is obtained by translating Theorem 4.4 from λ language to μ language. The last condition in (5.7) is equivalent to

$$\log \frac{1}{2\pi} \int_{-\infty}^{\infty} \frac{c(\mu)}{1 - i\mu} \, d\mu = \frac{1}{2\pi} \int_{-\infty}^{\infty} \frac{\log G'(\mu)}{1 + \mu^2} \, d\mu.$$

Corresponding to Theorem 4.2 we have

THEOREM 5.3 *Let $\{x(t), -\infty < t < \infty\}$ be a regular process. Then $x(t)$ can be written in the form*

$$x(t) = \int_{-\infty}^{0} c^*(s) \, d\xi(t + s) + v(t) = u(t) + v(t), \tag{5.9}$$

where c^* is Lebesgue measurable, $c^*(t) = 0$ for $t > 0$;

$$\int_{-\infty}^{0} |c^*(t)|^2 \, dt < \infty;$$

the $\xi(t)$ process has orthogonal increments with $\mathbf{E}\{|d\xi(t)|^2\} = dt$; every $\xi(t)$ increment is orthogonal to every $v(s)$; $\xi(t_2) - \xi(t_1) \in \mathfrak{M}_t$ if $t_1, t_2 \le t$; $v(t) \in \mathfrak{M}_{-\infty}$. Only the functions proportional (with constant of proportionality of modulus 1) to the function c^* of Theorem 5.2 (i) satisfy these conditions.

In this representation the $u(t)$ process is regular and the $v(t)$ process is deterministic. The prediction $\psi_{t-\tau,\tau}$ is given by

$$\psi_{t-\tau,\tau} = \int_{-\infty}^{-\tau} c^*(s) \, d\xi(t + s) + v(t), \tag{5.10}$$

and the prediction error is

$$\sigma_\tau^2 = \int_{-\tau}^{0} |c^*(s)|^2 \, ds, \tag{5.11}$$

which is the same as that for the $u(t)$ process. The spectral distribution of the $u(t)$ and $v(t)$ processes are respectively the absolutely continuous and singular components of the spectral distribution of the $x(t)$ process.

To derive this representation of $x(t)$ we use the spectral representation (5.1). Define G_v as the singular component of G, let S be the set of increase of G_v, of Lebesgue measure 0, and let S' be the complement of S. Define $u(t)$ and $v(t)$ by

$$u(t) = \int_{S'} e^{2\pi i t\mu} \, dy(\mu)$$

$$v(t) = \int_{S} e^{2\pi i t\mu} \, dy(\mu).$$

Then every $u(t)$ is orthogonal to every $v(t')$. The $u(t)$ and $v(t)$ processes are stationary in the wide sense, with respective spectral distributions the absolutely continuous and singular components of that of the $x(t)$ process. Since the $x(t)$ process is regular, Theorems 5.1 and 5.2 imply that there is a pair of functions c^*, c as described in Theorem 5.2. According to (5.8) the $u(t)$ process has spectral density $|c(\mu)|^2$, and according to X §8 this process can therefore be expressed as a process of moving averages of the form in (5.9). More specifically, we can write $u(t)$ in the form

$$u(t) = \int_{-\infty}^{\infty} e^{2\pi it\mu}\, c(\mu)\, d\xi^*(\mu) = \int_{-\infty}^{\infty} c^*(s)\, d\xi(t+s),$$

where the $\xi^*(\mu)$ and $\xi(t)$ processes have orthogonal increments with

$$\mathbf{E}\{|d\xi^*(\mu)|^2\} = d\mu, \qquad \mathbf{E}\{|d\xi(t)|^2\} = dt$$

and the $\xi(t)$ process is the Fourier transform of the ξ^* process. Here $\xi^*(\mu)$ is uniquely determined, neglecting values on a set of probability 0, by

$$\xi^*(\mu) = \int_{(-\infty,\,\mu]\cdot S'} \frac{dy(\mu')}{c(\mu')}.$$

Then $\xi^*(\mu)$ and $\xi(t)$ increments are orthogonal to every $v(s)$. For $s_1 < s_2 \leq t$ define Ψ by

$$\begin{aligned}(5.12)\qquad \Psi(\mu) &= \frac{e^{2\pi it\mu}}{c(\mu)} \int_{s_1-t}^{s_2-t} e^{2\pi i\mu\tau}\, d\tau, && \mu \in S' \\ &= 0, && \mu \in S.\end{aligned}$$

Then, by Theorem 5.2, $\Psi \in {}_t\mathfrak{N}$ and therefore the random variable

$$(5.13)\quad \psi = \int_{-\infty}^{\infty} \Psi(\mu)\, dy(\mu) = \int_{-\infty}^{\infty} \frac{e^{2\pi is_2\mu} - e^{2\pi is_1\mu}}{2\pi i\mu}\, d\xi^*(\mu) = \xi(s_2) - \xi(s_1)$$

belongs to $\mathfrak{N}_t$. If Γ is defined by

$$\begin{aligned}\Gamma(\mu) &= 0, && \mu \in S', \\ &= e^{2\pi it\mu}, && \mu \in S,\end{aligned}$$

$\Gamma \in {}_{-\infty}\mathfrak{N}$, and therefore

$$\int_{-\infty}^{\infty} \Gamma(\mu)\, dy(\mu) = \int_S e^{2\pi it\mu}\, dy(\mu) = v(t)$$

belongs to $\mathfrak{N}_{-\infty}$.

A function c^* satisfying the condition of the theorem must be proportional to the c^* of Theorem 5.2 because the manifold $\mathfrak{N}_{-\infty}$ will then be the closed linear manifold generated by the $v(t)$'s, $\mathfrak{N}_t$ will be that generated by $\mathfrak{N}_{-\infty}$ and by $\xi(s)$ increments with arguments $\leq t$, so that ${}_{-\infty}\mathfrak{N}$ and ${}_t\mathfrak{N}$ will be as described in Theorem 5.2 (i), and (i′) now implies the desired result.

The prediction $\psi_{t-\tau,}$ of $x(t)$ in terms of the past up to time $t-\tau$ is given by (5.10) because with this definition $\psi_{t-\tau,\tau} \in \mathfrak{N}_{t-\tau}$, and $x(t) - \psi_{t-\tau,\tau}$ is orthogonal to $\mathfrak{N}_{t-\tau}$ because this difference involves only ξ increments with arguments $\geq t-\tau$.

We have seen in (5.12) and (5.13) that $\psi = \xi(s_2) - \xi(s_1)$ corresponds in μ language to Ψ as defined by (5.12), in the correspondence (5.2). Then, since the prediction $\psi_{0,\tau}$ corresponds to the prediction function Ψ_τ, Ψ_τ is given by

$$\begin{aligned} \Psi_\tau(\mu) &= e^{2\pi i\tau\mu}\frac{\int_{-\infty}^{-\tau} e^{2\pi i\mu s}\, c^*(s)\, ds}{c(\mu)}, && \mu \in S' \\ &= e^{2\pi i\tau\mu}, && \mu \in S. \end{aligned}$$

This evaluation can of course also be checked directly from the fact that, as so defined, $\Psi_\tau \in {}_0\mathfrak{N}$ and $e^{2\pi i\tau\mu} - \Psi_\tau$ is orthogonal to ${}_0\mathfrak{N}$.

6. Generalization of §4 and §5

We now consider the following generalization of the prediction problem studied in §4. Let $\{x_n, -\infty < n < \infty\}$ be a stationary (wide sense) process, and let X be any random variable with $\mathbf{E}\{|X|^2\} < \infty$. We wish to approximate X by linear combinations of x_j's for $j \leq n$. More precisely, we wish to evaluate

$$\varphi_n(X) = \hat{\mathbf{E}}\{X \mid x_j, j \leq n\} = \hat{\mathbf{E}}\{X \mid \mathfrak{M}_n\},$$

the random variable in $\mathfrak{M}_n$ closest to X. In particular, if $X = x_{n+\nu}$, $\varphi_n(X) = \varphi_{n,\nu}$ and the problem of finding $\varphi_n(X)$ becomes exactly that studied in §4. If x is defined by

$$x = \hat{\mathbf{E}}\{X \mid x_j, j = 0, \pm 1, \cdots\} = \hat{\mathbf{E}}\{X \mid \mathfrak{M}_\infty\},$$

then $X - x$ is orthogonal to every x_j so that

$$\varphi_n(X) = \varphi_n(x) = \hat{\mathbf{E}}\{x \mid \mathfrak{M}_n\}$$

and in the following therefore we shall usually consider x rather than X. In λ language (see §4) the problem can be stated as follows: There is given a λ function f, measurable with respect to F, the spectral distribution function of the x_n process, with $\int_{-1/2}^{1/2} |f(\lambda)|^2 \, dF(\lambda) < \infty$. This function corresponds to x in the correspondence (2.2), so that

$$x = \int_{-1/2}^{1/2} f(\lambda) \, dy(\lambda). \tag{6.1}$$

The projection of f on ${}_n\mathfrak{M}$ is to be found, that is, the function in the closed linear manifold generated by the functions $\{e^{2\pi ij\lambda}, j \leq n\}$ [λ weighting $dF(\lambda)$] closest to f. The solution $\Phi_n(f)$ will be called the *nth approximation function* and $\varphi_n(X) = \varphi_n(x)$ will be called the *nth approximation.* Evidently

$$\varphi_n(X) = \varphi_n(x) = \int_{-1/2}^{1/2} \Phi_n(f, \lambda) \, dy(\lambda). \tag{6.2}$$

In particular, if $X = x_{n+\nu}$, $\Phi_n(f) = e^{2\pi in\lambda} \Phi_\nu$, where Φ_ν is the prediction function for lag ν discussed in §4. The nth mean square error is given by

$$\sigma_n^2(X) = \mathbf{E}\{|X - \varphi_n(X)|^2\} = \mathbf{E}\{|X - x|^2\} + \mathbf{E}\{|x - \varphi_n(x)|^2\}. \tag{6.3}$$

If the x_n process is non-regular, $\mathfrak{M}_n = \mathfrak{M}_\infty$ for all n. Then in this case $\varphi_n(x) = x$, $\Phi_n(f) = f$, and the nth mean square error is independent of n,

$$\sigma_n^2(X) = \mathbf{E}\{|X - x|^2\}.$$

In general it is clear that

$$\begin{aligned}
&\sigma_{n-1}^2(X) \geq \sigma_n^2(X), \\
&\lim_{n\to-\infty} \sigma_n^2(X) = \mathbf{E}\{|X - x|^2\} + \mathbf{E}\{|x - \hat{\mathbf{E}}\{x \mid \mathfrak{M}_{-\infty}\}|^2\} \\
&\lim_{n\to\infty} \sigma_n^2(X) = \mathbf{E}\{|X - x|^2\} \\
&\operatorname*{l.i.m.}_{n\to-\infty} \varphi_n(X) = \hat{\mathbf{E}}\{x \mid \mathfrak{M}_{-\infty}\} \\
&\operatorname*{l.i.m.}_{n\to\infty} \varphi_n(X) = x.
\end{aligned} \tag{6.4}$$

In the regular case, if we write the Fourier series for x in terms of the orthonormal sequence of the ξ_j's of the Wold decomposition theorem, Theorem 4.2,

$$x = \sum_{j=-\infty}^{\infty} \gamma_j \xi_j + x', \qquad \gamma_j = \mathbf{E}\{x\bar{\xi}_j\} = \mathbf{E}\{X\bar{\xi}_j\}, \qquad x' = \hat{\mathbf{E}}\{x \mid \mathfrak{M}_{-\infty}\}, \tag{6.5}$$

we can write $\varphi_n(x)$ in the form

$$\varphi_n(x) = \sum_{j=-\infty}^{n} \gamma_j \xi_j + x'. \tag{6.6}$$

In fact, as so defined, $\varphi_n(x) \in \mathfrak{M}_n$ and $x - \varphi_n(x)$ is orthogonal to $\mathfrak{M}_n$. We know from §4 that the λ function corresponding to ξ_n is $e^{2\pi in\lambda}/\sum_0^\infty c_j e^{-2\pi ij\lambda}$, where the c_j's are the constants of Theorems 4.2, 4.3, and 4.4.

It is easily verified that in the regular case we have the following correspondence between random variables and λ functions, under (2.2) (using the notation of §4):

$$\xi_n: \quad \begin{cases} \dfrac{e^{2\pi in\lambda}}{\sum_0^\infty c_j e^{-2\pi ij\lambda}}, & \lambda \notin S, \\ 0, & \lambda \in S \end{cases}$$

$$x': \quad \begin{cases} 0 & \lambda \notin S \\ f(\lambda) & \lambda \in S \end{cases}$$

$$x: \quad \begin{cases} \dfrac{\sum_{-\infty}^{\infty} \gamma_j e^{2\pi ij\lambda}}{\sum_{j=0}^{\infty} c_j e^{-2\pi ij\lambda}} = f(\lambda), & \lambda \notin S, \\ f(\lambda), & \lambda \in S, \end{cases}$$

$$\varphi_n(x): \quad \Phi_n(f) = \frac{\sum_{j=-\infty}^{n} \gamma_j e^{2\pi ij\lambda}}{\sum_{j=0}^{\infty} c_j e^{-2\pi ij\lambda}}, \quad \lambda \notin S,$$

$$= f(\lambda), \quad \lambda \in S,$$

where

$$\gamma_j = \mathbf{E}\{x\bar{\xi}_j\} = \int_{-1/2}^{1/2} e^{-2\pi ij\lambda} f(\lambda) \sum_0^\infty c_k e^{-2\pi ik\lambda}\, d\lambda.$$

The nth mean square error is

$$\sigma_n^2(X) = \mathbf{E}\{|X - x|^2\} + \sum_{j=n+1}^{\infty} |\gamma_j|^2. \tag{6.7}$$

The functions $\sum_{-\infty}^{\infty} \gamma_j e^{2\pi ij\lambda}$ and $\sum_{-\infty}^{n} \gamma_j e^{2\pi ij\lambda}$ are fundamental to this study. We observe that, by the above identification of the λ function corresponding to x,

$$\begin{aligned} \sum_{-\infty}^{\infty} \gamma_j e^{2\pi ij\lambda} &= f(\lambda) \sum_0^\infty c_j e^{-2\pi ij\lambda} \\ &= \frac{f(\lambda)F'(\lambda)}{\sum_0^\infty c_j e^{-2\pi ij\lambda}} \end{aligned} \tag{6.8}$$

for almost all λ. We have written the formula in the last form because of its connection with the covariances involved in the prediction problem,

$$\rho_n = \mathbf{E}\{X\bar{x}_n\} = \mathbf{E}\{x\bar{x}_n\} = \int_{-1/2}^{1/2} e^{-2\pi i n\lambda} f(\lambda)\, dF(\lambda). \tag{6.9}$$

This formula shows that the ρ_n's are the Fourier coefficients corresponding to a complex-valued function of bounded variation,

$$\int_{-1/2}^{\lambda} f(\lambda')\, dF(\lambda').$$

Thus the ρ_n's determine fF', F' determines the c_j's, and fF' together with the c_j's determines the γ_j's by (6.8).

In the continuous parameter case (using the notation of §5) there is a random variable X, and we wish to find

$$\psi_t(x) = \hat{\mathbf{E}}\{X \mid \mathfrak{N}_t\} = \hat{\mathbf{E}}\{x \mid \mathfrak{N}_t\},$$

where

$$x = \hat{\mathbf{E}}\{X \mid \mathfrak{N}_\infty\}.$$

We use the decomposition theorem, Theorem 5.3. Since $x \in \mathfrak{N}_\infty$ we can write it on the one hand in the form

$$x = \int_{-\infty}^{\infty} g(\mu)\, dy(\mu), \qquad \int_{-\infty}^{\infty} |g(\mu)|^2\, dG(\mu) < \infty, \tag{6.1'}$$

and on the other hand in the form

$$x = \int_{-\infty}^{\infty} \gamma^*(s)\, d\xi(s) + x', \qquad \int_{-\infty}^{\infty} |\gamma^*(s)|^2\, ds < \infty, \qquad x' = \hat{\mathbf{E}}\{x \mid \mathfrak{N}_{-\infty}\}. \tag{6.5'}$$

Then the prediction $\psi_t(x)$ can be written in the form

$$\psi_t(x) = \int_{-\infty}^{t} \gamma^*(s)\, d\xi(s) + x', \tag{6.6'}$$

and the prediction function $\Psi_t(g, \cdot)$, the μ function corresponding to $\psi_t(x)$, is given by

$$\Psi_t(g, \mu) = \frac{\int_{-\infty}^{t} \gamma^*(s) e^{2\pi i \mu s}\, ds}{c(\mu)} \qquad \mu \notin S$$

$$= g(\mu) \qquad \mu \in S.$$

The mean square error $\sigma_t^2(X) = \mathbf{E}\{|X - x|^2\} + \int_t^\infty |\gamma^*(s)|^2\, ds$. If ρ_t is defined by

$$\rho_t = \mathbf{E}\{X\overline{x(t)}\} = \mathbf{E}\{x\overline{x(t)}\} = \int_{-\infty}^{\infty} e^{-2\pi i t\mu} g(\mu)\, dG(\mu),$$

this covariance function determines gG' for almost all μ (Lebesgue measure), G' determines c, and [comparing (6.1′) and (6.5′)] γ^* is then determined by

$$\gamma^*(\mu) = \int_{-\infty}^{\infty} e^{-2\pi i\mu t}\gamma(t)\, dt,$$

$$\gamma(\mu) = g(\mu)c(\mu) = \frac{g(\mu)G'(\mu)}{\overline{c(\mu)}}. \tag{6.8′}$$

7. Multidimensional prediction

In this section we shall deal with N-dimensional random variables. We shall therefore adopt the convention that all random variables $x, y, z, \cdots$ are vector random variables unless specifically identified as scalars, and constants $a, b, c, \cdots$ are (N by N) matrices. It will be convenient to think of the random variables as (single-column) matrices. If M is a matrix, $\overline{M}$ will denote its conjugate transpose, $|M|^2$ will denote $M\overline{M}$, and $||M||^2$ will denote the sum of the elements of $M\overline{M}$ down the main diagonal, that is, the sum of the squares of the moduli of the elements of M. If M is a matrix of scalar random variables, $\mathbf{E}\{M\}$ is the matrix of their expectations.

According to our conventions, if x, y are random variables, xy is not defined unless $N = 1$, but $x\bar{y}$ and $|x|^2$ are defined. If $\mathbf{E}\{||x||^2\} < \infty$ and $\mathbf{E}\{||g||^2\} < \infty$, the distance between x and y is defined as $\mathbf{E}^{1/2}\{||x - y||^2\}$; x is said to be orthogonal to y if $\mathbf{E}\{x\bar{y}\} = 0$. If $\mathbf{E}\{|x|^2\} = I$ (identity matrix), x is said to be normed. Thus the concept of an orthonormal sequence of random variables is well defined, and in fact a sequence is orthonormal if and only if the sequence of scalar components is an orthonormal sequence. Observe that, if $x_1, \cdots, x_n$ is an orthonormal sequence,

$$\mathbf{E}\{|\sum_1^n a_j x_j|^2\} = \sum_1^n |a_j|^2.$$

The concepts of linear manifold and closed linear manifold are now defined as in IV §2 (note that the coefficients of combination are N by N matrices). There is a slight inconvenience in the orthogonalization of a sequence of random variables owing to the fact that cx may be the null vector even though c is not the null matrix and x is not the null vector.

A sequence $w_1, w_2, \cdots$ of random variables can be orthogonalized in the sense that there are linear combinations $y_1, y_2, \cdots$ of the w_j's such that each w_j is a linear combination of y_j's and conversely, and that either there are infinitely many y_j's which form an orthonormal sequence or there are finitely many y_j's, $y_1, \cdots, y_n$, such that the y_j's are mutually orthogonal, $y_1, \cdots, y_{n-1}$ are normed, and $\mathbf{E}\{|y_n|^2\}$ is a matrix with elements all 0 except for some 1's down the main diagonal. To see this simply orthogonalize the scalar components of the w_j's as in IV §2 to get an orthonormal sequence of scalar random variables. If there are finitely many in this orthonormal sequence, add zeros if necessary to obtain a multiple of N in the sequence. This sequence is then grouped in sets of N; each set is a y_j.

If x is a random variable with $\mathbf{E}\{||x||^2\} < \infty$ and if $y_1, y_2, \cdots$ is an orthonormal sequence, the series $\sum_j a_j y_j$ with $a_j = \mathbf{E}\{x\bar{y}_j\}$ is called the Fourier series of x with respect to the y_j's and the a_j's are called the Fourier coefficients. As in IV §3, it is proved that $\sum_j a_j y_j$ converges (in the sense of the distance used here) that $\sum_j ||a_j||^2 < \infty$, and that

$$\mathbf{E}\{||x||^2\} \geq \sum_j ||a_j||^2. \tag{7.1}$$

In the present case this means that $\sum_j |a_j|^2$ converges (that is, the matrix sum converges element by element) and that the Hermitian symmetric matrix $|x|^2 - \sum_j |a_j|^2$ is non-negative definite. Equality in (7.1) is equivalent to

$$|x|^2 = \sum_j |a_j|^2, \tag{7.1'}$$

and this is true if and only if x is in the closed linear manifold generated by the y_j's.

The projection $x_1 = \hat{\mathbf{E}}\{x \mid \mathfrak{M}\}$ of x on the closed linear manifold $\mathfrak{M}$ is as in IV §3 the closest random variable to x in $\mathfrak{M}$. Each scalar component of x_1 is the (scalar) projection of the corresponding scalar component of x on the closed linear manifold generated by the components of the random variables in $\mathfrak{M}$. The projection x_1 is characterized by the fact that $x_1 \in \mathfrak{M}$ and $x - x_1$ is orthogonal to $\mathfrak{M}$.

A process with orthogonal increments is defined exactly as in the case $N = 1$. It is proved that, if the random variables $\{y(t),\ a \leq t \leq b\}$ constitute such a process, there is a matrix function F such that, if $s < t$, $F(t) - F(s)$ is Hermitian symmetric, non-negative definite, and

$$\mathbf{E}\{|y(t) - y(s)|^2\} = F(t) - F(s).$$

It follows that the diagonal elements of F are monotone non-decreasing; all elements are of bounded variation. In the following, if M is Hermitian symmetric and non-negative definite, we shall denote the sum of the diagonal elements of M by $\Sigma(M)$. If Φ is an $N \times N$ matrix function whose elements are measurable with respect to $\Sigma(F(t))$, with

$$\Sigma\left(\int_a^b \Phi(t)\, dF(t)\overline{\Phi(t)}\right) < \infty,$$

the stochastic integral

$$\varphi = \int_a^b \Phi(t)\, dy(t)$$

is defined as in IX §2, and satisfies

$$\mathbf{E}\{\varphi_1\bar{\varphi}_2\} = \int_a^b \Phi_1(t)\, dF(t)\overline{\Phi_2(t)},$$

where $\Phi_i(t)$ is the integrand corresponding to φ_i. The random variables φ obtained in this way are exactly those in the closed linear manifold generated by the $y(t)$ increments. If $\varphi = 0$ with probability 1, the corresponding Φ satisfies

$$\mathbf{E}\{|\varphi|^2\} = \int_a^b \Phi(t)\, dF(t)\overline{\Phi(t)} = 0.$$

Any two functions Φ_1, Φ_2 differing by a function Φ satisfying this equation will have the same stochastic integral.

A process $\{x_n, -\infty < n < \infty\}$ is *stationary in the wide sense* if $\mathbf{E}\{||x_n||^2\} < \infty$ and if the *covariance (matrix) function*

$$R(n) = \mathbf{E}\{x_{m+n}\bar{x}_m\}$$

does not depend on m. The N component scalar processes are then also stationary in the wide sense. There is a spectral representation

$$x_n = \int_{-1/2}^{1/2} e^{2\pi in\lambda} I\, dy(\lambda) \qquad \mathbf{E}\{|dy(\lambda)|^2\} = dF(\lambda)$$

where the $y(\lambda)$ process has orthogonal increments and I is the identity. The function F is normalized as usual so that $F(-\infty) = 0$, $F(\lambda +) = F(\lambda)$; it is called the spectral distribution function of the process, and determines $R(n)$ by

$$R(n) = \int_{-1/2}^{1/2} e^{2\pi in\lambda}\, dF(\lambda).$$

From now on we shall use the notation of §4 for the corresponding concepts here. For example, σ_1^2 is the mean square error of prediction with lag 1,

$$\sigma_1^2 = \mathbf{E}\{|x_n - \varphi_{n-1,1}|^2\}.$$

The matrix σ_1^2 is Hermitian symmetric and non-negative definite. In the following we shall suppose that σ_1^2 is a non-singular matrix and derive the Wold decomposition, Theorem 4.2. This assumption on σ_1^2 excludes the possibility that some linear combination of components of x_n is completely predictable on the basis of past components of the x_j's. The matrix σ_1^2 is now positive definite, and there is a unique square root σ_1 which is Hermitian symmetric and positive definite. Define ξ_n by

$$\xi_n = \sigma_1^{-1}[x_n - \varphi_{n-1,\,1}].$$

Then

$$x_n - \varphi_{n-1,\,1} = \sigma_1 \xi_n, \qquad \mathbf{E}\{|\xi_n|^2\} = I, \qquad \xi_n \in \mathfrak{M}_n,$$

and the ξ_n's form an orthonormal sequence as in §4. We have the orthogonal development

$$x_n = \sum_{j=0}^{\infty} c_j \xi_{n-j} + v_n = u_n + v_n, \qquad c_j = \mathbf{E}\{x_n \bar{\xi}_{n-j}\},$$

and all the orthogonality and inclusion relations of §4 remain true. The spectral distribution function F_u is absolutely continuous with

$$F_u'(\lambda) = \left| \sum_{j=0}^{\infty} c_j e^{-2\pi i j \lambda} \right|^2$$

and

$$F = F_u + F_v.$$

If Φ is a λ function corresponding to ξ_n, we have, following §4,

$$\Phi(\lambda) \sum_{j=0}^{\infty} c_j e^{-2\pi i j \lambda} = e^{2\pi i n \lambda} I$$

for almost all λ and

$$\int_{-1/2}^{1/2} \Phi(\lambda)\, dF_v(\lambda) \overline{\Phi(\lambda)} = 0.$$

According to the first equation the series is a non-singular matrix for almost all λ, and we have

$$\Phi(\lambda) = e^{2\pi i n \lambda} \left[\sum_{j=0}^{\infty} c_j e^{-2\pi i j \lambda} \right]^{-1}.$$

Taking the absolutely continuous part in the second equation, we find that

$$\int_{-1/2}^{1/2} \Phi(\lambda)\, F_v'(\lambda)\overline{\Phi(\lambda)}\, d\lambda = 0,$$

where $F_v'(\lambda)$ is Hermitian symmetric and non-negative definite. Since the integrand is symmetric and non-negative definite, this means that $\Phi(\lambda)F_v'(\lambda)\overline{\Phi(\lambda)} = 0$ for almost all λ and, since $\Phi(\lambda)$ is almost never singular, we finally have that $F_v'(\lambda) = 0$ for almost all λ. Then F_v is the singular component of F just as in §4. The prediction functions and predictions are given by the formulas of §4. Since the analytic conditions determining the c_j's are not known, we shall not pursue this subject further.

SUPPLEMENT

The following is an outline of certain aspects of measure theory in the form needed in this book. Lengthy but standard and readily available proofs are omitted. Throughout the discussion, Ω is an abstract space of points ω.

1. Fields of point sets

DEFINITION *A class $\mathscr{F}$ of ω sets is called a field if it has the following properties:*

(i) $\Omega \in \mathscr{F}$;

(ii) if $\Lambda \in \mathscr{F}$, then $\Omega - \Lambda \in \mathscr{F}$;

(iii) *if* n *is any natural number, and if* $\Lambda_1, \cdots, \Lambda_n \in \mathscr{F}$, *then*

$$\bigcup_1^n \Lambda_j \in \mathscr{F}, \qquad \bigcap_1^n \Lambda_j \in \mathscr{F}.$$

A field is called a Borel field if it has the following additional property:

(iv) *if* $\Lambda_1, \Lambda_2, \cdots \in \mathscr{F}$, *then* $\bigcup_1^\infty \Lambda_j \in \mathscr{F}$ *and* $\bigcap_1^\infty \Lambda_j \in \mathscr{F}$.

THEOREM 1.1 *If $\mathscr{F}_0$ is a class of ω sets, there is a uniquely determined Borel field $\mathscr{F}$ of ω sets with the following two properties:*

(i) $\mathscr{F}_0 \subset \mathscr{F}$;

(ii) *if $\mathscr{F}_1$ is a Borel field of ω sets, and if $\mathscr{F}_0 \subset \mathscr{F}_1$, then $\mathscr{F} \subset \mathscr{F}_1$.*

The class $\mathscr{F}$ is the smallest Borel field of sets which includes all the sets of $\mathscr{F}_0$. There certainly is a Borel field of sets which includes all the sets of $\mathscr{F}_0$, for example, the Borel field of all ω sets. Define $\mathscr{F}$ as the class of sets each of which is in every Borel field of ω sets containing the sets of $\mathscr{F}_0$, that is, $\mathscr{F}$ is the intersection of all the Borel fields containing the sets of $\mathscr{F}_0$. Then $\mathscr{F}$ is a Borel field and has the two properties stated in the theorem. The uniqueness is a trivial consequence of these two properties.

The field $\mathscr{F}$ will be called the *Borel field generated by* $\mathscr{F}_0$, and will be denoted by $\mathscr{B}(\mathscr{F}_0)$.

THEOREM 1.2 *Let $\mathscr{F}_0$ be a field of ω sets. Then, if $\mathscr{G}$ is a class of ω sets with the following properties, $\mathscr{B}(\mathscr{F}_0) \subset \mathscr{G}$.*

(i) $\mathscr{F}_0 \subset \mathscr{G}$;

(ii) *If $\Lambda_j \in \mathscr{G}$, $j \geq 1$, and if either*

$$\Lambda_1 \subset \Lambda_2 \subset \cdots, \qquad \bigcup_1^\infty \Lambda_j = \Lambda$$

or

$$\Lambda_1 \supset \Lambda_2 \supset \cdots, \qquad \bigcap_1^\infty \Lambda_j = \Lambda,$$

then $\Lambda \in \mathscr{G}$.

The proof will be omitted. The property (ii) will be described by stating that $\mathscr{G}$ includes the limit of any monotone sequence of $\mathscr{G}$ sets.

If Ω is n-dimensional space, and if $\mathscr{F}_0$ is the class of open sets, the sets of $\mathscr{B}(\mathscr{F}_0)$ are called *n-dimensional Borel sets*. The same class of sets is obtained if $\mathscr{F}_0$ consists of the closed sets, or the (n-dimensional closed, or open) intervals. The same class of sets is also obtained if $\mathscr{F}_0$ is the class of finite unions of right semi-closed intervals, that is, of finite unions of sets of the form

$$\{\lambda_i < \xi_i \leq \mu_i, \qquad i = 1, \cdots, n\}.$$

Here the λ_i's and μ_i's are finite or infinite, except that, if $\mu_i = \infty$, "$\leq \mu_i$" is replaced by "$< \mu_i$" in the above definition, to reject points with infinite coordinates. This choice of $\mathscr{F}_0$ has the advantage that $\mathscr{F}_0$ is then a field.

In the following we shall discuss ω functions of various types. Any ω set defined by conditions on functions will be denoted by the defining conditions written between the braces. Thus, if x and y are ω functions, and if Y is a set of numbers, the ω set

$$\{x(\omega) < 3, y(\omega) \in Y\}$$

is the ω set on which $x(\omega) < 3$ and $y(\omega)$ is a number of the set Y.

If $\mathscr{F}$ is a Borel field of ω sets, and if x is an ω function, x is called *measurable with respect to* $\mathscr{F}$ if it is a real-valued function and if, for every real number c, $\{x(\omega) \leq c\} \in \mathscr{F}$, or if it is a complex-valued function whose real and imaginary parts are real measurable functions. It is sufficient in the above definition if the constant c ranges through a dense set of real numbers, rather than the whole class of real numbers. We note without proof the fact that linear combinations of functions measurable with respect to $\mathscr{F}$ are also measurable with respect to $\mathscr{F}$. If $\{x_n, n \geq 1\}$ is a sequence of ω functions measurable with respect to $\mathscr{F}$, the set of points of convergence of the sequence is an $\mathscr{F}$ set, and, if the sequence converges everywhere, its limit is a function measurable with respect to $\mathscr{F}$.

If $\mathscr{F}$ is the class of Borel measurable sets in n-dimensional space, the functions measurable with respect to $\mathscr{F}$ are called *Borel measurable* or *Baire* functions. A function of n complex variables is called a Borel measurable (or Baire) function if it is a Baire function of the $2n$ real and imaginary parts of the variables.

If A is a Borel set in n dimensions, the cylinder set it determines in $n' > n$ dimensions (by choosing the first n coordinates to determine points of A, and allowing the remaining coordinates to take arbitrary values) is a Borel set in n' dimensions. In fact, the class of n-dimensional Borel sets for which this assertion is true is obviously a Borel field which includes the n-dimensional intervals, and is therefore the class of n-dimensional

Borel sets. Corresponding to this fact, if x is a Baire function of n variables, it determines a function of $n' > n$ variables (whose function values are determined by the first n of the n' variables) and this function is a Baire function of n' variables.

THEOREM 1.3 *If $\mathscr{F}_0$ is a class of ω sets such that the class of finite unions of disjunct $\mathscr{F}_0$ sets is a field, and if $\mathscr{H}$ is a class of ω functions with the following properties, then $\mathscr{H}$ includes all ω functions measurable with the respect to $\mathscr{B}(\mathscr{F}_0)$.*

(i) *$\mathscr{H}$ includes every function which takes on the value 1 on an $\mathscr{F}_0$ set and 0 on its complement.*

(ii) *$\mathscr{H}$ includes every linear combination of a finite number of its functions.*

(iii) *If $x_n \in \mathscr{H}$, $n \geq 1$, and if $\lim_{n\to\infty} x_n(\omega) = x(\omega)$ exists and is finite for all ω, then $x \in \mathscr{H}$.*

In particular, if $\mathscr{H}$ is a class of functions of n variables, and if $\mathscr{F}_0$ is the class of right semi-closed intervals, then $\mathscr{H}$ includes every Baire function of n variables, if $\mathscr{H}$ has the above properties.

In the real [complex] case, the coefficients of combination in (ii) are real [complex]. We prove this theorem in the real case. Let $\mathscr{A}$ be the class of those $\mathscr{B}(\mathscr{F}_0)$ sets M with the property that, if $x(\omega) = 1$ on M and $x(\omega) = 0$ otherwise, then $x \in \mathscr{H}$. Then $\mathscr{A}$ includes the $\mathscr{F}_0$ sets, according to (i), finite unions of disjunct $\mathscr{F}_0$ sets, according to (ii), and the limits of monotone sequences of $\mathscr{A}$ sets, according to (iii). Since $\mathscr{B}(\mathscr{F}_0)$ is the Borel field generated by the field of finite unions of disjunct $\mathscr{F}_0$ sets, it follows from Theorem 1.2 that $\mathscr{A}$ includes all $\mathscr{B}(\mathscr{F}_0)$ sets. Now, if x is an arbitrary ω function measurable with respect to $\mathscr{B}(\mathscr{F}_0)$, define x_{jn} and x_n by

$$x_{jn}(\omega) = 1, \qquad \frac{j}{2^n} < x(\omega) \leq \frac{j+1}{2^n}$$

$$= 0 \text{ otherwise;}$$

$$x_n = \sum_{j=-n}^{n} \frac{j}{2^n} x_{jn}.$$

Then we have just seen that $x_{jn} \in \mathscr{H}$, and according to property (ii) of the theorem $x_n \in \mathscr{H}$ also. Finally $x \in \mathscr{H}$ by property (iii), because $x_n \to x$, as was to be proved.

THEOREM 1.4 *If $\mathscr{F}$ is a Borel field of ω sets, if $x_1, \cdots, x_n$ are ω functions measurable with respect to $\mathscr{F}$, and if F is a Baire function of n variables, then $F(x_1, \cdots, x_n)$ is an ω function measurable with respect to $\mathscr{F}$.*

This is a simple application of Theorem 1.3. In fact, if we define $\mathscr{H}$ as the class of Baire functions of n variables for which the assertion of the theorem is true (and $\mathscr{F}_0$ as the class of right semi-closed intervals), it is trivially verified that $\mathscr{H}$ has the properties described in Theorem 1.3, and therefore includes all Baire functions of n variables.

In the following we shall frequently consider ω sets of the form

$$\{[x_1(\omega), \cdots, x_n(\omega)] \in A\};$$

the indicated set is the ω set for which the n-tuple $x_1(\omega), \cdots, x_n(\omega)$ defines a point of the set A. If the x_j's are real, A is to be taken as a point set in n dimensions; if the x_j's are complex, A is to be taken as a point set in $2n$ dimensions, with the obvious conventions, that is, as a set of n-tuples each of whose coordinates is a complex number.

COROLLARY *If $\mathscr{F}$ is a Borel field of ω sets, if $x_1, \cdots, x_n$ are ω functions measurable with respect to $\mathscr{F}$, and if A is an n-dimensional Borel set ($2n$-dimensional if the x_j's are complex valued), then*

$$\{[x_1(\omega), \cdots, x_n(\omega)] \in A\} \in \mathscr{F}.$$

To prove this corollary, define F as the Baire function of n variables which is 1 on A and 0 otherwise. Then $F(x_1, \cdots, x_n)$ is measurable with respect to $\mathscr{F}$, according to Theorem 1.4, so that the ω set

$$\{F[x_1(\omega), \cdots, x_n(\omega)] = 1\} = \{[x_1(\omega), \cdots, x_n(\omega)] \in A\}$$

is in the class $\mathscr{F}$.

Let $\mathscr{F}$ be a Borel field of ω sets and let $\{x_t, t \in T\}$ be a family of ω functions measurable with respect to $\mathscr{F}$. Then we shall denote by $\mathscr{B}(x_t, t \in T)$ the Borel field of ω sets generated by the class of those of the form $\{x_t(\omega) \in A\}$ for $t \in T$ and A a right semi-closed interval. Obviously

$$\mathscr{B}(x_t, t \in T) \subset \mathscr{F}.$$

The Borel field $\mathscr{B}(x_t, t \in T)$ is the smallest Borel field of ω sets with respect to which the x_t's are all measurable. Then $\mathscr{B}(x_t, t \in T)$ can be substituted for $\mathscr{F}$ in the preceding corollary, and we find that, if $t_1, \cdots, t_n$ is a finite T set, and if A is an n-dimensional Borel set (or $2n$-dimensional if the x_t's are complex-valued),

$$\{[x_{t_1}(\omega), \cdots, x_{t_n}(\omega)] \in A\} \in \mathscr{B}(x_t, t \in T).$$

The class of ω sets of the type on the left forms a field which generates $\mathscr{B}(x_t, t \in T)$, and the same is true if A is restricted to be a finite union of right semi-closed intervals. In particular, if T is the set of integers $1, \cdots, n$, $\mathscr{B}(x_1, \cdots, x_n)$ is the class of sets of the form

$$\{[x_1(\omega), \cdots, x_n(\omega)] \in A\},$$

where A is an n-dimensional Borel set (or $2n$-dimensional if the x_j's are complex-valued).

THEOREM 1.5 *Let $\mathscr{F}$ be a Borel field of ω sets and let $x_1, \cdots, x_n$ be ω functions measurable with respect to $\mathscr{F}$. Then an ω function is measurable with respect to $\mathscr{B}(x_1, \cdots, x_n)$ if and only if it has the form $F(x_1, \cdots, x_n)$, where F is a Baire function of n variables.*

The "if" half of this theorem is a sharpening of the previous theorem, according to which $F(x_1, \cdots, x_n)$ is measurable with respect to $\mathscr{F}$ if F is a Baire function. Since $\mathscr{F}$ can be replaced by $\mathscr{B}(x_1, \cdots, x_n)$ in the statement of Theorem 1.4, however, that theorem asserts that $F(x_1, \cdots, x_n)$ is measurable with respect to $\mathscr{B}(x_1, \cdots, x_n)$. Conversely, let y be an ω function measurable with respect to $\mathscr{B}(x_1, \cdots, x_n)$. Suppose that y is real-valued, and define

$$\Lambda_{jm} = \left\{ \frac{j}{2^m} < y(\omega) \leq \frac{j+1}{2^m} \right\} \in \mathscr{B}(x_1, \cdots, x_n).$$

There is a Borel set A_{jm}' such that

$$\Lambda_{jm} = \{[x_1(\omega), \cdots, x_n(\omega)] \in A_{jm}'\}.$$

The Λ_{jm}'s (for fixed m) are disjunct. Hence, if we define A_{jm} by

$$A_{jm} = A_{jm}' - A_{jm}' \bigcup_{i \neq j} A_{im}',$$

the A_{jm}'s for fixed m are also disjunct, and

$$\Lambda_{jm} = \{[x_1(\omega), \cdots, x_n(\omega)] \in A_{jm}\}.$$

Define y_m by

$$y_m(\omega) = \frac{j}{2^m}, \qquad \omega \in \Lambda_{jm}, \qquad j = 0, \pm 1, \cdots.$$

Then we can write y_m in the form

$$y_m = F_m(x_1, \cdots, x_n),$$

where

$$F_m(\xi_1, \cdots, \xi_n) = \frac{j}{2^m}, \qquad (\xi_1, \cdots, \xi_n) \in A_{jm}, \qquad j = 0, \pm 1, \cdots$$

$$= 0 \text{ otherwise}.$$

Let R be the set of points in n-dimensional space determined by $[x_1(\omega), \cdots, x_n(\omega)]$ as ω varies. Then, since

$$\lim_{m\to\infty} F_m[x_1(\omega), \cdots, x_n(\omega)] = \lim_{m\to\infty} y_m(\omega) = y(\omega)$$

for all ω, it follows that

$$\lim_{m\to\infty} F_m(\xi_1, \cdots, \xi_n)$$

exists on R. Now the F_m's are Baire functions. Hence the set of points where the sequence of function values has a finite limit is a Borel set $R' \supset R$. Define the Baire function F as this limit on R' and 0 otherwise. Then

$$F(x_1, \cdots, x_n) = y,$$

so that y is a Baire function of the x_j's, as was to be proved. If y is complex-valued, the preceding result is applied to its real and imaginary parts.

THEOREM 1.6 *Let $\mathscr{F}$ be a Borel field of ω sets, and let $\{x_t, t \in T\}$ be a family of ω functions measurable with respect to $\mathscr{F}$. Let $\mathscr{F}_S = \mathscr{B}(x_t, t \in S)$, where $S \subset T$. Suppose that T is non-denumerable. Then, if $\Lambda \in \mathscr{F}_T$, there is a denumerable subset S (depending on Λ) of T, such that $\Lambda \in \mathscr{F}_S$. If x is an ω function measurable with respect to $\mathscr{F}_T$, there is a denumerable subset S (depending on x) such that x is measurable with respect to $\mathscr{F}_S$.*

Let $\mathscr{G}$ be the class of sets Λ with the property stated in the theorem. Then $\mathscr{G} \subset \mathscr{F}_T$ and we wish to prove that $\mathscr{G} = \mathscr{F}_T$. The class $\mathscr{G}$ includes every ω set $\{x_t(\omega) \in A\}$, where $t \in T$ and A is a Borel set. Moreover, it is easy to verify that $\mathscr{G}$ is a Borel field. Then $\mathscr{G} = \mathscr{F}_T$, because of the minimal property of $\mathscr{F}_T$, as was to be proved. Let x be any real ω function measurable with respect to $\mathscr{F}_T$. Then for each rational number r we have seen that there is a denumerable subset S_r of T such that

$$\{x(\omega) \leq r\} \in \mathscr{F}_{S_r}.$$

Let $S = \bigcup_r S_r$. Then S is a denumerable subset of T, and x is measurable with respect to $\mathscr{F}_S$. If x is a complex-valued function measurable with respect to $\mathscr{F}$, there are denumerable subsets S' and S'' of T with the property that the real [imaginary] part of x is measurable with respect to $\mathscr{F}_{S'}$ [$\mathscr{F}_{S''}$]. Then, if $S = S' \cup S''$, S is denumerable and x is measurable with respect to $\mathscr{F}_S$.

2. Set functions

DEFINITION *If $\mathscr{F}$ is a field of ω sets, a real finite numerically valued function q defined on the sets of $\mathscr{F}$ is called completely additive if, whenever $\Lambda_1, \Lambda_2, \cdots$ are finitely or denumerably infinitely many disjunct $\mathscr{F}$ sets, with $\bigcup_j \Lambda_j \in \mathscr{F}$ then,*

$$q\{\bigcup_j \Lambda_j\} = \sum_j q\{\Lambda_j\}.$$

If there are only finitely many Λ_j's, the hypothesis that $\bigcup_j \Lambda_j \in \mathscr{F}$ is implied by the hypothesis that the Λ_j's are $\mathscr{F}$ sets. If there are infinitely many Λ_j's, there is the same implication if the field is a Borel field.

It follows from this definition of complete additivity that, if $M_1, M_2, \cdots$ are $\mathscr{F}$ sets with either

$$M_1 \subset M_2 \subset \cdots, \qquad \bigcup_j M_j = M \in \mathscr{F}$$

or

$$M_1 \supset M_2 \supset \cdots, \qquad \bigcap_j M_j = M \in \mathscr{F},$$

then

$$\lim_{n\to\infty} q\{M_n\} = q\{M\}.$$

If $\Lambda = \{C\}$ is an ω set determined by conditions "C", we write $q\{\Lambda\}$ as $q\{C\}$ rather than as $q\{\{C\}\}$.

THEOREM 2.1 *Let $\mathscr{F}_0$ be a field of ω sets. Then, if q_1 and q_2 are completely additive set functions defined on the sets of $\mathscr{B}(\mathscr{F}_0)$, and if*

$$q_1\{\Lambda\} \leq q_2\{\Lambda\}, \qquad \Lambda \in \mathscr{F}_0,$$

it follows that

$$q_1\{\Lambda\} \leq q_2\{\Lambda\}, \qquad \Lambda \in \mathscr{B}(\mathscr{F}_0).$$

The theorem remains true if "$\leq$" is replaced by "$=$" in the hypothesis and conclusion.

To prove this theorem, we need only remark that the class of $\mathscr{B}(\mathscr{F}_0)$ sets on which q_1 and q_2 are in the stated relationship includes the $\mathscr{F}_0$ sets, and includes limits of monotone sequences of its sets. That is, this class satisfies the hypotheses of Theorem 1.2, and therefore is $\mathscr{B}(\mathscr{F}_0)$, as was to be proved.

DEFINITION *If $\mathscr{F}$ is a field of ω sets, a function q defined on the sets of $\mathscr{F}$ is called a measure if it is completely additive and non-negative, a probability measure if it is a measure and if $q\{\Omega\} = 1$.*

If q is a measure, the sets of the field on which it is defined are usually called *measurable.*

THEOREM 2.2 *Let $\mathscr{F}_0$ be a field of ω sets, and let q_0 be a measure defined on the sets of $\mathscr{F}_0$. Then there is a unique measure q, defined on the sets of $\mathscr{B}(\mathscr{F}_0)$, with*

$$q\{\Lambda\} = q_0\{\Lambda\}, \qquad \Lambda \in \mathscr{F}_0.$$

The uniqueness property is implied by Theorem 2.1. The existence proof will be omitted.

THEOREM 2.3 *Let $\mathscr{F}, \mathscr{F}'$ be Borel fields of ω sets, and let q be a measure defined on the sets of a Borel field $\mathscr{G} \supset \mathscr{F} \cup \mathscr{F}'$. Suppose that if $\Lambda' \in \mathscr{F}'$ there is a $\Lambda \in \mathscr{F}$ with the property that*

$$q\{\Lambda(\Omega - \Lambda') \cup (\Omega - \Lambda)\Lambda'\} = 0.$$

Then, if x' is an ω function measurable with respect to $\mathscr{F}'$, there is an ω function x, measurable with respect to $\mathscr{F}$, such that

$$q\{x(\omega) \neq x'(\omega)\} = 0.$$

Let $\mathscr{H}$ be the class of ω functions x' measurable with respect to $\mathscr{F}'$ for which the assertion is true. Then by hypothesis $\mathscr{H}$ includes every ω function which takes on the value 1 on an $\mathscr{F}'$ set and 0 on its complement. The class $\mathscr{H}$ obviously includes every linear combination of a finite number of its functions, and the limits of every convergent sequence of its functions. Hence $\mathscr{H}$ includes all ω functions measurable with respect to $\mathscr{F}'$, by Theorem 1.3, as was to be proved.

Let q be a measure defined on the sets of a Borel field $\mathscr{F}$, and consider all ω sets M with the property that there are sets M_1, $M_2 \in \mathscr{F}$, such that

$$M_1 \subset M \subset M_2, \qquad q\{M_2 - M_1\} = 0.$$

The class of sets M form a Borel field $\mathscr{F}^* \supset \mathscr{F}$, and $\mathscr{F}^* = \mathscr{F}$ if and only if $\mathscr{F}$ includes every subset of an $\mathscr{F}$ set of measure 0. If $M \in \mathscr{F}$ above, then the relations between M_1, M, and M_2 imply that

$$q\{M\} = q\{M_1\} = q\{M_2\}.$$

Even if $M \notin \mathscr{F}$, the second and third quantities here are equal. Then, if $M \in \mathscr{F}^* - \mathscr{F}$, we define $q\{M\}$ by this equation, and with this definition q becomes a measure defined on the sets of $\mathscr{F}^*$. Note that, if $M \in \mathscr{F}^*$, and if $q\{M\} = 0$, then the subsets of M are also $\mathscr{F}^*$ sets. Each set of $\mathscr{F}^* - \mathscr{F}$ differs from some $\mathscr{F}$ set by a subset of an $\mathscr{F}$ set of measure 0. The operation of extending the domain of the given measure from $\mathscr{F}$ to $\mathscr{F}^*$ is called *completing the measure*, and a measure with the property that $\mathscr{F} = \mathscr{F}^*$, that is, that subsets of sets of measure 0 are measurable (and have measure 0), is called *complete*. The operation of completion makes a measure a complete measure. We shall always use the notation $\mathscr{F}^*$ for the Borel field defined as just described. Note that $\mathscr{F}^*$ depends both on $\mathscr{F}$ and on the measure q.

THEOREM 2.4 *Let $\mathscr{F}_0$ be a field of ω sets, and suppose that $\mathscr{F}_1$ is a Borel field of sets with $\mathscr{F}_1 \supset \mathscr{B}(\mathscr{F}_0)$. Let q be a measure defined on $\mathscr{F}_1$, with the property that, if $\Lambda_1 \in \mathscr{F}_1$, there is a $\Lambda \in \mathscr{B}(\mathscr{F}_0)$ such that*

$$q\{\Lambda_1(\Omega - \Lambda) \cup (\Omega - \Lambda_1)\Lambda\} = 0.$$

Then:

(i) *If $\Lambda \in \mathscr{F}_1$, and if $\varepsilon > 0$, there is a $\Lambda_\varepsilon \in \mathscr{F}_0$ such that*

$$q\{\Lambda(\Omega - \Lambda_\varepsilon) \cup (\Omega - \Lambda)\Lambda_\varepsilon\} < \varepsilon.$$

(ii) *If $\Lambda \in \mathscr{B}(\mathscr{F}_0)$, there is a Λ' which is the intersection of denumerably many $\mathscr{F}_0$ sets, and a Λ'' which is the union of denumerably many $\mathscr{F}_0$ sets, such that*

$$\Lambda' \subset \Lambda \subset \Lambda'', \qquad q\{\Lambda'' - \Lambda'\} = 0.$$

(iii) *If x is an ω function measurable with respect to $\mathscr{F}_1$, and if $\varepsilon > 0$, there is a function x_ε, taking on a finite number of values, each on an $\mathscr{F}_0$ set, such that*

$$q\{|x(\omega) - x_\varepsilon(\omega)| \geq \varepsilon\} \leq \varepsilon$$

and (if x is integrable)

$$\int_\Omega |x - x_\varepsilon| \, dq < \varepsilon.$$

To prove (i) [(ii)], let $\mathscr{G}$ be the class of sets $\Lambda \in \mathscr{F}_1$ [$\Lambda \in \mathscr{B}(\mathscr{F}_0)$] for which the assertion is true. Then $\mathscr{G} \supset \mathscr{F}_0$, and it is easily shown that $\mathscr{G}$ includes limits of monotone sequences of its sets. But then $\mathscr{G} \supset \mathscr{B}(\mathscr{F}_0)$, by Theorem 1.2. In (i) it follows that $\mathscr{G} = \mathscr{F}_1$, because of the stated relation between $\mathscr{F}$ and $\mathscr{F}_1$ sets. Part (iii) is easily reduced to (i) by approximating x by a function taking on only finitely many values, each on an $\mathscr{F}_1$ set to which (i) is applicable.

Example 2.1 Let Ω be the real line, and let $\mathscr{F}_0$ be the class (field) of finite unions of right semi-closed intervals. Then $\mathscr{B}(\mathscr{F}_0)$ is the class of linear Borel sets. Let F be a bounded monotone non-decreasing function of ξ, continuous on the right, with $\lim\limits_{\xi \to -\infty} F(\xi) = 0$. If Λ is a finite union of right semi-closed intervals,

$$\Lambda = \bigcup_{j=1}^{n} (a_j, b_j], \quad \cdots < a_j < b_j < a_{j+1} < \cdots,$$

define

$$q\{\Lambda\} = \sum_{j=1}^{n} [F(b_j) - F(a_j)].$$

It can be shown that q is a measure on the sets of $\mathscr{F}_0$. Then by Theorem 2.2. the domain of definition of q can be extended to make q a measure of Borel sets, and the measure q can then be further extended by completion. The sets of the class $\mathscr{B}(\mathscr{F}_0)^*$ obtained by this completion are called Lebesgue-Stieltjes measurable, or measurable with respect to F, and q is called a Lebesgue-Stieltjes measure. We omit the easy generalization to unbounded functions F. The class of Lebesgue-Stieltjes measurable sets depends on the choice of F. For example, if F is constant except for jumps at finitely or enumerably infinitely many points, every ω set is Lebesgue-Stieltjes measurable. In general, however, $\mathscr{B}(\mathscr{F}_0)^*$ does not contain all ω sets. If δ is small but positive, the $\mathscr{F}_0$ set indicated above has measure near that of the larger open set $\bigcup_1^n (a_j, b_j + \delta)$ and near that of the smaller closed set $\bigcup_1^n [a_j + \delta, b_j]$. Thus Λ_ε in Theorem 2.4 (i) can be taken as a union of open, or closed, intervals losing of course the property that it is an $\mathscr{F}_0$ set, and Λ', Λ'' in Theorem 2.4 (ii) can be taken respectively as closed, and open, if convenient. If x is an ω function measurable with respect to F, that is, measurable with respect to the field $\mathscr{B}(\mathscr{F}_0)^*$, its integral over a set Λ is usually denoted by

$$\int_\Lambda x(\omega)\, dF(\omega).$$

In particular,

$$q\{\Lambda\} = \int_\Lambda dF(\omega), \qquad \Lambda \in \mathscr{B}(\mathscr{F}_0)^*.$$

Example 2.2 Let Ω be n-dimensional Cartesian space. The following is the generalization of the preceding example to n dimensions. Let $\mathscr{F}_0$ be the field of finite unions of right semi-closed n-dimensional intervals, so that $\mathscr{B}(\mathscr{F}_0)$ is the class of Borel sets in n dimensions. Let F be a bounded function of n variables, monotone non-decreasing and continuous on the right in each variable, with the following further properties:

(i) $$\lim_{\xi_j \to -\infty} F(\xi_1, \cdots, \xi_n) = 0, \qquad j = 1, \cdots, n;$$

(ii) if $a_i \leq b_i$, $i = 1, \cdots, n$, and, if S_k is the sum

$$\Sigma F(b_1, \cdots, b_n)$$

with the convention that in each summand k of the b_i's are replaced by the corresponding a_i's, and that the sum is taken over all $\binom{n}{k}$ possible replacements, then

$$S_0 - S_1 + \cdots + (-1)^n S_n \geq 0.$$

If Λ is a right semi-closed interval,

$$\Lambda = \{a_i < \xi_i \leq b_i, i = 1, \cdots, n\},$$

define

$$q\{\Lambda\} = S_0 - S_1 + \cdots + (-1)^n S_n,$$

and if Λ is a finite union of such (disjunct) intervals, define $q\{\Lambda\}$ as the corresponding sum. Then it is shown that q is a measure of $\mathscr{F}_0$ sets, and can therefore be extended to the Borel sets, and then completed. The measure obtained in this way is Lebesgue-Stieltjes measure in n dimensions; the measurable sets and functions are called measurable with respect to F. The remarks on the application of Theorem 2.4 made in the 1-dimensional case remain valid in n dimensions. The integral of a measurable and integrable ω function x over a measurable set Λ is denoted by

$$\int \cdot \underset{\Lambda}{\cdot} \cdot \int x(\xi_1, \cdots, \xi_n)\, d_{\xi_1, \cdots, \xi_n} F(\xi_1, \cdots, \xi_n),$$

and in particular

$$q\{\Lambda\} = \int \cdot \underset{\Lambda}{\cdot} \cdot \int d_{\xi_1, \cdots, \xi_n} F(\xi_1, \cdots, \xi_n).$$

Conversely, if q is a measure defined on the field of Borel sets of n-dimensional Cartesian space, and if F is defined by

$$F(\lambda_1, \cdots, \lambda_n) = q\{-\infty < \xi_i \leq \lambda_i, i = 1, \cdots, n\},$$

F has the properties described in Example 2.1 if $n = 1$, and in the present example if $n \geq 1$. The given measure then agrees with the Lebesgue-Stieltjes measure on finite unions of right semi-closed intervals, and therefore on all Borel sets, by Theorem 2.1.

If F_j for each $j = 1, \cdots, n$ has the properties described in Example 2.1, the function F defined by

$$F(\xi_1, \cdots, \xi_n) = \prod_1^n F_j(\xi_j)$$

satisfies the conditions on F in the present example. In this case we find that

$$S_0 - S_1 + \cdots + (-1)^n S_n = \prod_{j=1}^n [F_j(b_j) - F_j(a_j)],$$

and the integral with respect to F is usually written in the form

$$\int \cdots \int_\Lambda x(\xi_1, \cdots, \xi_n)\, dF_1(\xi_1) \cdots dF_n(\xi_n),$$

and evaluated by iterated integration. This n-dimensional measure is sometimes described as the product of the 1-dimensional measures on the coordinate axes determined by the F_j's.

Example 2.3 Let T be any infinite aggregate, and let Ω be the space of real-valued functions of $t \in T$. Then a point ω is a t function $\xi(\cdot)$. The following is the extension of the preceding examples to infinitely many dimensions. Let x_s be the ω function with the value $\xi(s)$ if ω is the function $\xi(\cdot)$, so that $x_s(\omega) = \xi(s)$. Let $t_1, \cdots, t_n$ be a finite subset of T, and let A be an n-dimensional Borel set. If the t function ω: $\xi(\cdot)$ has values $\xi(t_1), \cdots, \xi(t_n)$ at $t_1, \cdots, t_n$, the condition

$$[\xi(t_1), \cdots, \xi(t_n)] \in A$$

defines an ω set. This is the ω set

$$[x_{t_1}(\omega), \cdots, x_{t_n}(\omega)] \in A.$$

Let $\mathscr{F}_0$ be the class of all ω sets obtained in this way, for arbitrary n, $t_1, \cdots, t_n, A$. The class $\mathscr{F}_0$ is a field, but not a Borel field. Then $\mathscr{B}(\mathscr{F}_0) = \mathscr{B}(x_t, t \in T)$. Suppose that a set function q is defined on the sets of $\mathscr{F}_0$, with the following property: for each finite t set $t_1, \cdots, t_n$,

$$q\{[x_{t_1}(\omega), \cdots, x_{t_n}(\omega)] \in A\}$$

defines a function of the Borel set A which is a measure of n-dimensional Borel sets. This is obviously an additive function of $\mathscr{F}_0$ sets. One way of looking at this definition of q is that, for each finite t set $t_1, \cdots, t_n$, there is a function $F_{t_1, \cdots, t_n}$ of n variables, satisfying the conditions on such a function stated in Example 2.2, and

$$q\{[x_{t_1}(\omega), \cdots, x_{t_n}(\omega)] \in A\} = \int \cdots \int_A d_{\xi_1, \cdots, \xi_n} F_{t_1, \cdots, t_n}(\xi_1, \cdots, \xi_n).$$

The functions $\{F_{t_1, \cdots, t_n}\}$ must obey two consistency relations for this definition of q to be unique: if $\alpha_1, \cdots, \alpha_n$ is a permutation of the integers $1, \cdots, n$, then

$$F_{t_{\alpha_1}, \cdots, t_{\alpha_n}}(\xi_{\alpha_1}, \cdots, \xi_{\alpha_n}) = F_{t_1, \cdots, t_n}(\xi_1, \cdots, \xi_n)$$

and, if $m < n$,

$$F_{t_1, \cdots, t_m}(\xi_1, \cdots, \xi_m) = \lim_{\xi_{m+1}, \cdots, \xi_n \to \infty} F_{t_1, \cdots, t_n}(\xi_1, \cdots, \xi_n).$$

It is shown that q is a measure, and therefore can be extended to be a measure defined on the sets of $\mathscr{B}(\mathscr{F}_0)$, and then completed. We remark that $\mathscr{F}_0$ can be replaced throughout this discussion by the slightly smaller field (which generates the same Borel field) of ω sets of the form

$$[x_{t_1}(\omega), \cdots, x_{t_n}(\omega)] \in A,$$

where A is a finite union of n-dimensional right semi-closed intervals. We forbear extending the further remarks made in the n-dimensional case to the present case.

Leaving these examples, we make a few remarks on integration, and its relation to set functions. If $\mathscr{F}$ is a Borel field of ω sets, if q is a measure of $\mathscr{F}$ sets, and if x is an ω function measurable with respect to $\mathscr{F}$, we have supposed above that the integrability of x over a measurable set is defined in the usual way, and that the standard properties of the integral are known. We recall that, if x is a measurable ω function, it is integrable if and only if $|x|$ is. The integral of x over Λ is denoted by

$$\int_\Lambda x\, dq \quad \text{or} \quad \int_\Lambda x(\omega)\, dq.$$

If x is measurable and integrable over Ω, the set function f defined by

$$f(\Lambda) = \int_\Lambda x\, dq$$

is completely additive. Moreover, if $q\{|x(\omega)| > 0\} > 0, f(\Lambda)$ cannot vanish identically. That is, if $f(\Lambda) \equiv 0$, it follows that $x(\omega) = 0$ for almost all ω. Then two set functions f and g of this type are the same if and only if the corresponding integrands are equal for almost all ω. A completely additive function of $\mathscr{F}$ sets which vanishes whenever q vanishes is called *absolutely continuous* relative to q. The set function f is obviously absolutely continuous relative to q. Conversely (Radon-Nikodym theorem), if f is an absolutely continuous function of $\mathscr{F}$ sets, it can be expressed in the above integral form, with some x which is measurable relative to $\mathscr{F}$, and integrable. We have already remarked that x is then uniquely determined by the set function, disregarding values on sets of q measure 0. Theorem 2.5 will make this essentially unique determination of x explicit.

A completely additive function f of $\mathscr{F}$ sets is called *singular* relative to q if it does not vanish identically, and if there is an $\mathscr{F}$ set M such that

$$q(\mathrm{M}) = 0$$
$$f(\Lambda) = 0, \qquad \Lambda \subset \Omega - \mathrm{M}.$$

The set M is called the *singular set* of f relative to q, or, when f is non-negative, the *set of increase* of f. The set M is not uniquely determined, since every set of q measure 0 including a singular set of f relative to q is itself a singular set of f relative to q. According to a standard theorem of the theory, if f is completely additive, it can be written in the form $f = f_1 + f_2$, where f_1 is absolutely continuous relative to q, and f_2 is either singular relative to q or vanishes identically. This decomposition of f is unique, and, if f is non-negative, f_1 and f_2 are also non-negative. In the non-negative case, the condition for a singular set M of f relative to q becomes

$$q(\mathrm{M}) = 0$$
$$f(\Omega - \mathrm{M}) = 0.$$

Thus every completely additive f can be written in the form

$$f(\Lambda) = \int_{\Lambda} x\, dq + f_2(\Lambda),$$

where f_2 is the singular component, and x is determined uniquely, neglecting values on sets of q measure 0. It will be useful below to have an explicit representation of x in terms of f. The following discussion leads to x as a generalized derivative of f. For each n let $\mathrm{M}_1^{(n)}, \mathrm{M}_2^{(n)}, \cdots$ be finitely or denumerably infinitely many disjunct $\mathscr{F}$ sets, with union Ω. Suppose that each $\mathrm{M}_j^{(n+1)}$ is included in some $\mathrm{M}_k^{(n)}$. Define

$$\begin{aligned} x_n(\omega) &= \frac{f(\mathrm{M}_j^{(n)})}{q(\mathrm{M}_j^{(n)})}, \qquad \omega \in \mathrm{M}_j^{(n)}, \qquad q(\mathrm{M}_j^{(n)}) \neq 0 \\ &= 0, \qquad \omega \in \mathrm{M}_j^{(n)}, \qquad q(\mathrm{M}_j^{(n)}) = 0. \end{aligned}$$

Then x_n is a generalized difference quotient of f relative to q. If $\lim_{n\to\infty} x_n = x_\infty$ exists and is finite for almost all ω (q measure), x_∞ is called the *derivative of* f *with respect to* q *relative to the net of the* $M_j^{(n)}$*'s*. The derivative will, except in trivial cases, depend on the net. The following theorem is stated for future reference.

THEOREM 2.5 *If* f *is a completely additive function of* $\mathscr{F}$ *sets, and if* q *is a measure of* $\mathscr{F}$ *sets, then there is always a derivative* x_∞ *with respect to* q*, relative to a given net. The derivative is the density* x *of the absolutely continuous component of* f *relative to* q *if and only if the net satisfies the following two conditions, in which* $\mathscr{F}_\infty$ *is the Borel field generated by the class of* $M_j^{(n)}$*'s,* $j, n \geq 1$.

(i) x *is equal almost everywhere to a function which is measurable relative to* $\mathscr{F}_\infty$.

(ii) *The singular set of* f *is contained in some* $\mathscr{F}_\infty$ *set of measure* 0.

In most applications the net is chosen so that $\mathscr{F}_\infty = \mathscr{F}$, or at least that every $\mathscr{F}$ set is contained in some $\mathscr{F}_\infty$ set of the same measure. In this case the net satisfies the above conditions for every f. For example, the net has this property if Ω is a finite interval, if $\mathscr{F}$ is the class of its Lebesgue measurable subsets, if q measure is Lebesgue measure, and if the $M_j^{(n)}$'s are intervals (finitely many for each n) such that the maximum length of $M_1^{(n)}, M_2^{(n)}, \cdots$ goes to 0 when $n \to \infty$. Theorem 2.5 is proved by probability methods in VII §8 under the irrelevant assumption that q is a probability measure.

In Chapter V measures in abstract product spaces are used, and the following pages give the necessary background. They can be skipped by readers willing to accept the space X in Chapter V as the real line, or at least as finite dimensional Cartesian space.

Let X be any abstract space of points ξ, and let T be any aggregate. Let Ω be the space of functions ω of $t \in T$, with values in X. Then a point ω is a t function $\xi(\cdot)$. Let x_s be the ω function with the value $\xi(s)$ if ω is the function $\xi(\cdot)$, so that $x_s(\omega) = \xi(s)$. Throughout the following discussion it will be supposed that a Borel field $\mathscr{F}_X$ of X sets is given. We shall denote by $\mathscr{F}$ the Borel field of ω sets generated by the class of ω sets of the form $\{x_t(\omega) \in A\}$, where $t \in T$, $A \in \mathscr{F}_X$.

THEOREM 2.6 *If* $\Lambda \in \mathscr{F}$*, there is a sequence* $\{B_n\}$ *of* $\mathscr{F}_X$ *sets such that, if* $\mathscr{F}_X'$ *is the Borel field of* $\mathscr{F}_X$ *sets generated by* $\{B_n\}$*, and if* $\mathscr{F}'$ *is the Borel field of* $\mathscr{F}$ *sets generated by the class of sets of the form* $\{x_t(\omega) \in A\}$*, where* $t \in T$*,* $A \in \mathscr{F}_X'$*, then* $\Lambda \in \mathscr{F}'$.

Let $\mathscr{G}$ be the class of $\mathscr{F}$ sets Λ with the property stated in the theorem. Then $\mathscr{G}$ is obviously a Borel field, $\mathscr{G} \subset \mathscr{F}$, and, if $A \in \mathscr{F}_X$, then $\{x_t(\omega) \in A\} \in \mathscr{G}$. Hence $\mathscr{G} = \mathscr{F}$ by definition of $\mathscr{F}$.

Example 2.4 Let T be the class of integers $1, \cdots, n$. Then a point ω is an n-tuple $[\xi(1), \cdots, \xi(n)]$, and we shall denote $\mathscr{F}$ by $\mathscr{F}_X \times \cdots \times \mathscr{F}_X$ (n factors). It will be convenient to call the sets of this Borel field *generalized n-dimensional Borel sets*, and to call numerically valued functions measurable with respect to this Borel field *generalized Baire functions of n variables.* In particular, the generalized linear Borel sets are the $\mathscr{F}_X$ sets. If X is the real line, and if $\mathscr{F}_X$ is the class of linear Borel sets, the generalized concepts just defined reduce to the usual ones. For any T, we shall denote by $\mathscr{F}_0$ the field of ω sets of the form

$$\{[x_{t_1}(\omega), \cdots, x_{t_n}(\omega)] \in A\},$$

where n is an arbitrary positive integer, $t_1, \cdots, t_n$ are arbitrary points of T, and A is an arbitrary generalized n-dimensional Borel set. Then $\mathscr{F}_0 \subset \mathscr{F}$, $\mathscr{F}$ is the Borel field generated by $\mathscr{F}_0$, and $\mathscr{F} = \mathscr{F}_0$ if T is finite. We shall discuss measures of $\mathscr{F}$ sets in terms of measures of $\mathscr{F}_0$ sets.

Example 2.5 Let T be the class of integers $1, \cdots, n$. We shall set up a measure of $\mathscr{F} = \mathscr{F}_0$ sets of great importance in the theory of probability. Let p_0 be a measure of generalized linear Borel sets. For each $j < n$, let p_j be a function of $\xi_1, \cdots, \xi_j, A$, where $\xi_i \in X$ and A is a generalized linear Borel set, with the following properties.

(i) For fixed $\xi_1, \cdots, \xi_j$, $p_j(\xi_1, \cdots, \xi_j; \cdot)$ is a measure of generalized linear Borel sets.

(ii) For fixed A, $p_j(\cdot, \cdots, \cdot; A)$ is a generalized Baire function of $\xi_1, \cdots, \xi_j$.

Then we define a measure of $\mathscr{F}$ sets by the following iterated integral (evaluated from right to left)

$$\int p(d\xi_1) \int p(\xi_1;\, d\xi_2) \underset{\Lambda}{\cdots} \int p(\xi_1, \cdots, \xi_{n-1};\, d\xi_n) = q(\Lambda).$$

Example 2.6 This example generalizes the preceding one to infinitely many dimensions. Let T be the class of positive integers, and suppose that $p_0, p_1, \cdots$ have the properties described in the preceding example, except that each measure involved is to be a probability measure. Then the definition of $q(\Lambda)$ given above now defines q as an additive function of $\mathscr{F}_0$ sets, completely additive if the $\mathscr{F}_0$ sets are restricted to those defined, as above, by conditions on $x_1, \cdots, x_n$ for fixed n. The fact that the measures involved are probability measures implies that, if $\Lambda \in \mathscr{F}_0$, $q(\Lambda)$ is uniquely defined, that is, $q(\Lambda)$ is independent of the value of n used in the integration. Now, if X is the class of real numbers, and if $\mathscr{F}_X$ is the class of linear Borel sets, we have already discussed in Example 2.3 Kolmogorov's theorem, that every non-negative additive function of $\mathscr{F}_0$ sets, completely additive on each subclass of $\mathscr{F}_0$ determined by conditions on a fixed finite number of coordinate functions, is actually completely additive on $\mathscr{F}_0$, and therefore can be extended to be a measure of $\mathscr{F}$ sets. This theorem is not necessarily true for general X, $\mathscr{F}_X$, but *it is true when q is defined in terms of functions $p_0, p_1, \cdots$ as described above.* In probability theory, p_0 defines an initial probability distribution, $p_1, p_2, \cdots$ define the successive transition probability distributions, and these functions are frequently the given functions in terms of which the $\mathscr{F}$ probability measures are to be determined. The italicized statement thus asserts that such a determination is possible. To prove the statement it is sufficient to prove that, if

$$\Lambda_1 \supset \Lambda_2 \supset \cdots, \qquad \bigcap_n \Lambda_n = 0, \qquad \Lambda_n \in \mathscr{F}_0,$$

then

$$\lim_{n\to\infty} q(\Lambda_n) = 0.$$

In fact then, if $M_1, M_2, \cdots$ are disjunct $\mathscr{F}_0$ sets with union $M \in \mathscr{F}_0$, and if Λ_n is defined by

$$\Lambda_n = M - \bigcup_1^n M_j,$$

we have

$$\Lambda_1 \supset \Lambda_2 \supset \cdots, \qquad \bigcap_n \Lambda_n = 0, \qquad \Lambda_n \in \mathscr{F}_0,$$

$$q(\Lambda_n) = q(M) - \sum_1^n q(\Lambda_j),$$

so that the limit equation $q(\Lambda_n) \to 0$ implies the desired complete additivity. We can suppose that Λ_n has the form

$$\Lambda_n = \{[x_1(\omega), \cdots, x_{a_n}(\omega)] \in A_n\},$$

where A_n is a generalized a_n-dimensional Borel set. Then

$$\begin{aligned} q(\Lambda_n) &= \int p(d\xi_1) \underset{A_n}{\cdots} \int p(\xi_1, \cdots, \xi_{a_n-1};\, d\xi_{a_n}) \\ &= \int_X \Phi_n(\xi_1) p(d\xi_1), \end{aligned}$$

where Φ_n has the obvious definition. Here

$$\Phi_1 \geq \Phi_2 \geq \cdots,$$

so that the sequence $\{\Phi_n\}$ is convergent to some limit Φ, and we have

$$\lim_{n\to\infty} q(\Lambda_n) = \int_X \Phi(\xi_1) p(d\xi_1).$$

Now let δ be the limit on the left. We must prove that $\delta = 0$. If $\delta > 0$, it follows that there is an η_1, with $\Phi(\eta_1) > 0$. Now let $\Omega^{(1)}$ be the class of sequences $\omega^{(1)}$: $(\xi_2, \xi_3, \cdots)$, $\xi_j \in X$, and define a function of $\Omega^{(1)}$ sets by

$$q_1(\Lambda^{(1)}) = \int p(\eta_1;\, d\xi_2) \underset{A^{(1)}}{\cdots} \int p(\eta_1, \xi_2, \cdots, \xi_{n-1};\, d\xi_n),$$

where

$$\Lambda^{(1)} = \{[x_2^{(1)}(\omega^{(1)}), \cdots, x_n^{(1)}(\omega^{(1)})] \in A^{(1)}\},$$

$x_j^{(1)}$ is the obvious coordinate function, and $A^{(1)}$ is a generalized $(n-1)$-dimensional Borel set. Let $\Lambda_n^{(1)}$ be the $\omega^{(1)}$ set determined by the conditions

$$(\eta_1, \xi_2, \xi_3, \cdots) \in \Lambda_n.$$

Then

$$\Lambda_1^{(1)} \supset \Lambda_2^{(1)} \supset \cdots, \qquad \bigcap_n \Lambda_n^{(1)} = 0, \qquad q^{(1)}(\Lambda_n^{(1)}) \geq \Phi(\eta_1) > 0.$$

We are thus in a position to repeat the argument already given. In this way we find successively points $\eta_1, \eta_2, \cdots$ such that for every pair m, n there are values of $\xi_{m+1}, \xi_{m+2}, \cdots$ such that

$$(\eta_1, \cdots, \eta_m, \xi_{m+1}, \cdots) \in \Lambda_n.$$

Now consider the point ω_0: $(\eta_1, \eta_2, \cdots)$. For each n,

$$(\eta_1, \cdots, \eta_{a_n}, \xi_{a_n+1}{}^n, \xi_{a_n+2}{}^n, \cdots) \in \Lambda_n = \{[x_1(\omega), \cdots, x_{a_n}(\omega)] \in A_n\},$$

for properly chosen $\xi_{a_n+1}{}^n, \xi_{a_n+2}{}^n, \cdots$. That is,

$$(\eta_1, \cdots, \eta_{a_n}) \in A_n.$$

Then $\omega_0 \in \Lambda_n$, and therefore $\omega_0 \in \bigcap_n \Lambda_n$. This contradicts the hypothesis that the Λ_n's have a null intersection. Hence $\delta = 0$, as was to be proved.

Before applying Example 2.6 we remark that, although Ω was defined as an infinite product space all of whose factor spaces were the same, the factor spaces could have been taken to be different without altering the argument.

The first application is to independent measures. Suppose that $p_j(\cdot, \ldots, \cdot;\, A) = p_j(A)$ does not depend on $\xi_1, \cdots, \xi_j$. We have then set up the usual product measure in infinite dimensional space. The hypotheses have imposed no restriction on the p_j measures.

As a second application, let X be the real line, and let $\mathscr{F}_X$ be the class of linear Borel sets. Then, if q is any additive function of $\mathscr{F}_0$ sets, which, for each n, is a probability measure on the class of $\mathscr{F}_0$ sets of the form

$$\{[x_1(\omega), \cdots, x_n(\omega)] \in A\},$$

where A is an n-dimensional Borel set, q is a measure and can therefore be extended to $\mathscr{F}$ sets. This is actually a special case of the result we have just proved, because, according to I §9, and using probability language, there is a conditional distribution of x_n relative to $x_1, \cdots, x_{n-1}$ because the range of values of x_n is a Borel set, namely the line. That is, in the language we are using here, the $\mathscr{F}_0$ set function q is determined by functions $p_0, p_1, \cdots$ as described above. Of course this proof, as well as Kolmogorov's original one, is applicable more generally if X is a Borel set in n-dimensional space and if $\mathscr{F}_X$ is the class of Borel subsets of X. The obvious generalization to more general topological spaces will be omitted.

Finally we remark that we have assumed in this discussion that T is denumerable. This is not so strong a restriction as might appear at first. In fact, suppose that q is an additive function of $\mathscr{F}_0$ sets and that we are trying to prove that q is a measure by proving that if

$$\Lambda_1 \supset \Lambda_2 \supset \cdots \Lambda_n \in \mathscr{F}_0, \qquad \bigcap_n \Lambda_n = 0,$$

then

$$\lim_{n\to\infty} q(\Lambda_n) = 0.$$

Since each Λ_n is defined by conditions on finitely many x_t's, at most denumerably many parameter values are involved in all the Λ_n's and only these need be considered in the discussion. Thus the general case can be reduced to the case of denumerable T. Whether this reduction is useful or not depends on the hypotheses under examination.

Example 2.7 We now consider Example 2.5 in more detail, choosing $n = 2$ to avoid unessential complexities. We write $p(\xi; A)$ rather than $p_1(\xi; A)$. Suppose that a measure φ of $\mathscr{F}_X$ sets is given. Then for each $\xi_1 \in X$ we can write the measure $p(\xi_1; \cdot)$ as the sum of its absolutely continuous and singular components relative to φ, obtaining

$$p(\xi_1; A) = \int_A y(\xi_1, \xi_2)\varphi(d\xi_2) + \Delta(\xi_1; A).$$

It is desirable for many purposes to have y measurable in the pair ξ_1, ξ_2, that is, measurable with respect to the Borel field $\mathscr{F}_X \times \mathscr{F}_X = \mathscr{F}$. If y has this property, $\Delta(\cdot; A)$ will be a ξ_1 function measurable with respect to $\mathscr{F}_X$. According to our derivation, the ξ_2 function $y(\xi_1, \cdot)$ is measurable with respect to $\mathscr{F}_X$ and is uniquely determined only up to values on sets of φ measure 0. The problem is to choose this function for each ξ_1 to obtain ξ_1, ξ_2 measurability. This can be done as follows, under the additional hypothesis that *there is a sequence $\{B_j\}$ of $\mathscr{F}_X$ sets such that, if $\mathscr{G}$ is the Borel field generated by the class of B_j's, then $y(\xi_1, \cdot)$ is for each ξ_1 a ξ_2 function equal almost everywhere (φ measure) to a function measurable with respect to $\mathscr{G}$, and for each ξ_1 some $\mathscr{G}$ set is a singular set for the set function $\Delta(\xi_1; \cdot)$.* This hypothesis is satisfied in all cases of interest.

For example, if X is the real line, if $\mathscr{F}_X$ is the class of linear Borel sets, and if φ is any measure of Borel sets, completed or not, the B_j's of this condition can be taken as the open intervals with rational endpoints. Define

$$B_1^{(n)}, \cdots, B_{2^n}^{(n)}$$

as the intersections of the form $\bigcap_1^n A_j$, where A_j is either B_j or $X - B_j$. The $B_j^{(n)}$'s are disjunct and have union X, and the Borel field generated by the class of $B_j^{(n)}$'s, $j, n \geq 1$, is $\mathscr{G}$. Define

$$\begin{aligned} y_n(\xi_1, \xi_2) &= \frac{p(\xi_1; B_j^{(n)})}{\varphi(B_j^{(n)})}, \qquad \xi_2 \in B_j^{(n)}, \qquad \varphi(B_j^{(n)}) \neq 0, \\ &= 0 \qquad\qquad\qquad\quad \xi_2 \in B_j^{(n)}, \qquad \varphi(B_j^{(n)}) = 0. \end{aligned}$$

Then y_n is ξ_1, ξ_2 measurable, since it is a generalized Baire function of ξ_1 on each ξ_1, ξ_2 measurable set

$$\{\xi_1 \in X, \xi_2 \in B_j^{(n)}\}.$$

We have seen above that, for each ξ_1,

$$\lim_{n\to\infty} y_n(\xi_1, \xi_2) = y_\infty(\xi_1, \xi_2)$$

exists and is finite for almost all ξ_2 (φ measure). Here $y_\infty(\xi_1, \cdot)$ is the derivative of the measure $p(\xi_1; \cdot)$ with respect to φ measure relative to the net of $B_j^{(n)}$'s. Moreover, by hypothesis, $y(\xi_1; \cdot)$ is equal almost everywhere to a function measurable with respect to $\mathscr{G}$, and some singular set of $\Delta(\xi_1; \cdot)$ is a $\mathscr{G}$ set. Hence, by Theorem 2.5, $y_\infty(\xi_1; \cdot)$ is one version of the density of the absolutely continuous component of $p(\xi_1; \cdot)$ relative to φ measure. Moreover, y_∞ is ξ_1, ξ_2 measurable, as the limit of a sequence of ξ_1, ξ_2 measurable functions. The ξ_2 function $y_\infty(\xi_1; \cdot)$ may be undefined on an $\mathscr{F}_X$ set of φ measure 0. We define the function as 0 on this set, and the ξ_1, ξ_2 function obtained in this way is the desired function y.

3. Measure-preserving transformations

DEFINITION *Let Ω $[\tilde{\Omega}]$ be an abstract space of points ω $[\tilde{\omega}]$. Let q $[\tilde{q}]$ be a measure defined on the sets of a Borel field $\mathscr{F}$ $[\tilde{\mathscr{F}}]$ of ω $[\tilde{\omega}]$ sets. Let $\mathbf{T}$ be a single-valued transformation defined on the points of Ω, taking them into points of $\tilde{\Omega}$. If $\tilde{\Lambda} \in \tilde{\mathscr{F}}$, let $\mathbf{T}^{-1}\tilde{\Lambda}$ be the ω set $\{\mathbf{T}\omega \in \tilde{\Lambda}\}$. Then $\mathbf{T}$ is called a single-valued measure-preserving point transformation if the following two assertions are true:*

(i) *if $\tilde{\Lambda} \in \tilde{\mathscr{F}}$, then $\mathbf{T}^{-1}\tilde{\Lambda} \in \mathscr{F}$, and*

$$\tilde{q}\{\tilde{\Lambda}\} = q\{\mathbf{T}^{-1}\tilde{\Lambda}\};$$

(ii) *if $\Lambda \in \mathscr{F}$, there is a $\tilde{\Lambda} \in \tilde{\mathscr{F}}$ such that $\Lambda = \mathbf{T}^{-1}\tilde{\Lambda}$.*

According to this definition, $\mathbf{T}^{-1}\tilde{\Omega} = \Omega$, so that

$$\tilde{q}\{\tilde{\Omega}\} = q\{\Omega\}.$$

The same reasoning shows that, if $\tilde{R}$ is the range of $\mathbf{T}$, the $\tilde{\omega}$ set of images of ω points under $\mathbf{T}$, and, if $\tilde{R} \subset \tilde{\Lambda} \in \tilde{\mathscr{F}}$, it follows that

$$q\{\Omega\} = \tilde{q}\{\tilde{\Lambda}\} = \tilde{q}\{\tilde{\Omega}\}.$$

Thus $\tilde{R}$ has outer measure $\tilde{q}\{\tilde{\Omega}\}$ in the sense that every measurable $\tilde{\omega}$ set containing $\tilde{R}$ has measure $\tilde{q}\{\tilde{\Omega}\}$.

Each point $\tilde{\omega}$ corresponds to the ω set of points going into $\tilde{\omega}$ under $\mathbf{T}$. These ω sets will be called elementary sets below. Since every $\Lambda \in \mathscr{F}$ is the inverse image of some $\tilde{\Lambda}$, $\Lambda = \mathbf{T}^{-1}\tilde{\Lambda}$, it follows that every $\mathscr{F}$ set is the union of elementary ω sets, and that, if x is a measurable ω function, $x(\omega)$ is constant on each elementary ω set. Then every $\tilde{\omega}$ function $\tilde{x}$ determines a unique ω function x, denoted by $\mathbf{T}^{-1}\tilde{x}$, with

$$\tilde{x}(\tilde{\omega}) = x(\mathbf{T}^{-1}\tilde{\omega}).$$

Here $\mathbf{T}^{-1}\tilde{\omega}$ is to be interpreted as any one of the inverse images of $\tilde{\omega}$. If $\tilde{x}$ is $\tilde{\omega}$ measurable, $x = \mathbf{T}^{-1}\tilde{x}$ is ω measurable, because, if A is a Borel set, and *if* $\tilde{\Lambda}$ is defined by

$$\tilde{\Lambda} = \{\tilde{x}(\tilde{\omega}) \in A\},$$

then

$$\{x(\omega) \in A\} = \{\tilde{x}(\mathbf{T}\omega) \in A\} = \{\mathbf{T}\omega \in \tilde{\Lambda}\} = \mathbf{T}^{-1}\tilde{\Lambda},$$

so that the ω set on the left is measurable. Conversely, if x is a measurable ω function, there is a measurable $\tilde{\omega}$ function $\tilde{x}$, denoted by $\mathbf{T}x$ such that $x = \mathbf{T}^{-1}\tilde{x}$. In other words, the image function of x, which is defined by the above relation between x and $\tilde{x}$ only on $\tilde{R}$, can be defined on $\tilde{\Omega} - \tilde{R}$ to yield a measurable $\tilde{\omega}$ function. The extension on $\tilde{\Omega} - \tilde{R}$ is not unique, but two measurable image functions of x will agree on $\tilde{R}$ and therefore will agree on an $\tilde{\omega}$ set of measure $\tilde{q}\{\tilde{\Omega}\}$, since they agree on a measurable set containing $\tilde{R}$. We derive an image of x as follows. (We can assume that x is real.) Let $\Lambda_r = \{x(\omega) \leq r\}$ for rational r, and let $\tilde{\mathrm{M}}_r$ be an $\tilde{\omega}$ measurable set with

$$\Lambda_r = \mathbf{T}^{-1}\tilde{\mathrm{M}}_r.$$

Such a set $\tilde{\mathrm{M}}_r$ exists by hypothesis. Finally define

$$\tilde{\Lambda}_r = \bigcap_{s \geq r} \tilde{\mathrm{M}}_s.$$

Then $\tilde{\Lambda}_r$ is $\tilde{\omega}$ measurable, and

$$\tilde{\Lambda}_r \subset \tilde{\Lambda}_s, \qquad r < s$$

$$\Lambda_r = \mathbf{T}^{-1}\tilde{\Lambda}_r.$$

We have

$$x(\omega) = \underset{\omega \in \Lambda_r}{\text{G.L.B.}}\ r,$$

and we define

$$\begin{aligned} \tilde{x}(\tilde{\omega}) &= \underset{\tilde{\omega} \in \tilde{\Lambda}_r}{\text{G.L.B.}}\ r, \qquad && \tilde{\omega} \in \bigcup_r \tilde{\Lambda}_r, \\ &= 0, && \tilde{\omega} \in \tilde{\Omega} - \bigcup_r \tilde{\Lambda}_r. \end{aligned}$$

The transformation **T** operating on measurable ω functions is not necessarily single-valued, although it is single-valued if ω functions equal almost everywhere are considered identical. On the other hand, the inverse transformation $\mathbf{T}^{-1}$ (as applied to functions) is single-valued. In particular, applying the function transformation to functions taking on only the values 0 and 1, the transformation **T** induces a transformation operating on measurable ω sets, and taking them into measurable $\tilde{\omega}$ sets. This set transformation is single-valued if two $\tilde{\omega}$ sets differing only by a set of measure 0 are considered identical. The inverse set transformation is single-valued.

The following properties of the transformation **T**, considered as a transformation on functions, are now easily checked.

If $x_1, \cdots, x_n$ are measurable ω functions, if $x_j = \mathbf{T}^{-1}\tilde{x}_j$, and if Φ is a Baire function of n variables, then

$$\Phi(x_1, \cdots, x_n) = \mathbf{T}^{-1}\Phi(\tilde{x}_1, \cdots, \tilde{x}_n).$$

If $x_1, x_2, \cdots$ are measurable ω functions, and if $x_j = T^{-1}\tilde{x}_j$, then

$$\lim_{n\to\infty} x_n(\omega)$$

exists for almost all ω if and only if

$$\lim_{n\to\infty} \tilde{x}_n(\tilde{\omega})$$

exists for almost all $\tilde{\omega}$, and if these limits do exist they are transforms of each other under **T**, neglecting values on sets of measure 0. If x is a measurable ω function, and if $x = \mathbf{T}^{-1}\tilde{x}$, then x is integrable if and only if $\tilde{x}$ is, and, if these functions are integrable,

$$\int_\Omega x\, dq = \int_{\tilde{\Omega}} \tilde{x}\, d\tilde{q}.$$

The last point may deserve a few remarks. If $\tilde{x}(\tilde{\omega})$ takes on only the values 0 and 1, the same is true of $x(\omega)$, and the equality of integrals becomes the fact that **T** makes correspond measurable sets of the same measure. Then, since the class of functions x for which the statement is true includes linear combinations of functions in the class, it follows that the statement is true for functions $\tilde{x}$ which take on only finitely many values. The general case can obviously be reduced to the case of a non-negative $\tilde{x}$. Suppose then that $\tilde{x}$ is non-negative, and define

$$\begin{aligned} \tilde{x}_n(\tilde{\omega}) &= \frac{j}{2^n}, \quad \frac{j}{2^n} < \tilde{x}(\tilde{\omega}) \leq \frac{j+1}{2^n}, \quad j \leq n2^n - 1, \\ &= 0, \quad \tilde{x}(\tilde{\omega}) > n. \end{aligned}$$

Then $\lim_{n\to\infty} \tilde{x}_n(\tilde{\omega}) = \tilde{x}(\tilde{\omega})$, and, if $x_n = \mathbf{T}^{-1}\tilde{x}_n$, and $x = \mathbf{T}^{-1}\tilde{x}$, it follows that

$$\lim_{n\to\infty} x_n(\omega) = x(\omega)$$

for all ω. Moreover,

$$\tilde{x}_1(\tilde{\omega}) \leq \tilde{x}_2(\tilde{\omega}) \leq \cdots,$$

so that

$$x_1(\omega) \leq x_2(\omega) \leq \cdots.$$

Since x_n takes on only finitely many values,

$$\int_{\Omega} x_n \, dq = \int_{\tilde{\Omega}} \tilde{x}_n \, d\tilde{q}.$$

Then when $n \to \infty$ we obtain the desired inequality between the integrals of x and $\tilde{x}$ (in which infinite integrals are allowed).

Finally we observe that the essential starting point of the discussion is the existence of a single-valued transformation $\mathbf{T}^{-1}$, operating on $\tilde{\mathscr{F}}$ sets, and taking them into $\mathscr{F}$ sets in such a way that complements, unions, and intersections of $\tilde{\mathscr{F}}$ sets go into the corresponding complements, unions, and intersections of the image $\mathscr{F}$ sets. In fact, the original description of $\mathbf{T}$ as a single-valued point transformation served only to define $\mathbf{T}^{-1}$ as such a set transformation, and the whole discussion could have been carried through on the basis of the postulation of the desired properties of $\mathbf{T}^{-1}$. We have not done this because the greater generality achieved in this way is not required for the applications to be made in I (see the following examples and I §6; a more general approach is developed in X §1).

Example 3.1 Let Ω be an abstract space of points ω. Let q be a measure defined on the sets of a Borel field of ω sets, and let x be a real measurable ω function. Let $\mathscr{F} = \mathscr{B}(x)$ be the Borel field of ω sets of the form $\{x(\omega) \in \tilde{\Lambda}\}$, where $\tilde{\Lambda}$ is a linear Borel set. Let $\tilde{\Omega}$ be the real line, let $\tilde{\mathscr{F}}$ be the class of linear Borel sets, and let $\tilde{q}$ be the measure of Borel sets defined by

$$\tilde{q}\{\tilde{\Lambda}\} = q\{x(\omega) \in \tilde{\Lambda}\}.$$

For each ω define $\tilde{\omega} = \mathbf{T}(\omega) = x(\omega)$. Then $\mathbf{T}$ is a transformation taking a point ω into a real number $\tilde{\omega}$. Under this transformation, Ω goes into a linear set $\tilde{R}$. Note that $\mathbf{T}$ is single-valued, but that its inverse is multiple-valued, in general. If $\tilde{\Lambda}$ is a linear Borel set, we shall denote by $\mathbf{T}^{-1}\tilde{\Lambda}$ the set of all ω with $\mathbf{T}\omega \in \tilde{\Lambda}$. We remark that, if $\tilde{\Lambda}$ is a Borel set, and if $\Lambda = \mathbf{T}^{-1}\tilde{\Lambda}$, then

$$\tilde{q}\{\tilde{\Lambda}\} = \int_{\tilde{\Lambda}} dF(\tilde{\omega}), \qquad F(\tilde{\omega}) = q\{x(\omega) \leq \tilde{\omega}\},$$

where the integral is a Lebesgue-Stieltjes integral. In fact, we have seen in Example 2.2 that every measure of Borel sets can be expressed in this form, with suitable F, where F can be chosen to make the evaluation correct for $\tilde{\Lambda}$ a right semi-closed interval. In the present case the evaluation is correct for such a $\tilde{\Lambda}$ by definition of F. If we now restrict "measurable ω set" to mean "$\mathscr{F}$ set," the preceding results mean that $\mathbf{T}$ is a single-valued measure-preserving point transformation. Note that, if Φ is a Baire function of a single variable, then

$$\Phi(x) = \mathbf{T}^{-1}\Phi(\tilde{\omega}),$$

so that, by our general results,

$$\int_{\Omega} \Phi(x)\, dq = \int_{-\infty}^{\infty} \Phi(\tilde{\omega})\, dF(\tilde{\omega}).$$

In particular, we obtain the result

$$\int_{\Omega} x\, dq = \int_{-\infty}^{\infty} \tilde{\omega}\, dF(\tilde{\omega}),$$

a familiar result, which also follows trivially from the fact that the sums usually used in approximating the integral on the left are the Riemann-Stieltjes sums used in approximating the integral on the right.

Example 3.2 This example is the generalization of the previous one to an arbitrary dimensionality. Let Ω be an abstract space of points ω. Let q be a measure defined on the sets of a Borel field of ω sets, and let $\{x_t, t \in T\}$ be a family of real measurable functions. Let $\mathscr{F} = \mathscr{B}(x_t, t \in T)$. Let $\tilde{\Omega} : \{\tilde{\omega}\}$ be the class of real functions $\xi(\cdot)$ of $t \in T$. Let $\tilde{x}_s$ be the $\tilde{\omega}$ function with value $\xi(s)$ if $\tilde{\omega}$ is the function $\xi(\cdot)$, so that $\tilde{x}_s(\tilde{\omega}) = \xi(s)$. Let $t_1, \cdots, t_n$ be a finite subset of T, and let A be an n-dimensional right semi-closed interval, or a finite union of such intervals. Let $\tilde{\mathscr{F}}_0$ be the field of $\tilde{\omega}$ sets of the form

$$\{[\tilde{x}_{t_1}(\tilde{\omega}), \cdots, \tilde{x}_{t_n}(\tilde{\omega})] \in A\}.$$

Let $\tilde{\mathscr{F}} = \mathscr{B}(\tilde{x}_t, t \in T)$ be the Borel field of $\tilde{\omega}$ sets generated by $\tilde{\mathscr{F}}_0$. For each ω define $\tilde{\omega} = \mathbf{T}(\omega)$ as the function of t obtained from $x_t(\omega)$ when t varies. Then $\mathbf{T}$ is a single-valued point transformation taking Ω into $\tilde{R} \subset \tilde{\Omega}$. The inverse transformation is not single-valued, in general, and we define $\mathbf{T}^{-1}\tilde{\Lambda}$, for $\tilde{\Lambda}$ an $\tilde{\omega}$ set, as the ω set on which $\mathbf{T}\omega \in \tilde{\Lambda}$. Then, if $\tilde{\Lambda} \in \tilde{\mathscr{F}}$, it follows that $\mathbf{T}^{-1}\tilde{\Lambda} \in \mathscr{F}$. In fact, this is true by our definitions if $\tilde{\Lambda} \in \tilde{\mathscr{F}}_0$, and, since the $\tilde{\mathscr{F}}$ sets for which this assertion is true form a Borel field, the assertion is true for all $\tilde{\mathscr{F}}$ sets. If $\tilde{\Lambda} \in \tilde{\mathscr{F}}$, and if $\Lambda = \mathbf{T}^{-1}\tilde{\Lambda}$, define

$$\tilde{q}\{\tilde{\Lambda}\} = q\{\Lambda\}.$$

We remark that if T is infinite this $\tilde{q}$ measure is the measure determined from the measure of $\tilde{\mathscr{F}}_0$ sets by extension to $\tilde{\mathscr{F}}$, as discussed in Example 2.3, because the measures agree on $\tilde{\mathscr{F}}_0$ sets, and $\tilde{\mathscr{F}}$ is the Borel field generated by $\tilde{\mathscr{F}}_0$. The transformation $\mathbf{T}$ is a single-valued measure-preserving point transformation, if ω measurability of a set Λ is now taken to mean that $\Lambda \in \mathscr{F}$. According to our general results on this type of transformation,

$$q\{x_s(\omega) \leq c\} = \tilde{q}\{\tilde{x}_s(\tilde{\omega}) \leq c\}$$

and more generally, if $t_1, \cdot \cdot \cdot, t_n$ is a finite T set and if Φ is a Baire function of n variables, then

$$\int_{\Omega} \Phi(x_{t_1}, \cdot \cdot \cdot, x_{t_n})\, dq = \int_{\tilde{\Omega}} \Phi(\tilde{x}_{t_1}, \cdot \cdot \cdot, \tilde{x}_{t_n})\, d\tilde{q}.$$

That is to say, if either integral is defined, both are, and the two are equal. Thus for many purposes the family $\{x_t,\ t \in T\}$ of ω functions can be replaced by the family $\{\tilde{x}_t,\ t \in T\}$ of $\tilde{\omega}$ functions. The $\tilde{x}_t$'s are the coordinate variables in a space of dimensionality the cardinal number of the aggregate T.

CHAPTER III

§1

See Kolmogorov [5, 1933], Jessen [1, 1934] for various versions of the zero-one law.
imerous special cases had been observed before the general theorem was discovered.

§2

he strong half of Theorem 2.1, the inequality (2.1′), is due to Kolmogorov [1, 1928],
supposed that the y_j's were mutually independent. However, the fact that his
of uses only the hypotheses stated in Theorem 2.1, involving conditional expecta-
s, has been noted by several authors.
heorem 2.3 is due to Khintchine and Kolmogorov [1, 1924]. Theorems 2.4 and
re due to Kolmogorov [1, 1928, and the preceding reference]. The work of these
ors was developed further by Lévy [1, 1931], Jessen [1, 1934], Jessen and Wintner
935]. See also Marcinkiewicz [1, 1937; 2, 1938], Marcinkiewicz and Zygmund
937], van Kampen and Wintner [1, 1937], van Kampen [1, 1940], Wintner's books
938; 3, 1947], Lévy's book [4, 1937], Kunisawa [1, 1949]. Instead of dealing
ly with an infinite series of mutually independent random variables, one can treat
finite convolution of distribution functions of the random variables. This is the
ach used by van Kampen and Wintner. Lévy's approach stresses methods
ing explicitly the decreasing concentration of the successive partial sums of a
of mutually independent random variables. Kawata [1, 1941] simplifies this
ach somewhat by averaging Lévy's function of concentration of a distribution,
unisawa in the above reference gave a complete treatment based on such an
e. Kawata and Udagawa [1, 1949] proved the possibility in Theorem 2.7 of
involving the characteristic function on a set of positive Lebesgue measure,
ng slightly weaker results than those proved in that theorem.
proof of the corollary to Theorem 2.7 used the inequality

$$\left|\prod_j g_j - 1\right| \le \sum_j \left|g_j - 1\right|,$$

r any complex g_j's of modulus ≤ 1. It is sufficient to prove this inequality
ly many g_j's. It is trivially true for only one g_j. To finish the proof we need
ume that, if the inequality is true for $g_1, \cdots, g_n$, it is true for $g_1, \cdots, g_{n+1}$.
ows from

$$\left|\prod_1^{n+1} g_j - 1\right| = \left|\left(\prod_1^{n} g_j - 1\right) g_{n+1} + g_{n+1} - 1\right|$$

$$\le \left|\prod_1^{n} g_j - 1\right| + \left|g_{n+1} - 1\right|$$

$$\le \sum_1^{n+1} |g_j - 1|.$$

§3

y and sufficient conditions for the weak law of large numbers were found
orov [1, 1928], and in somewhat more general cases by Feller [4, 1937].

APPENDIX

CHAPTER I

§1–§5

The basic paper on probability as measure theory is Kolmogorov's [5, 1933].

The hypothesis that **P** measure is complete is used only when separability and measurability of a stochastic process (see II, §2) are of interest. Thus this hypothesis will never actually be needed when only finite or enumerably infinite collections of random variables (discrete parameter stochastic processes) are under consideration, and will not be needed for a considerable part of the study of non-denumerably infinite families of random variables (continuous parameter stochastic processes).

§6

Representation theory was stressed by Doob [4, 1938] as a device to reduce various probability theorems to standard measure theorems.

§7, §8

The measure theoretic definitions and basic properties of conditional probabilities and expectations were given by Kolmogorov [5, 1933].

§9

Theorem 9.4 was not originally in the text. Mrs. Shuh-teh Chen Moy pointed out to the author that it was contained in his proof of Theorem 9.5, and that it was worth separating out.

To avoid overburdening the text, Theorem 9.5 was not stated with maximum generality. In the first place, the only property of the range R actually used was its Lebesgue-Stieltjes measurability for every Lebesgue-Stieltjes measure. Since analytic sets have this property, the theorem remains true if R is only supposed analytic, rather than Borel. In the second place, the theorem is obviously true if its hypotheses are satisfied when Ω is replaced in the hypotheses by a subset Ω_1 of probability 1. We shall call the theorem, with R analytic and Ω replaced by Ω_1, the *generalized version of Theorem* 9.5. (It can be shown that, with the replacement of Ω by Ω_1, the theorem is no more general when R is only supposed analytic than when R is supposed Borel.) The hypotheses of the generalized version of Theorem 9.5 are satisfied (for every choice of $y_1, \cdots, y_n$) if Ω is a complete metric space, and if the given probability measure is a measure (completed or not) of Borel sets. The hypotheses are also satisfied (for every choice of $y_1, \cdots, y_n$) if Ω and the given probability measure have the property that, if x is any random variable, and if A is a linear set such that $\{x(\omega) \in A\}$ is a measurable ω set, then

$$\mathbf{P}\{x(\omega) \in A\} = \underset{B \supset A}{\text{G.L.B.}}\ \mathbf{P}\{x(\omega) \in B\},$$

where B is open. Gnedenko and Kolmogorov [1, 1949] impose the latter condition as part of their definition of a probability measure.

Note that, if the class $\mathscr{F}_1$ of measurable ω sets is the Borel field generated by a denumerable subclass, there is a random variable x such that the class of sets

$\{x(\omega) \in A\}$, for A a linear Borel set, is $\mathscr{F}_1$. Then, if the existence of a conditional probability distribution of x is assured by the generalized version of Theorem 9.5, there is a conditional distribution of $\mathscr{F}_1$ sets relative to any Borel field $\mathscr{F}$ of measurable ω sets.

Finally, consider the following example. The space Ω is the interval [0, 1]. Let $\mathscr{F}$ be the class of Borel subsets of Ω. Let A be a subset of Ω of outer Lebesgue measure 1 and inner Lebesgue measure 0, fixed throughout the following. Let $\mathscr{F}_1$ be the Borel field generated by A and the $\mathscr{F}$ sets, that is, $\mathscr{F}_1$ is the class of sets of the form $AB_1 \cup A'B_2$, where A' is the complement of A, and B_1, B_2 are $\mathscr{F}$ sets. A probability measure of $\mathscr{F}_1$ sets is defined by

$$\mathbf{P}\{AB_1 \cup A'B_2\} = \tfrac{1}{2}[m(B_1) + m(B_2)],$$

where $m(B)$ is the Lebesgue measure of B. It is easily verified that this definition is unique, and actually defines a probability measure of $\mathscr{F}_1$ sets. This probability measure reduces to Lebesgue measure on the Borel sets, and has value $\frac{1}{2}$ for the set A. It is easy to verify that there is no conditional probability distribution of $\mathscr{F}_1$ sets relative to $\mathscr{F}$. Let x be a function defined on this space, with the property that the class of sets $\{x(\omega) \in B\}$ for B Borel is $\mathscr{F}_1$. It is trivial to define such a function, and such a function cannot have a conditional probability distribution relative to $\mathscr{F}$. This example contradicts a theorem of Doob [4, 1938, Theorem 3.1] according to which a conditional probability measure of $\mathscr{F}_1$ sets relative to a field $\mathscr{F}$ always exists if $\mathscr{F}_1$ is the Borel field generated by a denumerable subclass of its sets. The incorrectness of this theorem and a related theorem [ibid., Theorem 1.1] were pointed out by Dieudonné and by Andersen and Jessen. A counterexample somewhat more special than that just given, and as such not quite contradicting the existence of a conditional probability distribution as defined in this book, is in Halmos [3, 1950, §48].

§10

The results in this section are due to Kolmogorov [5, 1933].

§11

The inequalities (11.8), (11.8′), (11.$\hat{8}$), (11.9), (11.$\hat{9}$), (11.10) are new as stated, but are implicit in the literature, at least in special cases. They were suggested to the writer on seeing (11.8) (with $a = 1/\mu$ and A the interval [0, a]), and a variation of (11.9), in Wintner [3, p. 18].

CHAPTER II

§1

For some purposes it is convenient to let the parameter of a stochastic process be a set in a certain additive family of sets. See Bochner [2, 1942] for a general discussion from this point of view. For example, the Brownian motion process, §9 Example 1, can be generalized by supposing that to each finite union I of k-dimensional intervals there corresponds a Gaussian random variable x_I, with expectation 0 and variance the k-dimensional volume of I, and that $x_{I_1}, \cdots, x_{I_n}$ have an n variate Gaussian distribution with $\mathbf{E}\{x_{I_r}x_{I_s}\}$ the volume of the intersection I_rI_s. When $k = 1$, if I_t is the interval [0, t], and if $y_t = x_{I_t}$, the stochastic process $\{y_t, 0 \leq t < \infty\}$ is a Brownian

motion process as defined in §9 Example 1. We observe that eve
with a member of a family of sets, as in this example, the existe
is guaranteed by the validity of the Kolmogorov consistency c
I §5, because the parameter was abstract-valued in that discussi

§2

For further discussion of the separability and measurability
and allied topics, see Doob [3, 1937; 5, 1940; 9, 1947], An
and Ambrose [1, 1940]. The point of view in §2, and used t
somewhat more general than that in the referenced papers i
basic ω space as function space or a slight modification ther
variables of the stochastic process under discussion become
now rejected except as an elegant special case. The phraseol
on separability and measurability therefore differs somewh
from that in §2. The relation between them is discusse
[1, 1940]. For a point of view similar to that in this book
If $\mathbf{P}$ measure had not been supposed complete, in the d
p. 51 the two ω sets on the last line would have been suppos
subset of Λ.

For an early treatment of

$$\int_a^b x(t, \omega)\, dt$$

not as an integral of the sample function $x(\cdot, \omega)$ but as a
mean limit of the Riemann sums, see Slutsky [1, 1928].

§3

The relation between strict and wide sense ideas is
various special cases. It has been defined more car
through more systematically than usual in the rest of
standing and organization of a large body of results.

§6

Markov processes were called *stochastically d*
[3, 1931]. The Markov property is sometimes car
that, if $s < t$, then the conditional probability $\mathbf{P}\{x_t($
for $r < s$. This definition is of course incorrect,
the Markov property or not) the indicated conditio
which is defined with no reference whatever to an

§7

The name *martingale* is due to Ville [1, 1939].
property E in Doob [5, 1940].

§9

Processes with independent increments were
[3, 1937], *homogeneous processes* in Cramér [1,
stationary increments), *integrals with random e*
additive processes in Lévy [7, 1948]. The s
initiated by de Finetti [1, 1929].

See also Marcinkiewicz [3, 1938], Doeblin [2, 1939], Gnedenko [1, 1939; 3, 1944], Kunisawa [1, 1949].

Theorem 3.4 is due to Kolmogorov [2, 1930].

See Loève [2, 1945] for these theorems for dependent random variables.

§4

The expression (4.6) for the characteristic function of an infinitely divisible law was discovered by Lévy [2, 1934]. A derivation in the particular case when the distribution has a second moment had already been given by Kolmogorov [4, 1932]. See also Cramér [1, 1937 Chapter VIII] for Kolmogorov's result. The first analytic derivations were given by Khintchine [3, 1937] and Feller [3, 1937]. The derivation given here is somewhat more straightforward than these because of the availability of the characteristic function inequalities of I §11.

For general discussions of limit laws of sums of mutually independent random variables see Doeblin [2, 1939], Gnedenko [1, 1939; 3, 1944], Khintchine [4, 1938], Gnedenko and Kolmogorov [1, 1949].

Necessary and sufficient conditions for the central limit theorem (essentially equivalent to Theorem 4.2) were obtained by Lévy [3, 1935] and in analytic form by Feller [1, 1935]. See also Doeblin [2, 1939], Gnedenko [1, 1939], Marcinkiewicz [3, 1938]. Theorem 4.3 is due to Lindeberg [1, 1922]. Theorem 4.4 is a classical version of the central limit theorem due to Liapounov. See Loève [2, 1945] for the extensions of these theorems to dependent random variables, and the references to the literature on these extensions.

§5

Theorem 5.1 is due to Kolmogorov [5, 1933]. The proof given is Kolmogorov's.

Theorem 5.2 is due to Doob [2, 1936].

CHAPTER IV

Since there are many good treatments of orthogonal functions, from various points of view, the treatment in IV is condensed. It cannot properly be omitted. For example, the Riesz-Fischer theorem that a sequence of functions converging in the mean according to the Cauchy criterion has a limit in the mean, that is, that L_2 is a complete space, is no less a probability theorem than the central limit theorem, and belongs in this book as much as the latter. This is not meant as an attempt to appropriate the theory of orthogonal functions for the theory of probability as distinguished from measure theory or Hilbert space theory. However, the relation of orthogonality between two functions is precisely the kind of relationship exploited by probability theory, and this is no less true because the theory of orthogonal functions was not developed in connection with what was considered the theory of probability at the time of the development. As a concession to tradition and the readability of texts on orthogonal functions the proof of the Riesz-Fischer theorem was omitted in I §4.

As a general reference to IV see books by Kaczmarz and Steinhaus [1, 1935] and by Stone [2, 1932]. In particular for §6 see Zygmund [1, 1935].

CHAPTER V

§1–§4

For general treatments of Markov chains, see books by Hostinsky [1, 1931], Fréchet [2, 1938] (in both of which detailed references to the literature are given), Romanovski [1, 1949], Feller [6, 1950]. Treatments of Markov chains with no restriction on the number of states have been given by Kolmogorov [6, 1936], Yosida and Kakutani [1, 1939], Doob [7, 1942], Feller [6, 1950]. The fundamental result in the case of finitely many states, that of §2 Case (b), goes back to Markov [1, 1906] and has been rediscovered frequently. The fundamental work in the case of infinitely many states is due to Kolmogorov [6, 1936].

§5

The treatment in §5 is essentially that of Doeblin [1, 1937], but generalized to an abstract state space, and amplified by an analysis of the possible classes of D triples $\varphi, \varepsilon, \nu$. Treatments of Markov processes with (possibly) continuous state spaces have been given in various degrees of generality, with many different methods, by Kolmogorov [3, 1931], Fréchet [1, 1934], Krylov and Bogolioubov [2, 3, 1937], Doob [4, 1938; 10, 1948], Doeblin [1, 1937; 5, 1940], Yosida and Kakutani [2, 1941], Beboutoff [1, 1942], Yaglom [1, 1947], Yosida [2, 1948]. The most far reaching of these is Doeblin's 1940 paper. The earlier and less definitive papers have been ignored in this listing.

§6

See also Doob [4, 1938; 10, 1948], Yosida [1, 1940], Kakutani [1, 1940] for the law of large numbers for Markov processes from a different point of view.

§7

The central limit theorem in the Markov chain case is due to Markov [2, 1924]. Theorem 7.5, under the additional assumption that f is bounded, is due to Doeblin [1, 1937].

For a discussion of the central limit theorem for a sequence whose elements are independent if sufficiently far apart, as in our application of Theorem 7.5′, see Hoeffding and Robbins [1, 1948].

CHAPTER VI

§1

For Markov chains with a continuous parameter see Kolmogorov [3, 1931], Krylov and Bogolioubov [1, 1936], Doeblin [4, 1940], Doob [7, 1942; 8, 1945]. See also Hille [1, 1948] for an analytic approach from the point of view of semi-groups. Theorem 1.1 is due to Doeblin [4, 1940]. Theorems 1.2, 1.3, and 1.4 are contained in much more general results of Doeblin [3, 1939]. (See also §2 below.)

§2

Pospišil [1, 1935–36] and Feller [2, 1936] treated the Markov processes considered in §2, obtaining existence and uniqueness theorems in the case of a bounded function q in (2.2). The treatment was purely analytic, considering the problem that of solving the Chapman-Kolmogorov integral equation. Doeblin [3, 1939] considered the problem probabilistically, treating the problem as that of the analysis of Markov processes under assumptions strong enough to assure that almost all sample functions are step functions. Feller [5, 1940] treated the problem as an analytic one, like Pospišil, weakening the hypotheses of Pospišil and Doeblin, and his own early ones. These authors all assumed, implicitly or explicitly, that the processes under consideration have the property that almost all the sample functions are step functions. In the special case of Markov chains, Doob [8, 1945] showed the possibility of dropping this assumption. The treatment in §2 follows that of the latter reference, appropriately generalized, and thus includes the results of Pospišil, Doeblin, and Feller, aside from the fact that those authors did not assume the stationarity of the stochastic transition functions. Lévy [8, 1951] gave a detailed analysis of the various possible types of sample functions in the chain case, including those whose discontinuities are not at a well-ordered set on the t-axis.

§3

The processes discussed in this section were first studied systematically by Kolmogorov [3, 1931], who established the fact that the transition probabilities satisfy the partial differential equations (3.4) and (3.4′). Feller [2, 1936] proved the appropriate existence and uniqueness theorems for solutions of these equations. Ito [3, 1946; 4, 1951] showed that these processes could be obtained constructively by solving stochastic differential equations. This work of Ito is given in §3, with modifications made possible by other work in this book. Fortet [1, 1943] used Feller's results to analyze the continuity and related properties of the sample functions of these processes. Kolmogorov, Feller, and Ito also discussed in the indicated references more general processes, combinations of those discussed in §2 and §3.

For a treatment of the solutions of the differential equations (3.4) and (3.4′) as limiting transition probabilities of sums of dependent random variables, in effect a study of one type of generalization of the central limit theorem, see, for example, Bernstein [2, 1938], Khintchine [1, 1933].

CHAPTER VII

§1

Martingales have been studied by many authors, referred to below. See particularly Lévy [5, 1937], Ville [1, 1939], Doob [5, 1940]. Semi-martingales are here introduced for the first time.

We recall that the random variables of a family $\{x_t,\ t \in T\}$ are said to be uniformly integrable if

$$\lim_{N\to\infty} \int_{\{|x_t(\omega)|>N\}} |x_t|\, d\mathbf{P} = 0$$

uniformly in t. A necessary and sufficient condition for uniform integrability is that $\mathbf{E}\{|x_t|\}$ be bounded in t, and that, if $\mathbf{P}\{\Lambda\} = \delta$,

$$\lim_{\delta\to 0} \int_{\Lambda} |x_t|\, d\mathbf{P} = 0.$$

It is sufficient for uniform integrability if $\mathbf{E}\{|x_t|^{\alpha}\}$ is bounded in t for some $\alpha > 1$. If $\{x_n, n \geq 1\}$ is a sequence of non-negative random variables converging with probability 1 to x, with expectations converging to the finite limit c, then (Fatou) $c \geq \mathbf{E}\{x\}$. There is equality if and only if the x_n's are uniformly integrable.

§2

Theorems 2.1 and 2.2 are new. Theorem 2.3 is a sharpening of a theorem of Halmos [1, 1939].

§3

Theorem 3.1, as applied to martingales, is due to Doob [5, 1940].

The theorem that any process which is a martingale in both parameter orders has the property that, for every pair of parameter values s, t,

$$\mathbf{P}\{x_s(\omega) = x_t(\omega)\} = 1$$

is new. The proof given in the text is due to J. R. Kinney and J. L. Snell.

Theorem 3.2 for martingales [in which case (3.4″) can be obtained from (3.4′) by applying (3.4′) to the process $\{-x_j, 1 \le j \le n\}$] was first used by Lévy and Ville.

Theorem 3.3 for martingales is due to Doob [13, 1951]. The fact that the theorem is also true for semi-martingales is due to J. L. Snell, and the proof in the text is his.

Theorem 3.4 is new.

§4

The martingale convergence theorems of this section are taken from Doob [5, 1940], with slightly strengthened subsidiary results. Various special cases had been discovered by other authors, as detailed below. The semi-martingale theorems are new (but see below the discussion of work of Andersen and Jessen, who obtained somewhat weaker theorems in a different formulation.

Theorem 4.1 (v) is due to Lévy [5, 1937, Theorem 68], under a regularity condition on $x_{n+1} - x_n$ slightly different from (4.2).

Theorem 4.1, Corollary 2, is due to Lévy [5, 1937, Corollary 68].

Theorem 4.3, Corollary 1, the second limit equation of (4.13′), is due essentially to Lévy [4, 1935; 5, 1937, Theorem 41], who proved the theorem in a slightly different form, in which z is a random variable which is the characteristic function of a point set, so that the conditional expectations become conditional probabilities.

Jessen [1, 1934] proved what is essentially Theorem 4.3, Corollary 1, in the case in which the y_j's are mutually independent with a common distribution, each distributed uniformly in the interval [0, 1] (see §7 for a statement of his result).

The following remarks are made to clarify the relation between the theorems of Andersen and Jessen [1, 1946; 3, 1948] and the martingale convergence theorems of §4. Let $\{x_n, \mathscr{F}_n, n \ge 1\}$ be a martingale, and let φ_n be the completely additive function of $\mathscr{F}_n$ sets defined by

(i) $$\varphi_n(\Lambda) = \int_\Lambda x_n \, d\mathbf{P}, \qquad \Lambda \in \mathscr{F}_n.$$

Then, if $m < n$,

$$\varphi_m(\Lambda) = \varphi_n(\Lambda), \qquad \Lambda \in \mathscr{F}_m.$$

Let $\mathscr{F}_\infty$ be the Borel field of ω sets generated by $\bigcup_n \mathscr{F}_n$. If there is a random variable x_∞ such that $\{x_n, \mathscr{F}_n, 1 \le n \le \infty\}$ is a martingale, as will be true for example if the x_n's are uniformly integrable, then (i) defines a completely additive absolutely continuous function φ_∞ of $\mathscr{F}_\infty$ sets when $n = \infty$. In this case φ_n is simply the function φ_∞ with domain of definition contracted to $\mathscr{F}_n$, and x_n is the density of φ_n relative to the given probability measure (also with domain contracted to $\mathscr{F}_n$). Suppose, however, that the hypothesis of the existence of x_∞ with the stated properties is replaced by the hypothesis that the x_n's are non-negative. Then, by Theorem 4.1, $\lim_{n\to\infty} x_n$ exists and is finite with probability 1, and we define x_∞ as this limit. Note that $\{x_n, 1 \le n \le \infty\}$ is now not necessarily a martingale, but that, using Fatou's lemma,

$$\lim_{n\to\infty} \int_\Lambda x_n \, d\mathbf{P} \ge \int_\Lambda x_\infty \, d\mathbf{P}, \qquad \Lambda \in \mathscr{F}_m,$$

so that $\{-x_n, 1 \le n \le \infty\}$ is a semi-martingale. If $\Lambda \in \bigcup_n \mathscr{F}_n$, $\varphi_n(\Lambda)$ is independent of n for large n, and we define

$$\varphi_\infty(\Lambda) = \lim_{n\to\infty} \varphi_n(\Lambda).$$

The set function φ_∞, defined on the field $\bigcup_n \mathscr{F}_n$, is not necessarily completely additive, as is seen by the following example, and complete additivity is a slightly weaker condition than the condition of uniform integrability of the x_n sequence, used above. In §8 an example is given of a martingale $\{x_n, n \ge 1\}$ with Ω the interval $[0, 1]$, having the following properties:

$$x_n \ge 0, \qquad \mathbf{E}\{x_n\} = 1, \qquad \lim_{n\to\infty} x_n(\xi) = 0, \qquad \xi \ne \tfrac{1}{2}.$$

The basic probability measure is Lebesgue measure. If we delete the point $\frac{1}{2}$ from Ω, $\lim_{n\to\infty} x_n = 0$ everywhere on Ω. Define $\mathscr{F}_n$ as the class of finite unions of the intervals

$$\left(\frac{j}{2^{n+1}}, \frac{j+1}{2^{n+1}}\right], \qquad j = 0, \cdots, 2^{n+1} - 1$$

except that 0 is included in the interval with $j = 0$ and $\frac{1}{2}$ is excluded from the interval with $j = 2^n - 1$. Then, if I_n is the interval with the latter value of j,

$$x_n(\xi) = 2^{n+1}, \qquad \xi \in I_n,$$

$$I_1 \supset I_2 \supset \cdots, \qquad \bigcap_1^\infty I_n = 0.$$

On the other hand,

$$1 = \int_{I_n} x_n \, d\mathbf{P} = \varphi_n(I_n) = \varphi_{n+1}(I_n) = \cdots = \varphi_\infty(I_n),$$

and it follows that φ_∞ is not completely additive, or we would have

$$\varphi(I_1) = \sum_1^\infty \varphi_\infty(I_n - I_{n+1}) = 0.$$

Conversely, let Ω be an abstract space, let $\mathscr{F}_1 \subset \mathscr{F}_2 \subset \cdots$ be a monotone sequence of Borel fields of ω sets, and let $\mathscr{F}_\infty$ be the Borel field of ω sets generated by $\bigcup_1^\infty \mathscr{F}_n$. Let $\mathbf{P}$ be a probability measure of $\mathscr{F}_\infty$ sets, let φ be a completely additive absolutely continuous (relative to $\mathbf{P}$) function of $\mathscr{F}_\infty$ sets, and let φ_n $[\mathbf{P}_n]$ be φ $[\mathbf{P}]$ with its domain of definition contracted to $\mathscr{F}_n$. Let x_n be the density of φ_n relative to $\mathbf{P}_n$. Andersen and Jessen [1, 1946] proved that $\lim_{n\to\infty} x_n = x_\infty$ with probability 1. Since, under the present hypotheses,

$$\varphi_n(\Lambda) = \int_\Lambda x_n \, d\mathbf{P} = \varphi_\infty(\Lambda) = \int_\Lambda x_\infty \, d\mathbf{P}, \qquad \Lambda \in \mathscr{F}_n,$$

we have

$$x_n = \mathbf{E}\{x_\infty \mid \mathscr{F}_n\}$$

with probability 1. Thus the Andersen-Jessen result becomes a special case of Theorem 4.3, according to which

$$\lim_{n\to\infty} \mathbf{E}\{x_\infty \mid \mathscr{F}_n\} = \mathbf{E}\{x_\infty \mid \mathscr{F}_\infty\}$$

with probability 1. More generally, Andersen and Jessen proved the existence of the limit x_∞ under the assumption that each φ_n is absolutely continuous, but that φ_∞ is completely additive, without necessarily being absolutely continuous. The limit is then the density of the absolutely continuous component of φ_∞. In this case,

$$\varphi_m(\Lambda) = \int x_m\, d\mathbf{P} = \varphi_n(\Lambda) = \int x_n\, d\mathbf{P} \qquad m < n, \qquad \Lambda \in \mathscr{F}_m,$$

so that the x_n process is a martingale. Moreover, if K is the variation of φ, and K_n that of φ_n, we have

$$K_1 \leq K_2 \leq \cdots \leq K, \qquad \mathbf{E}\{|x_n|\} = K_n,$$

so that the existence of the limit x_∞ is a consequence of Theorem 4.1. In a later paper [3, 1948], Andersen and Jessen assumed only that φ_∞ was completely additive, dropping the assumption of absolute continuity of the φ_n's. They defined x_n as the density of the absolutely continuous component of φ_n, $n < \infty$, and proved that in this case also $\lim_{n\to\infty} x_n$ exists and is finite with probability 1. To put this in the frame of martingale theory, suppose that φ is non-negative. (If this hypothesis is not true, φ can be expressed as the difference between two such φ's.) Suppose that $m < n$, and that $\Lambda \in \mathscr{F}_m$. Then, if the singular component of φ_m vanishes on Λ,

$$\varphi_m(\Lambda) = \int_\Lambda x_m\, d\mathbf{P} = \varphi_n(\Lambda) \geq \int_\Lambda x_n\, d\mathbf{P}.$$

Since the singular set of φ_m has probability 0, its introduction does not change the above integrals, and we have, with no restriction on the relation of Λ to the singular set of φ_m

$$\int_\Lambda x_m\, d\mathbf{P} \geq \int_\Lambda x_n\, d\mathbf{P}, \qquad \Lambda \in \mathscr{F}_m, \qquad m < n,$$

that is,

$$x_m \geq \mathbf{E}\{x_n \mid \mathscr{F}_m\}, \qquad m < n$$

with probability 1. In other words, the sequence $\{-x_n, \mathscr{F}_n, n \geq 1\}$ is a semi-martingale whose random variables are non-positive. We now conclude from Theorem 4.1s that $\lim_{n\to\infty} x_n$ exists and is finite with probability 1. We omit the discussion of the Andersen-Jessen identification of x_∞ with the density of the absolutely continuous component of φ_∞, but see §8, where such an identification is made. The discussion of derivatives in §8 is a special case of the work of Andersen and Jessen since a φ_∞ is given in each case. The specialization to derivatives with respect to nets in §8 was made in view of the applications to be made elsewhere in the book, and to avoid unnecessary abstraction in examples of martingale methodology. The situation treated by Andersen and Jessen is less general than that treated in §4 in that Andersen and Jessen always assume the existence of a completely additive φ_∞, whereas we have seen above that martingale theory treats cases in which the martingale is not derived from such a set function. On the other hand, Andersen and Jessen make a more complete identification of x_∞ with the density of the absolutely continuous component of φ_∞, obtaining thereby an explicit representation of the singular set of φ_∞.

We omit the case, also treated by Andersen and Jessen, in which the sequence of fields $\{\mathscr{F}_n\}$ is monotone non-increasing. In this case Andersen and Jessen obtain what are essentially the corresponding convergence theorems of §4 in a different language.

§5

The derivation of the zero-one law from martingale theory is due to Lévy [5, 1937]. The suggested application of martingale theory to prove many of the standard convergence theorems about infinite series of mutually independent random variables appears to be new.

Theorem 5.1 (and other results in this same area, proved by different methods) can be found in Marcinkiewicz and Zygmund [1, 1937], Marcinkiewicz [1, 1937]. The methods used in the text are taken from Doob [6, 1940].

§6

The application of martingale theory to derive the strong law of large numbers for mutually independent random variables with a common distribution function is taken from Doob [11, 1949].

§7

The theorems in this section are due to Jessen [1, 1934].

§8

The theorems of this section are easily derived from general theorems on derivatives of set functions due to de Possel [1, 1935]. They are implicit in Andersen and Jessen [1, 1946]. See the discussion of the Andersen-Jessen papers in the notes to §4.

§9

The application in §9 is given merely to clarify the significance of the likelihood ratio. See also Doob [13, 1951]. A deeper examination of the theory of statistical estimation from the point of view of martingales was given in Doob [11, 1949].

The *consistency* of the method of maximum likelihood, that is, the asymptotic correctness of the maximum likelihood estimates of the true parameter of a distribution in terms of a finite sample, was proved by Wald [2, 1949]. An earlier proof by Doob [1, 1934] was marred by an overenthusiastic use of the strong law of large numbers, but the slip is easily rectified, as noted by Doob in the Wald reference, to give the same results as Wald's method.

§10

The *fundamental theorem of sequential analysis* is due to Wald [1, 1944]. See Blackwell and Girshick [1, 1946] for another approach to this theorem in terms of martingale theory.

§11

The theorems in this section are new, except as indicated below. Some are, however, trivial generalizations of discrete parameter theorems.

In Doob [5, 1940] it was proved that the sample functions of a separable martingale with parameter set an interval almost all have left-hand limits at all points. In Doob [13, 1951] this result was strengthened to give Theorem 11.5 in the martingale case.

The treatment of the strong law of large numbers for processes with stationary independent increments is taken from Doob [6, 1940]. More precise results, involving upper limiting functions for x_t in (11.3) for $t \to \infty$, were obtained by Gnedenko [2, 1943].

Theorem 11.9 was stated with an indication of a (different) proof by Lévy [7, 1948, p. 78]. Lévy's statement is somewhat more general, but is easily reduced to Theorem 11.9.

For the central limit theorem for martingales see Lévy [3, 1935; 5, 1937].

§12

For further applications of martingale theory to the continuity of sample functions of Markov processes see Doob [7, 1942; 13, 1951].

CHAPTER VIII

§1

The study of processes with independent increments was initiated by de Finetti [1, 1929]. See Lévy's books [5, 1937; 7, 1948] for detailed treatments of these processes.

§2

The first rigorous study of the Brownian motion process was made by Wiener [1, 1923]. However, Bachelier [1, 1900] had already discovered many of the properties of this process. See Lévy's books [5, 1937; 7, 1948] for deep studies of the Brownian motion process and further references. Theorem 2.1 is due to Bachelier [1, 1900]. Theorem 2.2 is due to Wiener [1, 1923]. Theorem 2.3 is due to Lévy [6, 1940].

§3

For Einstein's work on the Brownian motion process see Einstein [1, 1906], and see Barnes and Silverman [1, 1934] for a discussion of the significance of this process in physical measurements.

§4, §5

The application of the Poisson process in §5 is new.

§6, §7

The centering of the general process of independent increments is due to Lévy [2, 1934], and §6 presumably carries out his ideas in somewhat more detail than he has given.

Theorem 7.1 is due to Lévy [2, 1934; 5, 1937], who proves also that (ii) (b) is implied by (ii) (a) even if there are fixed points of discontinuity.

Theorem 7.2 is due to Lévy [2, 1934]. See Lévy [5, 1937] for a detailed discussion of the significance of the Lévy formula (7.2) in terms of the sample function properties. Ito [1, 1942] has expressed the general process with independent increments as a kind of generalized integral of Poisson processes, thus exhibiting this relationship in an elegant way.

CHAPTER IX

§1, §2

Stochastic integrals of the type discussed in §2 were first discussed by Wiener, having been introduced by him, in a somewhat indirect form, in Wiener [1, 1923]. Such integrals are now a commonplace in Hilbert space discussions, in a somewhat different appearing form. For example, if, for each real number t, $\hat{\mathbf{E}}(t)$ is a projection operator acting on the L_2 space of measurable functions whose squares are integrable, projecting the space on the closed linear manifold $\mathfrak{M}_t$, if x is an element of L_2, and if $\mathfrak{M}_s \subset \mathfrak{M}_t$ when $s < t$, for example if the family of projections is a *canonical resolution of the identity* (see Stone [2, 1932]), then the family of elements of L_2 $\{\hat{\mathbf{E}}(t)x, -\infty < t < \infty\}$ is a stochastic process (assuming that the basic measure is a probability measure) with orthogonal increments. Integrals of the form

$$\int \Phi(t)\, d\hat{\mathbf{E}}(t)x$$

are standard tools in the Hilbert space operational calculus. (See the notes to X §3, and Stone [2, 1932].

For a very general approach to stochastic integrals see Bochner [2, 1942].

§3

This section is an adaptation of part of a paper by Khintchine [5, 1938], chosen for its general significance. See also Blanc-Lapierre [1, 1945] for more work in this same direction.

§5

The stochastic integral in §5 is a generalization of one defined by Ito [2, 1944], who treated the case in which the $y(t)$ process is the Brownian motion process. The use of martingale theory makes it possible to construct a closed system of these stochastic integrals, so that the integral with a variable upper limit defines a process of the same type as the process providing the original differential element.

CHAPTER X

§1

The discussion of measure-preserving transformations and the corresponding wide sense discussion are more general than usual in order to make it obvious that the theory of these transformations is exactly the same as that of stationary stochastic processes, although the two may not appear to be more than formally similar when stated at different levels of generality. For background on the subject see Hopf [1, 1937] and Halmos [2, 1949]. For general treatments (in probability language) of processes stationary in the wide sense see Cramér [2, 1940], Doob [12, 1949], Karhunen [1, 1946; 2, 1947], Lévy [7, 1948], Loève [1, 1945; 4, 1946], Maruyama [1, 1949], Slutsky [3, 1938], Wold [1, 1938]. In the following historical remarks no distinction is usually made between discrete and continuous parameter processes, or between real and complex cases where a proof in one case can be paraphrased into a proof for the other.

Theorem 1.1 is due to Doob [4, 1938].

§2

The ergodic theorem (Theorem 2.1) is due to G. D. Birkhoff [1, 1931]. The proof given here is taken, with insignificant modifications, from F. Riesz [1, 1945].

§3, §4

If $\{x_n, -\infty < n < \infty\}$ is a process which is stationary in the wide sense, it was shown in §1 that we can write $x_n = \mathbf{U}^n x_0$, where $\mathbf{U}$ is unitary, and that conversely such a formula always defines a process stationary in the wide sense. (Here we always take a unitary transformation as one defined on a specific space of integrable squared functions, rather than one defined on an abstract Hilbert space.) Now von Neumann [1, 1929] and Wintner [1, 1929] proved that, if $\mathbf{U}$ is a unitary transformation, with domain $\mathfrak{M}$, there is for each λ in the interval $[-\frac{1}{2}, \frac{1}{2}]$ a closed linear manifold $\mathfrak{M}(\lambda) \subset \mathfrak{M}$ such that:

(a) $\mathfrak{M}(-\frac{1}{2})$ is the manifold containing only the random variables vanishing almost everywhere, and $\mathfrak{M}(\frac{1}{2}) = \mathfrak{M}$;

(b) $\mathfrak{M}(\lambda) \subset \mathfrak{M}(\mu)$ if $\lambda < \mu$;

(c) $\mathfrak{M}(\lambda) = \bigcap_{\mu > \lambda} \mathfrak{M}(\mu), \qquad -\frac{1}{2} \le \lambda < \frac{1}{2}$,

and that, if $\hat{\mathbf{E}}(\lambda)$ is the projection of $\mathfrak{M}$ on $\mathfrak{M}(\lambda)$, then, for any $x, y \in \mathfrak{M}$,

$$\text{(i)} \qquad \mathbf{E}\{\mathbf{U}^n x)\bar{y}\} = \int_{-1/2}^{1/2} e^{2\pi i n\lambda}\, d\mathbf{E}\{[\hat{\mathbf{E}}(\lambda)x]\bar{y}\}.$$

The above properties of the projection operators imply that the process $\{\hat{\mathbf{E}}(\lambda)x, -\frac{1}{2} \le \lambda \le \frac{1}{2}\}$ is a process with orthogonal increments, and that $\mathbf{E}\{[\hat{\mathbf{E}}(\lambda)x]\bar{x}\}$ is real and monotone non-decreasing in λ. Thus, if $x = y$, (i) becomes another version of the expression (3.2) for $R(n)$. The equation (i) can be written in other ways, depending on the type of integral used, for example as

$$\text{(ii)} \qquad \mathbf{U}^n x = \int_{-1/2}^{1/2} e^{2\pi i n\lambda}\, d\hat{\mathbf{E}}(\lambda)x$$

and as

$$\text{(iii)} \qquad \mathbf{U}^n = \int_{-1/2}^{1/2} e^{2\pi i n\lambda}\, d\hat{\mathbf{E}}(\lambda).$$

In the form (ii) the von Neumann-Wintner representation becomes the spectral representation of a discrete parameter stationary process (wide sense), Theorem 4.1. We stress again that the probability situation is neither more nor less general than the Hilbert space situation (aside from the accident that in the probability discussion the measure space has measure 1). The two are the same, but in different language and with different emphasis. The reader is referred to Stone [2, 1932] for further details of the Hilbert space arguments. The proof of Theorem 4.1 given in the text follows a general method of Cramér [4, 1951].

The basic properties of wide sense stationary processes were given, in the continuous parameter case, by Khintchine [2, 1934], at a time when it was not yet clear that the theory was that of unitary translation groups in a different context. His work was translated into the discrete parameter case by Wold [1, 1938]. Khintchine proved Theorem 3.1 in the continuous parameter case, and then used the continuous parameter version of Theorem 3.2 to express the covariance function as a Fourier-Stieltjes transform. (See XI §3, §4.) The spectral representation theorem, Theorem 4.1, was published first by Cramér [2, 1942], but was apparently independently discovered at about the same time by Loève. See Lévy [7, 1948, pp. 123, 298] for a discussion of this matter. The Russian school had, however, already discovered the identity of the probability and Hilbert space problems. For example, Obukhoff [1, 1941] uses the spectral representation of Theorem 4.1 explicitly, referring as justification to earlier Hilbert space work of Kolmogorov (to be taken up in XI), who had mentioned the probability interpretation.

Many of the theorems on wide sense stationary processes, such as the representation of a covariance function as a Fourier-Stieltjes transform, are closely related to corresponding theorems on the harmonic analysis of individual functions, for which see Wiener [2, 1930].

Theorem 3.2 is due to Herglotz [1, 1911].

§6

Theorem 6.1, the law of large numbers for stationary processes (wide sense), also called the L_2 ergodic theorem, is due to von Neumann [3, 1932] in the language of Hilbert space transformations, to Khintchine [2, 1934] in the language of probability (both in the continuous parameter case). To follow through the full parallelism between strict and wide sense theorems, it would have been necessary to prove that, if $\mathbf{U}$ is isometric,

$$\underset{n\to\infty}{\text{l.i.m.}} \frac{1}{n+1} \sum_0^n \mathbf{U}^j x$$

exists for all x, and is the projection of x on the manifold of functions invariant under $\mathbf{U}$. This theorem was omitted because its proof would have taken the discussion too far afield.

Theorem 6.2 is due to Loève [1, 1945] (continuous parameter case), and was later proved independently by Blanc-Lapierre and Brard [1, 1946].

§7

Theorem 7.1 is new, but see also related work of Grenander [1, 1951] and Grenander and Rosenblatt [1, 1952]. Note that, according to Maruyama [1, 1949], a real stationary Gaussian process with zero means is metrically transitive if and only if its spectral distribution function is continuous.

§8, §9, §10

The material in these sections was obtained more or less independently by many authors. See the general references already given.

CHAPTER XI

(See also the notes to the corresponding sections of X.)

§1–§4

The continuous parameter theorem corresponding to that of von Neumann and Wintner on the form of the iterates of a unitary transformation is the following. Let $\{\mathbf{U}_t, -\infty < t < \infty\}$ be a family of unitary transformations with domain $\mathfrak{M}$, for which $\mathbf{U}_{s+t} = \mathbf{U}_s\mathbf{U}_t$, $-\infty < s, t < \infty$. Then (if an added continuity condition described below is imposed) there is for each real number λ a closed linear manifold $\mathfrak{M}(\lambda) \subset \mathfrak{M}$ such that

(a) $\bigcap_\lambda \mathfrak{M}(\lambda)$ is the manifold containing only the random variables which vanish almost everywhere, and $\bigcup_\lambda \mathfrak{M}(\lambda)$ is dense in $\mathfrak{M}$;

(b) $\mathfrak{M}(\lambda) \subset \mathfrak{M}(\mu)$ if $\lambda < \mu$;

(c) $\mathfrak{M}(\lambda) = \bigcap_{\mu>\lambda} \mathfrak{M}(\mu), -\infty < \lambda < \infty$;

and that, if $\hat{\mathbf{E}}(\lambda)$ is the projection of $\mathfrak{M}$ on $\mathfrak{M}(\lambda)$, then, for any $x, y \in \mathfrak{M}$,

$$\text{(i)} \qquad \mathbf{E}\{\mathbf{U}_t x)\bar{y}\} = \int_{-\infty}^{\infty} e^{2\pi i t\lambda}\, d\mathbf{E}\{[\hat{\mathbf{E}}(\lambda)x]\bar{y}\}$$

or, in alternative versions,

$$\text{(ii)} \qquad \mathbf{U}_t x = \int_{-\infty}^{\infty} e^{2\pi i t\lambda}\, d\hat{\mathbf{E}}(\lambda)x$$

$$\text{(iii)} \qquad \mathbf{U}_t = \int_{-\infty}^{\infty} e^{2\pi i t\lambda}\, d\hat{\mathbf{E}}(\lambda).$$

The first version, with $y = x$, yields the representation (3.2) of the covariance function as a Fourier-Stieltjes transform, the second yields the spectral representation (4.1) of a continuous parameter (wide sense) stationary process. The above result on unitary groups is due to Stone [1, 1930], under the hypothesis that, for all x, y, $\mathbf{E}\{(\mathbf{U}_t x)\bar{y}\}$ defines a continuous function of t. Von Neumann [2, 1932] proved that the measurability of this t function (for all x, y) implies its continuity if $\mathfrak{M}$ is separable.

Theorem 3.2 is due to Bochner [1, 1932].

§8

The fact that, if $\{x(t), -\infty < t < \infty\}$ is a Brownian motion process, $x'(t)$ is a (fictitious) stationary process with constant spectral density was first stressed by Wiener [see 3, 1930, as well as earlier papers].

§9

The French school describes linear operations as discussed in §9 as *filters*.

§11

The results of this section, aside from the part on stochastic integrals, are due to Kolmogorov [9, 1940; 10, 1940]. See also von Neumann and Schönberg [1, 1941].

CHAPTER XII

§1–§5

If F is monotone non-decreasing in the interval $[-\frac{1}{2}, \frac{1}{2}]$, Szegö [1, 1920] proved that

$$\lim_{n\to\infty} \operatorname*{Min}_{b_0,\cdots,b_{n-1}} \int_{-1/2}^{1/2} |e^{2\pi i n\lambda} - \sum_{j=0}^{n-1} b_j e^{2\pi i j\lambda}|^p\, dF(\lambda) = e^{\int_{-1/2}^{1/2} \log F'(\lambda)\, d\lambda},$$

with the obvious convention when the integral on the right is $-\infty$. Szegö treated this as an ordinary problem of polynomial approximation, proving the stated result for all $p > 0$, under the hypothesis that F was absolutely continuous. Wold [1, 1938] proved the fundamental decomposition result, Theorem 4.2. Kolmogorov [8, 1939; 11, 1941; 12, 1941] put the Wold decomposition theorem in its analytic setting, obtaining Theorems 4.1 and 4.3. The Kolmogorov results generalize the Szegö theorem, for the case $p = 2$, to arbitrary monotone F. The limit in Szegö's theorem is the mean square prediction error with lag 1. (See the general discussion of prediction theory in §1.) Wiener [3, 1942] obtained Kolmogorov's results for absolutely continuous F, independently, and solved the corresponding continuous parameter prediction problem, stressing the practical problem of finding the prediction explicitly in a useful form for electrical engineering. Krein [1, 1944; 2, 1945; 3, 1945] treated the discrete and continuous parameter problems in a somewhat more general form, with arbitrary F. Hanner [1, 1949] treated the continuous parameter problem in a more probabilistic manner than his predecessors, without the use of the spectral representation theorem, and was the first to obtain the continuous parameter analogue of the Wold decomposition theorem. Karhunen [3, 1950] obtained these results using the spectral representation theorem. Ahiezer [1, 1947] treated the Szegö problem with arbitrary F and $p \geq 1$, proving that the indicated limit is 0 if and only if the logarithmic integral is $-\infty$, proving also the corresponding result in the continuous parameter case. Loève ([3, 1946] and a section by Loève in Lévy [7, 1948]) obtained a version of the Wold decomposition theorem for non-stationary processes.

SUPPLEMENT

The reader is referred to Halmos [3, 1950] for general measure theoretic background material, and the proofs omitted in the Supplement.

§2

(Examples 2.3, 2.6.) The fact that finite dimensional measures of Borel sets can be extended to infinite dimensional measures as described in Example 2.3 is due to Daniell [1, 1918–1919; 2, 1919–1920] and Kolmogorov [5, 1933]. A proof that the theorem is applicable even if the factor spaces are abstract appeared in Doob [4, 1938]. The latter result is, however, incorrect in general. See the counterexample in Andersen and Jessen [2, 1948] or Halmos [3, 1950, §49]. The first proof that this result is correct at least in the case of independent factor measures was given by von Neumann [4, 1935] (see also Andersen and Jessen [2, 1948]). The fact that the result is correct whenever there are conditional probability distributions, the essential hypothesis of the text, is due to Ionescu Tulcea [1, 1949].

BIBLIOGRAPHY

N. I. AHIEZER

[1] Lectures on the theory of approximation. Moscow–Leningrad, 1947 (Russian).

WARREN AMBROSE

[1] On measurable stochastic processes. *Trans. Am. Math. Soc.* 47, 66–79 (1940).

ERIK SPARRE ANDERSEN, BØRGE JESSEN

[1] Some limit theorems on integrals in an abstract set. *Danske Vid. Selsk. Mat.-Fys. Medd.* 22, no. 14, 29 pp. (1946).

[2] On the introduction of measures in infinite product sets. *Danske Vid. Selsk. Mat.-Fys. Medd.* 25, no. 4, 8 pp. (1948).

[3] Some limit theorems on set-functions. *Danske Vid. Selsk. Mat.-Fys. Medd.* 25, no. 5, 8 pp. (1948).

L. BACHELIER

[1] Théorie de la speculation. *Ann. Sci. École Norm. Sup.* (3), 21–86 (1900).

R. B. BARNES, S. SILVERMAN

[1] Brownian motion as a natural limit to all measuring processes. *Revs. Modern Phys.* 6, 162–192 (1934).

M. BEBOUTOFF

[1] Markoff chains with a compact state space. *Rec. Math.* (*Mat. Sbornik*) N.S., 10 (52), 213–238 (1942).

SERGE BERNSTEIN

[1] Sur l'extension du théorème limite du calcul des probabilités aux sommes de quantités dépendantes. *Math. Ann.* 97, 1–59 (1927).

[2] Équations differentielles stochastiques. *Actualités Sci. Ind.* 738, 5–31 (1938).

GEORGE D. BIRKHOFF

[1] Proof of the ergodic theorem. *Proc. Natl. Acad. Sci. U.S.A.* 17, 656–660 (1931).

D. BLACKWELL, M. A. GIRSHICK

[1] On functions of sequences of independent chance vectors with applications to the problem of the "random walk" in k dimensions. *Ann. Math. Statistics* 17, 310–317 (1946).

A. BLANC-LAPIERRE

[1] Sur certaines fonctions aléatoires stationnaires. Applications à l'étude des fluctuations due à la structure de l'électricité. Thesis, Université de Paris, 1945, 80 pp.

A. Blanc-Lapierre, R. Brard

[1] Les fonctions aléatoires stationnaires et la loi des grands nombres. *Bull. Soc. Math. France* 74, 102–115 (1946).

S. Bochner

[1] Fouriersche Integrale. Leipzig 1932.
[2] Stochastic processes. *Ann. Math.* 48, 1014–1061 (1942).

K. L. Chung, W. H. J. Fuchs

[1] On the distribution of values of sums of random variables. *Mem. Am. Math. Soc.* no. 6, 12 pp. (1951).

H. Cramér

[1] Random variables and probability distributions. *Cambridge Tracts in Math.* no. 36 (1937).
[2] On the theory of stationary random processes. *Ann. Math.* 41, 215–230 (1940).
[3] On harmonic analysis in certain functional spaces. *Ark. Mat. Astr. Fys.* 28B, no. 12, 17 pp. (1942).
[4] A contribution to the theory of stochastic processes. *Proc. Sec. Berkeley Symp. Math. Statistics and Prob.* Berkeley 1951, 329–339.

P. J. Daniell

[1] Integrals in an infinite number of dimensions. *Ann. Math.* (2) 20, 281–288 (1918–1919).
[2] Functions of limited variation in an infinite number of dimensions. *Ann. Math.* (2) 21, 30–38 (1919–1920).

W. Doeblin

[1] Sur les propriétés asymptotiques de mouvement régis par certains types de chaines simples. *Bull. Math. Soc. Roum. Sci.* 39, no. 1, 57–115; no. 2, 3–61 (1937).
[2] Sur les sommes d'un grand nombre de variables aléatoires indépendantes. *Bull. Sci. Math.* 63, 23–64 (1939).
[3] Sur certains mouvements aléatoires discontinus. *Skand. Aktuarietidskr.* 22, 211–222 (1939).
[4] Sur l'équation matricielle $A^{(t+s)} = A^{(t)}A^{(s)}$ et ses applications aux probabilités en chaine. *Bull. Sci. Math.* (2), 62, 21–32 (1938); 64, 35–37 (1940).
[5] Éléments d'une théorie générale des chaines simple constantes de Markoff. *Ann. Sci. École Norm. Sup.* (3) 57, 61–111 (1940).

J. L. Doob

[1] Probability and statistics. *Trans. Am. Math. Soc.* 36, 759–775 (1934).
[2] Note on probability. *Ann. Math.* 37, 363–367 (1936).
[3] Stochastic processes depending on a continuous parameter. *Trans. Am. Math. Soc.* 42, 107–140 (1937).
[4] Stochastic processes with an integral-valued parameter. *Trans. Am. Math. Soc.* 44, 87–150 (1938).
[5] Regularity properties of certain families of chance variables. *Trans. Am. Math. Soc.* 47, 455–486 (1940).

[6] The law of large numbers for continuous stochastic processes. *Duke Math. J.* 6, 290–306 (1940).
[7] Topics in the theory of Markoff chains. *Trans. Am. Math. Soc.* 52, 37–64 (1942).
[8] Markoff chains—denumerable case. *Trans. Am. Math. Soc.* 58, 455–473 (1945).
[9] Probability in function space. *Bull. Am. Math. Soc.* 53, 15–30 (1947).
[10] Asymptotic properties of Markoff transition probabilities. *Trans. Am. Math. Soc.* 63, 393–421 (1948).
[11] Application of the theory of martingales. Le Calcul des Probabilités et ses Applications. Colloques Internationaux du Centre National de la Recherche Scientifique, Paris 1949, 23–27.
[12] Time series and harmonic analysis. *Proc. Berkeley Symp. Math. Statistics and Prob.* Berkeley 1949, 303–343.
[13] Continuous parameter martingales. *Proc. Sec. Berkeley Symp. Math. Statistics and Prob.* Berkeley 1951, 269–277.

J. L. Doob, Warren Ambrose

[1] On two formulations of the theory of stochastic processes depending upon a continuous parameter. *Ann. Math.* 41, 737–745 (1940).

A. Einstein

[1] Zur Theorie der Brownschen Bewegung. *Ann. Phys.* IV 19, 371–381 (1906).

William Feller

[1] Über den zentralen Grenzwertsatz der Wahrscheinlichkeitsrechnung. *Math. Z.* 40, 521–559 (1935); II 42, 301–312 (1937).
[2] Zur Theorie der stochastischen Prozesse (Existenz und Eindeutigkeitssätze). *Math. Ann.* 113, 113–160 (1936).
[3] On the Kolmogoroff-P. Lévy formula for infinitely divisible distribution functions. *Proc. Yugoslav Acad. Sci.* 82, 95–112 (1937).
[4] Uber das Gesetz der grossen Zahlen. *Acta Univ. Szeged* 8, 191–201 (1937).
[5] On the integro-differential equations of purely discontinuous Markoff processes. *Trans. Am. Math. Soc.* 48, 488–515 (1940). Errata. Ibid. 58, 474 (1945).
[6] An introduction to probability theory and its applications. Vol. 1. New York 1950.

Bruno de Finetti

[1] Sulle funzioni a incremento aleatorio. *Rend. Accad. Naz. Lincei Cl. Sci. Fis. Mat. Nat.* (6) 10, 163–168 (1929).

Robert Fortet

[1] Les fonctions aléatoires du type de Markoff associées a certaines equations linéaires aux derivées partielles du type paraboliques. *J. Math. Pures Appl.* 22, 177–243 (1943).

Maurice Fréchet

[1] Sur l'allure asymptotique de la suite des itérés d'un noyau de Fredholm. *Quart. J. Math. Oxford* Ser. 5, 106–144 (1934).
[2] Recherches théoriques modernes sur le calcul des probabilités II. Méthode des fonctions arbitraires. Théorie des événements en chaine dans le cas d'un nombre fini d'états possibles. Paris 1938.

B. V. GNEDENKO

[1] On the theory of limit theorems for sums of independent random variables. (Russian) *Bull. Acad. Sci. U.R.S.S.* 181–232, 643–647 (1939).
[2] Sur la croissance des processus stochastiques homogènes à accroissements indépendants. (Russian) *Izvestiya Akad. Nauk S.S.S.R.* 7, 89–110 (1943).
[3] Limit theorems for sums of independent random variables. (Russian) *Uspehi Matem. Nauk* 10, 115–165 (1944).

B. V. GNEDENKO, A. KOLMOGOROV

[1] Limit distributions for sums of independent random variables. (Russian) Moscow–Leningrad (1949).

PAUL R. HALMOS

[1] Invariants of certain stochastic transformations: the mathematical theory of gambling systems. *Duke Math. J.* 5, 461–478 (1939).
[2] Measurable transformations. *Bull. Am. Math. Soc.* 55, 1015–1034 (1949).
[3] Measure theory. New York 1950.

OLAF HANNER

[1] Deterministic and non-deterministic stationary random processes. *Ark. Mat.* 1, 161–177 (1949).

G. HERGLOTZ

[1] Uber Potenzreihen mit positivem reellen Teil im Einheitskreis. *Ber. Verh. Kgl. Sachs. Ges. Wiss. Leipzig Math.-Phys. Kl.* 63, 501–511 (1911).

EINAR HILLE

[1] Functional analysis and semi-groups. *Am. Math. Soc. Coll. Publ.* Vol. 31 1948.

W. HÖFFDING, H. ROBBINS

[1] The central limit theorem for dependent random variables. *Duke Math. J.* 15, 773–780 (1948).

EBERHARD HOPF

[1] Ergodentheorie. *Erg. Math.* 5, no. 2 (1937).

B. HOSTINSKY

[1] Méthodes générales du calcul des probabilités. *Mém. Sci. Math.* 52, 1931.

C. T. IONESCU TULCEA

[1] Mesures dans les espaces produits. *Atti Accad. Naz. Lincei Rend. Cl. Sci. Fis. Mat. Nat.* (8) 7 (1949), 208–211 (1950).

KIYOSI ITO

[1] On stochastic processes (I) (Infinitely divisible laws of probability). *Jap. J. Math.* 18, 261–301 (1942).
[2] Stochastic integral. *Proc. Imp. Acad. Tokyo* 20, 519–524 (1944).
[3] On a stochastic integral equation. *Proc. Jap. Acad.* nos. 1–4, 32–35 (1946).
[4] On stochastic differential equations. *Mem. Am. Math. Soc.* 4, 51 pp. (1951).

Børge Jessen

[1] The theory of integration in a space of an infinite number of dimensions. *Acta Math.* 63, 249–323 (1934).

Børge Jessen, Aurel Wintner

[1] Distribution functions and the Riemann zeta function. *Trans. Am. Math. Soc.* 38, 48–88 (1935).

Stefan Kaczmarz, Hugo Steinhaus

[1] Theorie der Orthogonalreihen. Warsaw–Lwow (1935).

Shizuo Kakutani

[1] Ergodic theorems and the Markoff process with a stable distribution. *Proc. Imp. Acad. Tokyo* 16, 49–54 (1940).

E. R. van Kampen

[1] Infinite product measures and infinite convolutions. *Am. J. Math.* 62, 417–448 (1940).

E. R. van Kampen, Aurel Wintner

[1] On divergent infinite convolutions. *Am. J. Math.* 59, 635–654 (1937).

Kari Karhunen

[1] Zur Spektraltheorie stochastischer Prozesse. *Ann. Acad. Sci. Fennicae Ser. A, I. Math. Phys.* 34, 7 pp. (1946).
[2] Uber lineare Methoden in der Wahrscheinlichkeitsrechnung. *Ann. Acad. Sci. Fennicae Ser. A, I. Math. Phys.* 37, 79 pp. (1947).
[3] Uber die Struktur stationärer zufälliger Funktionen. *Ark. Mat.* 1, 141–160 (1950).

Tatsuo Kawata

[1] The function of mean concentration of a chance variable. *Duke Math. J.* 8, 666–677 (1941).

Tatsuo Kawata, Masatomo Udagawa

On infinite convolutions. *Kodai Math. Sem. Rep.* no. 3, 15–22 (1949).

A. Khintchine

[1] Asymptotische Gesetze der Wahrscheinlichkeitsrechnung. *Erg. Math.* 4, 77 pp. (1933).
[2] Korrelationstheorie der stationäre stochastischen Prozesse. *Math. Ann.* 109, 604–615 (1934).
[3] Zur Theorie der unbeschränkt teilbaren Verteilungsgesetze. *Rec. Math.* (*Mat. Sbornik*) N.S. 2, 79–119 (1937).
[4] Limit laws for sums of independent random variables. (Russian) Moscow–Leningrad 1938.
[5] Theorie der abklingende Spontaneffekte. (Russian, German summary) *Izvestiya Akad. Nauk S.S.S.R. Ser. mat.* 3, 313-322 (1938).

A. Khintchine, A. Kolmogorov

[1] Uber Konvergenz von Reihen, deren Glieder durch den Zufall bestimmt werden. *Rec. Math.* (*Mat. Sbornik*) 32, 668–677 (1924).

A. Kolmogorov

[1] Über die Summen durch den Zufall bestimmter unabhängiger Grössen. *Math. Ann.* 99, 309–319 (1928); Bemerkungen zu meiner Arbeit "Über die Summen zufälliger Grössen." *Math. Ann.* 102, 484–488 (1930).
[2] Sur la loi forte des grands nombres. C. R. Acad. Sci. Paris 191, 910–912 (1930).
[3] Über die analytischen Methoden in der Wahrscheinlichkeitsrechnung. *Math. Ann.* 104, 415–458 (1931).
[4] Sulla forma generale di una processo stocastico omogeneo. (Una problema di Bruno de Finetti.) *Rend. R. Accad. Naz. Lincei Cl. Sci. Fis. Mat. Nat.* 15 (6), 805–808 (1932)
[5] Grundbegriffe der Wahrscheinlichkeitsrechnung. *Erg. Mat.* 2, no. 3 (1933).
[6] Anfangsgründe der Markoffschen Ketten mit unendlich vielen möglichen Zuständen. *Rec. Math. Moscou* (*Mat. Sbornik*) 1 (43), 607–610 (1936).
[7] Markov chains with a countable number of possible states. (Russian) *Bull. Math. Univ. Moscou* 1, no. 3, 16 pp. (1937).
[8] Sur l'interpolation et extrapolation des suites stationnaires. *C. R. Acad. Sci. Paris* 208, 2043–2045 (1939).
[9] Kurven in Hilbertschen Raum die gegenüber eine einparametrigen Gruppe von Bewegungen invariant sind. *C. R.* (*Doklady*) *Acad. Sci. U.R.S.S.* (*N.S.*) 26, 6–9 (1940).
[10] Wienersche Spiralen und einige andere interessante Kurven im Hilbertschen Raum. *C. R.* (*Doklady*) *Acad. Sci. U.R.S.S.* (*N.S.*) 26, 115–118 (1940).
[11] Stationary sequences in Hilbert space. (Russian) *Bull. Math. Univ. Moscou* 2, no. 6, 40 pp. (1941).
[12] Interpolation und Extrapolation von stationären zufälligen Folgen. *Bull. Acad. Sci. U.R.S.S. Ser. Math.* 5, 3–14 (1941).

M. Krein

[1] On the problem of continuation of helical arcs in Hilbert space. (Russian) *C. R.* (*Doklady*) *Acad. Sci. U.R.S.S.* 45, 139–142 (1944).
[2] On a generalization of some investigations of G. Szegö, V. Smirnoff, and A. Kolmogoroff. (Russian) *C. R.* (*Doklady*) *Acad. Sci. U.R.S.S.* (*N.S.*) 46, 91–94 (1945).
[3] On a problem of extrapolation of A. N. Kolmogoroff. (Russian) *C. R.* (*Doklady*) *Acad. Sci. U.R.S.S.* (*N.S.*) 46, 306–309 (1945).

N. M. Krylov, N. N. Bogolioubov

[1] Sur les propriétés ergodiques de l'équation de Smoluchowsky. *Bull. Soc. Math. France* 64, 49–56 (1936).
[2] Sur les probabilités en chaine. *C. R. Acad. Sci. Paris* 204, 1386–1388 (1937).
[3] Les propriétés ergodiques des suites de probabilités en chaine. *C. R. Acad. Sci. Paris* 204, 1454–1456 (1937).

Kiyonori Kunisawa

[1] On an analytical method in the theory of independent random variables. *Ann. Inst. Statist. Math. Tokyo* 1, 1–77 (1949).

Paul Lévy

[1] Sur les séries dont les termes sont des variables éventuelles indépendantes. *Studia Math.* 3, 119–155 (1931).

[2] Sur les intégrales dont les éléments sont des variables aléatoires indépendantes. *Ann. Scuola Norm. Sup. Pisa* (2) 3, 337–366 (1934); Observation sur un précédent mémoire de l'auteur. *Ibid.* 4, 217–218 (1935).

[3] Propriétés asymptotiques des sommes de variables aléatoires indépendantes ou enchainées. *J. Math. Pures Appl. Ser.* 8 14, 347–402 (1935).

[4] Propriétés asymptotiques des sommes de variables aléatoires enchainées. *Bull. Sci. Math.* (2) 59, 84–96, 109–128 (1935).

[5] Théorie de l'addition des variables aléatoires. Paris 1937.

[6] Le mouvement Brownien plan. *Am. J. Math.* 62, 487–550 (1940).

[7] Processus stochastiques et mouvement Brownien. Paris 1948.

[8] Systèmes markoviens et stationnaires. Cas dénombrable. *Ann. Sci. École Norm. Sup.* 68, 40–381 (1951).

J. W. Lindeberg

[1] Eine neue Herleitung des Exponentialgesetzes in der Wahrscheinlichkeitsrechnung. *Math. Zeitschr.* 15, 211–225 (1922).

Michel Loève

[1] Sur les fonctions aléatoires stationnaires de second ordre. *Rev. Sci.* 83, 297–303 (1945).

[2] Étude asymptotique de sommes de variables aléatoires liées. *J. Math. Pures Appl.* (9) 24, 249–318 (1945).

[3] Quelques propriétés des fonctions aléatoires de second ordre. *C. R. Acad. Sci. Paris* 222, 469–470 (1946).

[4] Fonctions aléatoires de second ordre. *Rev. Sci.* 84, 195–206 (1946).

J. Marcinkiewicz

[1] Quelques théorèmes sur les fonctions indépendantes. *Studia Math.* 7, 104–120 (1937).

[2] Sur les fonctions indépendantes I. *Fund. Math.* 30, 202–214 (1938).

[3] Sur les fonctions indépendantes II. *Fund. Math.* 30, 349–364 (1938).

J. Marcinkiewicz, A. Zygmund

[1] Sur les fonctions indépendantes. *Fund. Math.* 29, 60–90 (1937).

A. A. Markov

[1] Extension of the law of large numbers to dependent events. (Russian) *Bull. Soc. Phys. Math. Kazan* (2) 15, 135–156 (1906).

[2] Calculus of probability. (Russian) 4th ed. Moscow 1924.

Gisiro Maruyama

[1] The harmonic analysis of stationary stochastic processes. *Mem. Fac. Sci. Kyusyu Univ.* A 4, 45–106 (1949).

John von Neumann

[1] Allgemeine Eigenwerttheorie Hermitischer Funktionaloperatoren. *Math. Ann.* 102, 49–131 (1929).

[2] Uber einen Satz von Herrn M. H. Stone. *Ann. Math.* 33, 567–573 (1932).

[3] Proof of the quasi-ergodic hypothesis. *Proc. Natl. Acad. Sci. U.S.A.* 18, 70–82 (1932).

[4] Functional operators I. Measures and integrals. *Ann. Math. Studies* no. 21, Princeton, New Jersey. (Reprint of a multigraphed edition of 1935.)

John von Neumann, I. J. Schonberg

[1] Fourier integrals and metric geometry. *Trans. Am. Math. Soc.* 50, 226–251 (1941).

A. Obukhoff

[1] On the energy distribution in the spectrum of a turbulent flow. *C. R.* (*Doklady*) *Acad. Sci. U.R.S.S.* 32, 19–21 (1941).

Bedřich Pospišil

[1] Sur un problème de M. M. S. Bernstein et A. Kolmogoroff. *Časopis Pěst. Mat. Fys.* 65, 64–76 (1935–36).

Réné de Possel

[1] Sur la dérivation abstraite des fonctions d'ensemble. *C. R. Acad. Sci. Paris* 201, 579–581 (1935); *J. Math. Pures Appl.* 15, 391–409 (1936).

F. Riesz

[1] Sur la théorie ergodique. *Comm. Math. Helvetici* 17, 221–239 (1945).

V. I. Romanovski

[1] Discrete Markov chains. (Russian) Moscow–Leningrad 1949.

E. Slutsky

[1] Sur les fonctions éventuelles continues, intégrables et dérivables dans le sens stochastique. *C. R. Acad. Sci. Paris* 187, 878–880 (1928).

[2] Alcuni proposizioni sulla theoria degli funzioni aleatorie. *Giorn. Ist. Ital. Attuari* 8, 183–199 (1937).

[3] Sur les fonctions aléatoires presques periodiques et sur la decomposition des fonctions aléatoires stationnaires en composantes. *Actualités Sci. Ind.* 738, 35–55 (1938).

Marshall Harvey Stone

[1] Linear transformations in Hilbert space III. Operational methods and group theory. *Proc. Natl. Acad. Sci. U.S.A.* 16, 172–175 (1930).

[2] Linear transformations in Hilbert space and their applications to analysis. *Am. Math. Soc. Coll. Publ.* 15 (1932).

G. Szegö

[1] Beiträge zur Theorie der Toeplitzschen Formen. *Math. Zeitschr.* 6, 167–202 (1920).

Jean Ville

[1] Étude critique de la notion de collectif. Paris 1939.

A. Wald

[1] On cumulative sums of random variables. *Ann. Math. Statistics* 15, 283–296 (1944).
[2] Note on the consistency of the maximum likelihood estimate. *Ann. Math. Statistics* 20, 595–601 (1949).

Norbert Wiener

[1] Differential space. *J. Math. Phys. Math. Inst. Tech.* 2, 131–174 (1923).
[2] Generalized harmonic analysis. *Acta Math.* 55, 117–258 (1930).
[3] Extrapolation, interpolation, and smoothing of stationary time series. With engineering applications. Cambridge–New York 1949. (Reprinted from a publication issued with restricted circulation in 1942.)

Aurel Wintner

[1] Zur Theorie der beschränkten Bilinearformen. *Math. Zeitschr.* 30, 228–282 (1929).
[2] Asymptotic distributions and infinite convolutions. Princeton, 1938.
[3] The Fourier transforms of probability distributions. Baltimore 1947.

H. Wold

[1] A study in the analysis of stationary time series. Uppsala 1938.

A. M. Yaglom

[1] The ergodic principle for Markov processes with stationary distributions. (Russian) *Doklady Akad. Nauk S.S.S.R.* (*N.S.*) 56, 347–349 (1947).

Kosaku Yosida

[1] The Markoff process with a stable distribution. *Proc. Imp. Acad. Tokyo* 16, 43–48 (1940).
[2] Simple Markoff process with a locally compact phase space. *Math. Japonicae* 1, 99–103 (1948).

Kôsaku Yosida, Shizuo Kakutani

[1] Markoff process with an enumerable infinite number of possible states. *Jap. J. Math.* 16, 47–55 (1939).
[2] Operator theoretical treatment of Markoff's process and mean ergodic theorem. *Ann. Math.* 42, 188–228 (1941).

Antoni Zygmund

[1] Trigonometrical series. Warsaw-Lwow 1935.

Index

(This index includes the additional mathematical material, but not the names or historical remarks, in the Appendix.)